Phytochemicals for Health

Phytochemicals for Health

Edited by

Marcello Locatelli
Department of Pharmacy, University "G. d'Annunzio" of Chieti-Pescara, Chieti, Italy

Michał Tomczyk
Department of Pharmacognosy, Faculty of Pharmacy with the Division of Laboratory Medicine, Medical University of Bialystok, Bialystok, Poland

Laura Dugo
Department of Science and Technologies for Sustainable Development and One Health, University Campus Bio-Medico di Roma, Rome, Italy

Marina Russo
Messina Institute of Technology c/o Department of Chemical, Biological, Pharmaceutical and Environmental Sciences, former Veterinary School, University of Messina, Messina, Italy

ELSEVIER

Elsevier
Radarweg 29, PO Box 211, 1000 AE Amsterdam, Netherlands
125 London Wall, London EC2Y 5AS, United Kingdom
50 Hampshire Street, 5th Floor, Cambridge, MA 02139, United States

ISBN: 978-0-443-15366-2

For information on all Elsevier publications visit our website at https://www.elsevier.com/books-and-journals

Publisher: Candice Janco
Acquisitions Editor: Rafael Trombaco
Editorial Project Manager: Dan Egan
Production Project Manager: Paul Prasad Chandramohan
Cover Designer: Mark Rogers

Typeset by TNQ Technologies

Contents

PART 4 Inorganic compounds state-of-the-art

Contributors

Nurten Abaci Kaplan
Department of Pharmacognosy, Faculty of Pharmacy, Gazi University, Ankara, Turkey

G. Adami
University of Trieste, Department of Chemical and Pharmaceutical Sciences, Trieste, Italy

Imran Ali
Department of Chemistry, Jamia Millia Islamia (Central University), Delhi, New Delhi, India

Bashaer Alsharif
School of Pharmacy and Pharmaceutical Sciences and Trinity Biomedical Sciences, Institute, Trinity College Dublin, Dublin, Ireland; Department of Pharmaceutical Sciences, Pharmacy College, Umm Al-Qura University, Makkah, Saudi Arabia

Katia Arena
Messina Institute of Technology c/o Department of Chemical, Biological, Pharmaceutical and Environmental Sciences, Former Veterinary School, University of Messina, Messina, Italy

Adriana Arigò
University of Urbino Carlo Bo, Department of Pure and Applied Sciences, Urbino, Italy

P. Barbieri
University of Trieste, Department of Chemical and Pharmaceutical Sciences, Trieste, Italy

Fabio Boylan
School of Pharmacy and Pharmaceutical Sciences and Trinity Biomedical Sciences, Institute, Trinity College Dublin, Dublin, Ireland; Trinity Natural Products Research Centre, NatPro Centre, Trinity College Dublin, Dublin, Ireland

Francesco Cacciola
Messina Institute of Technology c/o Department of Chemical, Biological, Pharmaceutical and Environmental Sciences, Former Veterinary School, University of Messina, Messina, Italy

Giovanna Cafeo
Messina Institute of Technology c/o Department of Chemical, Biological, Pharmaceutical and Environmental Sciences, Former Veterinary School, University of Messina, Messina, Italy

Cecilia Cagliero
University of Turin, Department of Drug Science and Technology, Turin, Italy

Rosy Cavaliere
Department of Experimental and Clinical Medicine, University Magna Graecia of Catanzaro, Catanzaro, Italy

Luigi Ciriolo
Department of Pharmacy, University "G. d'Annunzio" of Chieti-Pescara, Chieti, Italy

Maria Chiara Cristiano
Department of Medical and Surgical Sciences, University Magna Graecia of Catanzaro, Catanzaro, Italy

M. Crosera
University of Trieste, Department of Chemical and Pharmaceutical Sciences, Trieste, Italy

Ersin Demir
Afyonkarahisar Health Sciences University, Faculty of Pharmacy, Department of Analytical Chemistry, Afyonkarahisar, Turkey

Laura Dugo
Department of Science and Technologies for Sustainable Development and One Health, University Campus Bio-Medico di Roma, Rome, Italy

Paola Dugo
Messina Institute of Technology c/o Department of Chemical, Biological, Pharmaceutical and Environmental Sciences, Former Veterinary School, University of Messina, Messina, Italy; Chromaleont s.r.l., c/o Department of Chemical, Biological, Pharmaceutical and Environmental Sciences, Former Veterinary School, University of Messina, Messina, Italy

Halina Ekiert
Department of Medicinal Plant and Mushroom Biotechnology, Faculty of Pharmacy, Jagiellonian University, Medical College, Kraków, Poland

Manon Ferrier
EA 2106 Biomolécules et Biotechnologies Végétales, UFR des Sciences Pharmaceutiques, Université de Tours, Tours, France

S. Fornasaro
University of Trieste, Department of Chemical and Pharmaceutical Sciences, Trieste, Italy

Nathalie Giglioli-Guivarc'h
EA 2106 Biomolécules et Biotechnologies Végétales, UFR des Sciences Pharmaceutiques, Université de Tours, Tours, France

Genny Grasselli
University of Urbino Carlo Bo, Department of Pure and Applied Sciences, Urbino, Italy

Christophe Hano
Institut de Chimie Organique et Analytique, Université d'Orléans-CNRS, UMR 7311, Orléans, France

Katarzyna Jakimiuk
Department of Pharmacognosy, Faculty of Pharmacy with the Division of Laboratory Medicine, Medical University of Bialystok, Bialystok, Poland

Abuzar Kabir
Department of Chemistry and Biochemistry, International Forensic Research Institute, Florida International Univeristy, Miami, FL, United States

Marta Klimek-Szczykutowicz
Department of Medicinal Plant and Mushroom Biotechnology, Faculty of Pharmacy, Jagiellonian University, Medical College, Kraków, Poland; Department of Pharmaceutical Sciences, Collegium Medicum, Jan Kochanowski University, Kielce, Poland

Arnaud Lanoue
EA 2106 Biomolécules et Biotechnologies Végétales, UFR des Sciences Pharmaceutiques, Université de Tours, Tours, France

Sabina Licen
University of Trieste, Department of Chemical and Pharmaceutical Sciences, Trieste, Italy

Clinio Locatelli
Department of Chemistry, University of Bologna, Bologna, Italy

Marcello Locatelli
Department of Pharmacy, University "G. d'Annunzio" of Chieti-Pescara, Chieti, Italy

Magdalena Anna Malinowska
Department of Organic Chemistry and Technology, Faculty of Chemical Engineering and Technology, Cracow University of Technology, Kraków, Poland

Antonia Mancuso
Department of Experimental and Clinical Medicine, University Magna Graecia of Catanzaro, Catanzaro, Italy

Fotouh R. Mansour
Department of Pharmaceutical Analytical Chemistry, Faculty of Pharmacy, Tanta University, Tanta, Egypt

Arianna Marengo
University of Turin, Department of Drug Science and Technology, Turin, Italy

Angela Marino
Department of Chemical, Biological, Pharmaceutical and Environmental Sciences, University of Messina, Messina, Italy

Laura Martín-Pozo
Department of Analytical Chemistry, Escuela Politécnica Superior, University of Seville, Seville, Spain

Dora Melucci
Department of Chemistry, University of Bologna, Bologna, Italy

Luigi Mondello
Messina Institute of Technology c/o Department of Chemical, Biological, Pharmaceutical and Environmental Sciences, Former Veterinary School, University of Messina, Messina, Italy; Chromaleont s.r.l., c/o Department of Chemical, Biological, Pharmaceutical and Environmental Sciences, University of Messina, Messina, Italy

Rossana Morabito
Department of Chemical, Biological, Pharmaceutical and Environmental Sciences, University of Messina, Messina, Italy

Mohd Mustaqeem
Department of Chemistry, Jamia Millia Islamia (Central University), Delhi, New Delhi, India

Ibukun O. Oresanya
Department of Pharmacognosy, Faculty of Pharmacy, Gazi University, Ankara, Turkey

Ilkay Erdogan Orhan
Department of Pharmacognosy, Faculty of Pharmacy, Lokman Hekim University, Ankara, Turkey; Full Member of Turkish Academy of Sciences (TÜBA), Vedat Dalokay Cad., Ankara, Turkey

Donatella Paolino
Department of Experimental and Clinical Medicine, University Magna Graecia of Catanzaro, Catanzaro, Italy

Miryam Perrucci
Department of Pharmacy, University "G. d'Annunzio" of Chieti-Pescara, Chieti, Italy

Alicja Piotrowska-Niczyporuk
Department of Plant Biology and Ecology, Faculty of Biology, University of Bialystok, Bialystok, Poland

Rossella Pirro
Department of Pharmacy, University "G. d'Annunzio" of Chieti-Pescara, Chieti, Italy

Alessia Remigante
Department of Chemical, Biological, Pharmaceutical and Environmental Sciences, University of Messina, Messina, Italy

Patrizia Rubiolo
University of Turin, Department of Drug Science and Technology, Turin, Italy

Marina Russo
Messina Institute of Technology c/o Department of Chemical, Biological, Pharmaceutical and Environmental Sciences, Former Veterinary School, University of Messina, Messina, Italy

Barbara Sgorbini
University of Turin, Department of Drug Science and Technology, Turin, Italy

Marta Sharafan
Department of Medicinal Plant and Mushroom Biotechnology, Medical College, Jagiellonian University, Kraków, Poland; Department of Organic Chemistry and Technology, Faculty of Chemical Engineering and Technology, Cracow University of Technology, Kraków, Poland

Elżbieta Sikora
Department of Organic Chemistry and Technology, Faculty of Chemical Engineering and Technology, Cracow University of Technology, Kraków, Poland

Ludovica Spagnuolo
Department of Science and Technologies for Sustainable Development and One Health, University Campus Bio-Medico di Roma, Rome, Italy

Sara Spinelli
Department of Chemical, Biological, Pharmaceutical and Environmental Sciences, University of Messina, Messina, Italy

Agnieszka Szopa
Department of Medicinal Plant and Mushroom Biotechnology, Faculty of Pharmacy, Jagiellonian University, Medical College, Kraków, Poland

Michał Tomczyk
Department of Pharmacognosy, Faculty of Pharmacy with the Division of Laboratory Medicine, Medical University of Bialystok, Bialystok, Poland

Monika Tomczykowa
Department of Organic Chemistry, Faculty of Medicine with the Division of Dentistry and Division of Medical Education in English, Medical University of Bialystok, Bialystok, Poland

Halil Ibrahim Ulusoy
Department of Analytical Chemistry, Faculty of Pharmacy, Cumhuriyet University, Sivas, Turkey

Roberto Laganà Vinci
Messina Institute of Technology c/o Department of Chemical, Biological, Pharmaceutical and Environmental Sciences, Former Veterinary School, University of Messina, Messina, Italy

About the editors

Marcello Locatelli is an Associate Professor of Analytical Chemistry and his research focuses on the development and validation of chromatographic methods for the qualitative and quantitative determination of biologically active molecules in human and animal, cosmetics, food, and environmental complex matrices. These procedures have been applied to different analytes and drug associations and to finding application in clinical and preclinical studies, to characterize new delivery systems of the active principle to improve their pharmacological properties. The development of the methods includes predictive models and chemometrics applied both for the optimization of extraction protocols and for final data processing. Particular attention is given to innovative (micro)-extraction techniques and new instrumental configurations for the quantitative analysis of complex matrices.

Michał Tomczyk is a Pharmacist, Full Professor of Pharmaceutical Sciences at the Department of Pharmacognosy, Faculty of Pharmacy with the Division of Laboratory Medicine, Medical University of Bialystok (Poland), member of the Senate of Medical University of Bialystok (2020–24), and an Editor-in-Chief of the international journal *Acta Poloniae Pharmaceutica—Drug Research.* He is also a member of the team for the Analysis of Pharmaceutical, Biomedical and Natural Products of the Committee of Analytical Chemistry of the Polish Academy of Sciences (PAN) since 2018, and various international scientific societies including the Society for Medicinal Plant and Natural Products Research (GA) and the Phytochemistry Society of Europe (PSE). His research focuses on unique natural products such as polyphenolic compounds and their applications in various areas of biomedicine; the natural products/compounds/extracts isolated by his laboratory are being examined for anticancer, anticaries, and antiinflammatory activities. His laboratory also investigates the different analytical methods of natural compounds using a combination of traditional and modern metabolomic methodologies (LC, HPTLC, and LC-MS). Additionally, he is the associate editor of *Frontiers in Pharmacology* and *Open Chemistry*, and editorial board member of *Planta Medica*, *Acta Pharmaceutica Hungarica*, and *Pharmacognosy Communications.*

Laura Dugo is an Associate Professor of Food Chemistry at the Department of Science and Technology for Sustainable Development and One Health, at University "Campus Bio-Medico di Roma". Her research activity focuses on food-derived compounds, their biological activity, and potential implication of diet composition on human health. Bioactive compounds are investigated through in vitro cellular models of oxidative stress, inflammation, and glycation. A specific focus of her research activity is the recovery of bioactive compounds from waste materials of the agri-food chain, with a circular economy perspective, aiming to identify bioactive molecule-rich matrices for nutraceuticals and food supplements manufacturing.

Marina Russo is an Associate Professor of Food Chemistry at the University of Messina. Her research activity primarily focuses on the development of analytical methods for characterization of nutraceuticals in functional foods. Her main topics of research are the development of HPLC and SFC analytical systems for the characterization of nutraceuticals, chiral and nonchiral oxygen heterocyclic compounds in functional foods, isolation and purification of phytochemicals from by-products and processing of agro-food industry, and development of on-line extraction techniques coupled with LC and SFC systems for the simultaneous extraction and analysis of bioactive compounds in vegetables; notably she received a postdoc grant for research activity for chemical characterization of by-products and processing of agro-food industry.

PART

Fundamentals in phytochemistry

1

CHAPTER

1 Introduction (general aspects of phytochemicals, definitions, classes, chemical structures, and properties)

Roberto Laganà Vinci[1], Francesco Cacciola[1] and Paola Dugo[1,2]

[1]*Messina Institute of Technology c/o Department of Chemical, Biological, Pharmaceutical and Environmental Sciences, Former Veterinary School, University of Messina, Messina, Italy;* [2]*Chromaleont s.r.l., c/o Department of Chemical, Biological, Pharmaceutical and Environmental Sciences, Former Veterinary School, University of Messina, Messina, Italy*

Phytochemicals are secondary metabolites produced by plant biosynthetic pathways as response to stress and pathogens microorganisms, while in humans play an important role in prevention of many disease and pathological conditions. Among these, terpenoids, phytosterols, carotenoids, alkaloids, capsacinoids, and phenolic compounds represent the most important families (Mendoza and Escamilla Silva, 2018). Despite their structural differences, some families can be grouped according to their metabolic pathways of synthesis. Phytosterols and carotenoids can be considered as terpenoids, which are synthesized through isoprenoid biosynthetic pathway that can follow the mevalonate pathway, responsible for the synthesis of phytosterols or the pathway of methylerythritol phosphate, responsible for the synthesis of carotenoids (Abdallah and Quax, 2017). Phenolic compounds are synthesized through the shikimate pathway, which leads to aromatic amino acids, followed by the phenylpropanoid pathway, which converts the aromatic amino acids into the different phenolic subclasses (Quideau et al., 2011). Capsacinoids share the phenylpropanoid pathway with phenolic compounds; however, even considering the presence of an aromatic ring with a hydroxyl group, they cannot be considered as phenolic compounds, because are synthetized through a combination of phenylpropanoid and fatty acids biosynthesis pathway (Reyes-Escogido et al., 2011). On the other hand, alkaloids, even sharing the shikimate pathway with phenolic compounds, are characterized by nitrogen heterocycles and can be synthesized through different pathways (Dewick, 2001).

Terpenoids are the largest family of natural compounds and are all derived from a five-carbon isoprene unit and can assume linear or cyclic forms, with varying degrees of oxidation. They can be principally divided, depending on the number of carbons that constitute them, into monoterpenes (C_{10}), sesquiterpenes (C_{15}), diterpenes (C_{20}), triterpenes (C_{30}), and tetraterpenes (C_{40}) (Fig. 1.1A). Among these, monoterpenes, sesquiterpenes, and diterpenes show anticancer, antiinflammatory, and antimicrobial properties, which seem to be enhanced when epoxy rings or hydroxyl groups are present (Abdallah and Quax, 2017; Brahmkshatriya and Brahmkshatriya, 2013).

Phytosterols, ascribable to triterpenes, are mainly represented by β-sitosterol, campesterol, and stigmasterol and possess a structure made by a total of 28–29 carbon atoms, with a polycyclic core

Phytochemicals for Health. https://doi.org/10.1016/B978-0-443-15366-2.00001-0

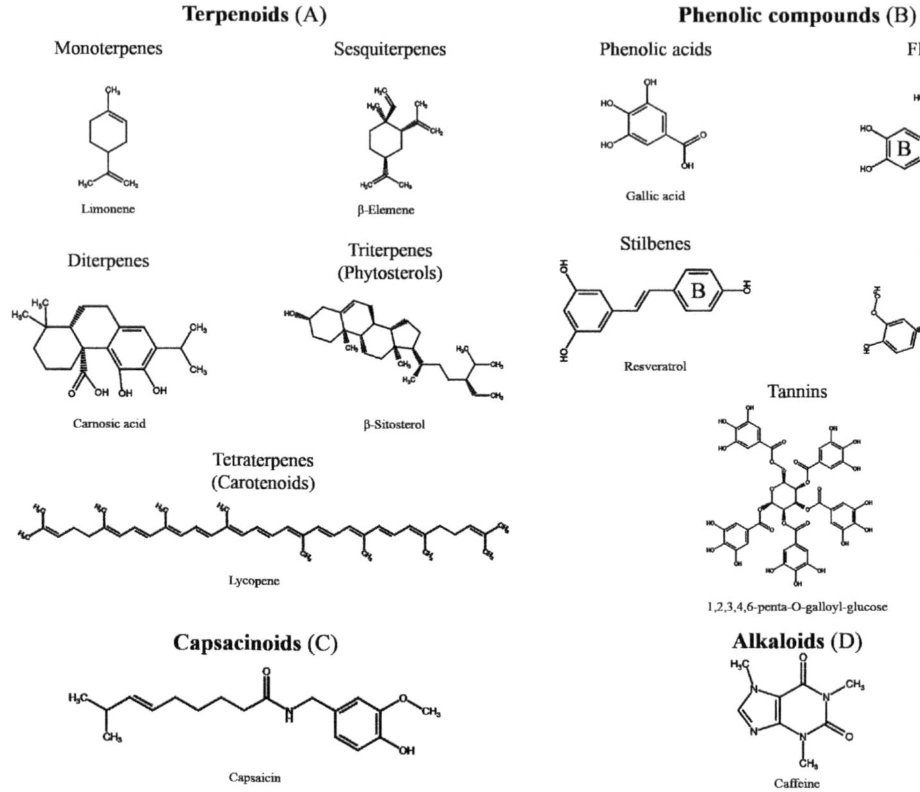

FIGURE 1.1

Examples of some of the main phytochemical compounds, divided by families and classes.

made of 17 carbon atoms and a carbon side chain. This makes them highly lipophilic molecules and integral part of the cell membranes of plants, contributing to the defense from plant stress. In humans, these are able to lower cholesterol levels by competing with it for absorption in the intestines. Phytosterols also show antiinflammatory and anticancer activities, with their biological activity increased by the presence of hydroxyl groups (Bouic, 2002; Du et al., 2022).

Among tetraterpenes, carotenoids, such as lutein and zeaxanthin, are the most important class, due to their role in plant photosynthesis and photo-protection and for their well-known antioxidant activity. This last property is strictly related to the conjugated double bonds, reaching its maximum in lycopene, which possesses 11 conjugated double bonds (Krinsky and Johnson, 2005).

Phenolic compounds (Fig. 1.1B) can be distinguished into five major classes, namely phenolic acids, flavonoids, stilbenes, lignans, and tannins, which in plants cover different roles, such as pigmentation and defense against ultraviolet radiation or pathogens. In general, number and position of hydroxyl groups strongly affect their antioxidant activity making polyhydroxy flavonoids powerful radical scavenging molecules; moreover, flavonoids and stilbenes show antiinflammatory and anticancer activities, with antiinflammatory effects strictly related to the presence of hydroxyl groups on the B-ring (Francenia Santos-Sánchez et al., 2019; Quideau et al., 2011; Rice-Evans et al., 1997).

Regarding capsacinoids (Fig. 1.1C), these compounds possess anticancer, antioxidant, and neuroprotective effects, which might be related to the presence of hydroxyl groups on the aromatic ring and to a fatty acid chain of 8/9 carbon atoms; the amide bond is essential for the interaction with receptors (Reyes-Escogido et al., 2011).

Finally, alkaloids (Fig. 1.1D) do present aromatic and heterocyclic structures, which allow them to form π-π stacking interactions with biological targets, giving them a wide range of biological activities, including analgesic, stimulant, and anticancer effects (Dewick, 2001).

References

Abdallah, I.I., Quax, W.J., 2017. A glimpse into the biosynthesis of terpenoids. KnE Life Sciences 3 (5), 81. https://doi.org/10.18502/kls.v3i5.981.

Bouic, P.J.D., 2002. Sterols and sterolins: new drugs for the immune system? Drug Discov. Today 7 (14), 775–778. https://doi.org/10.1016/S1359-6446(02)02343-7.

Brahmkshatriya, P.P., Brahmkshatriya, P.S., 2013. Terpenes: Chemistry, Biological Role, and Therapeutic Applications Natural Products: Phytochemistry, Botany and Metabolism of Alkaloids, Phenolics and Terpenes. Springer Berlin Heidelberg, India, pp. 2665–2691. https://doi.org/10.1007/978-3-642-22144-6_120.

Dewick, P.M., 2001. Medicinal Natural Products: A Biosynthetic Approach. Wiley. https://doi.org/10.1002/0470846275.

Du, Y., Fu, X., Chu, Y., Wu, P., Liu, Y., Ma, L., Tian, H., Zhu, B., 2022. Biosynthesis and the roles of plant sterols in development and stress responses. Int. J. Mol. Sci. 23 (4), 2332. https://doi.org/10.3390/ijms23042332.

Francenia Santos-Sánchez, N., Salas-Coronado, R., Hernández-Carlos, B., Villanueva-Cañongo, C., 2019. Shikimic Acid Pathway in Biosynthesis of Phenolic Compounds. IntechOpen. https://doi.org/10.5772/intechopen.83815.

Krinsky, N.I., Johnson, E.J., 2005. Carotenoid actions and their relation to health and disease. Mol. Aspect. Med. 26 (6), 459–516. https://doi.org/10.1016/j.mam.2005.10.001.

Quideau, S., Deffieux, D., Douat-Casassus, C., Pouységu, L., 2011. Plant polyphenols: chemical properties, biological activities, and synthesis. Angew. Chem. Int. Ed. 50 (3), 586–621. https://doi.org/10.1002/anie.201000044.

Reyes-Escogido, M.de L., Gonzalez-Mondragon, E.G., Vazquez-Tzompantzi, E., 2011. Chemical and pharmacological aspects of capsaicin. Molecules 16 (2), 1253–1270. https://doi.org/10.3390/molecules16021253.

Rice-Evans, C., Miller, N., Paganga, G., 1997. Antioxidant properties of phenolic compounds. Trends Plant Sci. 2 (4), 152–159. https://doi.org/10.1016/s1360-1385(97)01018-2.

Mendoza, N., Escamilla Silva, E.M., 2018. Introduction to Phytochemicals: Secondary Metabolites from Plants with Active Principles for Pharmacological Importance. IntechOpen. https://doi.org/10.5772/intechopen.78226.

CHAPTER

Role of active ingredients in human health

2

Sara Spinelli, Alessia Remigante, Angela Marino and Rossana Morabito
Department of Chemical, Biological, Pharmaceutical and Environmental Sciences, University of Messina, Messina, Italy

1. Introduction

Phytochemicals are plant based bioactive molecules commonly found in fruits, vegetables, seeds, and flours. More than 5000 dietary phytochemicals have been identified, but a large percentage of them remain unknown (R. H. Liu, 2013). Phytochemicals include a wide range of nonnutritive compounds synthesized by the cellular secondary metabolism of plants, with defensive, attractive, and protective functions (Kumar et al., 2023). In addition, they give color and flavor to plants and can play a key role in maintaining a good state of health in humans (Probst et al., 2017; Y. J. Zhang et al., 2015). The beneficial effects of many phytochemicals are achieved by their metabolites. Indeed, phytochemicals have low bioavailability due to their molecular structure, stability, and solubility (Aqil et al., 2013). In this context, colon microbiota plays a central role in mediating the intake of phytochemicals, converting them into easily assimilable compounds from colon epithelial cells (Arbo et al., 2020). Moreover, the close connection with the colon microbiota could produce interindividual differences in the effectiveness of phytochemicals (Ashcroft and Rorsman, 2012). The main catabolites generated by the colon microbiota are aromatic and phenolic acids up to three aromatic hydroxyl groups, or their mono- or dimethoxy analogs, which also contain a side chain of one to five carbon atoms with or without an aliphatic hydroxyl group (Aune, 2019). These free or conjugated catabolites are poured into the bloodstream and exert their protective actions at various levels. Indeed, several studies have shown that the intake of phytochemical containing foods is positively correlated with a lower occurrence of chronic diseases and mortality (Amer et al., 2017; Babu and Jayaraman, 2020; Bard et al., 2015). Among these, alkaloids, organosulfur compounds, phytosterols, carotenoids, and phenolic compounds represent the phytochemical classes that best support human health. Some examples of these phytochemicals are shown in Fig. 2.1.

Phytochemicals for Health. https://doi.org/10.1016/B978-0-443-15366-2.00002-2

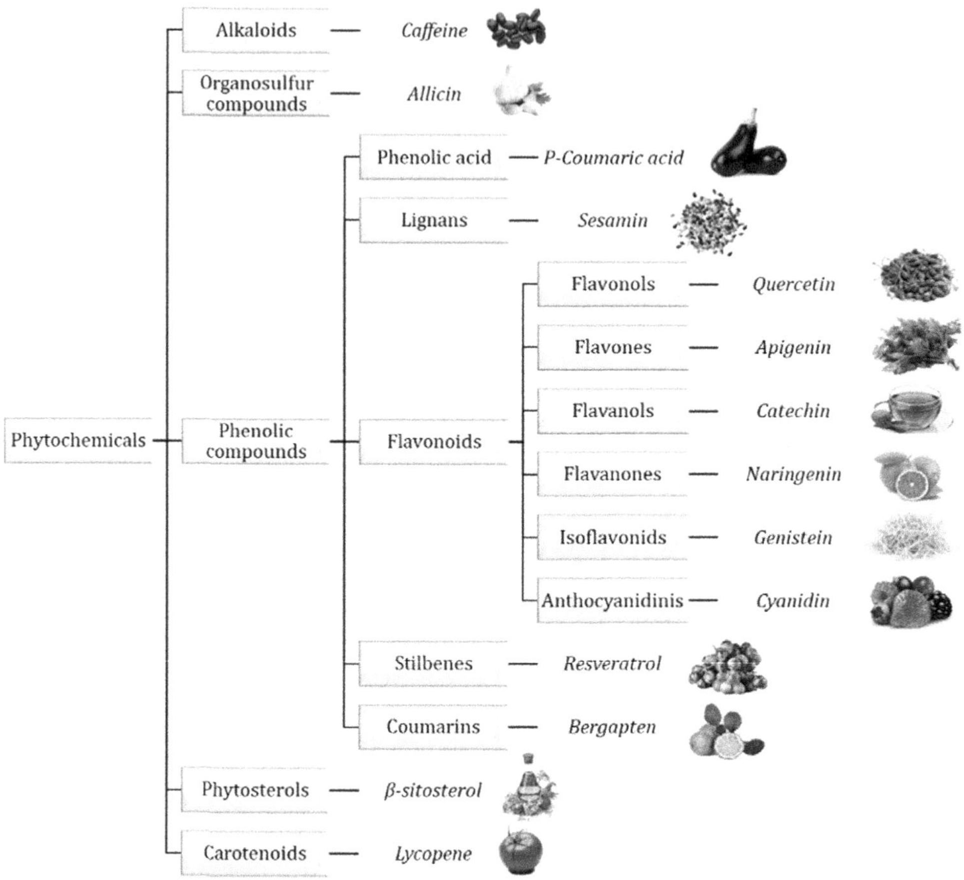

FIGURE 2.1

Classification of phytochemicals in the diet.

2. Classification

Phytochemicals classification is based on their chemical composition, Fig. 2.2, which is in close connection with their biological actions (Baskar et al., 2012) (Table 2.1).

2.1 Alkaloids

Alkaloids are nitrogen-containing compounds that possess different biological effects due to the different arrangement of atoms within their chemical structures. Alkaloids are produced by plants, bacteria, fungi, or animals and are synthesized from tyrosine, tryptophan, ornithine, histidine, or lysine, which form heterocyclic and nonheterocyclic compounds based on the position of the nitrogen atom (Lichman, 2021). In plants, alkaloids act as a nitrogen reserve useful to cope with the conditions

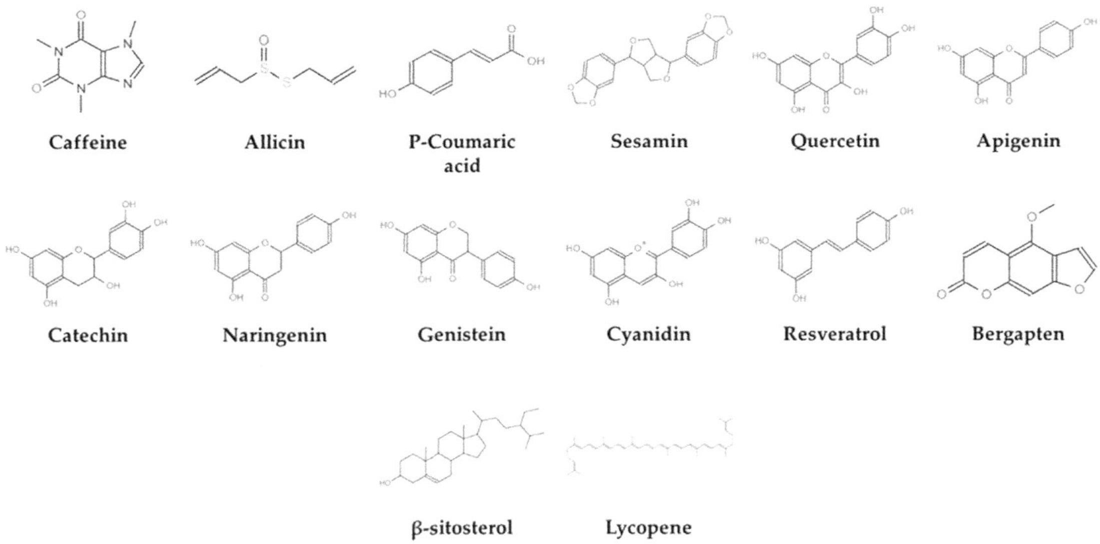

FIGURE 2.2

Chemical structures of the most important phytochemical molecules.

of environmental stress (Zuo et al., 2022). In human health, these compounds are effectively used due to their anticancer (Vincristine [Shirazi-Tehrani et al., 2022]; Vinblastine [Hoang et al., 2023]), antirheumatic (Colchicine [Dalbeth et al., 2014]), antiviral (Camptothecin [Yin et al., 2022]), anti-inflammatory (Tetrandrine [L. N. Gao et al., 2016]), antihypertensive and antipsychotic (Reserpine [Srivastava et al., 2008]), analgesic (Morphine [Klimas and Mikus, 2014; Nisar et al., 2015], Codeine [Derry et al., 2014], Caffeine [Pereira et al., 2021]) properties (Bhambhani et al., 2021).

2.2 Organosulfur compounds

Organosulfur compounds are widely spread in foods, such as vegetables (especially garlic, onion, and leek), cooked meat and fish (Sahu, 2002). Sulfur containing compounds comprise several classes of molecules with the same basic chemical structure (Miekus et al., 2020). They are responsible for pungent smells and flavors and play a pivotal role in protecting plants against different stresses, especially caused by pathogens and insects. With regard to their effects on human health, these compounds lead to several benefits, especially in terms of cancer prevention and treatment (alliin and allicin), liver detoxification (diallyl sulfide), and cardioprotection (allitridin).

2.3 Phytosterols

Phytosterols are fat-soluble compounds present in plant cells useful for maintaining membrane structure and stability (Jie et al., 2022). They are characterized by a tetracyclic structure comparable with cholesterol, which plays an identical structural role. The dietary content of phytosterols is higher in oleaginous fruits and seeds and oils obtained from them (Jiménez-Escrig et al., 2006), mainly

Table 2.1 Structures, properties, and action mechanisms of the main phytochemical substances.

Phytochemicals	Type	Subtype	Substance	Properties	Mechanism of action	References
Alkaloids	Xanthine alkaloid	Methylxanthine	Caffeine	Antioxidant	Mobilization of intracellular calcium	Amer et al. (2017), Herman and Herman (2012), Hosny et al. (2019), Karuppagounder et al. (2021), Reggiani (2021), and Ribeiro and Sebastio (2010)
				Neuroprotective	Competitive antagonism with adenosine receptors	
				Antiinflammatory	Inhibition of phosphodiesterases	
Organosulfur compounds	Oil soluble	Sulfoxide	Allicin	Antimicrobial	Breakage of bacterial DNA	X. Feng et al. (2017), Reiter et al. (2020), Sánchez-Gloria et al. (2021), Tao et al. (2023), and Toma et al. (2019)
				Immunomodulatory	Stimulation of B lymphocytes secretion	
				Antiinflammatory	Activation of Nrf2/Keap1 pathway	
Phenolic compounds	Phenolic acid	Not available	P-coumaric acid	Antioxidant	Scavenger of free radicals; increased activity of GPx, SOD and CAT	Bal et al. (2022), Daroi et al. (2022), Hakyemez et al. (2022), and X. Huang et al. (2020)
				Antiinflammatory	Reduction of IL-1β and NF-kB expression	
				Anticancer	Tumor cell cycle arrest (G2/M phase)	
	Lignans	Not available	Sesamin	Antioxidant	Activation of Nrf2/HO-1/NQO1 pathway;	Anju et al. (2021), Jeng et al. (2005), Kuo et al. (2020), Y. Wang et al. (2021) and Wen et al. (2022)
				Antiinflammatory	Reduction of IL-6 and TNFα production; inhibition of TRIM44 and NF-kB signaling pathways	

				Anticancer	Activation of p53/PTEN-mediated apoptosis	
				Antimicrobial	Inhibition of quorum-sensing and biofilm formation	
	Flavonoids	Flavonols	Quercetin	Antioxidant	Free radical scavenger; increase of endogenous antioxidant defenses	Abd El Mohsen et al. (2002), T. Chen et al. (2020), Di Petrillo et al. (2022), Qi et al. (2022), Remigante, Spinelli, Basile, et al. (2022), Reyes-Farias and Carrasco-Pozo (2019), Song et al. (2020), and Yang Wang et al. (2023)
				Antiviral	Inhibition of virus entry and replication	
				Antiinflammatory	Inhibition of TNF-α, NF-kB and AP-1 signaling pathways;	
				Antiaging	Protection against glycation and membrane oxidation	
				Anticancer	Modulation of PI3K/Akt/mTOR, Wnt/-catenin, and MAPK/ERK1/2 signaling pathways	
		Flavones	Apigenin	Anticancer	Inhibition of NF-kB and CK2 kinase signaling pathways; increased activity of caspase-3 and caspase-8	X. Feng et al. (2017), Imran et al. (2020), Sang et al. (2017), M. Wang et al. (2019)
				Antimicrobial	Inhibition of nucleic acid processing enzyme and c-Jun N-terminal kinase signaling pathway; alteration of the cell membrane potential	
				Antiaging	Increase of enzymatic endogenous	

Continued

Table 2.1 Structures, properties, and action mechanisms of the main phytochemical substances.—cont'd

Phytochemicals	Type	Subtype	Substance	Properties	Mechanism of action	References
					defenses; activation of the Nrf2 pathway.	
		Flavanols	Catechin	Antioxidant	Free radical scavenger; metal ion chelation; increase of enzymatic endogenous defenses; inhibition of prooxidant enzymes	Abd El Mohsen et al. (2002), Reygaert (2018), and Tarasova and Poroikov (2018)
				Antimicrobial	Bacterial cell membrane disruption; inhibition of bacterial DNA gyrase	
				Anticancer	Inhibition of metalloproteinases; DNA breakage; induction of apoptotic cell death and cell cycle arrest	
		Flavanones	Naringenin	Anticancer	Induction of cell cycle arrest and apoptotic cell death; metastasis regression; increased expression of IL-2 and INF-ɣ	Motallebi et al. (2022), Tutunchi et al. (2020), and Zaidun et al. (2018)
				Antiinflammatory	Reduction of TNF-α, IL-6 and IL-1β production; inhibition of NF-kB and MAPK signaling pathways	

				Antiviral	Inhibition of viral replication	
				Antidiabetic	Reduction of glucose absorption; increased glucose utilization by muscle and fat tissues	
				Antioxidant	Increase of enzymatic and nonenzymatic endogenous defenses; reduction of MDA levels	
		Isoflavonoids	Genistein	Anticancer	Cell cycle arrest (G2/M phases); induction of apoptotic cell death; antimetastatic and antiproliferative properties	Bhat et al. (2021), Mas-Bargues et al. (2022), and Rahman Mazumder and Hongsprabhas (2016)
				Antiaging	Free radical scavenger; increase of endogenous defenses; reduction of mitochondrial membrane potential; reduction of MDA and AGEs levels.	
		Anthocyanidins	Cyanidin	Anticancer	Reduction of cyclin-dependent kinase (CDK) signaling; induction of apoptotic cell death and chromatin condensation	Gopu and Shetty (2016), Safdar et al. (2023), and Y. Wang et al. (2016)
				Antidiabetic	Increased translocation of GLUT4; increased activity of PPARγ and secretion of	

Continued

Table 2.1 Structures, properties, and action mechanisms of the main phytochemical substances.—cont'd

Phytochemicals	Type	Subtype	Substance	Properties	Mechanism of action	References
					adiponectin and leptin	
				Antibacterial	Inhibition of biofilm formation and exopolysaccharide production	
				Antioxidant	Maintenance of mitochondrial membrane potential; free radical scavenger	
	Stilbenes	Phytoalexin	Resveratrol	Antiinflammatory	Increased activity of Sirt1, reduction of TNF-α, IL-1β, IL-6, IFN-γ, metalloproteases 1–3 and Cox-2 expression	D.-D. Huang et al. (2020), Huo et al. (2022), Malaguarnera (2019), and Meng et al. (2021)
				Antiaging	Increased activity of Nrf2, Sirt1 and antioxidant enzyme; maintenance of mitochondrial function; reduction of ROS and MDA levels	
				Anticancer	Induction of apoptotic cell death; modulation of PPAR, PGC1α, NF-kB, Nrf1-2, and p53 activity; reduction of several oncogenic miRNAs levels	
				Antidiabetic	Reduction of blood glucose levels; improvement of	

					insulin sensitivity and preservation of pancreatic β-cells	
				Antioxidant	Free radical scavenger; maintenance of mitochondrial function	
				Antiviral	Inhibition of virus entry and replication	
	Coumarins	Furocoumarin	Bergapten	Antioxidant	Free radical scavenger	S. Gao et al. (2022), Quetglas-Llabrés et al. (2022), and Zhou et al. (2017)
				Antiinflammatory	Inhibition of TNF-α, IL-6 and NF-κB signaling pathways	
				Antihemolytic	Maintenance of human erythrocyte membrane integrity	
				Anticancer	Induction of apoptotic cell death and cell cycle arrest	
Phytosterols	Not available	Not available	β-sitosterol	Antiviral	Inhibition of virus entry and replication	Babu and Jayaraman (2020), Baskar et al. (2012), C. Chen et al. (2022), Rajavel et al. (2018), and P. Zhang et al. (2023)
				Anticancer	Induction of apoptotic cell death; increased ROS production; DNA breakage, increased expression of p53	
				Antidiabetic	Inhibition of HbA1c formation; restoration of antioxidant enzymes	
				Antioxidant	Increase of enzymatic and nonenzymatic endogenous defenses; reduction of MDA levels	

Continued

Table 2.1 Structures, properties, and action mechanisms of the main phytochemical substances.—cont'd

Phytochemicals	Type	Subtype	Substance	Properties	Mechanism of action	References
Carotenoids	Tetraterpene	Not available	Lycopene	Anticancer	Induction of apoptotic cell death and cell cycle arrest; antimetastatic and antiangiogenic properties	Kapała et al. (2022), Leh and Lee (2022), and Mirahmadi et al. (2020)
				Antidiabetic and antiobesity	Modulation of AGE/RAGE, JNK/MAPK, PI3K/Akt, SIRT1/FoxO1/PPARγ signaling pathways and AchE activity	
				Antioxidant	Free radical scavenger; reduction of MDA levels, increase of enzymatic and nonenzymatic endogenous	
				Antiaging	Increased activity of Nrf2/HO-1 pathway; inhibition of apoptotic cell death; increase of antioxidant capacity	
				Antiinflammatory	Inhibition of TNF-α, NF-κB, JNK, IL-6, IL-1β, Cox2, and iNOS signaling pathways	

AchE, *Acetylcholinesterase;* AGEs, *Advanced Glycation end products;* AKT, *Protein Kinase B;* AP1, *Activating protein 1;* CAT, *Catalase;* c-Jun, *Activator protein 1;* CK2, *Casein kinase 2;* COX-2, *cycloxygenase inhibitor-2;* ERK1/2, *Extracellular signal-regulated Kinase 1/2;* FOXO1, *Forkhead box O1;* GPx, *Glutathione peroxidase;* HbA1c, *Glycated Hemoglobin;* HO-1, *Heme oxygenase;* IL-1β, *Interleukin-1β;* IL-2, *Interleukin-2;* IL-6, *Interleukin-6;* INF-ɣ, *Interferon-ɣ;* iNOS, *inducible nitric oxide synthase.;* JNK, *Jun N-terminal kinase;* MAPK, *Mitogen-activated protein kinase;* MAPK, *Mitogen-activated protein Kinase;* MDA, *Malondialdehyde;* mTOR, *mechanistic target of rapamycin;* NF-kB, *nuclear factor kappa-light-chain-enhancer of activated B cells;* NQO1, *NAD(P)H quinone oxido reductase;* Nrf2/Keap1, *Kelch-like ECH-associated protein 1-nuclear factor (erythroid-derived 2)-like 2;* PGC1-α, *Peroxisone proliferator activated receptor ɣ coactivator 1-α;* PI3K, *Phosphatidylinositol 3 Kinase;* PPAR-ɣ, *Peroxisome proliferator activated receptor ɣ;* PTEN, *Phosphatase and tensin homolog;* RAGE, *Receptor for advanced glycation end products;* SIRT-1, *NAD-dependent deacetylase sirtuin-1;* SOD, *Superoxide dismutase;* TNF-α, *Tumor necrosis factor-α;* TRIM44, *Tripartite motif containing 44;* Wnt, *Wingless-related integration site.*

rapeseed oil, wheat germ oil, and maize oil. Phytosterols are considered functional foods capable of lowering cholesterol levels (Bard et al., 2015). Phytosterols also have antiinflammatory and antioxidant effects (S. Feng et al., 2022) and have potential implications for the treatment of cancer and diabetes (Jones and Abumweis, 2009).

2.4 Carotenoids

Carotenoids are natural hydrophobic pigments composed of eight isoprenoid units that form acyclic or heterocyclic compounds. Carotenoids are widely distributed in plants, algae, fungi, birds, fish, crustaceans, and insects (Langi et al., 2018) The presence of double bonds in carotenoids structure is responsible for their color, ranging from yellow to red: fruits, yellow/orange vegetables, tomatoes, and egg yolk are the main carotenoids sources for humans (Cervantes-Paz et al., 2016). Carotenoids participate in several biological processes in plants, such as photosynthesis, photomorphogenesis, photoprotection, and development, but they even have a significant role in maintaining human health. In fact, they have antiinflammatory and antioxidant effects, thus counteracting aging, cancer, cardiovascular diseases, and/or arthritis (Soares et al., 2015).

2.5 Phenolic compounds

Polyphenols are found in plant-based foods and beverages, notably apples, berries, citrus fruit, plums, broccoli, cocoa, tea and coffee, and many others. There is substantial epidemiological evidence that a diet including high quantities of polyphenol rich fruit, vegetables, cocoa, and beverages protects against developing cardiovascular disease and type 2 diabetes. Considering the numerous beneficial effects reported so far and the growing interest of the scientific community, this chapter will shed light on phenolic compounds, or polyphenols, in order to highlight the potential mechanisms of action in the prevention and treatment of few examples of chronic diseases.

3. Phytochemicals combination: Synergistic or antagonistic effects

An interesting issue concerning the benefits of phytochemicals in human health is their combined administration. Combinations of several phytochemicals can cause a change in both final biological effects and bioavailability of each component. These combinations can improve or reduce the benefits conferred by individual bioactive compounds, as well as may induce facilitation/competition for cellular absorption and transport. This applies not only to the combination of pure phytochemicals but also to the association of different fruits, vegetables, and legumes, which can result in a synergistic, additive, or antagonist effect (S. Wang et al., 2011). Synergistic interactions between antioxidants can be explained by the theory of antioxidant regeneration: one antioxidant protects another antioxidant from oxidative degradation or isomerization by its own oxidation. An example is the synergy between β-carotene and lycopene: lycopene protects β-carotene from free radical-induced isomerization by its own oxidation. The ability of lycopene to protect β-carotene derives from its increased antioxidant scavenging activities (X.-D. Wang, 2012). In addition, it has been reported that β-carotene improves the absorption of lycopene (Phan et al., 2018).

Nevertheless, as said, the combination of several phytochemicals could have an antagonistic effect on the antioxidant capacity. For instance, the association between anthocyanins, such as pelargonidin-

3-glucoside, and flavonoid compounds (catechin, epicatechin, kaempferol, quercetin, quercetin-3-glucoside, and myricetin) seems to indicate activity a much weaker scavenger activity than individual compounds. Structural differences between these phytochemicals may be responsible for this effect. Indeed, the combination of structurally different substances could favor the interaction between hydrogen bonds, decreasing the availability of hydroxyl groups and their possibility of interaction with ROS. These findings suggest that phytochemicals interaction may elicit synergistic or antagonistic effects, which could explain the lower antioxidant yield of whole food extracts than the single compound. Thus, it is noticeable the importance of choosing the best combination of phytochemicals when new food supplements or fortified foods are designed.

4. Role of polyphenols in chronic diseases

4.1 Cancer

The onset of cancer is mainly related to the uncontrollable growth of tumor cells and has raised an increasing research interest in the latter years to investigate both cell mechanisms underlying tumor development and possible beneficial effects of drugs counteracting cancer cells proliferation (Y. J. Zhang et al., 2015). In this latter regard, plant polyphenols have been recognized as an effective resource for cancer therapy, as attested by several scientific evidence showing their potential anticancer properties for different types of cancer, i.e., breast, liver, stomach, colorectal, esophagus, lung, prostate, and uterine cervix (Reyes-Farias and Carrasco-Pozo, 2019). In particular, signaling pathways may induce tumor inhibitory and antiproliferative effects, when modulated by plant-derived compounds such as curcumin, 3-gallate epigallocatechin, resveratrol, and quercetin.

With regard to signaling, p53 deserves to be mentioned as the most important human tumor suppressor controlling cell cycle and apoptosis, which undergoes posttranslation modifications, including phosphorylation, acetylation, ubiquitination, and methylation, affecting its stability, localization, and function (Y. J. Zhang et al., 2015). The activity of p53 is reportedly regulated by flavanols as well as by quercetin by upregulating p53 mRNA, protein levels, and increasing caspase 3/7 activity in mesothelioma. In addition, curcumin has been recently described to promote hyperphosphorylation in Ser15 and, in turn, the expression of proapoptotic Bax genes in neuroblastoma cells (Sidhar and Giri, 2017).

A study performed by P. Zhang and Zhang, (2018) revealed that a combined treatment of curcumin and quercetin augments phosphorylation and acetylation levels in lung carcinogenesis by downregulating Bcl2 and upregulating p21 and BAX, leading thus to apoptosis. The anticancer effect of polyphenols may rely on their apparent affinity to cancer cells inducing regulatory effects, but not to normal cells, as demonstrated by Niedzwiecki and coauthors (2016). In this specific study, quercetin has been shown to provoke cytotoxic effects in breast cancer cells and leukemia cells, while not in normal cells, with quercetin directly interacting with DNA, thus increasing p53 and p-p53 levels, finally inducing apoptosis and stopping cell cycle in the S phase. The same authors also stated that quercetin reduced tumors, improved lifespan, with no detrimental effect in mice. More recently, the antioxidant capacity of red grapes has been investigated in human hepatocellular liver carcinoma (HepG2) cells, showing a positive correlation between total phenolic content and oxygen radical absorbance capacity (ORAC) values of the grape extracts (Cháirez-Ramírez et al., 2021). In this light, an increased fruit consumption may represent a good tool to counteract oxidative stress, hopefully reducing cancer risk, whose onset and development may be counteracted at level of PI3K/Akt/mTOR pathway, as reported by Z. Liu et al. (2020). The combined treatment achieved by mixing two polyphenols has been often demonstrated to elicit beneficial effects. This is the case of the isoflavone genistein and the polyphenol thearubigin, derived from black tea, whose combined in vitro application effectively inhibited prostate cancer cell growth. Moreover, thearubigin has been proven not to inhibit growth in the human prostate tumor cell line PC-3 when used alone, while, when combined with genistein at a 1:40 ratio, synergistically inhibited cell proliferation and induced a G2/M phase cycle arrest in a dose-dependent manner (Sakamoto, 2000). With regard to combined treatments, the ability of green tea polyphenols (GTPs) to synergize with chemotherapeutic agents to counteract lung cancer has raised much interest. In particular, [3H](−)-epigallocatechin gallate (EGCG) incorporation into human lung cancer cell line PC-9 was significantly enhanced by (−)-epicatechin, an inert tea polyphenol without a galloyl moiety. (−)-Epicatechin enhanced apoptosis, growth inhibition of PC-9 cells, and inhibition of tumor necrosis factor-α release from BALB/c-3T3 cells by EGCG and other tea polyphenols with a galloyl moiety in a dose-dependent manner. The effects of EGCG on apoptosis induction were also synergistically enhanced by other cancer-preventive agents, such as sulindac and tamoxifen (Suganuma et al., 1999). Therefore, on the basis of this and other evidences, vegetables, tomato sauce, fish, and vegetable fat introduced with diet are associated with a lower risk of PC progression, which may explain the strongly increasing scientific interest in the potential chemo-preventive role in PC of various phytochemicals present in food. The mechanisms underlying their anticancer activity have been shown to include inhibition of proliferation, induction of apoptosis, and arrest of the cell cycle. In particular, polyphenols such as lycopene, EGCG, and CUR exert their effect by downregulating different signal transduction pathways. With regard to green tea polyphenols containing EGCG, their anticancer properties have been shown by a recent evidence of epigenetic mechanisms as mainly responsible for the regulation of matrix metalloproteinases (MMP) and their tissue inhibitor (TIMP) associated with the progression of PC (Cabral-Pacheco et al., 2020).

Though many investigations provided several evidences for the anticancer effects of polyphenols, their introduction through the diet still remains a controversial issue, as in the case of breast cancer prevention. In this latter regard, high concentrations of polyphenols are able to block the proliferation of ERα (Estrogenic receptor alpha) and ERα^+ breast cancer cells, while lower concentrations can even stimulate the growth of ERα^+ cells. Specifically, carnosol inhibitory effect on breast cancer cells has been in vivo proven by observing a reduction of BC cell lines (MDA-MB-231) proliferation. Moreover, a significant inhibition of invasion and metastasis has been detected both in vitro and in vivo. The mechanism through which carnosol has been shown to act is the downregulation of MMP-9 activity and expression and the inhibition of the STAT3 signaling pathway through the ROS-dependent proteasome degradation of the STAT3 protein. With regard to skin tumors, their initiation can be attributed to various factors, with the excessive exposure to UV radiation as the main risk factor. The treatment of melanoma remains a challenging issue, due to its aggressive metastatic ability and resistance to current therapeutic approaches. Though no clinical studies are yet available assessing the efficacy of the most active polyphenols in skin cancer (SC), epidemiological evidence indicates that the poorly soluble polyphenols, such as CUR, RES, EGCG, coumarin, and quercetin, exert anticancer effects. Poorly soluble polyphenols, such as CUR, RES, quercetin, coumarin, and EGCG, may have significant benefits in the treatment of melanoma thanks to their antioxidant, antiinflammatory, antiproliferative, and chemoprotective properties. The combination of the carotenoid fucoxanthin and the polyphenol rosmarinic acid (RA) may function as a natural promising tool to prevent UVB-induced skin alterations as photoaging, skin inflammation, and its derivation to precancerous lesions and skin carcinomas (Rodríguez-Luna et al., 2019).

4.2 Cardiovascular diseases

Cardiovascular disease, the leading cause of death worldwide, is a chronic inflammatory condition contributing to atherosclerotic lesions with pathological consequences such as angina, myocardial infarction, or cerebrovascular attack. A major risk 90% of cardiovascular disease is currently associated with a poor diet quality, physical inactivity, and smoking (Tangney and Rasmussen, 2013). In this context, the beneficial effects of polyphenols, rising much interest of the researchers, have been based on epidemiological studies related to different dietary patterns and specific foods consumption. Nonetheless, the mechanisms underlying this correlation still remain not fully understood. However, one of the clear evidence is that polyphenols exert their action at molecular level, improving endothelial function and blocking platelet aggregation, which may explain their antithrombotic and antiaggregative effect. Based on this evidence, such compounds have an effective role in both cardiovascular disease prevention and treatment, corroborated by the association between polyphenols intake and a reduced risk of cardiovascular events, independently of other dietary and nondietary risk factors.

A further feature focused by researchers in this context is endothelial dysfunction and oxidative modification of low-density lipoprotein, which represent key factors in atherogenesis. Endothelial cells contribute to maintain the integrity of the vascular wall, mediate diffusion and active transport of numerous substances, and importantly, represent a nonthrombotic and nonadherent surface for platelets and leukocytes, which regulate vascular tone by producing nitric oxide (NO), prostaglandins, endothelins, and growth factors (Rajavel et al., 2018). An inverse relationship between grape products consumption and cardiovascular disease has been assessed, since clinical studies performed with grapes and grape juice have provided evidence for an improvement in endothelial function, a reduction in platelet aggregation, a positive influence on biomarkers such as LDL and HDL (high-density lipoprotein), and a reduction of blood pressure. Furthermore, the cardioprotective effect of grapes is achieved by prevention of LDL oxidation, improving antioxidant defenses and attenuating inflammatory responses (Iriti and Faoro, 2009). Among polyphenols, quercetin elicits favorable effects in hypertension conditions, as assessed by Serban et al. (2016) reviewing seven randomized controlled trials and pointing out the beneficial action of quercetin flavanol on hypertension. Apples, onions, and black tea are commonly considered the major sources of quercetin containing bioactive components similar to those of red grapes employed for red wine manufacturing as well as ingestion of black tea. Beneficial short- and long-term effects have been observed in coronary artery disease, attesting an improvement of endothelial vasomotor dysfunction. Moreover, quercetin has been shown to stimulate endothelial NO synthase (eNOS) phosphorylation and to enhance the production of NO bioavailability in endothelial cells, exhibiting thus vascular protective actions (Behl et al., 2020).

4.3 Aging

Single cells as well as the whole organism undergo morphological and biochemical changes with time, a process known as aging. It is a complex and tangled physiological phenomenon characterized by a diminished ability to face stress, because of a decline in the immune function, known as immune-senescence, resulting in a predisposition to infectious diseases and cancer. To explain the mechanisms of aging, many theories have been proposed. One of these is related to the molecular level, with oxidative stress or free radical hypothesis receiving wide support (Fulop et al., 2018). Therefore, ROS production and oxidative DNA damage can represent a favorable environment for age-related diseases development. As aged population is significantly and progressively increasing, with consequent strong impact of age-related disturbances on health and social life, a growing interest in identifying natural compounds for improving health, preventing chronic diseases, and extending longevity has been rising. In this regard, numerous polyphenols, including EGCG, curcumin, and quercetin, have been shown to extend lifespan in various model organisms (Fulop et al., 2018). In an old aluminum-induced neurotoxically rat model, oral gavage of 30 mg/mL/kg curcumin treatment significantly suppressed the activities of superoxide dismutase (SOD), glutathione peroxidase (GPx), glutathione-s-transferase (GST), protein kinase C (PKC), and Na^+/K^+-ATPase, thereby protecting the brain against aluminum toxicity in aging rats (Sharma et al., 2009). The mechanisms through which the antiaging effect of polyphenols may be achieved include antioxidant signaling, prevention of cellular senescence, microRNA targeting, influencing NO bioavailability, and promoting mitochondrial function (Luo et al., 2021).

A validated in vitro model of early aging has been recently performed in human erythrocytes by incubation with high doses of d-Gal (Remigante et al., 2021, 2020; Remigante, Spinelli, Trichilo, et al., 2022). The most obvious effects of aging on erythrocytes are, firstly, the increase in membrane lipid peroxidation and the reduction in the intracellular content of glutathione (GSH), and secondly, the glycation of hemoglobin, which is closely linked to Band 3 protein (B3p) function, an anion exchanger essential to gas exchange and erythrocytes deformability. In particular, different studies reported that aging process may affect the bicarbonate/chloride exchanger B3p in the erythrocyte membrane, which, as said, is involved in crucial physiological functions and, what is more, represents a major site of anchoring of the cytoskeleton to the erythrocyte membrane. Natural aging dependent oxidation of B3p increases the affinity for circulating anti-B3p antibodies, thus inducing modifications in the erythrocyte structure, including decreased deformability and increased sphericity (Perrone et al., 2023; Spinelli et al., 2023). Due to the close relationship between aging and oxidative stress, flavonoids have been considered as an excellent candidate to counteract age-related changes. Indeed, both the alteration of the B3p function and oxidative damage have been improved by pretreatment with quercetin, which preferentially ameliorated lipid peroxidation rather than protein oxidation. Moreover, quercetin prevented glycated A1c formation, while no protective effect on the endogenous antioxidant system (GSH/GSSG) was observed.

Functional foods and natural products with antioxidant properties can potentially improve the life functions of erythrocytes and, consequently, the homeostasis of the whole body. *Euterpe oleracea*, an Amazonian Brazilian fruit popularly known as Açaì (de Oliveira et al., 2021) presents several bioactive molecules with different properties, including antioxidant, antiinflammatory, and analgesic activities, and also modulates calcium homeostasis. The biological effects of Açaì are related to its chemical matrix, which includes numerous phytochemicals components, such as flavonoids. These molecules can directly neutralize reactive oxygen species (ROS) and/or inactivate molecules with a prooxidant capacity. Aging of circulating erythrocytes is closely related to oxidative stress, and Açaì extract has been shown to have beneficial effects against aging-related processes, such as oxidative damage and Hb glycation in human erythrocytes (Remigante, Spinelli, Straface, et al., 2022).

4.4 Neurodegenerative diseases

Neurodegenerative disorders, causing central nervous system cells death or dysfunction, usually get worse over time and have no cure. They can be genetic or be caused by cancer or a stroke and have aging as the primary risk factor for most of them, including Alzheimer's disease (AD) and Parkinson's disease (PD). The etiology of these diseases is further complicated by a number of proposed causative mechanisms, including oxidative stress, neuroinflammation, protein aggregation, iron toxicity, and mitochondrial dysfunction. Polyphenols are reported to improve many of these factors at a cellular level, which makes their use in complex neurodegenerative disorders compelling (Renaud and Martinoli, 2019). Although information on transport of polyphenols into the brain is limited compared with the measurement of plasma levels, an increasing number of studies have measured polyphenols and metabolites in the brains of rodents and pigs. Entry into the CNS of the most commonly studied polyphenols has been reported several times, for resveratrol (Vingtdeux et al., 2010), EGCG (Abd El Mohsen et al., 2002) and quercetin (Youdim et al., 2004). Some polyphenols are extensively transformed before they reach the brain, which may dampen their bioavailability. As an example, curcumin is highly lipophilic and, in theory, should easily gain entry to the brain (Yang et al., 2005). Therefore, polyphenols can be used to support neurodegenerative pathologies. Among fruits, grape provides a high content of polyphenols. In this context, different studies showed that grape-derived polyphenolic compounds inhibited amyloid fibril formation and, very recently, a grape seed polyphenolic extract has been demonstrated to significantly prevent Aβ protein oligomerization, by inhibiting the Aβ protein aggregation into high molecular weight oligomeric Aβ species, both in vitro and in Tg2576 mice (Tikhonova et al., 2020). There is also strong evidence for the utilization of natural polyphenols to counteract neurodegenerative disturbances due to their antioxidant and antiaging properties (Luo et al., 2021) and to significantly cause enhanced memory reconsolidation in transgenic mouse model of PD After dietary supplement of grape polyphenol (1.5 mL/kg) (Tikhonova et al., 2020).

4.5 Diabetes

Diabetes mellitus (DM) characterized by hyperglycemia, insulin resistance, pancreatic amyloid deposition, and β-cell dysfunction (Ashcroft and Rorsman, 2012) has a significant impact on quality of life worldwide with a high mortality for adults. Two consequences of diabetes critical for patients are oxidative stress and inflammation due to their detrimental impact on cell and tissue homeostasis and their role in the evolution of glucose metabolism pathology. Oxidative damage is exerted through different cellular mechanisms such as lipoperoxidation, protein oxidation, and reduction of enzyme antioxidants activity. Inflammation is mediated by an increase in the levels of circulating inflammatory cytokines as well as insulin resistance in liver, skeletal muscles, and vascular endothelium (Li et al., 2014). Therefore, the development of novel antidiabetic strategies is indeed a major challenge for researchers. Therefore, natural compounds are considered a primary resource for the discovery of potential therapeutics. As a matter of fact, many plants, in which main active constituents are polyphenols, have been shown to act as antidiabetic agents. In this regard, it has been proven that natural compounds, characterized by a high polyphenol content, can control carbohydrate metabolism by protecting and restoring β-cell integrity, enhancing insulin releasing activity, and increasing cellular glucose uptake. Among natural products, blackberries, red grapes, apricots, eggplants, and popular drinks such as coffee, cocoa, and green tea are all rich in polyphenols, which may dampen insulin resistance and be a valid and natural alternative in the treatment of diabetes (Solayman et al., 2016). Moreover, the intake of green tea and polyphenols contained in green tea exert a beneficial effect on insulin resistance and systemic inflammation. In an in vivo rat model of diabetes, induced by streptozotocin (STZ) application, orally ingested rutin decreased fasting plasma glucose levels (Kamalakkannan and Prince, 2006) and increased insulin and C-peptide levels. Both rutin and its aglycone quercetin restored impaired insulin and leptin signaling pathways in fructose-fed rats (Hu et al., 2012). Quercetin has shown similar effects to rutin. Oral quercetin supplementation in alloxan-induced diabetic mice, STZ rats, and rats fed a high cholesterol diet as well as intraperitoneal quercetin administration to STZ-rats lowered plasma glucose levels. Quercetin also increased insulin levels and protected β-cells from STZ-induced degeneration (Vessal et al., 2003). In rats receiving a high-cholesterol diet, quercetin prevented an increase in insulin levels. The myriad health-promoting effects of polyphenols can be attributed to their structural similarities and potential to modulate several cellular pathways common to many diseases, including diabetes. The beneficial effect of polyphenol in diabetes may include ROS scavenging, increased cellular antioxidant capability, diminished inflammation and cellular stress, finally inhibiting caspase activation and apoptosis (Reyes-Farias and Carrasco-Pozo, 2019). Polyphenolic compounds may also regulate insulin signaling, glucose metabolism, lipid levels, and other pathways. Taken together, these mechanisms may underlie the β-cell protective, insulin sensitizing, and hypoglycemic effects of polyphenols.

5. Conclusions and remarks

The current knowledge about biological effects of polyphenols and their positive impact on human health derive from plenty of investigations performed by both in vivo and in vitro approaches, which have explored the functional role of polyphenols at several levels. The activity of polyphenols and, in turn, their beneficial effects depend on the bioavailability due to their molecular structure, stability, and solubility, as well as the combination with other phytochemicals. A very wide body of evidence demonstrates that polyphenols significantly act on the redox balance offering insights into the mechanisms underlying their beneficial effects in cells and whole organisms. However, due to the high biological potential of polyphenols deserving much interest, further investigations are needed to corroborate what already known about their potential role in human health and to add more knowledge about their use to counteract diseases with high impact on human population.

References

Abd El Mohsen, M.M., Kuhnle, G., Rechner, A.R., Schroeter, H., Rose, S., Jenner, P., Rice-Evans, C.A., 2002. Uptake and metabolism of epicatechin and its access to the brain after oral ingestion. Free Radic. Biol. Med. 33 (12), 1693–1702. https://doi.org/10.1016/S0891-5849(02)01137-1.

Amer, M.G., Mazen, N.F., Mohamed, A.M., 2017. Caffeine intake decreases oxidative stress and inflammatory biomarkers in experimental liver diseases induced by thioacetamide: biochemical and histological study. Int. J. Immunopathol. Pharmacol. 30 (1), 13–24. https://doi.org/10.1177/0394632017694898.

Anju, V.T., Busi, S., Ranganathan, S., Rao Ampasala, D., Kumar, S., Suchiang, K., Kumavath, R., Dyavaiah, M., 2021. Sesamin and sesamolin rescues *Caenorhabditis elegans* from *Pseudomonas aeruginosa* infection through the attenuation of quorum sensing regulated virulence factors. Microb. Pathog. 155, 104912. https://doi.org/10.1016/j.micpath.2021.104912.

Aqil, F., Munagala, R., Jeyabalan, J., Vadhanam, M.V., 2013. Bioavailability of phytochemicals and its enhancement by drug delivery systems. Cancer Lett. 334 (1), 133–141. https://doi.org/10.1016/j.canlet.2013.02.032.

Arbo, B.D., André-Miral, C., Nasre-Nasser, R.G., Schimith, L.E., Santos, M.G., Costa-Silva, D., Muccillo-Baisch, A.L., Hort, M.A., 2020. Resveratrol derivatives as potential treatments for Alzheimer's and Parkinson's disease. Front. Aging Neurosci. 12. https://doi.org/10.3389/fnagi.2020.00103.

Ashcroft, F.M., Rorsman, P., 2012. Diabetes mellitus and the β cell: the last ten years. Cell 148 (6), 1160–1171. https://doi.org/10.1016/j.cell.2012.02.010.

Aune, D., 2019. Plant foods, antioxidant biomarkers, and the risk of cardiovascular disease, cancer, and mortality: a review of the evidence. Adv. Nutr. 10, S404. https://doi.org/10.1093/advances/nmz042.

Babu, S., Jayaraman, S., 2020. An update on β-sitosterol: a potential herbal nutraceutical for diabetic management. Biomed. Pharmacother. 131, 110702. https://doi.org/10.1016/j.biopha.2020.110702.

Bal, S.S., Leishangthem, G.D., Sethi, R.S., Singh, A., 2022. P-coumaric acid ameliorates fipronil induced liver injury in mice through attenuation of structural changes, oxidative stress and inflammation. Pestic. Biochem. Physiol. 180. https://doi.org/10.1016/j.pestbp.2021.104997.

Bard, J.M., Paillard, F., Lecerf, J.M., 2015. Effect of phytosterols/stanols on LDL concentration and other surrogate markers of cardiovascular risk. Diabetes Metabol. 41 (1), 69–75. https://doi.org/10.1016/j.diabet.2014.11.006.

Baskar, A.A., Al Numair, K.S., Gabriel Paulraj, M., Alsaif, M.A., Muamar, M.A., Ignacimuthu, S., 2012. β-sitosterol prevents lipid peroxidation and improves antioxidant status and histoarchitecture in rats with 1,2-dimethylhydrazine-induced colon cancer. J. Med. Food 15 (4), 335–343. https://doi.org/10.1089/jmf.2011.1780.

Behl, T., Bungau, S., Kumar, K., Zengin, G., Khan, F., Kumar, A., Kaur, R., Venkatachalam, T., Tit, D.M., Vesa, C.M., Barsan, G., Mosteanu, D.E., 2020. Pleotropic effects of polyphenols in cardiovascular system. Biomed. Pharmacother. 130, 110714. https://doi.org/10.1016/j.biopha.2020.110714.

Bhambhani, S., Kondhare, K.R., Giri, A.P., 2021. Diversity in chemical structures and biological properties of plant alkaloids. Molecules 26 (11). https://doi.org/10.3390/molecules26113374.

Bhat, S.S., Prasad, S.K., Shivamallu, C., Prasad, K.S., Syed, A., Reddy, P., Cull, C.A., Amachawadi, R.G., 2021. Genistein: a potent anti-breast cancer agent. Curr. Issues Mol. Biol. 43 (3), 1502–1517. https://doi.org/10.3390/cimb43030106.

Cabral-Pacheco, G.A., Garza-Veloz, I., Rosa, C.C.D.L., Ramirez-Acuña, J.M., Perez-Romero, B.A., Guerrero-Rodriguez, J.F., Martinez-Avila, N., Martinez-Fierro, M.L., 2020. The roles of matrix metalloproteinases and their inhibitors in human diseases. Int. J. Mol. Sci. 21 (24), 1–53. https://doi.org/10.3390/ijms21249739.

Cervantes-Paz, B., Victoria-Campos, C.I., Ornelas-Paz, J.d.J., 2016. Absorption of carotenoids and mechanisms involved in their health-related properties. Sub Cell. Biochem. 79, 415–454. https://doi.org/10.1007/978-3-319-39126-7_16.

Cháirez-Ramírez, M.H., de la Cruz-López, K.G., García-Carrancá, A., 2021. Polyphenols as antitumor agents targeting key players in cancer-driving signaling pathways. Front. Pharmacol. 12. https://doi.org/10.3389/fphar.2021.710304.

Chen, C., Shen, J.L., Liang, C.S., Sun, Z.C., Jiang, H.F., 2022. First discovery of beta-sitosterol as a novel antiviral agent against white spot syndrome virus. Int. J. Mol. Sci. 23 (18). https://doi.org/10.3390/ijms231810448.

Chen, T., Zhang, X., Zhu, G., Liu, H., Chen, J., Wang, Y., He, X., 2020. Quercetin inhibits TNF-α induced HUVECs apoptosis and inflammation via downregulating NF-kB and AP-1 signaling pathway in vitro. Medicine 99 (38), e22241. https://doi.org/10.1097/MD.0000000000022241.

Dalbeth, N., Lauterio, T.J., Wolfe, H.R., 2014. Mechanism of action of colchicine in the treatment of gout. Clin. Ther. 36 (10), 1465–1479. https://doi.org/10.1016/j.clinthera.2014.07.017.

Daroi, P.A., Dhage, S.N., Juvekar, A.R., 2022. p-Coumaric acid mitigates lipopolysaccharide induced brain damage via alleviating oxidative stress, inflammation and apoptosis. J. Pharm. Pharmacol. 74 (4), 556–564. https://doi.org/10.1093/jpp/rgab077.

de Oliveira, N.K.S., Almeida, M.R.S., Pontes, F.M.M., Barcelos, M.P., Silva, G.M., de Paula da Silva, C.H.T., Cruz, R.A.S., da Silva Hage-Melim, L.I., 2021. Molecular docking, physicochemical properties, pharmacokinetics and toxicity of flavonoids present in Euterpe oleracea Martius. Curr. Comput. Aided Drug Des. 17 (4), 589–617. https://doi.org/10.2174/1573409916666200619122803.

Derry, C.J., Derry, S., Moore, R.A., 2014. Caffeine as an analgesic adjuvant for acute pain in adults. Cochrane Database Syst. Rev. 2017 (10). https://doi.org/10.1002/14651858.CD009281.pub3.

Di Petrillo, A., Orrù, G., Fais, A., Fantini, M.C., 2022. Quercetin and its derivates as antiviral potentials: a comprehensive review. Phytother Res. 36 (1), 266–278. https://doi.org/10.1002/ptr.7309.

Feng, S., Wang, L., Shao, P., Sun, P., Yang, C.S., 2022. A review on chemical and physical modifications of phytosterols and their influence on bioavailability and safety. Crit. Rev. Food Sci. Nutr. 62 (20), 5638–5657. https://doi.org/10.1080/10408398.2021.1888692.

Feng, X., Yu, W., Li, X., Zhou, F., Zhang, W., Shen, Q., Li, J., Zhang, C., Shen, P., 2017. Apigenin, a modulator of PPARγ, attenuates HFD-induced NAFLD by regulating hepatocyte lipid metabolism and oxidative stress via Nrf2 activation. Biochem. Pharmacol. 136, 136–149. https://doi.org/10.1016/j.bcp.2017.04.014.

Fulop, T., Larbi, A., Dupuis, G., Page, A.L., Frost, E.H., Cohen, A.A., Witkowski, J.M., Franceschi, C., 2018. Immunosenescence and inflamm-aging as two sides of the same coin: friends or Foes? Front. Immunol. 8, 1960. https://doi.org/10.3389/fimmu.2017.01960.

Gao, L.N., Feng, Q.S., Zhang, X.F., Wang, Q.S., Cui, Y.L., 2016. Tetrandrine suppresses articular inflammatory response by inhibiting pro-inflammatory factors via NF-κB inactivation. J. Orthop. Res. 34 (9), 1557–1568. https://doi.org/10.1002/jor.23155.

Gao, S., Zou, X., Wang, Z., Shu, X., Cao, X., Xia, S., Shao, P., Bao, X., Yang, H., Xu, Y., Liu, P., 2022. Bergapten attenuates microglia-mediated neuroinflammation and ischemic brain injury by targeting Kv1.3 and Carbonyl reductase 1. Eur. J. Pharmacol. 933, 175242. https://doi.org/10.1016/j.ejphar.2022.175242.

Gopu, V., Shetty, P.H., 2016. Cyanidin inhibits quorum signalling pathway of a food borne opportunistic pathogen. J. Food Sci. Technol. 53 (2), 968–976. https://doi.org/10.1007/s13197-015-2031-9.

Hakyemez, I.N., Cevizci, M.N., Aksoz, E., Yilmaz, K., Uysal, S., Altun, E., 2022. Protective effects of p-coumaric acid against gentamicin-induced nephrotoxicity in rats. Drug Chem. Toxicol. 45 (6), 2825–2832. https://doi.org/10.1080/01480545.2021.1993703.

Herman, A., Herman, A.P., 2012. Caffeine's mechanisms of action and its cosmetic use. Skin Pharmacol. Physiol. 26 (1), 8–14. https://doi.org/10.1159/000343174.

Hoang, N.H., Huyen, N.T., Trang, D.T., Xuan Canh, N., Van Mao, C., Sopjani, M., Vuong, N.B., Xuan, N.T., 2023. Effects of Vinblastine and Vincristine on the function of chronic myeloid leukemic cells through expression of A20 and CYLD. Cell. Mol. Biol. 68 (10), 47–53. https://doi.org/10.14715/cmb/2022.68.10.7.

Hosny, E.N., Sawie, H.G., Elhadidy, M.E., Khadrawy, Y.A., 2019. Evaluation of antioxidant and anti-inflammatory efficacy of caffeine in rat model of neurotoxicity. Nutr. Neurosci. 22 (11), 789–796. https://doi.org/10.1080/1028415X.2018.1446812.

Hu, Q.H., Zhang, X., Pan, Y., Li, Y.C., Kong, L.D., 2012. Allopurinol, quercetin and rutin ameliorate renal NLRP3 inflammasome activation and lipid accumulation in fructose-fed rats. Biochem. Pharmacol. 84 (1), 113–125. https://doi.org/10.1016/j.bcp.2012.03.005.

Huang, D.-D., Shi, G., Jiang, Y., Yao, C., Zhu, C., 2020. A review on the potential of resveratrol in prevention and therapy of diabetes and diabetic complications. Biomed. Pharmacother. 125, 109767. https://doi.org/10.1016/j.biopha.2019.109767.

Huang, X., You, Y., Xi, Y., Ni, B., Chu, X., Zhang, R., You, H., 2020. p-Coumaric acid attenuates IL-1β-induced inflammatory responses and cellular senescence in rat chondrocytes. Inflammation 43 (2), 619–628. https://doi.org/10.1007/s10753-019-01142-7.

Huo, Y., Yang, D., Lai, K., Tu, J., Zhu, Y., Ding, W., Yang, S., 2022. Antioxidant effects of resveratrol in intervertebral disk. J. Invest. Surg. 35 (5), 1135–1144. https://doi.org/10.1080/08941939.2021.1988771.

Imran, M., Aslam Gondal, T., Atif, M., Shahbaz, M., Batool Qaisarani, T., Hanif Mughal, M., Salehi, B., Martorell, M., Sharifi-Rad, J., 2020. Apigenin as an anticancer agent. Phytother. Res. 34 (8), 1812–1828. https://doi.org/10.1002/ptr.6647.

Iriti, M., Faoro, F., 2009. Bioactivity of grape chemicals for human health. Nat. Prod. Commun. 4 (5), 611–634. https://doi.org/10.1177/1934578x0900400502.

Jeng, K.C.G., Hou, R.C.W., Wang, J.C., Ping, L.I., 2005. Sesamin inhibits lipopolysaccharide-induced cytokine production by suppression of p38 mitogen-activated protein kinase and nuclear factor-κB. Immunol. Lett. 97 (1), 101–106. https://doi.org/10.1016/j.imlet.2004.10.004.

Jie, F., Yang, X., Wu, L., Wang, M., Lu, B., 2022. Linking phytosterols and oxyphytosterols from food to brain health: origins, effects, and underlying mechanisms. Crit. Rev. Food Sci. Nutr. 62 (13), 3613–3630. https://doi.org/10.1080/10408398.2020.1867819.

Jiménez-Escrig, A., Santos-Hidalgo, A.B., Saura-Calixto, F., 2006. Common sources and estimated intake of plant sterols in the Spanish diet. J. Agric. Food Chem. 54 (9), 3462–3471. https://doi.org/10.1021/jf053188k.

Jones, P.J.H., Abumweis, S.S., 2009. Phytosterols as functional food ingredients: linkages to cardiovascular disease and cancer. Curr. Opin. Clin. Nutr. Metab. Care 12 (2), 147–151. https://doi.org/10.1097/MCO.0b013e328326770f.

Kamalakkannan, N., Prince, P.S.M., 2006. Antihyperglycaemic and antioxidant effect of rutin, a polyphenolic flavonoid, in streptozotocin-induced diabetic wistar rats. Basic Clin. Pharmacol. Toxicol. 98 (1), 97–103. https://doi.org/10.1111/j.1742-7843.2006.pto_241.x.

Kapała, A., Szlendak, M., Motacka, E., 2022. The anti-cancer activity of lycopene: a systematic review of human and animal studies. Nutrients 14 (23). https://doi.org/10.3390/nu14235152.

Karuppagounder, S.S., Uthaythas, S., Govindarajulu, M., Ramesh, S., Parameshwaran, K., Dhanasekaran, M., 2021. Caffeine, a natural methylxanthine nutraceutical, exerts dopaminergic neuroprotection. Neurochem. Int. 148, 105066. https://doi.org/10.1016/j.neuint.2021.105066.

Klimas, R., Mikus, G., 2014. Morphine-6-glucuronide is responsible for the analgesic effect after morphine administration: a quantitative review of morphine, morphine-6-glucuronide, and morphine-3-glucuronide. Br. J. Anaesth. 113 (6), 935–944. https://doi.org/10.1093/bja/aeu186.

Kumar, A., Nirmal, P., Kumar, M., Jose, A., Tomer, V., Oz, E., Proestos, C., Zeng, M., Elobeid, T., Sneha, K., Oz, F., Phytochemicals, M., 2023. Recent advances in health benefits and extraction method. Molecules 28 (2), 887. https://doi.org/10.3390/molecules28020887.

Kuo, T.N., Lin, C.S., Li, G.D., Kuo, C.Y., Kao, S.H., 2020. Sesamin inhibits cervical cancer cell proliferation by promoting p53/pten-mediated apoptosis. Int. J. Med. Sci. 17 (15), 2292–2298. https://doi.org/10.7150/ijms.48955.

Langi, P., Kiokias, S., Varzakas, T., Proestos, C., 2018. Carotenoids: from plants to food and feed industries. Methods Mol. Biol. 1852, 57–71. https://doi.org/10.1007/978-1-4939-8742-9_3.

Leh, H.E., Lee, L.K., 2022. Lycopene: a potent antioxidant for the amelioration of type II diabetes mellitus. Molecules 27 (7). https://doi.org/10.3390/molecules27072335.

Li, Z., Geng, Y.N., Jiang, J.D., Kong, W.J., 2014. Antioxidant and anti-inflammatory activities of Berberine in the treatment of diabetes mellitus. Evid. base Compl. Alternative Med. 2014. https://doi.org/10.1155/2014/289264.

Lichman, B.R., 2021. The scaffold-forming steps of plant alkaloid biosynthesis. Nat. Prod. Rep. 38 (1), 103–129. https://doi.org/10.1039/d0np00031k.

Liu, R.H., 2013. Health-promoting components of fruits and vegetables in the diet. Adv. Nutr. 4 (3), 384S. https://doi.org/10.3945/an.112.003517.

Liu, Z., Hong, Z., Qu, P., 2020. Proteomic analysis of human endometrial tissues reveals the roles of PI3K/AKT/mTOR pathway and tumor angiogenesis molecules in the pathogenesis of endometrial cancer. BioMed Res. Int. 2020. https://doi.org/10.1155/2020/5273969.

Luo, J., Si, H., Jia, Z., Liu, D., 2021. Dietary anti-aging polyphenols and potential mechanisms. Antioxidants 10 (2), 1–20. https://doi.org/10.3390/antiox10020283.

Malaguarnera, L., 2019. Influence of resveratrol on the immune response. Nutrients 11 (5). https://doi.org/10.3390/nu11050946.

Mas-Bargues, C., Borrás, C., Viña, J., 2022. The multimodal action of genistein in Alzheimer's and other age-related diseases. Free Radic. Biol. Med. 183, 127–137. https://doi.org/10.1016/j.freeradbiomed.2022.03.021.

Meng, T., Xiao, D., Muhammed, A., Deng, J., Chen, L., He, J., 2021. Anti-inflammatory action and mechanisms of resveratrol. Molecules 26 (1). https://doi.org/10.3390/MOLECULES26010229.

Miekus, N., Marszałek, K., Podlacha, M., Iqbal, A., Puchalski, C., Swiergiel, A.H., 2020. Health benefits of plant-derived sulfur compounds, glucosinolates, and organosulfur compounds. Molecules 25 (17). https://doi.org/10.3390/molecules25173804.

Mirahmadi, M., Azimi-Hashemi, S., Saburi, E., Kamali, H., Pishbin, M., Hadizadeh, F., 2020. Potential inhibitory effect of lycopene on prostate cancer. Biomed. Pharmacother. 129, 110459. https://doi.org/10.1016/j.biopha.2020.110459.

Motallebi, M., Bhia, M., Rajani, H.F., Bhia, I., Tabarraei, H., Mohammadkhani, N., Pereira-Silva, M., Kasaii, M.S., Nouri-Majd, S., Mueller, A.L., Veiga, F.J.B., Paiva-Santos, A.C., Shakibaei, M., 2022. Naringenin: a potential flavonoid phytochemical for cancer therapy. Life Sci. 305. https://doi.org/10.1016/j.lfs.2022.120752.

Niedzwiecki, A., Roomi, M.W., Kalinovsky, T., Rath, M., 2016. Anticancer efficacy of polyphenols and their combinations. Nutrients 8 (9). https://doi.org/10.3390/nu8090552.

Nisar, N., Li, L., Lu, S., Khin, N.C., Pogson, B.J., 2015. Carotenoid metabolism in plants. Mol. Plant 8 (1), 68–82. https://doi.org/10.1016/j.molp.2014.12.007.

Pereira, C.P.M., Souza, A.C.R., Vasconcelos, A.R., Prado, P.S., Name, J.J., 2021. Antioxidant and anti-inflammatory mechanisms of action of astaxanthin in cardiovascular diseases (review). Int. J. Mol. Med. 47 (1), 37–48. https://doi.org/10.3892/ijmm.2020.4783.

Perrone, P., Spinelli, S., Mantegna, G., Notariale, R., Straface, E., Caruso, D., Falliti, G., Marino, A., Manna, C., Remigante, A., Morabito, R., 2023. Mercury chloride affects band 3 protein-mediated anionic transport in red blood cells: role of oxidative stress and protective effect of olive oil polyphenols. Cells 12 (3). https://doi.org/10.3390/cells12030424.

Phan, M.A.T., Paterson, J., Bucknall, M., Arcot, J., 2018. Interactions between phytochemicals from fruits and vegetables: effects on bioactivities and bioavailability. Crit. Rev. Food Sci. Nutr. 58 (8), 1310–1329. https://doi.org/10.1080/10408398.2016.1254595.

Probst, Y.C., Guan, V.X., Kent, K., 2017. Dietary phytochemical intake from foods and health outcomes: a systematic review protocol and preliminary scoping. BMJ Open 7 (2). https://doi.org/10.1136/bmjopen-2016-013337.

Qi, W., Qi, W., Xiong, D., Long, M., 2022. Quercetin: its antioxidant mechanism, antibacterial properties and potential application in prevention and control of toxipathy. Molecules 27 (19). https://doi.org/10.3390/molecules27196545.

Quetglas-Llabrés, M.M., Quispe, C., Herrera-Bravo, J., Catarino, M.D., Pereira, O.R., Cardoso, S.M., Dua, K., Chellappan, D.K., Pabreja, K., Satija, S., Mehta, M., Sureda, A., Martorell, M., Satmbekova, D., Yeskaliyeva, B., Sharifi-Rad, J., Rasool, N., Butnariu, M., Bagiu, I.C., Bagiu, R.V., Calina, D., Cho, W.C., 2022. Pharmacological properties of bergapten: mechanistic and therapeutic aspects. Oxid. Med. Cell. Longev. 2022. https://doi.org/10.1155/2022/8615242.

Rahman Mazumder, M.A., Hongsprabhas, P., 2016. Genistein as antioxidant and antibrowning agents in in vivo and in vitro: a review. Biomed. Pharmacother. 82, 379–392. https://doi.org/10.1016/j.biopha.2016.05.023.

Rajavel, T., Packiyaraj, P., Suryanarayanan, V., Singh, S.K., Ruckmani, K., Pandima Devi, K., 2018. β-Sitosterol targets Trx/Trx1 reductase to induce apoptosis in A549 cells via ROS mediated mitochondrial dysregulation and p53 activation. Sci. Rep. 8 (1), 2071. https://doi.org/10.1038/s41598-018-20311-6.

Reggiani, C., 2021. Caffeine as a tool to investigate sarcoplasmic reticulum and intracellular calcium dynamics in human skeletal muscles. J. Muscle Res. Cell Motil. 42 (2), 281–289. https://doi.org/10.1007/s10974-020-09574-7.

Reiter, J., Hübbers, A.M., Albrecht, F., Leichert, L.I.O., Slusarenko, A.J., 2020. Allicin, a natural antimicrobial defence substance from garlic, inhibits DNA gyrase activity in bacteria. Int. J. Med. Microbiol. 310 (1), 151359. https://doi.org/10.1016/j.ijmm.2019.151359.

Remigante, A., Morabito, R., Marino, A., 2021. Band 3 protein function and oxidative stress in erythrocytes. J. Cell. Physiol. 236 (9), 6225–6234. https://doi.org/10.1002/jcp.30322.

Remigante, A., Morabito, R., Spinelli, S., Trichilo, V., Loddo, S., Sarikas, A., Dossena, S., Marino, A., 2020. D-galactose decreases anion exchange capability through band 3 protein in human erythrocytes. Antioxidants 9 (8), 1–18. https://doi.org/10.3390/antiox9080689.

Remigante, A., Spinelli, S., Basile, N., Caruso, D., Falliti, G., Dossena, S., Marino, A., Morabito, R., 2022. Oxidation stress as a mechanism of aging in human erythrocytes: protective effect of quercetin. Int. J. Mol. Sci. 23 (14), 7781. https://doi.org/10.3390/ijms23147781.

Remigante, A., Spinelli, S., Straface, E., Gambardella, L., Caruso, D., Falliti, G., Dossena, S., Marino, A., Morabito, R., 2022. Açaì (Euterpe oleracea) extract protects human erythrocytes from age-related oxidative stress. Cells 11 (15), 2391. https://doi.org/10.3390/cells11152391.

Remigante, A., Spinelli, S., Trichilo, V., Loddo, S., Sarikas, A., Pusch, M., Dossena, S., Marino, A., Morabito, R., 2022. d-Galactose induced early aging in human erythrocytes: role of band 3 protein. J. Cell. Physiol. 237 (2), 1586–1596. https://doi.org/10.1002/jcp.30632.

Renaud, J., Martinoli, M.G., 2019. Considerations for the use of polyphenols as therapies in neurodegenerative diseases. Int. J. Mol. Sci. 20 (8), 1883. https://doi.org/10.3390/ijms20081883.

Reyes-Farias, M., Carrasco-Pozo, C., 2019. The anti-cancer effect of quercetin: molecular implications in cancer metabolism. Int. J. Mol. Sci. 20 (13), 3177. https://doi.org/10.3390/ijms20123177.

Reygaert, W.C., 2018. Green tea catechins: their use in treating and preventing infectious diseases. BioMed Res. Int. 2018, 9105261. https://doi.org/10.1155/2018/9105261.

Ribeiro, J.A., Sebastio, A.M., 2010. Caffeine and adenosine. J. Alzheimer's Dis. ISSN: 1387-2877 20, S3–S15. https://doi.org/10.3233/JAD-2010-1379.

Rodríguez-Luna, A., Ávila-Román, J., Oliveira, H., Motilva, V., Talero, E., 2019. Fucoxanthin and rosmarinic acid combination has anti-inflammatory effects through regulation of NLRP3 inflammasome in UVB-exposed HaCaT keratinocytes. Mar. Drugs 17 (8), 451. https://doi.org/10.3390/md17080451.

Safdar, M.A., Aslam, R.M.N., Shakeel, A., Shiza, Waqar, M., Jmail, A., Mehmood, M.H., Gul, H., 2023. Cyanidin as potential anticancer agent targeting various proliferative pathways. Chem. Biol. Drug Des. 101 (2), 438–452. https://doi.org/10.1111/cbdd.14173.

Sahu, S.C., 2002. Dual role of organosulfur compounds in foods: a review. J. Environ. Sci. Health—Part C Environ. Carcinog. Ecotoxicol. Rev. 20 (1), 61–76. https://doi.org/10.1081/GNC-120005388.

Sakamoto, K., 2000. Synergistic effects of thearubigin and genistein on human prostate tumor cell (PC-3) growth via cell cycle arrest. Cancer Lett. 151 (1), 103–109. https://doi.org/10.1016/S0304-3835(99)00423-1.

Sánchez-Gloria, J.L., Martínez-Olivares, C.E., Rojas-Morales, P., Hernández-Pando, R., Carbó, R., Rubio-Gayosso, I., Arellano-Buendía, A.S., Rada, K.M., Sánchez-Muñoz, F., Osorio-Alonso, H., 2021. Anti-inflammatory effect of allicin associated with fibrosis in pulmonary arterial hypertension. Int. J. Mol. Sci. 22 (16). https://doi.org/10.3390/ijms22168600.

Sang, Y., Zhang, F., Wang, H., Yao, J., Chen, R., Zhou, Z., Yang, K., Xie, Y., Wan, T., Ding, H., 2017. Apigenin exhibits protective effects in a mouse model of d -galactose-induced aging via activating the Nrf2 pathway. Food Funct. 8 (6), 2331–2340. https://doi.org/10.1039/C7FO00037E.

Serban, M.C., Sahebkar, A., Zanchetti, A., Mikhailidis, D.P., Howard, G., Antal, D., Andrica, F., Ahmed, A., Aronow, W.S., Muntner, P., Lip, G.Y.H., Graham, I., Wong, N., Rysz, J., Banach, M., 2016. Effects of quercetin on blood pressure: a systematic review and meta-analysis of randomized controlled trials. J. Am. Heart Assoc. 5 (7), e002713. https://doi.org/10.1161/JAHA.115.002713.

Sharma, D., Sethi, P., Hussain, E., Singh, R., 2009. Curcumin counteracts the aluminium-induced ageing-related alterations in oxidative stress, Na+, K+ ATPase and protein kinase C in adult and old rat brain regions. Biogerontology 10 (4), 489–502. https://doi.org/10.1007/s10522-008-9195-x.

Shirazi-Tehrani, E., Vafadar, A., Keshavarzi, M., Firouzabadi, N., 2022. Anticancer properties of vincristine is modulated by microRNAs in acute lymphoblastic leukemia Nalm6 cell line. Anti Cancer Drugs 33 (1), e680. https://doi.org/10.1097/CAD.0000000000001234.

Sidhar, H., Giri, R.K., 2017. Induction of Bex genes by curcumin is associated with apoptosis and activation of p53 in N2a neuroblastoma cells. Sci. Rep. 7, 41420. https://doi.org/10.1038/srep41420.

Soares, N.d.C.P., Teodoro, A.J., Lotsch, P.F., Granjeiro, J.M., Borojevic, R., 2015. Anticancer properties of carotenoids in prostate cancer. A review. Histol. Histopathol. 30 (10), 1143–1154. https://doi.org/10.14670/HH-11-635.

Solayman, M., Ali, Y., Alam, F., Islam, M.A., Alam, N., Khalil, M.I., Gan, S.H., 2016. Polyphenols: potential future arsenals in the treatment of diabetes. Curr. Pharmaceut. Des. 22 (5), 549–565. https://doi.org/10.2174/1381612822666151125001111.

Song, X., Wang, Y., Gao, L., 2020. Mechanism of antioxidant properties of quercetin and quercetin-DNA complex. J. Mol. Model. 26 (6). https://doi.org/10.1007/s00894-020-04356-x.

Spinelli, S., Straface, E., Gambardella, L., Caruso, D., Falliti, G., Remigante, A., Marino, A., Morabito, R., 2023. Aging injury impairs structural properties and cell signaling in human red blood cells; Açaì berry is a keystone. Antioxidants 12 (4). https://doi.org/10.3390/antiox12040848.

Srivastava, D., Arya, U., SoundaraRajan, T., Dwivedi, H., Kumar, S., Subramaniam, J.R., 2008. Reserpine can confer stress tolerance and lifespan extension in the nematode *C. elegans*. Biogerontology 9 (5), 309–316. https://doi.org/10.1007/s10522-008-9139-5.

Suganuma, M., Okabe, S., Kai, Y., Sueoka, N., Sueoka, E., Fujiki, H., 1999. Synergistic effects of (−)-epigallocatechin gallate with (−)-epicatechin, sulindac, or tamoxifen on cancer-preventive activity in the human lung cancer cell line PC-9. Cancer Res. 59 (1), 44–47.

Tangney, C.C., Rasmussen, H.E., 2013. Polyphenols, inflammation, and cardiovascular disease. Curr. Atherosclerosis Rep. 15 (5). https://doi.org/10.1007/s11883-013-0324-x.

Tao, Z., Geng, D., Tao, J., Wang, J., Liu, S., Wang, Q., Xu, F., Xiao, S., Wang, R., 2023. Synergistic antibacterial effect and mechanism of allicin and an *Enterobacter cloacae* bacteriophage. Microbiol. Spectr. 11 (1). https://doi.org/10.1128/spectrum.03155-22.

Tarasova, O., Poroikov, V., 2018. HIV resistance prediction to reverse transcriptase inhibitors: focus on open data. Molecules 23 (4), 956. https://doi.org/10.3390/molecules23040956.

Tikhonova, M.A., Tikhonova, N.G., Tenditnik, M.V., Ovsyukova, M.V., Akopyan, A.A., Dubrovina, N.I., Amstislavskaya, T.G., Khlestkina, E.K., 2020. Effects of grape polyphenols on the life span and neuroinflammatory alterations related to neurodegenerative Parkinson disease-like disturbances in mice. Molecules 25 (22), 5339. https://doi.org/10.3390/MOLECULES25225339.

Toma, V.A., Tigu, A.B., Farcaş, A.D., Sevastre, B., Taulescu, M., Gherman, A.M.R., Roman, I., Fischer-Fodor, E., Pârvu, M., 2019. New aspects towards a molecular understanding of the allicin immunostimulatory mechanism via Colec12, MARCO, and SCARB1 receptors. Int. J. Mol. Sci. 20 (15), 3627. https://doi.org/10.3390/ijms20153627.

Tutunchi, H., Naeini, F., Ostadrahimi, A., Hosseinzadeh-Attar, M.J., 2020. Naringenin, a flavanone with antiviral and anti-inflammatory effects: a promising treatment strategy against COVID-19. Phytother Res. 34 (12), 3137−3147. https://doi.org/10.1002/ptr.6781.

Vessal, M., Hemmati, M., Vasei, M., 2003. Antidiabetic effects of quercetin in streptozocin-induced diabetic rats. Comp. Biochem. Physiol. C Toxicol. Pharmacol. 135 (3), 357−364. https://doi.org/10.1016/S1532-0456(03)00140-6.

Vingtdeux, V., Giliberto, L., Zhao, H., Chandakkar, P., Wu, Q., Simon, J.E., Janle, E.M., Lobo, J., Ferruzzi, M.G., Davies, P., Marambaud, P., 2010. AMP-activated protein kinase signaling activation by resveratrol modulates amyloid-β peptide metabolism. J. Biol. Chem. 285 (12), 9100−9113. https://doi.org/10.1074/jbc.M109.060061.

Wang, M., Firrman, J., Liu, L.S., Yam, K., 2019. A review on flavonoid apigenin: dietary intake, ADME, antimicrobial effects, and interactions with human gut microbiota. BioMed Res. Int. 2019. https://doi.org/10.1155/2019/7010467.

Wang, S., Meckling, K.A., Marcone, M.F., Kakuda, Y., Tsao, R., 2011. Synergistic, additive, and antagonistic effects of food mixtures on total antioxidant capacities. J. Agric. Food Chem. 59 (3), 960−968. https://doi.org/10.1021/jf1040977.

Wang, X.-D., 2012. Lycopene metabolism and its biological significance. Am. J. Clin. Nutr. 96 (5), 1214S−1222S. https://doi.org/10.3945/ajcn.111.032359.

Wang, Y., Fu, X.T., Li, D.W., Wang, K., Wang, X.Z., Li, Y., Sun, B.L., Yang, X.Y., Zheng, Z.C., Cho, N.C., 2016. Cyanidin suppresses amyloid beta-induced neurotoxicity by inhibiting reactive oxygen speciesmediated DNA damage and apoptosis in PC12 cells. Neural. Regen. Res. 11 (5), 795−800. https://doi.org/10.4103/1673-5374.182707.

Wang, Y., Wen, J., Almoiliqy, M., Wang, Y., Liu, Z., Yang, X., Lu, X., Meng, Q., Peng, J., Lin, Y., Sun, P., 2021. Sesamin protects against and ameliorates rat intestinal ischemia/reperfusion injury with involvement of activating Nrf2/HO-1/NQO1 signaling pathway. Oxid. Med. Cell. Longev. 2021. https://doi.org/10.1155/2021/5147069.

Wang, Y., Wan, R., Peng, W., Zhao, X., Bai, W., Hu, C., 2023. Quercetin alleviates ferroptosis accompanied by reducing M1 macrophage polarization during neutrophilic airway inflammation. Eur. J. Pharmacol. 938, 175407. https://doi.org/10.1016/j.ejphar.2022.175407.

Wen, L., Mao, W., Xu, L., Cai, B., Gu, L., 2022. Sesamin exerts anti-tumor activity in esophageal squamous cell carcinoma via inhibition of TRIM44 and NF-κB signaling. Chem. Biol. Drug Des. 99 (1), 118−125. https://doi.org/10.1111/cbdd.13937.

Yang, F., Lim, G.P., Begum, A.N., Ubeda, O.J., Simmons, M.R., Ambegaokar, S.S., Chen, P., Kayed, R., Glabe, C.G., Frautschy, S.A., Cole, G.M., 2005. Curcumin inhibits formation of amyloid β oligomers and fibrils, binds plaques, and reduces amyloid in vivo. J. Biol. Chem. 280 (7), 5892−5901. https://doi.org/10.1074/jbc.M404751200.

Yin, D., Yin, L., Wang, J., Shen, X., Dai, Y., Zhao, R., Hu, X., Hou, H., Zhang, D., Wang, G., Qi, K., Pan, X., 2022. Antiviral and virucidal activities of camptothecin on fowl adenovirus serotype 4 by blocking virus replication. Front. Cell. Infect. Microbiol. 12. https://doi.org/10.3389/fcimb.2022.823820.

Youdim, K.A., Qaiser, M.Z., Begley, D.J., Rice-Evans, C.A., Abbott, N.J., 2004. Flavonoid permeability across an in situ model of the blood-brain barrier. Free Radic. Biol. Med. 36 (5), 592−604. https://doi.org/10.1016/j.freeradbiomed.2003.11.023.

Zaidun, N.H., Thent, Z.C., Latiff, A.A., 2018. Combating oxidative stress disorders with citrus flavonoid: naringenin. Life Sci. 208, 111−122. https://doi.org/10.1016/j.lfs.2018.07.017.

Zhang, P., Liu, N., Xue, M., Zhang, M., Liu, W., Xu, C., Fan, Y., Meng, Y., Zhang, Q., Zhou, Y., 2023. Anti-inflammatory and antioxidant properties of β-sitosterol in copper sulfate-induced inflammation in zebrafish (*Danio rerio*). Antioxidants 12 (2). https://doi.org/10.3390/antiox12020391.

Zhang, P., Zhang, X., 2018. Stimulatory effects of curcumin and quercetin on posttranslational modifications of p53 during lung carcinogenesis. Hum. Exp. Toxicol. 37 (6), 618–625. https://doi.org/10.1177/0960327117714037.

Zhang, Y.J., Gan, R.Y., Li, S., Zhou, Y., Li, A.N., Xu, D.P., Li, H.B., Kitts, D.D., 2015. Antioxidant phytochemicals for the prevention and treatment of chronic diseases. Molecules 20 (12), 21138–21156. https://doi.org/10.3390/molecules201219753.

Zhou, Y., Wang, J., Yang, W., Qi, X., Lan, L., Luo, L., Yin, Z., 2017. Bergapten prevents lipopolysaccharide-induced inflammation in RAW264.7 cells through suppressing JAK/STAT activation and ROS production and increases the survival rate of mice after LPS challenge. Int. Immunopharm. 48, 159–168. https://doi.org/10.1016/j.intimp.2017.04.026.

Zuo, Y., Li, B., Guan, S., Jia, J., Xu, X., Zhang, Z., Lu, Z., Li, X., Pang, X., 2022. EuRBG10 involved in indole alkaloids biosynthesis in *Eucommia ulmoides* induced by drought and salt stresses. J. Plant Physiol. 278, 153813. https://doi.org/10.1016/j.jplph.2022.153813.

PART

2

Chemical and biological aspects

CHAPTER

From plants to extracts

3

Marcello Locatelli[1], Miryam Perrucci[1], Luigi Ciriolo[1], Rossella Pirro[1], Imran Ali[2], Fotouh R. Mansour[3], Halil Ibrahim Ulusoy[4] and Abuzar Kabir[5]

[1]*Department of Pharmacy, University "G. d'Annunzio" of Chieti-Pescara, Chieti, Italy;* [2]*Department of Chemistry, Jamia Millia Islamia (Central University), Delhi, New Delhi, India;* [3]*Department of Pharmaceutical Analytical Chemistry, Faculty of Pharmacy, Tanta University, Tanta, Egypt;* [4]*Department of Analytical Chemistry, Faculty of Pharmacy, Cumhuriyet University, Sivas, Turkey;* [5]*Department of Chemistry and Biochemistry, International Forensic Research Institute, Florida International Univeristy, Miami, FL, United States*

1. Introduction

Through years, medicinal plants have always been used, but during the past century, their importance has increasingly grown. It should be emphasized that plants produce different substances, primary metabolites with the aim of helping life's cells and secondary metabolites, also known as phytochemicals, which have biological effects (Manousi et al., 2019; Süntar, 2020).

Their biological activity, which includes antiinflammatory, antioxidant, antimicrobial, and anticancer properties, has fascinated both scientists and pharmaceutical company, especially for the attention of consumers for biological compounds, because they are considered more safe and reliable, and they are more cheap than synthesis products (Abubakar and Haque, 2020; Alamri et al., 2023; Manousi et al., 2019). It's also interesting to know the subdivision of main group of phytochemicals, respectively in order of quantities present: Phenolics, Terpenoids and Steroids, Alkaloids, and others (as illustrated in Fig. 3.1) (Saxena et al., 2013).

The hardest part is treating plants, as complex matrices in general, and it's one of the most interesting challenges for analytical chemists (Alampanos and Samanidou, 2021; Alamri et al., 2023; Diuzheva et al., 2020; Saxena et al., 2013). In addition, pretreatment needs to follow Green Analytical Chemistry (GAC) and Green Sample Preparation (GSP) principles, in general the last one acronym mentioned indicates preextraction and extraction methods with an eco-friendly and green approach (Alampanos and Samanidou, 2021; Alamri et al., 2023; Locatelli et al., 2023; Manousi et al., 2019). In Fig. 3.2, a schematic representation of GSP principles is reported.

Data from literature suggest that more than 80% of the total analytical time is due to sampling and pretreatment of the sample, and it (Martinho et al., 2019) also influences the quality of analysis and possibility of making errors (Bakarr Kanu, 2021). In fact, the main problem of extraction from complex matrices is the big quantity of different compound, for example, only from phenolic class

Phytochemicals for Health. https://doi.org/10.1016/B978-0-443-15366-2.00003-4

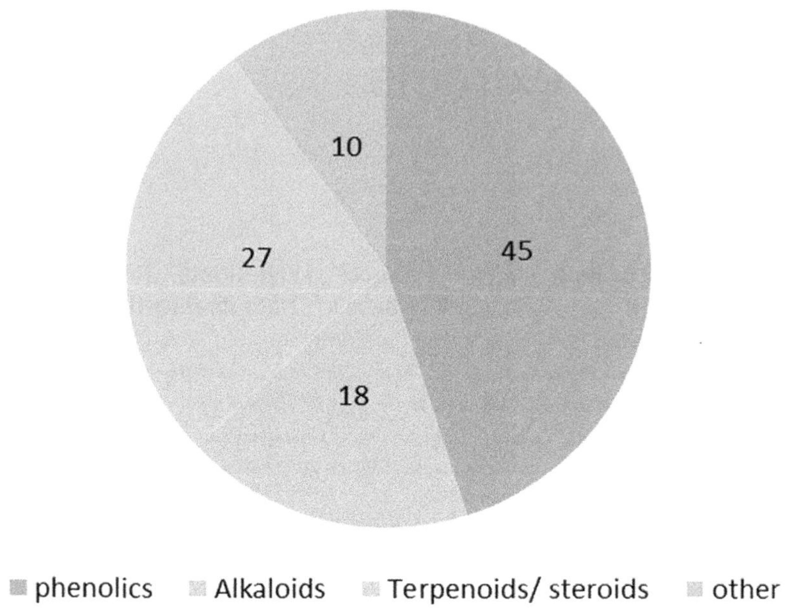

FIGURE 3.1

Main groups of phytochemicals.

there are Polyphenols, Phenolic Acid, and Others and for each subclasses various chemical species (Manousi et al., 2019).

Thus, extraction represents the first crucial step to obtain good results in the future analysis of separation or characterization (Manousi et al., 2019). Obviously, before starting with analytical procedures, researchers have to consider the nature of the drug, the concentration, and its therapeutic purpose (Manousi et al., 2019). Extraction can be divided based on the approach, classical or *conventional* and the newest or *nonconventional* (Manousi et al., 2019). With the first one term, we indicate extraction method as Soxhlet or Percolation, Digestion, instead nonconventional methods are Microwave-assisted extraction (MAE), Ultrasound-assisted extraction (UAE), Supercritical Fluid Extraction (SFE) (Abubakar and Haque, 2020; Manousi et al., 2019).

1.1 Preextraction methods

Before extraction, it is convenient to decide whether to use fresh or dried plants to preserve phytochemicals and to obtain the major extraction (Fonmboh et al., 2020; Manousi et al., 2019; Azwanida, 2015). The choice depends on thermal stability of the analyte(s) and time of Design of Experiment (DoE), because dried plants material can stay longer than fresh, and it also can avoid bacterial growth (Azwanida, 2015; Manousi et al., 2019; Markhali, 2021; Nortjie et al., 2022). Also the quantity recovered can change by choosing one type of drying instead of another, for example, Ahmad-Qasem demonstrated that olive leaves treated with hot air at 120°C had more phenolics recovery compared with freeze-drying (Ahmad-Qasem et al., 2013), instead Martino et al. found higher phenolics

FIGURE 3.2

A schematic representation of 10 principles of GSP.

compound from freeze-drying olive leaves than fresh (Martinho et al., 2019). The final aim is to always preserve phytocompounds, respecting each characteristic, from thermolable to light-sensitive. Freeze and drying constitute a method based on sublimation, instead microwave-drying used electric and magnetic radiation to arrive at the same result (Azwanida, 2015; Abubakar and Haque, 2020; Manousi et al., 2019). Another aspect to consider is uniformity and smallness of particle size, with the aim of increasing the contact surface between plants and solvent (Azwanida, 2015). Borhan et al. in their work tried with different sizes of grinded sample, obtaining best results with about 400 μm (Borhan et al., 2013).

2. Conventional extraction method

Different typologies of extraction techniques were developed during the year to isolate, preconcentrate, and clean-up analytes from matrix interferences before the analyses or other uses. In the context of the extraction of phytochemicals from different parts of plant, solvent-based techniques have always been most used. All these techniques are based on the physicochemical affinity of analytes to solvent used during the extraction.

Some of conventional solvent-based techniques to extract products from plants are:

- **Maceration:** is a very simple extraction technique that consists of soaking the material (after having fragmented or powdered it) into a large volume of the opportune solvent for 3 days at room temperature, then the solid parts are separated with filtration, and the extracted mixture can be concentrate promoting the solvent evaporation. This technique is not selective, shows low efficiency, and requires excessive use of solvents and time (Ćujić et al., 2016; Manousi et al., 2019).
- **Infusion:** the material is poured into a boiling solvent (particularly water) for times ranging from 2 to 30 min (Cittan et al., 2018).
- **Percolation:** in this case, the solvent passes through the material with a constant flow and gradually extracts the compounds from matrix. The most used solvent for percolation is ethyl alcohol (Paz et al., 2018).
- **Soxhlet:** is quite an efficient technique, based on the continuous flow of hot solvent through the material. The part of plants is placed into a porous bag obtained with filter paper while the solvent is placed into a bottom flask. When the solvent is heated, its vapor crosses the porous bag and starts to extract compounds before getting recondensed and returning on the bottom. This continuous cycle allows the exhaustive extraction of phytochemicals increase the efficiency of this technique compared with other conventional techniques (Alara et al., 2018; Manousi et al., 2019).

These conventional extraction techniques (Table 3.1) are still used in some cases due to their simplicity and high knowledge (Garcia-Vaquero et al., 2020).

The most used extraction techniques are:

- Solid-phase extraction (SPE): is a widely used procedure thanks to the high recovery and simple preparation. It is based on specific cartridge composed by organic materials that allowed us to separate analytes from other matrix interferents (Wang et al., 2020).
- Dispersive-solid phase extraction (dSPE): in this method, the solid sorbent is poured into the liquid sample ensuring high contact surface between analyte and sorbent (Ścigalski and Kosobucki, 2020).
- Matrix solid-phase dispersion (MSPD): a blending of sorbent and sample is manually prepared and then is compacted in a syringe barrel to carry out the selective elution of the analyte(s) with the appropriate solvent (Ramos, 2019).

3. Non-conventional extraction methods

In most cases, conventional extraction methods have been replaced by more modern techniques. These novel extraction techniques allowed to overcome the limitations of conventional techniques and to obtain the maximum recovery with a minimum use of toxic solvents, energy, and time (Garcia-Vaquero et al., 2020).

Often these cutting-edge methods of extraction combine the theories of the conventional techniques such as the affinity of compounds to the solvent, with innovative tools and technologies that increase their performances.

Table 3.1 Some examples of the conventional extraction method on plants.

Vegetable material	Time and temperature	Mass/solvent	References
***Psidium cattleianum* sabine**	120 minutes 40°C	1 g/10 mL EtOH 90%	Meregalli et al. (2020)
Brazilian olive leaves	1440 minutes at room temperature	0.5 g/25 mL of 70% EtOH	Rosa et al. (2021)
***Clitoria ternatea* L.**	150 minutes 50°C	1 g/15 mL H_2O	Mehmood et al. (2019)
Helichrysum leucocephalum	4320 minutes at room temperature	100 g/500 mL EtOH	Goldansaz et al. (2018)
Atriplex mollis	1440 minutes ×3 at room temperature	842 g/15 L EtOH: H2O (80:20 *v:v*) + LLE	Boutaoui et al. (2018)
Vernonia amygdalina	240 minutes	100 g/1000 mL EtOH 60% *v/v* (soxhlet)	Alara et al. (2019)
Thymus algeriensis	1440 minutes at room temperature	2000 g/15 L EtOH: H_2O (70:30 *v:v*)	(Boutaoui et al., 2018)
Cassia alata	120 minutes 60°C	1 g/20 mL EtOH	Yi Ling et al. (2019)
Calamintha vulgaris	– –	300 g/100 mL H_2O + 100 mL n-Hexane or chloroform or ethyl acetate.	Khan et al. (2018)
Rosmarinus Officinalis L.	1440 minutes at room temperature	0.5 g/10 mL H_2O/EtOH (95:5, 70:30, 50:50 *v:v*)	(Pontillo et al., 2021)
Vernonia cinerea	180 minutes at room temperature	10 g/200 mL EtOH 60% *v/v*	Alara et al. (2018)

EtOH, *ethanol;* H_2O, *water;* LLE, *liquid-liquid extraction.*

Among the most recent and used are:

- *Supercritical fluid extraction (SFE)*
  With specific condition of pressure and temperature, gas and liquid phases are indistinctive. On state diagram, this region is called critical region, and it is delimited by values of critical pressure and critical temperature. Supercritical fluid shows both fluids and gas properties when critical temperature and pressure values are reached (Hizir and Barron, 2016). These properties are exploited in these techniques, where the supercritical fluid flowing through the packed material dissolves chemicals. The obtained extract is dry, because by varying the temperature and pressure parameters, the solvent can be removed without leaving traces in the final product. One of the most supercritical fluids used is carbon dioxide (CO_2) in some cases, the extractive power and selectivity can be improved by adding little amounts of other solvents (Belo et al., 2019). SFE allows to achieve high yields and maintain the biological properties of the extract compounds, further is also compatible with the main analytical tools such as GC, HPLC, MS, and IR (Trentini et al., 2019).

- *Ultrasound-assisted extraction (UAE)*
Ultrasound waves at high intensity cause a phenomenon called cavitation, which consists of the formation and rapidly collapse of bubbles into the solvent. The energy of the cavitation converts into heat and promotes the disruption of biological matrix and enhances mass transfer.
This technique is cheap and environmentally friendly compared with the conventional technique and can be also boosted if coupled with other powerful extraction technologies, such as the use of deep eutectic solvents or ionic liquids (Murador et al., 2019).
- *Microwave-assisted extraction (MAE)*
Using microwaves frequency between 300 MHz and 300 GHz, the power of the solvent can be increased. These waves that propagate in the solvent generate heat, and facilitate the extraction process. This technique allowed to obtain the disruption of plants cells and consequently release of the bioactive compounds, enhancing the solvent penetration into cellular walls (Taşkın and Aksoylu Özbek, 2020).
- *Pressurized liquid extraction (PLE)*
Also known as accelerated solvent extraction, this technique uses high temperature (ranges 50–200°C) and high pressure (3.5–20 MPa), to reduce the viscosity of the extraction solvent and promote the matrix disintegration and therefore maximize the speed of the process (Teixeira et al., 2021). The efficiency yield is higher than that of the conventional techniques, and the possibility to use an automated equipment makes this technique very promising.
- *Enzyme-assisted extraction (EAE)*
This technology is particularly used for this type of extraction because the use of enzyme allowed to obtain the hydrolysis of cell wall components, avoiding the use of more stressing condition such as high temperature or pressure. This possibility is the main advantage of this technique that can be used for thermolabile compounds, maintaining the original efficacy of biological products (Catalkaya and Kahveci, 2019).

In addition to solvent-based extraction techniques, during the years were developed different techniques based on the use of different composed material able to selectively retain certain analytes, as reported in Table 3.2.

The use of sorbent-based techniques has increased the extraction from complex matrices, with the aim of drastically reducing the formation of artifacts and waste products (Greco et al., 2023, 2023b).

However, conventional techniques are very expensive, in terms of time and solvents used, and for this reason it is increasingly common to the use of microextraction techniques and the research of new advanced sorbent materials and technologies (Casado et al., 2020; Merone et al., 2020).

Among the microextraction techniques that have found wide use in recovery of phytocompounds from biological matrices include:

- Microextraction by packed sorbent (MEPS): this technique allows to extract, concentrate, and cleanup the analyte with a single device and in one step. The sorbent is placed directly inside a micro syringe used to withdraw the sample (Nahar et al., 2021).
- mSPEed: is like MEPS, but the presence of a one-way check valve allows a constant and high-pressure flow through the sorbent (González-Gómez et al., 2021, 2022).
- Solid-phase microextraction (SPME): is simpler than SPE and solvent-free technique, which uses a fused silica fiber coated with the sorbent material, which forms the stationary phase of the micro syringe. This technique can be applied on solid, liquid, and gas samples (Aresta et al., 2018).

Table 3.2 Some examples of some nonconventional extraction methods on plants.

Vegetable material	Time and temperature	Mass/solvent	Method	References
Psidium cattleianum **sabine**	90 minutes 40°C	1 g/10 mL EtOH 90%	UAE	Meregalli et al. (2020)
Brazilian olive leaves	3 minutes 86°C	0.5 g/25 mL of H_2O	MAE	Rosa et al. (2021)
Brazilian olive leaves	29 minutes 27°C	0.5 g/25 mL of H_2O	UAE	Rosa et al. (2021)
Clitoria ternatea **L.**	150 minutes 50°C	1 g/15 mL H_2O	UAE	Mehmood et al. (2019)
Atriplex mollis	60 minutes 40°C	20 g/180 mL CO_2	SFE	Boutaoui et al. (2018)
Atriplex mollis	10 minutes 80°C	1 g/20 mL H_2O	MAE	Boutaoui et al. (2018)
Monotheca buxifolia	30 minutes at room temperature	1 g/3 mL MeOH or $CHCl_3$	UAE	Ali et al. (2020)
Vernonia amygdalina	5 minutes 78°C	10 g/100 mL EtOH	MAE	Alara et al. (2019)
Thymus algeriensis	60 minutes 40°C	20 g/180 mL CO_2	SFE	(Boutaoui et al., 2018)
Thymus algeriensis	15 minutes 100°C	1 g/20 mL H_2O	MAE	(Boutaoui et al., 2018)
Cassia alata	5 minutes at room temperature	1 g/20 mL EtOH	UAE	Yi Ling et al. (2019)
Cassia alata	4 minutes at room temperature	1 g/20 mL EtOH	MAE	Yi Ling et al. (2019)
Herba artemisia scopariae	30 minutes at room temperature	2 g/10 mL NADES or EtOH or H_2O	US	Yue et al. (2020)
Rosmarinus Officinalis L.	60 minutes 50°C	0.5 g/10 mL H_2O/ EtOH (95:5, 70:30, 50:50 v:v)	MAE	(Pontillo et al., 2021)
Rosmarinus Officinalis L.	120 minutes 50°C	100 g/2.5 g enzyme in 20 mL buffer	EAE	(Pontillo et al., 2021)
Vernonia cinerea	2 minutes 60°C	1 g/14 mL EtOH 47% *v/v*	MAE	Alara et al. (2018)
Rhizoma smilacis glabrae	0.2 minutes at room temperature	1 g/20 mL VB3PILs	MAE	Fan et al. (2020)
Harpagophytum procumbens	8 minutes 80°C	500 mg/20 mL H_2O	MAE	(Diuzheva et al., 2018; Ferrante et al., 2017; Locatelli et al., 2017)

Continued

Table 3.2 Some examples of some nonconventional extraction methods on plants.—cont'd

Vegetable material	Time and temperature	Mass/solvent	Method	References
Harpagophytum procumbens	2 minutes at room temperature	50 mg/5 mL H_2O + 600 μL ethyl acetate + 500 μL ACN	DLLME	Diuzheva et al. (2018)
Harpagophytum procumbens	10 minutes at room temperature	50 mg/5 mL H_2O + 600 μL ethyl acetate + 500 μL ACN	UA-DLLME	Diuzheva et al. (2018)
Harpagophytum procumbens	1 minute at room temperature	50 mg/1 mL H_2O + 2 mL ACN +1 mL glucose solution (400 g/L)	SULLE	Diuzheva et al. (2018)

$CHCl_3$, *chloroform;* DLLME, *dispersive liquid-liquid microextraction;* EAE, *enzyme-assisted extraction;* MAE, *Microwave-assisted extraction;* MeOH, *methanol;* NADES, *natural deep eutectic solvent;* SFE, *supercritical fluid extraction;* SULLE, *sugaring-out liquid-liquid microextraction;* UA-DLLME, *ultrasound-assisted dispersive liquid-liquid microextraction;* UAE, *Ultrasound-assisted extraction;* VB3PILs, *vitamin B3-based protic ionic liquids.*

- Stir bar sorptive extraction (SBSE): based on the same principles of SPME but the amount of sorbent phase coating is larger on the stir bar, providing a higher extraction capacity. The most used stationary phase is dimethylpolysiloxane (PDMS), sometimes derivatized to increase selectivity toward specific analytes (Wang et al., 2020).
- Miniaturized conventional techniques: these techniques are modification of classical extraction techniques, reducing the amount of sorbent and using only few amounts of solvent. The main are dispersive micro solid-phase extraction (d-MSPE) and miniaturized matrix solid-phase extraction (m-MSPD) (Gao et al., 2023; Sajid et al., 2021).

However, in addition to new extraction techniques, the goal is continuously to develop new materials that can increase the performance of these technologies (e.g., novel silica-based sorbent materials, mesostructured silicas) or to study the use of new solvents that can reduce classic toxic solvents in according to the demands of the green analytical chemistry (e.g., deep eutectic solvents) (Ali Redha, 2021; González-Gómez et al., 2021; Imoisili et al., 2020).

4. Green approach in analytical procedures

Following the 12 principles of GAC (Fig. 3.3), it's preferable to use nonconventional methods for extraction of phytochemicals from plants. These principles wanted to be more eco-friendly, decreasing use of solvents and operating time (Manousi et al., 2019), confirming more green techniques than the nonconventional ones, because they use less hazardous solvents, less time of extraction procedure, and consequently, less energy. Using conventional methods of extraction has caused, during the years,

FIGURE 3.3

Schematic representation of 12 principles of GAC.

environmental problems related with pollution by toxic solvents and large use of them, and precisely because of the important use of these, the extraction must be followed by a process of concentration, and this means more energy and time-consuming process (Alara et al., 2021). To avoid these problems now, it is better using assisted extraction methods (MAE, UAE, and EAE) or SFC. These newest methods meet all principles of GSP and GAC: minor volume and less hazardous solvents, higher yields, better reproducibility, less extraction time, and less toxic waste.

For example, MAE is chosen for extraction of phenolics and flavonoids, because using microwave can destroy polyphenols with many substituents and high temperature sensible (Alara et al., 2021). In addition, using MAE generates heat, which improve the diffusivity of the solvent into the sample, obtaining better extraction, but in the same time it needs attention with some compound more thermolabile (Alara et al., 2018; Alara et al., 2019, 2021; Alara et al., 2018).

Another technique used requires less volume of solvent used, assuring, at the same time, good and fast extraction is UAE (Alara et al., 2021), indeed it is cheaper than the previous (Manousi et al., 2019; Shams et al., 2015). UAE is useable for laboratory or industry, small and large scale both, and it can be direct, when probes are directly immersed in the sample, or indirect, using an ultrasonic bath (Shams et al., 2015).

EAE is another green technique, because it reduces amount of solvents and time- and energy consumption, the only disadvantage can be the higher cost than MAE or UAE (Manousi et al., 2019). A method to limit costs can be the association with other nonconventional technique being cheaper, with the aim of increasing the yield of the extraction process (Puri et al., 2012).

Another important factor to consider is the solvent used in extraction, because plants material are rich of polar and nonpolar compounds and, consequently, an optimal extraction needs both type of solvents and two different methods to obtain a good and real extraction yield (Chuo et al., 2022).

As known for a long time, green solvents should follow 12 principles: availability, low price, recyclability, grade, synthesis, toxicity, biodegradability, performance, stability, flammability, storage, and renewability (Gu and Jérôme, 2013) and these to preserve humans' and environment's health (Chemat et al., 2019). Water is a perfect example of green solvent, because it presents each parameters request, but only without distillation (time and energy consumption) (Häckl and Kunz, 2018).

Other examples of green solvents are Supercritical fluids, Ionic Liquids, and Deep Eutectic Solvents (DESs), which in recent years are becoming more and more famous and known, and they represent the greenest class of solvents (Choi and Verpoorte, 2019). DESs are eutectic mixture of 2/3 components, bound together through hydrogen bonding. From this class, Natural Deep Eutectic Solvent (NADES) is born, called "natural" because one component of the eutectic mixture is a primary metabolite group (sugar, organic acids) (Hikmawanti et al., 2021).

Maceration, optimizing well ratio and type of solvents, can give results comparable to modern technique, but it produces bigger amount of waste than MAE or UAE and for this reason it does not match with green principles (Azwanida, 2015).

5. Comparison of the different procedures in term of yield

From a comparison between Tables 3.1 and 3.2, we can highlight immediately an important reduction of extraction time using nonconventional method, following GAC principles. Da Rosa et al. in their study reported how to have the major total phenolic compounds with three different extraction methods, maceration as conventional and MAE and UAE as nonconventional. Reduction of time is the first thing that stands out, from 24 h with maceration to 3 min for MAE and 29 for UAE and a yield better for MAE, this maybe because in a previous study was reported how MAE release faster cell compounds disrupting cell walls (Rosa et al., 2021; Zhang et al., 2008). Also temperature is a factor to consider during extraction of these phytochemical compounds, because often they are thermo-labels and microwave used during nonconventional extraction like these can increase the temperature, giving not optimal results. For these reasons, da Rosa et al. followed protocol already validated (Rafiee et al., 2011; Rosa et al., 2021).

Boutaoui et al. in their study have tried them too with three extraction methods, maceration (3 times, 24 h each ones), SFE (60 min), and MAE (10 min), which resulted in the best approach in terms of yield. Additionally, they reported not significant differences between time considering (5, 10, 15 min) at 80°C selected as right temperature. Another interesting particularity is that extraction with MAE is made with water, following GAC request of greenness (Boutaoui et al., 2018; Häckl and Kunz, 2018).

In the same year, this group conducted a similar study on Atriplex mollis, using maceration, SFE, and MAE as methods for extraction. Analysis on extract from maceration gave better results using ethyl acetate. More compounds were identified with SFE, but also in this case MAE was shown better results in terms of metabolites, yield and time-occurred, factor that should not be underestimated (Boutaoui et al., 2018).

Ling et al. investigated firstly the better particle size diameter of the powder to obtain bigger amount of total phenolic compounds. Also they emphasize an important reduction of time using non-conventional methods, such as UAE and MAE than maceration (respectively, 5, 4 and 120 min), giving also better yields. The results are the consequence of a better mass transfer due to sonication and microwave irradiation (Yi Ling et al., 2019).

Interesting was the study conducted by Pontillo et al. on rosemary leaves, which compared conventional and nonconventional extraction methods, using a classic hydroethanolic extraction, MAE, and EAE. Their conclusions about the EAE were fascinating, because the aforementioned method could be a promising pretreatment for a future conventional solid-liquid extraction, 1 h treatment with EAE and 24 h conventional extraction water/ethanol 50/50 *v/v*. Thus, the process gets longer, but results were the best one. EAE is an extraction method very promising, but many parameters affect the results. In addition, good results are always goals, and they can compensate high costs due to EAE (Pontillo et al., 2021). Many studies reported the high costs as important disadvantage and, consequently, this is a limitation for extraction of samples in large scale. Instead, many times EAE need not just one type of enzyme, but a combination of them to extract phytochemicals (Kumar et al., 2023).

Interesting is also a study of Alwazeer et al. done in 2023, which reported the use of two types of hydrogen-rich water for phytochemicals extraction from olive leaves. The different solvents used were: hydrogen-rich water, magnesium water, water/ethanol (50/50 v/v), and magnesium water/ethanol (50/50 v/v). The extraction consisted of leaving in agitation for 24 h the mixtures. Results reported that hydrogen-rich water was the better solvent to extract, giving a difference of 15% from simple water (Alwazeer and Allam Elnasanelkasim, 2023).

The study conducted by Fan et al. resulted innovative, because they coupled use of ionic liquid to MAE, obtaining results higher than those of conventional extraction methods (with ethanol, methanol, and β-cyclodextrins) (Fan et al., 2020). By modulating viscosity, hydrophobicity, hydrogen bonding, they tried with five different vitamin B3-based protic ionic liquids, obtaining good results. Additionally, the IL used can be recycled without losing extraction ability. Thus, this method matches perfectly the eco-friendly requirements of GAC (Fan et al., 2020).

Diuzheva et al. in their research compared DLLME, UA-DLLME, and SULLE with another nonconventional method, but yet validated, MAE. About SULLE, trying different sugars at various concentrations, they noted a peak split for each of the three sugars. DLLME was difficult due to the multiplicity of parameters to consider, from volume to type of extraction solvent and to obtain good yields of extraction, they had to use solvents with lighter density than water, for example, ethyl acetate.

However, the best recovery was obtained using MAE with sodium chloride and salt-aided DLLME (Diuzheva et al., 2018, 2020; Ferrante et al., 2017; Locatelli et al., 2017).

6. Selectivity of the different approach and how to modulate the extraction respect a specific target

Choosing extraction procedure for phytochemical from plants is crucial as choosing extraction solvent. Some basic questions that precede extraction are based on polar/nonpolar compounds, thermally stable or not, which instruments are present in laboratory, and how much time and cost the procedure needs (Tzanova et al., 2020).

About solvents, factors to consider are: selectivity, boiling point, reactivity, viscosity, safety, cost, vapor pressure, recovery. Selective for the phytochemical(s) to be extracted, the lowest possible boiling point to remove it easily, low reactivity and viscosity, safely as possible for human and environmental health, cheap, the same of reactivity for vapor pressure, and the term recovery indicates the ease with which to separate it from the extract (Manousi et al., 2019). Water is the major example that contains all these features (Manousi et al., 2019).

Ethanol and methanol are often used in both conventional and nonconventional methods, obtaining good yields. This is because these solvents destabilize phospholipids of the membrane, increasing cell permeability (Oreopoulou et al., 2019). Ethanol and mixtures of ethanol/water in different percentages are also most used, because in addition to optimal extraction, it prevents bacterial growth (da Silva et al., 2022; Manousi et al., 2019). In Table 3.3, there is a schematic representation of the main phytochemicals and solvents used for each compound.

Because of flammability and toxicity of many conventional solvents, there are other modern and more green solvents. One example is ionic liquids, which are organic salt in liquid state, and they present an organic cation coupled with organic or inorganic anion (Manousi et al., 2019). ILs can be divided in 11 categories: neutral ILs, acidic ILs, basic ILs, ILs with amphoteric anions, functionalized ILs, protic ILs, chiral ILs, supported ILs, bio-ILs, poly-ILs, and energetic ILs (Gong et al., 2023; Lim

Table 3.3 Schematic representation of the main phytochemicals and solvents used for each compound.

Water	Methanol	Ethanol	Acetone	Ethyl acetate
Anthocyanins	Anthocyanins	Anthocyanins	Phenolic diterpenes	Phenolic diterpenes
Phenolic acid glycosides	Phenolic acids	Phenolic acids	Flavonoids	Flavonoid aglycones
Rosmarinic acid	Flavonoids	Flavonoids	Tannins	
Saponins	Flavonoids glucosides	Flavonoids glucosides		
Terpenoids	Tannins	Tannins		
	Saponins	Terpenoids		
	Terpenoids	Alkaloids		

et al., 2022). Imidazolium ionic liquids are mostly used for extraction of phytochemicals from plants (Gong et al., 2023).

Another category of GAC-friendly solvents characterized by greenness is Deep Eutectic Solvents (DESs) and Natural DES (NADES). They are composed of hydrogen bond donors and hydrogen bond acceptors (where often choline chloride is reported). NADESs are natural because of being composed of sugar, organic acids, and amino acids. Also, DES and NADES match perfectly with criteria by GAC, because they're simple to find and product, nonvolatile, low toxicity, low energy consumption, nonflammable, stable at high temperature. Also NADES can be grouped in: ionic liquid NADES, neutral NADES, neutral NADES with acids, neutral NADES with bases, amino acids-based NADES. Thus, they are applicable in a wide range of compounds (Hikmawanti et al., 2021).

Also supercritical fluids are considered eco-friendly and used during extraction processes thanks to their characteristics and, for sure, CO_2 is the most supercritical fluid used. In Fig. 3.4, there is a schematic representation of physical states of CO_2 to have an easier comprehension.

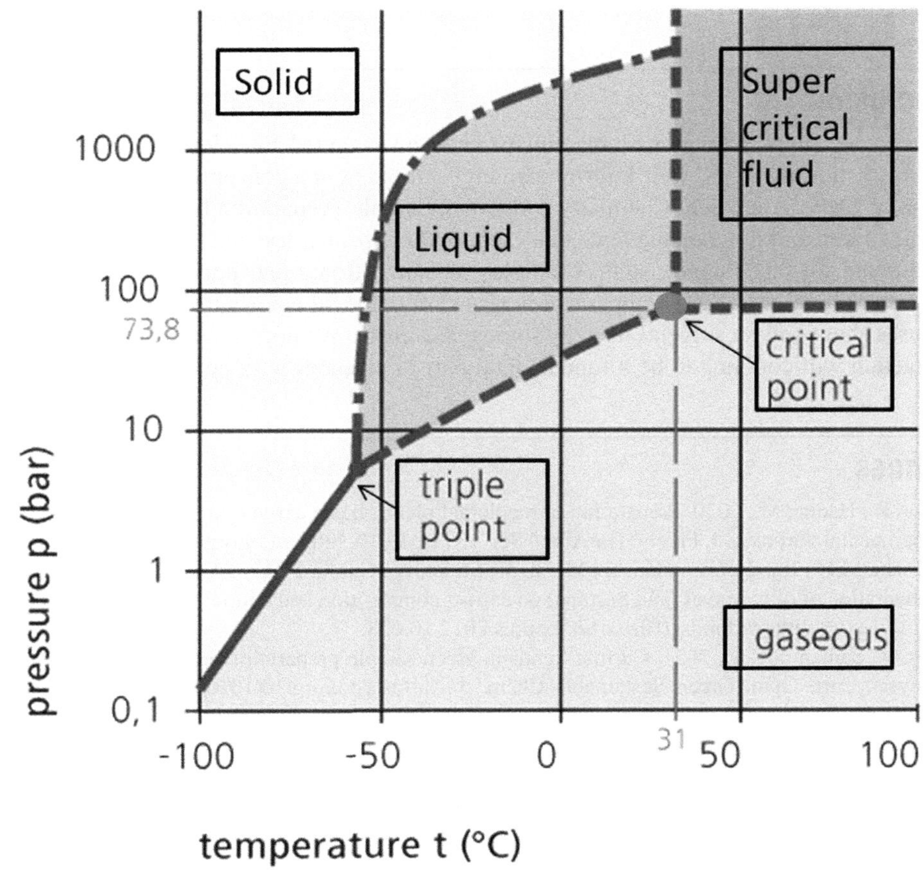

FIGURE 3.4

A schematic representation of physical states of CO_2.

Characteristics of a supercritical fluid are: low viscosity, high density and diffusivity, obtainable playing with temperature critical and pressure critical values. It's advisable to optimize some parameters before real samples extraction, because time extraction, temperature and pressure, type of matrix could affect the yield. Sometimes supercritical fluids are used also as cosolvent or modifier during extractions, to increase solubility of some target compounds poorly soluble in certain solvents (Uwineza and Waśkiewicz, 2020).

About MAE, one of the most used nonconventional extraction methods, radiations interact with polar compounds or solvents and heat is transferred by conduction, thus hydrogen bonding will break. As heat is product, thermolabile compounds can't be extracted with this method, so it's preferable extract small-compounds, as phenolic acid, quercetin, molecules that do not undergo destruction by heat. Phenolics and flavones cannot be subjected to repeated cycles due to possible oxidation, instead tannins and anthocyanins are thermosensible (Azwanida, 2015).

UAE, for its part, presents good result thanks to reduction of extraction time and solvents, but it's preferable using energy less than 20 kHz, because higher energy can lead to formation of free radicals (Azwanida, 2015).

7. Conclusion

During the years, plant extractions have always been investigated by scientists, starting from conventional extraction methods, well known, arriving to the new nonconventional extraction methods. The advent of Green Analytical Chemistry and Green Sample Preparation has changed the chards on table, turning more and more on methods that espouse these principles, and these methods are, in fact, nonconventional. In this chapter, some examples of conventional and nonconventional method extractions are provided with their comparison in terms of yield and environmental approach. There are some focuses about newest solvents used by some researchers. We hope that, with the research's help, plant extraction will continue to be a fundamental part of sciences with countless applications.

References

Abubakar, A.R., Haque, M., 2020. Preparation of medicinal plants: basic extraction and fractionation procedures for experimental purposes. J. Pharm. BioAllied Sci. 12 (1), 1–10. https://doi.org/10.4103/jpbs.JPBS_175_19.

Ahmad-Qasem, M.H., Barrajón-Catalán, E., Micol, V., Mulet, A., García-Pérez, J.V., 2013. Influence of freezing and dehydration of olive leaves (var. Serrana) on extract composition and antioxidant potential. Food Res. Int. 50 (1), 189–196. https://doi.org/10.1016/j.foodres.2012.10.028.

Alampanos, V., Samanidou, V., 2021. Current trends in green sample preparation before liquid chromatographic bioanalysis. Curr. Opin. Green Sustainable Chem. 31. https://doi.org/10.1016/j.cogsc.2021.100499. http://www.journals.elsevier.com/current-opinion-in-green-and-sustainable-chemistry.

Alamri, A.S., Saleem, H., Pervaiz, I., Khurshid, U., Butt, J., Chohan, T.A., Alhomrani, M., Locatelli, M., Zainal Abidin, S.A., H Alqahtani, Y.N., Rengasamy, K.R.R., Mahomoodally, M.F., 2023. Insight into the phytochemical, biological, and in silico studies of Erythrina suberosa roxb.: a source of novel therapeutic bioactive products from a medicinal plant. Food Biosci. 52. https://doi.org/10.1016/j.fbio.2023.102429. http://www.journals.elsevier.com/food-bioscience/.

Alara, O.R., Abdurahman, N.H., Ukaegbu, C.I., 2018a. Soxhlet extraction of phenolic compounds from Vernonia cinerea leaves and its antioxidant activity. J. Appl. Res. Med. Aromat. Plants 11, 12–17. https://doi.org/10.1016/j.jarmap.2018.07.003.

Alara, O.R., Abdurahman, N.H., Ukaegbu, C.I., Azhari, N.H., 2018b. Vernonia cinerea leaves as the source of phenolic compounds, antioxidants, and anti-diabetic activity using microwave-assisted extraction technique. Ind. Crop. Prod. 122, 533–544. https://doi.org/10.1016/j.indcrop.2018.06.034.

Alara, O.R., Abdurahman, N.H., Ukaegbu, C.I., Kabbashi, N.A., 2019. Extraction and characterization of bioactive compounds in Vernonia amygdalina leaf ethanolic extract comparing Soxhlet and microwave-assisted extraction techniques. J. Taibah Univ. Sci. 13 (1), 414–422. https://doi.org/10.1080/16583655.2019.1582460.

Alara, O.R., Abdurahman, N.H., Ukaegbu, C.I., 2021. Extraction of phenolic compounds: a review. Curr. Res. Food Sci. 4, 200–214. https://doi.org/10.1016/j.crfs.2021.03.011.

Ali, J.S., Saleem, H., Mannan, A., Zengin, G., Mahomoodally, M.F., Locatelli, M., Abidin, S.A.Z., Ahemad, N., Zia, M., 2020. Metabolic fingerprinting, antioxidant characterization, and enzyme-inhibitory response of Monotheca buxifolia (Falc.) A. DC. extracts. BMC Complem. Med. Thera. 20 (1). https://doi.org/10.1186/s12906-020-03093-1. https://link.springer.com/journal/12906/volumes-and-issues.

Ali Redha, A., 2021. Review on extraction of phenolic compounds from natural sources using green deep eutectic solvents. J. Agric. Food Chem. 69 (3), 878–912. https://doi.org/10.1021/acs.jafc.0c06641.

Alwazeer, D., Allam Elnasanelkasim, M., 2023. Hydrogen-rich water as a green solvent for the extraction of phytochemicals from agri-food wastes. Sustain. Chem. Pharm. 33, 101035. https://doi.org/10.1016/j.scp.2023.101035.

Aresta, A., Cotugno, P., Massari, F., Zambonin, C., 2018. Determination of trans-resveratrol in wines, spirits, and grape juices using solid-phase micro extraction coupled to liquid chromatography with UV diode-array detection. Food Anal. Methods 11 (2), 426–431. https://doi.org/10.1007/s12161-017-1013-0.

Azwanida, N.N., 2015. A review on the extraction methods use in medicinal plants, principle, Strength and limitation. Med. Aromatic Plants 04 (03). https://doi.org/10.4172/2167-0412.1000196.

Bakarr Kanu, A., 2021. Recent developments in sample preparation techniques combined with high-performance liquid chromatography: a critical review. J. Chromatogr. A 1654, 462444. https://doi.org/10.1016/j.chroma.2021.462444.

Belo, Y.N., Al-Hamimi, S., Chimuka, L., Turner, C., 2019. Ultrahigh-pressure supercritical fluid extraction and chromatography of Moringa oleifera and Moringa peregrina seed lipids. Anal. Bioanal. Chem. 411 (16), 3685–3693. https://doi.org/10.1007/s00216-019-01850-x.

Borhan, M.Z., Ahmad, R., Rusop, M., Abdullah, S., 2013. Impact of nanopowders on exraction yield of Centella asiatica. Adv. Mat. Res. 667, 246–250. https://doi.org/10.4028/www.scientific.net/AMR.667.246.

Boutaoui, N., Zaiter, L., Benayache, F., Benayache, S., Cacciagrano, F., Cesa, S., Secci, D., Carradori, S., Giusti, A.M., Campestre, C., Menghini, L., Locatelli, M., 2018. Atriplex mollis Desf. aerial parts: extraction procedures, secondary metabolites and color analysis. Molecules 23 (8). https://doi.org/10.3390/molecules23081962. http://www.mdpi.com/1420-3049/23/8/1962/pdf.

Casado, N., Morante-Zarcero, S., Pérez-Quintanilla, D., Câmara, J.S., Sierra, I., 2020. Two novel strategies in food sample preparation for the analysis of dietary polyphenols: micro-extraction techniques and new silica-based sorbent materials. Trends Food Sci. Technol. 98, 167–180. https://doi.org/10.1016/j.tifs.2018.06.020.

Catalkaya, G., Kahveci, D., 2019. Optimization of enzyme assisted extraction of lycopene from industrial tomato waste. Separ. Purif. Technol. 219, 55–63. https://doi.org/10.1016/j.seppur.2019.03.006.

Chemat, F., Abert-Vian, M., Fabiano-Tixier, A.S., Strube, J., Uhlenbrock, L., Gunjevic, V., Cravotto, G., 2019. Green extraction of natural products. Origins, current status, and future challenges. TrAC, Trends Anal. Chem. 118, 248–263. https://doi.org/10.1016/j.trac.2019.05.037.

Choi, Y.H., Verpoorte, R., 2019. Green solvents for the extraction of bioactive compounds from natural products using ionic liquids and deep eutectic solvents. Curr. Opin. Food Sci. 26, 87−93. https://doi.org/10.1016/j.cofs.2019.04.003.

Chuo, S.C., Nasir, H.M., Mohd-Setapar, S.H., Mohamed, S.F., Ahmad, A., Wani, W.A., Muddassir, M., Alarifi, A., 2022. A glimpse into the extraction methods of active compounds from plants. Crit. Rev. Anal. Chem. 52 (4), 667−696. https://doi.org/10.1080/10408347.2020.1820851.

Cittan, M., Altuntaş, E., Çelik, A., 2018. Evaluation of antioxidant capacities and phenolic profiles in Tilia cordata fruit extracts: a comparative study to determine the efficiency of traditional hot water infusion method. Ind. Crop. Prod. 122, 553−558. https://doi.org/10.1016/j.indcrop.2018.06.044.

Ćujić, N., Šavikin, K., Janković, T., Pljevljakušić, D., Zdunić, G., Ibrić, S., 2016. Optimization of polyphenols extraction from dried chokeberry using maceration as traditional technique. Food Chem. 194, 135−142. https://doi.org/10.1016/j.foodchem.2015.08.008.

da Silva, R.F., Carneiro, C.N., de Sousa, C.B. do C., Gomez, F.J.V., Espino, M., Boiteux, J., Fernández, M.de los Á., Silva, M.F., Dias, F.de S., 2022. Sustainable extraction bioactive compounds procedures in medicinal plants based on the principles of green analytical chemistry: a review. Microchem. J. 175, 107184. https://doi.org/10.1016/j.microc.2022.107184.

Diuzheva, A., Carradori, S., Andruch, V., Locatelli, M., De Luca, E., Tiecco, M., Germani, R., Menghini, L., Nocentini, A., Gratteri, P., Campestre, C., 2018. Use of innovative (Micro)Extraction techniques to characterise harpagophytum procumbens root and its commercial food supplements. Phytochem. Anal. 29 (3), 233−241. https://doi.org/10.1002/pca.2737. http://onlinelibrary.wiley.com/journal/10.1002/(ISSN)1099-1565.

Diuzheva, A., Locatelli, M., Tartaglia, A., Goga, M., Ferrone, V., Carlucci, G., Andruch, V., 2020. Application of liquid-phase microextraction to the analysis of plant and herbal samples. Phytochem. Anal. 31 (6), 687−699. https://doi.org/10.1002/pca.2939. http://onlinelibrary.wiley.com/journal/10.1002/(ISSN)1099-1565.

Fan, Y., Li, W., Zhang, S., Sun, S., Yang, L., 2020. Vitamin B3-based protic ionic liquids as green solvents for the isolation of astilbin from rhizoma smilacis glabrae. Ind. Crop. Prod. 152, 112563. https://doi.org/10.1016/j.indcrop.2020.112563.

Ferrante, C., Recinella, L., Locatelli, M., Guglielmi, P., Secci, D., Leporini, L., Chiavaroli, A., Leone, S., Martinotti, S., Brunetti, L., Vacca, M., Menghini, L., Orlando, G., 2017. Protective effects induced by microwave-assisted aqueous harpagophytum extract on rat cortex synaptosomes challenged with amyloid β-peptide. Phytother Res. 31 (8), 1257−1264. https://doi.org/10.1002/ptr.5850. http://onlinelibrary.wiley.com/journal/10.1002/(ISSN)1099-1573.

Fonmboh, D.J., Abah, E.R., Fokunang, T.E., Herve, B., Teke, G.N., Rose, N.M., Borgia, N.N., Fokunang, L.B., Andrew, B.N., Kaba, N., Bathelemy, N., Charles Ntungwen, F., 2020. An overview of methods of extraction, isolation and characterization of natural medicinal plant products in improved traditional medicine research. Asian J. Res. Med. Pharmaceut. Sci. 31−57. https://doi.org/10.9734/ajrimps/2020/v9i230152.

Gao, W., Qin, Y., Zang, Y., Zhu, J., Chen, W., Tong, S., Zeng, Y., Chu, C., 2023. Miniaturized matrix solid-phase dispersion and solid-phase clear-up combined with capillary electrophoresis for efficient determination of trace bioactive components in complicated sample matrix: take Wubi Shanyao Pill as an example. J. Separ. Sci. https://doi.org/10.1002/jssc.202300164. http://onlinelibrary.wiley.com/journal/10.1002/(ISSN)1615-9314.

Garcia-Vaquero, M., Rajauria, G., Tiwari, B., 2020. Conventional Extraction Techniques: Solvent Extraction Sustainable Seaweed Technologies: Cultivation, Biorefinery, and Applications. Elsevier, Ireland, pp. 171−189. https://doi.org/10.1016/B978-0-12-817943-7.00006-8. https://www.sciencedirect.com/book/9780128179437.

Goldansaz, S.M., Mahboubi, A., Yazdi-nejad, A., Jahanbakhshi, M., Mojab, F., 2018. Investigation on total phenolic content, antibacterial, and antioxidant activity of ethanolic extract of Helichrysum leucocephalum Boiss. America J Essen Oils Nat Pro. 6, 20–24.

Gong, J., Liang, C., Majeed, Z., Tian, M., Zhao, C., Luo, M., Li, C., 2023. Advances of imidazolium ionic liquids for the extraction of phytochemicals from plants. Separations 10 (3), 151. https://doi.org/10.3390/separations10030151.

González-Gómez, L., Morante-Zarcero, S., Pérez-Quintanilla, D., Sierra, I., 2021. Simultaneous determination of furanic compounds and acrylamide in insect-based foods by HPLC-QqQ-MS/MS employing a functionalized mesostructured silica as sorbent in solid-phase extraction. Foods 10 (7), 1557. https://doi.org/10.3390/foods10071557.

González-Gómez, L., Pereira, J.A.M., Morante-Zarcero, S., Câmara, J.S., Sierra, I., 2022. Green extraction approach based on µSPEed followed by HPLC-MS/MS for the determination of atropine and scopolamine in tea and herbal tea infusions. Food Chem. 394, 133512. https://doi.org/10.1016/j.foodchem.2022.133512.

Greco, V., Giuffrida, A., Locatelli, M., Savini, F., de Grazia, U., Ciriolo, L., Perrucci, M., Kabir, A., Ulusoy, H.I., D'Ovidio, C., Ali, I., 2023. Emerging trends in pharmacotoxicological and forensic sample treatment procedures. Appl. Sci. 13 (5), 2836. https://doi.org/10.3390/app13052836.

Greco, V., Giuffrida, A., Locatelli, M., Savini, F., de Grazia, U., Ciriolo, L., Perrucci, M., Kabir, A., Ulusoy, H.I., Ovidio, C., Catena, A.M., Ali, I., 2023b. Emerging procedures and solvents in biological sample pre-treatment. Adv. Sample Prep. 6, 100066.

Gu, Y., Jérôme, F., 2013. Bio-based solvents: an emerging generation of fluids for the design of eco-efficient processes in catalysis and organic chemistry. Chem. Soc. Rev. 42 (24), 9550–9570. https://doi.org/10.1039/c3cs60241a.

Häckl, K., Kunz, W., 2018. Some aspects of green solvents. Compt. Rendus Chem. 21 (6), 572–580. https://doi.org/10.1016/j.crci.2018.03.010.

Hikmawanti, N.P.E., Ramadon, D., Jantan, I., Mun'im, A., 2021. Natural deep eutectic solvents (Nades): phytochemical extraction performance enhancer for pharmaceutical and nutraceutical product development. Plants 10 (10). https://doi.org/10.3390/plants10102091. https://www.mdpi.com/2223-7747/10/10/2091.

Hizir, M., Barron, A.R., 2016. Basic Principles of Supercritical Fluid Chromatography and Supercritical Fluid Extraction.

Imoisili, P.E., Ukoba, K.O., Jen, T.C., 2020. Green technology extraction and characterisation of silica nanoparticles from palm kernel shell ash via sol-gel. J. Mater. Res. Technol. 9 (1), 307–313. https://doi.org/10.1016/j.jmrt.2019.10.059.

Khan, S., Khan, T., Shah, A.J., 2018. Total phenolic and flavonoid contents and antihypertensive effect of the crude extract and fractions of Calamintha vulgaris. Phytomedicine 47, 174–183. https://doi.org/10.1016/j.phymed.2018.04.046.

Kumar, A., Nirmal, P., Kumar, M., Jose, A., Tomer, V., Oz, E., Proestos, C., Zeng, M., Elobeid, T., Sneha, K., Oz, F., Phytochemicals, M., 2023. Recent advances in health benefits and extraction method. Molecules 28 (2), 887. https://doi.org/10.3390/molecules28020887.

Lim, J.R., Chua, L.S., Mustaffa, A.A., 2022. Ionic liquids as green solvent and their applications in bioactive compounds extraction from plants. Process Biochem. 122, 292–306. https://doi.org/10.1016/j.procbio.2022.10.024.

Locatelli, M., Ferrante, C., Carradori, S., Secci, D., Leporini, L., Chiavaroli, A., Leone, S., Recinella, L., Orlando, G., Martinotti, S., Brunetti, L., Vacca, M., Menghini, L., 2017. Optimization of aqueous extraction and biological activity of harpagophytum procumbens root on ex vivo rat colon inflammatory model. Phytother Res. 31 (6), 937–944. https://doi.org/10.1002/ptr.5821. http://onlinelibrary.wiley.com/journal/10.1002/(ISSN)1099-1573.

Locatelli, M., Kabir, A., Perrucci, M., Ulusoy, S., Ulusoy, H.I., Ali, I., 2023. Green profile tools: current status and future perspectives. Adv. Sample Prep. 6. https://doi.org/10.1016/j.sampre.2023.100068. https://www.sciencedirect.com/journal/advances-in-sample-preparation/issues.

Manousi, N., Sarakatsianos, I., Samanidou, V., 2019. Extraction Techniques of Phenolic Compounds and Other Bioactive Compounds from Medicinal and Aromatic Plants Engineering Tools in the Beverage Industry: Volume 3: The Science of Beverages. Elsevier, Greece, pp. 283–314. https://doi.org/10.1016/B978-0-12-815258-4.00010-X. https://www.sciencedirect.com/book/9780128152584.

Markhali, F.S., 2021. Roles of drying, size reduction, and blanching in sustainable extraction of phenolics from olive leaves. Processes 9 (9). https://doi.org/10.3390/pr9091662. https://www.mdpi.com/2227-9717/9/9/1662/pdf.

Martinho, D., Amin, K., Rosa, E., 2019. Extraction of phenolic compounds from olive leaf extracts and their effect on proliferation of human carcinoma cell lines. Agric. Sci. 10 (10), 1271–1285. https://doi.org/10.4236/as.2019.1010094.

Mehmood, A., Ishaq, M., Zhao, L., Yaqoob, S., Safdar, B., Nadeem, M., Munir, M., Wang, C., 2019. Impact of ultrasound and conventional extraction techniques on bioactive compounds and biological activities of blue butterfly pea flower (Clitoria ternatea L.). Ultrason. Sonochem. 51, 12–19. https://doi.org/10.1016/j.ultsonch.2018.10.013.

Meregalli, M.M., Puton, B.M.S., Camera, F.D.M., Amaral, A.U., Zeni, J., Cansian, R.L., Mignoni, M.L., Backes, G.T., 2020. Conventional and ultrasound-assisted methods for extraction of bioactive compounds from red araçá peel (Psidium cattleianum Sabine). Arab. J. Chem. 13 (6), 5800–5809. https://doi.org/10.1016/j.arabjc.2020.04.017. http://www.sciencedirect.com/science/journal/18785352.

Merone, G.M., Tartaglia, A., Locatelli, M., D'Ovidio, C., Rosato, E., Grazia, U.de, Santavenere, F., Rossi, S., Savini, F., 2020. Analytical chemistry in the 21st century: challenges, solutions, and future perspectives of complex matrices quantitative analyses in biological/clinical field. Analytica 1 (1), 44–59. https://doi.org/10.3390/analytica1010006.

Murador, D.C., Braga, A.R.C., Martins, P.L.G., Mercadante, A.Z., de Rosso, V.V., 2019. Ionic liquid associated with ultrasonic-assisted extraction: a new approach to obtain carotenoids from orange peel. Food Res. Int. 126. https://doi.org/10.1016/j.foodres.2019.108653. http://www.elsevier.com/inca/publications/store/4/2/2/9/7/0.

Nahar, L., Uddin, S.J., Alam, M.A., Sarker, S.D., 2021. Extraction of naturally occurring cannabinoids: an update. Phytochem. Anal. 32 (3), 228–241. https://doi.org/10.1002/pca.2987. http://onlinelibrary.wiley.com/journal/10.1002/(ISSN)1099-1565.

Nortjie, E., Basitere, M., Moyo, D., Nyamukamba, P., 2022. Extraction methods, quantitative and qualitative phytochemical screening of medicinal plants for antimicrobial textiles: a review. Plants 11 (15). https://doi.org/10.3390/plants11152011. http://www.mdpi.com/journal/plants.

Oreopoulou, A., Tsimogiannis, D., Oreopoulou, V., 2019. Extraction of Polyphenols from Aromatic and Medicinal Plants: An Overview of the Methods and the Effect of Extraction Parameters. Elsevier BV, pp. 243–259. https://doi.org/10.1016/b978-0-12-813768-0.00025-6.

Paz, J.E.W., Contreras, C.R., Munguía, A.R., Aguilar, C.N., Inungaray, M.L.C., 2018. Phenolic content and antibacterial activity of extracts of Hamelia patens obtained by different extraction methods. Braz. J. Microbiol. 49 (3), 656–661. https://doi.org/10.1016/j.bjm.2017.03.018.

Pontillo, A.R.N., Papakosta-Tsigkri, L., Lymperopoulou, T., Mamma, D., Kekos, D., Detsi, A., 2021. Conventional and enzyme-assisted extraction of rosemary leaves (rosmarinus officinalis L.): toward a greener approach to high added-value extracts. Appl. Sci. 11 (8), 3724. https://doi.org/10.3390/app11083724.

Puri, M., Sharma, D., Barrow, C.J., 2012. Enzyme-assisted extraction of bioactives from plants. Trends Biotechnol. 30 (1), 37–44. https://doi.org/10.1016/j.tibtech.2011.06.014.

Rafiee, Z., Jafari, S.M., Alami, M., Khomeiri, M., 2011. Microwave-assisted extraction of phenolic compounds from olive leaves; a comparison with maceration (Iran). J. Animal Plant Sci. 21 (4), 738–745.

Ramos, L., 2019. Use of new tailored and engineered materials for matrix solid-phase dispersion. TrAC, Trends Anal. Chem. 118, 751–758. https://doi.org/10.1016/j.trac.2019.07.006.

Rosa, G.S.d., Martiny, T.R., Dotto, G.L., Vanga, S.K., Parrine, D., Gariepy, Y., Lefsrud, M., Raghavan, V., 2021. Eco-friendly extraction for the recovery of bioactive compounds from Brazilian olive leaves. Sustain. Mat. Technol. 28. https://doi.org/10.1016/j.susmat.2021.e00276. https://www.journals.elsevier.com/sustainable-materials-and-technologies.

Sajid, M., Nazal, M.K., Ihsanullah, I., 2021. Novel materials for dispersive (micro) solid-phase extraction of polycyclic aromatic hydrocarbons in environmental water samples: a review. Anal. Chim. Acta 1141, 246–262. https://doi.org/10.1016/j.aca.2020.07.064.

Saxena, M., Saxena, J., Nema, R., Singh, D., Gupta, A., 2013. Phytochemistry of medicinal plants. J. Pharmacogn. Phytochem. 1, 168–182.

Ścigalski, P., Kosobucki, P., 2020. Recent materials developed for dispersive solid phase extraction. Molecules 25 (21), 4869. https://doi.org/10.3390/molecules25214869.

Shams, K.A., Abdel-Azim, N.S., Saleh, I.A., Hegazy, M.E.F., El-Missiry, M.M., Hammouda, F.M., 2015. Green technology: economically and environmentally innovative methods for extraction of medicinal and aromatic plants (MAP) in Egypt. J. Chem. Pharmaceut. Res. 7, 1050–1074.

Süntar, I., 2020. Importance of ethnopharmacological studies in drug discovery: role of medicinal plants. Phytochemistry Rev. 19 (5), 1199–1209. https://doi.org/10.1007/s11101-019-09629-9.

Taşkın, B., Aksoylu Özbek, Z., 2020. Optimisation of microwave effect on bioactives contents and colour attributes of aqueous green tea extracts by central composite design. J. Food Meas. Char. 14 (4), 2240–2252. https://doi.org/10.1007/s11694-020-00471-8.

Teixeira, R.F., Benvenutti, L., Burin, V.M., Gomes, T.M., Ferreira, S.R.S., Zielinski, A.A.F., 2021. An eco-friendly pressure liquid extraction method to recover anthocyanins from broken black bean hulls. Innovative Food Sci. Emerging Technol. 67. https://doi.org/10.1016/j.ifset.2020.102587. http://www.elsevier.com/locate/ifset.

Trentini, C.P., Cuco, R.P., Cardozo-Filho, L., Silva, C.d., 2019. Extraction of macauba kernel oil using supercritical carbon dioxide and compressed propane. Can. J. Chem. Eng. 97 (3), 785–792. https://doi.org/10.1002/cjce.23236. http://onlinelibrary.wiley.com/journal/10.1002/(ISSN)1939-019X.

Tzanova, M., Atanasov, V., Yaneva, Z., Ivanova, D., Dinev, T., 2020. Selectivity of current extraction techniques for flavonoids from plant materials. Processes 8 (10), 1222. https://doi.org/10.3390/pr8101222.

Uwineza, P.A., Waśkiewicz, A., 2020. Recent advances in supercritical fluid extraction of natural bioactive compounds from natural plant materials. Molecules 25 (17), 3847. https://doi.org/10.3390/molecules25173847.

Wang, M.Q., Ma, W.J., Shi, J., Zhu, Y., Lin, Z., Lv, H.P., 2020. Characterization of the key aroma compounds in Longjing tea using stir bar sorptive extraction (SBSE) combined with gas chromatography-mass spectrometry (GC–MS), gas chromatography-olfactometry (GC-O), odor activity value (OAV), and aroma recombination. Food Res. Int. 130. https://doi.org/10.1016/j.foodres.2019.108908. http://www.elsevier.com/inca/publications/store/4/2/2/9/7/0.

Wang, R., Li, C., Li, Q., Zhang, S., lv, F., Yan, Z., 2020. Electrospinning fabrication of covalent organic framework composite nanofibers for pipette tip solid phase extraction of tetracycline antibiotics in grass carp and duck. J. Chromatogr. A 1622, 461098. https://doi.org/10.1016/j.chroma.2020.461098.

Yi Ling, Y., Sook Fun, P., Yeop, A., Yusoff, M.M., Gimbun, J., 2019. Assessment of maceration, ultrasonic and microwave assisted extraction for total phenolic content, total flavonoid content and kaempferol yield from cassia alata via microstructures analysis. Mat. Today Proc. 19, 1273–1279. https://doi.org/10.1016/j.matpr.2019.11.133.22147853. In: https://www.sciencedirect.com/journal/materials-today-proceedings.

Yue, Y., Huang, Q., Fu, Y., Chang, J., 2020. A quick selection of natural deep eutectic solvents for the extraction of chlorogenic acid from herba artemisiae scopariae. RSC Adv. 10 (39), 23403–23409. https://doi.org/10.1039/d0ra03786a.

Zhang, B., Yang, R., Liu, C.Z., 2008. Microwave-assisted extraction of chlorogenic acid from flower buds of *Lonicera japonica* Thunb. Separ. Purif. Technol. 62 (2), 480–483. https://doi.org/10.1016/j.seppur.2008.02.013.

CHAPTER

4 From extracts to quantitative analyses: Method validation, analytical approaches (chromatographic, spectroscopic, and electrochemical procedures) and innovative instrumentations LC × LC

Laura Martín-Pozo[1], Katia Arena[2], Francesco Cacciola[2], Luigi Mondello[2,3] and Paola Dugo[2,3]

[1]*Department of Analytical Chemistry, Escuela Politécnica Superior, University of Seville, Seville, Spain;* [2]*Messina Institute of Technology c/o Department of Chemical, Biological, Pharmaceutical and Environmental Sciences, Former Veterinary School, University of Messina, Messina, Italy;* [3]*Chromaleont s.r.l., c/o Department of Chemical, Biological, Pharmaceutical and Environmental Sciences, University of Messina, Messina, Italy*

1. Introduction

Phytochemicals are plant-based bioactive compounds considered of great value by communities worldwide for their high antioxidant potential and well-known health-promoting properties. They encompass a wide range of structurally diverse classes of compounds including carotenoids e.g., carotene, lutein, or lycopene, phenolics, e.g., coumarins, flavonoids, phenolic acids, stilbenes, or tannins, phytosterols, e.g., campesterol, stigmastanol, and β-sitosterol, tocols, e.g., tocopherols and tocotrienols, fatty acids, volatile compounds, e.g., terpenes and aromatic compounds, and plant derived alkaloids, e.g., caffeine (Abdullah et al., 2022). Many of them are bioactive compounds and have been shown to possess a multitude of properties and beneficial health effects, such as antioxidant, anticancer, antimicrobial, antiinflammatory, antiviral, hypoglycemic, antihypertensive, hypocholesterolemic, antiaging, among others, which help prevent and mitigate various diseases (Bansal and Priyadarsini, 2022; Nwozo et al., 2023).

This has led to a gain in public interest and acceptance of medicinal plants and plant-based products. Phytochemicals can also be used in functional foods, beverages, and other food products of good nutritional value and economic significance and as precursors in the synthesis of new drugs (Bitwell et al., 2023). Nowadays, major concerns lie in the efficacy, safety, and quality of phytotherapeutics. The content of bioactive constituents is a key parameter to evaluate as a quality control,

Phytochemicals for Health. https://doi.org/10.1016/B978-0-443-15366-2.00004-6

for example, of herbal preparations intended for dietary supplementation (Indrayanto, 2022). In view of this, the development of rapid and simple quantitative methods becomes essential not only for the characterization of natural products but also to assess that formulations and plant-derived products comply with safety, control, and quality requirements.

Extraction of phytochemicals is an essential step in natural products research. Compounds must first be extracted from the matrix, isolated, and purified before being quantitatively analyzed (Mosić et al., 2020). There is a relentless drive to improve and develop enhanced extractive techniques with have greater efficiency and profitability. The physicochemical properties of phytochemicals must be considered, and the most appropriate technique and solvent selected to achieve maximum recovery (Kumar et al., 2023; Zhang et al., 2018). Over decades, a multitude of quantitative methods have been developed for the analysis of phytochemicals, most of them based on chromatography and spectroscopy. Liquid chromatography (LC) coupled with diode array detection (DAD) and mass spectrometry (MS) is the preferred technique for rapid quali–quantitative analysis of different phytocompound varieties, because it allows rapid screening of phytocompounds in addition to providing structural information with high sensitivity, selectivity, and precision (Chiriac et al., 2021).

In quantitative analysis, the amount of sample components is determined, and the reliability of the result is of great importance, especially in the context of quality control management. Validation plays a key role in producing reproducible results, and for this reason, all analytical procedures for phytochemical research must first be validated before their application, according to the latest official guidelines in force and the latest pharmacopoeias. Accordingly, these methods must be accurate, precise, and reproducible (Betz et al., 2011; Indrayanto, 2022).

The aim of this chapter is to provide a comprehensive review of the principal methodologies for the quantitative analysis of major phytochemicals. The most important aspects of the validation of methods to be applied in the analysis of natural products are described and discussed. The main techniques used in extraction and purification are described, as well as the most suitable techniques for their identification and quantification. Finally, comprehensive two-dimensional liquid chromatography (LC × LC) is proposed as an advanced and innovative technique for quali–quantitative purposes in field of phytochemical research.

2. Method validation

Various organizations have created guidelines for the validation of analytical methods. Among them, it has been demonstrated that it is adequate for the purpose for which it has been developed (Betz et al., 2011). The International Council for Harmonization of Technical Requirements for Pharmaceuticals for Human Use (ICH) proposed the ICH Q2(R1) guideline for method validation, which is in force since November 2005 (ICH Q2, R1, 2015). This guidance combines the parental Q2A guideline containing the analytical methods for validation of the required parameters and the Q2B methodological guideline.

The Food and Drug Administration (FDA) similarly combined its previously published guidelines, being the same as the ICH Q2(R1) guidance (ICH Q2, R1, 2015). FDA has also published two more industry-specific guidelines on the validation of drugs and biologics and the validation of bioanalytical methods, intended to complement the ICH guidance (Analytical Procedures and Methods Validation for Drugs and Biologics, 2015; FDA, 2018).

Nevertheless, in 2018, it was decided to develop a new ICH quality guideline based on ICH Q14 (analytical method development) and the revision of ICH Q2(R1) for better clarity and simplification. Both documents together address the harmonized scientific and technical basis for the proper development and validation of analytical methods throughout their entire life cycle. Drafts of the two documents can be consulted on the official website (published on March 24, 2022) (Analytical Procedure Development Q14, 2022; International Council for Harmonization, 2022).

Other guidelines for method validation are the AOAC guidelines (Appendix F, 2016) for Standard Method Performance, AOAC guidelines (Appendix K, 2023) for validation of supplements and botanicals, Eurachem Guide (Analytical Methods—A Laboratory Guide to Method Validation and Related Topics, 2014; Blanks in Method Validation, 2021) and the FDA ORA Laboratory Manual (n.d.), a reference for Office of Regulatory Affairs (ORA) laboratories (ORA, 2022).

Unfortunately, many of the published methods in phytochemical analysis fail to describe method validation. While accuracy, precision, and fitness for purpose are often assumed, they usually do not provide validation data, or if they do, only partial validation data are provided. Without proper validation, the reliability of the data cannot be confirmed and may not be replicated. Quantitative analytical methods for natural products should generally be validated following two steps. First, by evaluating the (linear) calibration model using the predicted coefficient of determination (R^2) and other statistical analysis to estimate the model fit. If the model meets the requirements, the remaining parameters, namely accuracy, precision, range, detection limits (DLs), quantification limits (QLs), and robustness must be evaluated following the directives and recommendations of official ongoing guidelines. Typical reliable validation characteristics that should be considered in quantitative assays are schematized in Fig. 4.1 and described further:

Specificity/selectivity: Ability to unequivocally determine the analyte in the presence of other analytes that may be present, e.g., impurities, degradation products, matrix. It can be demonstrated

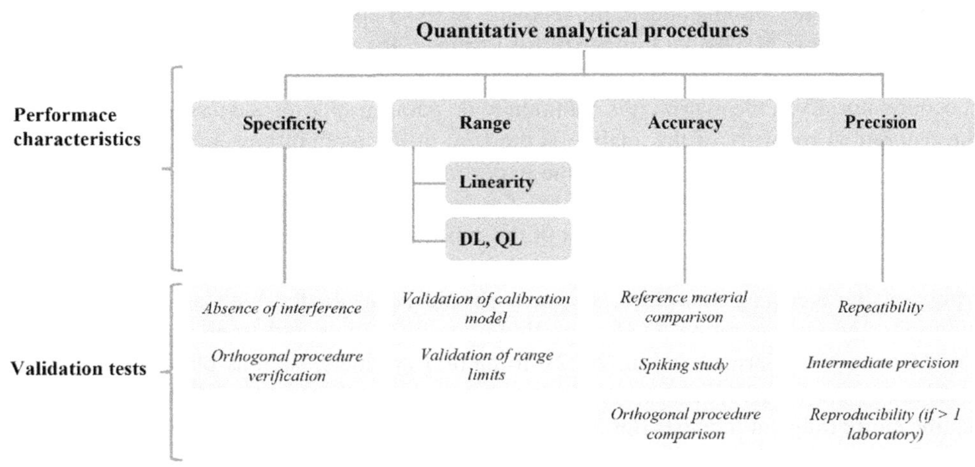

FIGURE 4.1

Validation data required in quantitative analytical procedures and possible tests that can be applied. *DL*, detection limit; *QL*, quantification limit.

Modified from ICH Q2(R2) guideline.

through the absence of interference substances in the retention time of the compounds to be determined or through the measurement with a second comparable analytical procedure. It is essential to minimize possible interferences during the sample treatment procedure. When the analytical method is not sufficiently selective, it is recommended to combine two or more procedures (International Council for Harmonization, 2022).

Linearity: Within a given range refers to the ability of the analytical process to obtain a response directly proportional to the amount of analyte in the sample. The range is the interval between the lowest and highest concentration in the sample at which the analytical process shows acceptable precision and accuracy. A minimum of five concentration levels are recommended, although additional concentrations may be required in complex models (International Council for Harmonization, 2022). The linearity of the calibration curve is usually expressed by the correlation coefficient (r) or the R^2. A correlation coefficient close to the unity is generally indicative of good linearity; however, it is not sufficient evidence to conclude that the calibration line is linear but the linear relationship should be evaluated with appropriate statistical methods (e.g., analysis of variance (ANOVA), lack-of-fit test, Mandel test), which indicate how well the data fit the linear model. In the case of analytical procedures showing nonlinear response other models or functions describing the relationship between response and concentration are required. For the case of the determination of phytochemicals that can be found at very different concentrations, it is sometimes necessary to dilute the final extracts to bring the components that exceed the linear dynamic range into the linear working range. In this situation, two analyses are required: the analysis of the undiluted sample and the analysis of the diluted sample, to ensure that all compounds of interest can be quantified with precision and accuracy (Betz et al., 2011).

Accuracy: Expresses the closeness of the experimental result to the real or accepted value. It is performed by recovery tests using certified reference material or analyte standards added to analyte-free matrices, the latter being the most widely used in studies of natural products. In a spike recovery assay, a certain amount of the compound of interest is added to the matrix and by comparing the amount found versus the amount added, the recovery of the method is calculated, which is an estimate of the accuracy (Betz et al., 2011). It is ideal, for instance, for the analysis of traces of contaminants such as pesticides in food; nevertheless, in the case of phytochemical studies where the components of interest occur naturally in the matrix, it is common to do parallel spiking and unspiking analysis. In this case, the theoretical recovery of the analyte is the sum of the amount present in the sample plus the amount added, and the difference between the experimental amount and the theoretical amount provides the estimate of accuracy (Betz et al., 2011). On the other hand, when certified reference materials are available (which contain a known amount of the substance of interest), these can be used instead or in addition to the spiking assay. Acceptable recovery values vary depending on the concentration and the purpose of the analysis. Therefore, accuracy should be evaluated with an adequate number of determinations and concentration levels e.g., three concentrations and three replicates of each (International Council for Harmonization, 2022). In natural products, such as plant extracts and foods, phytochemicals may be present in a broad range of concentrations, so recovery must be evaluated covering the entire range of interest (Betz et al., 2011). As a typical performance reference, about 95% of the results in a single laboratory at 1 $\mu g\ g^{-1}$ should fall within 80%–120% of the mean. If the recovery is below the accepted range, investigation should be done to improve the average recovery (extraction) while if the recovery is above the range, it is recommended to improve the separation (Appendix F, 2016; Appendix K, 2023).

Precision: It refers to the degree of agreement of the results when conditions are kept constant and is usually referred to as the relative standard deviation (RSDs) of simultaneous replicates. According to the ICH/FDA guideline, the precision is given by repeatability, intermediate precision, and reproducibility. *Repetitively* should be evaluated using a minimum of nine determinations, i.e., three concentration levels and three replicates of each or a minimum of six replicates at 100% concentration; *intermediate precision* should be assessed studying typical events or variations that could affect the precision (e.g., different days, environmental conditions, etc.); *reproducibility* is evaluated through interlaboratory tests, and although it is not usually mandatory for submission, it should be considered in the standardization of analytical procedures (International Council for Harmonization, 2022). Several factors can affect and reduce the accuracy of the data, such as the use of uncalibrated equipment, low chromatographic resolution, and handling errors. Usually, to obtain a representative value, replicates are performed on the same day (intraday precision) and over several consecutive days (interday precision). Repeatability varies with concentration (Appendix K, 2023).

Sensitivity: It is determined by the DLs and QLs, which are defined as the smallest amount of analyte that can be detected or quantitatively determined with sufficient precision and accuracy, respectively. They can be estimated in several ways, some of them are: I) on a signal-to-noise basis, with a signal-to-noise of 3:1 being considered acceptable for estimating the DL and a signal-to-noise of 10:1 for the QL; II) on the basis of the standard deviation of the response in the linear range and slope, expressed as 3. $3\sigma/S$ for the DL and $10\sigma/S$ for the QL, σ being the standard deviation and S the slope of the calibration curve; III) based on the standard deviation of the blank, measuring the background response by analyzing an adequate number of blanks; IV) based on its visual evaluation experimentally, by analyzing samples of known concentration establishing, which is the lowest concentration that can be detected or quantified (Analytical Procedure Development Q14, 2022; International Council for Harmonization, 2022; ICH Q2, R1, 2015).

Robustness: Measures the ability of the analytical procedure to remain unaffected by deliberate variations in parameters. It can be submitted as part of the development of the analytical procedure or should be available upon request.

It should be considered that during the life cycle of an analytical procedure, some changes may be necessary, requiring in those cases a total or partial revalidation (International Council for Harmonization, 2022). In addition, the matrix effect (ME) can also be studied, being defined as the matrix on the response enhancement or suppression of the analytes (Zhang et al., 2019). It is an important parameter to assess in the development of a method and can be measured in several ways; one of them is by considering the slopes of the calibration curves of the phytochemicals in synthetic matrix and in real matrix. To compensate for the matrix effect, a matrix match calibration approach is used (Pereira-Coelho et al., 2023).

3. Analysis of phytochemical compounds

The study of phytochemicals in materials of natural origin has been the focus of attention for decades due to the therapeutic properties that many of them have been shown to have (antioxidant, antiinflammatory, antimicrobial, etc.). At this point, a large quantity of literature is available in scientific databases on analysis techniques both qualitative and quantitative, experiencing an exponential growth (Fig. 4.1A). In particular, among the over 3350 document results found after searching TITLE-ABS-

KEY (phytochemicals AND quantitative AND analysis) in the Scopus database, 1614 have been published in just 5 years, since 2019 (accessed on May 22, 2023). Of these, the top subject areas have been Pharmacology, Toxicology, and Pharmaceutics, followed by Biochemistry, Genetics, and Molecular Biology (Fig. 4.2).

For the elaboration of this chapter, special emphasis was placed on scientific articles published during the last 5 years, in order to provide an up-to-date and current view of the state of the art in analytical procedures for the analysis of phytochemicals.

3.1 Chromatographic techniques

They are based on the physical separation of the components of a mixture by distribution between two phases, a stationary phase, and a mobile phase, based on the different interaction capacity of each component. Among the most widely used in the determination of the different classes of phytocompounds are gas chromatography (GC) and LC coupled to various detectors.

3.1.1 Gas chromatography

GC coupled to MS is suitable for the analysis of terpenoids and other volatile compounds. It is used for the characterization of components of the oily fraction (nonpolar or medium polarity compounds) or highly volatile compounds (including alkanes, esters, alcohols, aromatic hydrocarbons, terpene derivatives, and steroids). The use of the GC-MS technique for the analysis of nonvolatile or semivolatile compounds, such as polyphenols, is uncommon due to the prior derivatization step required to increase the volatility and stability of the compounds (Chiriac et al., 2021). Nonpolar capillary columns coated with 5% phenyl methylpolysiloxane or equivalent, of dimensions generally 30 m × 0.25 mm × 0.25 μm film thickness, are the most widely used in the separation, using helium as carrier gas usually at a flow rate of 1 mL min^{-1}. The injector temperature is around 250–270 °C, and temperature ramps are employed usually starting in the range 40–100 °C to a final temperature between 200–350 °C. In GC-MS, the mass spectrometer operates in electron impact (EI) mode at 70 eV.

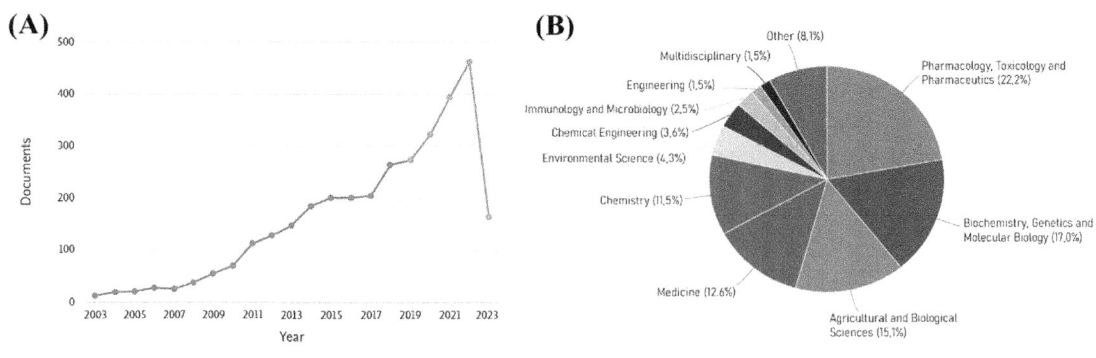

FIGURE 4.2

Number of papers published in the last two decades (A) and subject area of the papers results in the last 5 years (B).

Data from Scopus database.

3.1.2 Liquid chromatography

LC, especially in combination with UV/Vis spectrometry diode array detection (DAD) and tandem mass spectrometry (MS/MS), is typically used for the analysis of nonvolatile compounds, such as phenolic acids, anthocyanins, flavonoids, tocopherols, and carotenoids. Due to the high solubility of these classes of compounds, reversed phase (RP) LC is used, where a nonpolar packed column and mobile phases of polar nature are used as stationary phase. The most commonly employed analytical columns are C_{18} (silica bonded with alkyl chains), and to a lesser extent C_8, for the separation of these classes of compounds (Chiriac et al., 2021; Zhang et al., 2019) of variable length, diameter, and particle size; usually water (phase A) with organic solvents such as methanol or acetonitrile (B) is used as mobile phase composition; although C_{30} columns have been shown to be suitable for separating lipid-soluble vitamins and phytochemicals, such as carotenoids, chlorophylls, and tocopherols, using methanol and methyl tert-butyl ether as mobile phases (Cortés-Herrera et al., 2019; Rao et al., 2023). In order to improve the separation, phosphoric acid, acetic acid, trifluoroacetic acid, or formic acid is usually added to the mobile phase as additive, at varying percentages. Typically, the separation takes place using a gradient elution, as it achieves a better separation of the analytes, although the isocratic mode has provided good resolution in some studies. Normal phase (NP) LC is also used by some authors, although to a lesser extent than RP, and has been applied in the analysis of flavan-3-ols and procyanidins according to their degree of polymerization. In this type of technique, analytical columns with more polar packing such as dihydroxypropyl group bonded to silica gel and organic solvent mixtures are used as mobile phases with almost no water (Goto et al., 2021). Finally, an alternative increasingly used and applied in the analysis of phytochemicals is hydrophilic interaction liquid chromatography (HILIC) separation, which used in multidimensional techniques in the first dimension combined with RP in the second dimension has shown excellent results in the separation of polyphenols in a variety of matrices with a high-resolution capacity (Arena et al., 2023).

3.2 Spectroscopic techniques

These are instrumental methods based on the interaction of electromagnetic radiation with molecules to identify them or determine their concentration. In the quantitative analysis of phytochemical compounds, the most widely used are described as follows:

3.2.1 UV/vis spectroscopy

The region from 190 to 600 nm is of special interest in the determination of phytochemicals since most of them show their maximum absorption in this wavelength range. They can be used for identification and quantification by knowing the absorption maximum corresponding to their structural characteristics (e.g., total phenolic extract (280 nm), flavones (320 nm), etc.) (Arena et al., 2023; Lama-Muñoz and Contreras, 2022). UV/vis detectors and DAD are widely used together with LC because of their simplicity and low cost. The main disadvantage of UV/vis detection in the identification of natural compounds is its low sensitivity, resolution, and specificity, since when two compounds coelute it is not possible to distinguish them. Therefore, it is common to use MS detection in parallel to obtain complementary information (Lama-Muñoz and Contreras, 2022).

Spectrophotometric (colorimetric) methods are useful for the quantification of the total content of selected groups of phytochemicals and are the most extensively used for this purpose because of their simplicity and low cost. They are assays that determine specific functional groups and are based on different principles. Folin–Ciocalteu method is widely used to determine the total content of phenolic compounds (TPC); it is based on the ability of phenols to react with a reagent at basic pH, giving rise to an intense blue color (Folin and Ciocalteu, 1927; Georgiana Munteanu and Apetrei, 2021). The vanillin and proanthocyanins assay has been used to determine the total content of proanthocyanins (condensed tannins); it is based on the binding of the aldehydic group of vanillin to C-6 of the catechin A cycle, forming a red chromophore complex (Schofield et al., 2001); while complexation with aluminum trichloride ($AlCl_3$) has been used to determine total flavonoid content (TFC) (Shraim et al., 2021); the addition of $AlCl_3$ forms an acid complex with the oxo and hydroxyl groups of the flavonoids forming a yellow-colored Al(III)-flavonoid chelate complex that can be measured spectrophotometrically (Fig. 4.3). Spectrophotometric methods provide useful qualitative as well as quantitative information. The main disadvantage is that they provide an estimate of the total content and do not allow quantification of individual compounds (Ignat et al., 2011).

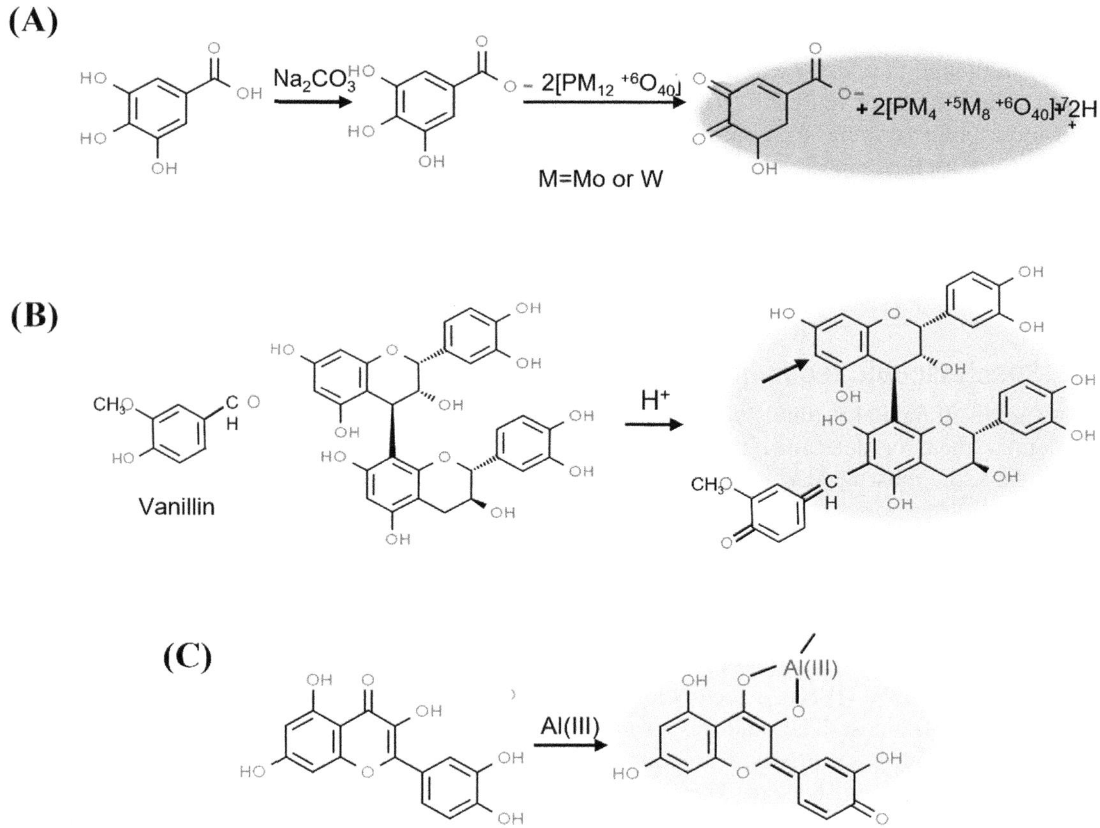

FIGURE 4.3

Reactions of selected spectrophotometric for total content determination: Folin–Ciocalteu method (A), vanillin method (B), and Al(III) quelate formation (C).

3.2.2 Infrared spectroscopy

The infrared spectroscopy (comprising the infrared region (IR) from 700 to 1 nm) has also been applied allowing rapid and nondestructive analysis. It relates to vibrational changes when molecules are excited with infrared radiation (Ben Salah and Allouche, 2019). It is divided into three types according to the wavelength range: near-IR (0.78−2.5 μm), mid-IR (2.5−50 μm), and far-IR (50−1000 μm) (NIR, MIR, FIR, respectively) (Johnson et al., 2023). IR spectroscopy is not only of great utility for structural elucidation of new molecules but also shows great potential for quantification purposes and relative prediction of phytochemical levels in natural matrices (Johnson et al., 2020). In particular, NIR and MIR have been applied to perform compositional predictions in a variety of matrices, including foods such as grapes, tomatoes, olive oil, cocoa bean, red wine, and tea, among others (Johnson et al., 2023; Lama-Muñoz and Contreras, 2022). Polyphenols, anthocyanins, carotenoids, and ascorbic acid have been the most studied bioactive compounds. Hernández-Hernández et al. (2021) demonstrated the feasibility of NIR in the analysis of bioactive compounds in intact cocoa bean husk; by means of regression models, they could relate the constituents with the Visible + NIR reflectance spectrum and determined the total content of sugars and phenols, and the content of individual components, such as catechin, theobromine, and epigallocatechin, obtaining very good correlations. A possible strategy is to couple GC or LC to Fourier Transform Infrared (FTIR) spectrometers (third generation of infrared spectrometers). This combination allows us to separate the components through the capillary or analytical column, before collecting the FTIR spectrum of each eluted compound, which provides detailed information on the functional groups of the compounds (Johnson et al., 2023). An interesting application is that of Lucarini et al. (2020) who employed FTIR in the mid-IR region to determine qualitatively and quantitatively the profile of phenolic compounds along with that of fatty acids in grape seeds by analyzing the main functional groups. The main advantage over other analytical methods is that it is nondestructive, and the major drawback is the complexity of the spectrum, which makes it difficult to interpret (Johnson et al., 2023).

3.2.3 Mass spectrometry

MS is a powerful and advanced detection technique that has high sensitivity and selectivity in quantitative analysis. It is broadly used as a detection method for LC and GC. There are two types of spectrometers: low-resolution and high-resolution spectrometers. Low-resolution spectrometry employs quadrupole (Q), ion traps (IT), and triple quadrupole (QqQ) and is used in selective screening methods, identification, and quantification of phytochemicals. For data acquisition, two or three transitions (product) resulting from the fragmentation of the parent precursor are generally monitored using multiple reaction monitoring (MRM) mode. Low-resolution spectrometry employs time-of-flight (ToF) and orbitrap and is popular in the characterization of natural compounds as it provides complete information on the exact mass, elemental composition, and chemical structure of the compounds. Hybrids such as quadrupole-ToF (QToF) and ion-trap Orbitrap (IT-Orbitrap) offer even greater confirmatory capability, as they allow the molecular formula and fragmentation pattern of the compounds to be generated (Arena et al., 2023). High-resolution spectrometry is especially advantageous in the identification of new compounds (comprehensive screenings) while low-resolution spectrometry is especially advantageous in the identification and quantification of target compounds (selective screenings), although the former can also be used for quantitative purposes. The ionization source generally applied to MS hyphenation with LC is electrospray ionization (ESI), while for GC-MS, EI is the source of choice (at 70 eV). However, other ionization sources such as atmospheric pressure chemical ionization (APCI) have been applied providing even more complementary information to the above (Chiriac et al., 2021; Muller and de Villiers, 2023; Pereira-Coelho et al., 2023; Zhang et al., 2019).

3.3 Other quantitative techniques

Capillary electrophoresis (CE) has been used as an alternative in the analysis of various classes of phytochemicals. It is an analytical technique that applies an electric field to separate sample components, and these are separated by their electrophoretic mobility. Only charged molecules migrate through the capillary. The velocity depends on the charge, size, shape, and resistance of the particle (Gackowski et al., 2021). The most commonly used detectors in phytochemical analysis are UV/Vis DAD, fluorescence, or amperometry detectors. It has been applied to the analysis of polyphenols, coumarins alkaloids, and amino acids in herbal raw materials (Gackowski et al., 2021). CE capillaries have very narrow diameters, which results in good and improved resolution compared with LC. The limitation is that pH control is critical, as small variations affect the charge of the molecules and have a greater impact than with LC (Przybylska et al., 2021).

Electrochemical techniques are those in which the potential, current, or charge in the electrochemical cell serves as the analytical signal. In this context, biosensors with electrochemical detection have been proposed as alternatives to chromatographic and spectrometric techniques, in order to avoid the often-tedious sample preparation step. They have proven to be reliable on-site and have been applied for the quantification of various phytochemicals, such as phenolic compounds (Verma et al., 2023). This type of sensor is composed of two components, a biological part (analyte and receptor) being the enzyme-based sensor one of the most explored, and a transducer. They electrochemically convert nonactive analytes into products by catalysis according to the enzyme that performs the oxidation and reduction on the surface of the working electrode. The advantages include high sensitivity, precision, reliability, and short response time (Bounegru and Apetrei, 2021; Verma et al., 2023). Although the main limitation is the leaching of the enzyme from the substrate, and its reduced activity when the enzymes are immobilized on the support. Bounegru and Apetrei (2021) developed a new electrochemical biosensor based on carbon nanofibers modified with cobalt phthalocyanine and laccase, to determine p-coumaric acid. The biosensor showed optimal linearity characteristics 0.4–6.4 μM and low DL and QL 4.83×10^{-7} M and 1.61×10^{-6} M, respectively, and was applied to different phytoproducts, obtaining p-coumaric acid concentrations comparable with those obtained with the standard FTIR method.

Fig. 4.4 outlines the workflow of the entire process seen in this chapter, from sample pretreatment, extraction, to quantification of principal phytocompounds. Many studies combine the implementation of two or more different analytical methodologies since they allow obtaining complementary and more complete compositional information, as well as more accurate qualitative and quantitative data. An interesting combined application was carried out by Cervantes-Paz et al. (2023) who used GC-MS, HPLC-MS, and FTIR for the identification and quantification of fatty acids and lipid-soluble phytochemicals in avocado during its ripening process. In this work, GC was used for the determination of fatty acids, HPLC-MS for the determination of carotenes, chlorophylls, and tocopherols, and FTIR to determine the degree of degradation of unsaturated fatty acids at different ripening stages.

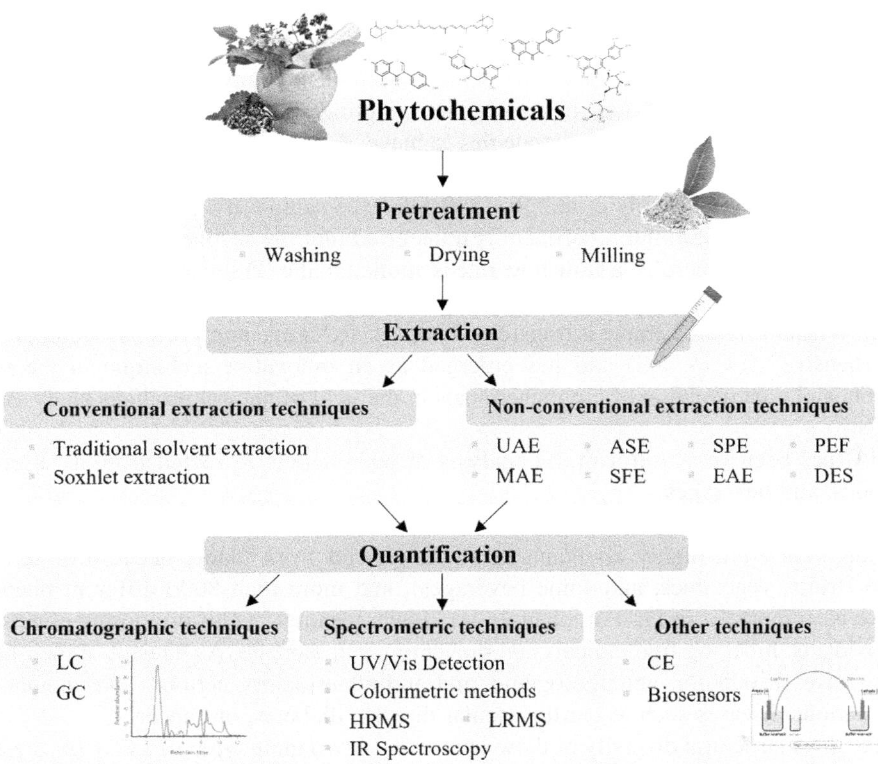

FIGURE 4.4

Summary representation of the possible workflow of the extraction and quantitative analysis of phytocompounds from natural products.

ASE, assisted solvent extraction; *CE*, capillary electrophoresis; *EAE*, enzymatic assisted extraction; *DES*, deep eutectic solvents; *GC*, gas chromatography; *HRMS*, high resolution mass spectrometry; *IR*, infrared; *LC*, liquid chromatography; *LRMS*, low resolution mass spectrometry; *MAE*, microwave assisted extraction; *PEF*, pulse electric field; *SFE*, supercritical fluids extraction; *SPE*, solid-phase extraction; *UAE*, ultrasound assisted extraction.

4. LC × LC: applications

Food-derived extracts and natural-real world samples are very complex matrices containing hundreds of chemical constituents of significantly varying polarity, structure, and concentration. Their separation and quantification are often a major challenge in the field of separative sciences. Given the physicochemical properties of most phytochemicals that we can find in nature, LC coupled to different detectors, especially to DAD and MS detectors, is considered the most popular and effective in the separation of these constituents (Ji et al., 2018). However, sometimes, when dealing with very complex samples, the resolving power of conventional LC is not sufficient and does not provide accurate quantification results because of analyte coelution and matrix effects; more innovative alternative analytical techniques with higher separation efficiency are required (Martín-Pozo et al., 2022). An interesting alternative to overcome this issue is the employment of multidimensional liquid chromatography (MD-LC) techniques, which involve the coupling of two or more orthogonal or near-orthogonal separation systems. These strategies achieve a considerable increase in peak capacity (several hundred can be achieved) which helps to solve component overlapping and the quantification of minority components (Cacciola et al., 2021), Briefly, the sample is initially separated in the first dimension (^{1}D) and afterward the ^{1}D effluent is transferred into the second dimension (^{2}D) through a modulator or interface. Generally, a slow flow rate is applied in the ^{1}D separation to reduce the fraction of solvent entering ^{2}D and increase the ^{1}D sampling rate, and microcolumns are used in the ^{1}D (Donato et al., 2012). When the entire sample is transferred from ^{1}D to ^{2}D and analyzed, the technique is known as "comprehensive" (LC × LC) and has emerged as an innovative technique in the separation, identification, and quantification of phytochemicals in the field of natural products analysis (Cacciola et al., 2017).

LC × LC has been very useful in the analysis of polyphenols extracted from a variety of plant extracts, foods, and beverages.

Polyphenols are a group of phytochemicals characterized by having more than one phenol group in their structure. They are highly abundant in foods derived from plants because of its secondary metabolism (fruits, vegetables, and some beverages), and more than 8000 different phenolic compounds have been reported so far. Polyphenols are known for their health-promoting effects, and they act as antioxidants, blocking free radicals and preventing cell damage. As a result of their antioxidant power, they have vasodilator, antiatherogenic, and antiinflammatory actions, and attenuate the risk factors for various diseases such as cardiovascular disease, diabetes, or cancer.

Given the great structural diversity of these compounds, two main types of LC × LC are applicable to the analysis of polyphenols contained in natural products and foods: RP-LC × RP-LC and HILIC × RP-LC. So far, most applications on polyphenols have been performed with RP-LC × RP-LC as it allows separating a large variety of molecules of similar molecular weights, structure, and polarity (Ji et al., 2018). As is well known, the combination of RP-LC mode in both dimensions does not pose a problem of mobile phase incompatibility, in contrast to other LC × LC modes, but suffers from very low orthogonality. At first, "full gradient" programs were applied and over time "tailored gradient" programs were investigated. Recently, applications employing HILIC × RP-LC setup have been emerging, which features two separation mechanisms with complementary selectivity for analyte retention and employs miscible and compatible mobile phases for common detectors, such as DAD and MS. However, here the weak solvent used in the ^{1}D is a strong solvent eluent for ^{2}D, leading to phase mismatch and poor focusing on the entrance of the second column (Cacciola et al., 2021). An interesting application of an RP-LC × RP-LC-PDA method was developed by Arena et al. (2019) for the characterization and quantification of polyphenols from *Pistacia vera* L. kernel samples. The authors validated the method in terms of linearity, sensitivity, and accuracy. They obtained satisfactory analytical characteristics, $R^2 > 0.99$, low DLs and QLs values, RSD <0.5% and accuracy <10% (as relative error deviation in a spiked sample), demonstrating the feasibility and reliability of the method for quantitative purposes. Also, Arena et al. (2022) employed a HILIC × RP-LC-PDA-ESI/MS method for the analysis of polyphenolic compounds in *Rhus coriaria* L. To overcome solvent incompatibility, they applied active modulation by introducing a flow makeup of water (weaker solvent) after separation on the ^{1}D column, thus reducing the solvent strength, as illustrated in Fig. 4.5.

Different standards of phenolic compounds were used to construct the calibration curves (gallic acid, protocatechuic acid, isoquercetin, myricetin, and cyanidin) and the amount of polyphenols was determined with a semiquantitative approach considering the classes of compounds. Gallic acid derivatives were significantly the most abundant class. Lately, the same research group, developed an LC × LC system, comprised of a ZIC-HILIC and C18 columns in the ^{1}D and ^{2}D, respectively, was tuned and employed for attaining high-resolution profiles of the polyphenolic pattern in seven commercial berry juices (Arena et al., 2023). The developed HILIC × RP-LC method was validated in terms of linearity range, correlation coefficients, DL, QL, precision (intra- and inter-day), and recovery. A total of 104 polyphenolic compounds belonging to different chemical classes (hydroxybenzoic and cinnamic acid derivatives, flavone glycosides, flavonols, flavonol glycosides, dihydroflavonols, and anthocyanin glycosides) have been characterized and quantified in the juices investigated. Fig. 4.6 reports the HILIC × LC-PDA plots ($\lambda = 280$ nm) of the elderberry, honeyberry, blueberry, and chokeberry juices.

Such a ZIC-HILIC × C18 platform based on focusing modulation, never employed so far for berry juices, showed a remarkable separation capability with high values of corrected peak capacity (up to 1372) and orthogonality (A_o up to 0.80), thus providing a great applicability to be advantageously employed for other complex food samples.

Although HILIC is almost exclusively used in the ^{1}D separation mode, Muller and de Villiers (2023) compared the two configurations for polyphenol analysis and found a marked improvement in

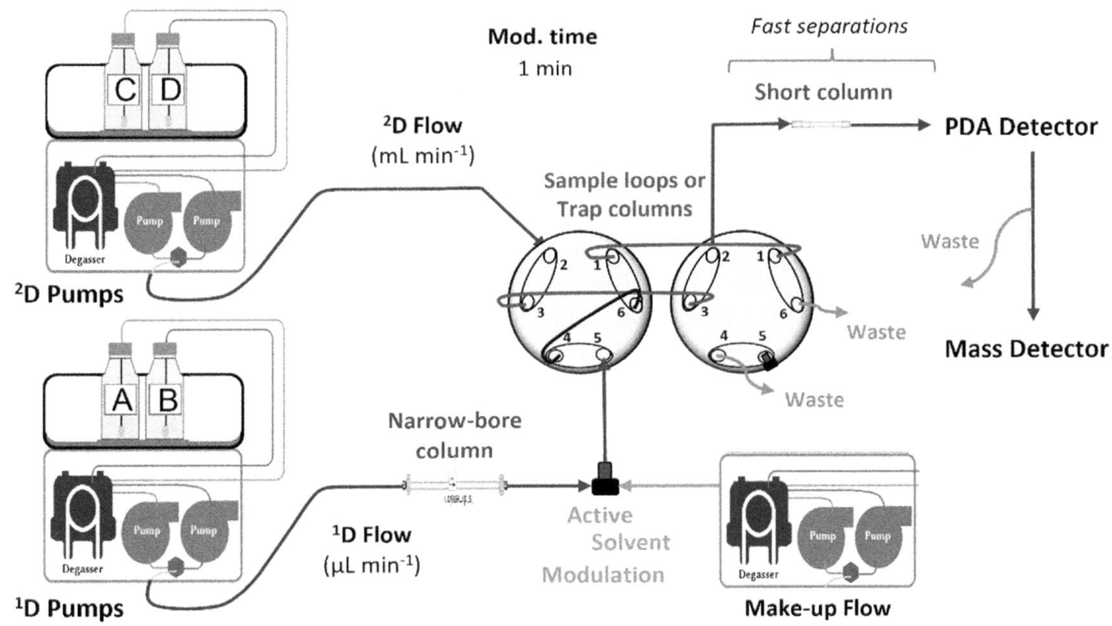

FIGURE 4.5

Typical configuration of an LC × LC-PDA-ESI/MS system interfacing two 6-port 2-position valves using active solvent modulation for solving solvent mismatching issue.

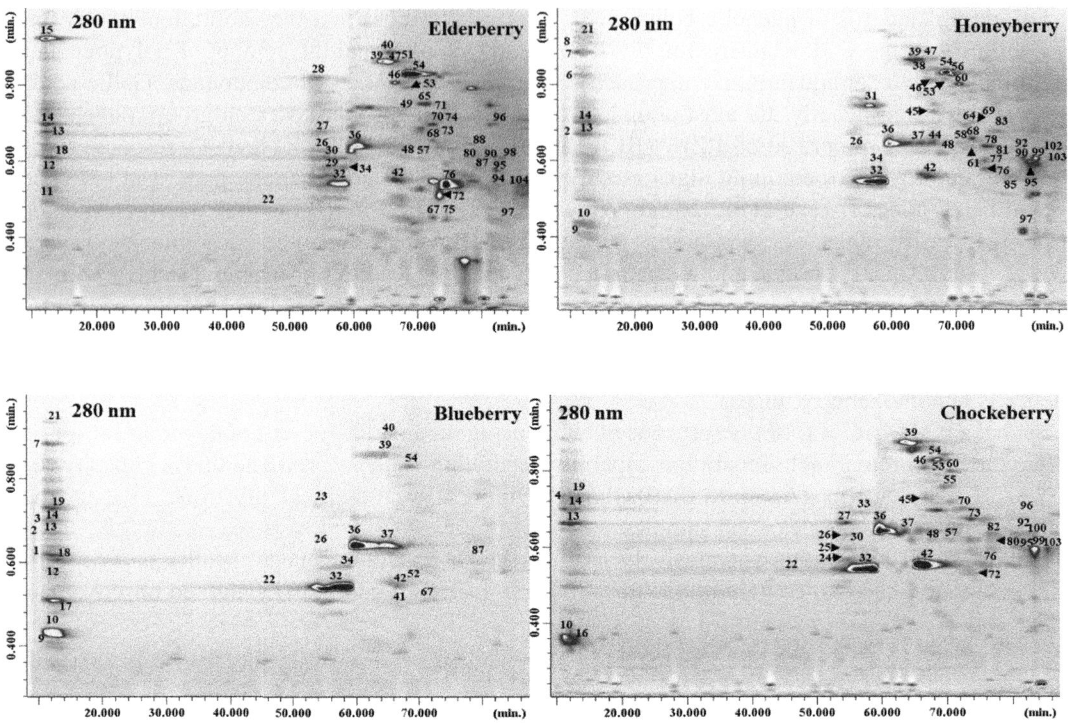

FIGURE 4.6

HILIC × LC-PDA plots (λ = 280 nm) of the elderberry, honeyberry, blueberry, and chokeberry.

Taken with kind permission of Arena *et al. From Arena, K., Cacciola, F., Miceli, N., Taviano, M.F., Cavò, E., Murphy, R.E., Dugo, P. & Mondello, L., 2023. Determination of the polyphenolic content of berry juices using focusing-modulated comprehensive two-dimensional liquid chromatography coupled to mass spectrometry detection. Analy. Bioanal. Chem. 415 (13), 2371–2382. https://doi.org/10.1007/s00216-022-04216-y.*

sensitivity using the RP-LC × HILIC configuration which, despite the detriment in resolution, could be a potential advantage for quantitative analysis.

LC × LC strategies aimed at lipid determination mainly involve RP-LC and HILIC mode coupling. The combination of these two separation mechanisms usually shows good orthogonality. An excellent application was reported by Navarro-Reig et al. (2018), who successfully used an LC × LC approach with chemometric tools to evaluate the effect of arsenic exposure on the lipidome of growing rice samples. In the ^{1}D, they used the RP-LC mode and in the ^{2}D an HILIC column. The chemometric-based data analysis was useful in the understanding, spectral and elution profile resolution as well as in the detection and statistical analysis of the method. On the other hand, a multitude of reported studies have analyzed triacylglycerols (TAGs) by Ag^+-LC × RP-LC combination in various vegetable oils such as rice (Mondello et al., 2005), linseed (Dugo et al., 2006), corn (Navarro-Reig et al., 2016; van der Klift et al., 2008), *Parinari curatellifolia* (Byrdwell, 2017), and borage oil (Arena et al., 2021). In the majority of applications, a silver-ion-coated cation exchange column (Ag^+ column) was employed for separation in the ^{1}D together with a C_{18} column in the ^{2}D. Furthermore, in all cases, the use of APCI-MS made identification easier, especially when commercial standards are not available; first, because of the absence of coelutions, and second, because it restricted the number of compounds that can be assigned to a well-defined spot position in the 2D chromatogram. These positive features make LC × LC methods not only highly useful in qualitative determination, the main goal of the reported works, but also in quantitative analysis of the lipid component.

Regarding the determination of carotenoids, the highly orthogonal NP-LC × RP-LC platforms have been the most extensively exploited (Cacciola et al., 2020). In NP-LC, mobile phases of a stronger nonpolar nature than those used in RP-LC are used, which leads to the downside of solvent incompatibility that must be overcome. In some applications, this limitation could be reduced by combining a microbore column in the ^{1}D with a conventional monolithic column in the ^{2}D (Dugo et al., 2008). Under NP-LC conditions, carotenoids are separated according to their degree of polarity, whereas in the RP-LC mode, they elute according to their different hydrophobicity. With regard stationary phases, taking into consideration the chemical and physical properties of carotenoids, in NP-LC mode cyanopropyl columns have been widely used while in RP-LC mode C_{30} and C_{18} columns have exhibited excellent results (Cacciola et al., 2016a; Cacciola et al., 2016b; Cacciola et al., 2012; Cacciola et al., 2020; Dugo et al., 2008). As for the detectors, DAD together with MS detection was the most applied, providing complementary information and being the most accurate for their unequivocal identification, taking into consideration that UV/vis absorption and MS spectrum of many carotenoids are similar, and it is difficult to dispose of commercial standards for all of them (Cacciola et al., 2020). A typical example of application of the NP-LC × RP-LC system was developed by Cacciola et al. (2012) for the analysis of carotenoid composition in red chili peppers. In this study they selected a microbore cyano column at the ^{1}D interfaced to a C_{18} at ^{2}D. A total of 33 compounds were separated and, as the authors could not quantitatively evaluate the components due to the lack of reference material, they successfully estimated the amount in (%) by carotenoid classes by plot integration. The most abundant carotenoids corresponded to di-ol-mono-keto-mono-ester class, which accounted for 36% of the total carotenoid content (Cacciola et al., 2012).

5. Conclusions

In this chapter, the main strategies for the determination of phytochemicals (e.g., phenolic compounds, carotenoids, and fatty acids) extracted from natural products have been described.

The most advanced chromatographic techniques including both GC and LC, coupled to different detectors, especially MS, are the favorite choice as they provide high sensitivity, selectivity, and speed for the identification and quantification of phytochemicals. In parallel, for a quantification of the total content of the phytochemical class, traditional colorimetric methods (or with some modifications) continue to be used, due to their simplicity and low cost. Studies combining the implementation of several techniques (e.g., FTIR, GC-MS, and LC-MS) are increasing, as more comprehensive qualitative and quantitative information can be obtained.

Unfortunately, developing methodologies fail to describe the validation of the method. They do not report a complete validation, and without validation data, the reliability of the data cannot be confirmed and may not be replicated. Quantitative analytical methods for natural products in general should be validated first by evaluating the calibration model and then, if the model meets the requirements, by evaluating the other validation parameters in terms of specificity, accuracy, precision, range, DLs, QLs, and robustness following the validation guidelines in force. In recent years, the scientific community is becoming more aware regarding the high importance of proper validation of the developed methods and the trend is changing favorably in recent years.

Finally, the search for innovative and more advanced methodologies that allow more accurate and reliable quantification at even lower levels is the order of the day and is a continuous challenge. In this sense, LC × LC has attracted attention as a novel technique for the analysis of all kinds of phytochemicals in a variety of matrices of natural origin, especially when dealing with highly complex samples. Thanks to its higher separation efficiency compared with one-dimensional LC, it solves component overlapping and reduces the matrix effect, allowing the quantification of minority components and providing more accurate results. Some limitations must also be considered, such as the inherent complexity of the multidimensional system and the dilution of the injected sample when subjected to two separations.

References

Abdullah, N.A., Tian, W., Zengliu, S., Zou, Y., Farooq, S., Huang, Q., Xiao, J., 2022. Recent advances in the extraction, chemical composition, therapeutic potential, and delivery of cardamom phytochemicals. Front. Nutr. 9. https://doi.org/10.3389/fnut.2022.1024820.

Analytical Methods—A Laboratory Guide to Method Validation and Related Topics, 2014.

Analytical procedure development Q14, International Council for Harmonisation of Technical Requirements for Pharmaceuticals for Human Use ICH Harmonized Guideline, 2022.

Analytical procedures and methods validation for drugs and biologics, U.S. Department of Health and Human Services Food and Drug Administration Center for Drug Evaluation and Research (CDER) Center for Veterinary Medicine (CVM), 2015.

Appendix F, 2016. Guidelines for Standard Method Performance Requirements.

Appendix K, 2023. Guidelines for Dietary Supplements and Botanicals.

Arena, K., Cacciola, F., Mangraviti, D., Zoccali, M., Rigano, F., Marino, N., Dugo, P., Mondello, L., 2019. Determination of the polyphenolic fraction of Pistacia vera L. kernel extracts by comprehensive two-dimensional liquid chromatography coupled to mass spectrometry detection. Anal. Bioanal. Chem. 411 (19), 4819–4829. https://doi.org/10.1007/s00216-019-01649-w. Available from: link.springer.de/link/service/journals/00216/index.htm.

Arena, P., Sciarrone, D., Dugo, P., Donato, P., Mondello, L., 2021. Pattern-type separation of triacylglycerols by silver thiolate × non-aqueous reversed phase comprehensive liquid chromatography. Separations 8 (6). https://doi.org/10.3390/separations8060088. www.mdpi.com/journal/separations.

Arena, K., Trovato, E., Cacciola, F., Spagnuolo, L., Pannucci, E., Guarnaccia, P., Santi, L., Dugo, P., Mondello, L., Dugo, L., 2022. Phytochemical characterization of Rhus coriaria L. Extracts by headspace solid-phase micro extraction gas chromatography, comprehensive two-dimensional liquid chromatography, and antioxidant activity evaluation. Molecules 27 (5), 1727. https://doi.org/10.3390/molecules27051727.

Arena, K., Cacciola, F., Miceli, N., Taviano, M.F., Cavò, E., Murphy, R.E., Dugo, P., Mondello, L., 2023. Determination of the polyphenolic content of berry juices using focusing-modulated comprehensive two-dimensional liquid chromatography coupled to mass spectrometry detection. Anal. Bioanal. Chem. 415 (13), 2371–2382. https://doi.org/10.1007/s00216-022-04216-y. https://www.springer.com/journal/216.

Bansal, A., Priyadarsini, C., 2022. Medicinal Properties of Phytochemicals and Their Production. IntechOpen. https://doi.org/10.5772/intechopen.98888.

Ben Salah, H., Allouche, N., 2019. Plant-Based Chemicals Extraction and Isolation. Springer Science and Business Media LLC, pp. 89–117. https://doi.org/10.1007/978-981-13-3810-6_4.

Betz, J.M., Brown, P.N., Roman, M.C., 2011. Accuracy, precision, and reliability of chemical measurements in natural products research. Fitoterapia 82 (1), 44–52. https://doi.org/10.1016/j.fitote.2010.09.011.

Bitwell, C., Indra, S.S., Luke, C., Kakoma, M.K., 2023. A review of modern and conventional extraction techniques and their applications for extracting phytochemicals from plants. Scientif. African 19. https://doi.org/10.1016/j.sciaf.2023.e01585. https://www.journals.elsevier.com/scientific-african.

Blanks in method validation, Supplement to Eurachem Guide, The Fitness for Purpose of Analytical Methods, 2021.

Bounegru, A.V., Apetrei, C., 2021. Development of a novel electrochemical biosensor based on carbon nanofibers–cobalt phthalocyanine–laccase for the detection of p-coumaric acid in phytoproducts. Int. J. Mol. Sci. 22 (17), 9302. https://doi.org/10.3390/ijms22179302.

Byrdwell, W.C., 2017. Comprehensive dual liquid chromatography with quadruple mass spectrometry (LC1MS2 × LC1MS2 = LC2MS4) for analysis of Parinari curatellifolia and other seed oil triacylglycerols. Anal. Chem. 89 (19), 10537–10546. https://doi.org/10.1021/acs.analchem.7b02753. http://pubs.acs.org/journal/ancham.

Cacciola, F., Donato, P., Giuffrida, D., Torre, G., Dugo, P., Mondello, L., 2012. Ultra high pressure in the second dimension of a comprehensive two-dimensional liquid chromatographic system for carotenoid separation in red chili peppers. J. Chromatogr. A 1255, 244–251. https://doi.org/10.1016/j.chroma.2012.06.076.

Cacciola, F., Giuffrida, D., Utczas, M., Mangraviti, D., Beccaria, M., Donato, P., Bonaccorsi, I., Dugo, P., Mondello, L., 2016a. Analysis of the carotenoid composition and stability in various overripe fruits by comprehensive two-dimensional liquid chromatography. LC-GC Eur. 29 (5), 252–256. http://www.chromatographyonline.com/lcgc/issue/issueList.jsp?id=2555.

Cacciola, F., Giuffrida, D., Utczas, M., Mangraviti, D., Dugo, P., Menchaca, D., Murillo, E., Mondello, L., 2016b. Application of comprehensive two-dimensional liquid chromatography for carotenoid analysis in red mamey (pouteria sapote) fruit. Food Anal. Methods 9 (8), 2335–2341. https://doi.org/10.1007/s12161-016-0416-7. http://www.springer.com/life+sci/food+science/journal/12161.

Cacciola, F., Donato, P., Sciarrone, D., Dugo, P., Mondello, L., 2017. Comprehensive liquid chromatography and other liquid-based comprehensive techniques coupled to mass spectrometry in food analysis. Anal. Chem. 89 (1), 414–429. https://doi.org/10.1021/acs.analchem.6b04370. http://pubs.acs.org/journal/ancham.

Cacciola, F., Rigano, F., Dugo, P., Mondello, L., 2020. Comprehensive two-dimensional liquid chromatography as a powerful tool for the analysis of food and food products. TrAC, Trends Anal. Chem. 127, 115894. https://doi.org/10.1016/j.trac.2020.115894.

Cacciola, F., Arena, K., Mandolfino, F., Donnarumma, D., Dugo, P., Mondello, L., 2021. Reversed phase versus hydrophilic interaction liquid chromatography as first dimension of comprehensive two-dimensional liquid chromatography systems for the elucidation of the polyphenolic content of food and natural products. J. Chromatogr. A 1645, 462129. https://doi.org/10.1016/j.chroma.2021.462129.

Cervantes-Paz, B., Yahia, E.M., Nuñez-Vilchis, A., 2023. Identification and quantification of fatty acids and lipid-soluble phytochemicals using GC-MS, HPLC-MS, and FTIR and their association with quality parameters during avocado ripening. J. Food Sci. 88 (1), 119–132. https://doi.org/10.1111/1750-3841.16390.

Chiriac, E.R., Chitescu, C.L., Geană, E.I., Gird, C.E., Socoteanu, R.P., Boscencu, R., 2021. Advanced analytical approaches for the analysis of polyphenols in plants matrices—a review. Separations 8 (5). https://doi.org/10.3390/separations8050065. https://www.mdpi.com/2297-8739/8/5/65/pdf.

Cortés-Herrera, C., Chacón, A., Artavia, G., Granados-Chinchilla, F., 2019. Simultaneous LC/MS analysis of carotenoids and fat-soluble vitamins in Costa Rican avocados (persea americana mill.). Molecules 24 (24), 4517. https://doi.org/10.3390/molecules24244517.

Donato, P., Cacciola, F., Tranchida, P.Q., Dugo, P., Mondello, L., 2012. Mass spectrometry detection in comprehensive liquid chromatography: basic concepts, instrumental aspects, applications and trends. Mass Spectr. Rev. 31 (5), 523–559. https://doi.org/10.1002/mas.20353.

Dugo, P., Kumm, T., Crupi, M.L., Cotroneo, A., Mondello, L., 2006. Comprehensive two-dimensional liquid chromatography combined with mass spectrometric detection in the analyses of triacylglycerols in natural lipidic matrixes. J. Chromatogr. A 1112 (1–2), 269–275. https://doi.org/10.1016/j.chroma.2005.10.070.

Dugo, P., Cacciola, F., Kumm, T., Dugo, G., Mondello, L., 2008a. Comprehensive multidimensional liquid chromatography: theory and applications. J. Chromatogr. A 1184 (1–2), 353–368. https://doi.org/10.1016/j.chroma.2007.06.074.

Dugo, P., Herrero, M., Giuffrida, D., Kumm, T., Dugo, G., Mondello, L., 2008b. Application of comprehensive two-dimensional liquid chromatography to elucidate the native carotenoid composition in red orange essential oil. J. Agric. Food Chem. 56 (10), 3478–3485. https://doi.org/10.1021/jf800144v.

FDA, 2018. Bioanalytical Method Validation, Guidance for Industry. Department of Health and Human Services Food and Drug Administration Center for Drug Evaluation and Research (CDER) Center for Veterinary Medicine (CVM).

Folin, O., Ciocalteu, V., 1927. On tyrosine and tryptophane determinations in proteins. J. Biol. Chem. 73 (2), 627–650. https://doi.org/10.1016/s0021-9258(18)84277-6.

Gackowski, M., Przybylska, A., Kruszewski, S., Koba, M., Mądra-Gackowska, K., Bogacz, A., 2021. Recent applications of capillary electrophoresis in the determination of active compounds in medicinal plants and pharmaceutical formulations. Molecules 26 (14), 4141. https://doi.org/10.3390/molecules26144141.

Georgiana Munteanu, I., Apetrei, C., 2021. Analytical methods used in determining antioxidant activity: a review. Int. J. Mol. Sci. 22 (7), 3380. https://doi.org/10.3390/ijms22073380.

Goto, T., Obara, M., Aoki, S., Okazawa, K., Konisho, K., Osakabe, N., Shoji, T., 2021. Evaluation of polyphenolic content and potential antioxidant activity of Japanese cultivars of peaches, prunes, and plums based on reversed- and normal-phase HPLC and principal component analyses. ACS Food Sci. Technol. 1 (10), 2019–2029. https://doi.org/10.1021/acsfoodscitech.1c00357. https://pubs.acs.org/page/afsthl/about.html.

Hernández-Hernández, C., Fernández-Cabanás, V.M., Rodríguez-Gutiérrez, G., Bermúdez-Oria, A., Morales-Sillero, A., 2021. Viability of near infrared spectroscopy for a rapid analysis of the bioactive compounds in intact cocoa bean husk. Food Control 120, 107526. https://doi.org/10.1016/j.foodcont.2020.107526.

ICH Q2, R1, 2015. Validation of Analytical Procedures: Text and Methodology—Guidance for Industry. U.S. Department of health and human services, food and drug administration. Biologics Evaluation and Research, p. 2015.

Ignat, I., Volf, I., Popa, V.I., 2011. A critical review of methods for characterisation of polyphenolic compounds in fruits and vegetables. Food Chem. 126 (4), 1821–1835. https://doi.org/10.1016/j.foodchem.2010.12.026.

Indrayanto, G., 2022. The importance of method validation in herbal drug research. J. Pharmaceut. Biomed. Anal. 214, 114735. https://doi.org/10.1016/j.jpba.2022.114735.

International Council for Harmonisation, 2022. of Technical Requirements for Pharmaceuticals for Human Use ICH Harmonized Guideline, p. 2022.

Ji, S., Wang, S., Xu, H., Su, Z., Tang, D., Qiao, X., Ye, M., 2018. The application of on-line two-dimensional liquid chromatography (2DLC) in the chemical analysis of herbal medicines. J. Pharmaceut. Biomed. Anal. 160, 301–313. https://doi.org/10.1016/j.jpba.2018.08.014. www.elsevier.com/locate/jpba.

Johnson, J., Mani, J., Ashwath, N., Naiker, M., 2020. Potential for Fourier transform infrared (FTIR) spectroscopy toward predicting antioxidant and phenolic contents in powdered plant matrices. Spectrochim. Acta Mol. Biomol. Spectrosc. 233, 118228. https://doi.org/10.1016/j.saa.2020.118228.

Johnson, J.B., Walsh, K.B., Naiker, M., Ameer, K., 2023. The use of infrared spectroscopy for the quantification of bioactive compounds in food: a review. Molecules 28 (7). https://doi.org/10.3390/molecules28073215. http://www.mdpi.com/journal/molecules.

Kumar, A., Nirmal, P., Kumar, M., Jose, A., Tomer, V., Oz, E., Proestos, C., Zeng, M., Elobeid, T., Sneha, V., Oz, F., 2023. Major phytochemicals: recent advances in health benefits and extraction method. Molecules 28 (2). https://doi.org/10.3390/molecules28020887. http://www.mdpi.com/journal/molecules.

Lama-Muñoz, A., Contreras, M.d.M., 2022. Extraction systems and analytical techniques for food phenolic compounds: a review. Foods 11 (22). https://doi.org/10.3390/foods11223671. www.mdpi.com/journal/foods.

Lucarini, M., Durazzo, A., Kiefer, J., Santini, A., Lombardi-Boccia, G., Souto, E.B., Romani, A., Lampe, A., Nicoli, S.F., Gabrielli, P., Bevilacqua, N., Campo, M., Morassut, M., Cecchini, F., 2020. Grape seeds: chromatographic profile of fatty acids and phenolic compounds and qualitative analysis by FTIR-ATR spectroscopy. Foods 9 (1). https://doi.org/10.3390/foods9010010. https://www.mdpi.com/2304-8158/9/1/10.

Martín-Pozo, L., Arena, K., Cacciola, F., Dugo, P., Mondello, L., 2022. Comprehensive two-dimensional liquid chromatography in food analysis. Is any sample preparation necessary? Green Analy. Chem. 3, 100025. https://doi.org/10.1016/j.greeac.2022.100025.

Mondello, L., Tranchida, P.Q., Stanek, V., Jandera, P., Dugo, G., Dugo, P., 2005. Silver-ion reversed-phase comprehensive two-dimensional liquid chromatography combined with mass spectrometric detection in lipidic food analysis. J. Chromatogr. A 1086 (1–2), 91–98. https://doi.org/10.1016/j.chroma.2005.06.017.

Mosić, M., Dramićanin, A., Ristivojević, P., Milojković-Opsenica, D., 2020. Extraction as a critical step in phytochemical analysis. J. AOAC Int. 103 (2), 365–372. https://doi.org/10.5740/jaoacint.19-0251.

Muller, M., de Villiers, A., 2023. A detailed evaluation of the advantages and limitations of online RP-LC × HILIC compared to HILIC × RP-LC for phenolic analysis. J. Chromatogr. A 1692, 463843. https://doi.org/10.1016/j.chroma.2023.463843.

Navarro-Reig, M., Jaumot, J., van Beek, T.A., Vivó-Truyols, G., Tauler, R., 2016. Chemometric analysis of comprehensive LC × LC-MS data: resolution of triacylglycerol structural isomers in corn oil. Talanta 160, 624–635. https://doi.org/10.1016/j.talanta.2016.08.005. https://www.journals.elsevier.com/talanta.

Navarro-Reig, M., Jaumot, J., Tauler, R., 2018. An untargeted lipidomic strategy combining comprehensive two-dimensional liquid chromatography and chemometric analysis. J. Chromatogr. A 1568, 80–90. https://doi.org/10.1016/j.chroma.2018.07.017. www.elsevier.com/locate/chroma.

Nwozo, O.S., Effiong, E.M., Aja, P.M., Awuchi, C.G., 2023. Antioxidant, phytochemical, and therapeutic properties of medicinal plants: a review. Int. J. Food Prop. 26 (1), 359–388. https://doi.org/10.1080/10942912.2022.2157425. https://www.tandfonline.com/loi/ljfp20.

ORA, 2022. ORA Laboratory Manual II Vol. I - Manual of quality policies for ORA regulatory laboratories (ORA-LAB.QM). Food Drug Adm. Office Regul. Aff.

Pereira-Coelho, M., Haas, I.C.d.S., Katrin Reinke, C., Dognini, J., Amboni, R.D.d.M.C., Vitali, L., Madureira, L.A.d.S., 2023. A green analytical method for the determination of polyphenols in wine by dispersive pipette extraction and LC-MS/MS. Food Chem. 405. https://doi.org/10.1016/j.foodchem.2022.134860. http://www.elsevier.com/locate/foodchem.

Przybylska, A., Gackowski, M., Koba, M., 2021. Application of capillary electrophoresis to the analysis of bioactive compounds in herbal raw materials. Molecules 26 (8), 2135. https://doi.org/10.3390/molecules26082135.

Rao, H., Ahmad, S., Aati, H.Y., Basit, A., Ahmad, I., Ahmad Ghalloo, B., Nadeem Shehzad, M., Nazar, R., Zeeshan, M., Nasim, M.J., ur Rehman Khan, K., 2023. Phytochemical screening, biological evaluation, and molecular docking studies of aerial parts of Trigonella hamosa (branched Fenugreek). Arab. J. Chem. 16 (7). https://doi.org/10.1016/j.arabjc.2023.104795. http://colleges.ksu.edu.sa/Arabic%20Colleges/CollegeOfScience/ChemicalDept/AJC/default.aspx. http://www.sciencedirect.com/science/journal/18785352.

Schofield, P., Mbugua, D.M., Pell, A.N., 2001. Analysis of condensed tannins: a review. Anim. Feed Sci. Technol. 91 (1–2), 21–40. https://doi.org/10.1016/S0377-8401(01)00228-0.

Shraim, A.M., Ahmed, T.A., Rahman, M.M., Hijji, Y.M., 2021. Determination of Total Flavonoid Content by Aluminum Chloride Assay: A Critical Evaluation. Academic Press, Qatar LWT, p. 150. https://doi.org/10.1016/j.lwt.2021.111932. https://www.journals.elsevier.com/lwt.

van der Klift, E.J.C., Vivó-Truyols, G., Claassen, F.W., van Holthoon, F.L., van Beek, T.A., 2008. Comprehensive two-dimensional liquid chromatography with ultraviolet, evaporative light scattering and mass spectrometric detection of triacylglycerols in corn oil. J. Chromatogr. A 1178 (1−2), 43−55. https://doi.org/10.1016/j.chroma.2007.11.039.

Verma, S., Thakur, D., Pandey, C.M., Kumar, D., 2023. Recent prospects of carbonaceous nanomaterials-based laccase biosensor for electrochemical detection of phenolic compounds. Biosensors 13 (3). https://doi.org/10.3390/bios13030305. http://www.mdpi.com/journal/biosensors/.

Zhang, Q.W., Lin, L.G., Ye, W.C., 2018. Techniques for extraction and isolation of natural products: a comprehensive review. Medicine 13 (1). https://doi.org/10.1186/s13020-018-0177-x. http://www.cmjournal.org/.

Zhang, L., Wang, S., Yang, R., Mao, J., Jiang, J., Wang, X., Zhang, W., Zhang, Q., Li, P., 2019. Simultaneous determination of tocopherols, carotenoids and phytosterols in edible vegetable oil by ultrasound-assisted saponification, LLE and LC-MS/MS. Food Chem. 289, 313−319. https://doi.org/10.1016/j.foodchem.2019.03.067. www.elsevier.com/locate/foodchem.

CHAPTER

5 From extract to chiral analyses of the natural products

Imran Ali[1], Mohd Mustaqeem[1], Ersin Demir[2] and Marcello Locatelli[3]

[1]*Department of Chemistry, Jamia Millia Islamia (Central University), Delhi, New Delhi, India;* [2]*Afyonkarahisar Health Sciences University, Faculty of Pharmacy, Department of Analytical Chemistry, Afyonkarahisar, Turkey;* [3]*Department of Pharmacy, University "G. d'Annunzio" of Chieti-Pescara, Chieti, Italy*

1. Introduction

Natural products have always played an important role in the discovery and development of therapeutic agents. These compounds have outstanding structural and functional diversity, resulting from a diverse range of biological sources. Chiral compounds stand out among natural products due to their specific biological activities that are unique to enantiomers. These tasks are critical to assuring the efficacy and safety of pharmaceutical medications. Chirality, a fundamental property in which molecules have nonsuperimposable mirror images, has a significant impact on the pharmacokinetics and pharmacodynamics of drugs. Chiral natural products are extremely important not only for therapeutic purposes but also for study into biological processes and the development of new synthetic technologies. The study of chiral natural compounds is fundamentally multidisciplinary, involving organic chemistry, pharmacology, and analytical chemistry. This chapter attempts to provide a complete review of the procedures involved in the extraction, separation, and analysis of chiral natural products, with a focus on both conventional and modern approaches. By investigating these methodologies, we hope to highlight both advances and current problems in this dynamic field, stressing the significance of chiral analysis in modern drug discovery and development.

Extraction is the initial critical step in the investigation of natural substances. The process involves extracting the target molecules from the complex mixtures found in biological sources. Conventional methods of extraction, such as solvent extraction and steam distillation, have been enhanced by more sophisticated techniques such as supercritical fluid extraction (SFE) and microwave-assisted extraction (MAE), which provide better efficiency and selectivity. The choice of extraction method depends on the nature of the source material and the target compounds, with a focus on preserving the integrity and chirality of the natural products (Zhang et al., 2018). After the extraction process, isolating chiral chemicals becomes challenging since natural sources include several stereoisomers. Advanced chromatographic methods, such as HPLC and supercritical fluid chromatography (SFC), are crucial for distinguishing these stereoisomers. Moreover, the utilization of crystallization techniques, specifically those that incorporate chiral auxiliaries or resolving agents, is essential in the acquisition of substances that are enantiomerically pure. The advancement of these isolation techniques has been driven by the

Phytochemicals for Health. https://doi.org/10.1016/B978-0-443-15366-2.00005-8

requirement for precise, sensitive, and consistent results in chiral analysis. Chiral analysis is crucial for understanding the pharmacological characteristics of natural products. It involves the determination of the absolute configuration and enantiomeric purity of chiral compounds. To attain this objective, a range of analytical methods, including chiral chromatography, electrophoresis, and polarimetry, is utilized. Each method has its strengths and limitations, and their application depends on the specific requirements of the analysis. Chiral chromatography gives excellent resolution and sensitivity, whereas polarimetry allows for fast and nondestructive examination.

The research is focused on chiral natural compounds due to their potential for transforming the discovery and development of pharmaceuticals. The objective of this chapter is to enhance understanding and utilization of these chemicals by providing comprehensive details on the techniques employed for extracting, isolating, and analyzing them. By using emerging natural resources and employing advanced analytical techniques, it is feasible to discover novel chiral compounds that exhibit distinct biological properties. The ongoing demand for new medicines to address various health issues indicates the significance of this research. Chiral natural products offer a diverse array of biological effects, making them a promising source for discovering new drug candidates. Studying these substances contributes to the field of medicinal chemistry by providing valuable insights into the relationship between molecular structure and biological function.

2. Natural products and their importance

Natural products, also referred to as secondary metabolites, are a significant and valuable source of potent medicinal discoveries that are produced by the diverse Earth's flora and fauna. Indisputably, over the course of millions of years in the history of life, both terrestrial and marine organisms have evolved to adapt to a variety of biotic and abiotic triggers, resulting in the development of natural products. They have been used as drugs for a very long time because they are bioactive by design, and they are still a source of potential treatments today (David et al., 2015). A chemical entity that is produced by a naturally occurring biological organism and has pharmacological properties is referred to as a "natural product". Natural items could be helpful in the development of necessary medications. The crude material that comes from the bodies of microbes, animals, plants, or fermentation broths of microorganisms is made up of separate and structurally diverse chemical components. In the fields of biotechnology and pharmaceuticals, natural products have been essential. This is since many modern medications are either chemicals that occur naturally or molecules that are generated from compounds found in nature. Typically, the medicinal substances that are injected, swallowed, and breathed are a combination of intricate medicinal substances (Mathur and Hoskins, 2017). For certain individuals, synthetic medications might result in adverse consequences and are costlier than traditional herbal treatments, even though being "natural" does not necessarily guarantee safety. Since ancient times, they have been extensively utilized as medicinal substances, food items, and nutritional supplements. Herbal medications include a high concentration of bioactive compounds, which have positive effects on human health. Both natural products and traditional medications hold significant value. For a long time, natural products and their derivatives have been acknowledged as a valuable source of medicinal agents and structural variety. Natural products exhibit a vast array of diverse and complex chemical structures. Additionally, their ability to modify biological functions has garnered significant interest (Amit Koparde et al., 2019a,b).

The presence of natural products has been crucial for the survival of humanity on Earth. The remarkable natural phenomena always serve as a significant milestone in the pursuit of herbal medication research. In previous eras, the planet harbored several species of therapeutic plants. Medicinal plants hold significant and essential value on a worldwide scale. The planet has been decorated with medicinal plants, which represent a valuable reservoir of resilience. Each plant possesses unique medicinal qualities attributed to its own bioactive compounds. Natural pharmacological compounds are considered essential and play significant roles in the contemporary field of medicine. Their therapeutic function was justified by the existence of their bioactive compounds. Because of their disease-inhibiting properties, they are very valuable as natural pharmaceuticals, offering essential bioactive chemicals that are less harmful and more efficient. Additionally, they give biological and chemical methods for modifying and extracting strong medications from natural sources. Ayurvedic medicines primarily utilize plant-derived raw ingredients in the form of crude pharmaceuticals, including dried herbal powders, extracts, and mixtures of products. across addition to these systems, there exists a diverse legacy of ethnobotanical utilization of plants by several vibrant tribal tribes across the nation (Kar, 2006). Approximately 75,000 species of higher plants are believed to exist on Earth, with barely 10% having been utilized in traditional medicine. Only a small fraction, ranging from 1% to 5%, has undergone scientific study and is recognized to possess therapeutic properties (RamaRao and Gurjar, 1990). Herbal medicine, practiced across the globe, is rooted in traditional medicine. According to a widespread oral survey conducted in many parts of the world, it has been asserted that traditional medicines possess their own significance and fundamental principles. Exploring the chemical components of plants and conducting pharmacological investigations can serve as a foundation for establishing a primary compound through the process of herbal medicine development. Herbs in contemporary medicine have given crucial life-saving medications. However, out of the estimated 400,000 plant species, only 6% have been examined for their activity, and a mere 20% have been subjected to phytochemical investigation. Therefore, it is necessary to examine the different bioactive components and conduct phytoanalysis and phytopharmacological evaluations of herbal medications to accomplish the goals of herbal drug development. The working function of each green herbal drug, derived from plants, involves the synthesis of bioactive compounds. These compounds are considered as natural gifts and are preserved within the plants. They can be extracted and used as raw materials whenever needed. This extraction process is carried out using various scientific methods for the purpose of conducting scientific investigations and studying the discovery of herbal drugs. Pharmaceutical substances often include secondary metabolites derived from plants, which play a crucial role in the development of drugs. For green plant developers to ensure a sufficient supply of source material for the pharmaceutical industry, they must consider factors such as environmental changes, diverse geographical distribution, labor cost, and the selection of superior plants. This will enable the development of high-quality herbal drugs (Koparde Akshada Amit and Doijad, 2019).

Medicinal and aromatic plants, especially those with ethnopharmacological significance, have served as a vital source of treatments and health care for millennia. In ancient times, these remedies were commonly prepared in simple forms such as powders, tinctures, macerations, teas, infusions, percolation products, poultices, decoctions, tinctures, inhalations, and various other herbal preparations (Orch et al., 2021; Benkhnigue et al., 2022). The exact dosage and methods for using these plants to treat various ailments were traditionally handed down orally from one generation to another. Over time, this accumulated knowledge was systematically documented in traditional pharmacopeias, preserving the extensive information on the medicinal uses of these plants. Drug discovery from medicinal plants incorporates several academic fields and a wide range of research techniques. Botanists, ethnobotanists, ethnopharmacologists, and plant ecologists frequently collect and classify species that are of particular interest. Advancements in technology now allow for the conversion of plants into "production industries" that produce organic medicinal chemicals for the manufacturing of biotech medicines, therapies, and treatments (Paul and Ma, 2011). Natural chemicals and their structural analogs have made major contributions to the field of pharmacotherapy throughout the course of history, notably in the treatment of infectious illnesses (Atanasov et al., 2021). In addition to the physiologically active bioactive molecules that are derived from plants and have been shown to have direct therapeutic value as drug substances, several other naturally occurring bioactive compounds have also been found to be useful as "leads" or model molecules for the synthesis or semi-synthesis of drugs. Recent developments in several scientific and technological fields, such as engineering techniques, genome mining, improved analytical tools, and breakthroughs in microbial culture, have contributed to the resolution of existing difficulties and led to the development of new opportunities. There is a clear indication of these developments in the complexity of bioassays that are utilized in the process of developing medicinal plant drugs. Additionally, they provide data on the "mode of action" at the molecular scale in a manner that is both immediate and accurate. Because of this, there has been a resurgence of interest in natural chemicals as possible therapeutic approaches, particularly for the treatment of antimicrobial resistance. It is possible for researchers that specialize in ethnobotany, pharmacognosy, pharmaceutics, medicinal chemistry, botany, taxonomy, organic chemistry, molecular biology, biochemistry, microbiology, pharmacology, and plant ecology to work together in order to uncover novel chemical components that are present in therapeutic plants. Species collection, extraction, chemical separation, structural identification, and bioassays are the five processes that may be separated into standard protocols in plant natural product separation chemistry for the purpose of drug development (Pezzuto, 1997). These procedures are implemented using combinatorial and synthetic chemistry techniques, computational simulation and chemical informatics research, and specialized high-throughput screening methods in well-equipped pharmaceutical and other facilities. Plant-derived natural products continue to play a vital role in the process of discovering and developing drugs, since they offer an abundant source of bioactive compounds with diverse features. Multiple compelling arguments underscore the enduring significance of plant-derived natural chemicals in this particular environment (Chaachouay and Zidane, 2024).

Natural products have distinct features compared with manufactured substances. They possess a wide range of structural variations, frequently have distinct biological functions, and showcase an impressive level of steric intricacy. This is a clear result of the fact that the biosynthesis enzymes possess an intrinsic three-dimensional (3D) structure and chirality. In addition, due to the restricted range of building blocks utilized for biosynthesis, nature must create new and unique structures by diverging common intermediates into various frameworks. Due to their accumulation in enzymatic pathways, natural products have demonstrated their capacity to attach to enzyme or protein targets, and thus, they can also interact with disease-relevant target proteins (Ortholand and Ganesan, 2004). Multiple statistical studies analyze and compare synthetic chemicals, combinatorial libraries, medicines, and natural products (Henkel et al., 1999; Lee and Schneider, 2001; Feher and Schmidt, 2003). Natural products generally have a lower average number of nitrogen, sulfur, or halogen atoms compared with medicines and manufactured chemicals. However, they are significantly richer in oxygen and have a higher number of hydrogen-bond donors. This discrepancy likely stems from the inherent disparities between biosynthesis and (combinatorial) synthetic chemistry. Several biosynthetic routes rely on reagents with comparable quantities of oxygen and nitrogen atoms, such as amino acids, or reagents with a higher proportion of oxygen than nitrogen, as seen in polyketide synthesis. Natural products often exhibit a higher number of rings, generally with sp^3-hybridized bridgehead atoms, and chiral centers, in comparison to synthesized compounds (Krastel et al., 2006).

3. Chirality in the natural products

Before the 1970s, chemists generally considered natural products to be enantiomerically pure, with a few exceptions. This might perhaps be attributed to the statements made by Louis Pasteur. He considered life processes to be asymmetrical. Advancements in enantioselective synthetic techniques throughout the late 1970s and early 1980s enabled the production of many natural compounds in a condition of high enantiomeric purity. Through the comparison of the optical rotations of the synthetic products with the reported optical rotations of the natural products, it was progressively realized that a significant number of natural compounds exhibit enantiomeric impurity (Mori, 2011). Life is chiral, meaning that its constituents are predominantly made up of optically active molecules, whether they are macromolecules (such as proteins and nucleic acids) or tiny biomolecules. Therefore, these molecules exhibit distinct interactions with various enantiomers of chiral substances, resulting in a selective preference for a certain enantiomer. The ability to distinguish between enantiomers is particularly significant in the field of medicinal chemistry, since several pharmacologically active substances are utilized in the form of racemates, which comprised equal amounts of two enantiomers. Each of these enantiomers may exhibit distinct characteristics in terms of pharmacodynamics, pharmacokinetics, and toxicity. Using a single enantiomer in a medicine can enhance its bioactivity and decrease the occurrence and severity of negative side effects. This is particularly important when considering the composition of natural products, as the bulk of these compounds consist of one or more chiral centers (Valentová et al., 2023). Stereoisomers are substances that exhibit differences mainly in the spatial arrangement of their component atoms. These isomers may be classified into two categories: enantiomers and diastereoisomers. Enantiomers are a pair of substances that are mirror images of each other and cannot be superimposed. They only differ in the capacity to rotate planar polarized light, but their physicochemical characteristics are generally identical. These isomers are known as chiral and are commonly referred to as optical isomers. Diastereoisomers are a kind of stereoisomer, which does not have mirror-image structure. They can exist as either chiral or geometric (cis/trans) isomers. A racemate, also known as a racemic mixture, is formed when equal amounts of two enantiomers are combined. Chirality is an intrinsic characteristic of all biological systems. Biomacromolecules, which are made up of smaller chiral components, such as amino acids, carbohydrates, and lipids, arrange themselves into intricate 3D structures, displaying supramolecular chirality. Chiral macromolecular scaffolds possess asymmetric binding sites and catalytic centers that may identify and differentiate between specific stereoisomers of other chiral compounds (Blackmond, 2019; Burton and Berger, 2018). Chirality commonly occurs due to the existence of asymmetric centers in organic compounds. Typically, these centers are tetracoordinate, meaning they are attached to four separate atoms or groups. An example of such a molecule is the chiral chemotherapeutic drug melphalan.

Chirality can also arise from atropisomerism, which occurs due to restricted rotation around a single bond. An example of this is ortho-substituted biphenyl derivatives, which exhibit chirality due to their restricted rotation. In pharmacology, the action of drugs often involves interactions with chiral biomolecules, resulting in a stereochemical preference in enzyme and receptor systems for one enantiomer over another. This preference can lead to significant differences in both the quantity and quality of biological activity between enantiomers. The Easson–Stedman hypothesis is frequently used to explain these variations in biological efficacy. It speculates that the differences arise from the differing affinities of the enantiomers for the same binding site (Easson and Stedman, 1933). Specifically, the more active enantiomer is capable of forming at least three intermolecular contacts with the receptor surface, while the less active enantiomer typically makes only two such contacts. In some cases, particularly in drug interactions with enzymes, it is suggested that a fourth site, such as a specific orientation or an additional binding site, is necessary to distinguish between the enantiomers and account for their differing biological activities (Valentová et al., 2023).

For several years, it has been recognized that the enantiomers of chiral pharmaceuticals exhibit different pharmacodynamic and toxicological properties. These differences affect nearly all categories of drugs. In some cases, one enantiomer may be entirely inactive, while in others, both enantiomers may have distinct biological activities. Sometimes, both enantiomers are active, but only one may be associated with adverse effects, or they may have opposing effects on the same biological target (Hutt, 2003). These stereoselective differences can arise from both pharmacokinetic processes and drug interactions at pharmacological receptors (Hutt, 2007). The active enantiomer, depicted on the left in Fig. 5.1, has substituents B, C, and D, which form three contacts with binding sites b, c, and d on the receptor, facilitating a proper fit and interaction. In contrast, the inactive enantiomer, shown on the right, has a spatial arrangement that prevents it from making all three necessary contacts with the binding sites, resulting in insufficient interaction (McConathy and Owens, 2003). This highlights the significance of stereochemistry in the binding affinity and activity of

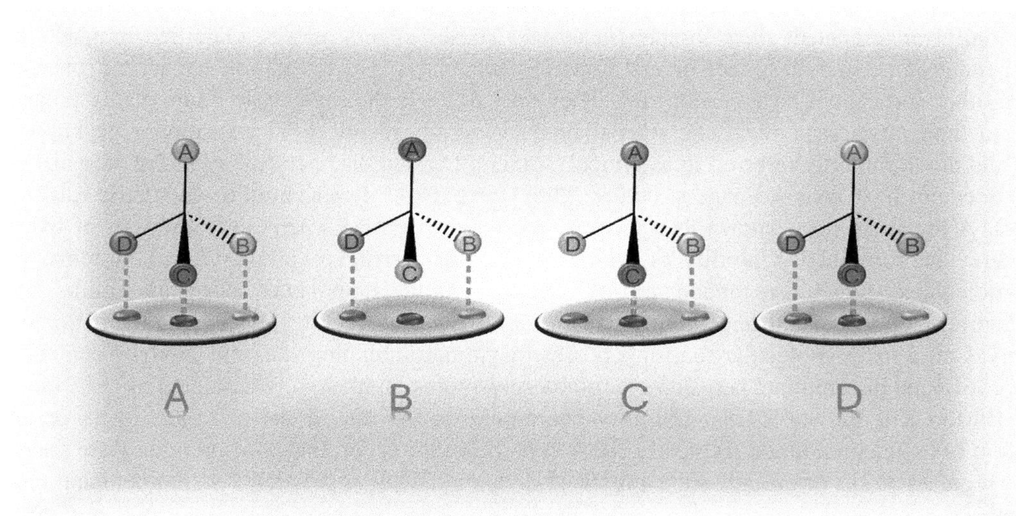

FIGURE 5.1

Interactions between an active and inactive enantiomer and a receptor.

enantiomers. Variations between enantiomers can occur during their absorption, distribution, metabolism, and excretion. As a result, after the administration of a racemic mixture, the enantiomers do not reach their site of action in equal concentrations. Advancements in chemical techniques, particularly stereoselective synthesis and stereospecific analysis methods, as well as regulatory requirements, have led to a significant increase in the number of chiral drugs submitted for approval as single enantiomers rather than racemates (Gandhi et al., 2020).

In contrast to the late 20th century, where around 55% of medications used in medical procedures were chiral, with half of them being utilized as racemates, the present trend in drug research mostly focuses on compounds that consist of only one enantiomeric form. Aside from newly developed chemical compounds, some existing racemic medications have undergone reassessment as prospective single enantiomer products, aiming to achieve an enhanced therapeutic profile or explore new therapeutic applications. As an illustration of the "chiral switching" notion, the first racemic nonsteroidal antiinflammatory medications (such as ibuprofen and ketoprofen) were commercially introduced as single (S)-enantiomers, as this particular enantiomer is primarily responsible for the antiinflammatory activity (Ali et al., 2021; Toledo and Briand, 2018). Multiple studies have shown a connection between these antiinflammatory medications and a reduction in both the occurrence and reappearance of cancer.

Remarkably, the (R)-enantiomer of flurbiprofen has demonstrated efficacy in treating colon and prostate cancer, as well as inhibiting the development of glioblastomas in both laboratory and animal models. Studying the impact of different arrangements at the chiral center on biological effects is quickly becoming an important aspect in the search for new chemotherapeutic drugs. Chirality is a crucial determinant of the structure of an anticancer medication and its interaction with molecular targets in cancer. This statement is true not just for medications derived from organic compounds but also for pharmaceuticals containing metals that have been specifically designed for use in cancer treatment (Wang et al., 2018). Pinto et al. (2021) conducted a literature study on the antitumor efficacy of both natural and synthesized chiral flavonoids.

The phenomenon of chirality is of utmost importance in determining the biological efficacy of naturally occurring substances. The enantiomers of natural products frequently display varying biological activity because of their unique interactions with biological systems. For example, limonene, a naturally occurring compound present in citrus fruits, occurs as a pair of enantiomers. The (R)-limonene possesses an aroma reminiscent of oranges, while its enantiomer presents a fragrance like lemons. Another instance is the alkaloid morphine, which is present in the poppy plant (Papaver somniferum). Morphine has five chiral centers, resulting in the molecule's ability to exist as 32 stereoisomers. The required narcotic analgesic effect is only shown by the levo isomer of morphine, which contains the stereodescriptors (5R, 6S, 9R, 13S, and 14R). Quinine possesses four chiral centers and can exist in 16 different stereo-isomeric forms. Among these, the levo-quinine with the specific configuration (3R, 4S, 8S, and 9R) exhibits the necessary antimalarial action. Unichirality is a characteristic feature of several natural compounds. An in-depth examination of chirality in natural products is crucial for comprehending their biological efficacy and advancing novel pharmaceuticals. Chiral natural compounds may be obtained from their natural sources, and their enantiomers can be distinguished by the application of different chromatographic methods. Subsequently, the separate enantiomers can undergo testing to determine their biological activity, potentially resulting in the identification of novel pharmaceuticals or drug development frameworks derived from natural products. The production of chiral natural compounds is a crucial component in

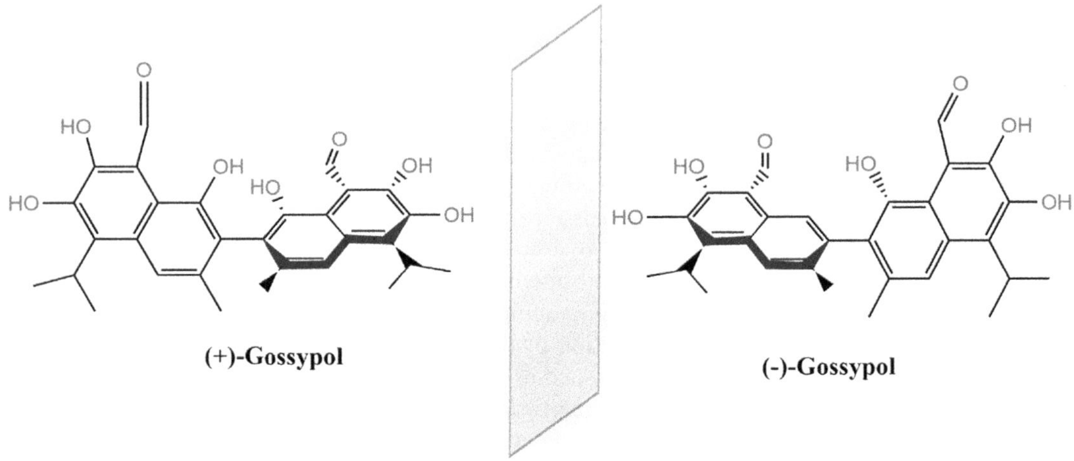

FIGURE 5.2

Structural representation gossypol mirror image.

the process of identifying and developing new drugs. The production of chiral molecules is a difficult task, and it is essential to establish effective methods for synthesizing enantiomerically pure substances. Fig. 5.2 presents the pair of Gossypol enantiomers, a chiral polyphenolic compound predominantly found in cotton plants. The illustration highlights the two-mirror image structure of Gossypol, making Gossypol an enantiomeric molecule with significant biological activities. Each enantiomer of Gossypol interacts differently with biological systems, influencing its pharmacological and toxicological properties.

4. Extraction techniques for the chiral natural products

Natural products possess a complex chemical structure and contain a small amount of active ingredients. Natural products are derived from natural sources. To effectively utilize natural product resources, it is imperative to do comprehensive research on the complex chemical composition of these substances. After the initial stages of research, which involve the processes of extraction and separation, the aim is to acquire raw extracts that contain active constituents obtained from natural sources. This will enable further research and development in the fields of separation, purification, and analysis. During the first phase, the process of extraction yields a mixture of desired and unidentified compounds. The extraction of natural chemicals from plants is mostly achieved by liquid extraction, steam distillation, sublimation, and expulsion. These procedures rely on the principle of extraction. The methods for extracting natural products can be categorized into traditional and modern techniques. Traditional approaches include maceration, percolation, decoction, reflux extraction, and Soxhlet extraction. On the other hand, modern methods incorporate technologies such as ultrasound and MAE.

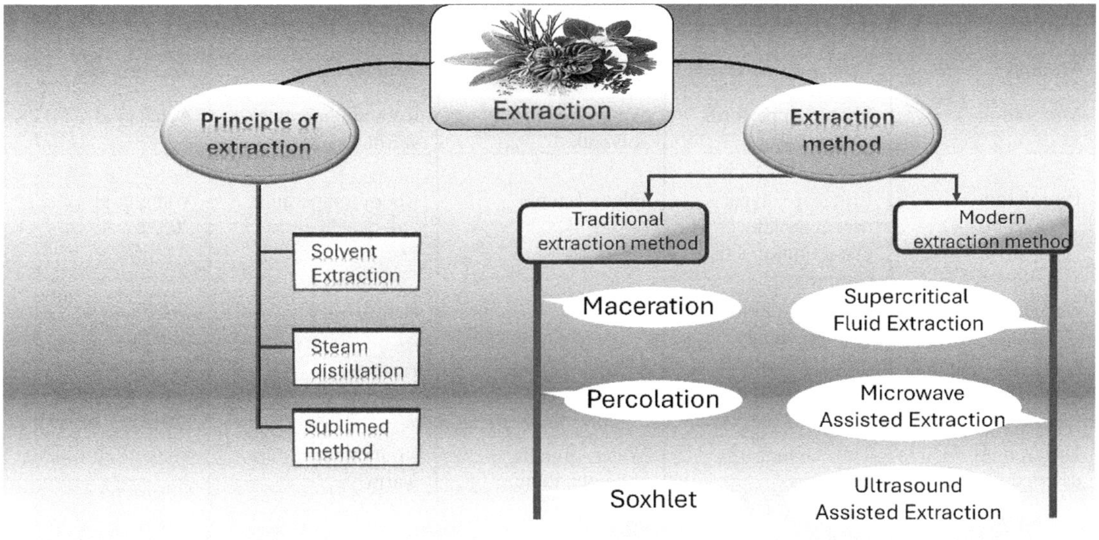

FIGURE 5.3

Classification of extraction methods.

Moreover, there exist environmentally sustainable and highly effective methods such as SFE and pressurized liquid extraction, which are gaining popularity in current research and applications. These diverse extraction methods offer researchers a range of choices to suit various needs in natural product isolation and analysis. Fig. 5.3 depicts these classifications. In contrast to conventional extraction procedures, the approaches provide several advantages, such as the conservation of the solvent and a decrease in the extraction time (Zhang et al., 2018). The extraction of natural products is essential for several purposes, such as the manufacturing of herbal medicines, the exploration of the biological properties of secondary metabolites, and the isolation of preexisting chemical compounds that are combined together (Talukdar and Choudhury, 2010).

4.1 Conventional or traditional extraction methods

Throughout the ages, humankind has depended on traditional extraction techniques to separate valuable compounds from different sources. These procedures, frequently uncomplicated and well established, have been crucial in the advancement of medications, scents, tastes, and a wide range of other things. This chapter examines the fundamental principles underlying classic methods such as maceration, percolation, and Soxhlet extraction. This chapter examines the advantages and disadvantages of each traditional extraction technique, as well as their suitability for different materials and desired compounds. Table 5.1 provides a comparison of conventional extraction methods. By studying these conventional methods, one can provide a solid basis for appreciating the advancements and elements of modern extraction technology.

Table 5.1 Traditional extraction methods: efficiency, solvents and temperatures.

Method	Characteristics	Solvent	Temperature	References
Maceration	Simple to perform, requires more time, lower yield	Water, organic solvents	Room temperature, warm	Achat et al. (2012)
Percolation	Efficient solvent use, complete extraction of active components	Ethanol, wine	Room temperature	Wang et al. (2020, 2023)
Decoction	Quick extraction, high efficiency	Water	Heat	Martins et al. (2015)
Soxhlet extraction	Solvent-saving, very efficient	Volatile organic solvents	Heat	Luque de Castro and Priego-Capote (2010)
Ultrasound-assisted extraction	Timesaving, enhances efficiency, reduces solvent use	Water, organic solvents, ionic liquid	Room temperature, warm	Song et al. (2023)
Microwave-assisted extraction	Timesaving, improves efficiency, reduces solvent use	Water, organic solvents, ionic liquid	Heat	Hu et al. (2019)
Supercritical fluid extraction	Environmentally friendly, efficient	Supercritical CO_2	Room temperature	Mehta et al. (2022)

4.1.1 Maceration

Maceration is a way to separate solids from liquids. In the past, maceration was used to get beneficial compounds out of plant or animal matter. This was done by mixing liquids with or without heat and agitating or shaking the mixture to help the compounds dissolve better. It involves letting a coarse powder made from natural materials soak in the right liquid at room temperature or with some light heat. The liquid is put into the covered container that has solid powder inside it. Extra-long periods of time are given for extraction, ranging from a few hours to several days. This is to give the solvent enough time to move through the plant cell wall or animal flesh and dissolve the chemicals that are there. The process only happens because molecules move around (Rasul, 2018). This technique is appropriate for extracting organic substances that are susceptible to thermal degradation and contain significant quantities of starch, mucilage, gum, and pectin. The efficiency of maceration is influenced by factors such as impregnation period, leaching solvent, material–liquid ratio, and raw material particle size (Zhang et al., 2023). Phenolic compounds were extracted from hazelnut shells by Di et al. (Di Michele et al., 2021) using several techniques such as maceration, ultrasonic bath, and high-power ultrasound. The isolated compounds were subsequently assessed for their antioxidant properties. The findings suggest that phenolic compounds produced by maceration extraction have more potent antioxidant properties. While maceration is a straightforward and efficient technique for extracting active constituents, it is laborious and yields very little quantities. Consequently, the process of maceration is frequently augmented by the utilization of ultrasonic technology to greatly improve its efficacy. In a conducted study aimed at improving the nutritional profile of olive oil, ultrasound-assisted maceration was employed to extract phenolic compounds from olive leaves. The findings demonstrated that olive oil extracted using ultrasound-assisted maceration exhibited a higher total phenol content compared with that obtained through traditional maceration methods. Additionally, this innovative approach significantly reduced the extraction time required for the process (Achat et al., 2012). In a separate investigation on the extraction of volatile compounds from orange peel, the utilization of microwave and ultrasonic aid has been found to greatly enhance the effectiveness of extraction and yield many volatile compounds (Chanivet et al., 2021). It is a relatively easy extraction procedure with the drawback of a long extraction period. Maceration is both an isocratic and a cold extraction process (Přichystal et al., 2016). It is ideal for the extraction of thermally liable chemicals. This approach extracts component elements from plants in solvent by immersing the plant sample in a specific solvent. It is done at room temperature in a constant condition.

4.1.2 Percolation

Percolation is a widely used method for extracting active compounds used in the production of tinctures and fluid extracts. It involves placing crushed herbs into a percolation cylinder and continuously adding solvent from the top to allow it to pass through the herbal material. Typically, a percolator shaped like a slender, cone-shaped container with open ends is used for this purpose. To begin, the powdered herbal ingredients are mixed with an appropriate amount of solvent and left to macerate in a tightly sealed container for several hours. The top of the percolator is then sealed, and additional solvent is added to form a layer above the material. This mixture is left to soak for an extended period within the sealed percolator. Once ready, the bottom of the percolator is unsealed, allowing the liquid extract to drip out gradually and in a controlled manner. More solvent is added as necessary until the percolate reaches about 75% of the desired volume. Afterward, the extract undergoes pressing to remove remaining liquid. The percolate is then clarified through filtration or settling processes before being collected for further use (Handa, 2008).

Percolation is a dynamic leaching technique that efficiently uses solvents, ensures thorough extraction of active chemicals, and directly collects the resulting leachates. Percolation extraction is performed at ambient temperature, offering a mild extraction method ideal for thermally delicate compounds. However, it is unsuitable for extracting natural materials that exhibit expansion tendencies and lack organized structures. The drawbacks of the percolation approach are the enormous amount of solvent required and the laborious nature of the extraction process. The percolation process is influenced by several factors, including the size of the powder, the composition of the solvent, the duration of extraction, the rate at which percolation occurs, and the amount of solvent used. The technique is frequently employed using various percentages of ethanol or white wine as a solvent. Therefore, it is crucial to take measures to prevent the solvent from evaporating. Wilson et al. (2022) obtained the whole cannabidiol from cannabis using both maceration and percolation methods. The findings indicated that percolation was more effective in extracting the cannabidiol compared to maceration. In a separate investigation concerning the extraction of phenolics from Allium sativum, it was shown that percolation yielded greater extraction and recovery rates compared with maceration (Ahmad et al., 2020). Similarly, while comparing the extraction of volatile components from grapeseed oil, it was found that Soxhlet extraction resulted in 67 volatile components, whereas percolation resulted in 60 volatile components (Sevindik et al., 2022).

4.1.3 Soxhlation

Soxhlet extraction, also known as continuous reflux extraction, utilizes the principle of solvent reflux to extract solid material efficiently. This method operates by continuously cycling pure solvent through the solid material. Compared with traditional reflux extraction, Soxhlet extraction offers higher efficiency and requires less solvent, addressing issues such as excessive solvent use and multiple extraction cycles (Luque de Castro and Priego-Capote, 2010). The process involves a Soxhlet extractor, which consists of a setup where the solvent boils and travels up a longer side arm. A concentrated droplet of solvent is periodically introduced into a porous container containing the solid mixture, where it dissolves and extracts the desired component. As the solvent fills the smaller side arm and exceeds its capacity, it initiates a siphoning action. The solvent, now containing dissolved components, is siphoned into the bottom flask below, leaving behind the solid material. Fresh solvent droplets continuously replace the spent solvent in the porous container, maintaining the extraction cycle. Soxhlet extraction is particularly suitable for extracting phenolic compounds and oils from plant materials. For example, Alara et al. (2018) employed Soxhlet extraction to extract phenolic compounds from Vernonia cinerea leaves. Their study investigated the effects of extraction time, solvent-to-feed ratio, and ethanol concentration on the yield of the product, as well as the total polyphenol and flavonoid content. They found that a 2-hour extraction time, a solvent-to-feed ratio of 1:20 g/mL, and a 60% v/v ethanol concentration yielded optimal results in terms of extraction efficiency and product quality.

Soxhlet extraction has both advantages and disadvantages. A significant quantity of medicine can be extracted using a considerably lesser amount of solvent. Significant efficiency in terms of time, energy, and eventually financial resources. The size of this method is minimal, and it is employed in batches. However, it becomes more cost-effective when it is transformed into a continuous extraction process. However, there are also some drawbacks to consider. Soxhlet extraction typically uses large volumes of organic solvents, which can be hazardous and expensive to dispose of. The process itself can be time-consuming, with a single extraction taking several hours.

4.1.4 Decoction

Water-soluble compounds can be extracted from medicinal plant material using a decoction, which is a water-based preparation. When working with plants that have tough and fibrous barks and roots, it is advisable to use the decoction method. To enhance the effectiveness of the dissolving process, it is common practice to finely fragment or pulverize the plant material. Boiling the plant material in water yields the liquid extract. Water can sometimes be substituted with glycerol or aqueous ethanol. The quantity of water used in this method depends on the level of hardness of the plant material (Zhang et al., 2018; Ponphaiboon et al., 2023). Delving deeper into these methods, Table 5.2 depicts the advantages and disadvantages of different traditional extraction techniques.

4.2 Nonconventional or modern extraction methods

4.2.1 Supercritical fluid extraction

A fluid beyond its critical point in terms of pressure and temperature is referred to as a "supercritical fluid." It is a highly dense fluid that exhibits properties that lie between those of a gas and a liquid. A supercritical fluid is a material that exhibits both liquid and gas properties when it reaches its critical point, which is defined by its critical temperature, critical pressure, and critical density. Currently, the

Table 5.2 Advantages and disadvantages of traditional extraction methods.

Extraction method	Advantages	Disadvantages
Maceration	Basic equipment, easy operation, operator expertise not required, energy efficient, suitable for high solubility substances, ideal for less potent, cost-effective extracts	Can be very time-consuming, incomplete extraction, high solvent requirement, sluggish process
Percolation	Less labor-intensive than maceration, efficient for thermolabile components, ideal for high-value crude drugs, thorough extraction	Requires skilled personnel, longer duration than Soxhlet, more solvent consumption, particle size consideration throughout
Soxhlet extraction	Simple technique, can handle large quantities simultaneously, no need for postextraction filtration, suitable for varying scales	Potential thermal degradation of heat-sensitive compounds, lengthy process, labor-intensive, requires large solvent volumes, limited variable control
Decoction	Suitable for heat-stable substances, no expensive equipment needed, easy to perform, no operator training required	Not suitable for heat-sensitive extracts, requires constant heating

fluid demonstrates a density and solubility comparable with that of a liquid, a viscosity that is about the same as that of a gas, and a diffusion coefficient that is 100 times greater than that of a liquid (Rozzi and Singh, 2002). Therefore, a supercritical fluid can act as a solvent like a liquid, but with improved mass transfer kinetics. SFE is a method that may be used instead of traditional solid liquid extraction. It requires less solvent and operates at lower temperatures. SFE is a type of liquid extraction that replaces the typical liquid solvent phase with a material that is above its critical point. Carbon dioxide (CO_2) is the most utilized supercritical extraction solvent due to its relatively low critical temperature, among the several options available.

To enhance the solvating properties of supercritical fluids like CO_2, organic solvents can be introduced as modifiers. CO_2, when used as a supercritical fluid, allows for extraction under moderate conditions, minimizing thermal degradation and improving the collection efficiency of volatile analytes. CO_2 is particularly effective in dissolving organic compounds with some level of lipophilicity, such as esters and lactones. Modifiers can be introduced into the CO_2 stream using a separate pump and mixing equipment, or they can be added directly to the sample matrix in the extraction cell before pressurization. Valves between the pump and extraction vessel, as well as between the vessel and restrictor, control the flow in this setup, allowing for static, dynamic, or hybrid extraction methods. The restrictor regulates pressure inside the extraction vessel to maintain optimal conditions. Supercritical fluid extraction (SFE) is widely employed in the food industry to extract fats and oils from seeds and foodstuffs. It is also utilized for extracting active compounds from medicinal plants, including steroids, terpenes, and alkaloids, various oxygen-containing heterocyclic compounds, aromatics, and phenolic compounds. This versatile technique finds applications across diverse industries due to its ability to selectively extract target compounds while maintaining product quality (Komal Patel and Panchal, 2019). Supercritical carbon dioxide ($SC-CO_2$) is widely recognized for its effectiveness in extracting nonpolar compounds due to its low polarity. However, its polarity can be enhanced by incorporating polar solvents such as hexane, methanol, ethanol, isopropanol, or dichloromethane as modifiers or cosolvents. Ethanol is often recommended as a cosolvent due to its lower toxicity and good miscibility with CO_2. $SC-CO_2$ is valued for its high selectivity, cost-effectiveness, safety (nonflammability and noncorrosiveness), inertness (which preserves the bioactivity of extracted substances), and ability to extract thermally labile compounds. It is extensively used in extracting a wide range of natural products, including nonpolar compounds such as lipids and volatile oils. Optimization of extraction conditions, particularly pressure and temperature, allows SC-CO_2 to achieve efficient extraction while maintaining the quality and integrity of the extracted compounds (Ponphaiboon et al., 2023).

SFE is a contemporary and environmentally favorable method that employs a unique state of matter known as a supercritical fluid as a solvent. CO_2 is a preferred option for several reasons. Initially, it is optimal for the extraction of heat-sensitive compounds due to its ability to achieve a supercritical state at temperatures that are close to ambient temperature. Secondly, the operation is simplified by the fact that the pressure necessary for this state is not excessively high. Thirdly, the density of CO_2 is highly responsive to fluctuations in temperature and pressure, and its capacity to dissolve substances increases proportionally within a specific range. This enables the precise regulation of the quantity of material extracted through the modification of these parameters. Lastly, CO_2 is nontoxic, nonflammable, chemically unreactive, affordable, readily available, and has a minimal environmental impact (Zhang et al., 2023). Zhang et al. (2022) extracted camphor tree essential oil using steam distillation and SC-CO_2, respectively. The results revealed that the SFE method produced a greater amount of essential oil than the steam distillation method that they used. In addition, the GC/MS analysis showed that the essential oils that were extracted using supercritical fluid had a more diverse composition. Considering that the temperature of the extraction and the pressure of the extraction are the primary elements that influence the extraction of supercritical fluid, the conditions of the extraction may be improved by combining the response surface components. The response surface approach was utilized by Zermane et al. (2014) in order to improve the supercritical extraction conditions for essential oil extracted from Algerian Myrtus communis L. leaves. The findings indicated that the maximum yield was achieved at a temperature of 313°C and a pressure of 30 MPa.

4.2.2 Microwave-assisted extraction

Microwaves are a kind of electromagnetic radiation that falls between the frequency ranges of 300 MHz to 300 GHz. They are classified as nonionizing radiation. The combined presence of ionic conduction and dipole rotation is the primary factor influencing microwave heating, as they directly affect polar materials and solvents. Ionic conduction refers to the movement of ions in response to an alternating electric field. The resistance of the solution to ion migration results in the generation of friction, leading to the eventual heating of the solution. Dipole rotation refers to the reorientation of a molecule's dipoles in response to a rapidly changing electric field. Microwaves generate heat by interacting with polar chemicals found in plant and animal matter, such as water and certain organic molecules, through processes of ionic conduction and dipole rotation. Both ethanol and methanol exhibit lower microwave absorption compared with water; however, their total heating efficiency surpasses that of water. However, hexane and other solvents with low polarity (low dielectric constant), such as chloroform, are transparent to microwaves and do not produce heat. Within the field of MAE, there is a simultaneous occurrence of heat and mass transfer, which results in a synergistic effect that enhances the extraction process and leads to higher extraction yield. The process of water evaporation leads to a rise in pressure inside the cells of plants and animals, causing them to enlarge and eventually burst. This promotes the extraction of components from damaged cells into the nearby solution. MAE can also be performed in the absence of a solvent. In a moisture-free or solvent-free MAE, the water present in the sample functions as a solvent and facilitates the breakdown of cells. Essential oils and antioxidants have been extracted from aromatic plants using a solvent-free technique called MAE. Several factors, including microwave power, time, dielectric properties of the sample combination, and type of solvent, can be adjusted to get the required extract characteristics (Zhang et al., 2018; Bachtler and Bart, 2021).

The microwave extraction method employs microwaves to disrupt the cell wall and cell membrane of plant cells, therefore facilitating the extraction of active substances contained inside the cell. Simultaneously, microwaves may quickly increase the temperature of the extraction system, therefore reducing both the time needed for extraction and the number of organic solvents needed. The efficiency of microwave extraction is influenced by several factors, including the duration of extraction, the temperature at which extraction takes place, the ratio of material to liquid, and the power of the microwave used. Hu et al. (2019) utilized microwaves to extract polysaccharides from Camptotheca acuminata fruits and improved the extraction process using response surface approach. The most favorable extraction circumstances were a material-liquid ratio of 1:40, a microwave power of 600 W, an extraction period of 14 minutes, and an extraction temperature of 70°C. Microwave extraction offers a higher level of efficiency as compared with traditional extraction procedures. In their study, Ding et al. (2014) conducted a comparison of ultrasound-assisted, microwave-assisted, and reflux extraction methods employing ionic liquids as extraction solvents. Their findings demonstrated that MAE exhibited the best level of extraction efficiency. Generally, MAE tends to have a higher yield compared with traditional extraction procedures.

Fernandez-Pastor et al. (2017) conducted a study comparing microwave-assisted extraction (MAE) with Soxhlet extraction methods for extracting triterpene acids from olive skins. Their findings showed that MAE not only significantly reduced extraction time but also improved the yield of triterpene acids compared with Soxhlet extraction. In another study focusing on lavender extraction using microwaves, essential oil obtained by MAE exhibited superior antibacterial activity compared with oil obtained through traditional hydrodistillation methods. This highlights MAE as an efficient and effective extraction method, capable of enhancing the bioactivity of extracted compounds due to its rapid and targeted heating effects.

4.2.3 Ultrasound-assisted extraction

Ultrasound is defined as sound waves or acoustic energy that has a frequency higher than the human hearing threshold of 20 kHz. Ultrasonication treatments for extracting natural compounds usually use frequencies ranging from 20 to 100 kHz and power densities between 10 and 1000 W/cm^2. Ultrasonic waves propagate through many materials by interacting with their mechanical and physical properties, including composition and structure. Ultrasonic waves cause molecules to move from their initial location due to a combination of compression and rarefaction (Mehta et al., 2022). The idea of the UAE is based on the use of cavitation phenomena, which increase the transfer of mass, and the intimate connection between the solvent and plant or animal tissues. Cavitation bubbles occur when the acoustic pressure generated by rarefaction cycles surpasses the cohesive forces between the molecules of the liquid media. If the acoustic pressure is insufficient, the cavitation bubbles will undergo expansion and contraction throughout the rarefaction and compression cycles. When the bubbles do not reach the necessary size for implosion or collapse, this is known as stable cavitation, which reduces the impact of cavitation. If the pressure reaches a sufficiently high level, transient cavitation produces a significant amount of energy, causing the microbubbles to keep expanding until they collapse or implode once they reach the critical size. The implosion of cavitation bubbles near tissue surfaces produces microjets, resulting in tissue rupture and the infiltration of solvent deep into the tissue matrix. Thermal energy release promotes the movement of mass. Ultrasonic energy during UAE leads to cavitation, thermal effects, and mechanical events that result in the breakage of cell walls, decrease in particle size, disintegration of the plant matrix, and improved mass transfer. Importantly, these processes do not alter the chemical structures or biological activity of the chemical contents present in the extracts. Ultrasound obviates the necessity of elevated temperatures in the extraction procedure, hence enabling the UAE to safeguard heat-sensitive compounds (Picot-Allain et al., 2021).

UAE harnesses the mechanical and cavitation effects of ultrasound to enhance the extraction of chemical components from natural products. The mechanical effect involves ultrasound waves transmitting through a substance, causing particles to vibrate and enhancing the dispersion of substances within the medium. On the other hand, the cavitation effect occurs when ultrasound induces microbubbles in the medium to vibrate and collapse, creating microjets and enhancing mass transfer. There are three primary types of ultrasonic devices used in UAE: the ultrasonic bath mode, sonotrode (ultrasonic probe) mode, and a special-design mode combining elements of both. The ultrasonic bath is widely used in small and laboratory-scale UAE setups, where a flask or tube is partially immersed in water to optimize cavitation. It is simple and accessible, comprising a transducer, a vessel, a timer, and often a heater to control temperature. However, for larger-scale extractions, a mechanical stirrer is typically employed to keep the matrix suspended and ensure uniform exposure to ultrasound. In contrast, the ultrasonic probe mode involves direct contact between an ultrasonic probe and the sample/solvent mixture. This setup allows for more targeted application of ultrasound energy but carries a risk of cross-contamination between samples due to direct contact with the probe. Special-design modes incorporate larger probes and strategic positioning of vessels to facilitate efficient extraction on a larger scale, often integrating stirrers and temperature control systems. Each mode offers advantages depending on the scale and specific requirements of the extraction process. Bath sonication is indirect, ensuring less contamination spread, while probe modes provide direct and intense ultrasound application but require careful handling to prevent sample contamination. These techniques collectively enhance the efficiency and effectiveness of extracting valuable compounds from natural sources for various applications in research and industry (Ponphaiboon et al., 2023).

Ultrasonic extraction is a more environmentally friendly and economically feasible way for extracting natural compounds, as compared with standard extraction methods. It significantly reduces the amount of time needed for extraction, improves the effectiveness of extraction, and at the same time minimizes the amount of solvent used. It functions as a more efficient and environmentally friendly substitute for conventional extraction techniques, resulting in energy conservation and reduced emissions. Lin et al. (2023) obtained Shatian pomelo peel polysaccharide using various extraction techniques. Their research indicated that ultrasonic extraction yielded higher amounts and exhibited more potent antioxidant activity. The primary determinants impacting the effectiveness of ultrasonic extraction are the ratio of solid to liquid, the strength of the ultrasonic waves, the duration of the ultrasonic treatment, and the temperature at which it is conducted. Most recent investigations have integrated response surface optimization with ultrasonic extraction settings. Achat et al. (2012) explored the feasibility of extracting phenolic compounds from olive leaves using ultrasonic maceration to enhance the nutritional content of olive oil. They found that optimal extraction conditions included an ultrasonic power of 60 W, an ultrasonic temperature of 16°C, and an ultrasonic duration of 45 minutes. Nurkhasanah et al. (Nurkhasanah et al. (2023) conducted an optimization study on the extraction of anthocyanins from pigmented corn using response surface methodology. Their research identified the best process conditions, which involved using a solvent composed of 36% methanol, maintaining an extraction pH of 7, applying an ultrasonic power of 73%, conducting the extraction for 10 min, and maintaining an extraction temperature of 70°C. In another study, the ultrasonic-assisted extraction (UAE) of clove oil and α-humulene demonstrated higher yields and shorter extraction times compared with conventional methods. This highlights UAE as an efficient and energy-saving technique for extracting these compounds (Wei et al., 2016).

5. Isolation techniques for the chiral natural products

Plants include phytochemicals, which have been demonstrated to be good for disease prevention and long-term health. Plants are commonly ingested for their important chemicals, which include saccharides, coumarins, lignans, flavonoids, terpenoids, and steroids. Plant health advantages and composition are becoming increasingly well-documented. Chiral natural products, a diverse and important family of chemicals, serve critical roles in biological processes and have enormous medicinal applications. However, separating these intriguing chemicals from their natural sources is typically difficult due to the occurrence of structurally identical but chemically different enantiomers. Due to the complexity of plant chemical compositions, the extraction and isolation of pure phytochemicals are essential steps before their structural identification and bioactivity screening. In recent years, advancements in extraction technologies and methodologies have significantly accelerated the extraction and analysis of phytochemicals. Extraction serves as the initial stage in phytochemistry research and is crucial for isolating desired components while minimizing undesired constituents. Isolation aims to extract maximum quantities of target chemical elements using physical and chemical processes, thereby purifying them into individual compounds or functional sections from plant extracts. Traditional isolation techniques such as solvent extraction, precipitation, crystallization, fractional distillation, salting-out, and dialysis remain widely used. Modern separation technologies including column chromatography, HPLC, ultrafiltration, and high-performance liquid drop countercurrent chromatography (HPLDCCC) have become indispensable for efficient and precise separation of phytochemicals. These methods play a critical role in enhancing the purity and efficacy of isolated compounds for further scientific investigation and application in various fields (Brusotti et al., 2014; Hosler and Mikita, 1987).

Despite significant advances in extraction and separation techniques, isolating natural products (NPs) from plants, marine species, or microbes remains a tough undertaking. Undoubtedly, hybrid approaches such as LC-NMR or LCMS enabled on-line structure elucidation and gave outstanding examples of NP identification without previous isolation (Sturm and Seger, 2012), but in many situations, the need to have purified compounds on hand remains. Full chemical structures, including stereochemistry, of novel NPs are expected to need isolated, highly pure compounds; nevertheless, the quantities required have greatly lowered, and 2D NMR spectra of tiny molecules may now be acquired with less than 100 mg within an acceptable period. Furthermore, biological activity must be tested in vitro and in vivo following a purification procedure to rule out interference from other chemicals. Last but not least, reference standards for the quality monitoring of herbal medicinal plants and herbal medicines rely heavily on isolated components of verified purity. In recent years, NPs have seen a resurgence in drug development programs, owing to their better chemical diversity over synthetic compound libraries (Rosén et al., 2009) and drug-like characteristics (Kingston, 2011). The traditional method of isolating NPs begins with the identification, collection, and processing of biological material, often by drying. Extraction with several solvents ranging from low to high polarity follows. Several purification processes are required before isolating pure chemicals, which is often done using (semi-)preparative HPLC or liquid−liquid chromatographic methods. In recent years, a trend toward separation techniques based on biological or pharmacological activity has been seen. However, by taking this method in its entirety, we will be able to restrict our understanding of the tremendous chemical variety of plants, microorganisms, and marine species, while also encouraging exploratory study on the chemistry of living organisms and the interaction changes in their metabolic profiles.

Identifying the chemical structures of plant compounds is crucial as it forms the basis for future research into their bioactivities, structure−activity relationships, metabolism in living organisms, structural modifications, and synthesis of active phytochemicals. However, conducting structural investigations using traditional chemical procedures such as chemical degradation and derivative synthesis can be challenging due to the limited quantities of substance obtained from plants. Traditional methods often require relatively large amounts of material, which may not be feasible when dealing with rare or low-yield plant compounds. Therefore, innovative approaches and sensitive analytical techniques, such as spectroscopic methods NMR, MS, chromatography coupled with spectroscopy LC-MS, LC-NMR, and computational chemistry, are increasingly employed to elucidate complex structures from small amounts of plant-derived material. These methods play a critical role in advancing our understanding of plant chemistry and facilitating the development of applications in various scientific and industrial fields. Thus, spectral analysis is mostly employed. In other words, the goal is to minimize the amount of sample used while maximizing the acquisition of structural information through the measurement and analysis of distinct spectra (Silverstein and Bassler, 1962). This section provides an overview of the typical techniques used to isolate phytochemicals and their unique uses.

5.1 Dialysis method

This technique involves substances selectively passing through natural or synthetic semipermeable membranes, such as dialysis bags, to achieve separation, classification, purification, or concentration. The process relies on creating concentration differences, pressure differentials, or potential variances.

For example, dialysis is used to remove inorganic salts, monosaccharides, and other impurities after separating and purifying substances such as saponins, proteins, polypeptides, and polysaccharides from plants. Conversely, the semipermeable membrane retains large molecular impurities while allowing small molecular substances to pass through into the solution on the other side. This method is crucial in biological and chemical applications for refining and isolating compounds based on their molecular size and characteristics (Tahara et al., 2017).

5.2 Fractional distillation method

Fractional distillation is a technique used to separate components within liquid compounds based on their different boiling points. This method is categorized into atmospheric, vacuum, molecular distillation, among others, and is particularly useful for separating volatile hydrocarbons and certain liquid alkaloids found in plants. For instance, in Cicuta virosa, the alkaloids coniine and conhydrine have distinct boiling points of 166–167 and 226°C, respectively, allowing them to be separated effectively using fractional distillation. In practice, compounds in liquid mixtures can be separated by repeatedly distilling the solution if their boiling point difference exceeds 100°C. When the boiling point difference is less than 25°C, a fractionation column is necessary to achieve separation. As the boiling point difference decreases, the fractionation apparatus must be more precise to effectively separate the components. Fractional distillation thus plays a crucial role in isolating and purifying compounds based on their boiling point differentials, making it a fundamental technique in chemical and pharmaceutical industries (Hanif et al., 2017).

5.3 Crystallization method

Crystallization is a highly effective technique for purifying substances by precipitating solutes from a mother liquor containing complex components. Recrystallization is often necessary when the initial crystals are impure and require further purification. This process exploits the differences in solubility of components in a solvent to separate them effectively from mixtures. Plant chemists frequently rely on crystallization as a primary method for obtaining purified compounds. Crystals can be obtained by freezing or concentrating the extract after extraction with a suitable solvent, particularly when the phytochemical content in a single plant is high (Escribà-Gelonch et al., 2018). The success of crystallization hinges on selecting an appropriate solvent that meets specific criteria: moderate boiling point, inertness toward the components, safety, affordability, and accessibility. Ideally, the solvent should be insoluble for impurities at both high and low temperatures or selectively soluble for the target compound. Common solvents used for crystallization include methanol, ethanol, acetone, ethyl acetate, acetic acid, and pyridine. Sometimes, a mixture of two or more solvents is employed when a single solvent fails to produce crystals. These mixed solvents typically consist of two miscible components: one with high solubility for the compound and another with low solubility. In practice, the crystallization process begins by heating the sample to dissolve it in the minimum amount of solvent with high solubility. The solution is then clouded by adding the second solvent with low solubility. After reaching saturation, the solution is cooled, leading to crystallization. Analyzing crystal form, color, melting point, melting range, and employing techniques such as thin-layer chromatography (TLC) or paper chromatography can provide initial insights into the purity of the crystallized product (Feng et al., 2019).

5.4 Classical chromatographic methods

Chromatography stands out as the most commonly used technique for separating chemical constituents in natural products, prized for its simplicity, speed, and high efficiency in separation. By leveraging different separation principles, operational modes, and chromatographic packings or even combining various chromatographic methods—it becomes possible to achieve the separation and purification of a diverse array of phytochemicals. Moreover, chromatography excels in identifying compounds, making it indispensable in both the analysis and purification processes of natural products.

5.4.1 Adsorption chromatography

Adsorption chromatography relies on the varying adsorptive capacities of adsorbents for different compounds. Commonly used adsorbents include silica gel, alumina, activated carbon, polyamide, among others. Silica gel is widely employed in adsorption chromatography for separating a wide range of plant chemical constituents. Alumina is preferred for separating alkaloids, steroids, and terpenoids, which are alkaline or neutral lipophilic components. Activated carbon is effective for separating water-soluble substances such as amino acids, carbohydrates, and certain glycosides. Polyamide is particularly useful for separating phenols, quinones, flavonoids, anthraquinones, and tannins due to its ability to form various hydrogen bonds with these compounds. Each adsorbent's specific properties make them suitable for different types of compounds in chromatographic separations (Poole, 2018).

5.4.2 Ion exchange chromatography

Ion exchange chromatography aims to separate chemical components based on their varying degrees of dissociation. In this technique, ion exchange resin acts as the stationary phase, while a mixture of water and solvent serves as the mobile phase. Ionic constituents in the mobile phase undergo adsorption onto the ion exchange resin through an ion exchange process. This method is particularly effective for separating ionic substances such as alkaloids, amino acids, organic acids, peptides, and flavonoids. The success of ion exchange chromatography depends largely on the compounds' degree of dissociation and their electric charges. Compounds with higher degrees of dissociation, whether acidic or alkaline, readily exchange with ions on the resin and can pose challenges during elution. Consequently, compounds with lower degrees of dissociation are eluted first, followed by those with higher degrees of dissociation, as they interact more strongly with the resin (Gerberding and Byers, 1998).

5.4.3 Partition chromatography

Partition chromatography separates components based on their differing partition coefficients between a stationary phase and a mobile phase, both immiscible liquids. Normal-phase chromatography uses a stationary phase more polar than the mobile phase, suitable for polar or moderately polar substances, employing carriers such as silica gel or cellulose powder. Silica gel with over 17% water content is favored due to reduced adsorption capacity. In contrast, reverse-phase chromatography employs a mobile phase more polar than the stationary phase, typically utilizing octadecylsilylated silica (ODS) as the stationary phase and methanol—water or acetonitrile—water mixtures as the mobile phase, ideal for separating nonpolar and moderately polar molecules. These techniques are crucial in analytical and preparative chromatography for their ability to effectively separate diverse chemical compounds based on their polarity characteristics (Feng et al., 2019).

5.5 New technologies and methods

5.5.1 High-performance liquid chromatography (HPLC)

HPLC is a fast and efficient technique for separating and analyzing substances, which was derived from standard column chromatography. The separation mechanism of this approach is identical to that of conventional column chromatography, encompassing adsorption chromatography, gel chromatography, partition chromatography, ion exchange chromatography, and other techniques. HPLC columns are manufactured using particle fillers with a diameter of 5–20 μm and high-pressure homogenate column loading method. The eluents are introduced into the column using a high-pressure infusion pump and are monitored by very sensitive detectors. The system also includes automatic recording and collecting devices. Consequently, it surpasses traditional column chromatography in terms of separation speed and efficiency. It possesses the attributes of exceptional efficiency, rapidity, and automation. Preparative HPLC is a technique that may be employed to produce a substantial quantity of samples with a high level of purity. HPLC has become more crucial in the separation, qualitative identification, and quantitative study of chemical components found in plants. When separating various chemical components of plants, it is important to remove small amounts of trace ingredients from a huge quantity of crude extracts. Typically, during the last phase of separation, samples with a high level of purity are generated using high or medium pressure liquid chromatography. Preparative HPLC mostly employs eluents with constant concentration. Gradient elution is occasionally employed for samples that provide challenges in separation. In addition, HPLC maintains the benefits of liquid chromatography, including its extensive variety of applications and the ability to easily replace the mobile phase. This technique can be utilized for the analysis of chemical components that are challenging to gasify, have large molecular weight, or exhibit thermal instability. The detectors frequently employed in HPLC are ultraviolet detectors and differential refractive index detectors; nevertheless, both exhibit certain limitations. Differential refractive index detectors exhibit sensitivity to variations in temperature, rendering them less than optimal for detecting tiny quantities of chemicals and precluding the use of gradient elution. Ultraviolet detectors are unable to detect samples that do not have ultraviolet absorption. Recently, the use of an evaporative light scattering detector (ELSD) has become prevalent in HPLC. It can not only identify materials that do not absorb ultraviolet light but also utilize gradient elution. It is appropriate for the majority of nonvolatile components (Ji et al., 2018).

5.5.2 Droplet countercurrent chromatography

DCCC is an advanced form of liquid–liquid partition chromatography that utilizes a countercurrent partition technique. In DCCC, the mobile phase flows through a column containing a liquid stationary phase. This setup necessitates the generation of droplets, where the mobile phase forms droplets that efficiently interact with the stationary phase. The key principle of DCCC lies in the continuous formation of new interfaces between the immiscible phases within thin partition extraction tubes. This dynamic process significantly enhances the separation of solutes dissolved in two-phase solvent systems. Components with different partition coefficients are isolated within their respective immiscible droplets, facilitating their separation. DCCC is particularly suitable for separating phytochemicals with high polarity. By leveraging the unique properties of droplet formation and continuous phase interaction, DCCC offers improved efficiency and resolution in the isolation of chemical components from complex mixtures. The separation efficiency of the separation effect is often superior to that of countercurrent partition chromatography, and it does not exhibit any emulsification phenomena. In addition, nitrogen is employed as the driving force for the mobile phase, therefore preventing oxidation of the separated material by atmospheric oxygen. However, this approach requires careful selection of a solvent solution capable of producing droplets. Additionally, the volume of material processed is limited, and specialized equipment is necessary. The DCCC method has excellent repeatability and is capable of handling crude extract samples ranging from milligram to gram quantities. It is applicable in both acidic and alkaline environments. By not employing solid separation carriers, it is possible to prevent the occurrence of irreversible adsorption and the widening of chromatographic peaks. When compared with preparative HPLC, DCCC utilizes a less amount of solvent, but, the separation process takes more time and the resolution is reduced (Ji et al., 2018).

5.5.3 High speed countercurrent chromatography

HSCCC is a type of liquid–liquid partition chromatography. Unlike traditional chromatography methods that utilize solid supports, HSCCC operates solely with solvents and Teflon tubing, preventing substrate loss due to binding with column materials. In HSCCC, compounds with higher partition coefficients in the mobile phase are eluted first, while those favoring the stationary phase are eluted later. This technique avoids the issues of irreversible adsorption and peak tailing commonly associated with solid-phase chromatography, as it eliminates the need for solid supports entirely. The chromatographic process achieves a sample recovery rate that is close to 100%. Additionally, it offers the benefits of reliable replication, exceptional purity of isolated molecules, and rapid processing. This method is suitable for separating and refining various types of phytochemicals, such as saponins, alkaloids, flavonoids, anthraquinones, lignans, triterpenes, proteins, and carbohydrates (Ji et al., 2018).

HSCCC is a type of CCC that has undergone significant advancements in terms of resolution, separation time, and sample loading capacity. It efficiently separates multigram quantities of materials within a few hours, making it highly effective for purifying natural products. However, using HSCCC requires a solid understanding of technical principles due to its unique parameters and separation process. A systematic approach to setting HSCCC conditions involves selecting appropriate two-phase solvent systems, determining the partition coefficient (K) of analytes, preparing the solvent system and sample solution, choosing the elution mode, flowrate, rotation speed, and monitoring the eluate in real time. These steps ensure optimal performance and reproducibility in HSCCC applications (Ito, 2005). Choosing the right solvent system is critical in developing the CCC (Countercurrent Chromatography) technique, accounting for about 90% of the overall effort. Unlike traditional liquid chromatography, CCC utilizes a two-phase solvent system comprising two immiscible solvents. One solvent acts as the stationary phase, while the other functions as the mobile phase. This approach offers a wide range of solvent combinations, enabling the separation of molecules with varying polarities. Several criteria must be met when selecting solvents for CCC:

1 **Analyte Stability and Solubility**: The solvents must dissolve the analytes effectively and maintain their stability throughout the separation process.
2 **Phase Separation**: The solvent system should readily separate into two distinct phases with a minimal volume ratio to minimize waste and optimize efficiency.
3 **Partition Coefficient (K) Range**: Ideally, the solvent system should exhibit partition coefficients (K values) suitable for the analytes being separated. For HSCCC, K values between 0.5 and 1.0 are generally optimal.

4 **Retention and Stability**: The solvent system should ensure that the stationary phase remains adequately retained within the column, preventing leakage or loss during the chromatographic process.

Careful consideration of these factors ensures the successful application of CCC techniques, enabling efficient and effective separation and purification of natural products and other analytes of interest (Sticher, 2008).

6. Importance of chiral analysis in natural products

The study of chiral substances has garnered significant interest due to the prevalence of chiral bioorganic molecules. Living organisms are composed of chiral biomolecules such as amino acids, carbohydrates, proteins, and nucleic acids. These biomolecules exist exclusively in one of two enantiomeric forms, for example, amino acids predominantly in the L-form and carbohydrates in the D-form within the human body. Enantiomers are molecular mirror images of each other that cannot be superimposed. The chirality, or handedness, of enantiomeric molecules arises from the presence of chiral elements in their structure, such as a chirality axis, chirality plane, or chirality center (like an asymmetric carbon atom). The chirality and optical activity of enantiomers are determined by their absolute configuration, which specifies the exact spatial arrangement of atoms in the molecule. Despite having identical physical and chemical properties, the two enantiomers of a compound interact differently with other chiral molecules due to variations in their atomic arrangements and thus their binding affinities. This characteristic plays a crucial role in biological interactions, as proteins, enzymes, and carbohydrates—all essential components of biological systems—exhibit chirality. Therefore, organisms may exhibit distinct responses to each enantiomer, a phenomenon that was first observed by Pasteur in 1857. This is especially significant when it comes to chemicals such as PACs and insecticides, which are specifically formulated to elicit biological responses (Purushoth Prabhu et al., 2016).

During the early 1980s, the field of analytical chiral separation faced significant challenges. The methods available for preparative synthesis and separation were notably less sophisticated compared to today's advanced techniques. However, it was evident that chiral medicines needed to undergo enantioseparation, with each enantiomer being used individually. In modern times, enantiomers are regarded as separate chemicals due to their potential to exhibit unique biological interactions, resulting in significantly different pharmacological, pharmacokinetic, or toxicological effects. The human body has a high degree of chiral selectivity, meaning that it interacts with each racemic medication in a distinct manner. Additionally, it metabolizes each enantiomer through a unique pathway, resulting in various pharmacological effects. One isomer may therefore exhibit the required therapeutic activity, whereas the other may be inert or have undesirable side effects (Li and Haynie, 2006). To illustrate the importance of chirality in drug development, Table 5.3 showcases some examples of chiral drugs and their effects.

Chirality in pharmacology can have profound effects on drug efficacy and safety. In many cases, one enantiomer of a drug may exhibit significantly stronger therapeutic effects than its counterpart. For example, one chiral form of naproxen demonstrates an antiinflammatory effect that is 28 times more potent than its enantiomer. Conversely, the other form of dopamine can be toxic to nerve cells, while the therapeutic form is essential in treating conditions such as Parkinson's disease by regulating

Table 5.3 Chiral drugs and their effects: some examples.

Chiral drug	Bioactivity Description
Albuterol	The D-isomer can lead to airway constriction, whereas the L-isomer prevents such side effects (Gad, 2024).
Ethambutol	The (S, S)-form acts as a tuberculostatic agent, while the (R, R)-form may cause optic neuritis and potential blindness.
Levodopa	L-dopa is used for Parkinson's disease treatment, but the D-form can cause severe side effects, including granulocytopenia.
Omeprazole	The S-enantiomer (Esomeprazole) is more effective at reducing stomach acid production compared to the R-enantiomer.
Penicillamine	The (S)-enantiomer is effective against arthritis, whereas the (R)-form is highly toxic.
Propoxyphene	The α-L-isomer serves as a cough suppressant, while the α-D-isomer is used for pain relief.
Thalidomide	The (S)-isomer is effective against nausea, but the (R)-isomer is teratogenic and can cause severe fetal deformities, such as underdeveloped limbs.
Ibuprofen	The S-enantiomer is responsible for the drug's antiinflammatory effects, while the R-enantiomer is largely inactive.

tremors. Historically, the implications of chirality in medicine are exemplified by thalidomide, a drug used in the 1960s to treat morning sickness. The racemic mixture, which includes both enantiomers, was marketed to pregnant women. Tragically, one enantiomer was found to cause severe birth defects, highlighting the critical importance of understanding and controlling chiral properties in pharmaceuticals to ensure safety and efficacy. However, only the (R)-enantiomer has therapeutic action. The teratogenic properties of the (S)-enantiomer were only identified after thalidomide had caused several hundred birth abnormalities. By then, chirality has been considered throughout the whole medication development and regulatory process. Using a medication's more active isomer often provides the following benefits: Fewer or less severe adverse effects that might arise from using the undesirable isomeric version. A patient's dosage is automatically cut in half. Waste is reduced because less unwanted isomer is manufactured. Racemic switching presents new business opportunities: a medication that was once sold as a racemate can be redeveloped and released as an enantiomerically pure form, which may help to prolong the patent protection of a crucial product. Studying the characteristics of a chiral drug molecule's enantiomers in terms of therapeutic effectiveness and safety has clear advantages. Because of this, before deciding whether to commercialize a medicine as one of the enantiomers or as a racemate, the FDA and the European Committee for Proprietary Medicinal Products have mandated since 1992 (FDA'S policy statement for the, 1992). that the qualities of each enantiomer be investigated independently. Strong methods for the identification and isolation of chiral drugs are needed for this. Furthermore, there is a growing recognition of the necessity to reassess the characteristics of enantiomers of racemic medicinal compounds that are presently on the market. The result has been a notable rise in the need for chiral analytical and separation techniques that are sensitive. Concurrently, there has been a notable surge in the quantity of novel compounds being developed in the pharmaceutical sector. Most pharmaceutical firms currently have explicit policies in place suggesting that all chiral bioactive compounds have their enantiomers isolated and examined. By using enantioselective synthesis, pure medication enantiomers may be obtained effectively. That said, this is nearly invariably costly, seldom feasible, and complex. Which chiral form of a chemical molecule will arise during a normal manufacturing process is mostly unpredictable. The many potential chiral forms of a chemical are often produced in equal proportions because of this lack of control. Therefore, it is frequently necessary to separate intermediates or end products from a racemic mixture. The development of techniques for sensitive, high throughput, and effective chiral separations, control over chemical synthesis, evaluation of enantiomeric purity, and pharmacodynamics calculation are all gaining attention. The development of strategies for chiral drug separation is strongly encouraged by the diverse instances of distinct pharmacokinetic, toxicological, and therapeutic features of the enantiomers of chiral medicines. The absorption, distribution, protein binding, and receptor affinity of enantiomers can all vary. These characteristics have been used to create effective methods for generating gram to multiton quantities of enantiomerically pure chiral pharmaceuticals, measuring small enantiomeric impurities in chiral medications, and accomplishing analytical-scale chiral separations. For chiral separations, chromatographic methods including TLC, SFE, GC, and HPLC have been developed. Effective substitutes for chromatographic methods include capillary zone electrophoresis, capillary gel electrophoresis, electrokinetic chromatography, and capillary electrochromatography (Li and Haynie, 2006).

In today's pharmaceutical sector, chirality is a big deal. The fact that enantiomers of a racemic medicine may exhibit distinct pharmacological activity, pharmacokinetic impacts, and pharmacodynamic properties has increased their level of attention. Due to its remarkable chiral selectivity, the human body responds differently to each racemic medication and uses various metabolic pathways to create distinct pharmacological effects from each enantiomer. Synthetic and chiral pharmaceuticals make up the bulk of the pharmaceutical industry at the moment. On the other hand, racemic combinations of several chiral medications are still available. It is critical to prepare and promote only the pure, therapeutically active version of a chiral medication to eliminate any potential side effects. As a result, research into methods for analyzing and isolating racemic medicines is critically important.

7. Chiral analysis of the chiral natural products

Chirality is a prevalent phenomenon in nature, where many biological systems exhibit a preference for specific enantiomers. For instance, proteins are predominantly composed of L-amino acids, and polysaccharides consist mainly of D-monosaccharides. This preference for single enantiomers is significant in biological functions and processes, highlighting the importance of chirality in biological systems (Richard et al., 2008). Enantiomers may have distinct biological activity, distinctive aromas or tastes, or various interactions with receptors, transport systems, or enzymes. The rising profile of chiral recognition has led to the exhaustive development of new analytical tools for the detection and quantification of enantiomeric forms. In addition, these techniques are employed to track the stereoselective behavior of enantiomers in many contexts, such as the manufacturing of drugs or agrochemicals, or the regulation of food components. In several fields, including the pharmacological, biological, environmental, and food sciences, chiral analysis has become an indispensable technique (Alvarez-Rivera et al., 2020). Chiral compounds are also employed in asymmetric synthesis, which involves the production of optically pure molecules. They are also employed in scientific investigations to ascertain response mechanisms and routes. Chiral compounds have a significant role in the agrochemical industry. Enantiomeric analysis is now performed using nonchromatographic techniques such as polarimetry, nuclear magnetic resonance (NMR), isotopic dilution, calorimetry, and enzyme approaches. The drawbacks of these approaches include the requirement for pure materials and the absence of enantiomer separation. Several factors influence the choice of a separation technique, including sample amount, cost, speed, sample state, and sensitivity. However, the most critical factor is the physical and chemical properties of the target enantiomers (mirror-image molecules) you want to separate (Singh et al., 2019). For instance, techniques such as Liquid Chromatography (LC), Capillary Electrophoresis (CE), or SFC are well-suited for analyzing heat-sensitive enantiomers. Conversely, Gas Chromatography (GC) is ideal for volatile ones (Fanali, 2017). Two-dimensional (2D) configurations can address challenges associated with complex sample matrices. For example, researchers have employed enantioselective 2D-GC to analyze essential oils, flavors, and polychlorinated biphenyls (PCBs), and amino acids. In recent years, chiral GC and GC-GC have become valuable tools for tasks such as quality control, origin determination, and detecting adulteration in food products. Additionally, they can track changes throughout food processing.

HPLC remains highly favored for separating compounds that are sensitive to heat and nonvolatile. Its widespread use is supported by a vast array of established methods. HPLC's adaptability for both analytical and preparative scale separations of enantiomers make it indispensable across various applications. Recent advancements include techniques to monitor changes in enantiomeric ratios during fermentation and for quality control and nutritional analysis of food. However, a significant area of focus for researchers employing chiral HPLC is the development of robust methods for food safety. This involves monitoring pesticide enantiomers to assess associated risks. Experimental designs that integrate statistical analyses to evaluate the impact of separation parameters, as well as advancements like 2D liquid chromatography, have proven invaluable in this field (Alvarez-Rivera et al., 2020).

SFC utilizes supercritical fluids—a state of matter where a solvent exists above its critical point, blurring the line between gas and liquid. These fluids offer advantages such as low viscosity and high diffusivity, enabling fast and efficient separations. Modifying temperature and pressure can effectively adjust the solvating capability of the mobile phase, thereby promoting the elution of specific compounds. Supercritical fluid chromatography (SFC) is particularly suitable for preparative separations due to its high capacity. CO_2 is commonly used as a mobile phase in SFC because of its ready availability, safety, and environmentally friendly characteristics. However, its low polarity necessitates the addition of organic solvents such as ethanol or methanol to enhance solubility. Recent research employing SFC has concentrated on separating chemical pesticide residues found in agricultural products. This focus is crucial due to the potential human health implications associated with these residues (Tao et al., 2018; Raimbault et al., 2019; D'Orazio et al., 2016).

CE and CEC are miniaturized techniques known for their low sample and solvent consumption. They offer fast, efficient enantioseparations and lower analysis costs due to minimal solvent usage. However, their microfluidic nature and limited sample capacity restrict them primarily to analytical applications. Examples include analyzing amino acids (protein and nonprotein) to glean information about food quality or biological processes, as many amino acids play crucial roles in metabolism. Cyclodextrins and macrocyclic antibiotics are the most common chiral selectors used in CE and CEC (Fanali, 2017).

7.1 Chromatography

Liquid chromatography plays a crucial function for medicinal chemists, serving as an essential technique for the separation of enantiomers in all phases of drug discovery and development. Various technical methods may be used to separate enantiomers on a large scale, such as traditional batch column chromatography (Samuelsson et al., 2018), recycling loop chromatography (Sousa et al., 2006; Lourenço et al., 2013), and VARICOL (multicolumn continuous chromatography, an enhanced version of SMB) technology (Lourenço et al., 2012), among others. LC has had significant progress over the past 4 decades, resulting in a wealth of technical expertise and a substantial amount of scientific data documented in the literature. From a theoretical perspective, the understanding of liquid/solid adsorption in physical chemistry is widely prevalent. In addition, the process of scaling up is simple. It may be done by following general guidelines, such as increasing the volumetric flow and injection volume in proportion to the square of the column radius ratio. One additional benefit of LC compared with other methods, such as SFC, is that the equipment required is less complex and more affordable. Additionally, there is no requirement for power to cool or heat substances such as CO_2 (Speybrouck and Lipka, 2016).

The direct separation and analysis of enantiomers using chiral HPLC (High-Performance Liquid Chromatography) have proven to be one of the most effective methods. It offers greater adaptability than chiral GC (Gas Chromatography), as it can separate a wide variety of nonvolatile compounds. This technique allows for the online detection and quantitation of both the mass and optical rotation of enantiomers, providing rapid and accurate chiral separation when appropriate detection devices are used. Chiral HPLC procedures can be categorized as either direct or indirect. Direct methods involve the use of chiral stationary phases (CSPs) and chiral additives in the mobile phase, while indirect methods involve the derivatization of samples. Direct separations using CSPs are more common and provide more predictable mechanistic outcomes compared to those using chiral additives in the mobile phase. Currently, there are nearly 100 HPLC CSPs available commercially. However, no single CSP can universally distinguish all classes of racemic compounds, making the selection of the appropriate CSP for enantioseparation a challenging task. This selection is primarily based on empirical data, intuition, trial-and-error, and sometimes luck. Another approach involves using predictive empirical principles based on the structures of the analytes and CSPs. Despite these methods, successful enantiomeric separation is not always guaranteed. Understanding the chemistry of the racemic analytes and the CSP is crucial, but interactions between the mobile phase, the analyte, and the CSP can complicate the process. When designing a chiral separation method, it is essential to consider all three components: the analyte, CSP, and mobile phase. Therefore, comprehending potential chiral recognition mechanisms is vital for achieving successful enantioseparation of specific classes of racemates on a specific CSP.

Bioactive compounds, such as active pharmaceutical ingredients (APIs), are frequently chiral molecules, with stereoisomers exhibiting distinct biological and therapeutic properties. However, the preparation of these molecules can result in racemic or scalemic compounds (it is not straightforward to generate the optically pure compound). The assessment of the enantiomeric purity of bioactive compounds, and consequently, their quality, is of paramount significance in the context of regulatory scopes. The distinct biological activities of various enantiomers of the same molecule are the reason why chiral analysis is so important in the field of natural products. This section examines the diverse analytical methods that are implemented to identify and characterize the chiral compounds present in natural products (Purushoth Prabhu et al., 2016).

Chromatographic methods are becoming increasingly successful due to the advancement of chromatographic instrumentation and the development of CSPs that are both highly efficient and versatile, enabling the resolution of a wide range of chemical classes of enantiomers which is shown in Table 5.4 Nevertheless, despite the fact that the development of new CSPs is a continuous and challenging issue that encompasses a variety of CSPs (Fernandes et al., 2018; Teixeira et al., 2019) and that significant progress has been made in understanding the fundamental principles of chiral recognition mechanisms (Scriba, 2016; Sardella et al., 2018) only a small number of CSPs possess the necessary characteristics and are effective for the preparative separation of enantiomers. For efficient enantioseparation on a preparative scale, CSPs must possess high-loading capabilities and strong enantiorecognition abilities. Additionally, CSPs suitable for preparative applications should be readily accessible and economically competitive compared with other enantioseparation techniques. When

Table 5.4 Chiral stationary phases for natural products separation.

CSP type	Representative Trade Names	Applications
Polysaccharide-based	CHIRALPAK IA CHIRALPAK IB CHIRALPAK IC CHIRALPAK ID CHIRALPAK IE CHIRALPAK IF CHIRALPAK IG	Alkaloids, amines, esters. Beta blockers, aromatics. Amines, esters, aromatics. Amino acids, esters. Alkaloids, amino acids. Beta blockers, chiral drugs. Amines, esters.
Synthetic polymer	Kromasil CHI-DMB Chiral column Kromasil CHI-TBB Chiral column	Acidic, neutral, basic compounds
Cyclodextrin	CYCLOBOND I column CYCLOBOND II column CYCLOBOND III column	Beta blockers
Protein phases	CHIRALPAK AGP (α 1-acid glycoprotein) CHIRALPAK HSA (human serum albumin) CHIRALPAK CBH (Cellobiohydrolase (stable enzyme))	Drugs, peptides Pharmaceuticals, steroids Glycosides, sugars
Chiral crown ether ((+) or (−) (18-Crown-6)-tetracarboxylic acid)	ChiroSil RCA (+) ChiroSil SCA (−) ChiralHyun-CR-1	Amino acids, esters, alcohols

these criteria are met, preparative chromatographic separations can successfully isolate desired enantiomers. Polysaccharide-based CSPs are particularly favored for these purposes due to their high-loading capacity, effective enantiodiscrimination performance, and versatility (Pinto et al., 2020). Despite certain limitations in the composition of the mobile phase, polysaccharide-based CSPs can be utilized in various elution modes, including normal, reversed, polar organic, and polar ionic. Additionally, immobilized polysaccharides have emerged as a reliable alternative, allowing for the use of a broader range of solvents as mobile phases (Padró and Keunchkarian et al., 2018; Grybinik and Bosakova, 2021; Aboul-Enein and Hassan, 2003). The structures of various polysaccharide-based Chiralpak columns, based on amylose and cellulose immobilized on silica gel, are depicted in Fig. 5.4.

Preparative enantioseparation often utilizes synthetic polymers such as polyacrylamide and cross-linked diallyltartardiamide, along with brush-type CSPs that incorporate neutral or anion exchange small molecules as chiral selectors. Cyclodextrin and macrocyclic antibiotic-based CSPs are also widely employed for this purpose (Sardella et al., 2017). In chromatography, besides the CSP, the mobile phase plays a critical role throughout the chromatographic process and requires careful optimization. It's essential to ensure that compounds are both soluble and stable in the mobile phase to achieve efficient separation. Additionally, an ideal mobile phase should be easy to remove, cost-effective, and environmentally friendly. From a green chemistry standpoint, the widespread use of certain organic solvents in LC poses environmental risks and potential hazards to personal health and safety. Therefore, there is a growing emphasis on developing alternative, more sustainable mobile phases that minimize these impacts while maintaining or improving separation performance (Armenta and de la Guardia, 2016; Shaaban, 2016). Nevertheless, normal and polar organic elution modes are

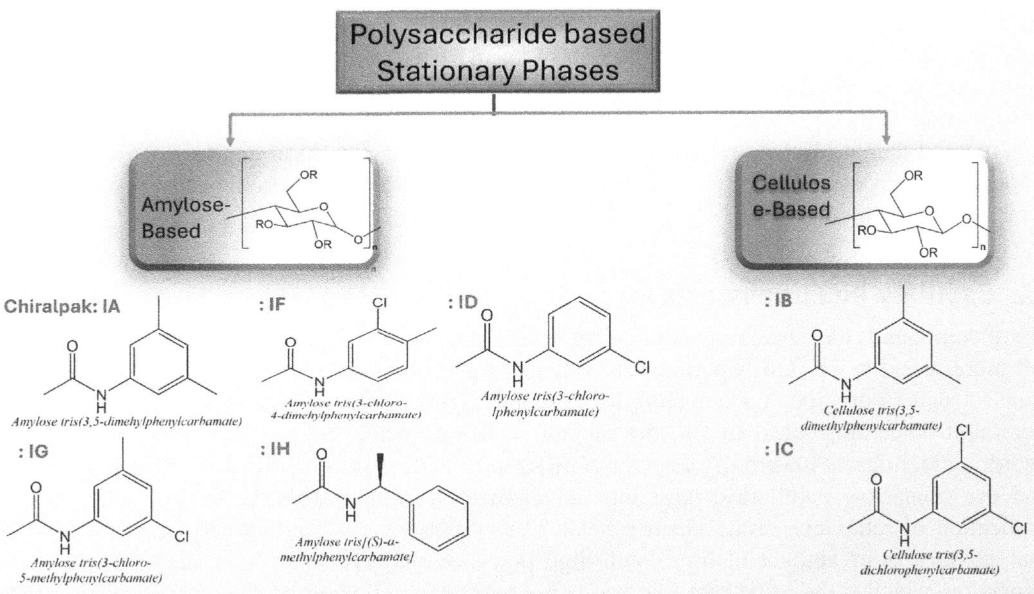

FIGURE 5.4

Structures of different polysaccharide-based Chiralpak columns.

more desirable than the reversed elution modes due to their greater ease of removal. The growing popularity of pure polar organic mobile phases, such as ethanol in LC, and the reduced usage or absence of organic solvents, as recommended by SFC (Speybrouck and Lipka, 2020), align with the move toward greener options. Improving efficiency and load capacity is important for enhancing the productivity of preparative enantioseparation. Additionally, faster flow rates and column equilibration can lead to shorter analysis and cycle durations without sacrificing resolution, which is also desired (Speybrouck and Lipka, 2016).

The chiral separation of ibuprofen (IBU) enantiomers was accomplished HPLC with a chiral stationary phase. The apparatus used included a Shimadzu LC-20AT HPLC system with a binary gradient pump model LC-20AT, an SPD-M20A diode-array detector, a CBM-20A communication bus module, and a DGU-20A5 degasser. The entire system was managed using LC solution software. The specific column employed was a Chiralcel OJ-H (cellulose tris(4-methylbenzoate)) coated on 5 mm silica-gel, 150 mm × 4.6 mm, 5 μm). The mobile phase for the separation was composed of n-hexane, 2-propanol, and trifluoroacetic acid in the ratio of 98:2:0.1 (v/v/v), with a flow rate maintained at 1.0 mL/minute. Detection was carried out using UV detection at a wavelength of 254 nm. The injection volume for the samples was set to 20 μL. For sample preparation, a stock solution of 1.0×10^{-2} mol/L (±)-ibuprofen was prepared by dissolving 0.02,063 g of the compound in 10 mL of n-hexane. This stock solution was then diluted to achieve a concentration range of 1.0×10^{-3} mol/L of ibuprofen enantiomers in 10-mL flasks for calibration and validation purposes.

The resulting chromatogram displayed distinct peaks for the (R)-IBU and (S)-IBU enantiomers in racemic IBU and pure (S)-IBU samples. The retention times for (R)-IBU and (S)-IBU were approximately 8.6 and 9.6 min, respectively, in racemic samples, and about 9.3 minutes for pure (S)-IBU samples. The chromatographic peaks were sharp and symmetrical with good baseline resolution, enabling precise quantification of each enantiomer (Valderrama et al., 2009). The method was validated over a concentration range of 50%–100% of (S)-IBU. A calibration curve was constructed by plotting the peak-area ratio of (S)-IBU against its nominal concentration, and linearity was confirmed by least squares linear regression analysis, which yielded a correlation coefficient greater than 0.99. This method demonstrates a robust and reliable approach for the chiral analysis of ibuprofen enantiomers, which is essential for pharmaceutical applications where the specific activity of enantiomers is of significant interest.

7.2 Capillary electrophoresis

Chiral compounds have been separated using several analytical procedures, such as HPLC, GC, TLC, and more recently, CE. CE is particularly valuable for the examination of several categories of substances, such as organic and inorganic ions, peptides, proteins, saccharides, medicines, optical isomers, and other molecules. In CE, the separation occurs when the analytes, driven by an applied electric field, migrate toward the detector at different speeds due to their distinct effective mobilities. The use of narrow capillaries, with internal diameters ranging from 10 to 100 μm, enables the application of relatively strong electric fields (200–1000 V/cm). This allows for the separation of analytes in a short amount of time, with high precision and efficiency. The major source of zone dispersion, which is the Joule heat, can be readily regulated in this setup. The many separation modes offered by CE, including zone electrophoresis (CZE), micellar electrokinetic chromatography (MEKC), gel electrophoresis (GE), and isotachophoresis (ITP), along with their high resolving power and efficiency, contribute to the competitiveness of the CE approach compared with other methods. CE has several extra benefits compared with other methods such as HPLC, including Only a minimal amount of sample and buffer (measured in nanoliters and microliters, respectively) is needed. Costly chiral columns may be circumvented by conveniently including the chiral selector within the BGE. The separation is very consistent since the buffer containing the chiral selector is refreshed after every run. The application of CE for the analysis of chiral substances has been extensively documented. Given the significant rise in publications, it is anticipated that chiral analysis using CE will soon be widely utilized for pharmaceutical, clinical, and environmental samples (Fanali and Salvatore, 1995).

There are two main approaches for separating diastereomers using CE: indirect and direct methods. The indirect method relies on the inherent physicochemical differences between diastereoisomers. Since diastereomers of the same compound have distinct properties, they can be separated by CE even with a nonchiral background electrolyte. However, this method doesn't achieve true enantiomeric separation as each peak represents a pair of unresolved enantiomers. To achieve true enantiomeric separation, the direct method utilizes a chiral selector. This selector interacts with the enantiomers during CE, forming temporary diastereomeric complexes. The strength of these interactions creates a mobility difference between the enantiomers, allowing them to be separated. The direct method is generally preferred due to its simplicity and efficiency. It avoids the need for pretreatment steps such as derivatization and uses readily available chiral selectors. Additionally, the method is less sensitive to the purity of the chiral selector compared to the indirect approach. CE has gained significant popularity in the last 10 years as a very effective and versatile method for separating enantiomers. It is considered a valuable supplement or alternative to HPLC in the field of pharmaceutical science and industry. Several chiral separation concepts that have been effectively utilized in HPLC have also been adapted for use in CE. The initial chiral separation using CE was conducted by Gassmann, Kuo, and Zare in 1985 (Gassmann et al., 1985). This methodology has many benefits, including high efficiency, the ability to use several chiral selectors to simplify method development, the use of tiny quantities of chiral selector and solvents, fast analysis, cheap overall cost, and no environmental effect. Utilizing low-UV wavelength (e.g., 200 nm) in CE enables the identification of contaminants that have weak chromophores, which might be challenging or unfeasible to detect using alternative techniques. CE is particularly effective for analyzing charged and polar substances that are not well-suited for traditional chromatographic procedures.

The enantioseparation of Warfarin (WF) was performed using capillary electrophoresis (CE) with various cyclodextrins (CD) as chiral selectors. The CE method allows for high flexibility in method development and optimization. The chiral selector is dissolved in the background electrolyte (BGE), which can be used in large excess to adjust the separation factor. This approach also allows for the combination of multiple chiral selectors to enhance the enantioseparation factor (Chankvetadze, 1997). The experimental setup included the use of native α, β, and γ-cyclodextrins (CDs), alongside various neutral and randomly substituted charged derivatives of CDs, serving as chiral selectors. The BGE composition and pH were optimized to achieve the best enantioseparation. A fused silica capillary with an effective length of 50 cm and an internal diameter of 50 μm was used. An applied voltage of 20 kV was used for the separation, with the capillary maintained at a constant temperature of 25°C. UV detection was performed at a wavelength of 200 nm.

A Chiralpak-AD column (Daicel Chemical Ind., Tokyo, Japan) was employed. The mobile phase consisted of a mixture of n-hexane and 2-propanol in a ratio of 90:10 (v/v). The flow rate was set to 1 mL/minute. The enantiomers were detected using an OR-990 optical rotation detector (JASCO,

Tokyo, Japan) connected online with a UV-970 ultraviolet detector (JASCO). The full HPLC chiral separation method involved the following details: racemic Warfarin was dissolved in the mobile phase to a final concentration of 1 mg/mL for sample preparation. An injection volume of 10 μL of the sample solution was injected into the HPLC system. The column temperature was maintained at room temperature (approximately 25°C). The total run time for the separation was 30 min (Chankvetadze et al., 2001). The final chromatogram shows effective separation of the Warfarin enantiomers. The resolution achieved between the enantiomers was sufficient to provide clear and distinct peaks, indicating successful enantioseparation. This method proved to be reliable and reproducible for the analysis of Warfarin enantiomers, making it suitable for both analytical and preparative applications. The use of CD as chiral selectors in CE, combined with the HPLC chiral separation method, provided a robust approach for the enantiomeric separation of Warfarin.

7.3 Polarimetry

Enantiomers of a chiral compound are challenging to distinguish because they share identical scalar physiochemical properties such as density, molecular weight, melting and boiling points, electronic and vibrational frequencies, reaction rates, and solubility. However, their vectorial properties differ; they are sign-inverted, meaning opposite enantiomers exhibit distinct behaviors when interacting with other chiral materials (e.g., biological materials, synthetic compounds, or drugs) or polarized light (Tang and Cohen, 2010). For instance, enantiomers rotate plane-polarized light in equal but opposite directions, producing optical rotation spectra that are mirror images of each other. Based on the direction of this rotation, a chiral molecule can be classified as dextrorotatory (positive, right-handed, clockwise) or levorotatory (negative, left-handed, counterclockwise). This characteristic, also known as circular birefringence (CB), arises from the different refractive indices for right circularly polarized (RCP) and left circularly polarized (LCP) light (Liehr, 1964). Similarly, chiral materials have distinct absorption cross-sections for RCP and LCP, known as circular dichroism. Molecular chirality can also manifest in other optical activities, such as Raman optical activity (ROA) and circularly polarized luminescence (CPL) (Gogoi et al., 2021).

Laser-based polarimeters with micro degree sensitivity in optical rotation (OR) have been developed, offering chromatographers a strong tool for analyzing chiral compounds (Yeung et al., 1980; Goodall, 1993). Polarimetry is an effective detection method for optically active substances, making it well-suited for analyzing chiral discrimination and determining enantiomeric purity. A polarimeter that uses lasers and operates at a certain wavelength might be considered a versatile device for detecting chiral compounds. This is because OR is a feature that can be measured at wavelengths that are far from the absorption band associated with optical activity. Mannschreck recently discussed lamp-based chiroptical detectors.

8. Selected examples of chiral analysis applied to specific natural products

Natural medicinal products are derived from naturally occurring substances that include active components. These components are extracted from various sources such as plants, microorganisms, minerals, and animals. Plants are the primary and most influential source of natural medicine, because to their wide range of chemical and structural variations and the many components found inside them. Some examples of pharmaceuticals generated from plants are TAX (obtained from the bark of willow trees) (Li et al., 2016), digoxin (extracted from the flower Digitalis lanata) (Gerwick, 2013), and morphine (derived from opium). It is estimated that in developed countries, more than 60% of all medications are either natural compounds or derivatives of secondary metabolites. Despite the challenges posed by several innovative drug development methodologies, natural products persist in generating new clinical candidates and medicinal molecules. These compounds continue to serve as crucial building blocks for the development of new drugs, especially in the field of antiinfective treatments such as azithromycin and amphotericin B, as well as acyclovir and romidepsin. Many of these have since progressed to clinical studies or been launched on the market. The integration of emerging technologies such as high throughput screening, automated separation techniques, and combinatorial chemistry has revolutionized the process of screening natural products in drug development. This is mostly owing to the wide range of chemical and structural diversity as well as biodiversity shown by these products (Mathur and Hoskins, 2017).

A significant number of medications currently available on the market are derived from natural sources. A prime example is aspirin (1), renowned for its analgesic efficacy and recognized as one of the most widely used medications globally. Its origins can be traced back to the plant genera Salix spp. and Populus spp., closely related to the compound salicin. Another notable example is the serendipitous discovery of the antibiotic penicillin (2) (Littleton, 2007) in the laboratory, derived from the fungus Penicillium notatum. Numerous other examples highlight the importance and value of natural products from plants and microbes in modern medicine. For instance, paclitaxel (Taxol) (3) (Littleton, 2007), originally obtained from the bark of the Pacific yew tree (Taxus brevifolia, Taxaceae), is a significant natural product with medical applications. Hypericin and pseudohypericin, derived from Hypericum perforatum (Guttiferae), exhibit antiviral activity against various retroviruses, including HIV. These compounds are crucial because they stabilize the HIV capsid structure, thereby inhibiting the release of reverse transcriptase and preventing the uncoating process (Amit Koparde et al., 2019a,b).

Natural compounds included in medicinal plants can effectively reduce the adverse effects of major diseases, for instance, plants can help with radiotherapy or oncochemotherapy. Because of their resistance qualities and cost-effectiveness, essential oils also provide substitutes for antibiotics and for costly therapies with unacceptable risk and advantages. In Brazil, a new antiinflammatory phytopharmaceutical, Acheflan, has been developed from an extract of Cordia verbenacea leaves, standardized in α-humulene. This natural product is now replacing conventional synthetic antiinflammatory drugs in the local market, reflecting the traditional use of this Brazilian medicinal plant backed by evidence-based phytotherapy. Several botanical products have received FDA approval and are sold globally. Fulyzaq, derived from the red sap of Croton lechleri, is used to treat HIV-related diarrhea, and Veregen, made from tea catechins, is used for treating external genital and perianal warts (Chen et al., 2008). Since 2005, Sativex, a titrated extract containing delta-9-tetrahydrocannabinol (THC) and cannabidiol (CBD), has been approved in multiple countries, including Canada, the United Kingdom, Germany, and New Zealand. In North America, Marinol (dronabinol) and Cesamet (nabilone) are used to treat nausea and vomiting associated with cancer chemotherapy. More recently, in 2012, the Dutch Medicines Evaluation Board approved a dry extract of Dioscorea nipponica, a traditional Chinese plant, for relieving headaches, muscle discomfort, and cramps (David et al., 2015).

The direct separation of thalidomide enantiomers was achieved using a commercially available cellulose tris-(4-methylphenyl benzoate) chiral stationary phase, specifically the Chiralcel OJ. The chromatographic conditions were meticulously optimized to ensure baseline separation of the enantiomers. The separation utilized a Chiralcel OJ column with dimensions of 250 × 4.6 mm I.D. and a particle size of 10 μm. The mobile phase consisted of a 50:50 (v/v) ratio of hexane and ethanol, with a flow rate maintained at 1.0 mL/minute. The process was carried out at a controlled temperature of 23°C, and UV detection was set at 240 nm. Injection volumes were precisely calibrated, with 2 μL for the racemic thalidomide mixture and 1 μL for each enantiomer (R-(+)-thalidomide and S-(−)-thalidomide) individually. In terms of chromatographic parameters, the capacity factor (k') for the first eluted enantiomer, R-(+)-thalidomide, was recorded at 9.67. The separation factor (α) was determined to be 1.54, and the resolution (R) achieved was 15.05. The procedure began with the preparation of the racemic mixture of thalidomide and the individual enantiomers. The Chiralcel OJ column was equilibrated with the mobile phase at a flow rate of 1.0 mL/minute until a stable baseline was achieved. Following this, 2 μL of the racemic mixture and 1 μL of each individual enantiomer were injected. The elution of the enantiomers was monitored at 240 nm, and their elution order was identified by comparing the retention times of the individual enantiomers under identical chromatographic conditions (Aboul-enein and Islam, 1991). The method described provides a straightforward and efficient approach for the enantiomeric separation of thalidomide. The use of cellulose tris-(4-methylphenyl benzoate) as a chiral stationary phase offers excellent stereochemical resolution. This method is suitable for both analytical and preparative scale separations, as evidenced by the baseline separation achieved. Consequently, it can be employed to determine the optical purity of thalidomide in both bulk and formulated dosage forms.

9. Automation and miniaturization in chiral analysis

The impact of a chiral substance in biological systems might vary significantly from its enantiomer due to the distinct relationship between structure and activity in chiral settings. Hence, the investigation of chiral substances with a focus on enantioselectivity is currently of significant interest. The various spectroscopic methods used in chiral recognition chromatography and electrophoresis are highly effective for separating chiral compounds. These methods have been continuously improved in terms of sensitivity and speed, thanks to the wide range of chiral stationary phases, chiral selectors, and chiral derivatization agents (CDAs) available for different analytes (Will et al., 2021). The subject of chiral analysis has experienced notable progress as a result of the incorporation of automation and downsizing, leading to a transformation of conventional approaches. Automation refers to the utilization of automated systems and equipment to carry out analytical operations with little human involvement, resulting in enhanced efficiency, accuracy, and consistency. Miniaturization is the process of decreasing the size of analytical machines and the quantities of samples and reagents used. This frequently results in quicker tests, lower expenses, and less harm to the environment.

The discipline of chiral analysis has significantly profited from the progress in automation and miniaturization, allowing more effective, precise, and high-capacity techniques. Technological advancements are crucial for analyzing natural products, as the existence of chiral centers can greatly impact the biological activity and pharmacokinetic aspects of the molecules.

9.1 Automation in chiral analysis

Automation is a crucial element of industrial operations. These procedures must be both productive and efficient, while also maintaining a high level of quality. Automation may enhance the efficiency of a laboratory by implementing parallel processing, integrating processes, extending operating hours, and reducing cycle times. Automation is often anticipated to liberate resources from repetitive work, generate idle time, and offer higher levels of consistency compared to manual procedures. Automation is crucial in the analytical laboratory due to the growing demand for enhanced productivity and cost reduction in the drug screening process (Laycock and Hartmann, 2005). The pharmaceutical and biotechnology industries have recently taken on the task of developing advanced methods to analyze and screen pharmaceutical natural products. These methods aim to generate potential medicinal products for the growing number of biological targets identified through the human genome initiative. The industries are focused on improving the speed and efficiency of this process. The conventional strategy of screening one compound against one target has been substituted by high throughput technologies, which include screening a large number of compounds against numerous targets. This enables the quick identification of 'hits' and the development of these hits into potential lead candidates. The progress in high throughput screening has mostly been achieved by incorporating automation/robotics into the laboratory. The introduction of combinatorial chemistry has led to significant advancements in high throughput synthesis. The introduction of combinatorial chemistry has equipped medicinal chemists with new methods to expedite the synthesis of pharmaceutical compounds. These methods include the ability to analyze extensive collections of compounds using automated parallel analysis and split-couple-recombine (or split/mix) methodologies (Zeng et al., 1998). Chiral analysis automation involves several elements such as preparing samples, collecting data, and processing data. Automated systems are specifically engineered to efficiently manage repeated operations with utmost accuracy, hence minimizing the risk of human mistakes and guaranteeing uniform outcomes. Robotic sample handlers have the capability to handle many samples at the same time, resulting in a substantial improvement in throughput and efficiency. These systems have the capability to carry out intricate processes including liquid–liquid extraction, solid-phase extraction (SPE), and derivatization with minimum oversight.

Automation significantly improves the ability to reproduce methods consistently and accurately. Automated systems guarantee the constant execution of every stage of the analytical process, resulting in outcomes that are more dependable and precise. Moreover, automation enables the uniformity of analytical techniques, simplifying the comparison of outcomes among various laboratories and research. Automated systems in chiral chromatography have the capability to precisely regulate parameters like flow rate, temperature, and solvent composition. State-of-the-art software enables the immediate monitoring and fine-tuning of these parameters, hence maximizing the separation conditions for various chiral substances. Automated data analysis systems can also efficiently detect and measure chiral analytes, hence minimizing the time needed for data interpretation. These automated devices enable the concurrent analysis of several samples, resulting in a substantial reduction in the time and effort required for human procedures.

9.1.1 Sample preparation

Automated sample preparation systems have been designed to process various biological matrices, such as plasma, serum, and tissues. These systems have the capability to carry out intricate preparatory procedures, such as protein precipitation, SPE, and liquid–liquid extraction (LLE), with minimum human involvement. Robotic liquid handlers and sophisticated software have enabled the preparation of sizable sample batches with exceptional accuracy and consistency. An example of this is the Hamilton MicroLab Star robotic system, which has the ability to automate various tasks such as preparing standard curves and quality control samples, diluting samples, transferring samples, and extracting proteins by precipitation (Zhang et al., 2011; Gu et al., 2006).

9.1.2 MS/MS method development

An analytical workflow is a sequential series of operations conducted on a sample to provide qualitative and/or quantitative data. MS workflow is an analytical process in which detection is carried out using Mass Spectrometry (MS). Automation of MS workflows involves the use of automated processes in various phases of handling and treating samples before MS detection, introducing samples, acquiring MS data, and processing MS data. It is important to acknowledge that, often, automation is only found in a single or a few phases of a process in the majority of circumstances. Therefore, if the phrases "automated MS" "MS automation" are used without a specific context, they might be deceptive. Automation is a vital component of contemporary analytical laboratories. The widespread adoption of automated analytical systems has been supported by advancements in microcomputer technology. Often, automated MS systems are just enhanced versions of conventional automated systems, where a mass spectrometer is incorporated instead of an optical detector. MS enhances the automated analytical procedures by delivering a high level of selectivity. Undoubtedly, automated MS systems have better throughputs compared with manual sample introduction methods. Automation enables analysts to conduct many and distant sample processing activities for the identification of analytes using MS. Automated MS systems possess the capacity to facilitate progress in several scientific fields (Elpa et al., 2020).

9.1.3 Sample analysis

Currently, chemical laboratories frequently employ autosamplers to provide unattended and accurate sample delivery to analytical equipment for extended periods of time. Autosamplers are commonly available as a package with hyphenated instruments such as GC or LC, although they may also be bought independently, depending on the user's taste. There are autosamplers available that have both linear and circular motions, like a carousel. Autosamplers that use robotic systems with 3D motions in space (X-YZ), known as cartesian coordinate robots, are connected to GC- or LC-MS instruments to provide versatile platforms for various applications (Elpa et al., 2020). Recent developments in ultrahigh-performance liquid chromatography (UHPLC) and fast scanning triple quadrupole mass spectrometers have made it possible to quickly analyze materials, often taking only 1–2 min per sample. Multiplexing approaches include the simultaneous operation of numerous LC systems in a single mass spectrometer, resulting in increased throughput while maintaining high separation quality. Automated systems have the capability to alternate between optimization and sample analysis modes, enabling the uninterrupted and efficient processing of extensive sample sets (Laycock and Hartmann, 2005).

9.1.4 Data processing and management

Automation is the utilization of integrated software solutions to facilitate data processing and management. These systems are capable of managing the complete process, starting from the receiving of samples to the generation of reports. They include functionalities such as automatic peak integration, calibration, and quantitation, along with data evaluation and quality control checks. Global databases provide the exchange of optimum SRM conditions among many instruments and locations, therefore minimizing duplication and enhancing uniformity (Shou and Zhang, 2012). Vendors have made substantial efforts and advancements in the past decade to develop software that is easy for users to navigate and that automates several mundane tasks. An example of this is the utilization of autotuning and calibration scripts to adjust and align with a set of reference peaks in a tuning standard.

9.2 Miniaturization in chiral analysis

Analytical methods commonly encompass several equally important stages, including sampling, sample preparation, isolation of the target substances, identification, quantification, and data management. Contemporary chromatographic and electrophoretic instrumental techniques have reached a level of development that allows for the combination of several separation techniques and detectors with a high capacity for providing detailed information (Sweedler, 2002). Currently, the phrase miniaturization is used in several fields of knowledge, such as engineering, physics, medicine, materials science, computer science, and chemistry. Miniaturization, in its most general meaning, refers to the formation of innovative systems that are significantly smaller than traditional systems. In the field of analytical chemistry, miniaturization refers not only to the reduction in size of analytical instruments, apparatus, and devices, but also to the use of significantly smaller quantities of components, such as chemicals and solvents, for performing analytical operations (Pena-Pereira, 2014). "Miniaturization" is a crucial concept in all analytical procedures, and sample preparation is an integral component of this process. The term encompasses the sample size, the number of procedures performed on the sample, the duration required to complete the full analytical process, the total expenditures associated with the process, the amount of hazardous organic solvents utilized, and the quantity of waste formed. Meeting these standards ensures improved process automation, increased throughput of analytical instruments, and, most importantly, enhanced quality of analytical findings. This quality is characterized by high accuracy and precise determination of compounds at low concentration levels. In response to these expectations, a diverse array of novel sample preparation procedures has been devised to tackle the growing complexity of real samples and the more challenging analytical tasks. It's important to note that in the field of analytical chemistry, the term "miniaturization" often refers specifically to micro-total analysis systems (μ-TAS) and lab-on-a-chip (LOC) devices. These terms highlight the intricate nature of reducing the size of analytical systems. However, miniaturization should be considered more broadly, encompassing improvements at every stage of the analytical process, not just the physical downsizing. This perspective allows us to appreciate the full scope of advancements made in the field (Pena-Pereira, 2014). A recent trend in analytical chemistry is the push toward miniaturizing analytical systems. This trend encompasses automation and portability across different stages of the analytical process, including sample preparation, analytical separation, and detection. The integration of these stages has led to the creation of fully miniaturized systems. The drive for miniaturization stems from the need to tackle challenges and meet the demands of new research areas. As a result, analytical microsystems are designed to handle very small sample volumes with high sensitivity, selectivity, and precision, while also reducing analysis time, reagent and solvent usage, and enabling fieldwork when possible.

Analytical separation techniques are implemented in analytical chemistry to separate target analytes before their detection. To separate and identify a large number of compounds, a variety of separation techniques have been devised. The most frequently employed analytical separation techniques are liquid chromatography (LC), GC, and CE. The development of compact, portable devices and the application of micro-scale techniques are all components of chiral analysis' miniaturization. Lab-on-a-chip systems, which are also referred to as microfluidic devices, have revolutionized the field by facilitating the precise manipulation of tiny sample volumes. These devices facilitate quicker and more efficient analyses by integrating multiple analytical stages on a single chip, such as sample injection, separation, and detection.

The utilization of microcolumns in chiral chromatography is a preeminent illustration of miniaturization. In comparison to conventional columns, microcolumns have smaller internal diameters, which results in a decrease in the quantity of samples and solvent necessary. This not only reduces the cost of analysis but also reduces refuse generation. Microcolumns also offer enhanced mass transfer and shorter diffusion paths, which result in higher separation efficiency and quicker analysis times. The potential for portability is another benefit of miniaturization. Compact analytical devices are applicable in a variety of environments, such as field analysis and point-of-care testing. Portable chiral analyzers facilitate the rapid screening of chiral compounds outside the laboratory, which is particularly advantageous in the areas of pharmaceutical quality control, environmental monitoring, and forensic analysis.

A demonstration of the separation process utilizing nano-liquid chromatography (nano-LC) and capillary electrophoresis (CE), with an emphasis on the examination of racemic substances including trans-stilbene oxide, lorazepam, and oxazepam. A capillary column loaded with silica and coated with 25% (w/w) cellulose tris(3-chloro-4-methylphenylcarbamate) was used in this research to achieve separation. The mobile phase for these experiments consisted of acetonitrile (ACN) and methanol (MeOH) mixtures, with buffer solutions added to maintain appropriate conductivity. Specifically, the mobile phases included 50%–100% methanol mixed with acetonitrile and 5 mM ammonium formate (pH 2.5 and 3.5) or ammonium acetate (pH 4.5, 5.5, and 6.5). The flow rate in nano-LC was 0.70 mm/second, whereas in CE, it was 0.81 mm/second. An applied voltage of +15 kV was used in CE, with an injection pressure of 10 bar for 0.2 minutes, followed by a 10 bar × 0.2-minute plug. The separations were carried out at a temperature of 20°C (Fanali et al., 2008).

The chromatograms in Fig. 5.5 of the document illustrate the separation of enantiomers, showing that increasing the polarity of the mobile phase resulted in higher retention factors for the studied compounds. Based on the study, capillary columns coated with cellulose tris(3-chloro-4-methylphenyl carbamate) on silica are viable for miniaturized separation of enantiomers using both pressure-driven CEC and nano-LC. These CSPs demonstrated excellent selectivity across various tested compounds, highlighting their efficacy in both analytical techniques.

In general, miniaturizing analytical separation techniques seeks to enhance the efficiency and speed of the separation process while also reducing costs and improving portability. It aims to minimize the consumption of samples, solvents, and reagents, as well as the waste produced. The initial move toward miniaturization arose from the practical need to conserve laboratory space. In reality, the initial analytical separation techniques that were available were primarily oversized, whereas analytical separation systems that are significantly smaller are presently available (Bartle and Myers, 2002). The appropriate search for solutions required for challenging research activities, in conjunction with advancements in related fields such as instrumental engineering and material

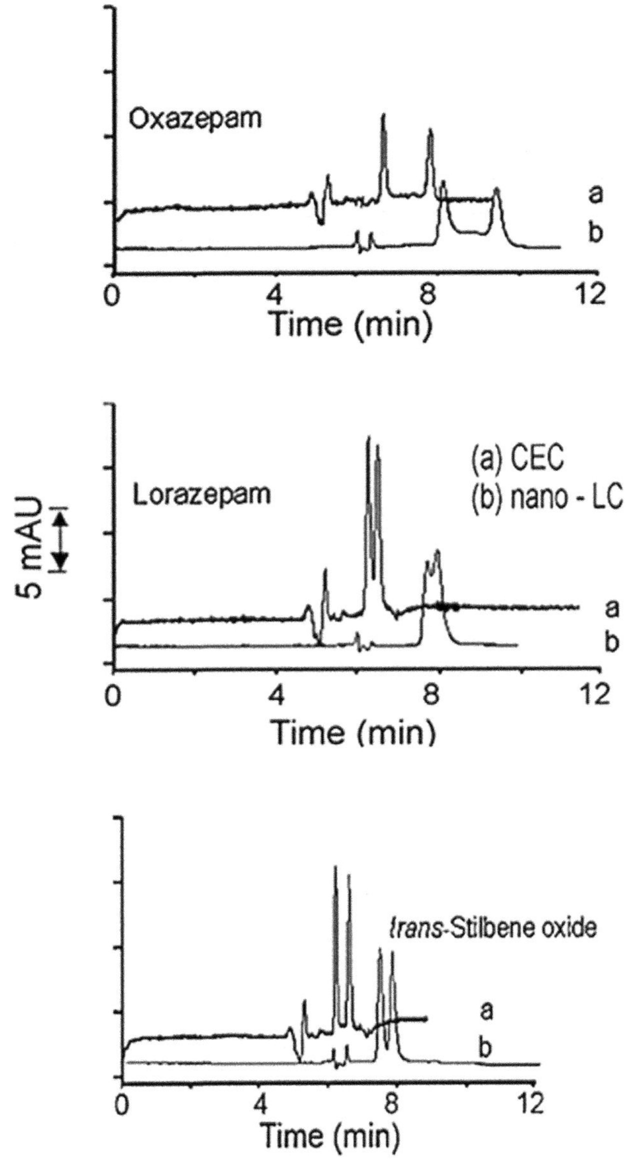

FIGURE 5.5

CEC and nano-LC separation of trans-stilbene oxide, lorazepam, and oxazepam enantiomers.

sciences, has resulted in further miniaturization. The miniaturization of analytical separation techniques has also been significantly influenced by the advancement of advanced fabrication technologies, including microelectromechanical systems (MEMS). Fig. 5.6 highlights the primary factors

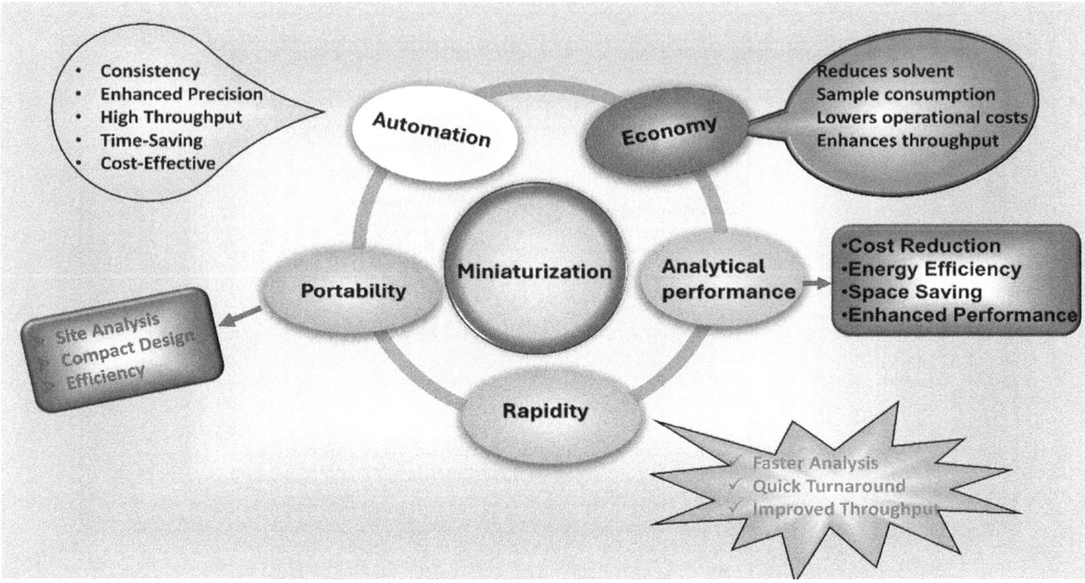

FIGURE 5.6

Key drivers of analytical systems miniaturization.

contributing to the trend toward smaller, more efficient analytical systems. It emphasizes the importance of cost reduction, energy efficiency, and space-saving, which are critical for improving overall system performance (Javier Pena-Pereira, 2014).

10. Challenges and future directions

The study of chiral natural products faces several challenges that prevent the complete utilization of the therapeutic properties of these compounds. One of the main challenges that has to be solved is the complexity of separating and characterizing chiral compounds from natural sources, which often consist of many stereoisomers. To accurately differentiate and measure these stereoisomers, sophisticated analytical techniques are required. The extraction and separation procedures face significant hurdles in terms of sustainability when working with limited quantities of natural materials. This holds particularly true when dealing with natural materials. Overcoming the obstacle of developing methods that are both efficient and cost-effective for the large-scale synthesis of single enantiomers is another challenge that must be addressed. The integration of automation and downsizing into analytical procedures offers several fascinating alternatives; nevertheless, it also requires a significant investment and the acquisition of new technical skills. In the future, research should prioritize the development of extraction technologies that are both environmentally sustainable and gentle, while also improving the sensitivity and specificity of chiral analytical procedures. Exploring undiscovered natural resources and employing biotechnological techniques may lead to the discovery of novel chiral compounds with unique pharmacological properties, which are not found in any other molecules. To fully harness the potential of chiral natural products in drug discovery and development, specialists from many disciplines need to collaborate and work together in a coordinated manner.

11. Conclusion

The study of chiral natural products is a pivotal area in the field of medicinal chemistry due to its promising potential for developing new and effective therapeutic treatments. This chapter provides a comprehensive evaluation of the approaches and techniques used in the extraction, separation, and analysis of chiral natural compounds. Researchers may optimize the utilization of chiral natural products in drug development by acquiring a deeper understanding of their unique features and characteristics. This information can lead to the development of drugs that are more powerful and have fewer side effects. The ongoing advancements in analytical technology and processes persistently promote advancement in this field, irrespective the obstacles that are now being tackled. There is a high level of confidence for the future of research on chiral natural products due to the potential to make significant contributions to the pharmaceutical industry and the overall improvement of global health. The complete therapeutic capacity of these fascinating substances can be realized through the continuous advancement of novel concepts and the cooperation of several persons.

References

Aboul-Enein, I.A., Hassan, Y., 2003. Chiral Separations by Liquid Chromatography and Related Technologies, vol 90. CRC Press.

Aboul-enein, H.Y., Islam, M.R., February 1991. Direct HPLC separation of thalidomide enantiomers using cellulose tris-4-methylphenyl benzoate chiral stationary phase. J. Liq. Chromatogr. 14 (4), 667–673. https://doi.org/10.1080/01483919108049279.

Achat, S., et al., July 2012. Direct enrichment of olive oil in oleuropein by ultrasound-assisted maceration at laboratory and pilot plant scale. Ultrason. Sonochem. 19 (4), 777–786. https://doi.org/10.1016/j.ultsonch.2011.12.006.

Ahmad, R., et al., November 2020. Optimization of extraction and quantification technique for phenolics content of garlic (<scp> *Allium sativum* </scp>): an application for comparative phytochemical evaluation based on cultivar origin. Biomed. Chromatogr. 34 (11). https://doi.org/10.1002/bmc.4942.

Alara, O.R., Abdurahman, N.H., Ukaegbu, C.I., December 2018. Soxhlet extraction of phenolic compounds from Vernonia cinerea leaves and its antioxidant activity. J. Appl. Res. Med. Aromat. Plants 11, 12–17. https://doi.org/10.1016/j.jarmap.2018.07.003.

Ali, I., Suhail, M., Aboul-Enein, H.Y., Kon'kova, T., June 2021. Recent trends in chiral separations by 2D-HPLC. Chromatographia 84 (6), 535–548. https://doi.org/10.1007/s10337-021-04030-1.

Alvarez-Rivera, G., Bueno, M., Ballesteros-Vivas, D., Cifuentes, A., February 2020. Chiral analysis in food science. TrAC Trends Anal. Chem. 123, 115761. https://doi.org/10.1016/j.trac.2019.115761.

Amit Koparde, A., Chandrashekar Doijad, R., Shripal Magdum, C., 2019. Natural products in drug discovery. In: Pharmacognosy - Medicinal Plants. IntechOpen. https://doi.org/10.5772/intechopen.82860.

Amit Koparde, A., Chandrashekar Doijad, R., Shripal Magdum, C., 2019. Natural Products in Drug Discovery. IntechOpen. https://doi.org/10.5772/intechopen.82860.

Armenta, S., de la Guardia, M., June 2016. Green chromatography for the analysis of foods of animal origin. TrAC Trends Anal. Chem. 80, 517–530. https://doi.org/10.1016/j.trac.2015.06.012.

Atanasov, A.G., Zotchev, S.B., Dirsch, V.M., Supuran, C.T., March 2021. Natural products in drug discovery: advances and opportunities. Nat. Rev. Drug Discov. 20 (3), 200–216. https://doi.org/10.1038/s41573-020-00114-z.

Bachtler, S., Bart, H.-J., January 2021. Increase the yield of bioactive compounds from elder bark and annatto seeds using ultrasound and microwave assisted extraction technologies. Food Bioprod. Process. 125, 1–13. https://doi.org/10.1016/j.fbp.2020.10.009.

Bartle, K.D., Myers, P., September 2002. History of gas chromatography. TrAC Trends Anal. Chem. 21 (9–10), 547–557. https://doi.org/10.1016/S0165-9936(02)00806-3.

Benkhnigue, O., Chaachouay, N., Khamar, H., El Azzouzi, F., Douira, A., Zidane, L., March 2022. Ethnobotanical and ethnopharmacological study of medicinal plants used in the treatment of anemia in the region of Haouz-Rehamna (Morocco). J. Pharm. Pharmacogn. Res. 10 (2), 279–302. https://doi.org/10.56499/jppres21.1196_10.2.279.

Blackmond, D.G., March 2019. The origin of biological homochirality. Cold Spring Harb. Perspect. Biol. 11 (3), a032540. https://doi.org/10.1101/cshperspect.a032540.

Brusotti, G., Cesari, I., Dentamaro, A., Caccialanza, G., Massolini, G., January 2014. Isolation and characterization of bioactive compounds from plant resources: the role of analysis in the ethnopharmacological approach. J. Pharm. Biomed. Anal. 87, 218–228. https://doi.org/10.1016/j.jpba.2013.03.007.

Burton, A.S., Berger, E.L., May 2018. Insights into abiotically-generated amino acid enantiomeric excesses found in meteorites. Life 8 (2), 14. https://doi.org/10.3390/life8020014.

Chaachouay, N., Zidane, L., February 2024. Plant-derived natural products: a source for drug discovery and development. Drugs Drug Candidates 3 (1), 184–207. https://doi.org/10.3390/ddc3010011.

Chanivet, M., Durán-Guerrero, E., Rodríguez-Dodero, M. del C., Barroso, C.G., Castro, R., April 2021. Application of accelerating energies to the maceration of sherry vinegar with citrus fruits. J. Sci. Food Agric. 101 (6), 2235–2246. https://doi.org/10.1002/jsfa.10843.

Chankvetadze, B., 1997. Capillary Electrophoresis in Chiral Analysis. John Wiley Sons.

Chankvetadze, B., Burjanadze, N., Crommen, J., Blaschke, G., January 2001. Enantioseparation of warfarin by capillary electrophoresis with UV and LIF detection using single and dual cyclodextrin type chiral selectors. Chromatographia 53 (S1), S296–S301. https://doi.org/10.1007/BF02490345.

Chen, S.T., Dou, J., Temple, R., Agarwal, R., Wu, K.-M., Walker, S., Oct. 2008. New therapies from old medicines. Nat. Biotechnol. 26 (10), 1077–1083. https://doi.org/10.1038/nbt1008-1077.

David, B., Wolfender, J.-L., Dias, D.A., April 2015. The pharmaceutical industry and natural products: historical status and new trends. Phytochem. Rev. 14 (2), 299–315. https://doi.org/10.1007/s11101-014-9367-z.

Di Michele, A., et al., Oct. 2021. Hazelnut shells as source of active ingredients: extracts preparation and characterization. Molecules 26 (21), 6607. https://doi.org/10.3390/molecules26216607.

Ding, X., Li, L., Wang, Y., Chen, J., Huang, Y., Xu, K., December 2014. Design of guanidinium ionic liquid based microwave-assisted extraction for the efficient extraction of Praeruptorin A from Radix peucedani. J. Sep. Sci. 37 (23), 3539–3547. https://doi.org/10.1002/jssc.201400828.

D'Orazio, G., Asensio-Ramos, M., Fanali, C., Hernández-Borges, J., Fanali, S., Sep. 2016. Capillary electrochromatography in food analysis. TrAC Trends Anal. Chem. 82, 250–267. https://doi.org/10.1016/j.trac.2016.06.012.

Easson, L.H., Stedman, E., January 1933. Studies on the relationship between chemical constitution and physiological action. Biochem. J. 27 (4), 1257–1266. https://doi.org/10.1042/bj0271257.

Elpa, D.P., Prabhu, G.R.D., Wu, S.-P., Tay, K.S., Urban, P.L., February 2020. Automation of mass spectrometric detection of analytes and related workflows: a review. Talanta 208, 120304. https://doi.org/10.1016/j.talanta.2019.120304.

Escribà-Gelonch, M., Hessel, V., Maier, M.C., Noël, T., Neira d'Angelo, M.F., Gruber-Woelfler, H., February 2018. Continuous-flow in-line solvent-swap crystallization of Vitamin D 3. Org. Process Res. Dev. 22 (2), 178–189. https://doi.org/10.1021/acs.oprd.7b00351.

Fanali, S., Aug. 2017. An overview to nano-scale analytical techniques: nano-liquid chromatography and capillary electrochromatography. Electrophoresis 38 (15), 1822–1829. https://doi.org/10.1002/elps.201600573.

Fanali, S., di Roma, A.D., 1995. An introduction to chiral analysis by capillary electrophoresis. In: Ist. di Cromatogr. del Cons. Naz. delle Ric.

Fanali, S., Dorazio, G., Lomsadze, K., Chankvetadze, B., November 2008. Enantioseparations with cellulose tris(3-chloro-4-methylphenylcarbamate) in nano-liquid chromatography and capillary electrochromatography. J. Chromatogr. B 875 (1), 296–303. https://doi.org/10.1016/j.jchromb.2008.07.017.

FDA'S policy statement for the development of new stereoisomeric drugs. Chirality 4 (5), January 1992, 338–340. https://doi.org/10.1002/chir.530040513.

Feher, M., Schmidt, J.M., January 2003. Property distributions: differences between drugs, natural products, and molecules from combinatorial chemistry. J. Chem. Inf. Comput. Sci. 43 (1), 218–227. https://doi.org/10.1021/ci0200467.

Feng, W., Li, M., Hao, Z., Zhang, Z., 2019. Analytical methods of isolation and identification. In: Phytochem. Hum. Heal.

Fernandes, C., Phyo, Y.Z., Silva, A.S., Tiritan, M.E., Kijjoa, A., Pinto, M.M.M., April 2018. Chiral stationary phases based on small molecules: an update of the last 17 years. Sep. Purif. Rev. 47 (2), 89–123. https://doi.org/10.1080/15422119.2017.1326939.

Fernandez-Pastor, I., et al., March 2017. Microwave-assisted extraction versus Soxhlet extraction to determine triterpene acids in olive skins. J. Sep. Sci. 40 (5), 1209–1217. https://doi.org/10.1002/jssc.201601130.

Gad, S.E., 2024. Albuterol. In: Encyclopedia of Toxicology. Elsevier, pp. 239–245. https://doi.org/10.1016/B978-0-12-824315-2.01100-3.

Gandhi, K., Shah, U., Patel, S., December 2020. Drug stereochemistry: a prodigy for pharmacology and drug development. Curr. Drug Discov. Technol. 17 (5), 565–573. https://doi.org/10.2174/1570163816666190502101803.

Gassmann, E., Kuo, J.E., Zare, R.N., November 1985. Electrokinetic separation of chiral compounds. Science 230 (4727), 813–814. https://doi.org/10.1126/science.230.4727.813.

Gerberding, S., Byers, C., May 1998. Preparative ion-exchange chromatography of proteins from dairy whey. J. Chromatogr. A 808 (1–2), 141–151. https://doi.org/10.1016/S0021-9673(98)00103-4.

Gerwick, W.H., 2013. Plant sources of drugs and chemicals. In: Encyclopedia of Biodiversity. Elsevier, pp. 129–139. https://doi.org/10.1016/B978-0-12-384719-5.00111-8.

Gogoi, A., Konwer, S., Zhuo, G.-Y., February 2021. Polarimetric measurements of surface chirality based on linear and nonlinear light scattering. Front. Chem. 8. https://doi.org/10.3389/fchem.2020.611833.

Goodall, D.M., April 1993. Chiral analysis based on polarimetric detection. TrAC Trends Anal. Chem. 12 (4), 177–184. https://doi.org/10.1016/0165-9936(93)87020-X.

Grybinik, S., Bosakova, Z., Sep. 2021. An overview of chiral separations of pharmaceutically active substances by HPLC (2018–2020). Monatshefte für Chemie - Chem. Mon. 152 (9), 1033–1043. https://doi.org/10.1007/s00706-021-02832-5.

Gu, H., Unger, S., Deng, Y., December 2006. Automated tecan programming for bioanalytical sample preparation with EZTecan. Assay Drug Dev. Technol. 4 (6), 721–733. https://doi.org/10.1089/adt.2006.4.721.

Handa, S.S., 2008. An overview of extraction techniques for medicinal and aromatic plants. Extr. Technol. Med. Aromat. plants 1 (1), 21–40.

Hanif, M.A., et al., July 2017. Raman spectroscopy for the characterization of different fractions of hemp essential oil extracted at 130 °C using steam distillation method. Spectrochim. Acta Part A Mol. Biomol. Spectrosc. 182, 168–174. https://doi.org/10.1016/j.saa.2017.03.072.

Henkel, T., Brunne, R.M., Müller, H., Reichel, F., March 1999. Statistical investigation into the structural complementarity of natural products and synthetic compounds. Angew. Chemie Int. Ed. 38 (5), 643–647. https://doi.org/10.1002/(SICI)1521-3773(19990301)38:5<643::AID-ANIE643>3.0.CO;2-G.

Hosler, D.M., Mikita, M.A., April 1987. Ethnobotany: the chemist's source for the identification of useful natural products. J. Chem. Educ. 64 (4), 328. https://doi.org/10.1021/ed064p328.

Hu, W., et al., July 2019. Microwave-assisted extraction, physicochemical characterization and bioactivity of polysaccharides from Camptotheca acuminata fruits. Int. J. Biol. Macromol. 133, 127–136. https://doi.org/10.1016/j.ijbiomac.2019.04.086.

Hutt, J.V.A.J., 2003. The chiral switch: the development of single enantiomer drugs from racemates. Univ. Comeniana. Acta Fac. Pharm. 50, 7–23.

Hutt, A.J., June 2007. Chirality and pharmacokinetics: an area of neglected dimensionality? Drug Metabol. Drug Interact. 22 (2–3), 79–112. https://doi.org/10.1515/DMDI.2007.22.2-3.79.

Ito, Y., February 2005. Golden rules and pitfalls in selecting optimum conditions for high-speed counter-current chromatography. J. Chromatogr. A 1065 (2), 145–168. https://doi.org/10.1016/j.chroma.2004.12.044.

Javier Pena-Pereira, F., 2014. Miniaturization in Sample Preparation. DE GRUYTER OPEN. https://doi.org/10.2478/9783110410181.

Ji, S., et al., Oct. 2018. The application of on-line two-dimensional liquid chromatography (2DLC) in the chemical analysis of herbal medicines. J. Pharm. Biomed. Anal. 160, 301–313. https://doi.org/10.1016/j.jpba.2018.08.014.

Kar, A., 2006. Pharmacognosy and Pharma Biotechnology. New age Int. Ltd, New Delhi, pp. 5–11.

Kingston, D.G.I., March 2011. Modern natural products drug discovery and its relevance to biodiversity conservation. J. Nat. Prod. 74 (3), 496–511. https://doi.org/10.1021/np100550t.

Komal Patel, D.P.I., Panchal, N., 2019. Techniques adopted for extraction of natural products extraction methods: maceration, percolation, soxhlet extraction, Turbo distillation, supercritical fluid extraction. Int. J. Adv. Res. Chem. Sci. 6 (4). https://doi.org/10.20431/2349-0403.0604001.

Koparde Akshada Amit, C.S.M., Doijad, R.C., 2019. Natural products in drug discovery. In: Pharmacogn. Plants. IntechOpen.

Krastel, P., Petersen, F., Roggo, S., Schmitt, E., Schuffenhauer, A., 2006. Aspects of chirality in natural products drug discovery. Chiral. Drug Res. 67–94. https://doi.org/10.1002/9783527609437.ch3.

Laycock, J.D., Hartmann, T., 2005. Automation. In: Integrated Strategies for Drug Discovery Using Mass Spectrometry. Wiley, pp. 511–542. https://doi.org/10.1002/0471721034.ch18.

Lee, M.-L., Schneider, G., May 2001. Scaffold architecture and pharmacophoric properties of natural products and trade drugs: application in the design of natural product-based combinatorial libraries. J. Comb. Chem. 3 (3), 284–289. https://doi.org/10.1021/cc000097l.

Li, B., Haynie, D., 2006. Chiral drug separation. In: Encycl. Chem. Process, pp. 449–458. https://doi.org/10.1081/E-ECHP-120039232.

Li, J., Larregieu, C.A., Benet, L.Z., December 2016. Classification of natural products as sources of drugs according to the biopharmaceutics drug disposition classification system (BDDCS). Chin. J. Nat. Med. 14 (12), 888–897. https://doi.org/10.1016/S1875-5364(17)30013-4.

Liehr, A.D., April 1964. Interaction of electromagnetic radiation with matter. I. Theory of optical rotatory power: topic A. Trigonal dihedral compounds 1a. J. Phys. Chem. 68 (4), 665–722. https://doi.org/10.1021/j100786a001.

Lin, B., Wang, S., Zhou, A., Hu, Q., Huang, G., Aug. 2023. Ultrasound-assisted enzyme extraction and properties of Shatian pomelo peel polysaccharide. Ultrason. Sonochem. 98, 106507. https://doi.org/10.1016/j.ultsonch.2023.106507.

Littleton, J., May 2007. The future of plant drug discovery. Expert Opin. Drug Discov. 2 (5), 673–683. https://doi.org/10.1517/17460441.2.5.673.

Lourenço, T.C., Bósio, G.C., Cassiano, N.M., Cass, Q.B., Moreau, R.L.M., January 2013. Chiral separation of 3,4-methylenedioxymethamphetamine (MDMA) enantiomers using batch chromatography with peak shaving recycling and its effects on oxidative stress status in rat liver. J. Pharm. Biomed. Anal. 73, 13–17. https://doi.org/10.1016/j.jpba.2012.01.025.

Lourenço, T.C., et al., March 2012. Albendazole sulfoxide enantiomers: preparative chiral separation and absolute stereochemistry. J. Chromatogr. A 1230, 61–65. https://doi.org/10.1016/j.chroma.2012.01.070.

Luque de Castro, M.D., Priego-Capote, F., April 2010. Soxhlet extraction: past and present panacea. J. Chromatogr. A 1217 (16), 2383–2389. https://doi.org/10.1016/j.chroma.2009.11.027.

Martins, N., Barros, L., Santos-Buelga, C., Silva, S., Henriques, M., Ferreira, I.C.F.R., January 2015. Decoction, infusion and hydroalcoholic extract of cultivated thyme: antioxidant and antibacterial activities, and phenolic characterisation. Food Chem. 167, 131–137. https://doi.org/10.1016/j.foodchem.2014.06.094.

Mathur, S., Hoskins, C., June 2017. Drug development: Lessons from nature. Biomed. Rep. 6 (6), 612–614. https://doi.org/10.3892/br.2017.909.

McConathy, J., Owens, M.J., April 2003. Stereochemistry in drug action. Prim. Care Companion CNS Disord 5 (2). https://doi.org/10.4088/PCC.v05n0202.

Mehta, N., et al., Sep. 2022. Ultrasound-assisted extraction and the encapsulation of bioactive components for food applications. Foods 11 (19), 2973. https://doi.org/10.3390/foods11192973.

Mori, K., July 2011. Bioactive natural products and chirality. Chirality 23 (6), 449–462. https://doi.org/10.1002/chir.20930.

Nurkhasanah, A., Fardad, T., Carrera, C., Setyaningsih, W., Palma, M., July 2023. Ultrasound-assisted anthocyanins extraction from pigmented corn: optimization using response surface methodology. Methods Protoc 6 (4), 69. https://doi.org/10.3390/mps6040069.

Orch, H., Chaachouay, N., Douiri, E.M., Faiz, N., Zidane, L., Douira, A., 2021. Use of medicinal plants in dermato-cosmetology: an ethnobotanical study among the population of izarène. Jordan J. Pharm. Sci. 14 (3), 323–340.

Ortholand, J.-Y., Ganesan, A., June 2004. Natural products and combinatorial chemistry: back to the future. Curr. Opin. Chem. Biol. 8 (3), 271–280. https://doi.org/10.1016/j.cbpa.2004.04.011.

Padró, J.M., Keunchkarian, S., July 2018. State-of-the-art and recent developments of immobilized polysaccharide-based chiral stationary phases for enantioseparations by high-performance liquid chromatography (2013–2017). Microchem. J. 140, 142–157. https://doi.org/10.1016/j.microc.2018.04.017.

Paul, M., Ma, J.K., January 2011. Plant-made pharmaceuticals: leading products and production platforms. Biotechnol. Appl. Biochem. 58 (1), 58–67. https://doi.org/10.1002/bab.6.

Pena-Pereira, F., 2014. 1 from conventional to miniaturized analytical systems. In: Miniaturization in Sample Preparation. De Gruyter Open, Poland, pp. 1–28. https://doi.org/10.2478/9783110410181.1.

Pezzuto, J.M., January 1997. Plant-derived anticancer agents. Biochem. Pharmacol. 53 (2), 121–133. https://doi.org/10.1016/S0006-2952(96)00654-5.

Picot-Allain, C., Mahomoodally, M.F., Ak, G., Zengin, G., Aug. 2021. Conventional versus green extraction techniques — a comparative perspective. Curr. Opin. Food Sci. 40, 144–156. https://doi.org/10.1016/j.cofs.2021.02.009.

Pinto, M.M.M., Fernandes, C., Tiritan, M.E., April 2020. Chiral separations in preparative scale: a medicinal chemistry point of view. Molecules 25 (8), 1931. https://doi.org/10.3390/molecules25081931.

Pinto, C., Cidade, H., Pinto, M., Tiritan, M.E., December 2021. Chiral flavonoids as antitumor agents. Pharmaceuticals 14 (12), 1267. https://doi.org/10.3390/ph14121267.

Ponphaiboon, J., Krongrawa, W., Aung, W.W., Chinatangkul, N., Limmatvapirat, S., Limmatvapirat, C., July 2023. Advances in natural product extraction techniques, Electrospun fiber fabrication, and the integration of experimental design: a comprehensive review. Molecules 28 (13), 5163. https://doi.org/10.3390/molecules28135163.

Poole, C.F., Aug. 2018. Chromatographic test methods for characterizing alkylsiloxane-bonded silica columns for reversed-phase liquid chromatography. J. Chromatogr. B 1092, 207–219. https://doi.org/10.1016/j.jchromb.2018.06.011.

Přichystal, J., Schug, K.A., Lemr, K., Novák, J., Havlíček, V., November 2016. Structural analysis of natural products. Anal. Chem. 88 (21), 10338–10346. https://doi.org/10.1021/acs.analchem.6b02386.

Purushoth Prabhu, T., Shinesudev, kurian, M., Manjushsa, P., Minnu, P.S., Mohamed Hisham, C., Shrikumar, S., 2016. Pharmaceutical review and its importance of chiral chromatography. Int. J. Res. Pharm. Chem. 6 (3), 476–484.

Raimbault, A., Dorebska, M., West, C., July 2019. A chiral unified chromatography–mass spectrometry method to analyze free amino acids. Anal. Bioanal. Chem. 411 (19), 4909–4917. https://doi.org/10.1007/s00216-019-01783-5.

RamaRao, A.V., Gurjar, M.K., 1990. Drugs from plant resources: an overview. Pharmatimes 22 (5), 19–20.

Rasul, M.G., 2018. Conventional extraction methods use in medicinal plants, their advantages and disadvantages. Int. J. Basic Sci. Appl. Comput. 2 (6).

Richard, G.I., et al., May 2008. Chiral recognition of amino acids by use of a fluorescent resorcinarene. Appl. Spectrosc. 62 (5), 476–480. https://doi.org/10.1366/000370208784344514.

Rosén, J., Gottfries, J., Muresan, S., Backlund, A., Oprea, T.I., April 2009. Novel chemical space exploration via natural products. J. Med. Chem. 52 (7), 1953–1962. https://doi.org/10.1021/jm801514w.

Rozzi, N.L., Singh, R.K., April 2002. Supercritical fluids and the food industry. Compr. Rev. Food Sci. Food Saf. 1 (1), 33–44. https://doi.org/10.1111/j.1541-4337.2002.tb00005.x.

Samuelsson, J., Leśko, M., Enmark, M., Högblom, J., Karlsson, A., Kaczmarski, K., Jun. 2018. Optimizing column length and particle size in preparative batch chromatography using enantiomeric separations of Omeprazole and Etiracetam as models: feasibility of Taguchi empirical optimization. Chromatographia 81 (6), 851–860. https://doi.org/10.1007/s10337-018-3519-z.

Sardella, R., Ianni, F., Marinozzi, M., Macchiarulo, A., Natalini, B., April 2017. Laboratory-scale preparative enantioseparations of pharmaceutically relevant compounds on commercially available chiral stationary phases for HPLC. Curr. Med. Chem. 24 (8), 796–817. https://doi.org/10.2174/0929867323666160907111107.

Sardella, R., Ianni, F., Macchiarulo, A., Pucciarini, L., Carotti, A., Natalini, B., January 2018. Elucidation of the chromatographic enantiomer elution order through computational studies. Mini-Rev. Med. Chem. 18 (2). https://doi.org/10.2174/1389557516666161018143629.

Scriba, G.K.E., Oct. 2016. Chiral recognition in separation science – an update. J. Chromatogr. A 1467, 56–78. https://doi.org/10.1016/j.chroma.2016.05.061.

Sevindik, O., Kelebek, H., Rombolà, A.D., Selli, S., May 2022. Grape seed oil volatiles and odour activity values: a comparison with Turkish and Italian cultivars and extraction methods. J. Food Sci. Technol. 59 (5), 1968–1981. https://doi.org/10.1007/s13197-021-05212-3.

Shaaban, H., Oct. 2016. New insights into liquid chromatography for more eco-friendly analysis of pharmaceuticals. Anal. Bioanal. Chem. 408 (25), 6929–6944. https://doi.org/10.1007/s00216-016-9726-2.

Shou, W.Z., Zhang, J., May 2012. Recent development in software and automation tools for high-throughput discovery bioanalysis. Bioanalysis 4 (9), 1097–1109. https://doi.org/10.4155/bio.12.51.

Silverstein, R.M., Bassler, G.C., November 1962. Spectrometric identification of organic compounds. J. Chem. Educ. 39 (11), 546. https://doi.org/10.1021/ed039p546.

Singh, A., Kaur, N., Kumar Chopra, H., November 2019. Chiral recognition methods in analytical chemistry: role of the chiral ionic liquids. Crit. Rev. Anal. Chem. 49 (6), 553–569. https://doi.org/10.1080/10408347.2019.1565985.

Song, Z., Xiong, X., Huang, G., May 2023. Ultrasound-assisted extraction and characteristics of maize polysaccharides from different sites. Ultrason. Sonochem. 95, 106416. https://doi.org/10.1016/j.ultsonch.2023.106416.

Sousa, M.E., et al., July 2006. Multimilligram enantioresolution of low-solubility xanthonolignoids on polysaccharide chiral stationary phases using a solid-phase injection system. J. Chromatogr. A 1120 (1–2), 75–81. https://doi.org/10.1016/j.chroma.2005.12.071.

Speybrouck, D., Lipka, E., Oct. 2016. Preparative supercritical fluid chromatography: a powerful tool for chiral separations. J. Chromatogr. A 1467, 33–55. https://doi.org/10.1016/j.chroma.2016.07.050.

Speybrouck, D., Lipka, E., January 2020. Productivity and solvent waste in supercritical fluid chromatography for preparative chiral separations: a guide for a convenient strategy. J. Chromatogr. A 1610, 460549. https://doi.org/10.1016/j.chroma.2019.460549.

Sticher, O., 2008. Natural product isolation. Nat. Prod. Rep. 25 (3), 517. https://doi.org/10.1039/b700306b.

Sturm, S., Seger, C., Oct. 2012. Liquid chromatography–nuclear magnetic resonance coupling as alternative to liquid chromatography–mass spectrometry hyphenations: Curious option or powerful and complementary routine tool? J. Chromatogr. A 1259, 50–61. https://doi.org/10.1016/j.chroma.2012.05.032.

Sweedler, J.V., July 2002. The continued evolution of hyphenated instruments. Anal. Bioanal. Chem. 373 (6), 321–322. https://doi.org/10.1007/s00216-002-1358-z.

Tahara, S., Yamamoto, S., Yamajima, Y., Miyakawa, H., Uematsu, Y., Monma, K., 2017. A rapid dialysis method for analysis of artificial sweeteners in foods (2nd report). Food Hyg. Saf. Sci. (Shokuhin Eiseigaku Zasshi) 58 (3), 124–131. https://doi.org/10.3358/shokueishi.58.124.

Talukdar, D., Choudhury, D., 2010. Phytochemical screening and TLC profiling of plant extracts of Cyathea gigantea (Wall. Ex. Hook.) Haltt. and Cyathea brunoniana. Wall. ex. Hook. (Cl. and Bak.). Asian Univ. Sci. Technol. 5 (1), 70–74.

Tang, Y., Cohen, A.E., April 2010. Optical chirality and its interaction with matter. Phys. Rev. Lett. 104 (16), 163901. https://doi.org/10.1103/PhysRevLett.104.163901.

Tao, Y., et al., February 2018. Supercritical fluid chromatography–tandem mass spectrometry-assisted methodology for rapid enantiomeric analysis of fenbuconazole and its chiral metabolites in fruits, vegetables, cereals, and soil. Food Chem. 241, 32–39. https://doi.org/10.1016/j.foodchem.2017.08.038.

Teixeira, J., Tiritan, M.E., Pinto, M.M.M., Fernandes, C., February 2019. Chiral stationary phases for liquid chromatography: recent developments. Molecules 24 (5), 865. https://doi.org/10.3390/molecules24050865.

Toledo, M.V., Briand, L.E., July 2018. Relevance and bio-catalytic strategies for the kinetic resolution of ketoprofen towards dexketoprofen. Crit. Rev. Biotechnol. 38 (5), 778–800. https://doi.org/10.1080/07388551.2017.1399249.

Valderrama, P., Romero, A.L., Imamura, P.M., Poppi, R.J., Oct. 2009. Figures of merit in the quantification of ibuprofen enantiomers by chiral HPLC. J. Chromatogr. Sci. 47 (9), 777–780. https://doi.org/10.1093/chromsci/47.9.777.

Valentová, J., Lintnerová, L., Mikľášová, N., Oboňová, B., Habala, L., March 2023. Analogues of anticancer natural products: chiral aspects. Int. J. Mol. Sci. 24 (6), 5679. https://doi.org/10.3390/ijms24065679.

Wang, Y., Huang, H., Zhang, Q., Zhang, P., 2018. Chirality in metal-based anticancer agents. Dalt. Trans. 47 (12), 4017–4026. https://doi.org/10.1039/C8DT00089A.

Wang, W.-Y., Qu, H.-B., Gong, X.-C., March 2020. [Research progress on percolation extraction process of traditional Chinese medicines]. Zhongguo Zhongyao Zazhi 45 (5), 1039–1046. https://doi.org/10.19540/j.cnki.cjcmm.20191221.305.

Wang, W., Ding, F., Qu, H., Gong, X., May 2023. Mechanism modeling and application of Salvia miltiorrhiza percolation process. Sci. Rep. 13 (1), 8311. https://doi.org/10.1038/s41598-023-35529-2.

Wei, M.-C., Xiao, J., Yang, Y.-C., November 2016. Extraction of α-humulene-enriched oil from clove using ultrasound-assisted supercritical carbon dioxide extraction and studies of its fictitious solubility. Food Chem. 210, 172–181. https://doi.org/10.1016/j.foodchem.2016.04.076.

Will, J.M., Behrens, A., Macke, M., Quarles, C.D., Karst, U., January 2021. Automated chiral analysis of amino acids based on chiral derivatization and trapped ion mobility–mass spectrometry. Anal. Chem. 93 (2), 878–885. https://doi.org/10.1021/acs.analchem.0c03481.

Wilson, J., Simpson, T., Spelman, K., Oct. 2022. Total cannabidiol (CBD) concentrations and yields from traditional extraction methods: percolation vs. maceration. Front. Pharmacol. 13. https://doi.org/10.3389/fphar.2022.886993.

Yeung, E.S., Steenhoek, L.E., Woodruff, S.D., Kuo, J.C., Aug. 1980. Detector based on optical activity for high performance liquid chromatographic detection of trace organics. Anal. Chem. 52 (9), 1399–1402. https://doi.org/10.1021/ac50059a006.

Zeng, L., Burton, L., Yung, K., Shushan, B., Kassel, D., January 1998. Automated analytical/preparative high-performance liquid chromatography–mass spectrometry system for the rapid characterization and purification of compound libraries. J. Chromatogr. A 794 (1–2), 3–13. https://doi.org/10.1016/S0021-9673(97)01008-X.

Zermane, A., Larkeche, O., Meniai, A.-H., Crampon, C., Badens, E., January 2014. Optimization of essential oil supercritical extraction from Algerian Myrtus communis L. leaves using response surface methodology. J. Supercrit. Fluids 85, 89–94. https://doi.org/10.1016/j.supflu.2013.11.002.

Zhang, J., Wei, S., Ayres, D.W., Smith, H.T., Tse, F.L., Sep. 2011. An automation-assisted generic approach for biological sample preparation and LC–MS/MS method validation. Bioanalysis 3 (17), 1975–1986. https://doi.org/10.4155/bio.11.178.

Zhang, Q.-W., Lin, L.-G., Ye, W.-C., December 2018. Techniques for extraction and isolation of natural products: a comprehensive review. Chin. Med. 13 (1), 20. https://doi.org/10.1186/s13020-018-0177-x.

Zhang, H., et al., Aug. 2022. Extraction of camphor tree essential oil by steam distillation and supercritical CO2 extraction. Molecules 27 (17), 5385. https://doi.org/10.3390/molecules27175385.

Zhang, M., Zhao, J., Dai, X., Li, X., December 2023. Extraction and analysis of chemical compositions of natural products and plants. Separations 10 (12), 598. https://doi.org/10.3390/separations10120598.

CHAPTER

Antidiabetic potential of capsaicinoids

6

Alicja Piotrowska-Niczyporuk[1], Katarzyna Jakimiuk[2], Monika Tomczykowa[3] and Michał Tomczyk[2]
[1]*Department of Plant Biology and Ecology, Faculty of Biology, University of Bialystok, Bialystok, Poland;* [2]*Department of Pharmacognosy, Faculty of Pharmacy with the Division of Laboratory Medicine, Medical University of Bialystok, Bialystok, Poland;* [3]*Department of Organic Chemistry, Faculty of Medicine with the Division of Dentistry and Division of Medical Education in English, Medical University of Bialystok, Bialystok, Poland*

1. Introduction

Capsaicinoids constitute a family of natural products (phenolic alkaloids) isolated from the dried fruits of plants from the *Capsicum* genus, e.g., chili peppers. The genus *Capsicum* (Solanaceae family) consists of approximately 37 wild and semidomesticated species and five well-known domesticated species, namely, *C. annuum* L., *C. baccatum* L., *C. chinense* Jacq., *C. frutescens* L. and *C. pubescens* Ruiz and Pav, with more than 200 varieties (Bosland and Votava, 2012). *Capsicum* species are native to Central and Southern America, with Brazil the center for most wild peppers and Peru and Bolivia the primary centers of cultivated *Capsicum* diversity (van Zonneveld et al., 2015).

The chemical structure of each individual capsaicinoid contains a vanilloid group (an aromatic ring with a hydroxyl and a methyl group) attached to a long hydrocarbon chain and a polar amide group (Blum et al., 2003; Contreras-Padilla and Yahia, 1998). The variability of their content in fresh peppers is related to the pungency of the pepper. Capsaicin ($C_{18}H_{27}NO_3$, *trans*-8-methyl-*N*-vanillyl-6-nonenamide) is a crystalline compound found most abundantly (46%) among other capsaicinoids in *Capsicum* plants. Other capsaicinoids (Fig. 6.1) include dihydrocapsaicin (41%), nordihydrocapsaicin (7%), norcapsaicin (7%), homocapsaicin (3%), homodihydrocapsaicin (2%), and others (Antonio et al., 2018; Bennett and Kirby, 1968; Wesolowska et al., 2011).

The level of capsaicinoids ranges greatly among *Capsicum* species, from 1.2 to 6580 $\mu g\ g^{-1}$ of plant fresh biomass (Victoria-Campos et al., 2015). Capsaicinoids contain approximately 30% of the total capsaicinoid mixture and are mostly responsible for the pungency of capsaicin (Nelson et al., 2004). Nevertheless, the definite percentage of capsaicin and other capsaicinoids varies depending on the pepper source and extraction procedure (Reddy and Sasikala, 2013). These compounds are available in relatively large amounts in fruits of various pepper variants from the genus *Capsicum* L., and they are responsible for the spicy character of pepper plants (Adaszek et al., 2017). The pungency of all these molecules emphasizes that this activity is defined mainly by the benzene ring region; however, the length of the acyl chain can modify it (Reyes-Escogido Mde et al., 2011).

Phytochemicals for Health. https://doi.org/10.1016/B978-0-443-15366-2.00006-X

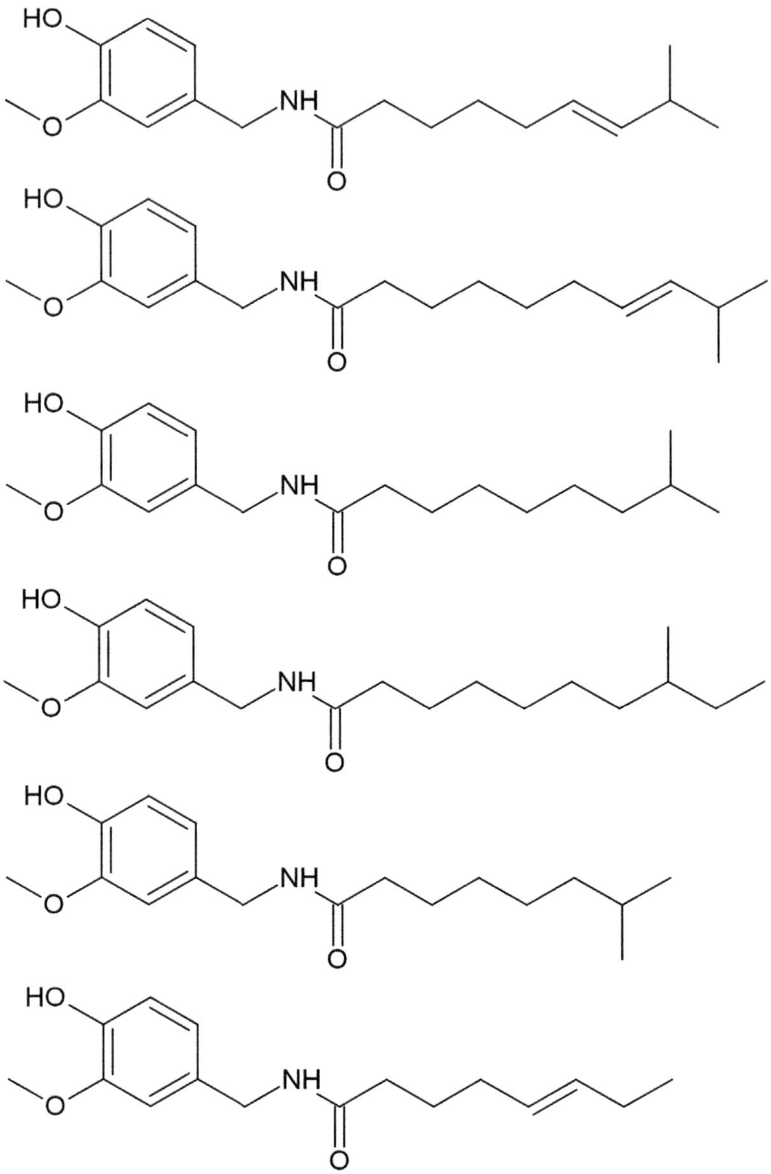

capsaicin

homocapsaicin

dihydrocapsaicin

homodihydrocapsaicin

nordihydrocapsaicin

norcapsaicin

FIGURE 6.1

Principal capsaicinoids found within *Capsicum* genus, (A) capsaicin, (B) homocapsaicin, (C) dihydrocapsaicin, (D) homodihydrocapsaicin, (E) nordihydrocapsaicin, and (F) norcapsaicin.

Capsaicinoids show evidence of a wide variety of biological effects; thus, they have been the targets of extensive research since their identification (Hassan et al., 2019). They possess beneficial effects on human health and could play a role in the prevention of many diseases. Among all capsaicinoids, capsaicin is the most thoroughly studied in experimental and clinical investigations (Díaz-Laviada, 2010). For example, capsaicin is used in medicine as a chemical agent for treating rheumatoid arthritis, postoperative pain, acute/chronic neuropathic pain, and musculoskeletal pain (Vadivelu et al., 2010). Moreover, all capsaicinoids present therapeutic properties to treat pain and have been used to study the pathophysiology of pain and diseases (Walpole et al., 1996). Capsaicinoids can bind to transient receptor potential vanilloid (TRPV) ion-channel receptors, which are expressed predominantly by sensory neurons (Cortright and Szallasi, 2004). This group of compounds functions as a high affinity agonist of the TRPV1 receptor (O'Neill et al., 2012). Moreover, capsaicinoids are responsible for releasing and depleting substance P from afferent nociceptive-specific neurons (Bevan and Szolcsa'nyi, 1990). Capsaicinoids may activate TRPV receptors in sensory nerves and cause sensations that range from hotness to burning. Following this, analgesia resulting from the depletion of substance P leads to the desensitization of small sensory neurons to noxious heat and chemical stimuli and the loss of their ability to release mediators involved in neurotransmission and inflammation. Thus, these effects of capsaicinoids have been used in some chronic pain syndromes, such as postherpetic neuralgia, musculoskeletal pain, and bladder dysfunction (Caterina and Julius, 2001; Hayman and Kam, 2008). Moreover, the thermogenic effect of capsaicinoids can both promote weight loss and help maintain body mass (Fan et al., 2019). However, some of the medical activities of capsaicinoids (e.g., antineoplastic and cardioprotective effects) have been found to be independent of the TRPV1 receptor (Rollyson et al., 2014).

Recent studies have indicated that capsaicinoids possess antioxidant, anticarcinogenic, and anti-inflammatory properties (Giuffrida et al., 2013; Lu et al., 2017; de Sá Mendes and de Andrade Gonçalves, 2020). These compounds have been found to induce apoptosis in different cancerous cells and inhibit carcinogenesis in the prostate, skin, breast, colon, lung, and human urinary bladder. On the other hand, excessive exposure to capsaicin can be toxic, causing local irritation, respiratory problems, and an increased potential for certain types of cancer (Fernández-Bedmar and Alonso-Moraga, 2016; Lu et al., 2017; de Sá Mendes and de Andrade Gonçalves, 2020). Capsaicin also has the potential to treat nervous system disorders, including arthritis, cystitis, and human immunodeficiency virus (HIV). This organic compound can also be applied as a natural inhibitor to pathogenic microorganisms in food due to its antibacterial and antifungal activity against a number of pathogenic microorganisms (e.g., *Pseudomonas aeruginosa*, *Klebsiella pneumoniae*, *Staphylococcus aureus*, and *Candida albicans*) (Topuz et al., 2011; Nascimento et al., 2014).

Evidence has shown that capsaicinoids are promising candidates that may be targeted under different medicinal strategies in the treatment of obesity (Kantar et al., 2016; Urbina et al., 2017; Fan et al., 2019). The potential mechanisms underlying the antidiabetic properties of capsaicinoids include (1) the inhibition of enzymes involved in sugar metabolism; (2) effects on glucose uptake enhancers; (3) the modulation of insulin secretagogues; and (4) the inhibition of adipogenesis and activation of brown adipose tissue (BAT) activity (Zheng et al., 2017).

1.1 Enzyme inhibitors: α-amylase and α-glucosidase

Diabetes mellitus is a chronic metabolic disorder related to blood glucose levels (Hameed et al., 2015; Malik and Hu, 2012). Diabetes can be classified into two types: insulin-dependent (type 1, T1D) and non-insulin-dependent (type 2, T2D) (Heise et al., 2014). The release of glucose from carbohydrates by the carbohydrate hydrolytic enzymes α-glucosidase and α-amylase may cause hyperglycemia in blood plasma (Chakrabarti and Rajagopalan, 2002; Singh et al., 2018). The role of α-amylases is to catalyze the hydrolysis of starch. They are endo-acting enzymes since they attack the interior parts of starch chains. Starch is the most prevalent storage polysaccharide in plants and a macromolecule composed of two types of polysaccharides, amylose (a linear α-(1 $\rightarrow$ 4)-linked glucan) and amylopectin (α-(1 $\rightarrow$ 6)-branched linkages). Specifically, α-amylases hydrolyze the-(1 $\rightarrow$ 4)-glycosidic linkages in starch molecules, leading to the production of maltose, maltotriose, maltotetraose, maltodextrins, and glucose. In humans, α-amylases are found in the salivary glands, which secrete the enzyme into the mouth, and the pancreas, which secretes the enzyme into the small intestine (Robyt, 2008). α-Glucosidases are carbohydrate hydrolytic enzymes found in the luminal surface of enterocytes and are secreted in the small intestine. α-Glucosidase is a key enzyme catalyzing the hydrolytic cleavage of disaccharides (maltose and sucrose) into monosaccharides (glucose and fructose). Hence, the inhibition of the activity of enzymes involved in sugar metabolism can retard the elevation of blood sugar and suppress postprandial hyperglycemia. Some studies indicate that capsaicinoids may modulate the activity of both α-glucosidase and α-amylase and can be used for the therapeutic treatment of diabetes mellitus type 2 (Robyt, 2008; Kim et al., 2014).

Various extracts from the whole fruit and pure capsaicinoids have been reported to inhibit the activities of both α-amylase and α-glucosidase based on in vitro studies, which are potential targets for the treatment of T2D, a metabolic disorder characterized by chronic insulin resistance and loss of functional β-cells due to relative (rather than absolute) deficiency of insulin (Tundis et al., 2013). In the study of diabetes, the application of capsaicin inhibited the activity of α-amylase and α-glucosidase, which are the key enzymes that affect the increase in blood glucose levels according to a docking study. The obtained results revealed that capsaicin had better inhibition potential on α-glucosidase activity than α-amylase activity. Other capsaicinoids, such as dihydrocapsaicin, have also been reported to improve human health, and they are effective compounds in the treatment of diabetes (Thongnum and Chanthai, 2018). For example, Nanok and Sansenya (2020) indicated that the IC_{50} values of capsaicin and dihydrocapsaicin on the activity of both α-glucosidase and α-amylase were 1.60–6.61 mM. Moreover, capsaicin has also been reported to inhibit α-glucosidase and α-amylase enzymes by molecular docking, and its inhibitory efficacy seems to have higher potential on α-glucosidase than α-amylase (Thongnum and Chanthai, 2018). The kinetic results revealed that inhibitory potential of both compounds on each enzyme was similar to the IC_{50} values. The inhibition constant (Ki) showed that capsaicin has similar inhibitory efficiency on both α-glucosidase and α-amylase, with Ki values of 0.32 and 0.33 mM, respectively. Dihydrocapsaicin had higher inhibitory potential on α-amylase than α-glucosidase, with Ki values of 0.30 and 0.77 mM, respectively. However, capsaicin and dihydrocapsaicin had lower inhibitory potentials than the commercial diabetes drug acarbose (Nanok and Sansenya, 2020). These results indicated that capsaicin was more likely to bind at the active site of the free enzyme–inhibitor complex (EI) than the enzyme–substrate–inhibitor complex (ESI) for both α-amylase and α-glucosidase. The binding efficiency of capsaicin with these enzymes is associated with the lower Ki of the EI complex than the ESI complex. The binding efficiency of dihydrocapsaicin with α-amylase and α-glucosidase was also similar to that of the enzyme–capsaicin complex. Dihydrocapsaicin exhibited higher inhibitory activity on both α-glucosidase and α-amylase enzymes than capsaicin (Nanok and Sansenya, 2020). Furthermore, the effects of chilli on diabetes have been investigated by Islam and Choi (2008), who found that red chili extract induced an increase in insulin and reduced hyperglycemia in T2D.

Extracts from pepper plants have also been investigated regarding their inhibitory effect on α-amylase and α-glucosidase activities. Tundis et al. (2013) examined the antidiabetic properties of four different pepper cultivars *(C. annuum)* harvested at two different maturity stages using α-amylase and α-glucosidase in vitro inhibition assays. The cultivar, maturity stage, and polarity of the solvent significantly affected the impact of the extracts on α-amylase and α-glucosidase inhibition. In general, peppers harvested at the immature stage showed the highest inhibitory influence on the activity of both α-amylase and α-glucosidase. Moreover, the lipophilic fraction showed significantly higher α-amylase and lower α-glucosidase inhibition activities than the crude extract. The authors speculated that the α-amylase inhibition activity of the pepper extracts was due to the fatty acid (e.g., myristic acid, methyl stearate, and methyl linoleate), flavonoid (e.g., luteolin), and capsaicin contents while the α-glucosidase inhibition activity of the extracts was due to the luteolin contents.

Di Sotto et al. (2018) noted that the ethanolic extracts of *C. annuum* peel and seeds decreased the activity of α-amylase, with IC_{50} values of 164.7 and 109.9 μg mL^{-1}, respectively. The inhibition activities of the extracts were lower than that of acarbose (IC_{50} of 52.2 μg mL^{-1}). However, the peel and seed extracts were reported to possess very low α-glucosidase inhibition activities. The α-amylase and α-glucosidase inhibition activities of the pulp and edible parts were negligible (Di Sotto et al., 2018). The authors suggested that enzyme inhibition activities could be attributed to multiple bioactive compounds present in *C. annuum* extracts, such as fatty acids, polyphenols, carotenoids, and flavonoids. Different classes of bioactive compounds present in conjunction with capsaicin may contribute to the inhibition activity of the pepper extracts. The obtained results also indicated that pepper extracts showed higher α-amylase inhibition activities than α-glucosidase. Several parameters are responsible for the efficiency of the inhibition activity of the extracts, including the cultivar, maturity stage at harvest, part of the fruit, and type of solvent used for extraction (Di Sotto et al., 2018).

Watcharachaisoponsiri et al. (2016) studied the α-amylase and α-glucosidase inhibitory activity of different commercial chili pepper extracts, including yellow pepper, green pepper, cayenne pepper, red chinda pepper, green chinda pepper, young pepper, chili spur pepper, sweet pepper, and bird chili, and reported that chili peppers (5 mg mL^{-1}) extracted with 70% (v/v) aqueous ethanol exhibited anti-α-glucosidase activities within the range of 23%–66% inhibition, with sweet pepper and Cayenne pepper providing the highest and the lowest inhibitory activities, respectively. All chili peppers exhibited anti-α-amylase activities within the range of 27%–58% inhibition, with green China pepper and sweet pepper providing the highest and the lowest inhibitory activities, respectively. Both enzyme inhibitions were dose-dependent, in which inhibition was elevated as the concentration of chili pepper extract increased. Thus, chili peppers might be potential sources of natural products that can be used to combat diabetes and obesity through the inhibition of α-glucosidase and α-amylase and could thus be used for the future development of functional foods, nutraceuticals, or dietary supplements (Watcharachaisoponsiri et al., 2016; Narang et al., 2017).

2. Glucose uptake enhancers

2.1 Effect on glucose transporters (GLUTs)

Glucose uptake is dependent on a family of eight developmentally regulated glucose transporter (GLUT) membrane proteins, each of which has a specific tissue distribution. GLUT1 has a high affinity for glucose and is insulin-independent, and it is responsible for basal glucose uptake. GLUT1 is found in most tissues and maintains glucose transport across the blood−brain barrier. After birth, GLUT1 levels decrease while the other isoforms increase, e.g., GLUT2 increases in the liver, GLUT3 increases in the brain, and GLUT4 increases in the muscle. The action of insulin on glucose uptake is predominantly mediated by GLUT4, which is found in the muscle, fat, and heart tissue, where glucose uptake is insulin-mediated. GLUT2 and GLUT5 contain different subunits, with the former transporting high concentrations of glucose and the latter transporting fructose. GLUTs play an important role in the active absorption of monosaccharides by intestinal mucosa cells (Burcelin and Thorens, 2001; Chen et al., 2016).

The function of capsaicinoids is beneficial for people with diabetes through the modulation of sugar absorption. In a study by Zhang et al. (2017), the weight of model rats was significantly lower than that in the normal group and the weights of the rats in the capsaicin-treated groups increased significantly. Such a weight increase was due to improved glycometabolism and typical diabetes symptoms (increased food and water intake, diuresis, and decreased body weight) by capsaicin, although the risk of overeating was reduced (Zhang et al., 2017). Other studies indicated the significant effect of capsaicinoids on glycometabolism because improved glucose metabolism and increased insulin levels and glycogen content were observed in T1D rats treated with capsaicin (Bady et al., 2006).

Some studies have also shown that hepatic glycogen content is related to the development of diabetes. Capsaicin can restrain glycogen decomposition under stressful conditions to effectively use fatty acids as an energy source (Oh and Ohta, 2003). In this study, capsaicin significantly increased insulin levels and glycogen content, and this finding was validated by an analysis of liver-related genes (Zhang et al., 2017). Two key genes, namely liver X receptor (LXR) and pancreatic and duodenal homeobox 1 (PDX-1), associated with glycometabolism were abundantly expressed in the liver and played a crucial role in the treatment of diabetes. LXR is the key gene that activates the inhibition of liver sugar dysplasia and reduces serum glucose levels. This gene also inhibits the expression of phosphoenolpyruvate carboxykinase (PEPCK) and glucose-6-phosphatase (G6pase), thereby suppressing sugar dysplasia (Commerford et al., 2007). These results revealed that LXR can also stimulate islet cells to secrete insulin, and this activity is connected with glucokinase (GK) and GLUT2, which can stimulate insulin secretion. GK and GLUT2 are considered glucose sensors in β cells and can modulate insulin secretion by affecting the concentrations of blood glucose (Bae et al., 2010). PDX-1 can directly regulate not only the expression of GK and GLUT2 but also the expression of other genes (amylin and synaptophysin), thereby stimulating islet cells to secrete insulin (Kulkarni et al., 2011). In addition, PDX-1 can induce liver cells to develop into pancreatic endocrine and exocrine cells, which induce liver cells to secrete insulin (Koizumi et al., 2004).

Zhang et al. (2017) showed that capsaicin significantly increased the expression of the PDX-1, LXR, GLUT2, and GK genes in the liver. Interestingly, the expression of LXR was increased in the model group, which was perhaps due to a self-regulation mechanism when exposed to hyperglycemic stress in the body (Laffitte et al., 2003; Zhang et al., 2017). The role of the pancreas in T1D rats is very important. Related research suggests that the insulin receptor substrate (IRS) family is closely related to the incidence of type 2 diabetes mellitus because knockout of the IRS1 gene in rats was related to insulin resistance (Danielsson et al., 2005) and knockout of IRS2 in rats was related to severe diabetes symptoms (Morino et al., 2005). In addition, studies have shown that capsaicin activates TRPV1 ion channels and stimulates islet β cells to secrete insulin, thus leading to a reduction in blood glucose levels in rats. The results obtained from Zhang et al. (2017) showed that capsaicin significantly increased the expression levels of PDX-1, IRS1, IRS2, and GLUT2 in the pancreas of T1D rats. The experimental results indicated that capsaicin treatment suppressed elevated blood glucose levels by three pathways: stimulation of insulin secretion, amelioration of pancreatic dysfunction, and improvement of glycometabolism (Zhang et al., 2017).

TRPV1 receptors play a crucial role in the development of T1D and are considered the main channel through which capsaicin elicits its physiological functions in the body (Sharma et al., 2013; Tsurugizawa et al., 2013). The expression of TRPV1 was significantly increased in the liver after treatment with capsaicin, and TRPV1 protein expression in the capsaicin group was 1.56 times (pancreas) that in the capsiate group (Zhang et al., 2017). These results indicate that the pungent ingredient of chili peppers is mainly responsible for the activation of TRPV1 receptors and elicits hypoglycemic effects. More importantly, the expression levels of GLUT2, IRS1, and IRS2 were associated with the TRPV1 expression level. These findings were verified by the interesting phenomenon wherein the expression of TRPV1 was upregulated by capsiate but remained low and the expressions of GLUT2, IRS1, and IRS2 were also downregulated. On the basis of the aforementioned analysis and experimental results, the authors indicate that the signaling pathway TRPV1−(nPDX-1)−(GLUT2/GK) or TRPV1−(PDX-1)−(IRS1/2) may exist in the liver or pancreas (Zhang et al., 2017). In general, the hypoglycemic effect of pepper could be due to the synergistic effect of several factors. Generally, capsaicin can activate the TRPV1 ion channel in the liver and increase the expression of TRPV1, LXR, and PDX-1. LXR and PDX-1 are responsible for controlling the glycometabolism balance by regulating the expression levels of GK, GLUT2, PEPCK, and G6pase, thereby inhibiting gluconeogenesis and promoting glycogen synthesis. The upregulated expression of PDX-1 in the presence of capsaicin stimulated insulin secretion by regulating the expression levels of GLUT2 and cooperated with IRS1/2 to improve glucose metabolism. However, in the intestine, the expression of SGTL1, GLUT2, and GLUT5 was inhibited by capsaicin, thereby leading to an apparent reduction in the digestibility rate of total sugar and an increase in the excretion of sugar in the feces. These results proved that capsaicin possesses high antidiabetic actions in T1D (Zhang et al., 2013). However, for diabetic patients (70 kg), the consumption of approximately 54−80 g of red pepper (dry weight) every day can achieve the effect of the treatment (Ben-Chaim et al., 2006; Lee et al., 2016).

2.2 Effects on sodium-dependent glucose cotransporters

Type 2 diabetes mellitus (T2D) and associated obesity are health problems throughout the developed world. Some studies indicate that sodium-glucose cotransporter (SGLT1) is dysregulated in obesity and T2D (Stearns et al., 2010). Intestinal mucosal cells are generally known to contain significant amounts of sodium-dependent glucose cotransporters (SGLTs) and GLUTs, which synergistically work to complete the transport of glucose in the body. SGLT coordinates the secondary active transport of glucose across the apical membrane of enterocytes throughout the small bowel. SGLT1 is upregulated approximately 3- to 4-fold in both patients and animal models with T2D, while murine knockout of the SGLT1 regulatory subunit RS1, which overexpresses SGLT1, leads to profound obesity (Dyer et al., 2002; Osswald et al., 2005). Therefore, the suppression of SGLT1-mediated glucose absorption will slow intestinal glucose uptake, thereby reducing postprandial hyperglycemia and influencing the metabolic state.

The SGLT family has become one of the current research hotspots in the pathogenesis of diabetes (Zhang et al., 2018). The hypoglycemic effects of hypoglycemic drugs (dapagliflozin and canagliflozin) may result from their competitive inhibition of SGLT in the mucosa of the small intestine (SGTL1) or proximal renal tubule of the kidney (SGTL2). These drugs can selectively inhibit SGLT2 expression in the kidney, resulting in the removal of excess glucose and its associated calories through urine and decreased blood glucose levels (Manova et al., 2014; Mori et al., 2016). Thus, inhibiting the expression of related genes in the intestines may be one of the hypoglycemic mechanisms of capsaicinoids. Zhang et al. (2018) demonstrated that capsaicinoid treatment significantly downregulated the expression levels of GLUT2, GLUT5, and SGLT1 in diabetic rats. The reliability of this mechanism was confirmed by the increase in fecal sugar excretion (Zhang et al., 2018).

Capsaicin can enhance first- and second-phase insulin secretion during hyperglycemic clamps (Kwon et al., 2013). Improved hepatic insulin sensitivity was also associated with reduced hepatic glucose output and increased hepatic glycogen storage. These changes were related to enhanced pAkt/PEPCK and pAMPK signaling pathways (Kwon et al., 2013). These changes suggest inhibition of gluconeogenesis and activation of glycogen synthesis by capsaicin (Zhang et al., 2017; Panchal et al., 2018).

In the study by Stearns et al. (2010), to determine the role of vagal afferents in SGLT1 control, animals were pretreated with capsaicin to selectively deafferent the vagus. Infusion of D-glucose led to a 2.9-fold upregulation in SGLT1 compared with water, and this effect was replicated by D-fructose or saccharin. This response was strongest after isolated infusions of the duodenum and proximal jejunum, and a distal blunted effect was observed; moreover, the topography matched the expression profile of the sweet taste receptor. The reflex was abolished by capsaicin pretreatment (Stearns et al., 2008, 2010).

3. Insulin secretagogues

Capsaicinoids could significantly decrease the concentration of blood glucose by increasing the levels of insulin and glycogen, and these effects might be associated with the regulation of several key glycometabolism-related genes (Panchal et al., 2018).

Incretins are hormones that are released from the gut into the bloodstream in response to food ingestion, and they then modulate the insulin secretory response to the products within the nutrients in the food. The insulin secretory response of incretins, called the incretin effect, accounts for at least 50% of the total insulin secreted after oral glucose. Therefore, incretin hormones are insulinotropic because they induce insulin secretion at the usual physiological concentrations seen in the plasma after ingestion (Kim and Egan, 2008).

Because insulin signaling is already desensitized in T2D and injection of insulin loses its effectiveness over time, diabetes researchers are investigating different strategies for improving blood glucose levels. Different signaling pathways also modulate blood glucose levels and activate a similar second messenger signaling cascade, in particular, glucagon-like polypeptide 1 (GLP-1), although others are under investigation, such as GLP-2, oxyntomodulin or glucose-dependent insulinotropic polypeptide (GIP) (Kim and Egan, 2008).

GLP-1 is an endogenous 30-amino acid peptide hormone. It has several physiological roles in different tissues to control cell metabolism. GLP-1 receptor (GLP-1R) belongs to the class B family of G-protein coupled receptors. The receptors for glucagon, GLP-2 and GIP, also belong to this group. Activation of the receptor activates an adenylate cyclase, increases inositol triphosphate (IP3) levels and gene expression, and has a range of other downstream effects (Anand et al., 2018). Some studies have shown that the application of capsaicinoids significantly improves the plasma levels of ghrelin, GIP, and GLP-1, which are comprehensively involved in energy homeostasis and metabolic regulation (Clarke et al., 2014).

Another study revealed that dietary capsaicinoids increased the *Firmicutes/Bacteroidetes* ratio and *Faecalibacterium* abundance, accompanied by increased plasma levels of GLP-1, GIP, and gastric inhibitory polypeptide and decreased plasma ghrelin levels. Further enterotype analysis revealed that these subjects could be clustered into the *Bacteroides* enterotype (E1) and *Prevotella* enterotype (E2), and the aforementioned beneficial effects were mainly obtained in E1 subjects. Moreover, E1 subjects had significantly higher fecal *Faecalibacterium* abundance and butyrate concentration after capsaicinoid application than E2 subjects (Kang et al., 2016). The increased *Firmicutes/Bacteroidetes* ratio and *Faecalibacterium* abundance coincided with the increase in plasma levels of GLP-1 and GIP and the decrease in plasma ghrelin levels in healthy Chinese adults. Strikingly, further gut enterotype stratification analysis revealed that the aforementioned benefits were mainly obtained in E1 subjects, who also had significantly lower plasma lipopolysaccharide-binding protein and fecal Gram-negative bacteria levels and higher fecal butyrate concentrations than E2 subjects (Kang et al., 2016). The E1 group plays important roles in protecting mucosal barrier function and stimulating the secretion of GI hormones (Tolhurst et al., 2012). The potent insulinotropic action of the gut incretin GLP-1 that induces satiety is considered to be at least partly mediated by sensory afferents that innervate the gut because sensory denervation by neonatal capsaicin administration in mice impaired the insulinotropic effects of GLP-1 (Ahren, 2004; Holst, 2007).

Song et al. (2017) indicated that a low-capsaicin (0.01%) or high-capsaicin (0.02%) diet for 6 weeks significantly increased fecal butyrate and plasma total GLP-1 levels but decreased plasma total ghrelin, TNF-α, IL-1β, and IL-6 levels compared with a normal diet in mice. The beneficial effects of dietary capsaicin on glucose homeostasis are likely associated with alterations in specific bacteria at the genus level. These alterations in bacteria induced by dietary capsaicin contribute to improved glucose homeostasis by increasing short-chain fatty acids, regulating gastrointestinal hormones, and inhibiting proinflammatory cytokines (Song et al., 2017).

Other studies revealed that capsaicin increased GLP-1 and decreased ghrelin secretion, indicating a possible interaction between TRPV1 and GLP-1, which is an incretin hormone that induces expansion of insulin-secreting β-cell mass, and this change augments glucose-stimulated insulin secretion (Smeets and Westerterp-Plantenga, 2009). Moreover, GLP-1 may enhance insulin sensitivity in humans and rodents (Parlevliet et al., 2010; Panchal et al., 2018). Capsaicin administration through injection to Zucker diabetic fatty rats prevented increased glucose-induced insulin secretion and improved glucose metabolism. These changes were supported by the loss of TRPV1- and calcitonin gene-related (CGRP) peptide-coexpressing nerve fibers in the pancreas (Gram et al., 2007). Acute capsaicin treatment increased GLP-1 and insulin secretion in wild-type mice in a TRPV1-dependent manner (Wang et al., 2012). Chronic dietary capsaicin increased plasma GLP-1 and insulin concentrations, with improved glucose tolerance in wild-type mice; these responses were not seen in TRPV1−/− mice. In db/db mice, TRPV1 activation by capsaicin improved glucose homeostasis, increased GLP-1 production in the distal ileum, and increased plasma GLP-1 concentrations (Wang et al., 2012; Panchal et al., 2018).

GLP-1 is also synthesized in the brain and acts as a neurotransmitter (Amouyal and Andreelli, 2016; Anand et al., 2018). Various peripheral signals (circulating concentrations of substrates or hormones, fasting, and meals) modulate the synthesis and release of GLP-1 in the brain. GLP-1 may also act as a neurotransmitter at a distance. This information is then relayed to the brain, which in turn may act on peripheral tissues through efferent vagal nerves, resulting in GLP-1-mediated effects through the nervous system (Holst, 2007). This loop enables GLP-1 to act on target tissue through the vagus nerve without the need to physically reach the target tissue. This mode of action is consistent with the very limited half-life of the hormone and is made possible by the substantial innervations of the gut. This action has been extensively demonstrated in various murine models, including mice denervated at birth following the early postnatal injection of capsaicin. In this model, physiological doses of GLP-1 do not stimulate the glucose-dependent secretion of insulin, thus reflecting the need for maintaining the integrity of nerve pathways to mediate the effects of exogenous GLP-1 on insulin (Ahren, 2004). An increase in GLP-1 secretion by the nervous system is observed after oral administration of glucose (usually between 10 and 15 min in healthy individuals), which is incompatible with physical contact between food and intestinal L cells (Holst, 2007; Amouyal and Andreelli, 2016). Anand et al. (2018) showed that the nociceptive capsaicin/heat pain receptor TRPV1 was not affected by acute application of these gut hormones, indicating that GLP-1R (receptor GLP-1) may not participate in noxious signaling involving TRPV1. A clinical trial of the GLP1 analogue ROSE-010 reported a reduction in pain during acute exacerbations of irritable bowel syndrome and increased colonic transit, thus providing relief of constipation. These results also indicated enhanced ATP signaling in the presence of exendin-4 in human DRG neurons and by others in rodent neurons (Richards et al., 2014; Anand et al., 2018), suggesting that GLP-1R may participate in modulating gut motility rather than pain signaling. Nerve growth factor (NGF) is currently regarded as the major regulator of TRPV1 expression and hypersensitivity in nociceptors and inflamed bowel (Anand et al., 2006). The neurite-promoting effects combined with the lack of sensitization of capsaicin responses suggest that GLP-1R agonists may be beneficial in other conditions, such as diabetic neuropathy, and have the potential to provide neurite regeneration without the burden of nociceptor sensitization or enhancement of pain. These studies showed that increased GLP-1R innervations may be involved in inflammatory bowel disease and afferent signaling and potentially provide a peripheral therapeutic target for satiety and weight loss (Anand et al., 2018).

4. Peroxisome proliferator-activated receptor (PPARγ) ligands

4.1 Effects on PPARγ gene

Peroxisome proliferator-activated receptors (PPARs) belong to a subfamily of the nuclear receptor superfamily of ligand-inducible transcription factors. PPARα, PPARβ/δ, and PPARγ isotypes of PPARs encoded by separate genes have been identified (Wang et al., 2014).

PPARs mainly control the expression of gene networks involved in adipogenesis, lipid metabolism, inflammation, and metabolic homeostasis maintenance. PPARs can be activated by dietary fatty acids and their metabolites, and they act as lipid sensors that are able to markedly redirect metabolism (Evans et al., 2004). The gene transcription process was identical in all three PPAR subtypes. After ligand binding, PPARs form heterodimers with another ligand-activated nuclear receptor, the retinoid X receptor (RXR). The PPAR-RXR heterodimer binds to peroxisome proliferator response elements (PPREs) in the promoter region of the respective target genes. The transcription process is then initiated upon recruitment of different transcriptional cofactors (Park et al., 2014; Wang et al., 2014).

PAPRγ activation results in growth arrest and/or apoptosis in a variety of cancer cells. The potential of capsaicin to induce apoptotic cell death in human colon cancer cells and the association of PPARγ with capsaicin action were investigated (Kim et al., 2004). Treatment with capsaicin or the PPARγ ligand troglitazone induced apoptotic cell death in a dose-dependent manner in HT-29 human colon cancer cells. Capsaicin-induced cell death was completely blocked by bisphenol diglycidyl ether, a specific PPARγ antagonist. Capsazepine, a specific antagonist of the vanilloid receptor, did not inhibit capsaicin-induced apoptosis. The authors suggest that capsaicin-induced apoptotic cell death in HT-29 human colon cancer cells could be associated with the PPARγ pathway without the involvement of the vanilloid receptor. Therefore, capsaicin may have a beneficial effect for the treatment of colon cancer (Kim et al., 2004).

PPARγ plays an important role in the transcriptional regulation of genes involved in adipogenesis, fat deposition, insulin sensitivity, inflammatory response, and hepatic fibrosis. In the liver, PPAR-γ induces a phenotypic switch in hepatic stellate cells (HSCs), the main fibrogenic cell type, from activated to quiescent, and it has been shown that PPARγ agonists/ligands inhibit both liver fibrosis and HSC activation by activating the PPARγ pathway (Choi et al., 2017). The administration of capsaicin restored the induction of dimethylnitrosamine (DMN) PPARγ expression during hepatic fibrosis in rats. Moreover, the ability of capsaicin to inhibit fibrogenic transforming growth factor-beta 1 (TGF-β1) secretion as well as collagen secretion and expression in HSCs by activating PPAR-γ expression was also reported (Bitencourt et al., 2012). The findings thus demonstrate that capsaicin prevents DMN-induced liver fibrosis by acting on several of the responsible pathways (Choi et al., 2017). A clinical study reported that capsaicin-induced satiety may be involved in pain from gastrointestinal stress, which indicates that determining the appropriate dosing of capsaicinoids is important for the treatment of patients with liver fibrosis. Therefore, capsaicin should be further studied for its proper management and use as a candidate antifibrotic agent before clinical application for patients with liver fibrosis (Choi et al., 2017).

Decreases in body weight, body fat, and serum lipids, such as triglycerides, low-density lipoprotein, and high-density lipoprotein, in obese rats treated with capsaicin were reported by Tan et al. (2014). In these studies, capsaicin upregulated the expression of both PPARα and PPARγ, the uncoupling protein-2 (UCP2) gene, and adiponectin. This compound also downregulated the expression of the leptin gene (Narang et al., 2017).

Kang et al. (2010) investigated the effect of 0.015% capsaicin in dietary supplements for 10 weeks in obese mice fed a high-fat diet. The inhibition of insulin resistance and fat accumulation in hepatocytes was reported after applying this compound. Capsaicin also decreased the levels of fasting glucose, insulin, triglycerides, leptin, tumor necrosis factor-a, interleukin-6, monocyte chemoattractant protein-1, and TRPV1. Increases in the mRNA expression of adiponectin, PPARα, and PPARγ coactivator-1α were also observed in mice after supplementation with capsaicin (Kang et al., 2010; Narang et al., 2017).

Moreover, Lee et al. (2013) showed that the topical application of 0.075% capsaicin to male mice fed a high-fat diet significantly decreased weight gain and lipid accumulation in mesenteric and epididymal adipose tissues. Capsaicin decreased serum levels of glucose, cholesterol, and triglycerides and was able to induce the expression of adiponectin and adipokines, including both PPARα, PPARγ, visfatin, and adipsin and decrease the expression of tumor necrosis factor-a and interleukin-6. These results indicated that topical application of capsaicin can reduce lipid accumulation in adipose tissues and decrease the inflammatory process in an animal model (Lee et al., 2013; Narang et al., 2017).

The results obtained from the studies of Szoka and Palka (2020) indicated that capsaicin stimulated intrinsic and extrinsic pathways of apoptosis in a dose-dependent manner in the human glioblastoma LN-18 cell line and that this phenomenon was not dependent on TRPV1. Activation of PPARγ, a ligand-dependent transcription factor, also induced apoptosis in glioblastoma cells. Although PPARγ ligands (thiazolidinediones, such as rosiglitazone and pioglitazone) promoted apoptosis in LN-18 cells, capsaicin augmented this effect. These findings suggest that capsaicin-dependent upregulation of PPARγ represents the mechanism for augmentation of cell death by thiazolidinediones (Szoka and Palka, 2020).

Hepatic vagotomy involves dissection of both afferent and efferent vagal branches innervating the liver (Uno et al., 2012). Therefore, to confirm the involvement of afferent vagal signals originating in the liver, capsaicin was applied to the hepatic branch of the vagus in 8-week-old mice. Previous studies (Uno et al., 2006) reported that the expression of calcitonin gene-related peptide, a sensory neuropeptide, was markedly decreased in capsaicin-treated vagal nerves while immunoreactivity for S100 proteins was similar in vehicle- and capsaicin-treated nerves, indicating selective deafferentation of the vagal nerve by capsaicin treatment. The authors (Uno et al., 2012) reported that perivagal capsaicin treatment did not affect endogenous PPARγ expression in the livers of mice. Under these conditions, blockade of afferent vagal signals from the liver prevented systolic blood pressure elevation during obesity development in mice. Thus, hepatic PPARγ upregulation associated with obesity is involved in the development of hypertension through the afferent vagal pathway from the liver (Uno et al., 2012).

PPARγ is a nuclear receptor that regulates intestinal inflammation. PPARγ is highly expressed in the colon and can be activated by various dietary ligands. A number of fatty acids, such as polyunsaturated fatty acids or eicosanoids, are considered endogenous PPARγ activators. Nevertheless, other nutrients, such as glutamine, flavonoids, and spicy food containing capsaicinoids, are also able to activate PPARγ, which plays a key role in bacterial-induced inflammation. Thus, the antiinflammatory properties of probiotics may be mediated through PPARγ (Marion-Letellier et al., 2009).

Capsaicin, the spicy component of hot pepper, has been found to lower glucose, insulin, and leptin levels and reduce the impairment of glucose tolerance in obese mice. Capsaicin is the archetypal agonist of TRPV1, and the aforementioned effect can be modulated by the expression/activation of this receptor. However, luciferase assays revealed that capsaicin is capable of binding PPAR*s* and PPAR-target gene levels were higher in the livers of obese mice supplemented with dietary capsaicin than in those of obese controls (Kang et al., 2010; Rigano et al., 2017).

4.2 Effects on PGC-1α gene

Peroxisome proliferator-activated receptor-γ coactivator-1α (PGC-1α) is a principal regulator of the expression of genes involved in mitochondrial biogenesis, respiratory function, and carbohydrate and lipid metabolism. It also participates in the pathogenesis of obesity, diabetes, and cardiomyopathy (Luo et al., 2012). Increased expression of PGC-1α results in increased mitochondrial function and oxidative muscle fibers and protects against metabolic disease during aging (Wenz et al., 2009).

Some studies have reported that capsaicin, an agonist of TRPV1, acts through TRPV1-mediated Ca^{2+} mobilization (Caterina, 2003). Other results (Luo et al., 2012) confirmed that capsaicin induced a significant increase in Ca^{2+} in myotubes, which was blocked by the specific TRPV1 antagonist 5′-iodoresiniferatoxin (iRTX). Ca^{2+}- and calmodulin-dependent protein kinases can activate the PGC-1α gene promoter and induce PGC-1α expression, thus contributing to mitochondrial biogenesis in myocytes (Yoshioka et al., 2009). Capsaicin exposure resulted in a significant increase in PGC-1α expression, mitochondrial biogenesis, and ATP production in cultured myotubes, which can be abolished by inhibition of TRPV1 or depletion of intracellular Ca^{2+} (Luo et al., 2012). For that reason, a TRPV1-mediated Ca^{2+} increase is critical for capsaicin-induced PGC-1α expression in myotubes (Luo et al., 2012). Several earlier studies showed that a single dose of capsaicin temporarily promoted oxygen consumption and lipid utilization and enhanced swimming capability in rodents, which was associated with sympathetic activation (Oh and Ohta, 2003).

The studies of Luo et al. (2012) showed that chronic dietary capsaicin continuously improved exercise endurance with no effect on food intake. First, in vitro studies on cultured myotubes showed that capsaicin induced PGC-1α expression, mitochondrial biogenesis, and ATP production through TRPV1-mediated Ca^{2+} signals. Second, dietary capsaicin administration in mice increased the expression of TRPV1 and PGC-1α in skeletal muscles and consequently upregulated the PGC-1α target genes involved in fatty acid oxidation and mitochondrial respiration and downregulated the genes involved in glycolysis. Third, dietary capsaicin increased mitochondrial mass and respiratory function and promoted fiber transformation from glycolytic type II fibers to more oxidative type I fibers, which contain more mitochondria and produce more ATP from lipid oxidation but less ATP from glycolysis than type II fibers, thus providing stable energy for a longer time with less lactic acid accumulation (Joyner and Coyle, 2008; Wenz et al., 2009; Luo et al., 2012). Thus, the benefits of capsaicin on exercise endurance and energy metabolism might be attributed to the restoration of mitochondrial function and lipid oxidative capacity through TRPV1-mediated PGC-1α upregulation in skeletal muscles (Luo et al., 2012). In addition, the authors (Luo et al., 2012) showed that TRPV1 is upregulated by capsaicin treatment in vitro and in vivo, and they also found that exercise and a high-fat diet affect the protein expression of TRPV1 in muscles. These findings also suggest that dietary capsaicin supplementation can be a promising lifestyle modification strategy to improve skeletal muscle function and prevent cardiometabolic diseases (Luo et al., 2012).

Recent studies also indicate that capsaicin acts by activating the sympathetic system to induce brown adipose tissue (BAT) thermogenesis and reduce fat accumulation (Azhar et al., 2016). The application of capsaicin in mice has been shown to induce thermogenesis through the activation of BAT, as evidenced by the increase in markers related to mitochondrial biogenesis, such as PPARγ, PGC-1α, and uncoupling protein 1 (UCP1). Capsaicin also induced the development of beige adipocytes at an early stage of adipogenesis (Kida et al., 2016). The effect of capsaicin treatment has been compared with that of chronic cold exposure, in which sympathetic stimulation results in the activation of BAT (Saito, 2015). Capsaicin binds to the intestinal TRPV1 receptor, also referred to as the capsaicin receptor, thereby initiating the sympathetic response observed with treatment and increasing thermogenesis and lipid metabolism in adipose tissues responsible for the decrease in body fat and obesity (Caterina et al., 2000; Yiangou et al., 2001; Dinis et al., 2005; Baskaran et al., 2016).

The effect of capsaicin on selected thermogenic and related genes, such as PGC-1α, PPARα, UCP1, brain-derived neurotrophic factor (BDNF), nuclear receptor interacting protein 1 (NRIP1), nuclear receptor coactivator 1 (NCOA1), cell death activator (CIDEA), mitogen-activated protein kinase 14 (MAPK14), sirtuin 1 (SIRT1), and prostaglandin-endoperoxide synthase 2 (PTGS2), in BAT and white adipose tissue (WAT) (visceral and subcutaneous) was confirmed in the studies of Baboota et al. (2014). In BAT, a high-fat diet significantly upregulated gene expression compared with control animals, and these genes were further significantly upregulated upon capsaicin administration relative to the high-fat diet group. Similarly, in vWAT, these genes were significantly increased in high-fat-diet-fed mice, but no significant difference was observed in the capsaicin-treated group compared with the control group. Different trends in the expression levels of these genes were observed for sWAT, BDNF, NCOA1, PPARα and PTGS2, which showed similar patterns as BAT, i.e., a significant increase in expression in the high-fat-diet group and further upregulation in the high-fat-diet group exposed to capsaicin, whereas UCP1, NRIP1, CIDEA, PGC1α, MPAK14, and SIRT2 were significantly down-regulated in the high-fat-diet group relative to the control but significantly upregulated in the diet group treated with capsaicin (Baboota et al., 2014). The authors suggest that the differential effect in the two types of WAT may be because brown-like transformation is more prominent in sWAT than vWAT, which is less susceptible to browning (Kajimura and Saito, 2014). The obtained results from this study indicate that oral administration of capsaicin modulates the hypothalamic satiety-associated genotype, alters the gut microbial composition, increases the expression of BAT-associated genes in subcutaneous WAT, and increases the expression of thermogenesis and mitochondrial biogenesis genes in BAT. These results provide interesting explanations for the antiobesity effect of capsaicin (Baboota et al., 2014).

The application of capsaicin at low (0.1 μM) and high (10 μM) concentrations increased the expression of the PPARγ2 and PGC-1α genes when cells were treated during the late stage of brown adipogenesis (Kida et al., 2016). PPARγ2 promotes a transcriptional cascade involved in the function of adipocytes, while PGC-1α stimulates mitochondrial biogenesis as well as brown adipocyte function, including the transcriptional activation of UCP1. Based on the results, the authors indicate that capsaicin stimulates the differentiation of brown preadipocytes present in the late stage of brown adipogenesis and increases expression levels in brown adipocytes or both. Compared with the treatment with capsaicin during the late stage of brown adipogenesis, PPARγ2 and PGC-1α were not upregulated when cells were treated with capsaicin during the early stage of brown adipogenesis. This differential response to capsaicin may be related to increases in the expression level of TRPV1 during brown adipogenesis (Kida et al., 2016).

Mostafavian et al. (2020) showed that the expression of the PGC-1α and UCP1 genes was decreased in high-fat diet rats compared with the normal diet group. Additionally, the expression of PGC-1α and UPC1 in the High-Fat-Training (HFDT), High-Fat-Capsaicin (HFDCap), and High-Fat-Training-Capsaicin (HFDTCap) groups was significantly increased compared with that in the high-fat diet rats. The expression of PGC-1α and UPC-1 in the HFDTCap group was also significantly increased compared with that in the high-fat diet and HFDCap groups. The authors speculate that 8 weeks of progressive training combined with capsaicin administration had an effect on the browning of visceral adipose tissue in high-fat diet rats by increasing the expression of PGC-1α and UCP1 genes (Mostafavian et al., 2020).

To investigate whether the combined effects of capsaicin on promoting lipid metabolism were related to the induction of browning in 3T3-L1 cells (the mouse embryonic fibroblast cell line possessing the ability to differentiate into mature adipocytes), the levels of beige-specific genes and proteins in these cells were investigated (Fan et al., 2019). Increases in the expression of beige-specific genes, including T-Box transcription factor 1 (Tbx1), transmembrane protein 26 (TMEM26), cluster of differentiation 40 (CD40), UCP1, and PR domain containing 16 (PRDM16), were observed in cells treated with capsaicin. Furthermore, capsaicin treatment increased the protein levels of UCP1, PRDM16, and PGC-1α in 3T3-L1 cells, which are markers of beige adipocyte biogenesis in 3T3-L1 cells. The results indicated that the application of capsaicin could effectively induce beige adipocyte biogenesis in 3T3-L1 white adipocytes (Fan et al., 2019). PGC-1α is also a beige-specific marker, and the increase in both gene expression and protein synthesis in capsaicin-treated cells suggests that this protein is more resistant to degradation or possesses more stability in the presence of this compound (Fan et al., 2019). The authors speculate that the potential mechanism of inducing browning of white adipocytes by capsaicinoids may be associated with the activation of the PPARγ/β_3-AR (beta-3 adrenergic receptor) signaling pathways. The study supported the potential therapeutic utility of capsaicin to treat obesity (Fan et al., 2019).

In addition to the above findings, Jeong et al. (2014) reported that 0.1, 1, 5, and 10 μmol of capsaicin can decrease lipid accumulation during adipocyte differentiation of bovine bone marrow mesenchymal stem cells for 2, 4, and 6 days. Capsaicin also inhibited the expression of PPARγ, fatty acid binding protein, and stearoyl-CoA desaturase (SCD). The number of apoptotic cells increased in dose- and time-dependent manners. Furthermore, capsaicin increased the expression levels of apoptotic genes, such as B-cell lymphoma 2-associated X protein (BAX) and caspase 3 (CASP3). Overall, capsaicin inhibited fat deposition by triggering apoptosis (Jeong et al., 2014). Moreover, Ibrahim et al. (2015) reported that capsaicin inhibited the adipogenic differentiation of mesenchymal stem cells by repressing PPARγ, CCAAT/enhancer-binding protein alpha (C/EBPa), fatty-acid-binding protein (FABP4), and SCD-1 gene expression (Ibrahim et al., 2015; Narang et al., 2017). Transdifferentiation of mesenchymal stem cells into adipocytes is an added detrimental factor causing the intensification of obesity; therefore, capsaicin-induced inhibition of the differentiation of mesenchymal stem cells into adipocytes may provide benefits for human health and antiobesity effects (Ibrahim et al., 2015).

4.3 Effects on adiponectin

Adipocytes are highly specialized cells that affect energy homeostasis by secreting a number of biologically active adipokines (Lee et al., 2013). Research into adipokines has strongly influenced our understanding of the relationship between increased adipose tissue and obesity. Adipokines, which are biologically active molecules secreted from adipose tissues or adipocytes, play major roles in the regulation of food intake, insulin sensitivity, energy metabolism, and the vascular microenvironment (Fan et al., 2019).

Among adipokines, tumor necrosis factor-alpha (TNF-α), interleukin-6 (IL-6), monocyte chemoattractant protein-1 (MCP-1), and adiponectin are strongly associated with obesity-induced inflammation and obesity-related pathologies. The circulation levels of TNF-α, IL-6, and MCP-1 are positively correlated with the levels of inflammatory markers, such as C-reactive protein, whereas adiponectin is negatively correlated with these levels (Kang et al., 2007). Adiponectin improves insulin sensitivity by increasing energy expenditure and fatty acid oxidation through activation of adenosine monophosphate-activated protein kinase (AMPK) and by increasing the expression of target genes of PPARα, such as fatty acid translocase/CD36, acyl-CoA oxidase, and UCP2 (Kadowaki and Yamauchi, 2005). Adiponectin is thought to improve insulin sensitivity by upregulating the expression of insulin receptor substrate-1 in skeletal muscles and to attenuate the development of atherosclerosis by suppressing the expression of adhesion molecules in vascular endothelial cells and inhibiting the accumulation of monocyte/macrophage-derived foam cells in the vascular wall (Kadowaki et al., 2006). Studies have shown that capsaicin has the potential to serve as a fat-reducing drug, possibly by increasing the expression of adiponectin, thus reducing fat accumulation in the adipose tissue of obese mice (Lee et al., 2013).

Capsaicin has also been identified as a stimulator of adiponectin production in mouse and human adipocytes. This compound effectively lowers blood glucose and triglyceride levels in mice (Yamazaki et al., 2008).

Topical application of capsaicin has been shown to increase the expression of adiponectin and other adipokines, thus reducing fat accumulation in the adipose tissue of obese mice (Kang et al., 2007). In addition, capsaicin treatment decreased the levels of IL-6 and MCP-1 and increased the level of adiponectin released from obese fat tissues and fat cells. This dual action of capsaicin may be favorable for improving the dysregulation of adipokine release through the normalization of the balance between the secretion of proinflammatory and antiinflammatory adipokines in insulin resistance or atherosclerosis Kang et al. (2007); Kang et al. (2007) showed that intraperitoneal injection of capsaicin (6 mg/kg, BW) in vivo altered lipid energy metabolism, whereas Kawada et al. (1986) indicated that dietary capsaicin supplementation (0.014%) lowered the perirenal adipose tissue weight and the levels of serum triglycerides in obese rats. Additionally, Kang et al. (2007) demonstrated that intraperitoneal injection of capsaicin (2 mg/kg, BW) significantly decreased the expression levels of MCP-1, IL-6, and cyclooxygenase-2 (COX-2) genes in the adipose tissue of obese mice but increased the expression level of the adiponectin gene. Moreover, there was a significant decrease in macrophage infiltration into the adipose tissue of obese mice. These results indicated that the beneficial dual action of capsaicin in vitro can be relevant in vivo (Kang et al., 2007).

Capsaicin treatment (0.015% in food for 10 weeks) in mice fed a high-fat diet for 10 weeks lowered obesity, fasting glucose, insulin, leptin, inflammatory markers in adipose tissue and liver and hepatic triglycerides (Kang et al., 2010). The application of this compound increased adiponectin mRNA/protein in the adipose tissue as well as PPARα/PGC-1α mRNA in the liver. Thus, the improved metabolic and inflammatory status in adipose tissue and liver suggest that dietary capsaicin may reduce insulin resistance. These effects were associated with its dual action on PPARα and TRPV1 expression/activation (Kang et al., 2010). In pancreatectomized diabetic rats exposed to capsaicin, a reduction in body weight gain, visceral fat accumulation, and serum leptin and higher glucose tolerance without modulating energy intake were observed (Kwon et al., 2013; Panchal et al., 2018).

Substantial upregulation of the adiponectin gene was observed under topical capsaicin treatment in obese mice (Lee et al., 2013). Since plasma adiponectin levels are reduced in obesity, the findings suggested that the reduced visceral fat accumulation resulting from topical application of capsaicin led to adiponectin secretion from visceral adipose tissue. The authors (Lee et al., 2013) speculated that topical application of capsaicin increased energy metabolism through the sensory-central sympathetic nerve axis and suppressed the accumulation of visceral fat. These results indicated that topical application of capsaicin to obese mice effectively limited fat accumulation in adipose tissues, reduced inflammation, and increased insulin sensitivity (Lee et al., 2013).

Visceral obesity is the main risk factor for nonalcoholic fatty liver disease (NAFLD), and inappropriate storage of triglycerides in adipocytes and higher concentrations of free fatty acids may add to increased hepatic lipid storage and progressive liver damage (Shin et al., 2020). Capsaicin attenuated inflammation in liver tissue, which might be induced by phosphorylation of AMPK signaling through the TRPV1 receptor (Kang et al., 2011). This hepatic AMPK activation is mediated by systemic increases in the concentration of adiponectin as the mechanism of action of capsaicin (Shin et al., 2020). Immunohistochemical analysis revealed elevated levels of adiponectin in the liver tissue of capsaicin-treated mice. Adiponectin increased fatty acid combustion and energy consumption, which led to decreased triglyceride content in the liver through activation of the AMPK and PPARα pathways. An increase in AMPK phosphorylation and PPARα protein levels was observed in high-fat-diet-induced obese mice treated with capsaicin (Shin et al., 2020). These data suggest that lower fatty liver symptoms induced by topical application of capsaicin are associated with AMPK-mediated adiponectin signaling, β-oxidation upregulation, and de novo lipogenesis in high-fat-diet-induced NAFLD mice. Therefore, adiponectin seems to be involved in the capsaicin-induced inhibition of hepatic lipid accumulation (Shin et al., 2020). The elevation of adiponectin levels by natural product supplementation might benefit patients with metabolic syndrome and its animal model, and the topical application of capsaicin could potentially be beneficial in preventing the complications of NAFLD (Shin et al., 2020).

Adipogenesis is the critical and original process of fatty adipose accumulation, which suggests that decreased preadipocyte differentiation, proliferation, and lipogenesis have the potential to reduce obesity (Zheng et al., 2017). Hsu and Yen (2007) demonstrated that capsaicin inhibited the expression of PPARγ, C/EBP-α, and leptin but induced the upregulation of adiponectin at the protein level. Thus, it efficiently induced apoptosis and inhibited adipogenesis in 3T3-L1 preadipocytes and adipocytes in vitro (Hsu and Yen, 2007). Zhang et al. (2007) found that capsaicin treatment prevented adipogenesis of 3T3-L1 preadipocytes in vitro, with increased intracellular calcium. Male C57BL/6 obese mice fed a high-fat diet for 10 weeks that received a supplement of 0.015% capsaicin showed decreased fasting glucose, insulin, and leptin concentrations and markedly improved glucose intolerance in obese mice, accompanied by decreased TRPV1 expression in adipose tissue, increased adiponectin expression in adipose tissue, and increased PPARα and PGC-1α expression in the liver (Kang et al., 2010). Ohnuki et al. (2001) demonstrated that a $10\ mg\ kg^{-1}$ body weight capsaicin treatment could markedly suppress body fat accumulation and promote energy metabolism in mice. Hence, these studies suggested that capsaicin could decrease adipogenesis and regulate gene function related to lipid metabolism, thus indicating its potential to promote weight loss.

In KKAy genetically obese and diabetic mice, capsaicin supplementation in the diet lowered the concentrations of blood glucose, insulin, and triglycerides through a reduction in hepatic triglyceride deposition (Kang et al., 2011). Furthermore, inflammation in adipose tissue and liver was reduced by capsaicin, with no changes in body weight or adiposity. Hepatic AMPK activation, systemic increases in the concentration of adiponectin, and increased adiponectin receptor 2 (AdipoR2) in the liver were suggested as the mechanisms of action of capsaicin in reducing these metabolic complications (Kang et al., 2011; Panchal et al., 2018).

In rats fed a high-fat diet, the application of capsaicin reduced the weight of perirenal adipose tissue and serum triglyceride concentrations and topical capsaicin application reduced body weight and fat gain and lowered fat mass in mesenteric and epididymal adipose tissues (Lee et al., 2016). Capsaicin-chitosan microspheres in obese rats fed a high-fat diet showed more effective obesity reductions than capsaicin while decreasing the proportion of body fat. It was proposed that these changes were caused by increases in mRNA and protein expression of PPARα, UCP2, and adiponectin genes along with decreases in leptin expression (Tan et al., 2014; Panchal et al., 2018).

Mun et al. (2014) studied the effect of the combination of corn gluten hydrolyzate and capsaicin on rats fed high-fat diets. They observed that this combination reduced body weight, fasting plasma glucose, insulin and leptin levels, and Homeostatic Model Assessment for Insulin Resistance (HOMA-IR) values while increasing plasma adiponectin. Plasma triglyceride, total cholesterol, HDL cholesterol, and LDL cholesterol concentrations were not changed by the combination of corn gluten hydrolyzate and capsaicin treatment, whereas hepatic total lipid, triglyceride, and total cholesterol concentrations were decreased. It has been suggested that BAT activation, adrenosympathetic nervous system activation for catecholamine production, and increased energy expenditure are responsible for the thermogenic activity of capsaicin (Mun et al., 2014; Panchal et al., 2018).

5. Synergistic or additive effects of capsaicinoids with commercially available antidiabetic drugs

Interactions with concomitant treatment of *C. annuum* and aspirin and other salicylic compounds were documented. The combined ASA-capsaicin therapies reached equivalent effectiveness in platelet aggregation inhibition as ASA monotherapy. Sandor and coauthors proved that capsaicin did not influence the inhibitory effect of ASA on platelet aggregation; thus, capsaicin-ASA treatment would combine the antiplatelet effect of ASA with the possible gastroprotection effect of capsaicin (Sandor et al., 2014).

6. Conclusions

This chapter aims to provide an overview of the mechanisms underlying the antidiabetic action of capsaicinoids. Most of the results were obtained from studies on animal systems, and since many of these mechanisms are poorly understood, further investigation is required in human models. The limited number of human studies has not provided a basis for the development of capsaicinoids as commercial antiobesity drugs. In addition, more advanced clinical studies are required to establish the efficacy of capsaicinoids in the treatment of obesity.

References

Adaszek, Ł., Gadomska, D., Staniec, M., Gołyński, M., Łyp, P., Zietek, J., Różańska, D., Orzelski, M., Śmiech, M., Winiarczyk, S., 2017. Clinical assessment of the anti-cancer activity of the capsaicin-containing habanero pepper extract in dogs - preliminary study. Med. Weter. 73, 404–411.

Ahren, B., 2004. Sensory nerves contribute to insulin secretion by glucagon-like peptide 1 in mice. Am. J. Physiol. Regul. Integr. Comp. Physiol. 286, R269–R272.

Amouyal, C., Andreelli, F., 2016. Increasing GLP-1 Circulating levels by bariatric surgery or by GLP-1 receptor agonists therapy: why are the clinical consequences so different? J. Diabetes Res. 5908656, 10.

Anand, U., Otto, W.R., Casula, M.A., Day, N.C., Davis, J.B., Bountra, C., et al., 2006. The effect of neurotrophic factors on morphology, TRPV1 expression and capsaicin responses of cultured human DRG sensory neurons. Neurosci. Lett. 399, 51–56.

Anand, U., Yiangou, Y., Akbar, A., Quick, T., MacQuillan, A., Fox, M., et al., 2018. Glucagon-like peptide 1 receptor (GLP-1R) expression by nerve fibres in inflammatory bowel disease and functional effects in cultured neurons. PLoS One 13, e0198024.

Antonio, A.S., Wiedemann, L.S.M., Veiga Junior, V.F., 2018. The genus *Capsicum*: a phytochemical review of bioactive secondary metabolites. RSC Adv. 8, 25767–25784.

Azhar, Y., Parmar, A., Miller, C.N., Samuels, J.S., Rayalam, S., 2016. Phytochemicals as novel agents for the induction of browning in white adipose tissue. Nutr. Metab. 13, 89.

Baboota, R.K., Murtaza, N., Jagtap, S., Singh, D.P., Karmase, A., Kaur, J., Bhutani, K.K., Boparai, R.K., Premkumar, L.S., Kondepudi, K.K., Bishnoi, M., 2014. Capsaicin-induced transcriptional changes in hypothalamus and alterations in gut microbial count in high fat diet fed mice. J. Nutr. Biochem. 25, 893–902.

Bady, I., Marty, N., Dallaporta, M., Emery, M., Gyger, J., Tarussio, D., Foretz, M., Thorens, B., 2006. Evidence from glut2-null mice that glucose is a critical physiological regulator of feeding. Diabetes 55, 988–995.

Bae, J.S., Kim, T.H., Kim, M.Y., Park, J.M., Ahn, Y.H., 2010. Transcriptional regulation of glucose sensors in pancreatic beta cells and liver: an update. Sensors 10, 5031–5033.

Baskaran, P., Krishnan, V., Ren, J., Thyagarajan, B., 2016. Capsaicin induces browning of white adipose tissue and counters obesity by activating TRPV1 channel dependent mechanisms. Br. J. Pharmacol. 173, 2369–2389.

Ben-Chaim, A., Borovsky, Y., Falise, M., Mazourek, M., Kang, B.C., Paran, I., Jahn, M., 2006. QTL analysis for capsaicinoid content in capsicum. Theor. Appl. Genet. 113, 1481–1490.

Bennett, D.J., Kirby, G.W., 1968. Constitution and biosynthesis of capsaicin. J. Chem. Soc. C Org. 0, 442–446.

Bevan, S., Szolcsányi, J., 1990. Sensory neuron-specific actions of capsaicin: mechanisms and applications. Trends Pharmacol. Sci. 11, 330–333.

Bitencourt, S., de Mesquita, F.C., Caberlon, E., da Silva, G.V., Basso, B.S., Ferreira, G.A., de Oliveira, J.R., 2012. Capsaicin induces de-differentiation of activated hepatic stellate cell. Biochem. Cell. Biol. 90, 683–690.

Blum, E., Mazoure, M., O'Connell, M., Curry, J., Thorup, T., Liu, K., Jahn, M., Paran, I., 2003. Molecular mapping of capsaicinoid biosynthesis genes and quantitative trait loci analysis for capsaicinoid content in capsicum. Theor. Appl. Genet. 108, 79–86.

Bosland, P.W., Votava, E.J.P., 2012. Vegetable and Spice Capsicum, second ed. CABI Publishing, Wallingford, UK.

Burcelin, R., Thorens, B., 2001. Evidence that extrapancreatic GLUT2-dependent glucose sensors control glucagon secretion. Diabetes 50, 1282–1289.

Caterina, M.J., 2003. Vanilloid receptors take a TRP beyond the sensory afferent. Pain 105, 5–9.

Caterina, M.J., Julius, D., 2001. The vanilloid receptor: a molecular gateway to the pain pathway. Annu. Rev. Neurosci. 24, 487–517.

Caterina, M.J., Leffler, A., Malmberg, A.B., Martin, W.J., Trafton, J., Petersen-Zeitz, K.R., Koltzenburg, M., Basbaum, A.I., Julius, D., 2000. Impaired nociception and pain sensation in mice lacking the capsaicin receptor. Science 288, 306–313.

Chakrabarti, R., Rajagopalan, R., 2002. Diabetes and insulin resistance associated disorders: disease and the therapy. Curr. Sci. 83, 533–1538.

Chen, L.H., Tuo, B.G., Dong, H., 2016. Regulation of intestinal glucose absorption by ion channels and transporters. Nutrients 8, 1–11.

Choi, J.H., Jin, S.W., Choi, C.Y., Kim, H.G., Lee, G.H., Kim, Y.A., Chung, Y.C., Jeong, H.G., 2017. Capsaicin inhibits dimethylnitrosamine-induced hepatic fibrosis by inhibiting the TGF-β1/Smad pathway *via* peroxisome proliferator-activated receptor gamma activation. J. Agric. Food Chem. 65, 317–326.

Clarke, G., Stilling, R.M., Kennedy, P.J., Stanton, C., Cryan, J.F., Dinan, T.G., 2014. Minireview: gut microbiota: the neglected endocrine organ. Mol. Endocrinol. 28, 1221–1238.

Commerford, S.R., Vargas, L., Dorfman, S.E., Mitro, N., Rocheford, E.C., Mak, P.A., Li, X., Kennedy, P., Mullarkey, T.L., Saez, E., 2007. Dissection of the insulin-sensitizing effect of liver x receptor ligands. Mol. Endocrinol. 21, 3002–3012.

Contreras-Padilla, M., Yahia, E.M., 1998. Changes in capsaicinoids during development, maturation, and senescence of Chile peppers and relation with peroxidase activity. J. Agric. Food Chem. 46, 2075–2079.

Cortright, D.N., Szallasi, A., 2004. Biochemical pharmacology of the vanilloid receptor TRPV1: an update. Eur. J. Biochem. 271, 1814–1819.

Danielsson, A., Ost, A., Lystedt, E., Kjolhede, P., Gustavsson, J., Nystrom, F.H., Stralfors, P., 2005. Insulin resistance in human adipocytes occurs downstream of IRS1 after surgical cell isolation but at the level of phosphorylation of IRS1 in type 2 diabetes. FEBS J. 272, 141–151.

de Sá Mendes, N., de Andrade Gonçalves, E.C.B., 2020. The role of bioactive components found in peppers. Trends Food Sci. Technol. 99, 229–243.

Di Sotto, A., Vecchiato, M., Abete, L., Toniolo, C., Giusti, A.M., Mannina, L., Di Giacomo, S., 2018. *Capsicum annuum* L var Cornetto di Pontecorvo PDO: Polyphenolic profile and *in vitro* biological activities. J. Funct.Foods 40, 679–691.

Díaz-Laviada, I., 2010. Effect of capsaicin on prostate cancer cells. Future Oncol. 6, 1545–1550.

Dinis, P., Charrua, A., Avelino, A., Nagy, I., Quintas, J., Ribau, U., Cruz, F., 2005. The distribution of sensory fibers immunoreactive for the TRPV1 (capsaicin) receptor in the human prostate. Eur. Urol. 48, 162–167.

Dyer, J., Wood, I.S., Palejwala, A., Ellis, A., Shirazi-Beechey, S.P., 2002. Expression of monosaccharide transporters in intestine of diabetic humans. Am. J. Physiol. Gastr. L 282, G241–G248.

Evans, R.M., Barish, G.D., Wang, Y.X., 2004. PPARs and the complex journey to obesity. Nat. Med. 10, 355–361.

Fan, L., Xu, H.Y., Yang, R.G., Zang, Y.F., Chen, J.F., Qin, H., 2019. Combination of capsaicin and capsiate induces browning in 3T3-L1 white adipocytes via activation of the peroxisome proliferator-activated receptor gamma/ beta(3)-adrenergic receptor signaling pathways. J. Agric. Food Chem. 67, 6232–6240.

Fernández-Bedmar, Z., Alonso-Moraga, A., 2016. *In vivo* and *in vitro* evaluation for nutraceutical purposes of capsaicin, capsanthin, lutein and four pepper varieties. Food Chem. Toxicol. 98, 89–99.

Giuffrida, D., Dugo, P., Torre, G., Bignardi, C., Cavazza, A., Corradini, C., et al., 2013. Characterization of 12 capsicum varieties by evaluation of their carotenoid profile and pungency determination. Food Chem. 140, 794–802.

Gram, D.X., Ahren, B., Nagy, I., Olsen, U.B., Brand, C.L., Sundler, F., Tabanera, R., Svendsen, O., Carr, R.D., Santha, P., et al., 2007. Capsaicin-sensitive sensory fibers in the islets of Langerhans contribute to defective insulin secretion in Zucker diabetic rat, an animal model for some aspects of human type 2 diabetes. Eur. J. Neurosci. 25, 213–223.

Hameed, I., Masoodi, S.R., Mir, S.A., Nabi, M., Ghazanfar, K., Ganai, B.A., 2015. Type 2 diabetes mellitus: from a metabolic disorder to an inflammatory condition. World J. Diabetes 6, 598–612.

Hassan, N.M., Yusof, N.A., Yahaya, A.F., Rozali, N.N.M., Othman, R., 2019. Carotenoids of capsicum fruits: pigment profile and health-promoting functional attributes. Antioxidants 8, 469.

Hayman, M., Kam, P.C.A., 2008. Capsaicin: a review of its pharmacology and clinical applications. Curr. Anaesth. Crit. Care 19, 338–343.

Heise, T., Nosek, L., Rønn, B.B., Endahl, L., Heinemann, L., Kapitza, C., Draeger, E., 2014. Lower within-subject variability of insulin detemir in comparison to NPH insulin and insulin glargine in people with type 1 diabetes. Diabetes 53, 1614–1620.

Holst, J.J., 2007. The physiology of Glucagon-like peptide 1. Physiol. Rev. 87, 1409–1439.

Hsu, C.L., Yen, G.C., 2007. Effects of capsaicin on induction of apoptosis and inhibition of adipogenesis in 3T3-L1 cells. J. Agric. Food Chem. 55, 1730–1736.

Ibrahim, M., Jang, M., Park, M., Gobianand, K., You, S., Yeon, S.H., et al., 2015. Capsaicin inhibits the adipogenic differentiation of bone marrow mesenchymal stem cells by regulating cell proliferation, apoptosis, oxidative and nitrosative stress. Food Funct. 6, 2165–2178.

Islam, M.S., Choi, H., 2008. Dietary red chilli (*Capsicum frutescens* L) is insulinotropic rather than hypoglycemic in type 2 diabetes model of rats. Phytother Res. 22, 1025–1029.

Jeong, J.Y., Suresh, S., Park, M.N., Jang, M., Park, S., Gobianand, K., You, S., Yeon, S.H., Lee, H.J., 2014. Effects of capsaicin on adipogenic differentiation in bovine bone marrow mesenchymal stem cell. Asian-Australas. J. Anim. Sci. 27, 1783–1793.

Joyner, M.J., Coyle, E.F., 2008. Endurance exercise performance: the physiology of champions. J. Physiol. 586, 35–44.

Kadowaki, T., Yamauchi, T., 2005. Adiponectin and adiponectin receptors. Endocr. Rev. 26, 439–451.

Kadowaki, T., Yamauchi, T., Kubota, N., Hara, K., Ueki, K., Tobe, K., 2006. Adiponectin and adiponectin receptors in insulin resistance, diabetes, and the metabolic syndrome. J. Clin. Invest. 116, 1784–1792.

Kajimura, S., Saito, M., 2014. A new era in brown adipose tissue biology: molecular control of brown fat development and energy homeostasis. Annu. Rev. Physiol. 76, 225–249.

Kang, C., Zhang, Y., Zhu, X., Liu, K., Wang, X., Chen, M., Wang, J., Chen, H., Hui, S., Huang, L., Zhang, Q., Zhu, J., Wang, B., Mi, M., 2016. Healthy subjects differentially respond to dietary capsaicin correlating with specific gut enterotypes. J. Clin. Endocrinol. Metab. 101, 4681–4689.

Kang, J., Tsuyoshi, G., Han, I.S., Kawada, T., Kim, Y.M., Yu, R., 2010. Dietary capsaicin reduces obesity-induced insulin resistance and hepatic steatosis in obese mice fed a high-fat diet. Obesity 18, 780–787.

Kang, J.H., Kim, C.S., Han, I.S., Kawada, T., Yu, R., 2007. Capsaicin, a spicy component of hot peppers, modulates adipokine gene expression and protein release from obese-mouse adipose tissues and isolated adipocytes, and suppresses the inflammatory responses of adipose tissue macrophages. FEBS Lett. 581, 4389–4396.

Kang, J.H., Tsuyoshi, G., Le Ngoc, H., Kim, H.M., Tu, T.H., Noh, H.J., Kim, C.S., Choe, S.Y., Kawada, T., Yoo, H., et al., 2011. Dietary capsaicin attenuates metabolic dysregulation in genetically obese diabetic mice. J. Med. Food 14, 310–315.

Kantar, M.B., Anderson, J.E., Lucht, S.A., Mercer, K., Bernau, V., Case, K.A., et al., 2016. Vitamin variation in *Capsicum* spp provides opportunities to improve nutritional value of human diets. PLoS One 11, 1–12.

Kawada, T., Hagihara, K., Iwai, K., 1986. Effects of capsaicin on lipid metabolism in rats fed a high fat diet. J. Nutr. 116, 1272–1278.

Kida, R., Yoshida, H., Murakami, M., Shirai, M., Hashimoto, O., Kawada, T., Matsui, T., Funaba, M., 2016. Direct action of capsaicin in brown adipogenesis and activation of brown adipocytes. Cell Biochem. Funct. 34, 34–41.

Kim, C.S., Park, W.H., Park, J.Y., Kang, J.H., Kim, M.O., Kawada, T., Yoo, H., Han, I.S., Yu, R., 2004. Capsaicin, a spicy component of hot pepper, induces apoptosis by activation of the peroxisome proliferator-activated receptor gamma in HT-29 human colon cancer cells. J. Med. Food 7, 267–273.

Kim, K.-T., Rioux, L.-E., Turgeon, S.L., 2014. Alpha-amylase and alpha-glucosidase inhibition is differentially modulated by fucoidan obtained from *Fucus vesiculosus* and *Ascophyllum nodosum*. Phytochemistry 98, 27–33.

Kim, W., Egan, J.M., 2008. The role of incretins in glucose homeostasis and diabetes treatment. Pharmacol. Rev. 60, 470–512.

Koizumi, M., Doi, R., Toyoda, E., Tulachan, S.S., Kami, K., Mori, T., Ito, D., Kawaguchi, Y., Fujimoto, K., Gittes, G.K., 2004. Hepatic regeneration and enforced PDX-1 expression accelerate transdifferentiation in liver. Surgery 136, 449–457.

Kulkarni, R.N., Jhala, U.S., Winnay, J.N., Krajewski, S., Montminy, M., Kahn, C.R., 2011. PDX-1 haploinsufficiency limits the compensatory islet hyperplasia that occurs in response to insulin resistance. J. Clin. Invest. 114, 828–836.

Kwon, D.Y., Kim, Y.S., Ryu, S.Y., Cha, M.R., Yon, G.H., Yang, H.J., Kim, M.J., Kang, S., Park, S., 2013. Capsiate improves glucose metabolism by improving insulin sensitivity better than capsaicin in diabetic rats. J. Nutr. Biochem. 24, 1078–1085.

Laffitte, B.A., Chao, L.C., Li, J., Walczak, R., Hummasti, S., Joseph, S.B., Castrillo, A., Wilpitz, D.C., Mangelsdorf, D.J., Collins, J.L., Saez, E., Tontonoz, P., 2003. Activation of liver X receptor improves glucose tolerance through coordinate regulation of glucose metabolism in liver and adipose tissue. Proc. Natl. Acad. Sci. U.S.A. 100, 5419–5424.

Lee, G.R., Mi, K.S., Yoon, D.J., Kim, A.R., Yu, R., Park, N.H., Han, I.S., 2013. Topical application of capsaicin reduces visceral adipose fat by affecting adipokine levels in high-fat diet-induced obese mice. Obesity 21, 115–122.

Lee, J., Park, S.J., Hong, S.C., Han, J.H., Choi, D., Yoon, J.B., 2016. QTL mapping for capsaicin and dihydrocapsaicin content in a population of *Capsicum annuum* 'NB1' × *Capsicum chinense* 'Bhut Jolokia. Plant Breed. 135, 376–383.

Lu, M., Ho, C., Huang, Q., 2017. Extraction, bioavailability, and bioefficacy of capsaicinoids. J. Food Drug Anal. 25, 27–36.

Luo, Z., Ma, L., Zhao, Z., He, H., Yang, D., Feng, X., Ma, S., Chen, X., Zhu, T., Cao, T., et al., 2012. TRPV1 activation improves exercise endurance and energy metabolism through PGC-1α upregulation in mice. Cell Res. 22, 551–564.

Malik, V.S., Hu, F.B., 2012. Sweeteners and risk of obesity and type 2 diabetes: the role of sugar-sweetened beverages. Curr. Diabetes Rep. 12, 95–203.

Manova, M., Petkova, E., Yordanova, S., Petkova, V., Petrova, G., 2014. Budget impact analysis of adding dapaglifozin to the therapy of diabetes mellitus type 2 in Bulgaria. Value Health 17, A337.

Marion-Letellier, R., Déchelotte, P., Iacucci, M., Gosh, S., 2009. Dietary modulation of peroxisome proliferator-activated receptor gamma. Gut 58, 586–593.

Mori, K., Saito, R., Nakamaru, Y., Shimizu, M., Yamazaki, H., 2016. Physiologically based pharmacokinetic-pharmacodynamic modeling to predict concentrations and actions of sodium-dependent glucose transporter 2 inhibitor canagliflozin in human intestines and renal tubules. Biopharm Drug Dispos. 37, 491–506.

Morino, K., Petersen, K.F., Dufour, S., 2005. Reduced mitochondrial density and increased IRS-1 serine phosphorylation in muscle of insulin-resistant offspring of type 2 diabetic parents. J. Clin. Invest. 115, 3587–3593.

Mostafavian, M., Abdi, A., Mehrabani, J., Barari, A., 2020. Effect of eight weeks of aerobic progressive training with capsaicin on changes in PGC-1α and UPC-1 expression in visceral adipose tissue of obese rats with diet. CMJA 10, 106–117.

Mun, J.M., Ok, H.M., Kwon, O., 2014. Corn gluten hydrolysate and capsaicin have complimentary actions on body weight reduction and lipid-related genes in diet-induced obese rats. Nutr. Res. 34, 458–465.

Nanok, K., Sansenya, S., 2020. α-Glucosidase, α-amylase, and tyrosinase inhibitory potential of capsaicin and dihydrocapsaicin. J. Food Biochem. 44, e13099.

Narang, N., Jiraungkoorskul, W., Jamrus, P., 2017. Current understanding of antiobesity property of capsaicin. Pharm. Rev. 11, 23–26.

Nascimento, P.L.A., Nascimento, T.C.E.S., Ramos, N.S.M., Silva, G.R., Gomes, J.E.G., Falcão, R.E.A., et al., 2014. Quantification, antioxidant and antimicrobial activity of phenolics isolated from different extracts of *Capsicum frutescens* (pimenta malagueta). Molecules 19, 5434–5447.

Nelson, A.J., Ragan, B.G., Bell, G.W., Ichiyama, R.M., Iwamoto, G.A., 2004. Capsaicin-based analgesic balm decreases pressor responses evoked by muscle afferents. Med. Sci. Sports Exerc. 36, 444–450.

Oh, T.W., Ohta, F., 2003. Capsaicin increases endurance capacity and spares tissue glycogen through lipolytic function in swimming rats. J. Nutr. Sci. Vitaminol. 49, 107–111.

Ohnuki, K., Haramizu, S., Oki, K., Watanabe, T., Yazawa, S., Fushiki, T., 2001. Administration of capsiate, a non-pungent capsaicin analog, promotes energy metabolism and suppresses body fat accumulation in mice. Biosci. Biotechnol. Biochem. 65, 2735–2740.

O'Neill, J., Brock, C., Olesen, A.E., Andresen, T., Nilsson, M., Dickenson, A.H., 2012. Unravelling the mystery of capsaicin: a tool to understand and treat pain. Pharmacol. Rev. 64, 939–971.

Osswald, C., Baumgarten, K., Stumpel, F., et al., 2005. Mice without the regulator gene Rsc1A1 exhibit increased Na+-D-glucose cotransport in small intestine and develop obesity. Mol. Cell Biol. 25, 78–87.

Panchal, S.K., Bliss, E., Brown, L., 2018. Capsaicin in metabolic syndrome. Nutrients 10, 630.

Park, S.Y., Kim, J.Y., Lee, S.M., Jun, C.H., Cho, S.B., Park, C.H., Joo, Y.E., Kim, H.S., Choi, S.K., Rew, J.S., 2014. Capsaicin induces apoptosis and modulates MAPK signaling in human gastric cancer cells. Mol. Med. Rep. 9, 499–502.

Parlevliet, E.T., de Leeuw van Weenen, J.E., Romijn, J.A., Pijl, H., 2010. GLP-1 treatment reduces endogenous insulin resistance via activation of central GLP-1 receptors in mice fed a high-fat diet. Am. J. Physiol. Endocrinol. Metab. 299, E318–E324.

Reddy, D.M., Sasikala, P., 2013. Capsaicin and color extraction from different verities of green and red chilli peppers of Andhra Pradesh. Int. J. Adv. Sci. Tech. Res. 2, 554–572.

Reyes-Escogido Mde, L., Gonzalez-Mondragon, E.G., Vazquez-Tzompantzi, E., 2011. Chemical and pharmacological aspects of capsaicin. Molecules 16, 1253–1270.

Richards, P., Parker, H.E., Adrienssens, A.E., Hodgson, J.M., Cork, S.C., Trapp, S., et al., 2014. Identification and characterisation of glucagon-like peptide 1 receptor expressing cells using a new transgenic mouse model. Diabetes 63, 1224–1233.

Rigano, D., Sirignano, C., Taglialatela-Scafati, O., 2017. The potential of natural products for targeting PPARα. Acta Pharm. Sin. B 7, 427–438.

Robyt, J.F., 2008. Starch: structure, properties, chemistry, and enzymology. In: Fraser-Reid, B.O., Tatsuta, K., Thiem, J. (Eds.), Glycoscience: Chemistry and Chemical Biology. Springer Berlin Heidelberg, Heidelberg, pp. 1437–1472.

Rollyson, W.D., Stover, C.A., Brown, K.C., Perry, H.E., Stevenson, C.D., McNees, C.A., Ball, J.G., Valentovic, M.A., Dasgupta, P., 2014. Bioavailability of capsaicin and its implications for drug delivery. J. Contr. Release 196, 96–105.

Saito, M., 2015. Capsaicin and related food ingredients reducing body fat through the activation of TRP and brown fat thermogenesis. Adv. Food Nutr. Res. 76, 1–28.

Sandor, B., Papp, J., Mozsik, G., Szolcsanyi, J., Keszthelyi, Z., Juricskay, I., Toth, K., Habon, T., 2014. Orally given gastroprotective capsaicin does not modify aspirin-induced platelet aggregation in healthy male volunteers (human phase I examination). Acta Physiol. Hung. 101, 429–437.

Sharma, S.K., Vij, A.S., Sharma, M., 2013. Mechanisms and clinical uses of capsaicin. Eur. J. Pharmacol. 720, 55–62.

Shin, M.K., Yang, S.-M., Hana, I.-S., 2020. Capsaicin suppresses liver fat accumulation in high-fat diet-induced NAFLD mice. Anim. Cell Syst. 24, 214–219.

Singh, G., Singh, A., Verma, R.K., Mall, R., Azeem, U., 2018. Synthesis, biological evaluation and molecular docking studies of novel benzimidazole derivatives. Comput. Biol. Chem. 72, 45–52.

Smeets, A.J., Westerterp-Plantenga, M.S., 2009. The acute effects of a lunch containing capsaicin on energy and substrate utilisation, hormones, and satiety. Eur. J. Nutr. 48, 229–234.

Song, J.-X., Ren, H., Gao, Y.-F., Lee, C.-Y., Li, S.-F., Zhang, F., Li, L., Chen, H., 2017. Dietary capsaicin improves glucose homeostasis and alters the gut microbiota in obese diabetic ob/ob mice. Front. Physiol. 8, 602.

Stearns, A.T., Balakrishnan, A., Rhoads, D.B., Tavakkolizadeh, A., 2010. Rapid upregulation of sodium-glucose transporter SGLT1 in response to intestinal sweet taste stimulation. Ann. Surg. 251, 865–871.

Stearns, A.T., Balakrishnan, A., Rounds, J., et al., 2008. Capsaicin-sensitive vagal afferents modulate post-transcriptional regulation of the rat Na+/glucose cotransporter SGLT1. Am. J. Physiol. Gastrointest. Liver Physiol. 294, G1078–G1083.

Szoka, L., Palka, J., 2020. Capsaicin up-regulates pro-apoptotic activity of thiazolidinediones in glioblastoma cell line. Biomed. Pharmacother. 132, 110741.

Tan, S., Gao, B., Tao, Y., Guo, J., Su, Z.Q., 2014. Antiobese effects of capsaicin-chitosan microsphere (CCMS) in obese rats induced by high fat diet. J. Agric. Food Chem. 62, 1866–1874.

Thongnum, K., Chanthai, S., 2018. Inhibitory reactivity of capsaicin with α-amylase and α-glucosidase related to antidiabetes using molecular docking and quantum calculation methods. Orient. J. Chem. 34, 2211–2228.

Tolhurst, G., Heffron, H., Lam, Y.S., et al., 2012. Short-chain fatty acids stimulate glucagon-like peptide-1 secretion *via* the G-protein-coupled receptor FFAR2. Diabetes 2012, 364–371.

Topuz, A., Dincer, C., Özdemir, K.S., Feng, H., Kushad, M., 2011. Influence of different drying methods on carotenoids and capsaicinoids of paprika (Cv, Jalapeno). Food Chem. 129, 860–865.

Tsurugizawa, T., Nogusa, Y., Ando, Y., Uneyama, H., 2013. Different TRPV1-mediated brain responses to intragastric infusion of capsaicin and capsiate. Eur. J. Neurosci. 38, 3628–3635.

Tundis, R., Menichini, F., Bonesi, M., Conforti, F., Statti, G., Menichini, F., Loizzo, M.R., 2013. Antioxidant and hypoglycaemic activities and their relationship to phytochemicals in *Capsicum annuum* cultivars during fruit development. LWT–Food Sci. Technol. 53, 370–377.

Uno, K., Katagiri, H., Yamada, T., Ishigaki, Y., Ogihara, T., Imai, J., Hasegawa, Y., Gao, J., Kaneko, K., Iwasaki, H., Ishihara, H., Sasano, H., Inukai, K., Mizuguchi, H., Asano, T., Shiota, M., Nakazato, M., Oka, Y., 2006. Neuronal pathway from the liver modulates energy expenditure and systemic insulin sensitivity. Science 312, 1656–1659.

Uno, K., Yamada, T., Ishigaki, Y., Imai, J., Hasegawa, Y., Gao, J., Kaneko, K., Matsusue, K., Yamazaki, T., Oka, Y., Katagiri, H., 2012. Hepatic peroxisome proliferator-activated receptor-γ-fat-specific protein 27 pathway contributes to obesity-related hypertension *via* afferent vagal signals. Eur. Heart J. 33, 1279–1289.

Urbina, S.L., Roberts, M.D., Kephart, W.C., Villa, K.B., Santos, E.N., Olivencia, A.M., et al., 2017. Effects of twelve weeks of capsaicinoid supplementation on body composition, appetite and self-reported caloric intake in overweight individuals. Appetite 113, 264–273.

Vadivelu, N., Mitra, S., Narayan, D., 2010. Recent advances in postoperative pain management. Yale J. Biol. Med. 83, 11–25.

van Zonneveld, M., Ramirez, M., Williams, D.E., Petz, M., Meckelmann, S., Avila, T., Bejarano, C., Ríos, L., Peña, K., Jäger, M., et al., 2015. Screening genetic resources of *Capsicum peppers* in their primary center of diversity in Bolivia and Peru. PLoS One 10, e0134663.

Victoria-Campos, C.I., Ornelas-Paz, J.D.J., Ramos-Aguilar, O.P., Failla, M.L., Chitchumroonchokchai, C., Ibarra-Junquera, V., et al., 2015. The effect of ripening, heat processing and frozen storage on the *in vitro* bio-accessibility of capsaicin and dihydrocapsaicin from Jalapeno peppers in absence and presence of two dietary fat types. Food Chem. 181, 325–332.

Walpole, C.S., Bevan, S., Bloomfield, G., Breckenridge, R., James, I.F., Ritchie, T., Szallasi, A., Winter, J., Wrigglesworth, R., 1996. Similarities and differences in the structure-activity relationships of capsaicin and resiniferatoxin analogues. J. Med. Chem. 39, 2939–2952.

Wang, L., Waltenberger, B., Pferschy-Wenzig, E.-M., Blunder, M., Liu, X., Malainer, C., Blazevic, T., Schwaiger, S., Rollinger, J.M., Heiss, E.H., Schuster, D., Kopp, B., Bauer, R., Stuppner, H., Dirsch, V.M., Atanasov, A.G., 2014. Natural product agonists of peroxisome proliferator-activated receptor gamma (PPARγ): a review. Biochem. Pharmacol. 92, 73–89.

Wang, P., Yan, Z., Zhong, J., Chen, J., Ni, Y., Li, L., Ma, L., Zhao, Z., Liu, D., Zhu, Z., 2012. Transient receptor potential vanilloid 1 activation enhances gut glucagon-like peptide-1 secretion and improves glucose homeostasis. Diabetes 61, 2155–2165.

Watcharachaisoponsiri, T., Sornchan, P., Charoenkiatkul, S., Suttisansanee, U., 2016. The α-glucosidase and α-amylase inhibitory activity from different chili pepper extracts. Int. Food Res. J. 23, 1439–1445.

Wenz, T., Rossi, S.G., Rotundo, R.L., Spiegelman, B.M., Moraes, C.T., 2009. Increased muscle PGC-1alpha expression protects from sarcopenia and metabolic disease during aging. Proc. Natl. Acad. Sci. U.S.A. 106, 20405–20410.

Wesolowska, A., Jadczak, D., Grzeszczuk, M., 2011. Chemical composition of the pepper fruit extracts of hot cultivars *Capsicum annuum* L. Acta Sci. Pol. Hortorum Cultus 10, 171–184.

Yamazaki, Y., Kawano, Y., Uebayasi, M., 2008. Induction of adiponectin by natural and synthetic phenolamides in mouse and human preadipocytes and its enhancement by docosahexaenoic acid. Life Sci. 82, 290–300.

Yiangou, Y., Facer, P., Dyer, N.H., Chan, C.L., Knowles, C., Williams, N.S., Anand, P., 2001. Vanilloid receptor 1 immunoreactivity in inflamed human bowel. Lancet 357, 1338–1339.

Yoshioka, T., Inagaki, K., Noguchi, T., et al., 2009. Identification and characterization of an alternative promoter of the human PGC-1alpha gene. Biochem. Biophys. Res. Commun. 381, 537–543.

Zhang, L., Zhou, M., Fang, G., Tang, Y., Chen, Z., Liu, X., 2013. Hypocholesterolemic effect of capsaicinoids by increased bile acids excretion in ovariectomized rats. Mol. Nutr. Food Res. 57, 1080–1088.

Zhang, L.L., Yan Liu, D., Ma, L.Q., Luo, Z.D., Cao, T.B., Zhong, J., et al., 2007. Activation of transient receptor potential vanilloid type-1 channel prevents adipogenesis and obesity. Circ. Res. 100, 1063–1070.

Zhang, S., Ma, X., Zhang, L., Sun, H., Liu, X., 2017. Capsaicin reduces blood glucose by increasing insulin levels and glycogen content better than capsiate in streptozotocin-induced diabetic rats. J. Agric. Food Chem. 65, 2323–2330.

Zhang, S., You, Y., Liud, J., Qin, X., Liu, X., 2018. Hypoglycaemic effect of capsaicinoids via elevation of insulin level and inhibition of glucose absorption in streptozotocin-induced diabetic rats. J. Funct.Foods 51, 94–103.

Zheng, J., Zheng, S., Feng, Q.Y., Zhang, Q., Xiao, X.H., 2017. Dietary capsaicin and its anti-obesity potency: from mechanism to clinical implications. Biosci. Rep. 37, BSR20170286.

CHAPTER

7

Qualitative and quantitative determination of the main components of *Capparis cartilaginea* Decne. infusion by HPLC-UV

Bashaer Alsharif[1,2] and Fabio Boylan[1,3]

[1]*School of Pharmacy and Pharmaceutical Sciences and Trinity Biomedical Sciences, Institute, Trinity College Dublin, Dublin, Ireland;* [2]*Department of Pharmaceutical Sciences, Pharmacy College, Umm Al-Qura University, Makkah, Saudi Arabia;* [3]*Trinity Natural Products Research Centre, NatPro Centre, Trinity College Dublin, Dublin, Ireland*

1. Introduction

The genus *Capparis* (Capparaceae), commonly called Caper, consists of about 250–400 species (Inocencio et al., 2006). *Capparis* are native to the Mediterranean basin and distributed in Europe, Africa, Madagascar, Asia, Australia, and tropical and subtropical areas of Southern America. As a spontaneous plant, capers have a large natural distribution. They grow in arid conditions, in rocky silty soils, stones, and depressions along roadsides, and can sustain adverse climatic conditions without showing any signs of stress. They have been known to human beings since the early days of civilization, and it is stated that the word *Capparis* was coined in the 4th century B.C. by Theophrastus (Inocencio et al., 2006). Pickles made from the flower buds, unripe fruit, or young twigs of Capers with vinegar and salt are sold commercially to be used as condiments (Rivera et al., 2003).

The roots, stems, leaves, buds, fruits, barks, and seeds of *Capparis* are used in traditional medicine to treat diseases such as rheumatism, stomach problems, headache, and toothache. These plants also have immunomodulatory, antioxidant, antibacterial, antifungal, and diuretic properties. In addition, the genus *Capparis* has pharmacological properties and is used to treat diabetes, sclerosis, allergy, as well as fever, and inflammatory diseases (Tlili et al., 2011).

The phytochemistry of the genus *Capparis* is characterized by the presence of a variety of biologically active compounds such as alkaloids, terpenoids, flavonoids, and phenolic compounds. These compounds are responsible for the diverse pharmacological activities exhibited by the plants in this genus. Alkaloids are one of the major classes of phytochemicals found in *Capparis* species. Some of the alkaloids identified in Caper plants include Cappariloside A, Cappariloside B, Cadabicine, Isocodonocarpine, Codonocarpine, and Capparispine (Fig. 7.1). These alkaloids have been shown to exhibit antifungal, antibacterial, antiviral, and antiinflammatory activities.

Phytochemicals for Health. https://doi.org/10.1016/B978-0-443-15366-2.00007-1

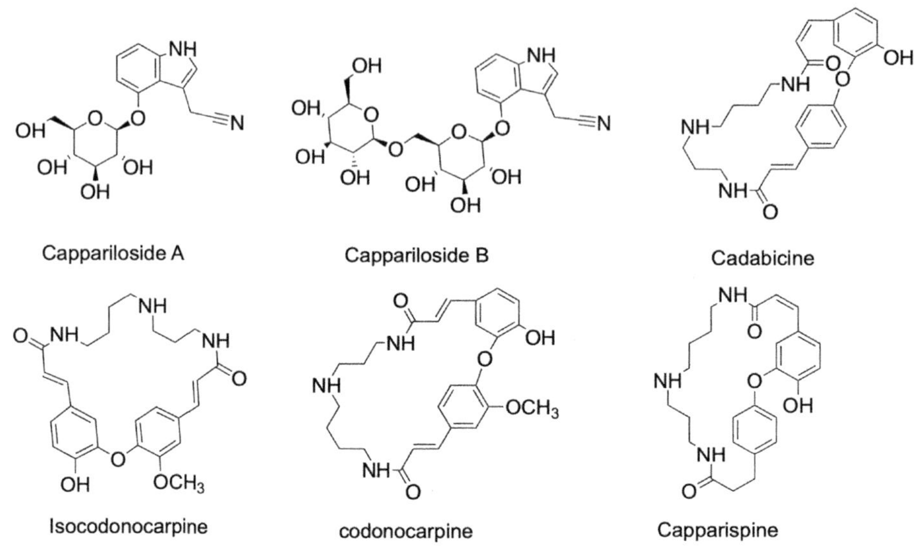

FIGURE 7.1

Chemical structures of alkaloids from the genus *Capparis*.

Terpenoids are also present in *Capparis* species and include compounds such as betulinic acid, oleanolic acid, ursolic acid, and phaseic acid (Fig. 7.2). They have been shown to exhibit antitumor, antiviral, and antiinflammatory activities (Annaz et al., 2022).

Phenolic compounds, including gallic acid, ellagic acid, and caffeic acid, have also been identified in *Capparis* fruits and leaves. Flavonoids represent the largest group of polyphenolic compounds occurring in *Capparis* species. Review papers on the phytochemistry and pharmacology of the *Capparis* genus have been published (Annaz et al., 2022; Gull et al., 2015; Shahrajabian et al., 2021). Three different subclasses of flavonoids have been described from the genus *Capparis*, including flavonols, flavones, and bioflavonoids. Compounds with a flavonol skeleton are the most frequently isolated ones. Among them quercetin, kaempferol, and their glycosides were found in higher numbers in *Capparis* species. Several papers on the isolation of flavonoids from the genus *Capparis have* been published so far, and all of these compounds are documented in Table 7.1.

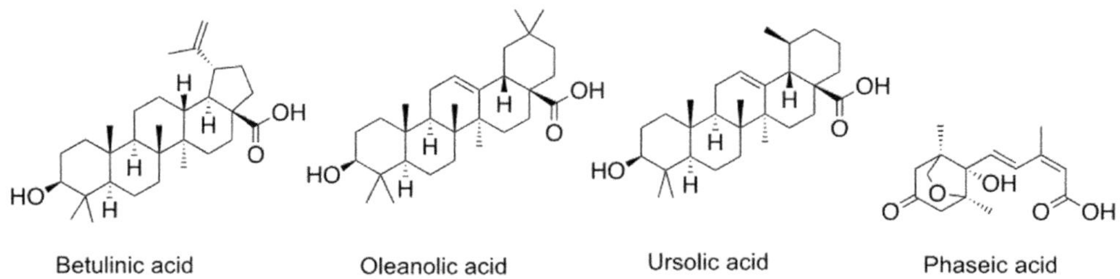

FIGURE 7.2

Chemical structures of terpenoids from the genus *Capparis*.

Table 7.1 Flavonoids isolated from *Capparis* species.

Compound name and structure	Plant species	References
Flavonols		
Kaempferol **R1** = H, **R2** = OH, **R3,4** = H	*C. atamisquea, C. cynophallophora, C. humilis, C. retusa, C. speciosa, C. tweediana, C. himalayensis*	Li et al. (2008); Pelotto and Del Pero Martinez (1998)
Quercetin **R1,2** = OH, **R3,4** = H	*C. atamisquea, C. cynophallophora, C. retusa, C. speciosa, C. tweediana, C. himalayensis*	Li et al. (2008); Pelotto and Del Pero Martinez (1998)
Isorhamnetin **R1** = OH, **R2** = OCH_3, **R3,4** = H	*C. atamisquea, C. retusa, C. salicifolia, C. speciosa, C. tweediana*	Pelotto and Del Pero Martinez (1998)
Rhamnocitrin **R1,4** = OH, **R2** = H, **R3** = OCH_3	*C. tweediana*	Pelotto and Del Pero Martinez (1998)
Rhamnetin **R1,2,4** = OH, **R3** = OCH_3,	*C. tweediana*	Sharaf et al. (2000)
Rhamnazin **R1,4** = OH, **R2,3** = OCH_3	*C. tweediana*	Sharaf et al. (2000)
Flavonol glycosides		
Quercetin -O-3-rutinoside **R1** = O-Rut, **R2-5** = OH	*C. spinosa, C. moonii, C. ovata, C. cartilaginea, C. deserti, C. spinosa* subsp. *C. spinosa* subsp. *rupestris, C. sinaica Veill*	Ahmed et al. (1972a,b); Argentieri et al. (2012); Ibrahim et al. (2013); Sharaf et al. (1997)
Kaempferol-O-3-rutinoside **R1** = O-Rut, **R2** = OH, **R3,4,5** = H	*C. spinosa, C. cartilaginea, C. decidua, C. sicula, C. orientalis*	Inocencio et al. (2000); Sharaf et al. (1997)
Quercetin -O-7-rutinoside **R1** = OH, **R2,3,5** = OH, **R4** = O-Rut,	*C. cartilaginea, C. decidua, C. spinosa*	Sharaf et al. (1997)
Quercetin 3-O-glucoside-7-O-rhamnoside **R1** = O-Glc, **R2,3,5** = OH, **R4** = O-Rha	*C. cartilaginea, C. decidua, C. spinosa*	Sharaf et al. (2000)

Continued

Table 7.1 Flavonoids isolated from *Capparis* species.—cont'd

Compound name and structure	Plant species	References
Quercetin 3-O-glucoside **R1** = O-Glc, **R2-5** = OH,	*C. spinose*	Sharaf et al. (2000)
Quercetin-3-O-[6‴-α-L-rhamnosyl-6″-β-D-glucosyl] -β-D-glucoside **R1** = O-Rha, **R2-5** = OH	*C. spinose*	Sharaf et al. (2000)
Flavones		
Apigenin **R1,3,5** = H, **R2,4,6** = OH	*C. himalayensis*	Li et al. (2008)
Luteolin **R1,2,3,5** = OH, **R4,6** = H	*C. himalayensis*	Zhou et al. (2011)
Wogonin **R1,2,5** = H, **R4,R6** = OH, **R3** = OCH_3	*C. himalayensis*	Zhou et al. (2011)
Oroxylin A **R1,2,3** = H, **R5** = OCH_3,**R4,** **R6** = OH	*C. himalayensis*	Zhou et al. (2011)
Sakuranetin **R1,3,5** = H, **R2,6** = OH, **R4** = OCH_3	*C. spinose*	Zhou et al. (2011)
Flavone glycosides		
Luteolin 7-O-glucoside **R1,2,6** = OH, **R3,5** = H, **R4** = O-Glc	*C. decidua*	Saxena and Goutam (2008)
Acacetin 7-rutinoside **R1,3,5** = H, **R2** = OCH_3, **R4** = O-Rut, **R6** = OH	*C. tener*	Su et al. (2008)
Bioflavonoids		
Iso-ginkgetin **R1** = OH, **R2** = OCH_3	*C. spinose*	Zhou et al. (2011)
Ginkgetin **R1** = OCH_3, **R2** = OH	*C. spinose*	Zhou et al. (2011)

1.1 *Capparis cartilaginea* Decne

Capparis cartilaginea Decne *or Capparis spinosa* subsp. *cartilaginea Botanist (C. cartilaginea)* is a member of the family Capparaceae. The plant has been reported from various locations around the globe. These include French southern territories near Madagascar, Somalia, Yemen, Kenya, Palestine, Djibouti, Oman, Jordan, Chad, Iran, Ethiopia, Egypt, Seychelles, Tanzania, and Saudi Arabia (Fici et al., 2020). According to Inocencio et al. (2006), *C. cartilaginea* Decne is a representative species of the sect Galeatae (Infrageneric rank), characterized by capers having flowers that are always solitary at the leaf axils. In traditional Arabian medicine, *C. cartilaginea* herb is used as a decoction for treating inflammation, pain in the ear, headache, healing of bruises, snakebite, and childbirth (Hamed et al., 2007). Asian literature on indigenous medicine states that it has been traditionally used in paralysis, gout, rheumatism, tuberculosis, and splenomegaly (Said, 1997). It has also been used as an expectorant, anthelmintic, diuretic, and laxative to relieve constipation. In Saudi Arabia, in the Al Qassim region, the *C. cartilaginea* tree is called by the local name (Youssef, 2013). It is traditionally utilized for seven medicinal purposes: antiinflammatory, purgative, tonic, antitumor, wound wash, disinfectant, and sudorific. In the Tabuk region of Saudi Arabia, *C. cartilaginea* has been reported to be utilized traditionally for the treatment of diabetes and rheumatism and also as an anthelminthic (Latif et al., 2014). In Yemen, *C. cartilaginea* has been traditionally used to treat tumors, heat cold, shortness of breath, and itching (Mothana et al., 2009).

Previous phytochemical studies on *C. cartilaginea* have reported the presence of flavonoid glycosides. Four flavonoids were isolated from *C. cartilaginea*, namely kaempferol-3-O-rutinoside, quercetin-3-O-rutinoside, quercetin-7-O-rutinoside, and quercetin-3-O-glucoside-7-O-rhamnoside (Ahmed et al., 1972a, 1972b). GC/MS analysis of glucosinolate aglycones obtained from *C. cartilaginea* from Egypt revealed four types of Isothiocyanates, butyl isothiocyanate (65.03%), 6-methylsulfonylhexyl isothiocyanate (29.86%), 7-methylsulfonylheptyl isothiocyanate (0.066%), and bezylsulfonyl-4bentenyl isothiocyanate (0.914%) (20). GC and GC-MS analysis of the essential oil obtained from the leaves of *C. cartilaginea* Decne from Saudi Arabia identified the major compounds as isopropyl isothiocyanate (31.0%), 2-methylbutanenitrile (21.4%), 2-butyl isothiocyanate (18.1%), isobutyronitrile (15.4%), and 3-methylbutanenitrile (8.2%) (Alsharif et al., 2022).

Several biological activities have been reported for *C. cartilaginea*, including antioxidant, anticancer, antimicrobial, osteoporotic, and larvicidal properties (Abutaha and Al-Mekhlafi, 2014; Hamed et al., 2007; Abdulkarim Moharram et al., 2018). *C. cartilaginea* extracts have also been found to exhibit hypotensive and spasmolytic activities (Gilani and Aftab, 1994). The pharmacological activities of *Capparis* species are attributed mainly to polyphenolic compounds in the genus (Tlili et al., 2011). The lead compounds must be identified and quantified to provide a deeper understanding of the pharmacological activities and further exploit the valuable plant resource. So in this work, the major phenolic compounds of *C. cartilaginea,* including gallic acid **(1)**, quercetin 3-(2G-rhamnosylrutinoside) (Manghaslin) **(2),** kaempferol 3-(2G-rhamnosylrutinoside) (Clitorin) **(3)**, quercetin 3-neohesperidoside **(4)**, kaempferol 3-neohesperidoside **(5)**, quercetin-3-O-rutinoside (rutin) **(6)**, kaempferol-3-O-rutinoside (nicotiflorin) **(7)**, isorhamnetin-3-o-rutinoside (Narcissin) **(8)** (Fig. 7.3) were determined and quantified simultaneously with a simple, rapid, and accurate analysis by reversed-phase liquid chromatography.

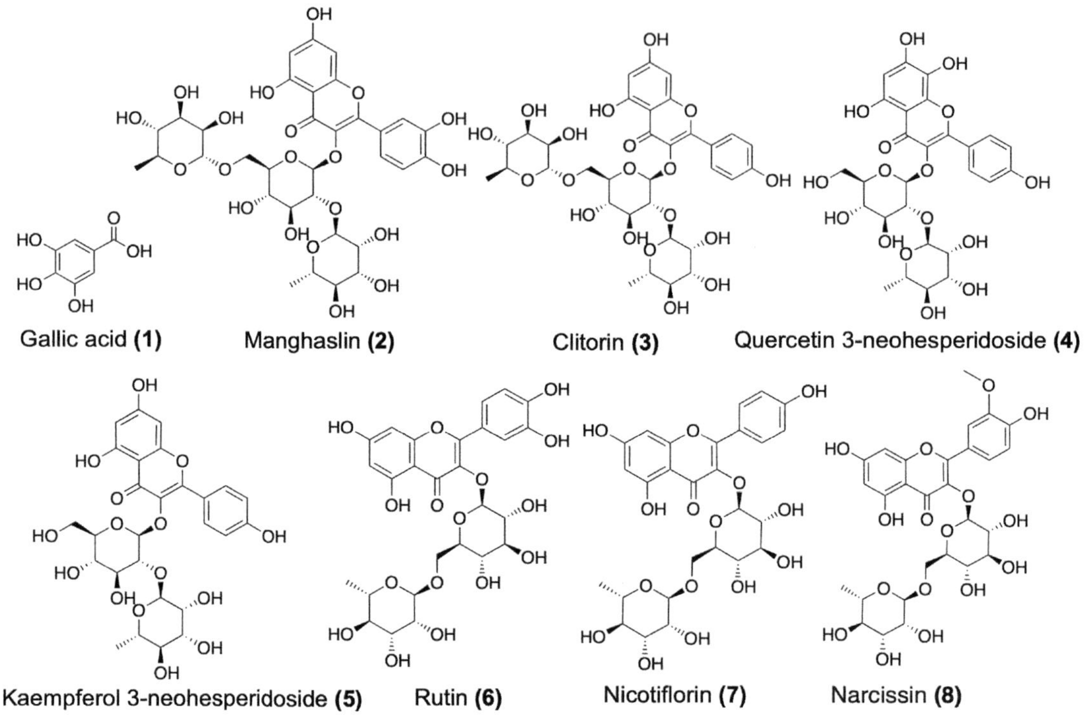

FIGURE 7.3

Chemical structures of phenolic compounds isolated from *C. cartilaginea*.

2. Experimental

2.1 Chemicals and reagents

The standard compounds: rutin, nicotiflorin, and gallic acid, were purchased from Sigma Aldrich (purity ≥95.0%), and narcissin was purchased from extrasynthese (purity ≥99.0%). Manghaslin, clitorin, quercetin 3-neohesperidoside, and kaempferol 3-neohesperidoside were obtained from the leaves of *C. cartilaginea* that were separated from the organic extracts of *C. cartilaginea* leaves infusion using High-Speed Counter-Current Chromatography (HSCCC) previously, and their levels of purity were above 95% by HPLC analysis (Alsharif et al., 2023). All solvents used for the extraction and chromatographic procedures were HPLC-grade solvents purchased from Trinity College Dublin HMF facilities (Dublin, Ireland). Deionized water was prepared using a Millipore water deionizing system.

2.2 Plant material and preparation of *C. cartilaginea* infusion and extract

The fresh leaves of *C. cartilaginea* were collected from Al-Taif region, Saudi Arabia, in June 2019. Professor Ammar Bader identified the plant, and voucher specimens (SA-UK, 2019−12) were deposited in the pharmacognosy lab's herbarium at Umm Al-Qura University. To prepare the infusion, 500 gm of the dried leaves were powdered and infused in 2.5 L of boiling water. Then the infusion was filtered, concentrated under reduced pressure and kept below 4°C for further analysis. The extract was prepared by infusing 500 gm of the dried leaves in 2.5 L hot water. Then, the infusion was filtered and partitioned with ethyl acetate (100 mL) 10 times and subsequently with butanol (100 mL) 10 times. The extracts obtained (5.2 g of ethyl acetate extract and 9.8 g of butanol extract) were combined, concentrated with a rotary evaporator, and kept below 4°C for further analysis.

2.3 Preparation of standard and sample solutions

A stock solution was prepared using 2 mg/mL of each compound and then diluted to the following concentrations: 1280, 640, 320, 160, 80, and 40 (μg/mL). The sample solutions used here were the extract and the infusion in a 1280 μg/mL concentration. All standard and sample solutions were filtered through a 0.45 mm (Minisart CA 0.45 μm 28 mm) before injection into the HPLC equipment.

2.4 Apparatus and conditions

HPLC analyses were performed on a Waters HPLC system with Waters Alliance 2695 fitted with Waters 2487 dual λ_absorbance detector, a 717plus autosampler, and an in-line degasser system. Data were collected using Empower Software. The analytical column used was LiChrospher RP-18 HPLC Column (25 cm × 4.6 mm). Mobile Phase 0.25% (v/v) phosphoric acid in water (solvent A) and 100% methanol (solvent B); Gradient: 0—5 minutes - 40% B, 5—10 minutes - 55% B, 10—15 minutes - 65% B, 15—20 minutes - 50% B, 20—25 minutes - 30% B, 25—30 minutes - 40% B; Flow Rate: 1 mL/minutes; Time of Analysis: 50 minutes; Injector Volume: 10 μL, and detection wavelength at 254 and 280 nm.

2.5 Method validation

To assess the validity of the method, tests were performed following the International Committee of Harmonization (ICH guidelines) regulation for the validation of analytical methods in terms of linearity of the calibration curves, accuracy, precision, stability, limit of detection (LOD), and limit of quantification (LOQ).

The linearity of the standard curves of each analyte was evaluated by plotting the peak areas (y) versus the corresponding concentrations (x, mg/mL). Method accuracy was calculated by spiking triplicate samples of the extract with standard stock solutions of standard compounds of one to eight at the expected analyte concentration, i.e., high, medium, and low. Three injections were made, and the theoretical amount of analyte in the sample and the average percentage of analyte recovered in the spiked solutions were calculated.

The method's precision was evaluated by analyzing known concentrations of the eight compounds in three replicates on the same day (intraday repeatability) and by duplicating the experiments on three consecutive days (interday reproducibility), respectively. The relative standard deviation (% RSD) values for t_R and peak area of each compound were calculated. To evaluate the stability of the solution extracted from *Capparis*, one of the sample solutions was stored at 4°C and analyzed at 0 hours, 6 hours, 12 hours, and 24 hours, respectively. To confirm the repeatability, six different sample solutions were prepared and analyzed, and the RSD values of the peak areas were calculated. Linear regression analysis using the least squares method was applied to the calibration curve of each standard analyte as a function of its concentration. The LOD and LOQ were calculated from the residual standard deviation of the regression (σ) line and the slope of the calibration curve (S) as follows: LOD $= 3.3\ \sigma/S$; LOQ $= 10\ \sigma/S$.

2.6 Statistics

All tests were performed in triplicates. Data are expressed as mean $\pm$ relative standard deviation. Slopes and intercepts of calibration graphs were calculated by linear regression. Correlation coefficient (r2) values were obtained by using the MS Office Excel program.

3. Results and discussion

A simple, rapid, and selective HPLC-UV method was developed to determine gallic acid and seven flavonoids from the extract and infusion of *C. cartilaginea*. To obtain the most efficient separation, the chromatographic conditions were systematically optimized. The mobile phase compositions, the gradient elution, and the detection wavelength were optimized. The results showed that MeOH and 25% phosphoric acid in a water solution with an optimal gradient system (described earlier) narrowed peak shape and improved tailing. Additionally, three UV wavelengths (254 and 280 nm) were selected to modify the separation conditions because most flavonoids can be detected by UV in the 200–300 nm range. The result indicated that the flavonoids from *C. cartilaginea* showed strong UV absorbance at 254 nm. The eight phenolic compounds present in the extract and infusion were effectively separated without obvious interference within 30 minutes (Figs. 7.4 and 7.5).

The quantitative method was validated in terms of linearity of the calibration curves, LOD, LOQ, precision, stability, repeatability, and accuracy. As shown in Table 7.2, all calibration curves exhibited excellent linear regression with coefficients of determination ranging from 0.9901 to 0.9998. The tested concentration ranges adequately covered variations in the amounts of the compounds investigated in the samples. The overall LODs and LOQs were less than 0.99 μg/mL and 3.01 μg/mL, respectively. Intraday and interday precision, repeatability, stability, and recovery of the investigated compounds are shown in Table 7.3. The results demonstrated that the RSD precision values were all $<2.36\%$, and stability was $<3.24\%$. The RSD values of repeatability were less than 1.82%, and below the acceptance criteria value for repeatability (RSD $\leq 2\%$). These results showed excellent repeatability of the analytical method. The overall recoveries lay between 97.51% and 100.19%, accompanied by the RSD values $<3.39\%$, per acceptance criteria of recovery of about 80%–120%. These results suggest that the established method using a simple and easy HPLC–UV is suitable for simultaneously determining the eight compounds in *C. cartilaginea*.

Previous studies on the analysis of Capparis' flavonoids, and polyphenolic compounds, have been reported using HPLC and LC-MS. In agreement with our results (Table 7.4), previous studies on Caper indicated that rutin is overall the main representative flavonoid (Akgül and Özcan, 1999; Giuffrida et al., 2002; Inocencio et al., 2000; Maldini et al., 2016; Tlili et al., 2011). HPLC analysis of *C. spinosa* from Iran showed high levels of rutin in the leaves and fruits; rutin was 16,939.2 $\pm$ 0.01 μg/g in leaves, and in fruit, 1019.52 $\pm$ 0.01 μg/g (Mohebali et al., 2018). Moreover, in *C. spinosa*, infusion rutin was

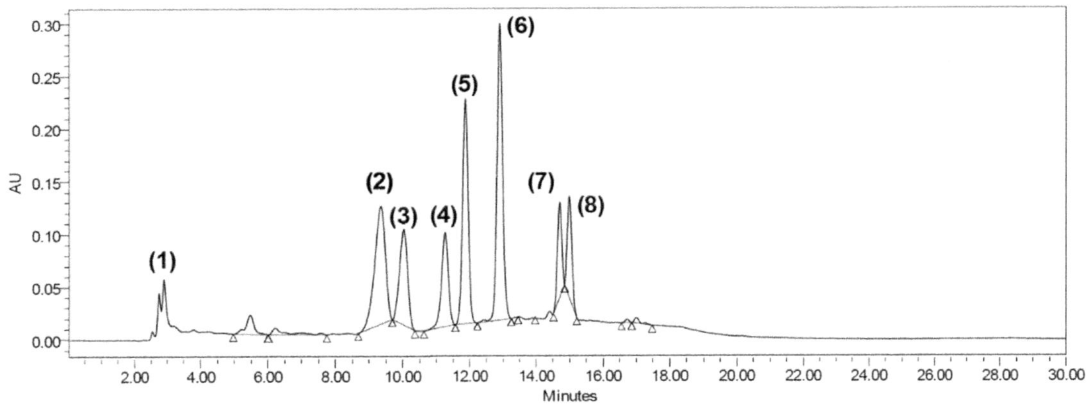

FIGURE 7.4

HPLC chromatogram of *C. cartilaginea* extract at 254 nm.

identified as the dominant flavonoid and represented nearly half of the total phenolics, followed by nicotiflorin and narcissin (Siracusa et al., 2011). Leaves and flower buds of *C. spinosa* from different locations in Tunisia were very rich in rutin (with an average of 1352.71 ± 254.93 and 693.14 ± 281.63 mg/100 g FW, respectively) (Tlili et al., 2010). Rutin was also the major compound in Moroccan, Italian, and Turkish samples of *C. spinosa* buds and aerial parts of *C. sinaica Veill* aerial parts and *C. deserti* from Egypt (Hassan et al., 2020; Hussein et al., 2022; Stefanucci et al., 2018). This compound has remarkable pharmacological and biological properties that have been exploited in human medicine and nutrition, such as antioxidant, antihypertensive, antiapoptotic, antiautophagic, and neuroprotective activities owing to its action against oxidative stress and inflammation (Hassan et al., 2020).

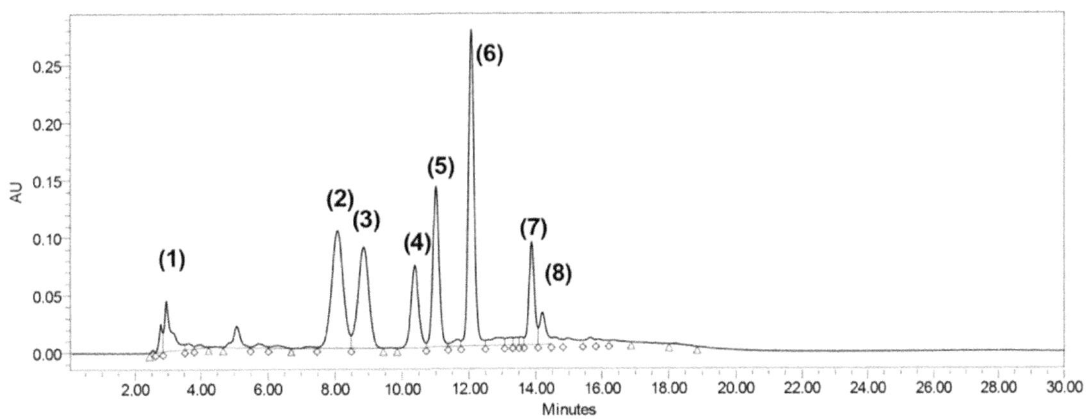

FIGURE 7.5

HPLC chromatogram of *C. cartilaginea* infusion at 254 nm.

Table 7.2 Calibration curve parameters, (LOD), (LOQ) for the standards.

Compounds	Wavelength (nm)	Regression equation[a]	Linear range (μg/mL)	R^2 value	LOD (μg/mL)	LOQ (μg/mL)
1	254	$y = 13331x + 179395$	1280–40	$R^2 = 0.9998$	0.72	2.18
2	254	$y = 9962.6x - 315198$	1280–40	$R^2 = 0.9954$	0.67	2.05
3	254	$y = 8160.2x - 35450$	1280–40	$R^2 = 0.9995$	0.21	1.58
4	254	$y = 9882.7x + 63857$	1280–40	$R^2 = 0.9983$	0.38	1.18
5	254	$y = 3090.6x + 92428$	1280–40	$R^2 = 0.9992$	0.46	1.39
6	254	$y = 21302x - 196240$	1280–40	$R^2 = 0.9998$	0.26	0.79
7	254	$y = 10388x - 75597$	1280–40	$R^2 = 0.9901$	0.99	3.01
8	254	$y = 8933.6x + 556576$	1280–40	$R^2 = 0.9920$	0.96	2.91

[a]*y is the value of the peak area, and x is the value of the investigated compound's concentration (μg/mL).*

Table 7.3 Precision, repeatability, stability, and recovery of the analyzed compounds.

Compounds	Precision[a] (RSD, %)		Repeatability[a] (RSD, %, n = 3)	Stability[a] (RSD, %, n = 3)	Recovery[a] (n = 3)	
Intra-day (n = 3)	Inter-day (n = 3)	Mean, %	RSD, %			
1	0.56	0.86	1.68	2.08	98.79 ± 2.03	1.65
2	1.36	1.04	1.24	1.27	111.97 ± 6.42	0.35
3	0.92	1.26	1.02	0.49	99.58 ± 5.48	2.34
4	1.12	0.85	1.53	2.93	99.61 ± 2.37	0.52
5	1.39	1.26	1.26	1.44	99.45 ± 1.48	1.22
6	2.04	2.07	1.78	1.12	105.59 ± 10.63	0.92
7	1.68	1.57	1.22	1.75	102.80 ± 6.77	1.29
8	0.88	2.36	1.82	3.24	99.65 ± 3.68	1.03

[a]*All Values are mean ± S.D. obtained by triplicate analyses.*

Two new flavonol triglycosides were detected and quantified in *C. cartilaginea* infusion and its extract, named manghaslin and clitorin. The content of manghaslin in the infusion and extract was found to be 170.23 ± 0.44 and 171.71 ± 0.65 μg/g, whereas clitorin was 165.52 ± 0.21 and 165.73 ± 0.32 μg/g, respectively. These compounds have been reported to exert antioxidant, antiinflammatory, and cholinesterase inhibitory effects (Chevalley et al., 2000; Kırmızıbekmez et al., 2019). Flavonol-O-triglycosides such as kaempferol-3-O-(2-rhamnoside)-rutinoside, quercetin-3-*O*-rhamnosyl-rhamnosyl-glucoside, and quercetin-3-O-rutinoside-7-O-rhamnoside were previously identified in different parts of *C. spinosa* plant, such as leaves, seeds, and flower buds (Bakr and El Bishbishy, 2016; Inocencio et al., 2000; Tagnaout et al., 2016). Similarly, glycosides of neohesperidoside were detected and quantified in *C. cartilaginea* for the first time; the content of

Table 7.4 Contents of the investigated compounds in the extract and infusion of *C. cartilaginea*.

Compound[a]	Extract (μg/g)	Infusion (μg/g)
1	14.56 ± 1.12	16.32 ± 0.96
2	171.71 ± 0.65	170.23 ± 0.44
3	165.73 ± 0.32	163.52 ± 0.21
4	84.56 ± 0.52	82.56 ± 0.72
5	201.84 ± 0.68	198.25 ± 0.33
6	248.75 ± 0.49	246.75 ± 0.17
7	123.63 ± 0.21	121.98 ± 0.63
8	67.99 ± 0.32	21.86 ± 0.12

[a]*RSD relative standard deviation.*

quercetin-3-neohesperidoside was 82.56 ± 0.72 μg/g in the infusion and 84.56 ± 0.52 μg/g in its extract, whereas kaempferol 3-neohesperidoside was 198.25 ± 0.33 μg/g in the infusion and 201.84 ± 0.68 μg/g in its extract.

Nicotiflorin and narcissin were reported as major flavonoids in *Capparis* (Bakr and El Bishbishy, 2016; Siracusa et al., 2011). Nicotiflorin was detected in both infusion and extract of *C. cartilaginea* with a content of 121.98 ± 0.63 μg/g and 123.63 ± 0.21 μg/g, respectively (Sharaf et al., 1997). In 1997, this compound was isolated for the first time from the leaves of Egyptian *C. cartilaginea* and reported as one of the main flavonoids in this species (Sharaf et al., 1997). It has been documented to exert diverse biological activities such as α-glucosidase inhibitory, antiglycation, and hepatoprotective activities (Altay et al., 2022).

The detection and quantification of narcissin (8, Rt = 14.19 minutes) in both *C. cartilaginea* infusion and its extract represent the first documentation of an O-methylated flavonoid in this species. In earlier studies, narcissin was reported as one of the dominant flavonoids in the aqueous infusions of *C. spinosa* (Siracusa et al., 2011). Nonetheless, our findings indicate that narcissin was present in a markedly lower amount in *C. cartilaginea* infusion (21.86 ± 0.12 μg/g) relative to the extract (67.99 ± 0.32 μg/g) and was generally the least prevalent flavonoid identified in *C. cartilaginea*.

The presence of gallic acid was determined in low concentrations in both *C. cartilaginea* infusion and its extract. The amount of gallic acid in the infusion was 16.32 ± 0.96 μg/g, while in the extract, it was 14.56 ± 1.12 μg/g. Studies have reported the presence of gallic acid in different parts of Caper plants, such as the fruit, leaf, and stem. For instance, a study analyzed the chemical composition of the fruit of *C. decidua*, and identified gallic acid as one of the major phenolic compounds present (Neeraj and Jangir, 2019). Another study evaluated the antioxidant and antidiabetic potential of methanolic extract of *C. spinosa* leaves and reported the presence of gallic acid in the extract (26.74 mg/100 g) (26.74 mg/100 g) (Tlili et al., 2017). In contrast, no gallic acid was detected in another study that analyzed the methanolic extract of C. spinosa leaves from Turkey (Aliyazicioglu et al., 2013). Gallic acid has been reported to have potential pharmacological properties, such as antioxidant, anticancer, antiinflammatory, and antimicrobial. Moreover, gallic acid can also protect biological cells, tissues, and organs from damage caused by oxidative stress (Bai et al., 2021).

4. Conclusion

The phenolic compounds of *Capparis* serve as significant chemical markers and contribute to the biological potential of this medicinal plant. In the present study, a simple, accurate, and reliable HPLC–UV method was established to determine and quantify the phenolic compounds in *C. cartilaginea* infusion and extract. This analytical procedure would be potentially helpful to improve the quality control of the leaves of *C. cartilaginea* for further development of phytopharmaceuticals with the analyzed plant.

References

Abdulkarim Moharram, B., Al-mahbashi, H.M., Ali, R.S., Ali Aqlan, F., 2018. PhytochemicaL, anti-inflammatory, antioxidant, cytotoxic and antibacterial study of Capparis cartilaginea decnefrom Yemen. Int. J. Pharm. Pharmaceut. Sci. 10 (6), 38. https://doi.org/10.22159/ijpps.2018v10i6.22905.

Abutaha, N., Al-Mekhlafi, A.F., 2014. Evaluation of the safe use of the larvicidal fraction of *Capparis cartilaginea* Decne. against Aedes caspius (Pallas) (Diptera: Culicidae) larvae. Afri. Entomol. 22 (4), 838–846. https://doi.org/10.4001/003.022.0402.

Ahmed, Z.F., Rizk, A.M., Hammouda, F.M., Seif El-Nasr, M.M., 1972a. Glucosinolates of Egyptian Capparis species. Phytochemistry 11 (1), 251–256. https://doi.org/10.1016/S0031-9422(00)89999-9.

Ahmed, Z., Rizk, A., Hammouda, F., El–Nasr, M., 1972b. Phytochemical investigation of Egyptian Capparis species. Planta Med. 21 (02), 156–160. https://doi.org/10.1055/s-0028-1099537.

Akgül, A., Özcan, M., 1999. Some compositional characteristics of capers (*Capparis* spp.) seed oil. Grasas y Aceites 50 (1), 49–52. https://doi.org/10.3889/gya.1999.v50.i1.635.

Aliyazicioglu, R., Eyupoglu, O.E., Sahin, H., Yildiz, O., Baltas, N., 2013. Phenolic components, antioxidant activity, and mineral analysis of *Capparis spinosa* L. Afr. J. Biotechnol. 12 (47), 6643–6649. https://doi.org/10.5897/AJB2013.13241.

Alsharif, B., Babington, G.A., Radulović, N., Boylan, F., 2022. Volatiles of Capparis cartilaginea Decne. from Saudi Arabia. Plants 11 (19), 2518. https://doi.org/10.3390/plants11192518.

Alsharif, B., Hante, N., Govoni, B., Verli, H., Kukula-Koch, W., Jose Santos-Martinez, M., Boylan, F., 2023. *Capparis cartilaginea* Decne (Capparaceae): isolation of flavonoids by high-speed countercurrent chromatography and their anti-inflammatory evaluation. Front. Pharmacol. 14. https://doi.org/10.3389/fphar.2023.1285243.

Altay, A., Yeniçeri, E.K.K., Taslimi, P., Taskin-Tok, T., Yılmaz, M.A., Köksal, E., 2022. Phytochemical analysis and biological evaluation of Hypericum linarioides bosse: in vitro and in silico studies. ChemistrySelect 7 (23). https://doi.org/10.1002/slct.202200039.

Annaz, H., Sane, Y., Bitchagno, G.T.M., Ben Bakrim, W., Drissi, B., Mahdi, I., El Bouhssini, M., Sobeh, M., 2022. Caper (*Capparis spinosa* L.): an updated review on its phytochemistry, nutritional value, traditional uses, and therapeutic potential. Front. Pharmacol. 13. https://doi.org/10.3389/fphar.2022.878749.

Argentieri, M., Macchia, F., Papadia, P., Fanizzi, F.P., Avato, P., 2012. Bioactive compounds from *Capparis spinosa* subsp. *rupestris*. Ind. Crop. Prod. 36 (1), 65–69. https://doi.org/10.1016/j.indcrop.2011.08.007.

Bai, J., Zhang, Y., Tang, C., Hou, Y., Ai, X., Chen, X., Zhang, Y., Wang, X., Meng, X., acid, G., 2021. Pharmacological activities and molecular mechanisms involved in inflammation-related diseases. Biomed. Pharmacother. 133, 110985. https://doi.org/10.1016/j.biopha.2020.110985.

Bakr, R.O., El Bishbishy, M.H., 2016. Profile of bioactive compounds of *Capparis spinosa* var. *aegyptiaca* growing in Egypt. Rev. Brasi. Farmacogn. 26 (4), 514–520. https://doi.org/10.1016/j.bjp.2016.04.001.

Chevalley, I., Marston, A., Hostettmann, K., 2000. New phenolic radical scavengers from saxifraga cuneifolia. Pharmaceut. Biol. 38 (3), 222–228. https://doi.org/10.1076/1388-0209(200007)3831-SFT222.

Fici, S., Souvannakhoummane, K., Lanorsavanh, S., Lamxay, V., 2020. Studies on the genus *Capparis* L. (Capparaceae) in Lao PGR. IV: a new species from the Khammouan Province. Phytotaxa 429 (1), 73–79. https://doi.org/10.11646/phytotaxa.429.1.6.

Gilani, A.H., Aftab, K., 1994. Hypotensive and spasmolytic activities of ethanolic extract of *Capparis cartilaginea*. Phytother Res. 8 (3), 145–148. https://doi.org/10.1002/ptr.2650080305.

Giuffrida, D., Salvo, F., Ziino, M., Toscano, G., Dugo, G., 2002. Initial investigation on some chemical constituents of capers (*Capparis spinosa* L.) from the island of Salina. Ital. J. Food Sci. 14 (1), 25–33.

Gull, T., Anwar, F., Sultana, B., Alcayde, M.A.C., Nouman, W., 2015. Capparis species: a potential source of bioactives and high-value components: a review. Ind. Crop. Prod. 67, 81–96. https://doi.org/10.1016/j.indcrop.2014.12.059.

Hamed, A.R., Abdel-Shafeek, K.A., Abdel-Azim, N.S., Ismail, S.I., Hammouda, F.M., 2007. Egypt chemical investigation of some *Capparis* species growing in Egypt and their antioxidant activity. Evid. Based Complemen. Alternat. Med. 4 (1), 25–28. https://doi.org/10.1093/ecam/nem110.

Hassan, H., Rawlinson, C., Morgan, D., Chen, R., 2020. Pharmacological properties of rutin and its potential uses for Alzheimer's disease. J. Exp. Stroke Transl. Med. 13 (2), 1–12.

Hussein, S.A.M., Mousa, M.A., Ewase, A.S.S., Soliman, A.M., 2022. Studying the chemical composition and hepatoprotective activity of Capparis sinaica veill towards CCl4 injury in Albino. Egypt. J. Chem. 65 (12), 761–767. https://doi.org/10.21608/EJCHEM.2022.143584.6268.

Ibrahim, A.K., Youssef, A.I., Arafa, A.S., Ahmed, S.A., 2013. Anti-H5N1 virus flavonoids from *Capparis sinaica* Veill. Nat. Prod. Res. 27 (22), 2149–2153. https://doi.org/10.1080/14786419.2013.790027.

Inocencio, C., Rivera, D., Alcaraz, F., Tomás-Barberán, F.A., 2000. Flavonoid content of commercial capers (*Capparis spinosa*, *C. sicula* and *C. orientalis*) produced in mediterranean countries. Eur. Food Res. Technol. 212 (1), 70–74. https://doi.org/10.1007/s002170000220.

Inocencio, C., Rivera, D., Obón, M.C., Alcaraz, F., Barreña, J.A., 2006. A systematic revision of *Capparis* section *Capparis* (Capparaceae). Ann. Mo. Bot. Gard. 93 (1), 122–149. https://doi.org/10.3417/0026-6493(2006)93[122:ASROCS]2.0.CO;2.

Kırmızıbekmez, H., İnan, Y., Reis, R., Sipahi, H., Gören, A.C., Yeşilada, E., 2019. Phenolic compounds from the aerial parts of *Clematis viticell*a L. and their in vitro anti-inflammatory activities*. Nat. Prod. Res. 33 (17), 2541–2544. https://doi.org/10.1080/14786419.2018.1448815.

Latif, A., Amer, H.M., Hamad, M.E., Alarifi, S.A.R., Almajhdi, F.N., 2014. Medicinal plants from Saudi Arabia and Indonesia: in vitro cytotoxicity evaluation on Vero and Hep-2 cells. J. Med. Plants Res. 8, 1065–1073. https://doi.org/10.5897/JMPR2014.5481.

Li, Y.Q., Yang, S.L., Li, H.R., Xu, L.Z., 2008. Two new alkaloids from *Capparis himalayensis*. Chem. Pharm. Bull. 56 (2), 189–191. https://doi.org/10.1248/cpb.56.189China.

Maldini, M., Foddai, M., Natella, F., Addis, R., Chessa, M., Petretto, G.L., Tuberoso, C.I.G., Pintore, G., 2016. Metabolomic study of wild and cultivated caper (*Capparis spinosa* L.) from different areas of Sardinia and their comparative evaluation. J. Mass Spectrom. 51 (9), 716–728. https://doi.org/10.1002/jms.3830.

Mohebali, N., Shahzadeh Fazeli, S.A., Ghafoori, H., Farahmand, Z., MohammadKhani, E., Vakhshiteh, F., Ghamarian, A., Farhangniya, M., Sanati, M.H., 2018. Effect of flavonoids rich extract of *Capparis spinosa* on inflammatory involved genes in amyloid-beta peptide injected rat model of Alzheimer's disease. Nutr. Neurosci. 21 (2), 143–150. https://doi.org/10.1080/1028415X.2016.1238026.

Mothana, R.A., Lindequist, U., Gruenert, R., Bednarski, P.J., 2009. Studies of the in vitro anticancer, antimicrobial and antioxidant potentials of selected Yemeni medicinal plants from the island Soqotra. BMC Compl. Altern. Med. 9 (1). https://doi.org/10.1186/1472-6882-9-7.

Neeraj, N.V., Jangir, S.S.B.L., 2019. Quality assessment and phytochemical profile of *Capparis decidua* (Forssk) Edgew. Pharma Innov. 8, 424–430.

Pelotto, J.P., Del Pero Martínez, M.A., 1998. Flavonoid aglycones from Argentinian *Capparis* species (Capparaceae). Biochem. Systemat. Ecol. 26 (5), 577–580. https://doi.org/10.1016/S0305-1978(97)00122-1.

Rivera, D., Inocencio, C., Obón, C., Alcaraz, F., 2003. Review of food and medicinal uses of *Capparis* L. Subgenus *Capparis* (Capparidaceae). Econ. Bot. 57 (4), 515–534. https://doi.org/10.1663/0013-0001(2003)057[0515:ROFAMU]2.0.CO;2.

Said, M., 1997. Hamdard pharmacopoeia of eastern medicine. In: Indian Medical Science Series. Sri Satguru Publications, Delhi.

Saxena, V., Goutam, A., 2008. Isolation and study of the flavone glycoside; luteolin-7-O-β-D-glucopyranoside from the seeds of the *Capparis decidua* (Forsk.). Int. J. Chem. Sci. 6 (1), 7–10.

Sharaf, M., El-Ansari, M.A., Saleh, N.A.M., 1997. Flavonoids of four Cleome and three Capparis species. Biochem. Systemat. Ecol. 25 (2), 161–166. https://doi.org/10.1016/S0305-1978(96)00099-3.

Sharaf, M., El-Ansari, M.A., Saleh, N.A.M., 2000. Quercetin triglycoside from *Capparis spinosa*. Fitoterapia 71 (1), 46–49. https://doi.org/10.1016/S0367-326X(99)00116-1.

Shahrajabian, M.H., Sun, W., Qi, C., 2021. Plant of the millennium, Caper (*Capparis spinosa* L.), chemical composition and medicinal uses. Bull. Natl. Res. Cent. 45 (1). https://doi.org/10.1186/s42269-021-00592-0.

Siracusa, L., Kulisic-Bilusic, T., Politeo, O., Krause, I., Dejanovic, B., Ruberto, G., 2011. Phenolic composition and antioxidant activity of aqueous infusions from *Capparis spinosa* L. and *Crithmum maritimum* L. before and after submission to a two-step in vitro digestion model. J. Agric. Food Chem. 59 (23), 12453–12459. https://doi.org/10.1021/jf203096q.

Stefanucci, A., Zengin, G., Locatelli, M., Macedonio, G., Wang, C.K., Novellino, E., Mahomoodally, M.F., Mollica, A., 2018. Impact of different geographical locations on varying profile of bioactives and associated functionalities of caper (*Capparis spinosa* L.). Food Chem. Toxicol. 118, 181–189. https://doi.org/10.1016/j.fct.2018.05.003.

Su, D.M., Tang, W.Z., Yu, S.S., Liu, Y.B., Qu, J., Yu, D.Q., 2008. Water-soluble constituents from roots of Capparis tenera. Zhongguo Zhongyao Zazhi 33 (9), 1021–1023.

Tagnaout, I., Zerkani, H., Mahjoubi, M., Bourakhouadar, M., Alistiqsa, F., Bouzoubaa, A., Zair, T., 2016. Phytochemical study, antibacterial and antioxidant activities of extracts of *Capparis spinosa* L. Int. J. Pharmacogn. Pharm. Res. Morocco 8 (12), 1993–2005.

Tlili, N., Khaldi, A., Triki, S., Munné-Bosch, S., 2010. Phenolic compounds and vitamin antioxidants of caper (*Capparis spinosa*). Plant Foods Hum. Nutr. 65 (3), 260–265. https://doi.org/10.1007/s11130-010-0180-6.

Tlili, N., Elfalleh, W., Saadaoui, E., Khaldi, A., Triki, S., Nasri, N., 2011. The caper (*Capparis* L.): ethnopharmacology, phytochemical and pharmacological properties. Fitoterapia 82 (2), 93–101. https://doi.org/10.1016/j.fitote.2010.09.006.

Tlili, N., Feriani, A., Saadoui, E., Nasri, N., Khaldi, A., 2017. Capparis spinosa leaves extract: source of bioantioxidants with nephroprotective and hepatoprotective effects. Biomed. Pharmacother. 87, 171–179. https://doi.org/10.1016/j.biopha.2016.12.052.

Youssef, R.S., 2013. Medicinal and non-medicinal uses of some plants found in the middle region of Saudi Arabia. J. Med. Plants Res. 7 (34), 2501–2513.

Zhou, H.F., Xie, C., Jian, R., Kang, J., Li, Y., Zhuang, C.L., Yang, F., Zhang, L.L., Lai, L., Wu, T., Wu, X., 2011. Biflavonoids from caper (*Capparis spinosa* L.) Fruits and their effects in inhibiting NF-kappa B activation. J. Agric. Food Chem. 59 (7), 3060–3065. https://doi.org/10.1021/jf105017j.

CHAPTER 8

Naturally occurring coumarins and derivatives may have potential for the treatment of Alzheimer's disease

Ibukun O. Oresanya[1], Nurten Abaci Kaplan[1] and Ilkay Erdogan Orhan[2,3]

[1]*Department of Pharmacognosy, Faculty of Pharmacy, Gazi University, Ankara, Turkey;* [2]*Department of Pharmacognosy, Faculty of Pharmacy, Lokman Hekim University, Ankara, Turkey;* [3]*Full Member of Turkish Academy of Sciences (TÜBA), Vedat Dalokay Cad., Ankara, Turkey*

1. Introduction

Coumarins are important class of secondary metabolites, which were firstly isolated from *Dipteryx odoranta* Wild (tonka beans, Fabaceae) (Ehlers et al., 1995). Coumarins are known for their pleasant vanilla-like odor and are naturally present in varying concentrations in many higher plants such as liquorice, cinnamon, apricot, sweet clover, strawberry, lavender, and cherry (Akeson et al., 1963; Blahová and Svobodová, 2012; Brown, 1962; Luo et al., 2016; Miller et al., 1996). They are found in quite a number of plant families, but most abundant in Apiaceae and Rutaceae families. In plants, coumarins function as a hormone and are involved in the plant stress-response system like some other secondary metabolites. Animals are also discouraged from eating plants containing coumarins because of the bitter taste (Bourgaud et al., 2006; Menezes and Diederich, 2019).

Generally, coumarins are unsaturated lactones, where a benzene ring is fused to an α-pyrone ring. They have a basic 2*H*-1-benzopyran-2-one skeleton and synthesized through phenylpropanoid pathway (Vogt, 2010). The subclasses of coumarins are simple coumarins, which can be hydroxylated, methoxylated, alkylated, or alkoxylated (e.g., umbelliferone, scopoletin, esculetin, scoparone, fraxetin, and daphnetin); furanocoumarins, where a five-membered furan ring attached to the coumarin nucleus can either be linear (e.g., psoralen, xanthotoxin, bergapten and isopimpinellin) or angular (e.g., angelicin, sphondin, and pimpinellin); dihydrofuranocoumarins (e.g., felamidine, rutaretin, marmesin); biscoumarin (e.g., dicoumarol); and pyranocoumarins (e.g., xanthyletin, decursin, decursinol), which are derived from the same pathway, whereas the most common phenylcoumarins (e.g., dispardiol, disparinol D, mammea A) are derived from the metabolism of isoflavone (Flores-Morales et al., 2023; Skalicka-Woźniak et al., 2016). First, phenylalanine, which is formed through the shikimic acid pathway, is converted by phenylalanine ammonia lyase (PAL) to give *trans*-cinnamic acid, which in turn leads to the formation of 4′-coumaroyl-S-CoA, an important intermediate in this pathway (Vogt, 2010). This crucial intermediate can then be transformed to many classes of

Phytochemicals for Health. https://doi.org/10.1016/B978-0-443-15366-2.00008-3

phenylpropanoids. The biosynthesis of coumarin from 4′-coumaroyl-S-CoA occurs through 6′-hydroxylation, *trans* > *cis* isomerization of the exocyclic double bond, and the final lactonization/cyclization reaction. A first and crucial step of coumarin biosynthesis is the 6′ (ortho) hydroxylation catalyzed by the 2-oxoglutarate-dependent dioxygenase F6′H1 (Bourgaud et al., 2006; Shimizu, 2014). Cytochromes P450 (CYPs) play an important role in the metabolic biotransformation of coumarins. For example, CYP2A6 is involved in the metabolic transformation of coumarin to 7-hydroxycoumarin in human liver microsomes (Stefanachi et al., 2018).

Coumarins have been studied for their numerous pharmacological activities such as antibacterial, anticoagulant, anticancer, antiviral, antidiabetic, antiinflammatory (Al-Warhi et al., 2020; Kontogiorgis et al., 2012), which is attributed to its oxaheterocyclic ring that allows easy binding to many protein targets. The lactone group present in coumarins also gives it the ability to make strong polar binding, such as hydrogen bonds and dipole–dipole interactions, and, sometimes, to acylate target proteins, as claimed for the covalent mechanism of inhibition of some enzymes. Coumarins have been reported to inhibit monoamine oxidase and cholinesterase enzymes, which are some key enzymes involved in neurodegenerative diseases. In various studies, coumarins were demonstrated to be involved in multifaceted mechanisms in the pathogenesis of Alzheimer's Disease (AD), especially the inhibition of cholinesterase and amyloid-beta aggregation. Therefore, this chapter focuses on recent studies on coumarins involving strategies to ameliorate AD as well as molecular docking studies.

2. Prevalence and pathogenesis of AD

2.1 Prevalence

AD is a progressive and irreversible neurological disease that is fatal in the course of time. AD is a broad term for a loss of memory and other mental faculties severe enough to interfere with everyday living. Alois Alzheimer, a German neuropathologist, diagnosed AD in 1906, which was later named after his surname. AD has been recognized as the most frequent kind of dementia among older adults since the beginning of the 21st century, accounting for 60%–70% of dementia cases worldwide. According to a 2015 Alzheimer's Association report, AD is classified into three types, which are autosomal-dominant AD (ADAD), early-onset AD (EOAD), and late-onset AD (LOAD). The majority of AD cases are caused by LOAD, which develops after the age of 65. EOAD occurs in 5% of instances when the illness manifests itself before the age of 65. ADAD is an uncommon variant of AD, in which the illness is inherited in an autosomal dominant way at a young age (Anand et al., 2017; Bateman et al., 2011; Chatila and Bradshaw, 2023; Duan et al., 2021; Klyucherev et al., 2022; Patil et al., 2020). Although AD is not an age-related dementia, it generally comes up after the age of 65 and has a 5% prevalence among individuals under the age of 65 (Torok et al., 2022). In 2020, healthcare expenses for dementia care reached up to approximately $300 billion. AD currently costs the United States (US) economy more than cancer or cardiovascular disease. Spending on AD is expected to reach $2 trillion per year in the USA by 2030. The high incidence of AD imposes a significant social cost, particularly on carers, and raises medical health expenditures; as a result, it is recognized as a significant economic concern, particularly in developing nations.

According to World Health Organization (WHO) data from 2016, it was predicted that more than 47.5 million individuals worldwide are living with dementia in 2016. According to the same analysis, this amount is expected to climb to 75.6 million by 2030. According to 2017 data from the Centers for

Disease Control and Prevention, AD is one of the top five causes of mortality among the elderly in the USA. According to the Alzheimer's Association's 2014 prevalence investigation, an additional case of AD is predicted to develop every 33 s by 2050. Moreover, India, Brazil, and China account for more than two-thirds of all dementia sufferers. This rate is predicted to grow to three-quarters by 2050. Therefore, more research is needed to understand the cause of the high prevalence of the disease in this location, which is currently unclear (Stanciu et al., 2019).

Despite the fact that women live longer than males, the disease is significantly more frequent among women (Soto-Lara et al., 2023). According to the findings of research looking into why AD is more frequent in women, it is believed that hormonal changes in women through events such as menopause and delivery raise the risk (Cui et al., 2023). A study found that women with AD had considerably quicker hippocampus shrinkage rates than males (Ardekani et al., 2016). Another study that looked at gender variations in the development of AD found that yearly atrophy rates in male patients with the disease were substantially slower than in females (Sampedro et al., 2015). Another study discovered that the degree of cerebral white matter lesion in women with AD was considerably higher than in males (Sawada et al., 2000). Estrogen increases the production and deposition of beta amyloid plaques by causing tau phosphorylation and accumulation. Women have more significant neuroinflammation, mitochondrial dysfunction, oxygen stress, apoptosis, autophagy, and blood–brain barrier (BBB) failure as a result of all of these factors. As a result, the severity and prevalence of AD in women rise (Cui et al., 2023). Gender determines illness progression as well as comorbidities. Women with AD are more likely to have hypertension, diabetes, depression, stroke or transient ischemic attack, and heart disease (Svob Strac et al., 2021).

To put a stop to the fast rise of AD, researchers are looking for novel chemicals that can be utilized in treatment of the disease. Natural coumarins obtained from plants are also regarded as a natural source that may be employed in the treatment of AD and the alleviation of its symptoms. The actual reasons of progressive nature of AD are yet unknown. However, some factors, such as abnormal tau protein phosphorylation, metal dyshomeostasis, calcium metabolism changes, oxidative stress, neuroinflammation, abnormal energy metabolism, and protein processing, i.e., unwanted amyloid beta (Aβ formation and aggregation), are reported to be imperative in AD pathogenesis (Anand et al., 2017; Patil et al., 2020). In AD patients, the production of cytokines that impact neuroinflammation such as tumor necrosis factor (TNF)-α, interleukin (IL)-1β, and nitric oxide (NO) rises, and so inflammation develops in neurons (Svob Strac et al., 2021).

2.2 Pathogenesis of AD

The etiology of AD is not clearly known; however, multiple hypotheses have been proposed concerning how it progresses. These hypotheses are now practiced in the treatment of AD. The hypotheses including the cholinergic hypothesis, amyloid-beta (Aβ) hypothesis, tau hypothesis, and oxidative stress hypothesis are all theories on the molecular mechanism of AD (Anand et al., 2017; Nascimento et al., 2022; Torok et al., 2022; Verma et al., 2022). One of the most accepted explanations describing the development of AD is the cholinergic hypothesis. According to cholinergic hypothesis, acetylcholine (ACh) and butyrylcholine (BCh) levels in the brain have been found to be lower in individuals with AD. It is a notion based on research done in the 1970s on the brains of patients with AD after they passed. The amount of choline acetyltransferase (CAT), which is responsible for ACh or BCh synthesis, was found to be lower in the brains of patients with AD. ACh is a neurotransmitter present in the brain that controls memory, learning, and cognitive functioning. In AD patients, unbalanced ACh release, transport, and related receptor malfunction induce behavioral abnormalities and cognitive impairment. These findings imply that cholinergic neuron ablation causes cholinergic function loss. As a result, a cholinergic replacement might be the initial focus for overcoming AD. Currently, the most commonly used drugs in clinic for the treatment of AD are based on the inhibition of the related enzymes, which are acetylcholinesterase (AChE) and butyrylcholinesterase (BChE). Donepezil and rivastigmine are the current drugs for the management of AD, and they increase ACh levels in the brain's nervous tissue (Anand et al., 2017; Dwyer et al., 2009; Pejchal et al., 2011; Wilcock et al., 1982).

Another explanation for the genesis of AD is Aβ hypothesis. According to this hypothesis, AD is characterized by a dense accumulation of neurotoxic Aβ plaques and tau protein neurofibril tangles in the brain's nerve tissue. Aβ plaques are primarily generated by two enzymes, e.g., β-secretase (BACE1) and γ-secretase, consecutively breaking down a big precursor protein such as APP (amyloid precursor protein). Another hypothesis is tau hypothesis, where tau protein plays a crucial role in microtubule stability by assisting in cell integrity. It has six isoforms, each with three major binding sites: An *N*-terminal bulge, a C-terminal microtubule binding domain, and a shorter tail sequence. It is discovered phosphorylated in the axonal membrane in its natural condition. Tau protein hyperphosphorylation induces microtubule instability and neurofibrillary tangles (NFT) formation. In AD, factors involving tau protein hyperphosphorylation include the production of nuclear factor-kappaB (NF-kB), tumor necrosis factor-alpha (TNF-α), and interleukins (ILs), all of which induce neurotic damage in the brain. Tau is converted into insoluble intracellular NFTs after being hyperphosphorylated, eventually leading to neurodegeneration. NFTs are made up mostly of tau protein, a microtubule-associated protein (MAP) that binds microtubules in cells to enable neuronal trafficking. Tau separates from microtubules and collects into tangles throughout the development of AD, limiting transport and inducing microtubule fragmentation (Anand et al., 2017; Dwyer et al., 2009; Maccioni et al., 2010; Nisbet et al., 2015; Pejchal et al., 2011). The oxidative stress theory also explains the genesis of AD. The levels of free oxygen radicals (ROS) and free nitrogen radicals (RNS) in the brains of patients with AD were discovered to be extremely high. After entering the mitochondria, it disrupts the electron transport chain (ETC) mechanism, resulting in free radical formation, oxidative stress, and, ultimately, neurodegeneration (Butterfield and Boyd-Kimball, 2004). Excessive oxygen consumption in the brain increases free radical formation, and decreasing levels of antioxidant enzymes such as superoxide dismutase (SOD), catalase (CAT), glutathione peroxidase (GPx), and glutathione reductase (GR) can additionally contribute to development of neurodegeneration (Verma et al., 2022).

Monoamine oxidases (MAOs) are a class of enzymes that are responsible for the oxygen-mediated deamination of endogenous and exogenous amine groups. MAO inhibitors are gaining popularity not only for their functions in monoamine neurotransmitter metabolism and oxidative stress, but also for their extra neuroprotective and neurorescue actions against AD (Jeong et al., 2006). MAO-A and treatment of neuropsychiatric illnesses such as resistant depression and atypical depression. MAO-B inhibitors (such as selegiline) are used to treat neurodegenerative illnesses such as Parkinson and AD. MAO-A selective inhibitors are used to treat mental disorders including anxiety and depression, whereas MAO-B selective inhibitors are frequently used to treat neurodegenerative illnesses such as AD and Parenchyma disease. Because depression is common in Alzheimer's patients, there are therapeutic options that include both inhibitors (Behl et al., 2021; Carreiras and Marco, 2004; Verma et al., 2022). MAO-B expression rises about 4-fold in neural tissues of the aged, where MAO-B catalytic activity increases in the brains of AD patients, increasing neurotransmitter consumption and neuronal degeneration. As a result, MAO-B inhibition is especially addressed in AD. Monoamine neurotransmitters (i.e., serotonin, dopamine, etc.) stored in neuron terminals rise, when MAO-B is inhibited (Anand et al., 2017; Belov Kirdajova et al., 2020; Klyucherev et al., 2022; Nascimento et al., 2022; Stanciu et al., 2019; Svob Strac et al., 2021; Torok et al., 2022; Verma et al., 2022; Zhang et al., 2021).

Another neurochemical explanation describing the emergence of AD is *N*-methyl-D- aspartate (NMDA)-mediated glutamatergic hypothesis, which states that glutamate is a neurotransmitter that activates NMDA receptors and plays an active role in memory and cognition. However, in some circumstances, such as AD, glutamate overstimulation of NMDA receptors causes neurons to malfunction due to excitotoxicity (Anand et al., 2017; Francis, 2005; Verma et al., 2022). Memantine, a glutamatergic antagonist used clinically to treat AD, was developed based on this hypothesis and protects the brain tissue from glutamate-mediated excitotoxicity.

Microglia are the brain's major innate immune cells that respond nonspecifically against any foreign antigen and are abundant in the brains of persons with AD. It is also hypothesized to play a function in the early stages of AD pathogenesis. Microglia have both proinflammatory and antiinflammatory properties. Although antiinflammatory microglial phenotype of microglia predominates during the outset of AD, it promotes a switch to the proinflammatory microglial type as the disease progresses. Recent research suggests that in the treatment of AD, particular microglia toll-like receptor (TLR) agonists and antagonists are coupled to enhance microglial activation. Microglia are the brain's primary innate immune cells that react nonspecifically to any foreign antigen and are numerous in Alzheimer's patients' brains. It is also thought to be involved in the early stages of AD pathogenesis. Microglia possess both pro- and antiinflammatory characteristics. Although the antiinflammatory microglial phenotype of microglia predominates at the start of the disease, as the illness advances, it encourages a flip to the proinflammatory microglial type. According to recent study, certain TLR agonists and antagonists are paired to promote microglial activation in the treatment of AD (Fiebich et al., 2018; Svob Strac et al., 2021).

Generally, Aβ plaque aggregation, tau hyperphosphorylation, overexpression of cyclase enzymes, lipid dyshomeostasis, NMDA receptor hyperexcitation, MAO-B hyperactivation, hypometabolism, disruption of the blood—brain barrier (BBB), oxidative stress, hyperactivation of AChE, neurovascular dysfunction, reduced synaptic ACh levels, and neuronal degeneration are all pathophysiological aspects that contribute to the course of AD.

The current medications such as donepezil, rivastigmine, galantamine used for the treatment of AD have several negative effects. It induces bradycardia, for example, by enhancing vagal activation in the sinoatrial and atrioventricular nodes. By increasing stomach acid output, it can induce peptic ulcers. It is especially significant in people who are on nonsteroidal antiinflammatory medicines. Patients may also experience nausea, vomiting, lack of appetite, and dehydration. Drugs such as rivastigmine can increase bronchial secretions, making breathing harder for asthma sufferers (Anand et al., 2017; Dorababu, 2022; Husna Ibrahim et al., 2020; Nascimento et al., 2022; Svob Strac et al., 2021). This is another reason that has necessitated the search for alternative medicines, especially from natural origins, for the treatment of AD.

3. Natural coumarins with cholinesterase and MAO inhibitory activities through in vitro studies

Researchers have continued to look for phytochemicals with phenolic structures or nonalkaloid AChE inhibitors, such as terpenes, coumarins, and other phenolic substances because of the severe toxicity of alkaloidal inhibitors (i.e., rivastigmine, galantamine). In a study conducted by Dall'Acqua et al. (2010), three coumarin-type substances showing inhibition on AChE were isolated from dichloromethane extract of the root of *Ferulago campestris* (Besser) Grecescu (Apiaceae) and their structures were identified as umbelliprenin, coladonin, and coladin. IC_{50} values of each on AChE were 380.1 ± 16.1 μM, 1171.1 ± 8.7 μM, and 602.0 ± 4.2 μM, respectively. In the same study, IC_{50} value of galantamine, which was used as the reference drug, was determined as 6.7 ± 0.2 μM. The obtained coumarins had a mild inhibitory impact on AChE (Dall'Acqua et al., 2010). In another study, coumarins and furanocoumarins (citrusoside A, citrusoside B, citrusoside D, oxypeucedanin, oxypeucedanin hydrate, isoimperatorin, 6′-hydroxy-7-methoxy bergamottin, 6′,7′-dihydroxybergamottin, and bergomottin) were isolated from the hexane and dichloromethane extracts of the fruit peels of the plant *Citrus hystrix* DC. (Rutaceae), which is common in Southeast Asia. The inhibitory effects of isolated coumarins and furanocoumarins on AChE are given in Table 8.1. IC_{50} value of galantamine in the same study was 3.2 ± 0.2 μM. The highest AChE inhibitory activity was observed in 6′-hydroxy-7′-methoxybergamottin and 6′,7′-dihydroxybergamottin (Youkwan et al., 2010).

AChE inhibition by scopolin (IC_{50}: 65.2 ± 1.2 μM) and scopoletin (IC_{50}: 87.1 ± 0.7 μM), which were isolated as coumarin derivatives from the methanolic extract of the underground parts of *Scopolia carniolica* Jaqc. (Solanaceae) using different chromatographic methods. In the same study, galantamine's IC_{50} value was 3.6 ± 0.2 μM. Both isolated coumarins displayed moderate, concentration-dependent, and long-lasting inhibitory activities (Rollinger et al., 2004). In a study on *Mesua elegans* (King) Kosterm (Calophyllaceae), hexane extract prepared from its bark exhibited a significant AChE inhibition. Isolation from the hexane extract of the plant afforded a total of nine different 4-phenylcoumarins, three of which had not been isolated before. Mesuagenins A, C, and D were the newly isolated compounds. Other coumarins isolated from the same plant are isomammeigin, 5,7-dihydroxy-8-(2-methylbutanoyl)-6-[(E)-3,7-dimethylocta-2,6-dienyl]-4-phenyl-2H-chromen-2-one, 5,7-dihydroxy-8-(3-methylbutanoyl)-6-[(E)-3,7-dimethylocta-2,6-dienyl]-4-phenyl-2H-chromen-2-one, mammea A/BA cyclo F, mammea A/BA, whose inhibition values are tabulated in Table 8.1. Among them, mesuagenin B inhibited AChE at the strongest level. In the same study, IC_{50} value of tacrine on AChE as reference drug was determined as 0.074 ± 0.012 μM (Awang et al., 2010).

AChE inhibitory activity directed fractionation was performed on the methanolic extract prepared from the underground parts of *Angelica gigas* Nakai. (Apiaceae). Peucedanone, a new coumarin derivative, was isolated for the first time in addition to 11 known coumarin derivatives. Among the isolated coumarins, decursinol showed the highest inhibitory activity against AChE in vitro with IC_{50} value of 2.8×10^{-5} M (Table 8.1) (Kang et al., 2001). Lee et al. (2004) investigated AChE inhibition of methanolic extract from the powdered branches of *Vaccinium oldhamii* Miq. (Ericaceae) through activity-directed fractionation. As a result of different chromatographic methods, taraxerol (IC_{50}: 79 μM) and scopoletin (IC_{50}: 52 μM) were isolated from the extract as the active substances. Both compounds showed concentration-dependent AChE inhibition. In vitro AChE inhibition of xanthotoxin and isopimpinellin, the furanocoumarins isolated from the chlorophosome extract of *Angelica*

Tables 8.1 Inhibitory effects of natural coumarins on MAO-B.

Natural coumarins	Plants obtained	Inhibition effect on MAO-B (IC_{50} ± S.D.[a] or inhibition% ± S.D.)	References
7-[(R,E)-6′-hydroxy-3′,7′-dimethylocta-2′,7′-dienyloxy]coumarin	*Dictamnusalbus* L.	IC_{50}: 0.5 μM	Jeong et al. (2006)
((7E)-3′,7′-dimethylocta-2′,6′-dienyloxy)coumarin		IC_{50}: 0.6 μM	
Lacinartin	*Zanthoxylum schinifolium* Siebold & Zucc.	IC_{50}: 28.6 μM	Genovese et al. (2009)
Bergapten	*Peucedanum japonicum* Thunb.	IC_{50}: 13.8 μM	Huong et al. (1999)
Psoralen		IC_{50}: 35.8 μM	
Xanthotoxin		IC_{50}: 40.7 μM	
Praeruptorin A		IC_{50}: 27.4 μM	
7-(6′R -hydroxy-3′,7′-dimethyl-2′E, 7′-octadienyloxy) coumarin	*Dictamnus albus* L.	IC_{50}: 0.5 μM	Jeong et al. (2006); Kong et al. (2001)
Auraptene		IC_{50}: 0.6 μM	
Psoralen	*Psoralea corylifolia* L.	IC_{50}: 61.8 μM	Kong et al. (2001)
Isopsoralen		IC_{50}: 12.8 μM	
Osthenol	*Angelica pubescens* Maxim.	IC_{50}: >60.0 μM	Baek et al. (2019)
Bakuchicin	*Psoralea corylifolia* L.	IC_{50}: >60.0 μM	
Scopoletin		IC_{50}: >60.0 μM	
Isoimperatorin		IC_{50}: >60.0 μM	
Umbelliferone	*Angelica decursiva (Miq.)* Franch. & Sav.	IC_{50}: 147.37 ± 3.65 μM	Seong et al. (2019)
6-Formyl umbelliferone		IC_{50}: 15.31 ± 0.68 μM	
8-Formyl umbelliferone		IC_{50}: 81.35 ± 2.59 μM	
Umbelliferone-6-carboxylic acid		IC_{50}: 44.27 ± 5.52 μM	
Esculetin		IC_{50}: >300 μM	
Geiparvarin	*Geijera parviflora* Lindl.	3.65%	Carotti et al. (2002)

[a]*Standard deviation.*

acutiloba (Siebold & Zucc.) Kitag (Apiaceae), was determined. Xanthotoxin (IC_{50}: 0.58 μmol/mL) and isopimpinellin (IC_{50}: 0.32 μmol/mL) inhibited AChE at concentrations of 1.00 μmol/mL by 55.8% and 80.7%, respectively (Miyazawa et al., 2004). 2′-O-Ethylmurrangatin, murranganone, and paniculatin are coumarin derivatives obtained from the methanolic extract of the dried leaves of *Murraya paniculata* (L.) Jack (Rutaceae), where 2′-*O*-ethylmurrangatin was inactive on AChE

(Choudhary et al., 2002). IC_{50} values of murranganone and paniculatin were determined to be 79.14 ± 0.096 μM and 31.65 ± 0.036 μM, respectively. IC_{50} value of the positive control on AChE was 0.041 ± 0.001 μM. In the same study, IC_{50} values were 842.85 ± 13.04 μM, 74.36 ± 0.64 μM, and >100 μM for murranganone and paniculatin, respectively, in BChE inhibition assay. IC_{50} value of eserin used as positive control was 0.857 ± 0.008 μM on BChE. According to the findings, paniculatin had the most inhibitory impact on AChE, whereas murranganone had the greatest effect on BChE.

Orhan et al. (2021) investigated BChE inhibition of coumarin derivatives isolated from the roots of some Apiaceae species. From *Heptaptera anatolica* (Boiss.) Tutin (Apiaceae), badrakemin, 14′-acetoxybadrakemin, badrakemone, colladonin, karatavicinol, and umbelliprenin were isolated. Among them, 14′-acetoxybadrakemone, colladonin acetate, and 14′-acetoxycholladonine were isolated from the *Heptaptera cilicica* (Boiss. & Bal.) Tutin. Osthol, oxypeucedanin, oxypeucedanin hydrate, isoimperatorin, and scopoletin from *Neocryptodiscus papillaris* (Boiss.) Herrnst. & Heyn; deltoin, smyrnioridin, and marmesin from *Petroedmondia syriaca* (Boiss.) Tamirsch were isolated. Their IC_{50} values are given in Table 8.1, while IC_{50} values of galanthamine used as positive control in the same study on AChE and BChE were determined as 4.9 and 60.2 μM, respectively.

MAO-A and B are the enzymes important in the metabolism of intracellular neurotransmitters in central nervous system (CNS). MAO inhibitors have demonstrated clinical usefulness as possible therapeutic targets for a number of neurological and neurodegenerative illnesses for decades. MAO-B inhibitory capacity of coumarins isolated from methanol extract from the aerial parts of *Dictamnus albus* L. (Rutaceae) was examined in vitro (Jeong et al., 2006). The results are presented in Table 8.1. Iproniazid (IC_{50}: 20.8 μM) was used as a positive control. A number of coumarins were isolated from the methanol extract obtained from the roots of *Artemisia capillaris* Thunb. (Asteraceae), and their AChE and BChE inhibitory effects were determined as tabulated in Tables 8.1 and 8.2. Berberine used as a positive control in the same study had an IC_{50} of 0.717 ± 0.01 μM and 7.01 ± 0.31 μM on AChE and BChE, respectively (Ali et al., 2016). IC_{50} values on AChE and BChE of 8-acetyl-7-hydroxyycoumarin isolated from the rhizomes of *Nardostachys jatamansi* (D.Don) DC are given in Tables 8.1 and 8.2 (Sharma et al., 2022).

4. In vivo studies on coumarins against AD

Over the years, researchers have tried to develop animal models to closely represent the human AD. Mouse models are one of the most important research tools for finding new treatments for AD (Hall and Roberson, 2012). Actually, no single mouse model exhibits all features of AD currently due to the complexity of the pathophysiology of AD, which includes cognitive and behavioral deficits, amyloid plaques and neurofibrillary tangles, gliosis, synapse loss, axonopathy, neuron loss, neuroinflammation, and neurodegeneration. However, various mouse models have been useful and were able to develop AD phenotypes to varying degrees and in different combinations. More also, AD mouse model may be a better model for early preclinical stages of AD than the later dementia stages, which suggests that early treatment would be beneficial clinically (Zahs and Ashe, 2010). AD rat models are also used in *in vivo* AD studies and offer some advantages over mice in their display of better social behavior and more accurate motor coordination (Whishaw et al., 2001). Transgenic AD animal models have been found very convenient as the animals are made to reflect various aspects of AD such as cognitive deficit, inflammation, amyloid, and tau aggregation (Do Carmo and Cuello, 2013). Interestingly,

Table 8.2 Natural coumarins tested in *in vivo* studies against AD.

Natural coumarins	Plant source	Method/effect	Doses/animals	References
Imperatorin	*Angelica officinalis* fruits	Imperatorin (5 and 10 mg/kg) improved memory and learning in PA test in scopolamine-induced cognitive impairment. The activity of antioxidant enzymes was also increased by imperatorin.	1, 5, and 10 mg/kg (*i.p.*) for 7 days/ male Swiss mice	Budzynska et al. (2015)
Auraptene	*Citrus* species	Auraptene improved spatial memory in $A\beta_{1\text{-}42}$ peptide-induced AD model. It was also shown to prevent neuronal death by reducing the Bax/ Bcl-2 ratio in rat hippocampus.	5, 10, and 25 mg/ kg (*i.p.*) for 12 days/male Wistar rats	Joghataee et al. (2020)
		Auraptene enhanced memory in MWM test after induction of cerebral hypoperfusion common using carotid ligation or two vessel occlusions. Auraptene also decreased MDA and increased glutathione (GSH) content as well as antioxidant enzymes in cortex and hippocampus tissues.	4, 8 and 25 mg/kg (p.o.) for 20 days/ male Wistar rats	Ghanbarabadi et al. (2016)
		Auraptene reversed memory deficit induced by scopolamine from PA test.	50, 75, 100 mg/kg (*p.o.*) for 7 days/ adult male mice	Tabrizian et al. (2015)
Xanthotoxin (XAN)	*Pastinaca sativa* (fruits)	XAN (15 mg/kg) enhanced spatial and short-term memory in STZ-induced AD, it significantly decreased AChE activity, and	5, 10, and 15 mg/ kg (*i.p.*) adult male Wistar rats	Hindam et al. (2020)

Continued

Table 8.2 Natural coumarins tested in *in vivo* studies against AD.—cont'd

Natural coumarins	Plant source	Method/effect	Doses/animals	References
		significantly mitigated the oxidative stress observed in STZ-treated rats as demonstrated by decreased MDA levels along with GSH increase.		
		XAN improved cognition and showed neuroprotective effect from PA test after scopolamine-induced cognitive impairment.	1, 2.5, and 5 mg/kg/male Swiss mice	Skalicka-Wozniak et al. (2018)
Umbelliferone (UMB)	*Heracleum leskovii* (fruits)	UMB (15 mg/kg) prevented STZ-induced impairments in spatial and short-term memory, it also increased the level of Nrf2 and HO-1 thereby mitigating oxidative stress in the hippocampus of rats.	5, 10, and 15 mg/kg (*i.p.*)/adult male Wistar rats	Hindam et al. (2020)
	Commercial UMB	UMB showed neuroprotective effect in rats with cerebral ischemic-reperfusion induced by middle cerebral artery occlusion.	15 and 30 mg/kg (*i.p.*) for 7 days/ Sprague–Dawley male rats	Wang et al. (2015)
Nodakenin	*Angelica decursiva*	Nodakenin (1 mg/kg) enhanced cognition in PA test, increased proliferation of new cells, and enhanced hippocampal neurogenesis.	0.3 or 1 mg/kg (*p.o.*) for 14 days male ICR (CD-1) mice	Gao et al. (2015)
Esculetin		Esculetin can attenuate oxidative damage and neuroinflammation by elevating Neuronal antioxidant enzyme activities, decreasing oxidative	10, 20, and 30 mg/kg (*p.o.*) for 6 weeks/ Sprague–Dawley rats	Yao et al. (2022)

Table 8.2 Natural coumarins tested in *in vivo* studies against AD.—cont'd

Natural coumarins	Plant source	Method/effect	Doses/animals	References
		damage Biomarkers, reducing proinflammatory cytokines' levels, and triggering Cholinergic mechanisms in D-gal provoked AD animals.		
Scopoletin	*Artemisia umbelliformis* (aerial parts)	Scopoletin can cross BBB and significantly increased brain dopamine level.	8 mg/kg (*i.p.*) for 21 days/male Swiss albino mice	Basu et al. (2016)
Pteryxin (PTE)	*Mutellina purpurea* (fruits)	PTE at 16 mg/kg showed significant improvement in learning and memory of 5xFAD mice and regulated. Expression of proteins associated with cognition, oxidative stress, synaptic plasticity and Aβ-related pathway.	8, 16 mg/kg (*i.p.*) for 7 days/5xFAD mice	Kiris et al. (2022)
Fraxetin (FRA)	*Fraxinus rhynchophylla* (bark)	FRA improved memory and behavioral deficits, decreased elevated serum corticosterone levels, and increased serotonin and dopamine levels.	20, 40, and 60 mg/kg (*p.o.*)/adult male BALB/c mice	Ahmed et al. (2023)
Scoparone	*Artemisia umbelliformis*	Scoparone at 2.5, 5, and 12.5 mg/kg significantly improved learning or memory acquisition in PA test.	2.5, 5, 12.5, 25 mg/kg (*i.p.*)/Swiss albino mice	Kowalczyk et al. (2022)
Osthol (OST)		OST ameliorated cognitive impairments in APP/PS1 mice, inhibited neuronal apoptosis, promoted neurogenesis and proliferation of NSCs.	20 mg/kg twice a day for 6 weeks (*i.g.*)/APP/PS1 mice	Xiao et al. (2021)

Continued

Table 8.2 Natural coumarins tested in *in vivo* studies against AD.—cont'd

Natural coumarins	Plant source	Method/effect	Doses/animals	References
Daphnetin (DAP)	*Daphne odora*	DAP showed neuroprotective effect against cerebral ischemia/reperfusion injury in mice induced middle cerebral artery occlusion.	5, 10, and 20 mg/kg (*i.p.*)/C57BL/6 mice	Liu et al. (2016)
		DAP improved spatial learning and memory in a double-transgenic mouse model of AD. DAP also reduced Aβ aggregation.	100 mg/kg (*i.g.*) (APP)/presenilin 1 (PS1) double-transgenic male mice	Gao et al. (2022)
Bergapten (BER)	*Heracleum leskovii* L. (fruits)	BER improved learning and memory in scopolamine-induced memory impairment using PA test.	12.5, 25, 50, and 100 mg/kg (*i.p.*) male Swiss mice	Kowalczyk et al. (2020)
	Heracleum leskovii	BER elicited neuroprotective effect in STZ-induced AD, it improved cognitive function in Morris water maze and object recognition tests as well as reduction in Aβ and tau hyperphosphorylation expression by 89% and 61%, respectively.	25 mg/kg (*i.p.*) given to male Swiss Albino mice for 21 days	Salem et al. (2021)
Angelicin		ANG showed neuroprotective effects against isoflurane-induced neurotoxicity by alleviating cognitive deficit in Morris water maze test and reducing inflammation.	2.5 mg/kg (*i.p.*) administered to pregnant C57BL/6 J mice for 3 days	Liu et al. (2023)
Anomalin	*Angelica anomala* (roots)	ANO suppressed neuropathic pain in mice induced by streptozotocin as well as sodium-nitroprusside (SNP)-induced neuroinflammation.	50 mg/kg (*i.p.*) was given to male ICR mice for 5 days	Khan et al. (2016)

Table 8.2 Natural coumarins tested in *in vivo* studies against AD.—cont'd

Natural coumarins	Plant source	Method/effect	Doses/animals	References
Herniarin	*Matricaria chamomilla*	HER showed neuroprotective effect in ischemic stroke induced by middle cerebral artery occlusion (MCAO). Antioxidant enzymes were increased and pro-inflammatory cytokines were reduced in HER-treated groups.	10, 20 mg/kg (*i.p.*) was administered to adult male rats for 7 days	Asgharzade et al. (2021)
Fraxin	*Cortex Fraxini*	Fraxin exerted neuroprotective effect in rats with ischemic stroke induced by MCAO.	5, 10, 20 mg/kg (*i.p.*) was given to Sprague–Dawley rats	Yao et al. (2022)

animal model for sporadic AD, which represents 95% of the AD types, is also available (Zhang et al., 2020).

Different coumarins have been tested in *in vivo* studies using different AD animal models. For instance, the effect of imperatorin, a naturally occurring furanocoumarin, on scopolamine-induced cognitive deficit and oxidative stress was investigated by Budzynska et al. (2015). Imperatorin was isolated from *Angelica officinalis* Moench (syn. *Angelica archangelica* L.). fruits (Apiaceae), whose identity and purity were confirmed through RP-HPLC and ^{1}H NMR analyses. Passive avoidance (PA) test was applied in this study to evaluate learning and memory in male Swiss mice impaired by scopolamine (1 mg/kg) and treated with imperatorin at the doses of 1, 5, and 10 mg/kg, while rivastigmine (0.5 mg/kg, *i.p.*) was used as the reference drug. At the end of the experiment (7 days), cerebral cortex and hippocampus tissues were excised and the level of antioxidant enzymes as well as malondialdehyde (MDA) as an oxidative stress marker, in both tissues, was determined. The results of the study indicated that the administration of imperatorin injections (5 and 10 mg/kg) improved memory acquisition and consolidation as well as locomotion activity in the mice. Additionally, the data revealed that imperatorin was able to prevent prooxidant effects of scopolamine in the brain. There was a significant increase in antioxidant enzymes (SOD, GPx, and GR) activities and a decrease in MDA concentration in the prefrontal cortex and hippocampus. The outcomes suggested that imperatorin may become a new treatment strategy for AD treatment after evaluation in further preclinical and clinical studies.

Joghataee et al. (2020) investigated neuroprotective effect of auraptene, a simple coumarin derivative, in Aβ peptide-induced AD. Auraptene is the main coumarin in the peel of oranges and also common in genus *Citrus* (Rutaceae) (Furukawa et al., 2012). In this study, AD was induced in male Wistar rats by injecting $A\beta_{1\text{-}42}$ into the hippocampus of the rats. Morris water maze test was used to determine the effect of auraptene (5, 10, and 25 mg/kg/day) administered intraperitoneally for 12 days on learning and spatial memory. Concentrations of Bax and Bcl-2 proteins, known mediators of apoptosis, were also evaluated in the brain's hippocampus. The outcomes disclosed that auraptene enhanced the spatial memory performance in Morris water maze test by significantly reducing the time spent in finding the platform and increasing the distance traveled in the target quadrant by auraptene-treated animals, when compared with the disease groups. More so, treatment with auraptene decreased the Bax/Bcl-2 ratio in the hippocampus in the treated rats (Wang et al., 2016), while injection of Aβ was reported to result into neuronal death (Jazvinšćak Jembrek et al., 2015). The mentioned study, therefore, exhibited that auraptene apart from improving memory can also protect neuronal cells from apoptosis. Several other studies have also shown the potency of auraptene in the treatment of AD through various in vivo models (Ghanbarabadi et al., 2016; Tabrizian et al., 2015) as described in Table 8.1.

Xanthotoxin (8-methoxypsoralen), a photoactive furanocoumarin, was isolated from *Pastinaca sativa* L. (Apiaceae) fruit, while umbelliferone (7-hydroxycoumarin) was isolated from the fruit of *Heracleum leskovii* Grossch (Apiaceae). Hindam et al. (2020) evaluated the neuroprotective effects of xanthotoxin and umbelliferone in streptozotocin (STZ)-induced cognitive dysfunction in rats and investigated the possible role of the Janus kinase/signal transducers and activators of transcription (JAK2/STAT3) and nuclear factor erythroid-derived 2 and heme oxygenase-1 [nuclear related factor 2 (Nrf2)/HO-1] signaling pathways. JAK/STAT signaling pathway mediates interferon signaling and adaptive immunity (O'Shea and Plenge, 2012). Activation of this pathway also leads to inflammation of CNS; therefore, regulation of JAK/STAT pathway may have neuroprotective effect by preventing neuroinflammation in AD (Yan et al., 2018). On the other hand, activation of the Nrf2/HO-1 pathway inhibits oxidative stress, tau phosphorylation, and Ab deposition, while causes improvements in learning and memory functions (Ali et al., 2018). Sporadic AD was induced by intracerebroventricular (ICV) injection of STZ (3 mg/kg), after which xanthotoxin and umbelliferone were administered to different animal groups at doses of 5, 10, 15 mg/kg. The results indicated that xanthotoxin and umbelliferone (15 mg/kg) prevented STZ-induced impairments in spatial and short-term memory, significantly decreased AChE activity, and mitigated the oxidative stress observed in STZ-treated rats as demonstrated by decreased MDA levels along with GSH increase. Xanthotoxin protected neurons from inflammation-related injury induced by STZ, inhibits the inflammatory response through downregulating the JAK2/STAT3 signaling pathway, while umbelliferone increased the level of Nrf2 and HO-1 in the hippocampus of rats, suggesting that the neuroprotective effect of umbelliferone might rely partly on the Nrf2/HO-1 pathway and improved neuron cell viability. This study suggested that xanthotoxin and umbelliferone might be promising candidates for the management of AD after being tested in clinical studies.

Gao et al. (2015) investigated the effect of nodakenin (NOD), a coumarin present in many Apiaceae species (Jung and Huneck, 1991), on scopolamine-induced memory impairment. This group investigated the neurogenesis of adult hippocampal cells neurogenesis by NOD (0.3 and 1 mg/kg) administration in mice. 5-Bromo-2-deoxyuridine (BrdU) immuno-staining was used to measure the proliferation, differentiation, and survival of newborn cells. Studies have shown the importance of neurogenesis in cognition and learning (Deng et al., 2010; Morgan, 2007), and in AD patients, hippocampal neurogenesis is reduced (Boekhoorn et al., 2006). Therefore, proneurogenesis substances such as NOD can be helpful in the management of AD. The results of the study demonstrated that subchronic administration of nodakenin (1 mg/kg) to mice enhanced cognition in PA test, increased proliferation of new cells and their survival, and enhanced hippocampal neurogenesis. The results suggested that NOR might be a promising natural agent for future AD treatment.

Esculetin is a coumarin derivative found in many plant species such as *Citrus x limonia* Osbeck (Rutaceae), *Fraxinus rhynchophylla* Hance, and *Fraxinus chinensis* Roxb. from Oleaceae. Yao et al. (2022) investigated the effect of esculetin on D-galactose (D-gal)-induced AD model and the mechanism involved. Previous studies have shown that D-gal can be used to induce AD in animal models. Exposure of animals to mild doses of D-gal causes aging-related cognitive dysfunction, oxidative stress, reduced immune responses, neurotoxicity, increase in AChE activity, and decrease in activity of antioxidant enzymes (Gao et al., 2015; Wei et al., 2005). Subcutaneous administration of D-gal at 150 mg/kg was used to initiate AD in Sprague−Dawley rats, after which esculetin (10, 20, and 30 mg/kg) was given orally for 6 weeks. Morris water maze test was employed to assess memory and learning, while antioxidant and AChE enzymatic activities were determined in the brain tissues and other organs. The outcomes displayed that esculetin significantly improved the spatial learning and memory indices as well as antioxidant enzyme activity. Additionally, esculetin reduced the activity of AChE and inflammatory markers. The authors concluded that esculetin can alleviate inflammation, oxidative stress, and cognitive dysfunction in D-gal-induced AD rats.

Kiris et al. (2022) investigated the therapeutic potential of two coumarins; pteryxin (PTE), a dihydropyranocoumarin, and scopoletin (SCO), a simple coumarin, in a 5xFAD mouse model of AD. 5xFAD mouse is a transgenic mouse widely used to mimic AD, because of its similar biochemical and behavioral phenotypes to those of human AD (Jawhar et al., 2012; Maarouf et al., 2013). It has five familial AD mutations, it presents early and aggressive Aβ-plaques, gliosis, synaptic degeneration, neuronal loss, and progressive cognitive deficits (Bouter et al., 2014; Sasaguri et al., 2017). Scopoletin and pteryxin were isolated from aerial parts of *Artemisia umbelliformis* Lam. (Asteraceae) and fruits of *Mutellina purpurea* (Poir.) Reduron, Charpin & Pimenov (Apiaceae), respectively. Then, 12-month-old 5xFAD mice were intraperitoneally given PTE and SCO (8 and 16 mg/kg) for 7 days, and Morris water maze test was performed; afterward, brain tissues were collected after the behavioral test for proteome analysis. The results revealed that PTE at 16 mg/kg showed significant improvement in learning and memory of the mice better than SCO, PTE also regulated expression of proteins associated with cognition, oxidative stress, synaptic plasticity, and Aβ-related pathway. This study disclosed the therapeutic potential of PTE in slowing the progression of AD.

The potential of fraxetin (FRA), a simple coumarin compound, was evaluated on the alleviation of chronic stress by Ahmed et al. (2023). Exposure to chronic stress may lead to decrease in neurotransmitter levels in the prefrontal cortex and hippocampus of the brain, leading to cognitive dysfunction (Bondi et al., 2008). Fraxetin were orally administered at doses of 20, 40, and 60 mg/kg to adult male BALB/c mice, after induction of chronic unpredictable stress by exposure to series of mild stressors for 14 days. Behavioral tests were performed and postmortem frontal cortical, hippocampal, and striatal tissues were analyzed through high-performance liquid chromatography (HPLC) for neurochemical changes. Administration of FRA improved memory and behavioral deficits, decreased elevated serum corticosterone levels, and increased serotonin and dopamine levels, which are important neurotransmitters in the brain. In another study by Zhang et al. (2022), FRA was shown to inhibit neurotoxicity induced by isoflurane in hippocampal neuronal cells and suppressed ROS-dependent autophagy. These studies suggested the importance of FRA in the management of AD.

Xiao et al. (2021) investigated the potentials of osthol (OST) in promoting endogenous neural stem cells (NSCs) proliferation and neurogenesis in AD mice. Neurogenesis is known to alter the progression of AD (Zhao et al., 2008); also mobilization of endogenous NSCs to generate sufficient neurons is also an interesting alternative to prevent neurodegeneration (Uliassi et al., 2017). In this study, APP/PS1 mice were treated with OST, and Morris water maze test was used to assess cognition. Proliferation of NSCs in hippocampal regions, level of the proteins related to neuronal survival and neurogenesis were monitored by Nissl staining, immunofluorescence analysis, and western blot. The findings revealed that OST could improve the learning and memory function, inhibit neuronal apoptosis, elevate the expression of glial cell line derived neurotrophic factor, promote neurogenesis, and proliferation of NSCs. It was concluded that OST could serve as a therapeutic agent to reverse or slow down the progress of AD.

Different methods including chemicals have been so far employed to induce AD-like symptoms. The methods and chemicals include Aβ peptide, aluminum chloride ($AlCl_3$), D-gal, heavy metals (e.g., Cu, Zn), scopolamine, streptozotocin, and lipopolysaccharide, among others (Rapaka et al., 2022). All the aforementioned studies depicted the potential of different types of natural coumarins in the treatment of AD; however, more preclinical studies need to be conducted on coumarins. There are many coumarins yet to be tested in vivo.

5. Molecular docking of naturally occurring coumarins against AD-related enzymes

Molecular modeling techniques help to understand the binding and orientation of a drug candidate, by giving deeper insight of the molecular mechanisms, responsible for the observed biological effects. Takomthong et al. (2020) reported the molecular docking studies and neuroprotective effect of nine coumarins (phellopterin, isoimpinellin, toddalolactone, toddaculin, toddacoumaquinone, toddalenone, toddanone, artanin, and fraxinol) isolated from *Toddalia asiatica* (L.) Lam. (Rutaceae). All tested coumarins inhibited AChE moderately, where fraxinol possessed the highest inhibitory potential (IC_{50}: 18 μM). The tested coumarins also inhibited $A\beta_{1-42}$-aggregation induced by AChE apart from isopimpinellin and fraxinol. The group also investigated the binding interactions of the mentioned coumarins with AChE and $A\beta_{1-42}$ as well as their structure—activity relationship. Seven out of nine coumarins were identified as multifunctional compounds inhibiting the pathogenesis of AD. Phellopterin, the most potent multifunctional coumarin, exhibited an ability to protect neuronal cell damage induced by H_2O_2 and $A\beta_{1-42}$. The structure—activity relationship and molecular docking studies suggested that the substituents on positions 6, 7, and 8 of coumarin ring influenced AChE and $A\beta_{1-42}$ aggregation inhibition. The results suggested that coumarins with multitarget ligands could be useful for a multifactorial disease such as AD. Many cellular processes are involved in the pathogenesis of AD, such as oxidative stress, cholinergic deficit, inflammation, metal imbalance, protein misfolding, aggregation, and monoaminergic disturbances that have been observed in brains (Deng et al., 2019; Ekströ;m et al., 2022; Kowalczyk et al., 2020). All the aforementioned mechanisms contribute to neuronal apoptosis and memory impairments in AD and make it challenging to find a cure for AD despite huge efforts engaging several potential drug targets. In recent times, drugs with multitarget strategies have been proposed, and they might be more promising (Cummings et al., 2019). Many evidences indicate that multitarget-directed ligands/molecules may improve the therapeutic efficacy of AD drugs (Pravin and Jozwiak, 2022).

Khalid et al. (2023) investigated the AChE and BChE inhibition by three natural coumarins [2′-*O*-ethylmurrangatin, murranganone (MUR), and paniculatin (PAN)] isolated from *M. paniculata* leaves and performed molecular docking studies to understand their interactions with active gorges of AChE and BChE. Absorption, distribution, metabolism, and excretion (ADME) profiles of coumarins were predicted using Swiss ADME, a web-based server tool. The ADME test is used to confirm the drug likeliness of a molecule that would make it an orally active drug in humans. These properties include good membrane absorption and permeation, hydrogen donors and acceptors, good molecular weight, cell permeability, and ability to cross BBB among others. The results of the studies presented that only MUR and PAN inhibited both enzymes in vitro. On the other hand, 2′-O-ethylmurrangatin did not show any inhibition, which was consistent with the molecular docking results, where it had low binding affinity for AChE and BChE. The findings from molecular docking simulations revealed that these three coumarins gained access to the binding sites of AChE and BChE. However, MUR and PAN showed higher binding affinity and binding energy toward both enzymes. Docking simulations revealed that hydrogen bonding, mediated by ketone and hydroxyl functionalities in various positions, significantly contributed to the binding of the tested coumarins to the receptor and ligand-AChE complex for PAN, the most active compound was comparable with that of tacrine, the reference drug. Additionally, ADME test revealed good physiochemical properties including lipophilicity, size, polarity, solubility, saturation, and flexibility of the tested coumarins. This study suggested the drug-likeness of MUR and PAN for AD.

Anupama et al. (2023) performed molecular docking studies on five pyranocoumarins (e.g., seselin, jatamansine, jatamansinol, jatamansinone, and dihydrojatamansin) isolated from *Nardostachys jatamansi* (D.Don) DC. (Caprifoliaceae), a memory-enhancing plant in Ayurveda medicine. The potential binding modes of the pyranocoumarins and reference drugs (galanthamine and rivastigmine) were docked into AChE, BChE, glycogen synthase kinase 3β (GSK3β), and kelch-like ECH-associating protein 1 (Keap1). Cholinesterases (AChE and BChE) are involved in the cholinergic system in the brain, causing decline in the level of ACh in AD brains. GSK3β is linked to tau protein phosphorylation (Mondragón-Rodríguez et al., 2012), while Keap1 in the cytoplasm leads to the activation of Nrf2 and its associated genes (Brandes and Gray, 2020), which are associated with oxidative stress also implicated in AD pathogenesis (Kerr et al., 2017). Therefore, a molecule that can bind to AChE, BChE, GSK3β, and Keap1 would be a multitarget therapy for AD treatment. The molecular docking score of the pyranocoumarins, especially dihydrojatamansin (−9.14 kcal/mol), was higher than that of rivastigmine (−4.65 kcal/mol) and galanthamine (−7.14 kcal/mol), when docked with AChE. For BChE, the docking score of jatamansinol (−7.80 kcal/mol) was comparable with that of rivastigmine (−7.69 kcal/mol) and galanthamine (−8.63 kcal/mol). More so, the docking score of those five pyranocoumarins, e.g., jatamansine and jatamansinol in particular, was better than those of the reference drugs, when docked with GSK3β and Keap1. ADME properties of the tested pyranocoumarins, such as brain/blood partition coefficient, Madin−Darby canine kidney (MDCK) cell permeability, Caco-2 cell permeability, human oral absorption, and predicted values for CNS activity, were within the acceptable range in accordance with Lipinski's rule of five, which predicts the likelihood of oral activeness of a drug in humans. It considers that a druggable molecule with good membrane absorption and permeation would have molecular weight around 500 Da, hydrogen donors not more than 5, hydrogen acceptors not more than 10, and octanol/water partition coefficient not more than 5 (Lipinski et al., 1997). This study confirmed the druggability of the tested pyranocoumarins from *N. jatamansi* and may be potential AD multitarget (AChE, BChE, GSK3b, and Keap1) drug candidates.

Gay et al. (2020) examined the neuroprotective effect of scopoletin among other phenolic compounds against neurotoxicity and SH-SY5Y cell apoptosis induced by H_2O_2. Scopoletin prevented cell death, decreased the levels of apoptotic cells as well as ROS. The group also performed molecular docking calculations for scopoletin for its possible binding with sirtuin-1 (SIRT1) protein that is involved in the modulation of neurodegeneration by extending the survival of neuronal cells (Hubbard and Sinclair, 2014). AutoDock 4.2.6 program was used, and atomic coordinate of scopoletin was obtained from PubChem database, docking of scopoletin was compared with that of resveratrol, a known activator of SIRT1 (Allard et al., 2009). Docking analyses pointed out to the presence of tight interactions between the activator-binding sites of SIRT1 and scopoletin, among other compounds tested, was similar to those of resveratrol. The findings indicated that scopoletin might be the potential candidate for prevention and/or treatment of neurodegenerative diseases for further investigations.

Orhan et al. (2019) investigated the anticholinesterase activity and performed molecular docking analyses of hyperforin and hyuganin C. Hyperforin is a prenylatedphloroglucinol derivative from *Hypericum perforatum* L. (Clusiaceae), while hyuganin C is a rare coumarin derivative from *Mutellina purpurea*. Hyuganin C showed a remarkable stronger BChE inhibitory activity than that of galanthamine, the reference drug. From the docking simulations, hyuganin C exhibited decent interactions at the active site of BChE. It displayed a strong network of hydrogen bonds and hydrophobic interactions with BChE active site. Hyuganin C blocked the access to key residues involved in the normal functioning of BChE such as parts of the catalytic triad and anionic and peripheral anionic sites. At the same time, they complemented some of the hydrophobic residues of the cavity. In this report, hyuganin C was therefore suggested as a hopeful agent for AD drug development afterward further toxicity studies.

Umbelliferone and pimpinellin isolated from *Zosima absinthifolia* Link (Apiaceae) exhibited high inhibition on AChE and BChE among other coumarins tested and were docked with AChE and BChE, respectively, by Karakaya et al. (2019). Umbelliferone possessed a good docking score, binding energy (−7.46 kcal/mol), and molecular interactions with AChE, when compared with donepezil as the reference drug. Pimpinellin also showed good docking score binding energy (−5.78 kcal/mol) and molecular interactions with BChE. Kozioł et al. (2020) isolated rutamarin, a dihydrofuranocoumarin from *Ruta graveolens* L. (Rutaceae), and analyzed its molecular interactions with human MAO-B (*h*MAO-B). The group firstly tested the *h*MAO-A and B inhibitory potential of rutamarin and *R. graveolens* extracts. MAO-A and B are implicated in the etiology of AD. MAO B is elevated in plaque-associated glia in the brains of AD patients. Inhibitors of MAO-A and B are also known to exert neuroprotective effect by reduction of oxidative stress (Kennedy et al., 2003). Rutamarin showed a weak (25.2%) and strong inhibition (95.3%) against *h*MAO-A and *h*MAO-B, respectively. Docking results recommended that rutamarin could bind to the active site of *h*MAO-B binding pocket.

Molecular docking analyses have shown that the tested natural coumarins have stable molecular interactions with several proteins including AChE, BChE, MAO-B, SIRT1, GSK3β, Keap1, and $A\beta_{1-42}$, all of which are associated with AD. In silico ADME analysispredicted good oral bioavailability and drug-likeness property for murranganone, paniculatin, seselin, jatamansine, jatamansinol, jatamansinone, and dihydrojatamansin. Generally, more in silico studies are still needed to identify potential multitarget inhibitors with good pharmacokinetic profile for AD treatment.

6. Why are there not many clinical trials yet on coumarins for AD?

Quite a number of natural coumarins have given promising results in both in vitro and in vivo preclinical studies for AD pathogenesis. Nevertheless, very few natural coumarins have proceeded into clinical trials. Recently, Galluzzi et al. (2022) designed and registered a 36-week long, randomized, double-blinded, and placebo-controlled clinical trial, to test the effect of auraptene and naringenin mostly found in the peel of *Citrus* species on subjective cognitive decline (SCD) in 80 older adults between 65 and 75 years old. The study proposed to recruit the elderly with SCD and administered 400 mg of citrus peel extract containing 0.1 mg of auraptene and 3 mg of naringenin or the placebo at a 1:1 ratio for 36 weeks. This trial is still ongoing and expected to receive the results soon.

The paucity of clinical trials for coumarins in AD treatment could be due to the assumed possible toxicity of coumarins. One of the earliest uses of coumarin in humans was for the treatment of lymphedema, but its use was discontinued in some countries based on some hepatotoxicity reports (Casley-Smith and Casley-Smith, 1997; Farinola and Piller, 2005). Coumarins were also used as an anticoagulant and to treat asthma and breast cancer (Marshall et al., 1994; Musa et al., 2008). Despite the promising health benefits of coumarins, its hepatotoxic effect was observed in animal studies (Born et al., 2000). It was also observed to induce tumor in rodents (Lake, 1999). Pitaro et al. (2022) presented a narrative review on the hepatotoxicity of coumarins in preclinical and clinal trials. The report mentioned that the cases of hepatotoxicity from coumarin administration are rare in humans. But some studies have shown coumarin-induced transaminase elevation, which is an indicator of drug-induced hepatotoxicity (Abraham et al., 2010; Arora and Goldhaber, 2006). An important finding from the review is that coumarin-induced hepatotoxicity is restricted to a small subset of patients, and further studies on the safety of coumarins are still needed. Normally, coumarin core structure is metabolized by cytochrome (CYP) P450-linked mono-oxygenase enzyme system (CYP2A6) in liver microsomes and hydroxylated to form 7-hydroxycoumarin, which is nontoxic (Vassallo et al., 2004). However, in some cases, ingested coumarins go through another pathway involving other CYP450 isoforms (i.e., CYP1A1, CYP1A2, and CYP2E1), where an unstable 3,4-epoxycoumarin is formed and spontaneously rearranged to form O-hydroxyphenylacetaldehyde, which is a hepatotoxic aldehyde (Egan et al., 1990). Factors such as genetic variability, especially in the expression of CYP2A6, and environmental factors can significantly induce interindividual variations in the metabolism of coumarin and influence the responses of individuals to coumarin treatment (Barot et al., 2015). Therefore, more studies focusing on susceptibility factors of individuals to coumarin-induced hepatotoxicity should be carried out. The study by Hu and Piller (2017) identified pharmacogenetic strategies among others to avoid hepatotoxicity induced by coumarins. This will encourage the less vulnerable individuals to enjoy the numerous health benefits of coumarins.

Again despite the neuroprotective effects of furanocoumarins, they are also considered as a phototoxic class of coumarins, which can yield mutagenic and carcinogenic products in the presence of long-wave UV light (Peroutka et al., 2007). In order to resolve these concerns, many more toxicity studies are needed so as to harness the beneficial potentials inherent in this important class of metabolite.

Natural coumarins are present in a diverse human diet ranging from fruits, vegetables, nuts, wine, coffee to tea; therefore, humans are highly exposed to coumarins through diet. It is believed that dietary coumarins poses little or no health risk to humans and human data indicated a tolerable dose intake (TDI) of coumarin equal to 0.1 mg/kg body. Huge evidences suggest that toxic effects are strongly dose-dependent, and only high doses of coumarins may produce dangerous effects (Stefanachi et al., 2018).

7. Conclusions

Coumarins are naturally occurring substances present in a wide range of plants and often come crossed through daily diet. The highlighted in vitro and in vivo studies in this review depicted the neuroprotective potential of various coumarin derivatives in the treatment of AD. However, there are still a lot of natural coumarins yet to be tested for their potentials in the treatment of AD, especially using in vivo methods. More in vivo studies are still needed in this area to determine the mechanisms involved in anti-AD activity of coumarins. Additionally, more toxicological, bioavailability and pharmacokinetic studies are still needed. Despite significant potential of coumarins to improve cognition and memory, there is no report yet on clinical trials for natural coumarins with anti-AD potential, which is probably due to lack of adequate information on the toxicity of the bioactive coumarins. Besides, molecular docking and simulation studies can be desired as well to identify more druggable coumarin candidates with multitarget effect toward AD.

References

Abraham, K., Wöhrlin, F., Lindtner, O., Heinemeyer, G., Lampen, A., 2010. Toxicology and risk assessment of coumarin: focus on human data. Mol. Nutr. Food Res. 54 (2), 228–239. https://doi.org/10.1002/mnfr.200900281.

Ahmed, Z., Tokhi, A., Arif, M., Rehman, N.U., Sheibani, V., Rauf, K., Sewell, R.D., 2023. Fraxetin attenuates disrupted behavioral and central neurochemical activity in a model of chronic unpredictable stress. Front. Pharmacol. 14. https://doi.org/10.3389/fphar.2023.1135497.

Akeson, W.R., Gorz, H.J., Haskins, F.A., 1963. Effect of genotype and growth stage on distribution of melilotic acid, *o*-coumaric acid, and coumarinic acid in *Melilotus alba* Desr. Crop Sci. 3 (2), 167–171.

Al-Warhi, T., Sabt, A., Elkaeed, E.B., Eldehna, W.M., 2020. Recent advancements of coumarin-based anticancer agents: an up-to-date review. Bioorg. Chem. 103, 104163. https://doi.org/10.1016/j.bioorg.2020.104163.

Ali, M.Y., Jannat, S., Jung, H.A., Choi, R.J., Roy, A., Choi, J.S., 2016. Anti-Alzheimer's disease potential of coumarins from *Angelica decursiva* and *Artemisia capillaris* and structure-activity analysis. Asian Pac. J. Tropical Med. 9 (2), 103–111. https://doi.org/10.1016/j.apjtm.2016.01.014.

Ali, T., Kim, T., Rehman, S.U., Khan, M.S., Amin, F.U., Khan, M., Ikram, M., Kim, M.O., 2018. Natural dietary supplementation of anthocyanins *via* PI3K/Akt/Nrf2/HO-1 pathways mitigate oxidative stress, neurodegeneration, and memory impairment in a mouse model of Alzheimer's disease. Mol. Neurobiol. 55, 6076–6093. https://doi.org/10.1007/s12035-017-0798-6.

Allard, J.S., Perez, E., Zou, S., De Cabo, R., 2009. Dietary activators of SIRT1. Mol. Cell. Endocrinol. 299 (1), 58–63. https://doi.org/10.1016/j.mce.2008.10.018.

Anand, A., Patience, A.A., Sharma, N., Khurana, N., 2017. The present and future of pharmacotherapy of Alzheimer's disease: a comprehensive review. Eur. J. Pharmacol. 815, 364–375. https://doi.org/10.1016/j.ejphar.2017.09.043.

Anupama, K.P., Shilpa, O., Antony, A., Gurushankara, H.P., 2023. Jatamansinol from *Nardostachys jatamansi*: a multi-targeted neuroprotective agent for Alzheimer's disease. J. Biomol. Struct. Dyn. 41 (1), 200–220. https://doi.org/10.1080/07391102.2021.2005681.

Ardekani, B.A., Convit, A., Bachman, A.H., 2016. Analysis of the MIRIAD data shows sex differences in hippocampal atrophy progression. J. Alzheim. Dis. 50 (3), 847–857. https://doi.org/10.3233/JAD-150780.

Arora, N., Goldhaber, S.Z., 2006. Anticoagulants and transaminase elevation. Circulation 113 (15), e698–e702. https://doi.org/10.1161/CIRCULATIONAHA.105.603100.

Asgharzade, S., Khorrami, M.B., Forouzanfar, F., 2021. Neuroprotective effect of herniarin following transient focal cerebral ischemia in rats. Metab. Brain Dis. 36, 2505–2510. https://doi.org/10.1007/s11011-021-00841-1.

Awang, K., Chan, G., Litaudon, M., Ismail, N.H., Martin, M.-T., Gueritte, F., 2010. 4-Phenylcoumarins from *Mesua elegans* with acetylcholinesterase inhibitory activity. Bioorg. Med. Chem. 18 (22), 7873–7877. https://doi.org/10.1016/j.bmc.2010.09.044.

Baek, S.C., Kang, M.-G., Park, J.-E., Lee, J.P., Lee, H., Ryu, H.W., Park, C.M., Park, D., Cho, M.-L., Oh, S.-R., 2019. Osthenol, a prenylated coumarin, as a monoamine oxidase A inhibitor with high selectivity. Bioorg. Med. Chem. Lett 29 (6), 839–843. https://doi.org/10.1016/j.bmcl.2019.01.016.

Barot, K.P., Jain, S.V., Kremer, L., Singh, S., Ghate, M.D., 2015. Recent advances and therapeutic journey of coumarins: current status and perspectives. Med. Chem. Res. 24, 2771–2798. https://doi.org/10.1007/s00044-015-1350-8.

Basu, M., Mayana, K., Xavier, S., Balachandran, S., Mishra, N., 2016. Effect of scopoletin on monoamine oxidases and brain amines. Neurochem. Int. 93, 113–117. https://doi.org/10.1016/j.neuint.2016.01.001.

Bateman, R.J., Aisen, P.S., De Strooper, B., Fox, N.C., Lemere, C.A., Ringman, J.M., Salloway, S., Sperling, R.A., Windisch, M., Xiong, C., 2011. Autosomal-dominant Alzheimer's disease: a review and proposal for the prevention of Alzheimer's disease. Alzheimer's Res. Ther. 3 (1), 1–13. https://doi.org/10.1186/alzrt59.

Behl, T., Kaur, D., Sehgal, A., Singh, S., Sharma, N., Zengin, G., Andronie-Cioara, F.L., Toma, M.M., Bungau, S., Bumbu, A.G., 2021. Role of monoamine oxidase activity in Alzheimer's disease: an insight into the therapeutic potential of inhibitors. Molecules 26 (12), 3724. https://doi.org/10.3390/molecules26123724.

Belov Kirdajova, D., Kriska, J., Tureckova, J., Anderova, M., 2020. Ischemia-triggered glutamate excitotoxicity from the perspective of glial cells. Front. Cell. Neurosci. 14, 51. https://doi.org/10.3389/fncel.2020.00051.

Blahová, J., Svobodová, Z., 2012. Assessment of coumarin levels in ground cinnamon available in the Czech retail market. Sci. World J. 2012, 263851. https://doi.org/10.1100/2012/263851.

Boekhoorn, K., Joels, M., Lucassen, P.J., 2006. Increased proliferation reflects glial and vascular-associated changes, but not neurogenesis in the presenile Alzheimer hippocampus. Neurobiol. Dis. 24 (1), 1–14. https://doi.org/10.1016/j.nbd.2006.04.017.

Bondi, C.O., Rodriguez, G., Gould, G.G., Frazer, A., Morilak, D.A., 2008. Chronic unpredictable stress induces a cognitive deficit and anxiety-like behavior in rats that is prevented by chronic antidepressant drug treatment. Neuropsychopharmacology 33 (2), 320–331. https://doi.org/10.1038/sj.npp.1301410.

Born, S.L., Caudill, D., Smith, B.J., Lehman-McKeeman, L.D., 2000. *In vitro* kinetics of coumarin 3, 4-epoxidation: application to species differences in toxicity and carcinogenicity. Toxicol. Sci. 58 (1), 23–31. https://doi.org/10.1093/toxsci/58.1.23.

Bourgaud, F., Hehn, A., Larbat, R., Doerper, S., Gontier, E., Kellner, S., Matern, U., 2006. Biosynthesis of coumarins in plants: a major pathway still to be unravelled for cytochrome P450 enzymes. Phytochemistry Rev. 5, 293–308. https://doi.org/10.1007/s11101-006-9040-2.

Bouter, Y., Kacprowski, T., Weissmann, R., Dietrich, K., Borgers, H., Brauß, A., Sperling, C., Wirths, O., Albrecht, M., Jensen, L.R., 2014. Deciphering the molecular profile of plaques, memory decline and neuron loss in two mouse models for Alzheimer's disease by deep sequencing. Front. Aging Neurosci. 6, 75. https://doi.org/10.3389/fnagi.2014.00075.

Brandes, M.S., Gray, N.E., 2020. NRF2 as a therapeutic target in neurodegenerative diseases. ASN Neuro 12, 1759091419899782. https://doi.org/10.1177/1759091419899782.

Brown, S.A., 1962. Biosynthesis of coumarin and herniarin in lavender. Science 137 (3534), 977–978.

Budzynska, B., Boguszewska-Czubara, A., Kruk-Slomka, M., Skalicka-Wozniak, K., Michalak, A., Musik, I., Biala, G., 2015. Effects of imperatorin on scopolamine-induced cognitive impairment and oxidative stress in mice. Psychopharmacology 232, 931–942. https://doi.org/10.1007/s00213-014-3728-6.

Butterfield, D.A., Boyd-Kimball, D., 2004. Amyloid β-peptide (1-42) contributes to the oxidative stress and neurodegeneration found in Alzheimer disease brain. Brain Pathol. 14 (4), 426–432. https://doi.org/10.1111/j.1750-3639.2004.tb00087.x.

Carotti, A., Carrieri, A., Chimichi, S., Boccalini, M., Cosimelli, B., Gnerre, C., Carotti, A., Carrupt, P.-A., Testa, B., 2002. Natural and synthetic geiparvarins are strong and selective MAO-B inhibitors. Synthesis and SAR studies. Bioorg. Med. Chem. Lett 12 (24), 3551–3555. https://doi.org/10.1016/s0960-894x(02)00798-9.

Carreiras, M., Marco, J., 2004. Recent approaches to novel anti-Alzheimer therapy. Curr. Pharmaceut. Des. 10 (25), 3167. https://doi.org/10.2174/1381612043383421.

Casley-Smith, J., Casley-Smith, J., 1997. Coumarin in the treatment of lymphoedema and other high-protein oedemas. In: Coumarins: Biology, Applications and Mode of Action. John Wiley & Sons, Inc., New York, NY, p. 348.

Chatila, Z.K., Bradshaw, E.M., 2023. Alzheimer's disease genetics: a dampened microglial response? Neuroscientist 29 (2), 245–263. https://doi.org/10.1177/10738584211024531.

Choudhary, M.I., Khalid, A., Sultani, S.Z., 2002. A new coumarin from *Murraya paniculata*. Planta Med. 68 (01), 81–83. https://doi.org/10.1055/s-2002-19874.

Cui, S.-S., Jiang, Q.-W., Chen, S.-D., 2023. Sex difference in biological change and mechanism of Alzheimer's disease: from macro-to micro-landscape. Ageing Res. Rev. 101918. https://doi.org/10.1016/j.arr.2023.101918.

Cummings, J., Feldman, H.H., Scheltens, P., 2019. The "rights" of precision drug development for Alzheimer's disease. Alzheimer's Res. Ther. 11 (1), 1–14. https://doi.org/10.1186/s13195-019-0529-5.

Dall'Acqua, S., Maggi, F., Minesso, P., Salvagno, M., Papa, F., Vittori, S., Innocenti, G., 2010. Identification of non-alkaloid acetylcholinesterase inhibitors from *Ferulago campestris* (Besser) Grecescu (Apiaceae). Fitoterapia 81 (8), 1208–1212. https://doi.org/10.1016/j.fitote.2010.08.003.

Deng, G., Wu, C., Rong, X., Li, S., Ju, Z., Wang, Y., Ma, C., Ding, W., Guan, H., Cheng, X., 2019. Ameliorative effect of deoxyvasicine on scopolamine-induced cognitive dysfunction by restoration of cholinergic function in mice. Phytomedicine 63, 153007. https://doi.org/10.1016/j.phymed.2019.153007.

Deng, W., Aimone, J.B., Gage, F.H., 2010. New neurons and new memories: how does adult hippocampal neurogenesis affect learning and memory? Nat. Rev. Neurosci. 11 (5), 339–350. https://doi.org/10.1038/nrn2822.

Do Carmo, S., Cuello, A.C., 2013. Modeling Alzheimer's disease in transgenic rats. Mol. Neurodegener. 8 (1), 1–11. https://doi.org/10.1186/1750-1326-8-37.

Dorababu, A., 2022. Pharmacological report of recently designed multifunctional coumarin and coumarin–heterocycle derivatives. Arch. Pharmazie 355 (2), 2100345. https://doi.org/10.1002/ardp.202100345.

Duan, W., Zhou, G.D., Balachandrasekaran, A., Bhumkar, A.B., Boraste, P.B., Becker, J.T., Kuller, L.H., Lopez, O.L., Gach, H.M., Dai, W., 2021. Cerebral blood flow predicts conversion of mild cognitive impairment into Alzheimer's disease and cognitive decline: an arterial spin labeling follow-up study. J. Alzheim. Dis. 82 (1), 293–305. https://doi.org/10.3233/JAD-210199.

Dwyer, B.E., Zacharski, L.R., Balestra, D.J., Lerner, A.J., Perry, G., Zhu, X., Smith, M.A., 2009. Getting the iron out: phlebotomy for Alzheimer's disease? Med. Hypotheses 72 (5), 504–509. https://doi.org/10.1016/j.mehy.2008.12.029.

Egan, D., O'kennedy, R., Moran, E., Cox, D., Prosser, E., Thornes, R.D., 1990. The pharmacology, metabolism, analysis, and applications of coumarin and coumarin-related compounds. Drug Metabol. Rev. 22 (5), 503–529. https://doi.org/10.3109/03602539008991449.

Ehlers, D., Pfister, M., Bork, W.-R., Toffel-Nadolny, P., 1995. HPLC analysis of tonka bean extracts. Z. für Lebensmittel-Untersuchung und -Forsch. A 201, 278–282. https://doi.org/10.1007/BF01193004.

Ekström, F., Gottinger, A., Forsgren, N., Catto, M., Iacovino, L.G., Pisani, L., Binda, C., 2022. Dual reversible coumarin inhibitors mutually bound to monoamine oxidase B and acetylcholinesterase crystal structures. ACS Med. Chem. Lett. 13 (3), 499–506. https://doi.org/10.1021/acsmedchemlett.2c00001.

Farinola, N., Piller, N., 2005. Pharmacogenomics: its role in re-establishing coumarin as treatment for lymphedema. Lymphatic Res. Biol. 3 (2), 81–86. https://doi.org/10.1089/lrb.2005.3.81.

Fiebich, B.L., Batista, C.R.A., Saliba, S.W., Yousif, N.M., De Oliveira, A.C.P., 2018. Role of microglia TLRs in neurodegeneration. Front. Cell. Neurosci. 12, 329. https://doi.org/10.3389/fncel.2018.00329.

Flores-Morales, V., Villasana-Ruíz, A.P., Garza-Veloz, I., González-Delgado, S., Martinez-Fierro, M.L., 2023. Therapeutic effects of coumarins with different substitution patterns. Molecules 28 (5), 2413. https://doi.org/10.3390/molecules28052413.

Francis, P.T., 2005. The interplay of neurotransmitters in Alzheimer's disease. CNS Spectr. 10 (S18), 6–9. https://doi.org/10.1017/s1092852900014164.

Furukawa, Y., Watanabe, S., Okuyama, S., Nakajima, M., 2012. Neurotrophic effect of citrus auraptene: neuritogenic activity in PC12 cells. Int. J. Mol. Sci. 13 (5), 5338–5347. https://doi.org/10.3390/ijms13055338.

Galluzzi, S., Zanardini, R., Ferrari, C., Gipponi, S., Passeggia, I., Rampini, M., Sgrò, G., Genovese, S., Fiorito, S., Palumbo, L., 2022. Cognitive and biological effects of citrus phytochemicals in subjective cognitive decline: a 36-week, randomized, placebo-controlled trial. Nutr. J. 21 (1), 64. https://doi.org/10.1186/s12937-022-00817-6.

Gao, J., He, H., Jiang, W., Chang, X., Zhu, L., Luo, F., Zhou, R., Ma, C., Yan, T., 2015. Salidroside ameliorates cognitive impairment in a d-galactose-induced rat model of Alzheimer's disease. Behav. Brain Res. 293, 27–33. https://doi.org/10.1016/j.bbr.2015.06.045.

Gao, P., Wang, Z., Lei, M., Che, J., Zhang, S., Zhang, T., Hu, Y., Shi, L., Cui, L., Liu, J., 2022. Daphnetin ameliorates Aβ pathogenesis *via* STAT3/GFAP signaling in an APP/PS1 double-transgenic mouse model of Alzheimer's disease. Pharmacol. Res. 180, 106227. https://doi.org/10.1016/j.phrs.2022.106227.

Gao, Q., Jeon, S.J., Jung, H.A., Lee, H.E., Park, S.J., Lee, Y., Lee, Y., Ko, S.Y., Kim, B., Choi, J.S., 2015. Nodakenin enhances cognitive function and adult hippocampal neurogenesis in mice. Neurochem. Res. 40, 1438–1447. https://doi.org/10.1007/s11064-015-1612-3.

Gay, N.H., Suwanjang, W., Ruankham, W., Songtawee, N., Wongchitrat, P., Prachayasittikul, V., Prachayasittikul, S., Phopin, K., 2020. Butein, isoliquiritigenin, and scopoletin attenuate neurodegeneration *via* antioxidant enzymes and SIRT1/ADAM10 signaling pathway. RSC Adv. 10 (28), 16593–16606. https://doi.org/10.1039/c9ra06056a.

Genovese, S., Epifano, F., Curini, M., Dudra-Jastrzebska, M., Luszczki, J.J., 2009. Prenyloxyphenylpropanoids as a novel class of anticonvulsive agents. Bioorg. Med. Chem. Lett 19 (18), 5419–5422. https://doi.org/10.1016/j.bmcl.2009.07.110.

Ghanbarabadi, M., Iranshahi, M., Amoueian, S., Mehri, S., Motamedshariaty, V.S., Mohajeri, S.A., 2016. Neuroprotective and memory enhancing effects of auraptene in a rat model of vascular dementia: experimental study and histopathological evaluation. Neurosci. Lett. 623, 13–21. https://doi.org/10.1016/j.neulet.2016.04.047.

Hall, A.M., Roberson, E.D., 2012. Mouse models of Alzheimer's disease. Brain Res. Bull. 88 (1), 3–12. https://doi.org/10.1016/j.brainresbull.2011.11.017.

Hindam, M.O., Sayed, R.H., Skalicka-Woźniak, K., Budzyńska, B., El Sayed, N.S., 2020. Xanthotoxin and umbelliferone attenuate cognitive dysfunction in a streptozotocin-induced rat model of sporadic Alzheimer's disease: the role of JAK2/STAT3 and Nrf2/HO-1 signalling pathway modulation. Phytother Res. 34 (9), 2351–2365. https://doi.org/10.1002/ptr.6686.

Hu, M., Piller, N.B., 2017. Strategies for avoiding benzopyrone hepatotoxicity in lymphedema management—the role of pharmacogenetics, metabolic enzyme gene identification, and patient selection. Lymphatic Res. Biol. 15 (4), 317–323. https://doi.org/10.1089/lrb.2017.0020.

Hubbard, B.P., Sinclair, D.A., 2014. Small molecule SIRT1 activators for the treatment of aging and age-related diseases. Trends Pharmacol. Sci. 35 (3), 146–154. https://doi.org/10.1016/j.tips.2013.12.004.

Huong, D.T.L., Choi, H.C., Rho, T.C., Lee, H.S., Lee, M.K., Kim, Y.H., 1999. Inhibitory activity of monoamine oxidase by coumarins from *Peucedanum japonicum*. Arch Pharm. Res. (Seoul) 22, 324–326. https://doi.org/10.1007/BF02976373.

Husna Ibrahim, N., Yahaya, M.F., Mohamed, W., Teoh, S.L., Hui, C.K., Kumar, J., 2020. Pharmacotherapy of Alzheimer's disease: seeking clarity in a time of uncertainty. Front. Pharmacol. 11, 261. https://doi.org/10.3389/fphar.2020.00261.

Jawhar, S., Trawicka, A., Jenneckens, C., Bayer, T.A., Wirths, O., 2012. Motor deficits, neuron loss, and reduced anxiety coinciding with axonal degeneration and intraneuronal Aβ aggregation in the 5XFAD mouse model of Alzheimer's disease. Neurobiol. Aging 33 (1). https://doi.org/10.1016/j.neurobiolaging.2010.05.027, 196. e129-196. e140.

Jazvinšćak Jembrek, M., Hof, P.R., Šimić, G., 2015. Ceramides in Alzheimer's disease: key mediators of neuronal apoptosis induced by oxidative stress and Aβ accumulation. Oxid. Med. Cell. Longev. 2015, 346–783. https://doi.org/10.1155/2015/346783.

Jeong, S.H., Han, X.H., Hong, S.S., Hwang, J.S., Hwang, J.H., Lee, D., Lee, M.K., Ro, J.S., Hwang, B.Y., 2006. Monoamine oxidase inhibitory coumarins from the aerial parts of *Dictamnus albus*. Arch Pharm. Res. (Seoul) 29, 1119–1124. https://doi.org/10.1007/BF02969302.

Joghataee, S., Mohammad-zadeh, M., Amin, B., Jafari, F., Tondar, M., Gholami, O., 2020. Auraptene has neuroprotective and memory enhancing effects in a rat model of Alzheimer's disease. Neurol. Asia 25 (3), 353–360.

Jung, P.A.D.J., Huneck, S., 1991. Gigasol and other coumarins from *Angelica gigas*. Phytochemistry 30 (2), 710–712. https://doi.org/10.1016/0031-9422(91)83763-B.

Kang, S.Y., Lee, K.Y., Sung, S.H., Park, M.J., Kim, Y.C., 2001. Coumarins isolated from *Angelica gigas* inhibit acetylcholinesterase: structure– activity relationships. J. Nat. Prod. 64 (5), 683–685. https://doi.org/10.1021/np000441w.

Karakaya, S., Koca, M., Yılmaz, S.V., Yıldırım, K., Pınar, N.M., Demirci, B., Brestic, M., Sytar, O., 2019. Molecular docking studies of coumarins isolated from extracts and essential oils of *Zosima absinthifolia* Link as potential inhibitors for Alzheimer's disease. Molecules 24 (4), 722. https://doi.org/10.3390/molecules24040722.

Kennedy, B., Ziegler, M., Alford, M., Hansen, L., Thal, L., Masliah, E., 2003. Early and persistent alterations in prefrontal cortex MAO A and B in Alzheimer's disease. J. Neural. Transm. 110 (7), 789–801. https://doi.org/10.1007/s00702-003-0828-6.

Kerr, F., Sofola-Adesakin, O., Ivanov, D.K., Gatliff, J., Gomez Perez-Nievas, B., Bertrand, H.C., Martinez, P., Callard, R., Snoeren, I., Cocheme, H.M., 2017. Direct Keap1-Nrf2 disruption as a potential therapeutic target for Alzheimer's disease. PLoS Genet. 13 (3), e1006593. https://doi.org/10.1371/journal.pgen.1006593.

Khalid, A., Khan, W., Zia, K., Ahsan, W., Alhazmi, H., Abdalla, A.N., Najmi, A., Khan, A., Bouyahya, A., Ul-Haq, Z., 2023. Natural coumarins from *Murraya paniculata* as mixed-type inhibitors of cholinesterases: *in vitro* and *in silico* investigations. Front. Pharmacol. 14, 585. https://doi.org/10.3389/fphar.2023.1133809.

Khan, S., Choi, R.J., Lee, J., Kim, Y.S., 2016. Attenuation of neuropathic pain and neuroinflammatory responses by a pyranocoumarin derivative, anomalin in animal and cellular models. Eur. J. Pharmacol. 774, 95–104. https://doi.org/10.1016/j.ejphar.2016.02.008.

Kiris, I., Skalicka-Wozniak, K., Basar, M.K., Sahin, B., Gurel, B., Baykal, A.T., 2022. Molecular effects of pteryxin and scopoletin in the 5xFAD Alzheimer's disease mouse model. Curr. Med. Chem. 29 (16), 2937–2950. https://doi.org/10.2174/0929867328666210827152914.

Klyucherev, T.O., Olszewski, P., Shalimova, A.A., Chubarev, V.N., Tarasov, V.V., Attwood, M.M., Syvänen, S., Schiöth, H.B., 2022. Advances in the development of new biomarkers for Alzheimer's disease. Transl. Neurodegener. 11 (1), 1–24. https://doi.org/10.1186/s40035-022-00296-z.

Kong, L.D., Tan, R.X., Woo, A.Y.H., Cheng, C.H.K., 2001. Inhibition of rat brain monoamine oxidase activities by psoralen and isopsoralen: implications for the treatment of affective disorders. Pharmacol. Toxicol. 88 (2), 75–80. https://doi.org/10.1034/j.1600-0773.2001.d01-86.x.

Kontogiorgis, C., Detsi, A., Hadjipavlou-Litina, D., 2012. Coumarin-based drugs: a patent review (2008–present). Expert Opin. Ther. Pat. 22 (4), 437–454. https://doi.org/10.1517/13543776.2012.678835.

Kowalczyk, J., Budzyńska, B., Kurach, Ł., Pellegata, D., El Sayed, N.S., Gertsch, J., Skalicka-Woźniak, K., 2022. Neuropsychopharmacological profiling of scoparone in mice. Sci. Rep. 12 (1), 822. https://doi.org/10.1038/s41598-021-04741-3.

Kowalczyk, J., Kurach, Ł., Boguszewska-Czubara, A., Skalicka-Woźniak, K., Kruk-Słomka, M., Kurzepa, J., Wydrzynska-Kuźma, M., Biała, G., Skiba, A., Budzyńska, B., 2020. Bergapten improves scopolamine-induced memory impairment in mice *via* cholinergic and antioxidative mechanisms. Front. Neurosci. 14, 730. https://doi.org/10.3389/fnins.2020.00730.

Kozioł, E., Luca, S.V., Ağalar, H.G., Sağlık, B.N., Demirci, F., Marcourt, L., Wolfender, J.-L., Jóźwiak, K., Skalicka-Woźniak, K., 2020. Rutamarin: efficient liquid–liquid chromatographic isolation from *Ruta graveolens* L. and evaluation of its *in vitro* and *in silico* MAO-B inhibitory activity. Molecules 25 (11), 2678. https://doi.org/10.3390/molecules25112678.

Lake, B., 1999. Coumarin metabolism, toxicity and carcinogenicity: relevance for human risk assessment. Food Chem. Toxicol. 37 (4), 423–453. https://doi.org/10.1016/s0278-6915(99)00010-1.

Lee, J.H., Lee, K.T., Yang, J.H., Baek, N.I., Kim, D.K., 2004. Acetylcholinesterase inhibitors from the twigs of *Vaccinium oldhami* Miquel. Arch Pharm. Res. (Seoul) 27, 53–56. https://doi.org/10.1007/BF02980046.

Lipinski, C.A., Lombardo, F., Dominy, B.W., Feeney, P.J., 1997. Experimental and computational approaches to estimate solubility and permeability in drug discovery and development settings. Adv. Drug Deliv. Rev. 23 (1–3), 3–25. https://doi.org/10.1016/s0169-409x(00)00129-0.

Liu, J., Chen, Q., Jian, Z., Xiong, X., Shao, L., Jin, T., Zhu, X., Wang, L., 2016. Daphnetin protects against cerebral ischemia/reperfusion injury in mice *via* inhibition of TLR4/NF-κB signaling pathway. BioMed Res. Int. 2016, 2816056. https://doi.org/10.1155/2016/2816056.

Liu, J., Miao, M., Wei, F., 2023. Angelicin alleviates maternal isoflurane exposure-induced offspring cognitive defects through the carbonic anhydrase 4/aquaporin-4 pathway. Mol. Biotechnol. https://doi.org/10.1007/s12033-023-00723-0.

Luo, K., Jahufer, M., Wu, F., Di, H., Zhang, D., Meng, X., Zhang, J., Wang, Y., 2016. Genotypic variation in a breeding population of yellow sweet clover (*Melilotus officinalis*). Front. Plant Sci. 7, 972. https://doi.org/10.3389/fpls.2016.00972.

Maarouf, C.L., Kokjohn, T.A., Whiteside, C.M., Macias, M.P., Kalback, W.M., Sabbagh, M.N., Beach, T.G., Vassar, R., Roher, A.E., 2013. Molecular differences and similarities between Alzheimer's disease and the 5XFAD transgenic mouse model of amyloidosis. Biochem. Insights 6. https://doi.org/10.4137/BCI.S13025. BCI. S13025.

Maccioni, R.B., Farías, G., Morales, I., Navarrete, L., 2010. The revitalized tau hypothesis on Alzheimer's disease. Arch. Med. Res. 41 (3), 226–231. https://doi.org/10.1016/j.arcmed.2010.03.007.

Marshall, M., Mohler, J., Edmonds, K., Williams, B., Butler, K., Ryles, M., Weiss, L., Urban, D., Bueschen, A., Markiewicz, M., 1994. An updated review of the clinical development of coumarin (1, 2-benzopyrone) and 7-hydroxycoumarin. J. Cancer Res. Clin. Oncol. 120, S39–S42. https://doi.org/10.1016/j.arcmed.2010.03.007.

Menezes, J.C., Diederich, M.F., 2019. Natural dimers of coumarin, chalcones, and resveratrol and the link between structure and pharmacology. Eur. J. Med. Chem. 182, 111637. https://doi.org/10.1016/j.ejmech.2019.111637.

Miller, K.G., Poole, C., Pawlowski, T., 1996. Classification of the botanical origin of cinnamon by solid-phase microextraction and gas chromatography. Chromatographia 42, 639−646.

Miyazawa, M., Tsukamoto, T., Anzai, J., Ishikawa, Y., 2004. Insecticidal effect of phthalides and furanocoumarins from *Angelica acutiloba* against *Drosophila melanogaster*. J. Agric. Food Chem. 52 (14), 4401−4405. https://doi.org/10.1021/jf0497049.

Mondragón-Rodríguez, S., Perry, G., Zhu, X., Moreira, P.I., Williams, S., 2012. Glycogen synthase kinase 3: a point of integration in Alzheimer's disease and a therapeutic target? Int. J. Alzheimer's Dis. 2012, 276803. https://doi.org/10.1155/2012/276803.

Morgan, D., 2007. Amyloid, memory and neurogenesis. Exp. Neurol. 205 (2), 330−335. https://doi.org/10.1016/j.expneurol.2007.03.006.

Musa, M.A., Cooperwood, J.S., Khan, M.O.F., 2008. A review of coumarin derivatives in pharmacotherapy of breast cancer. Curr. Med. Chem. 15 (26), 2664−2679. https://doi.org/10.2174/092986708786242877.

Nascimento, A., Fernandes, R., Carvalho, A., Frigieri, I., Alves, R., Chorilli, M., 2022. Insights for Alzheimer's disease pharmacotherapy and current clinical trials. Neurochem. Int. 159, 105401. https://doi.org/10.1016/j.neuint.2022.105401.

Nisbet, R.M., Polanco, J.-C., Ittner, L.M., Götz, J., 2015. Tau aggregation and its interplay with amyloid-β. Acta Neuropathol. 129, 207−220. https://doi.org/10.1007/s00401-014-1371-2.

O'Shea, J.J., Plenge, R., 2012. JAK and STAT signaling molecules in immunoregulation and immune-mediated disease. Immunity 36 (4), 542−550. https://doi.org/10.1016/j.immuni.2012.03.014.

Orhan, I.E., Senol Deniz, F.S., Trædal-Henden, S., Cerón-Carrasco, J.P., den Haan, H., Peña-García, J., Pérez-Sánchez, H., Emerce, E., Skalicka-Wozniak, K., 2019. Profiling auspicious butyrylcholinesterase inhibitory activity of two herbal molecules: hyperforin and hyuganin C. Chem. Biodivers. 16 (5), e1900017. https://doi.org/10.1002/cbdv.201900017.

Orhan, I.E., Tosun, F., Deniz, F.S.S., Eren, G., Mıhoğlugil, F., Akalgan, D., Miski, M., 2021. Butyrylcholinesterase-inhibiting natural coumarin molecules as potential leads. Phytochem. Lett. 44, 48−54. https://doi.org/10.1016/j.phytol.2021.05.001.

Patil, P., Thakur, A., Sharma, A., Flora, S.J.S., 2020. Natural products and their derivatives as multifunctional ligands against Alzheimer's disease. Drug Dev. Res. 81 (2), 165−183. https://doi.org/10.1002/ddr.21587.

Pejchal, V., Stepankova, S., Padelkova, Z., Imramovsky, A., Jampilek, J., 2011. 1, 3-Substituted imidazolidine-2, 4, 5-triones: synthesis and inhibition of cholinergic enzymes. Molecules 16 (9), 7565−7582. https://doi.org/10.3390/molecules16097565.

Peroutka, R., Schulzová, V., Botek, P., Hajšlová, J., 2007. Analysis of furanocoumarins in vegetables (Apiaceae) and citrus fruits (Rutaceae). J. Sci. Food Agric. 87 (11), 2152−2163. https://doi.org/10.1002/jsfa.2979.

Pitaro, M., Croce, N., Gallo, V., Arienzo, A., Salvatore, G., Antonini, G., 2022. Coumarin-induced hepatotoxicity: a narrative review. Molecules 27 (24), 9063. https://doi.org/10.3390/molecules27249063.

Pravin, N., Jozwiak, K., 2022. Effects of Linkers and substitutions on multitarget directed ligands for Alzheimer's disease: emerging paradigms and strategies. Int. J. Mol. Sci. 23 (11), 6085. https://doi.org/10.3390/ijms23116085.

Rapaka, D., Adiukwu, P.C., Bitra, V.R., 2022. Experimentally induced animal models for cognitive dysfunction and Alzheimer's disease. MethodsX, 101933. https://doi.org/10.1016/j.mex.2022.101933.

Rollinger, J.M., Hornick, A., Langer, T., Stuppner, H., Prast, H., 2004. Acetylcholinesterase inhibitory activity of scopolin and scopoletin discovered by virtual screening of natural products. J. Med. Chem. 47 (25), 6248−6254. https://doi.org/10.1021/jm049655r.

Salem, M.A., Budzyńska, B., Kowalczyk, J., El Sayed, N.S., Mansour, S.M., 2021. Tadalafil and bergapten mitigate streptozotocin-induced sporadic Alzheimer's disease in mice *via* modulating neuroinflammation, PI3K/Akt, Wnt/β-catenin, AMPK/mTOR signaling pathways. Toxicol. Appl. Pharmacol. 429, 115697. https://doi.org/10.1016/j.taap.2021.115697.

Sampedro, F., Vilaplana, E., De Leon, M.J., Alcolea, D., Pegueroles, J., Montal, V., Carmona-Iragui, M., Sala, I., Sánchez-Saudinos, M.-B., Antón-Aguirre, S., 2015. APOE-by-sex interactions on brain structure and metabolism in healthy elderly controls. Oncotarget 6 (29), 26663. https://doi.org/10.18632/oncotarget.5185.

Sasaguri, H., Nilsson, P., Hashimoto, S., Nagata, K., Saito, T., De Strooper, B., Hardy, J., Vassar, R., Winblad, B., Saido, T.C., 2017. APP mouse models for Alzheimer's disease preclinical studies. EMBO J. 36 (17), 2473–2487. https://doi.org/10.15252/embj.201797397.

Sawada, H., Udaka, F., Izumi, Y., Nishinaka, K., Kawakami, H., Nakamura, S., Kameyama, M., 2000. Cerebral white matter lesions are not associated with apoE genotype but with age and female sex in Alzheimer's disease. J. Neurol. Neurosurg. Psychiatr. 68 (5), 653–656. https://doi.org/10.1136/jnnp.68.5.653.

Seong, S.H., Ali, M.Y., Jung, H.A., Choi, J.S., 2019. Umbelliferone derivatives exert neuroprotective effects by inhibiting monoamine oxidase A, self-amyloidβ aggregation, and lipid peroxidation. Bioorg. Chem. 92, 103293. https://doi.org/10.1016/j.bioorg.2019.103293.

Sharma, A., Nuthakki, V.K., Gairola, S., Singh, B., Bharate, S.B., 2022. A coumarin–donepezil hybrid as a blood–brain barrier permeable dual cholinesterase inhibitor: isolation, synthetic modifications, and biological evaluation of natural coumarins. ChemMedChem 17 (18), e202200300. https://doi.org/10.1002/cmdc.202200300.

Shimizu, B.-I., 2014. 2-Oxoglutarate-dependent dioxygenases in the biosynthesis of simple coumarins. Front. Plant Sci. 5, 549. https://doi.org/10.3389/fpls.2014.00549.

Skalicka-Wozniak, K., Budzynska, B., Biala, G., Boguszewska-Czubara, A., 2018. Scopolamine-induced memory impairment is alleviated by xanthotoxin: role of acetylcholinesterase and oxidative stress processes. ACS Chem. Neurosci. 9 (5), 1184–1194. https://doi.org/10.1021/acschemneuro.8b00011.

Skalicka-Woźniak, K., Orhan, I.E., Cordell, G.A., Nabavi, S.M., Budzyńska, B., 2016. Implication of coumarins towards central nervous system disorders. Pharmacol. Res. 103, 188–203. https://doi.org/10.1016/j.phrs.2015.11.023.

Soto-Lara, M., Silva-Loredo, M., Monroy-Córdoba, J.R., Flores-Ordoñez, P., Cervera-Delgadillo, N.G., Carrillo-Mora, P., 2023. Alternative medicine therapies in neurological disorders: prevalence, reasons and associated factors. A systematic review. Compl. Ther. Med. 73, 102932. https://doi.org/10.1016/j.ctim.2023.102932.

Stanciu, G.D., Luca, A., Rusu, R.N., Bild, V., Beschea Chiriac, S.I., Solcan, C., Bild, W., Ababei, D.C., 2019. Alzheimer's disease pharmacotherapy in relation to cholinergic system involvement. Biomolecules 10 (1), 40. https://doi.org/10.3390/biom10010040.

Stefanachi, A., Leonetti, F., Pisani, L., Catto, M., Carotti, A., 2018. Coumarin: a natural, privileged and versatile scaffold for bioactive compounds. Molecules 23 (2), 250. https://doi.org/10.3390/molecules23020250.

Svob Strac, D., Konjevod, M., Sagud, M., Nikolac Perkovic, M., Nedic Erjavec, G., Vuic, B., Simic, G., Vukic, V., Mimica, N., Pivac, N., 2021. Personalizing the care and treatment of Alzheimer's disease: an overview. Pharmacogenomics Personalized Med. 14, 631–653. https://doi.org/10.2147/PGPM.S284615.

Tabrizian, K., Yaghoobi, N.S., Iranshahi, M., Shahraki, J., Rezaee, R., Hashemzaei, M., 2015. Auraptene consolidates memory, reverses scopolamine-disrupted memory in passive avoidance task, and ameliorates retention deficits in mice. Iran. J. Basic Med. Sci. 18 (10), 1014.

Takomthong, P., Waiwut, P., Yenjai, C., Sripanidkulchai, B., Reubroycharoen, P., Lai, R., Kamau, P., Boonyarat, C., 2020. Structure–activity analysis and molecular docking studies of coumarins from *Toddalia asiatica* as multifunctional agents for Alzheimer's disease. Biomedicines 8 (5), 107. https://doi.org/10.3390/biomedicines8050107.

Torok, J., Anand, C., Verma, P., Raj, A., 2022. Connectome-based biophysics models of Alzheimer's disease diagnosis and prognosis. Transl. Res. 254, 13–23. https://doi.org/10.1016/j.trsl.2022.08.008.

Uliassi, E., Gandini, A., Perone, R.C., Bolognesi, M.L., 2017. Neuroregeneration *versus* neurodegeneration: toward a paradigm shift in Alzheimer's disease drug discovery. Future Med. Chem. 9 (10), 995–1013. https://doi.org/10.4155/fmc-2017-0038.

Vassallo, J.D., Hicks, S.M., Daston, G.P., Lehman-McKeeman, L.D., 2004. Metabolic detoxification determines species differences in coumarin-induced hepatotoxicity. Toxicol. Sci. 80 (2), 249–257. https://doi.org/10.1093/toxsci/kfh162.

Verma, A., Waiker, D.K., Bhardwaj, B., Saraf, P., Shrivastava, S.K., 2022. The molecular mechanism, targets, and novel molecules in the treatment of Alzheimer's disease. Bioorg. Chem. 119, 105562. https://doi.org/10.1016/j.bioorg.2021.105562.

Vogt, T., 2010. Phenylpropanoid biosynthesis. Mol. Plant 3 (1), 2–20. https://doi.org/10.1093/mp/ssp106.

Wang, Q., Zhang, L., Yuan, X., Ou, Y., Zhu, X., Cheng, Z., Zhang, P., Wu, X., Meng, Y., Zhang, L., 2016. The relationship between the Bcl-2/Bax proteins and the mitochondria-mediated apoptosis pathway in the differentiation of adipose-derived stromal cells into neurons. PLoS One 11 (10), e0163327. https://doi.org/10.1371/journal.pone.0163327.

Wang, X., Li, R., Wang, X., Fu, Q., Ma, S., 2015. Umbelliferone ameliorates cerebral ischemia–reperfusion injury *via* upregulating the PPAR gamma expression and suppressing TXNIP/NLRP3 inflammasome. Neurosci. Lett. 600, 182–187. https://doi.org/10.1016/j.neulet.2015.06.016.

Wei, H., Li, L., Song, Q., Ai, H., Chu, J., Li, W., 2005. Behavioural study of the D-galactose induced aging model in C57BL/6J mice. Behav. Brain Res. 157 (2), 245–251. https://doi.org/10.1016/j.bbr.2004.07.003.

Whishaw, I.Q., Metz, G.A., Kolb, B., Pellis, S.M., 2001. Accelerated nervous system development contributes to behavioral efficiency in the laboratory mouse: a behavioral review and theoretical proposal. Dev. Psychobiol. 39 (3), 151–170. https://doi.org/10.1002/dev.1041.

Wilcock, G., Esiri, M., Bowen, D., Smith, C., 1982. Alzheimer's disease: correlation of cortical choline acetyltransferase activity with the severity of dementia and histological abnormalities. J. Neurol. Sci. 57 (2–3), 407–417. https://doi.org/10.1016/0022-510x(82)90045-4.

Xiao, H., Wang, Y., Wu, Y., Li, H., Liang, X., Lin, Y., Kong, L., Ni, Y., Deng, Y., Li, Y., 2021. Osthol ameliorates cognitive impairments *via* augmenting neuronal population in APP/PS1 transgenic mice. Neurosci. Res. 164, 33–45. https://doi.org/10.1016/j.neures.2020.04.001.

Yan, Z., Gibson, S.A., Buckley, J.A., Qin, H., Benveniste, E.N., 2018. Role of the JAK/STAT signaling pathway in regulation of innate immunity in neuroinflammatory diseases. Clin. Immunol. 189, 4–13. https://doi.org/10.1016/j.clim.2016.09.014.

Yao, H., Zhao, J., Song, X., 2022. Protective effects of fraxin on cerebral ischemia-reperfusion injury by mediating neuroinflammation and oxidative stress through PPAR-γ/NF-KB pathway. Brain Res. Bull. 187, 49–62. https://doi.org/10.1016/j.brainresbull.2022.06.010.

Yao, X., Yu, M., Zhou, Y., Guan, H., 2022. Esculetin ameliorates cognitive impairments in D-galactose-induced Alzheimer's disease rats by inhibiting inflammation and oxidative stress. Phcog. Mag. 18 (79).

Youkwan, J., Sutthivaiyakit, S., Sutthivaiyakit, P., 2010. Citrusosides A–D and furanocoumarins with cholinesterase inhibitory activity from the fruit peels of *Citrus hystrix*. J. Nat. Prod. 73 (11), 1879–1883. https://doi.org/10.1021/np100531x.

Zahs, K.R., Ashe, K.H., 2010. 'Too much good news'—are Alzheimer mouse models trying to tell us how to prevent, not cure, Alzheimer's disease? Trends Neurosci. 33 (8), 381–389. https://doi.org/10.1016/j.tins.2010.05.004.

Zhang, C., Lv, Y., Bai, R., Xie, Y., 2021. Structural exploration of multifunctional monoamine oxidase B inhibitors as potential drug candidates against Alzheimer's disease. Bioorg. Chem. 114, 105070. https://doi.org/10.1016/j.bioorg.2021.105070.

Zhang, L., Chen, C., Mak, M.S., Lu, J., Wu, Z., Chen, Q., Han, Y., Li, Y., Pi, R., 2020. Advance of sporadic Alzheimer's disease animal models. Med. Res. Rev. 40 (1), 431–458. https://doi.org/10.1002/med.21624.

Zhang, T., Zhou, B., Sun, J., Song, J., Nie, L., Zhu, K., 2022. Fraxetin suppresses reactive oxygen species-dependent autophagy by the PI3K/Akt pathway to inhibit isoflurane-induced neurotoxicity in hippocampal neuronal cells. J. Appl. Toxicol. 42 (4), 617–628. https://doi.org/10.1002/jat.4243.

Zhao, C., Deng, W., Gage, F.H., 2008. Mechanisms and functional implications of adult neurogenesis. Cell 132 (4), 645–660. https://doi.org/10.1016/j.expneurol.2020.113178.

CHAPTER 9

Increasing of biologically active metabolite production using plant biotechnology techniques in *Nasturtium officinale* (watercress) microshoot cultures

Marta Klimek-Szczykutowicz[1,2], Halina Ekiert[1] and Agnieszka Szopa[1]

[1]*Department of Medicinal Plant and Mushroom Biotechnology, Faculty of Pharmacy, Jagiellonian University, Medical College, Kraków, Poland;* [2]*Department of Pharmaceutical Sciences, Collegium Medicum, Jan Kochanowski University, Kielce, Poland*

Abbreviations

C	Control cultures
DW	Dry weight
ETH	Ethephon
F–C	Folin–Ciocalteu
GAL	Gallic acid equivalent
Gi	Growth index
GLP	Good laboratory practice
GMP	Good manufacturing practice
GSLs	Glucosinolates
HPLC-DAD	High-performance liquid chromatography with diode-array detection
IUCN	Union for Conservation of Nature
JA	Jasmonic acid
MeJA	Methyl jasmonate
MS	Murashige and Skoog
NaSA	Sodium salicylate
Phe	Phenylalanine
RE	Rutoside equivalent
SA	Salicylic acid
SIN	Sinigrin equivalent
TE	Trolox equivalent
TIS	Temporary immersion system
Trp	Tryptophan

Phytochemicals for Health. https://doi.org/10.1016/B978-0-443-15366-2.00009-5

UHPLC-DAD-ESI-MS	Ultra-high-performance liquid chromatography diode array detector electrospray ionization tandem mass spectrometry
UHPLC-DAD-MS/MS	Ultra-high-performance liquid chromatography-tandem mass spectrometry
YeE	Yeast extract

1. *Nasturtium officinale*—chemical, medicinal, nutritional, and cosmetological characteristics

1.1 Botanical characteristics

Nasturtium officinale R. Br. (Robert Brown) – watercress, is an aquatic or semiaquatic species belonging to the Brassicaceae family. This species is very rare in natural habitats in Europe. In Poland, it has been classified by the International Union for Conservation of Nature (IUCN) (The IUCN Red List of Threatened Species, 2023) as partially endangered. The synonymous names of this species are *Nasturtium aquaticum* Wahlenb., *Rorippa nasturtium-aquaticum* (L.) Hayek and *Rorippa officinalis* (W.T. Aiton) P. Royen. In foreign languages, this species is known as: watercress (English); Brunnenkresse, Wasserkresse, Brunnkresse (German); cresson de fontaine (French); berro (Spanish); dou ban cai and 豆瓣菜 (Chinese) (IUCN).

N. officinale is a perennial aquatic or semiaquatic species with creeping or floating stems. The stems are 10–70 cm long, hollow, irregularly shaped. The feathery leaves are dark green. The inflorescence is a raceme. The flowers are small, and white with four petals. The fruit is a silique containing small, red-brown seeds (Manton, 1935; World Flora Online, 2023).

N. officinale has natural habitats in Europe, North Africa and Asia. This species has also settled in North America, South America and Australia. In its natural environment, it usually grows in shallow streams with a gentle current, ditches, pond edges and swampy areas. Commercial cultivation began in southern England in 1808 (Klimek-Szczykutowicz et al., 2018a,b). Currently, artificial breeding is the main source of obtaining this plant, because the number of its natural sites is decreasing. According to the IUCN, the species is classified as "Least Concern" (IUCN). Each European country has a separate classification of *N. officinale*, e.g., in Estonia this species has the status of an extinct or probably extinct species, in Austria and Sweden the status of an endangered species, and in Poland is partially endangered (Klimek-Szczykutowicz et al., 2018a,b; IUCN).

1.2 Traditional use

Herb of *N. officinale* (*Nasturtii herba*) is a raw material used in medicine in Europe and in other countries around the world. *N. officinale* has long been used in the form of extracts from fresh or dried herbs in traditional Persian medicine, Iran, Azerbaijan, Morocco, and also in Mauritius. The leaves have been used in traditional Moroccan medicine to improve overall health in pediatric medicine (Teixidor-Toneu et al., 2016). In Iran and Azerbaijan, this raw material was recommended in the treatment of e.g., hypertension, hyperglycemia and renal colic (Asadi-Samani et al., 2017). On the other hand, in Mauritius, it was used in the treatment of asthma, cough and bronchial problems (Suroowan and Mahomoodally, 2016).

Nowadays, the *N. officinale* herb is an officially accepted plant material for use in phytotherapy in Germany. According to the German Commission Monograph E, *Nasturtii herba* can be used in the form of fresh or dried aerial parts of the plant for the treatment of rhinitis. *N. officinale* also has an important position in the German Commission Monograph D on homeopathic medicines. Indications for its use are the treatment of inflammation of the urinary tract (Blumenthal et al., 2000).

1.3 Chemical composition

The conducted phytochemical studies showed differences in qualitative and quantitative content between different parts of the plant. Above-ground parts are characterized by the greatest amount of compounds and diversity (Afsharypuor and Salehi, 2008; Jeon et al., 2017). The main groups of secondary metabolites, characteristic of *N. officinale*, are: glucosinolates (GSLs), isothiocyanates and polyphenolic compounds (Fig. 9.1).

GSLs are a special group of compounds found mainly in the Brassicaceae family. Most GSLs are thermally and chemically stable. Plants containing GSLs produce a special enzyme − myrosinase (β-thioglucosidase), which hydrolyzes the molecules of these glucosides. As a result of crushing the fresh raw material and destroying the tissue structure, enzymatic decomposition of GSLs occurs. Hydrolysis products include: isothiocyanates, nitriles, elemental sulfur, thiocyanates, epithionitriles, oxazolidine-2-thiones and/or indole compounds (Fig. 9.1) (Holst and Williamson, 2004).

According to the professional classification, GSLs are divided into three groups due to the amino acid side chain: aliphatic GSLs, aromatic GSLs and indole GSLs. Aliphatic GSLs are derived from alanine (Ala), leucine (Leu), isoleucine (Ile), methionine (Met) or valine (Val). Aromatic GSLs are derivatives of phenylalanine (Phe) or tyrosine (Tyr), while indole GSLs are derivatives of tryptophan

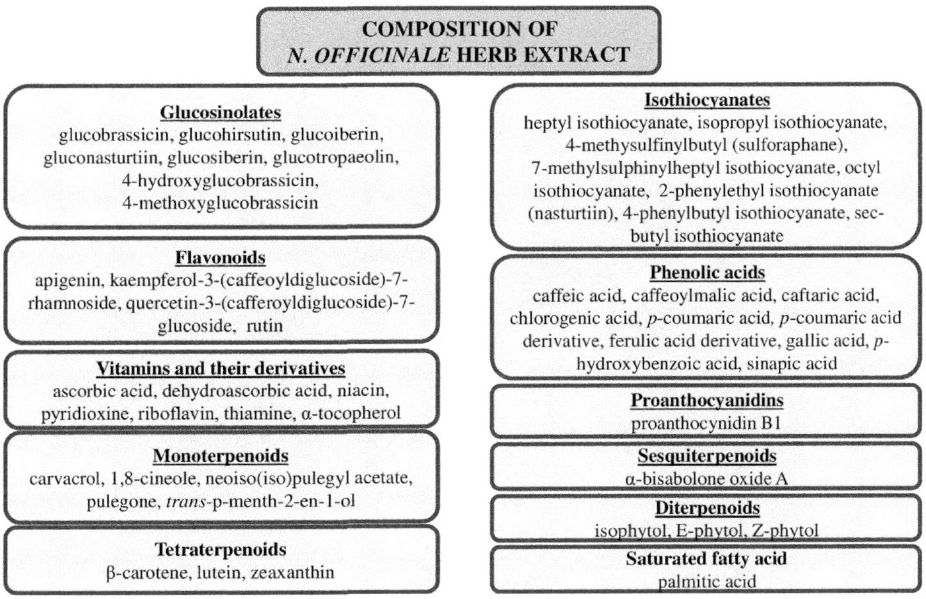

FIGURE 9.1

Chemical profile of *N. officinale* herb extract.

(Trp) (Blažević et al., 2020; Holst and Williamson, 2004). The most important metabolites responsible for the biological activity of *N. officinale* are GSLs. There are eight specific GSLs in the species: Phe derivatives – gluconasturtiin and glucotropaeolin, Trp derivatives – glucobrassicin, 4-hydroxyglucobrassicin, 4-methoxyglucobrassicin, and Met derivatives - glucohirsutin, glucoiberin, and glucosiberin (Fig. 9.2). The presence of these compounds has been found in extracts from leaves, stems, roots, flowers and seeds. The highest content of GSLs was found in flowers and herbs. The dominant compounds in all parts of the plant are gluconasturtiin (2-phenylethyl GSL), and the hydrolysis product of this glucoside – isothiocyanate 2-phenylethyl (nasturtiin) (Fig. 9.3) (Afsharypuor and Salehi, 2008; Jeon et al., 2017).

Another important group of secondary metabolites found in *N. officinale* are polyphenols, such as flavonoids, phenolic acids and proanthocyanidins. In leaf and herb extracts, apart from glycosidic compounds of quercetin and kaempferol, the presence of apigenin and rutoside was also confirmed. Among the phenolic acids, the presence of, among others, chlorogenic acid, gallic acid, caffeic acid, caffeoylmalonic acid, caftaric acid, *p*-coumaric acid and sinapic acid as well as derivatives of ferulic acid, has been documented (Boligon et al., 2013; Martínez-Sánchez et al., 2008; Zeb, 2015).

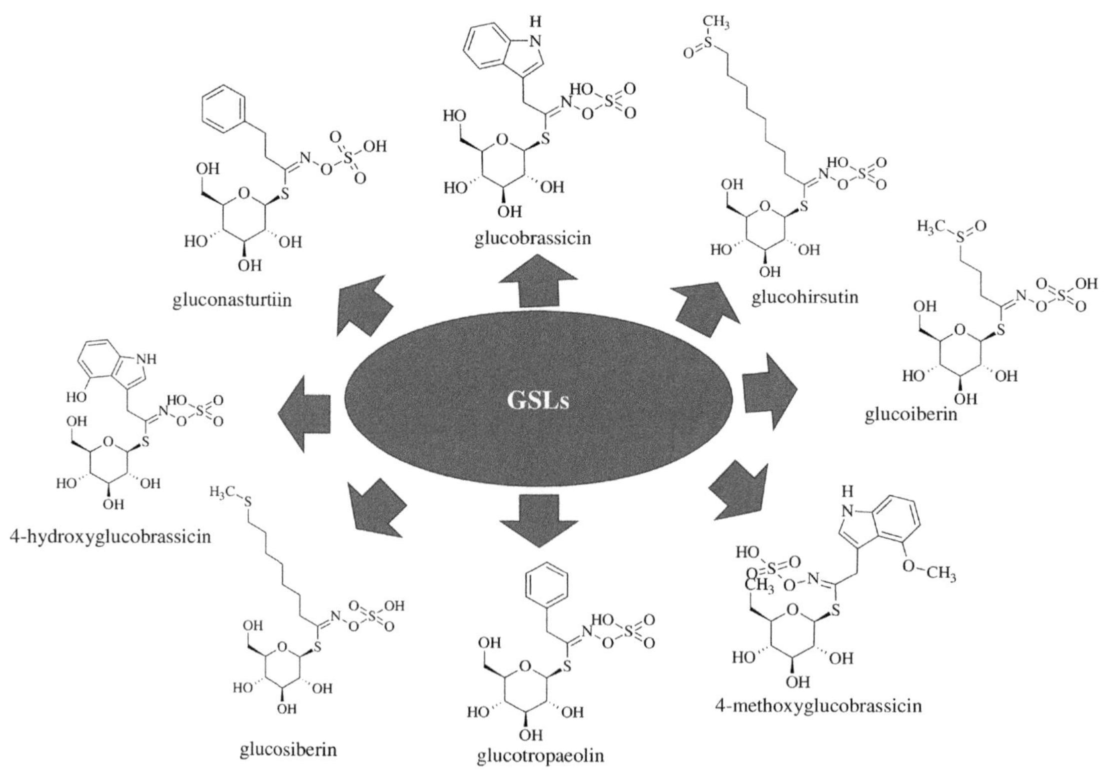

FIGURE 9.2

Chemical structure of GSLs characteristic for *N. officinale*.

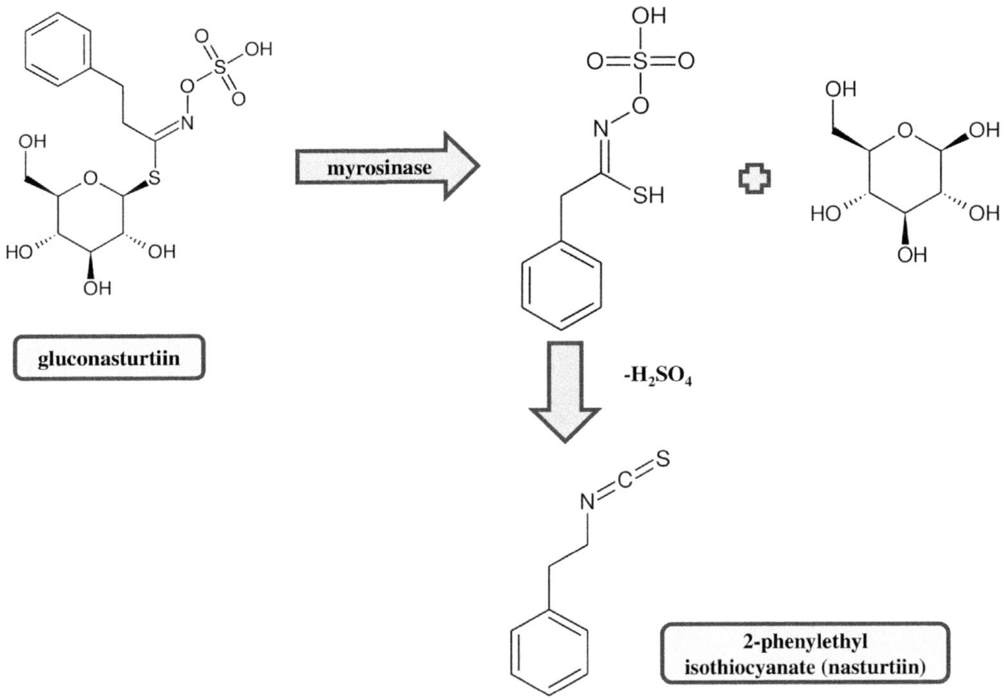

FIGURE 9.3

Reaction of gluconasturtiin decomposition to 2-phenylethyl isothiocyanate (nasturtiin) and glucose as a result of the myrosinase activity.

1.4 Biological activity of *N. officinale* extracts

The biological activity of *N. officinale* extracts (Fig. 9.4) is mainly determined by the presence of GSLs and isothiocyanates in their composition. Specific activity profiles have been proven for some of these compounds. Modern research has proven their strong anticancer and antioxidant effects (Amiri, 2012; Casanova et al., 2013; De Souza et al., 2016; Giallourou et al., 2019; Li et al., 2016; Martínez-Sánchez et al., 2008; Shahani et al., 2017; Yuan et al., 2016). The latest studies prove anticancer effects in the treatment of lung, colon, breast, liver and bile duct cancer. The main compounds responsible for this activity are isothiocyanates (Arumugam et al., 2020; Boyd et al., 2006; De Souza et al., 2016; Li et al., 2016; Schuchardt et al., 2019; Yuan et al., 2016). Other studies have also shown the antiinflammatory and antipsoriatic effects of sulforaphane – an isothiocyanate isolated from the *N. officinale* herb (Camponogara et al., 2019; Yehuda et al., 2012). Extracts from the herb also have scientifically proven effects, e.g., antiallergic, antibacterial and hypolipemic, as well as lower glucose levels and can inhibit the enzyme – angiotensin-converting enzyme, thanks to which they also affect the functioning of the circulatory system (Bahramikia and Yazdanparast, 2008; Hoseini et al., 2009; Mahdavi et al., 2019; Mohammadi et al., 2017; Quezada-Lázaro et al., 2016; Sadeghi et al., 2014; Schuchardt et al., 2019; Yaricsha et al., 2017; Yazdanparast et al., 2008; Zafar et al., 2017).

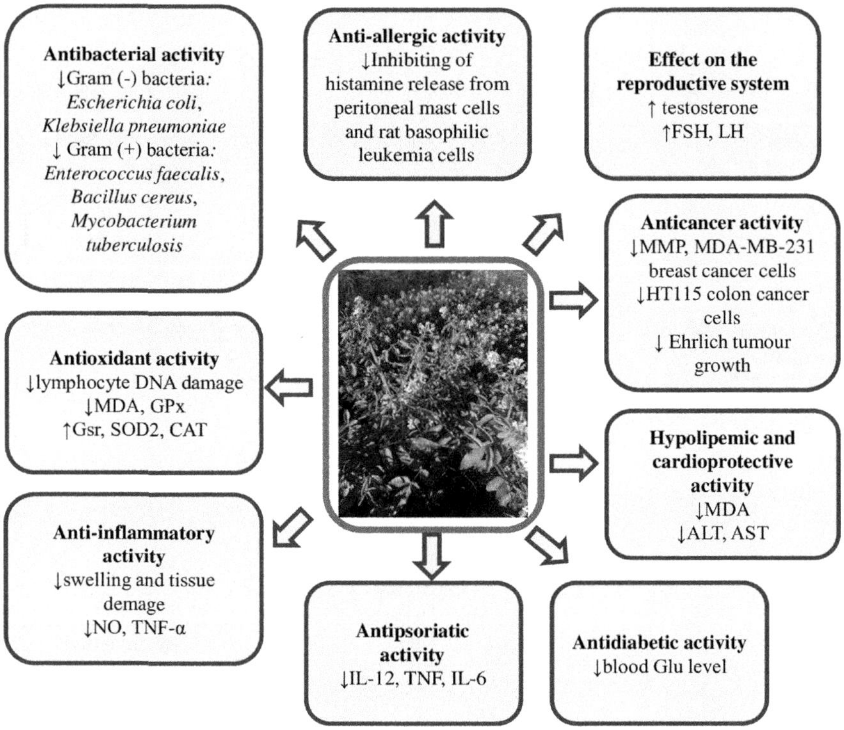

FIGURE 9.4

Biological activities of *N. officinale* herb extracts.

Extracts from the herb of *N. officinale* are successfully used in cosmetic preparations, as well as in patented cosmetic compositions. The first mention of the cosmetic use of this raw material comes from the textbook of the Italian doctor and writer - Trota from Salerno, entitled "De Ornatu Mulierum" from the 12th century (Głusiuk, 2018). Nowadays, extracts have scientifically documented, valuable from the point of view of cosmetology, antioxidant, antiaging, skin brightening, antiacne, antiinflammatory, antibacterial, and tyrosinase-inhibiting properties (Klimek-Szczykutowicz et al., 2018a,b; Thibane et al., 2019).

2. Plant biotechnology techniques important in pharmaceutical and cosmetic research development

Biotechnology of medicinal plants is an important field of science, especially for pharmaceutical and cosmetic utility. In vitro plant cultures can be a potential source of many valuable, biologically active compounds used in the treatment of numerous diseases (Mohaddab et al., 2022; Smetanska, 2008). Due to the lack of contamination of the raw material, e.g., with heavy metals, pesticides, microorganisms or aflatoxins, biomass from in vitro cultures could be an attractive innovative source of raw material for pharmaceutical, food and cosmetic industries. Another advantage of in vitro plant cultures is the possibility of obtaining plant material from rare and protected species, regardless of climatic conditions and seasons (Ekiert, 2009; Karuppusamy, 2009; Kayser and Müller, 2003; Malepszy, 2012). A valued advantage of plant in vitro cultures used in cosmetology is also the fact that they can be a source of plant stem cells (Ekiert, 2009; Karuppusamy, 2009).

Medicinal plants are a very interesting subject of research in biotechnology (25% of drugs used in Europe are drugs of natural origin). Research conducted in the world's laboratories has shown that in vitro plant cultures can produce therapeutically important compounds in quantities many times higher than in parent plants; these include rosmarinic acid, shikonin, ginsenosides, berberine and paclitaxel. Industrial biotechnological procedures for obtaining the above-mentioned compounds have been implemented (or are being implemented) mainly by companies from Japan (e.g., Mitsui, Nitto-Denki), Germany (e.g., Diversa, Phyton), America (ESCA genetics) and Korea (e.g., Samyang Genex Corp.) (Ekiert, 2009; Karuppusamy, 2009).

Plant in vitro cultures of various medicinal plant species enable the production of renewable biomass, usually with a high biosynthetic capacity of the desired compounds. Biomass, which is a raw material, is obtained under constant controlled conditions in accordance with Good Laboratory Practice (GLP) and Good Manufacturing Practice (GMP), which ensures high quality and high value of potential biotechnological raw materials. The advantage of in vitro cultures is the ability to intensify biosynthesis and increase the accumulation of desired metabolites in biomass, using various biotechnological methods. The most common are: optimization of the mineral and hormonal composition of the medium, selection of highly productive lines and the use of elicitors and precursors. It is possible to obtain in vitro-grown biomass in a continuous production process on a large scale (Filová, 2014; Mohaddab et al., 2022; Smetanska, 2008).

Plant in vitro cultures allow the use of the biochemical potential of cells in various growth systems and become an alternative and current source of production of secondary metabolites of commercial importance. The rapid production of homogeneous and renewable biomass capable of biosynthesis of substances with pharmacological properties, as well as the possibility of constant control of this process and influencing the increase in the accumulation of these compounds indicate the possibility of a wide application of biotechnology methods in the pharmaceutical and cosmetology industries (Filová, 2014; Mohaddab et al., 2022; Smetanska, 2008).

Many research centers around the world and pharmaceutical concerns conduct studies on the possibility of obtaining compounds from raw materials of medicinal plants obtained using biotechnological methods, as well as research programs for the industrial production of secondary metabolites with therapeutic properties in various plant in vitro systems, mainly in cell cultures. Plant in vitro cultures and the available tools of modern biotechnology can be a valuable method leading to the protection of genetic resources and help to multiply rare, endangered and protected plants with unchanged biological value (Filová, 2014; Smetanska, 2008).

The production of secondary metabolites in plant cultures can be stimulated by the use of various elicitors. It is one of the most promising methods used in plant biotechnology to increase productivity. Its mechanism is based on inducing stress in in vitro plant cultures. In plant physiology, "stress" is defined by a factor (biotic or abiotic) that can modify plant growth, reproduction, and function. Abiotic elicitors are both physical factors (e.g., ultraviolet radiation, temperature, mechanical damage) and chemical substances, such as ethephon (ETH), salicylic acid (SA), jasmonic acid (JA) and methyl jasmonate (MeJA). Biotic elicitors are obtained from microbial, plant or fungal sources, e.g., yeast extract (YeE), chitosan or scleroglucan. The exogenous addition of an elicitor to the culture medium of in vitro cultures stimulates the production of secondary metabolites (Narayani and Srivastava, 2017). Studies have shown, among others, that MeJA induces a higher accumulation of resveratrol in *Vitis vinifera* cell suspensions (Santamaria et al., 2012), gymnemic acid in suspension cultures of *Gymnema sylvestre* (Chodisetti et al., 2015), ginsenosides in suspension cultures of *Panax ginseng* (Thanh et al., 2005), and rosmarinic acid in suspension cultures of *Mentha piperita* (Krzyzanowska et al., 2012). Other scientific studies have also confirmed the stimulating effect of SA on the production of phenolic acids in *Salvia miltiorrhiza* cell cultures (Dong et al., 2010), and jaceosidins and syringins in suspension cultures of *Saussurea medusa* (Yu et al., 2006). Other studies have shown a stimulating effect of ETH on anthocyanins production in root cultures of *Daucus carota* ssp. *sativus* var. *atrorubens* (Barba-Espín et al., 2020), and pseudohypericins in shoot cultures of *Hypericum perforatum* (Rao et al., 2010). Research with YeE confirmed, among others, increased production of dibenzocyclooctadiene lignans in microshoot cultures of *Schisandra chinensis* (Szopa et al., 2018).

Another method of stimulating the production of secondary metabolites in vitro plant cultures is the exogenous supplementation of precursors of these compounds to culture media. The most commonly used precursor to increase the production of flavonoids, phenolic acids, lignans and coumarins is phenylalanine (Phe). Phe is a substrate in the first step of the biosynthetic phenylpropanoid pathway (Smetanska, 2008). Due to the structure of GSLs and the variable side chain derived from different amino acids, Phe can also be used to increase the production of aromatic GSLs, such as gluconasturtiin (Sánchez-Pujante et al., 2017). The use of Phe increased e.g., the production of flavonoids (cinnaroside, rutoside, casticin) and phenolic acids (neochlorogenic acid, *p*-hydroxybenzoic acid, caffeic acid and *p*-coumaric acid) in *Vitex agnus castus* microshoot cultures (Skrzypczak-Pietraszek et al., 2018). In other studies, Szopa et al. (Szopa et al., 2020) confirmed the positive effect of Phe supplementation on the production of phenolic acids in shoot cultures of *Aronia arbutifolia* and *Aronia melanocarpa*. Phe also stimulated flavonoid production in callus cultures of *Citrullus colocynthis* and *Abutilon indicum* (Sajjalaguddam and Paladugu, 2015).

Under the series of works conducted on microshoot cultures of *N. officinale* the optimization of secondary metabolites production was performed based on the use of biotic and abiotic elicitors in agitated cultures and biogenetic precursors in high-production Plantform bioreactors. The results proved the effectiveness of applied techniques in the production of GSLs and polyphenolic compounds and indicated the biological activity of studied extracts (inhibition of tyrosinase and the effect of antibacterial extracts).

3. Effect of elicitor treatments on metabolites production and biological activity of *N. officinale* agitated microshoot cultures

3.1 Research project and methodology

Under the biotechnological studies performed in the Department of Medicinal Plant and Mushroom Biotechnology (former Department of Pharmaceutical Botany), Faculty of Pharmacy, Jagiellonian University, Medical College, in vitro cultures of *N. officinale* were initiated and increasing the production of metabolites was examined the effect of elicitation treatments. As part of the experiments, the effect of elicitation on the antioxidant activity of biomass extracts and the ability to inhibit tyrosinase was also assessed (Klimek-Szczykutowicz et al., 2022).

The study was carried out on agitated microshoot cultures of *N. officinale* (Fig. 9.5) grown on Murashige and Skoog (MS) medium containing 1 mg/L 6-benzyladenine (BA) and 1 mg/L 1-naphthaleneacetic acid (NAA), under continuous white LED light (2.75 W/m^2, PPFD 40 $\mu mol/m^2/s$)

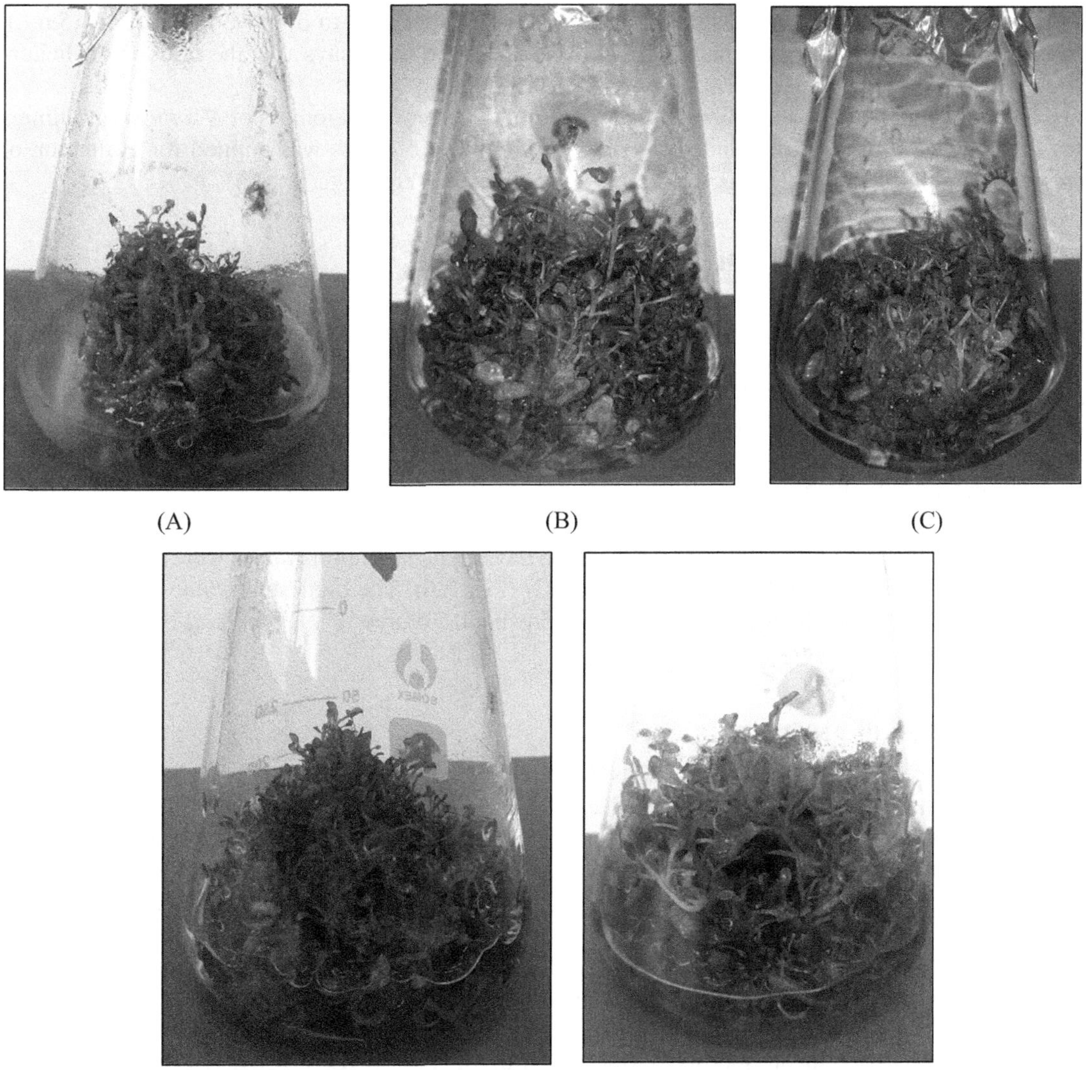

(A) (B) (C)

(D) (E)

FIGURE 9.5

Morphological appearance of experimental *N. officinale* microshoot cultures; (A) control cultures; and elicited cultures on 10th day of cultivation; (B) 50 μM ETH, (C) 100 μM MeJA, (D) 100 μM NaSA, (E) 1 mg/mL YeE (harvest time - 48 hours after adding elicitors).

at 25 ± 2°C. Three abiotic elicitors were tested: ethephon (ETH), methyl jasmonate (MeJA) and sodium salicylate (NaSA), and one biotic elicitor-yeast extract (YeE). Elicitors at concentrations of 25 and 50 μM (ETH), 50 and 100 μM (MeJA), 50 and 100 μM (NaSA) and 1 and 3 mg/mL (YeE) were supplemented to the experimental culture media on the 10th day of culture. Microshoots which were grown without elicitors were established as control (C), and on 10 day of the growth period, redistilled sterile H_2O was added to the flask in the amount corresponding to that used to dissolve the elicitors (5 mL). Experimental and C media and biomass were collected 1, 2, 4, 6 and 8 days after the addition of elicitors (Klimek-Szczykutowicz et al., 2022).

The effect of the culture conditions used on the appearance of the biomass of *N. officinale* cultures was found. The biomass growth index (Gi) for fresh and dry biomass was applied for evaluation of microshoots multiplication (Grzegorczyk and Wysokińska, 2011).

3.2 Biomass increments and biomass maintenance

The appearance of *N. officinale* microshoots was affected by the applied elicitors and their concentrations, and also the day of growth after elicitor addition. Microshoots growing in the media with the addition of NaSA were characterized by a dark green color and a large number of shoots. Microshoots that were grown with 25 μM ETH were characterized by a green color throughout the entire growth period. After 6 and 8 days, dying cultures were observed for 50 μM ETH. In both MeJA concentrations used, the microshoots were smaller and lighter in color 6 and 8 days after elicitor addition (Klimek-Szczykutowicz et al., 2022).

The Gi for the tested variants ranged from 6.47 (25 μM ETH, after 1 day) to max. 31.04 (50 μM NaSA after 2 days) and 31.59 (100 μM MeJA after 6 days). The maximum values were respectively 1.2 and 1.1 times higher than in the control cultures (C) (Fig. 9.6). The highest Gi values obtained were comparable to those obtained for 20-day-old microshoot cultures of *N. officinale* grown in RITA bioreactors (Gi = 31.71).

3.3 Production of glucosinolates

The analysis of the total content of GSLs was performed using the spectrophotometric method (Gallaher et al., 2012; Klimek-Szczykutowicz et al., 2019, 2020). Their contents depended on the elicitor used, its concentration and biomass harvesting time. The results of the analysis of the GSLs total content gave grounds for selecting the most productive variants for each of the tested elicitors; 25 μM ETH (4 days), 50 μM MeJA (1 day), 50 μM NaSA (4 days) and 3 mg/mL YeE (6 days) (Table 9.1). Biomass extracts from these variants were subjected to a detailed qualitative and quantitative analysis of GSLs using the UHPLC-DAD-MS/MS (ultra-high-performance liquid chromatography-tandem mass spectrometry) method (Blažević et al., 2019; Grosser and van Dam, 2017). In total, the presence of nine GSLs was confirmed in the analyzed extracts, such as 6-(methylsulfinyl)hexyl GSL (glucohesperin), 7-(methylsulfinyl)heptyl GSL, 8-(methylsulfinyl)octyl GSL (glucohirsutin), 7-(methylsulfanyl)heptyl GSL, 8-(methylsulfanyl)octyl GSL, 2-phenylethyl GSL (gluconasturtiin), 4-hydroxyindol-3-ylmethyl GSL (4-hydroxyglucobrassicin), indol-3-ylmethyl GSL (glucobrassicin) and 4-methoxyindol-3-ylmethyl GSL (4-methoxyglucobrassicin). The content of individual compounds ranged from 0.55 to 152.91 mg/100 g DW (dry weight). Quantitatively dominant GSLs were: 4-methoxyglucobrassicin (max. 152.91 mg/100 g DW, C, 4 days

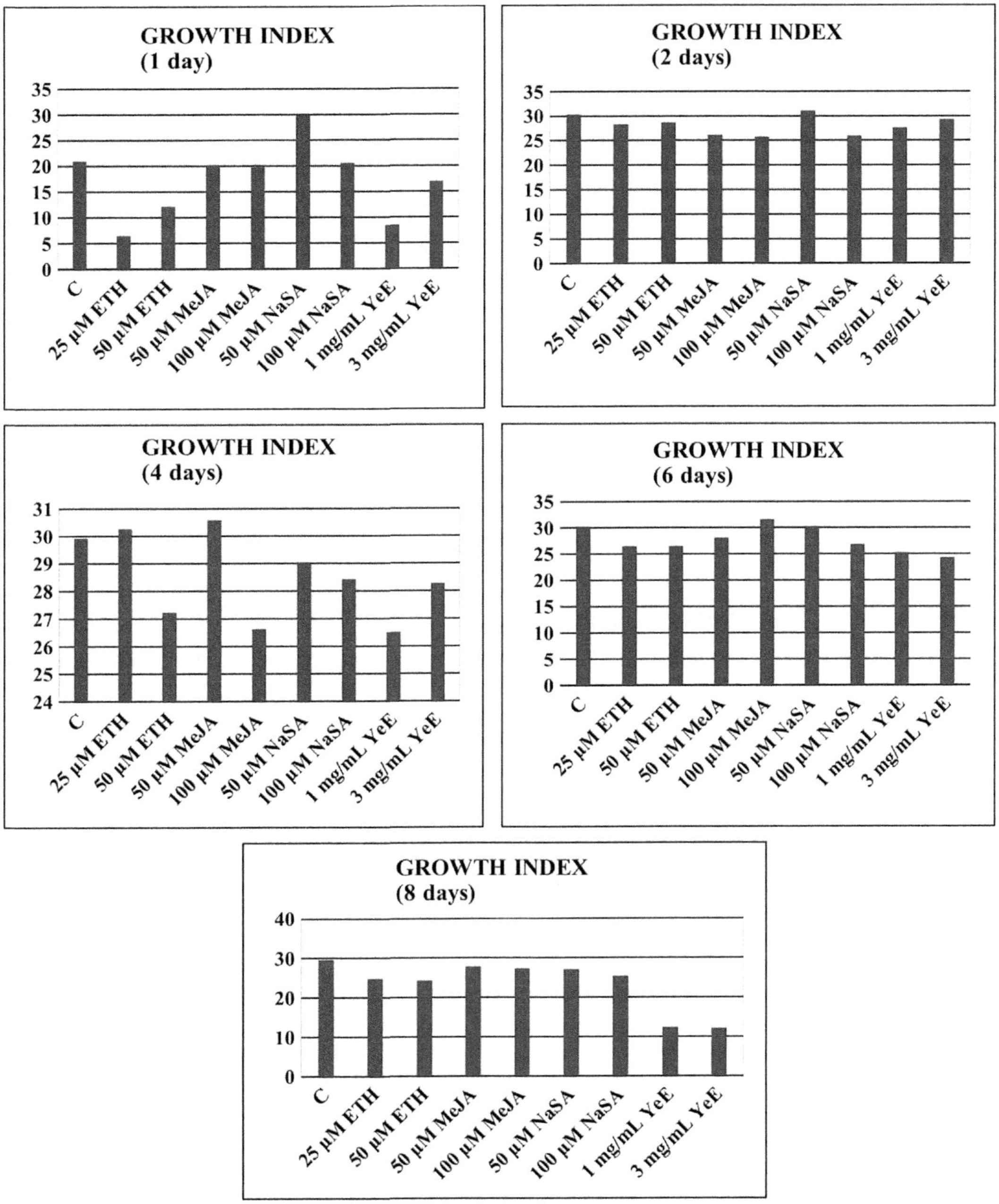

FIGURE 9.6

Growth index (Gi ± SD) values of *N. officinale* microshoot agitated cultures: Control (C) and experimental in vitro cultures after 1, 2, 4, 6, and 8–days after treatment with different elicitors ($P < 0.05$ *vs.* control, $n = 6$).

Table 9.1 Minimal and maximal total GSLs content (mg sinigrin equivalent (SIN)/100 g DW ±SD) for *N. officinale* control (C) and experimental in vitro cultures reached after elicitation.

Treatment		Min.–Max. total content	Condition for maximal content: Elicitor concentration	Condition for maximal content: Harvest time after treatment
C		117.75 ± 13.23–260.66 ± 14.57	nt[a]	4 days
Elicitors	ETH	109.99 ± 24.97–226.31 ± 13.70	25 μM	4 days
	MeJA	149.16 ± 28.27–298.36 ± 15.58	50 μM	1 day
	NaSA	167.32 ± 9.32–320.60 ± 68.51	50 μM	4 days
	YeE	94.99 ± 26.01–274.11 ± 10.72	3 mg/mL	6 days

[a]nt, *no treatment.*

after elicitation), glucobrassicin (max. 65.95 mg/100 g DW, 50 μM MeJA, 1 day after elicitation) and gluconasturtiin (max. 68.34 mg/100 g DW, 50 μM MeJA, 1 day after elicitation). The assays showed a clear, stimulating effect of 50 μM MeJA on the production of gluconasturtiin and glucobrassicin, their contents were 3.0 and 1.4-fold higher, respectively than in control (C).

3.4 Production of polyphenol compounds

Studies have confirmed the influence of the type of elicitor, its concentration and time of harvesting on the total content of flavonoids (Ramos et al., 2017). The total content of flavonoids ranged from 305.86 to 1131.33 mg rutoside equivalent (RE)/100 g DW. The highest contents of flavonoids were obtained 1 day after the elicitation of 100 μM NaSA (1131.33 mg RE/100 g DW), and after 8 days after the elicitation of 50 μM NaSA (1129.14 mg RE/100 g DW). They were 1.4 and 2.4 times higher, respectively than in C (Table 9.2) (Klimek-Szczykutowicz et al., 2022).

The total content of polyphenols determined by the Folin-Ciocalteu (F–C) method (Singleton et al., 1999) in the experimental microshoot cultures ranged from 131.89 to 336.89 mg gallic acid equivalent (GAL)/100 g DW. The highest content of polyphenols was obtained for C harvested after 6 days. Elicitation treatments did not increase overall polyphenol production (Table 9.2) (Klimek-Szczykutowicz et al., 2022).

As part of HPLC-DAD (high-performance liquid chromatography with diode-array detection) determinations (Ellnain-Wojtaszek and Zgórka, 1999; Sułkowska-Ziaja et al., 2017), a quantitative analysis of the content of two phenolic acids (*p*-coumaric acid and ferulic acid) and one flavonoid (rutoside) was performed, the presence of which was found by UHPLC-DAD-ESI-MS (ultra-high-performance liquid chromatography diode array detector electrospray ionization tandem mass spectrometry) in a previous study (Table 9.3) (Klimek-Szczykutowicz et al., 2020a,b). The content of *p*-coumaric acid ranged from 1.80 to 64.38 mg/100 g DW. The highest content was confirmed for C collected after 8 days. Ferulic acid content ranged from 1.39 to 17.76 mg/100 g DW. The highest content was obtained for microshoot cultures after elicitation of 50 μM NaSA and harvested after 8 days. The content of rutoside ranged from 1.14 to 21.17 mg/100 g DW; the maximum content was confirmed for C collected after 6 days (Table 9.3) (Klimek-Szczykutowicz et al., 2022).

Table 9.2 Minimal and maximal total flavonoid (mg of RE/100 g DW ±SD) and total polyphenol (mg of GAL/100 g DW±SD) contents in experimental *N. officinale* in vitro cultures reached after elicitation.

Treatment		Flavonoids			Polyphenols		
			Condition for maximal content			Condition for maximal content	
		Min. – Max. Total content	Elicitor concentration	Harvest time after elicitor treatment	Min. -Max. Total content	Elicitor concentration	Harvest time after elicitor treatment
C		305.86 ± 31.76–984.83 ± 26.72	nt[a]	4 days	167.22 ± 2.56–336.89 ± 8.03	nt	6 days
Elicitors	ETH	549.27 ± 29.88–1060.80 ± 90.75	50 μM	2 days	187.95 ± 5.16–314.27 ± 19.20	25 μM	1 day
	MeJA	580.00 ± 179.98–1037.47 ± 26.80	100 μM	1 day	224.75 ± 0.12–276.29 ± 19.66	100 μM	1 day
	NaSA	685.75 ± 26.66–1131.33 ± 9.85	100 μM	1 day	227.66 ± 2.50–293.21 ± 13.55	50 μM	8 days
	YeE	115.35 ± 35.98–824.18 ± 64.57	1 mg/mL	8 days	131.89 ± 5.16–265.89 ± 7.61	1 mg/mL	4 days

[a]nt, *no treatment.*

Table 9.3 Minimal - maximal amounts of the main phenolic compounds (mg/100 g DW±SD) in extracts from *N. officinale* control (C) and experimental in vitro cultures after elicitation.

		Condition for maximal content			
Polyphenol compounds	**Min. – Max. Content**	**Elicitor**	**Elicitor concentration (μM)**	**Harvest time after elicitor treatment**	**C**
p-Coumaric acid	1.80 ± 0.39–38.76 ± 0.72	MeJA	100	6 days	64.38 ± 9.26
Ferulic acid	1.72 ± 0.01–17.76 ± 2.07	NaSA	50	8 days	16.46 ± 0.09
Rutoside	1.14 ± 0.23–17.00 ± 1.45	NaSA	100	8 days	21.17 ± 2.67

3.5 Antioxidant activities

The assessment of the antioxidant potential of biomass extracts was performed using the CUPRAC (Biesaga-Kościelniak et al., 2014; Özyürek et al., 2007), DPPH (Blios, 1958) and FRAP (Benzie and Strain, 1996) methods. The antioxidant potential of the extracts determined by the CUPRAC method ranged from 0.84 to 2.97 mmol Trolox equivalent (TE)/100 g DW (Table 9.4). The maximum antioxidant potential was found for microshoot cultures elicited with 100 μM MeJA, collected after 1 day. It was 1.8 times higher than in C. The antioxidant potential determined by the DPPH method ranged from 0.53 to 2.30 mmol TE/100 g DW. The maximum antioxidant potential was documented for microshoot cultures elicited with 50 μM ETH, harvested after 6 days. It was 2.2 times higher than in C. The antioxidant potential of biomass extracts

Table 9.4 Minimal - maximal antioxidant activity, estimated by the CUPRAC, DPPH, and FRAP methods (expressed in mmol TE/100 g DW±SD), of extracts from *N. officinale* control (C) and experimental in vitro cultures after elicitation.

		Condition for maximal content			
Assay	**Min. – Max. Antioxidant activity**	**Elicitor**	**Elicitor concentration (μM)**	**Harvest time after elicitor treatment**	**C**
CUPRAC	1.02 ± 0.01–2.97 ± 0.03	MeJA	100	1 day	1.63 ± 0.06
DPPH	0.53 ± 0.05–2.30 ± 2.09	ETH	50	6 days	1.04 ± 0.01
FRAP	0.15 ± 0.01–0.83 ± 0.02	MeJA	100	1 day	0.40 ± 0.01

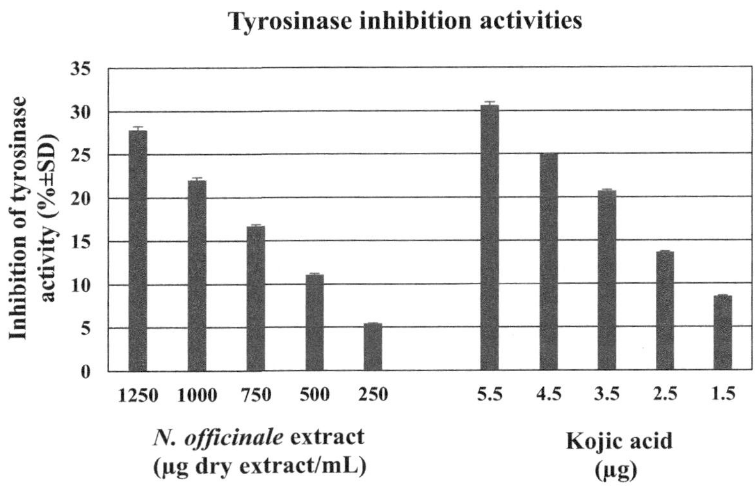

FIGURE 9.7

Inhibition of tyrosinase activity (%±SD) of extract from microshoots of *N. officinale* agitated cultures harvested after 1 day of elicitation with 100 µM MeJA in comparison to kojic acid.

was determined by the FRAP method and ranged from 0.15 to 0.83 mmol TE/100 g DW. The maximum antioxidant potential was confirmed for cultures elicited with 100 µM MeJA, collected after 1 day. It was 2.1 times higher compared to C (Table 9.4) (Klimek-Szczykutowicz et al., 2022).

3.6 Tyrosinase inhibition activities

One of the elements of the research was to assess the possibility of using the tested extracts from the microshoot cultures of *N. officinale* in cosmetic preparations and the effectiveness of eliminating skin discoloration. For extracts from microshoot cultures with a high antioxidant potential and a high content of flavonoids and polyphenols (agitated cultures elicited with 100 µM MeJA and collected after 1 day), were performed studies on tyrosinase inhibition activity. The results showed a linear dependence of tyrosinase inhibition with increasing concentration of the tested extract. The highest tyrosinase inhibition activity (27.84%) was found for 1250 µg/mL concentration of dry extract and was similar to tyrosinase inhibition for 5.5 µg/mL kojic acid, which was used as a reference substance (Fig. 9.7) (Klimek-Szczykutowicz et al., 2022).

4. Effect of exogenous supplementation of culture media with precursor compounds on metabolites production and biological activity of *N. officinale* microshoot agitated cultures and cultures grown in Plantform bioreactors

4.1 Research project and methodology

The exogenous supplementation of culture media with selected biosynthetic precursors was to examine the production increasing of bioactive metabolites in in vitro cultures of *N. officinale*.

Experiments at increasing the production scale of biomass and metabolites acquisition were carried out in TIS (temporary immersion system) – Plantform bioreactors (PlantForm, Sweden). The study also assessed the effect of supplementation of cultures with precursors on antioxidant and antibacterial activity. The studies were carried out in microshoot cultures of *N. officinale* grown in Plantform bioreactors on MS medium containing 1 mg/L BA and 1 mg/L NAA, under white LED light (2.75 W/ m^2, PPFD 40 μmol/m^2/s) in temp. 25 ± 2°C. The immersion cycle was 5 minutes every 1.5 hour at an aeration rate of 1.0 vvm. Two precursors were tested; amino acids - Phe (L-phenylalanine) and Trp (L-tryptophan). Phe is a substrate in the first stage of the phenylpropanoid pathway and can be used to increase the production of aromatic GSLs (Phe derivatives). On the other hand, Trp may affect the content of indole GSLs (Trp derivatives). Sterile Phe and Trp solutions were added to the experimental cultures at the time of culture establishment ("day 0") or on 10th day of culture growth. The following precursor concentrations were obtained in the culture media: 0.05; 0.1; 0.5; 1.0 and 3.0 mM. Supplementation with sterile redistilled water was used as C conditions (Klimek-Szczykutowicz et al., 2021).

4.2 Biomass increments and biomass maintenance

Microshoot cultures of *N. officinale* carried out in Plantform bioreactors were characterized by the production of numerous green microshoots for all tested variants of Phe and Trp supplementation (Fig. 9.8). Higher concentrations of precursors influenced the color of the microshoots. Gi ranged from 5.47 to 13.66 (Fig. 9.9). Precursors added at the time of breeding establishment (day 0) caused a slight decrease in biomass increments, while precursors added on the 10th day of cultivation favored higher Gi values. In the case of the highest concentrations of precursors (3.0 mM) supplemented on the 10th day of culture, biomass gains of 11.48 (for Phe) and 13.66 (for Trp) were obtained, and compared to C they were 1.4 and 1.6-fold higher, respectively (Fig. 9.9). In previous studies, the highest Gi index (31.71) was obtained for microshoot cultures grown in the RITA bioreactor after 20 days of cultivation and it was 2.3 times higher than for the highest results obtained in this study (3.0 mM Trp, 10th day) (Klimek-Szczykutowicz et al., 2021). In contrast, the microshoot cultures grown in the Plantform

(A) (B) (C)

FIGURE 9.8

Morphological appearance of selected, experimental microshoot cultures of *N. officinale* grown in Plantform bioreactors supplemented with biosynthetic precursors at the time of establishing the culture; (A) C (control), (B) 1.0 mM Phe, and (C) 1.0 mM Trp.

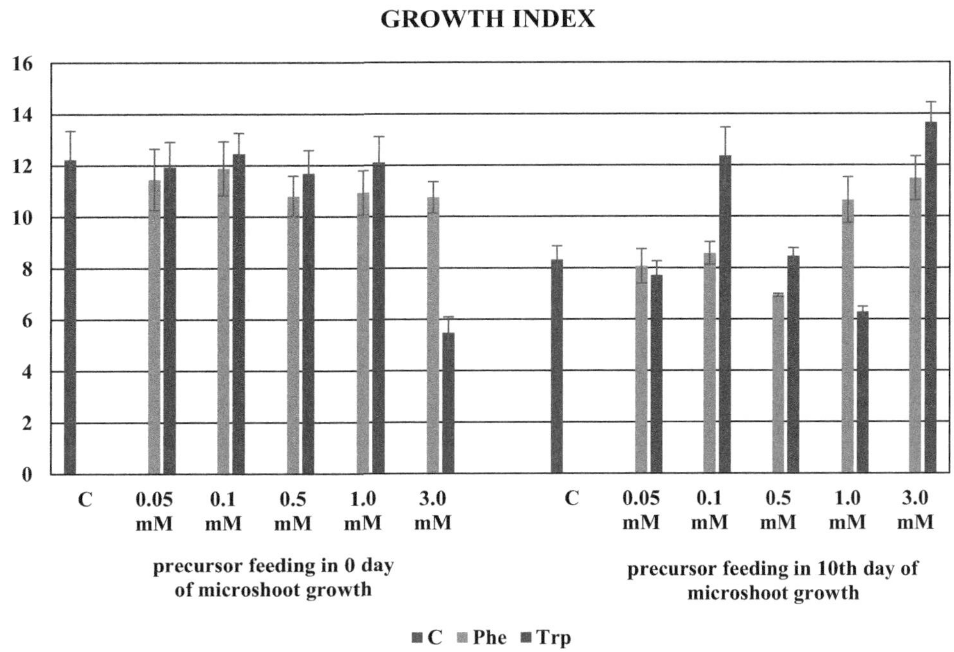

FIGURE 9.9

Growth index (Gi ± SD) values of bioreactor-grown *N. officinale*: control (C) and experimental in vitro cultures reached after precursor feeding in 0 and 10th day of microshoot growth ($P < 0.05$ vs. control, $n = 6$).

bioreactor showed a higher Gi compared to the agitated cultures. The maximum Gi (13.66) was 1.3 times higher than the maximum Gi for agitated cultures on MS medium containing 1 mg/L zeatin (Zea) and 1 mg/L NAA (Fig. 9.9) (Klimek-Szczykutowicz et al., 2020a,b).

4.3 Production of glucosinolates

The conducted studies included an analysis of the impact of the Phe and Trp precursors used, as well as the concentration and time of their supplementation on the total content of GSLs. The content of GSLs, determined spectrophotometrically (Gallaher et al., 2012), as a result of Phe supplementation, ranged from 141.06 to 269.20 mg SIN/100 g DW (Table 9.5). The highest content of GSLs was found after supplementation with 3.0 mM Phe on day 0 and it was 3.3 times higher than in the C. The total content of GSLs as a result of Trp supplementation ranged from 140.09 to 200.97 mg SIN/100 g DW (Table 9.5). The highest content after Trp supplementation was obtained for the concentration of 1.0 mM (day 10). The conducted biotechnological studies have shown that the time of supplementation of the substrate with precursors has a large impact on the production of GSLs. The highest results were obtained for precursors added at the time of culture establishment. It can be concluded that longer contact of plant tissue with precursors stimulates the production of GSLs better (Klimek-Szczykutowicz et al., 2021).

Table 9.5 Minimal and maximal GSLs total contents (mg SIN/100 g DW ±SD) for bioreactor-grown *N. officinale* control (C) and experimental in vitro cultures reached after precursor feeding.

Treatment		Min. – Max. Total content (mg/100 g DW±SD)	Condition for maximal content	
			Precursor concentration	Day of supplementation
	C	125.04 ± 5.05	nt[a]	nt
Precursors	Phe	141.06 ± 10.65–269.20 ± 15.78	3.0 mM	0
	Trp	140.09 ± 20.40–200.97 ± 5.17	1.0 mM	10

[a]nt, *no treatment.*

Based on the determination of the total content of GSLs, the most favorable concentration for each of the precursors (3.0 mM Phe and 3.0 mM Trp supplemented on day 0) was selected for the qualitative and quantitative analysis of these compounds by UHPLC-DAD-MS/MS (Blažević et al., 2020). For comparative purposes, extracts from C cultures were also analyzed. Significant differences in the qualitative composition and quantitative content of GSLs were found. Five GSLs were confirmed in extracts from C samples: Met derivatives: 7-(methylsulfinyl)heptyl GSL and 8-(methylsulfinyl)octyl GSL (glucohirsutin); Phe derivative: 2-phenylethyl GSL (gluconasturtiin) and Trp derivatives: indol-3-ylmethyl GSL (glucobrassicin) and 4-methoxyindol-3-ylmethyl GSL (4-methoxyglucobrassicin). In extracts from microshoot cultures supplemented with 3.0 mM Phe, the presence of six GSLs was confirmed: Met derivatives: 7-(methylsulfinyl)heptyl GSL and glucohirsutin; Phe derivative: gluconasturtiin and Trp derivatives: 4-hydroxyglucobrassicin, glucobrassicin and 4-methoxyglucobrassicin. In extracts from microshoot cultures supplemented with 3.0 mM Trp, four GSLs were found: Phe derivatives: gluconasturtiin; and Trp derivatives: 4-hydroxyglucobrassicin, glucobrassicin and 4-methoxyglucobrassicin. The content of individual compounds in the tested extracts ranged from 3.78 to 149.99 mg/100 g DW. The main GSLs were: 4-methoxyglucobrassicin and gluconasturtiin. For 4-methoxyglucobrassicin, the highest content (149.99 mg/100 g DW) was found for extracts supplemented with 3.0 mM Phe and it was as much as 11.7 times higher than in the C. The maximum content of gluconasturtiin (36.71 mg/100 g DW) was found for extracts from cultures supplemented with 3.0 mM Phe on day 0 and it was 2.3 times higher than in C (Klimek-Szczykutowicz et al., 2021).

Comparing studies with microshoot cultures of *N. officinale* carried out in two types of commercially available bioreactors - RITA and Plantform, significant differences were observed in GSLs research using the UHPLC-DAD-MS/MS method (Klimek-Szczykutowicz et al., 2020a,b, 2021). For non-elicited microshoot cultures grown in RITA bioreactors on MS medium with 1 mg/L BA and 1 mg/L NAA, gluconasturtiin content was 3.6 and 1.5 times higher after 20 days of growth compared to C and supplemented cultures, respectively 3.0 mM Phe conducted in Plantform bioreactors. The content of 4-methoxyglucobrassicin was 3.0 and 1.5-fold higher in cultures supplemented with 3.0 mM Phe and 3.0 mM Trp, respectively, compared to 20-day-old cultures grown in RITA bioreactors (Klimek-Szczykutowicz et al., 2020, 2021).

In the study by Wielanek et al. (2009), gluconasturtiin content was determined in hairy root cultures of *N. officinale* grown with the addition of 0.5 mM precursors (Phe, Cys, Met, and Ser). Biomass was harvested 10 days after supplementation. Studies have confirmed that Phe stimulates the production of gluconasturtiin more than other amino acids. Callus and suspension cultures of *Nasturtium montana* were also treated with precursors — Cys, Met and Trp in order to study their effect on the production of indole acetonitrile compounds, which are products of GSLs hydrolysis. Studies have confirmed the stimulating effect of supplementation with these compounds. The studies confirmed for the first time the effect of Phe and Trp supplementation on the total production of GSLs and the content of individual compounds from this group in microshoot cultures of *N. officinale* (Klimek-Szczykutowicz et al., 2021).

4.4 Production of polyphenol compounds

Spectrophotometric assays confirmed the effect of the addition of precursors, their concentration and supplementation time on the total content of flavonoids in *N. officinale* microshoot cultures grown in Plantform bioreactors (Ramos et al., 2017). The total content of flavonoids ranged from 739.22 to 1364.38 mg RE/100 g DW. The highest results were obtained for 3.0 mM Phe and 1.0 mM Trp supplemented on day 0. They were 1364.38 and 1324.14 mg RE/100 g DW and were respectively 2.4- and 2.2-times higher than in C (Table 9.6) (Klimek-Szczykutowicz et al., 2021).

The total content of polyphenols determined by the F—C assay in the tested extracts ranged from 151.43 to 1062.76 mg GAL/100 g DW (Singleton et al., 1999). The highest content of polyphenols was found as a result of supplementation with 3.0 mM Trp on the 10th day. It was 4.3 times higher than in C. High content of polyphenols — 873.11 mg GAL/100 g DW was also obtained after supplementation with 3.0 mM on day 0. It was 4.6-fold higher than in C cultures. Phe had a smaller effect on the content of polyphenols compared to Trp. The highest total polyphenol content was obtained after supplementation with 0.1 mM Phe at the time of culture establishment. It was 1.7 times higher than in the C (Table 9.6) (Klimek-Szczykutowicz et al., 2021).

Using the HPLC-DAD method (Ellnain-Wojtaszek and Zgórka, 1999; Sułkowska-Ziaja et al., 2017), the extracts from the tested *N. officinale* microshoot cultures grown in Plantform bioreactors were determined for the content of two phenolic acids (*p*-coumaric acid and ferulic acid) and one flavonoid (rutoside), the presence of which was demonstrated by the UHPLC-DAD-ESI-MS method in a previous study (Table 9.7) (Klimek-Szczykutowicz et al., 2020a,b). The highest content of *p*-coumaric acid (29.11 mg/100 g DW) was documented in microshoot cultures supplemented with 0.5 mM Trp on the 10th day of the culture growth period. It was 2.6 times higher than in the C samples. The maximum content of ferulic acid (27.76 mg/100 g DW) was confirmed in cultures supplemented with 3.0 mM Phe on day 0; this content was 7.6 times higher compared to the C (Table 9.7). The highest amount of rutoside (16.03 mg/100 g DW) was obtained in cultures supplemented with 0.1 mM Trp on the 10th day; this content was 4.2 times higher than in the C (Table 9.7) (Klimek-Szczykutowicz et al., 2021).

4.5 Antioxidant potential

In extracts from microshoot cultures, the antioxidant activity was tested using the CUPRAC (Biesaga-Kościelniak et al., 2014; Özyürek et al., 2007), DPPH (Blios, 1958) and FRAP (Benzie and Strain, 1996)

Table 9.6 Minimal and maximal total flavonoid (mg of RE/100 g DW±SD) and total polyphenol (mg of GAL/100 g DW±SD) contents in experimental *N. officinale* in vitro cultures reached after precursor feeding.

Treatment		Flavonoids			Polyphenols		
			Condition for maximal content			Condition for maximal content	
		Min. – Max. Total content	Precursor concentration	Day of supplementation	Min. -Max. Total content	Precursor concentration	Day of supplementation
C		863.71 ± 49.96	nt[a]	nt	248.02 ± 4.55	nt	nt
Precursors	Phe	808.38 ± 45.35–1364.38 ± 80.14	3.0 mM	0	228.29 ± 9.34–325.55 ± 7.99	0.1 mM	0
	Trp	739.22 ± 55.75–1324.14 ± 123.67	1.0 mM	0	151.43 ± 14.05–1062.76 ± 28.77	3.0 mM	10

[a]nt, *no treatment.*

Table 9.7 Minimal and maximal amounts of the main phenolic compounds (mg/100 g DW±SD) in extracts of bioreactor-grown *N. officinale* in vitro cultures reached after precursor feeding.

Polyphenol compounds	Min. – Max. Content	Condition for maximal content			C
		Precursor	Precursor concentration (mM)	Day of supplementation	
p-Coumaric acid	2.51 ± 0.20–29.11 ± 3.70	Trp	0.5	10	4.90 ± 0.35
Ferulic acid	2.87 ± 0.11–27.76 ± 2.12	Phe	3.0	0	3.66 ± 0.70
Rutoside	1.26 ± 0.32–11.66 ± 1.44	Trp	0.05	0	3.82 ± 0.60

assays. The metabolite determinations made it possible to select the supplementation variant that gave the most favorable results – 3.0 mM Phe added on day 0. The antioxidant activity of extracts from these cultures, as determined by the CUPRAC, DPPH, FRAP tests, was respectively 2.0, 1.8 and 3.0 times higher than in C cultures (Table 9.8). For this variant of supplementation, significant contents of flavonoids, polyphenolic compounds and carotenoids were also confirmed. The results suggest that the high content of these compounds may affect the confirmed antioxidant activity (Klimek-Szczykutowicz et al., 2021).

4.6 Antibacterial activities

Due to the high content of metabolites and antioxidant potential, the extract from cultures supplemented with 3.0 mM Phe on day 0 was selected for further studies on antibacterial activity on Gram-positive bacterial strains: *Staphylococcus aureus, Staphylococcus epidermidis, Propionibacterium*

Table 9.8 Minimal and maximal antioxidant activity, estimated with the CUPRAC, DPPH, and FRAP assays (expressed in mmol TE/100 g DW±SD), of extracts from bioreactor-grown *N. officinale* in vitro cultures reached after precursor feeding.

Assay	Min. – Max. Antioxidant activity	Condition for maximal content			C
		Precursor	Precursor concentration (mM)	Day of supplementation	
CUPRAC	1.70 ± 0.07–3.05 ± 0.18	Phe	0.1	0	1.54 ± 0.06
DPPH	0.54 ± 0.02–0.90 ± 0.03	Trp	0.5	10	0.69 ± 0.09
FRAP	0.31 ± 0.11–0.94 ± 0.26	Phe	0.1	0	0.24 ± 0.01

acnes, *Propionibacterium granulosum* and Gram-negative: *Escherichia coli*, *Pseudomonas aeruginosa*. The diffusion test in a solid medium showed the strongest inhibitory activity against *Propionibacterium* spp. strains (20–21 mm). The extract was not active against Gram-negative bacteria. The determined MIC (minimum inhibitory concentration) and MBC (minimum bactericidal concentration) values indicated significant activity against acne-causing bacteria - *P. acnes* (MIC in the range of 250–500 μg/mL, MBC in the range of 4–8). For the bacteria *S. aureus* and *S. epidermidis*, the MIC was 1000 μg/mL, and the MBC was 16 and 8, respectively. Cytotoxicity studies of the extract from microshoot cultures of *N. officinale* were also performed. After 2 days of incubation, it was shown that the extract at concentrations of 1.95–125.0 μg/mL did not significantly inhibit the viability of fibroblasts compared to the C (0 μg/mL medium). In turn, a slight decrease in cell viability to about 92% was observed at the highest concentration tested (250 μg/mL) (Klimek-Szczykutowicz et al., 2021).

5. Conclusions

The *N. officinale* herb (*Nasturtii herba*) has long been used in traditional medicine of Iran, Azerbaijan, Morocco, and Mauritius. The main groups of compounds determining the biological activity of *Nasturtii herba* are: GSLs, isothiocyanates, polyphenolic compounds, and vitamins. Scientific researchers have confirmed many valuable pharmacological properties of *Nasturtii herba* extracts, such as anticancer, antioxidant, antiinflammatory, and antibacterial properties. *N. officinale* today has the status of a "pro-health" plant used not only in phytotherapy but also in modern dietetics and cosmetology.

Increasing the production of biologically active metabolites using plant biotechnology techniques in *N. officinale* microshoot cultures is important due to the partial threat of this species. One of the methods of stimulating the production of active compounds in in vitro cultures is elicitation procedures, as well as adding precursors to media.

The studies on elicitation (with: ETH: ethephon, MeJA: methyl jasmonate, NaSA: sodium salicylate, and YeE: yeast extract) showed interesting differences in the accumulation of individual GSLs, and the stimulating influence of 50 μM MeJA on the production of gluconasturtiin (max. 68.34 mg/100 g DW) and glucobrassicin (max. 65.95 mg/100 g DW). Elicitation increased the total amounts of GSLs (max. 320.60 mg SIN/100 g DW, 50 μM NaSA). Moreover, the tested elicitors were confirmed to have a stimulating influence on the production of individual polyphenolic compounds, including *p*-coumaric acid (max. 38.76 mg/100 g DW, 100 μM MeJA), and rutoside (max 17.00 mg/100 g DW, 100 μM NaSA) (Table 9.9).

Stimulation of metabolite production *via* supplementation of the growth media in Plantform bioreactors with exogenous biosynthetic precursors (Phe: phenylalanine, and Trp: tryptophan) confirmed their stimulating influence on the production of flavonoids (max. 1364.38 mg RE/100 g DW, 3.0 mM Phe) and polyphenols (max. 1062.76 mg GAL/100 g DW, 3.0 mM Trp). The high biomass production (max. Gi = 10.74) and the efficient accumulation of GSLs achieved under these conditions helped to obtain applicative results of productivity determinations for 4-methoxyglucobrassicin (max. 149.99 mg/100 g DW, productivity = 398.77). Moreover, the tested precursors were confirmed to have a stimulating influence on the production of individual polyphenolic compounds, including ferulic acid (max. 27.76 mg/100 g DW, 3.0 mM Phe) (Table 9.9).

Table 9.9 Comparison of the maximal content of secondary metabolite in extracts from *N. officinale* in vitro cultures and from the herb of the parent plant.

Analyzed metabolites		In vitro cultures				Parent plant material (herb)
		Maximal content	Type of cultures	Variant MS medium	Duration of cultivation	
Total contents	Total content of GSLs (mg SIN/100 g DW ± SD)	320.60 ± 28.51	Agitated cultures	1 mg/L BA and 1 mg/L NAA, 50 μM NaSA, elicitation on the 10th day of cultivation	4 days after elicitation	499.89 ± 34.59
	Total content of flavonoids (mg RE/100 g DW ± SD)	1364.38 ± 80.14	Plantform bioreactor	1 mg/L BA and 1 mg/L NAA, 3 mM Phe adding day 0	20 days	1153.88 ± 50.21
	Total content of polyphenols (mg GAL/100 g DW ± SD)	1062.76 ± 28.77	Plantform bioreactor	1 mg/L BA and 1 mg/L NAA, 3 mM Trp adding day 10	20 days	459.32 ± 15.18
GSLs (mg/100 g DW ± S D)	Gluconasturtiin	68.34 ±5.14	Agitated cultures	1 mg/L BA and 1 mg/L NAA, 50 μM MeJA, elicitation on the 10th day of cultivation	1 day after elicitation	640.94 ± 2.96
	4-Hydroxyglucobrassicin	5.26 ± 1.57	Agitated cultures	1 mg/L BA and 1 mg/L NAA, 50 μM MeJA, elicitation on the 10th day of cultivation	1 day after elicitation	nd[a]
	Glucobrassicin	65.95 ± 0.49	Agitated cultures	1 mg/L BA and 1 mg/L NAA, 50 μM MeJA, elicitation on the 10th day of cultivation	1 day after elicitation	tr[a]
	4-Metoxyglucobrassicin	149.99 ± 17.44	Plantform bioreactor	1 mg/L BA and 1 mg/L NAA, 3 mM Phe adding day 0	20 days	23.47 ± 0.96

Continued

Table 9.9 Comparison of the maximal content of secondary metabolite in extracts from *N. officinale* in vitro cultures and from the herb of the parent plant.—cont'd

Analyzed metabolites		In vitro cultures				
		Maximal content	Type of cultures	Variant MS medium	Duration of cultivation	Parent plant material (herb)
Polyphenol compounds (mg/100 g DW±SD)	*p*- Coumaric acid	38.76 ± 0.72	Agitated cultures	1 mg/L BA and 1 mg/L NAA, 100 μM MeJA, elicitation on the 10th day of cultivation	6 days after elicitation	3.10 ± 0.21
	Ferulic acid	27.76 ± 2.12	Plantform bioreactor	1 mg/L BA and 1 mg/L NAA, 3 mM Phe adding day 0	20 days	12.69 ± 1.13
	Rutoside	17.00 ± 1.45	Agitated cultures	1 mg/L BA and 1 mg/L NAA, 100 μM NaSA, elicitation on the 10th day of cultivation	8 days after elicitation	7.20 ± 0.67

[a]*nd, not detected, tr—<0.1 μmol/g DW.*

In vitro cultures showed 1.6 times lower maximum total contents (for 50 μM NaSA, 4 days after elicitation) compared to the parent plant. On the other hand, in vitro cultures yielded 2.3 times higher maximum total polyphenol content (for 3 mM Trp adding day 10) and 1.2 times higher maximum total flavonoid content (3 mM Phe adding day 0) compared to the parent plant herb. The maximum content of the main GSL - gluconasturtiin was 9.4 times lower than in the *N. officinale* herb. In in vitro cultures, 6.4 times higher content of 4-methoxyglucobrassicin was obtained compared to the *N. officinale* herb, and higher content of 4-hydroxyglucobrassicin and glucobrassicin were detected, respectively not detected or present in trace amounts in the parent plant herb.

In addition, in vitro cultures compared to the herb were characterized by 12.5, 2.2 and 2.4 times higher content of *p*-coumaric acid, ferulic acid and rutoside, respectively.

The results obtained were compared with those of the parent plant material, which is presented in Table 9.9.

Productivity was also calculated for in vitro cultures. The highest productivity for GSLs, flavonoids and 4-methoxyglucobrassicin was obtained in cultures grown in the Plantform bioreactor with the addition of 3.0 mM Phe on day 0. For gluconasturtiin, glucobrassicin, *p*-coumaric acid, ferulic acid and rutoside, the highest productivity was obtained for cultures grown with 25 μmol ETH (4 days after elicitation). The highest productivity for 4-hydroxyglucobrassicin was obtained for cultures grown with 50 μmol MeJA (1 day after elicitation). The highest productivity of polyphenols was found in cultures growing in the Plantform bioreactor with 3.0 mM Trp added on day 0. Productivity values for the analyzed metabolites are presented in Table 9.10.

Table 9.10 Comparison of the productivity[a] of estimated metabolites in extracts from *N. officinale* in vitro cultures grown in the optimized cultivation conditions.

Analyzed metabolites		The best conditions for *N. officinale* microshoot cultivation			
		Elicitation 50 μmol MeJA 1 day	Elicitation 25 μmol ETH 4 days	Plantform 3.0 mM Phe 0 days	Plantform 3.0 mM Trp 0 days
Total contents	Total content of GSLs	73.11	671.44	715.70	292.15
	Total content of flavonoids	297.37	2730.53	3627.34	1848.68
	Total content of polyphenols	90.81	785.84	751.53	1299.71
GSLs	Gluconasturtiin	23.66	126.95	97.59	13.14
	4-Hydroxyglucobrassicin	6.21	2.38	nd[a]	nd
	Glucobrassicin	22.83	101.94	18.03	5.63
	4-Metoxyglucobrassicin	24.80	214.21	398.77	113.31
Polyphenol compounds	*p*- Coumaric acid	18.66	20.83	8.23	3.41
	Ferulic acid	7.19	11.25	9.77	1.91
	Rutoside	4.31	13.84	3.34	0.79

[a]*Productivity = max. concentration for each molecule (mg/100 g DW) × culture biomass (g/L)/100 g DW; nd, not detected*

In addition, for extracts from in vitro cultures, antioxidant activity was demonstrated by the CUPRAC, DPPH and FRAP assays. Extracts from in vitro cultures may be a potential ingredient of cosmetics due to the demonstrated tyrosinase-inhibiting properties and inhibiting the growth of acne-causing bacteria such as *P. acnes* and *P. granulosum.* The tested extracts did not show any cytotoxic effect (they did not significantly inhibit the viability of fibroblasts compared to the C in the tested concentration range).

To sum up, our studies on the example of *N. officinale* species, proved that plant in vitro cultured biomass could be a good and effective alternative source of raw material to the field-grown plants.

Acknowledgments

Funding: The study was realized as a part of the research projects: N/42/DBS/000122, and N42/DBS/000427 supported by the Polish Ministry of Science and Higher Education and SUPB.RN.23.002 supported by Jan Kochanowski University.

Author contribution statement: All the authors read and approved the manuscript in its final form. All authors declare that they have no conflict of interest.

References

Afsharypuor, S., Salehi, M., 2008. Volatile constituents of leaves and stems of *Nasturtium officinale* R. Br. J. Essent. Oil Res. 20 (6), 517–518. https://doi.org/10.1080/10412905.2008.9700076.

Amiri, H., 2012. Volatile constituents and antioxidant activity of flowers, stems and leaves of *Nasturtium officinale* R. Br. Nat. Prod. Res. 26 (2), 109–115. https://doi.org/10.1080/14786419.2010.534998.

Arumugam, A., Din Ibrahim, M., Kntayya, S.B., Ain, N.M., Iori, R., Galletti, S., Ioannides, C., Razis, A.F.A., 2020. Induction of apoptosis by gluconasturtiin-isothiocyanate (GNST-ITC) in human hepatocarcinoma HepG2 cells and human breast adenocarcinoma MCF-7 cells. Molecules 25 (5), 1240. https://doi.org/10.3390/molecules25051240.

Asadi-Samani, M., Moradi, M.T., Mahmoodnia, L., Alaei, S., Asadi-Samani, F., Luther, T., 2017. Traditional uses of medicinal plants to prevent and treat diabetes; an updated review of ethnobotanical studies in Iran. J. Nephropathol. 6 (3), 118–125. https://doi.org/10.15171/jnp.2017.20.

Bahramikia, S., Yazdanparast, R., 2008. Effect of hydroalcoholic extracts of *Nasturtium officinale* leaves on lipid profile in high-fat diet rats. J. Ethnopharmacol. 115 (1), 116–121. https://doi.org/10.1016/j.jep.2007.09.015.

Barba-Espín, G., Chen, S.T., Agnolet, S., Hegelund, J.N., Stanstrup, J., Christensen, J.H., Müller, R., Lütken, H., 2020. Ethephon-induced changes in antioxidants and phenolic compounds in anthocyanin-producing black carrot hairy root cultures. J. Exp. Bot. 71 (22), 7030–7045. https://doi.org/10.1093/jxb/eraa376.

Benzie, I.F.F., Strain, J.J., 1996. The ferric reducing ability of plasma (FRAP) as a measure of 'antioxidant power. Anal. Biochem. 239 (1), 70–76. https://doi.org/10.1006/abio.1996.0292.

Biesaga-Kościelniak, J., Dziurka, M., Ostrowska, A., Mirek, M., Kościelniak, J., Janeczko, A., 2014. Brassinosteroid improves content of antioxidants in seeds of selected leguminous plants. Aust. J. Crop. Sci. 8 (3), 378–388.

Blažević, I., Đulović, A., Čulić, V.Č., Burčul, F., Ljubenkov, I., Ruščić, M., Generalić Mekinić, I., Bunias, E.L., 2019. Glucosinolate profile and in vitro biological potential. Molecules 24 (4), 741. https://doi.org/10.3390/molecules24040741.

Blažević, I., Montaut, S., Burčul, F., Olsen, C.E., Burow, M., Rollin, P., Agerbirk, N., 2020. Glucosinolate structural diversity, identification, chemical synthesis and metabolism in plants. Phytochemistry 169, 112100. https://doi.org/10.1016/j.phytochem.2019.112100.

Blios, M.S., 1958. Antioxidant determinations by the use of a stable free radical. Nature 26, 1199–1200.

Blumenthal, Busse, W.R., Goldberg, A., Gruenwald, J., Hall, Riggins, C.W., Rister, R.S., 2000. The Complete German Commission E Monographs. Therapeutic Guide to Herbal Medicines. Elsevier.

Boligon, A.A., Janovik, V., Boligon, A.A., Pivetta, C.R., Pereira, R.P., Rocha, J.B.T.D., Athayde, M.L., 2013. HPLC analysis of polyphenolic compounds and antioxidant activity in *Nasturtium officinale*. Int. J. Food Prop. 16 (1), 61–69. https://doi.org/10.1080/10942912.2010.528111.

Boyd, L.A., McCann, M.J., Hashim, Y., Bennett, R.N., Gill, C.I.R., Rowland, I.R., 2006. Assessment of the anti-genotoxic, anti-proliferative, and anti-metastatic potential of crude watercress extract in human colon cancer cells. Nutr. Cancer 55 (2), 232–241. https://doi.org/10.1207/s15327914nc5502_15.

Camponogara, C., Silva, C.R., Brusco, I., Piana, M., Faccin, H., de Carvalho, L.M., Schuch, A., Trevisan, G., Oliveira, S.M., 2019. *Nasturtium officinale* R. Br. effectively reduces the skin inflammation induced by croton oil via glucocorticoid receptor-dependent and NF-κB pathways without causing toxicological effects in mice. J. Ethnopharmacol. 229, 190–204. https://doi.org/10.1016/j.jep.2018.10.011.

Casanova, N.A., Ariagno, J.I., López Nigro, M.M., Mendeluk, G.R., Gette, M.d.l.A., Petenatti, E., Palaoro, L.A., Carballo, M.A., 2013. In vivo antigenotoxic activity of watercress juice (*Nasturtium officinale*) against induced DNA damage. J. Appl. Toxicol. 33 (9), 880–885. https://doi.org/10.1002/jat.2746.

Chodisetti, B., Rao, K., Gandi, S., Giri, A., 2015. Gymnemic acid enhancement in the suspension cultures of *Gymnema sylvestre* by using the signaling molecules—methyl jasmonate and salicylic acid. Vitro Cell Dev. Biol. Plant 51 (1), 88–92. https://doi.org/10.1007/s11627-014-9655-8.

De Souza, D.A., Costa, P.M., Ribeiro, R.I.M.A., Vidigal, P.V.T., Pinto, F.C.H., 2016. Daily intake of watercress causes inhibition of experimental Ehrlich tumor growth. J. Bras. Patol. Med. Lab. 52 (6), 393–399. https://doi.org/10.5935/1676-2444.20160063.

Dong, J., Wan, G., Liang, Z., 2010. Accumulation of salicylic acid-induced phenolic compounds and raised activities of secondary metabolic and antioxidative enzymes in *Salvia miltiorrhiza* cell culture. J. Biotechnol. 148 (2–3), 99–104. https://doi.org/10.1016/j.jbiotec.2010.05.009.

Ekiert, H., 2009. Pharmaceutical aspects of plant biotechnology. Part 1. Introduction – methods and main research directions. Farm. Pol. 65 (1), 69–77.

Ellnain-Wojtaszek, M., Zgórka, G., 1999. High-performance liquid chromatography and thin-layer chromatography of phenolic acids from *Ginkgo biloba* L. leaves collected within vegetative period. J. Liq. Chromatogr. Relat. Technol. 22 (10), 1457–1471. https://doi.org/10.1081/JLC-100101744.

Filová, A., 2014. Production of secondary metabolities in plant tissue cultures. Research J. Agricult. Sci. 46 (1), 236–245.

Gallaher, C.M., Gallaher, D.D., Peterson, S., 2012. Development and validation of a spectrophotometric method for quantification of total glucosinolates in cruciferous vegetables. J. Agric. Food Chem. 60 (6), 1358–1362. https://doi.org/10.1021/jf2041142.

Giallourou, N.S., Rowland, I.R., Rothwell, S.D., Packham, G., Commane, D.M., Swann, J.R., 2019. Metabolic targets of watercress and PEITC in MCF-7 and MCF-10A cells explain differential sensitisation responses to ionising radiation. Eur. J. Nutr. 58 (6), 2377–2391. https://doi.org/10.1007/s00394-018-1789-8.

Głusiuk, A., 2018. Beauty care in th treaties pf Metrodora from Byzantine and Trotula from Salerno. Seminare. Poszukiwania naukowe 39 (1), 145–154. https://doi.org/10.21852/sem.2018.1.12.

Grosser, K., van Dam, N.M., 2017. A straightforward method for glucosinolate extraction and analysis with high-pressure liquid chromatography (HPLC). JoVE 121. https://doi.org/10.3791/55425.

Grzegorczyk, I., Wysokińska, H., 2011. Liquid shoot culture of *Salvia officinalis* L. for micropropagation and production of antioxidant compounds; effect of triacontanol. Acta Soc. Bot. Pol. 77 (2), 99–104. https://doi.org/10.5586/asbp.2008.013.

Holst, B., Williamson, G., 2004. A critical review of the bioavailability of glucosinolates and related compounds. Nat. Prod. Rep. 21 (3), 425–447. https://doi.org/10.1039/b204039p.

Hoseini, H.F., Gohari, A.R., Saeidnia, S., Majd, N.S., Hadjiakhoondi, A., 2009. The effect of *Nasturtium officinale* on blood glucose level in diabetic rats. Pharmacologyonline 3, 866–871.

International Union for Conservation of Nature (IUCN). The IUCN Red List of Threatened Species. *Nasturtium officinale*. https://www.iucnredlist.org. (Accessed 15 January 2023).

Jeon, J., Bong, S.J., Park, J.S., Park, Y.K., Arasu, M.V., Al-Dhabi, N.A., Park, S.U., 2017. De novo transcriptome analysis and glucosinolate profiling in watercress (*Nasturtium officinale* R. Br.). BMC Genom. 18 (1). https://doi.org/10.1186/s12864-017-3792-5.

Karuppusamy, S., 2009. A review on trends in production of secondary metabolites from higher plants by in vitro tissue, organ and cell cultures. J. Med. Plants Res. 3 (13), 1222–1239.

Kayser, O., Müller, R.H., 2003. Biotechnologia Farmaceutyczna. PZWL, p. 2003.

Klimek-Szczykutowicz, M., Szopa, A., Ekiert, H., 2018a. *Nasturtium officinale* (watercress) – a source of plant raw materials with valuable cosmetic , medicinal and pro-health benefits. Pol. J. Cosmetol. 21 (4), 290–299.

Klimek-Szczykutowicz, M., Szopa, A., Ekiert, H., 2018b. Chemical composition, traditional and professional use in medicine, application in environmental protection, position in food and cosmetics industries, and biotechnological studies of *Nasturtium officinale* (watercress) – a review. Fitoterapia 129, 283–292. https://doi.org/10.1016/j.fitote.2018.05.031.

Klimek-Szczykutowicz, M., Szopa, A., Blicharska, E., Dziurka, M., Komsta, Ł., Ekiert, H., 2019. Bioaccumulation of selected macro- and microelements and their impact on antioxidant properties and accumulation of glucosinolates and phenolic acids in in vitro cultures of *Nasturtium officinale* (watercress) microshoots. Food Chem. 300, 125184. https://doi.org/10.1016/j.foodchem.2019.125184.

Klimek-Szczykutowicz, M., Dziurka, M., Blažević, I., Đulović, A., Granica, S., Korona-Glowniak, I., Ekiert, H., Szopa, A., 2020a. Phytochemical and biological activity studies on *Nasturtium officinale* (watercress) microshoot cultures grown in RITA® temporary immersion systems. Molecules 25 (22), 5257. https://doi.org/10.3390/molecules25225257.

Klimek-Szczykutowicz, M., Szopa, A., Dziurka, M., Komsta, Ł., Tomczyk, M., Ekiert, H., 2020b. The influence of *Nasturtium officinale* R. Br. agar and agitated microshoot culture media on glucosinolate and phenolic acid production, and antioxidant activity. Biomolecules 10 (9), 1216. https://doi.org/10.3390/biom10091216.

Klimek-Szczykutowicz, M., Dziurka, M., Blažević, I., Đulović, A., Miazga-Karska, M., Klimek, K., Ekiert, H., Szopa, A., 2021. Precursor-boosted production of metabolites in *Nasturtium officinale* microshoots grown in Plantform bioreactors, and antioxidant and antimicrobial activities of biomass extracts. Molecules 26 (15), 4660. https://doi.org/10.3390/molecules26154660.

Klimek-Szczykutowicz, M., Dziurka, M., Blažević, I., Đulović, A., Apola, A., Ekiert, H., Szopa, A., 2022. Impacts of elicitors on metabolite production and on antioxidant potential and tyrosinase inhibition in watercress microshoot cultures. Appl. Microbiol. Biotechnol. 106 (2), 619–633. https://doi.org/10.1007/s00253-021-11743-8.

Krzyzanowska, J., Czubacka, A., Pecio, L., Przybys, M., Doroszewska, T., Stochmal, A., Oleszek, W., 2012. The effects of jasmonic acid and methyl jasmonate on rosmarinic acid production in *Mentha* × *piperita* cell suspension cultures. Plant Cell Tissue Organ Cult. 108 (1), 73–81. https://doi.org/10.1007/s11240-011-0014-8.

Li, Q., Zhan, M., Chen, W., Zhao, B., Yang, K., Yang, J., Yi, J., Huang, Q., Mohan, M., Hou, Z., Wang, J., 2016. Phenylethyl isothiocyanate reverses cisplatin resistance in biliary tract cancer cells via glutathionylation-dependent degradation of Mcl-1. Oncotarget 7 (9), 10271–10282. https://doi.org/10.18632/oncotarget.7171.

Mahdavi, S., Kheyrollahi, M., Sheikhloei, H., Isazadeh, A., 2019. Antibacterial and antioxidant activities of *Nasturtium officinale* essential oil on food borne bacteria. Open Microbiol. J. 13 (1), 81−85. https://doi.org/10.2174/1874285801913010081.

Malepszy, S., 2012. Biotechnologia Roślin. PWN, p. 2012.

Manton, I., 1935. The cytological history of Watercress (*Nasturtium officinale* R. Br.). Z. für Indukt. Abstammungs- Vererbungslehre 69 (1), 132−157. https://doi.org/10.1007/bf01762872.

Martínez-Sánchez, A., Gil-Izquierdo, A., Gil, M.I., Ferreres, F., 2008. A comparative study of flavonoid compounds, vitamin C, and antioxidant properties of baby leaf Brassicaceae species. J. Agric. Food Chem. 56 (7), 2330−2340. https://doi.org/10.1021/jf072975+.

Mohaddab, M., El Goumi, Y., Gallo, M., Montesano, D., Zengin, G., Bouyahya, A., Fakiri, M., 2022. Biotechnology and in vitro culture as an alternative system for secondary metabolite production. Molecules 27 (22), 8093. https://doi.org/10.3390/molecules27228093.

Mohammadi, J., Motlagh, F.T., Mohammadi, N., 2017. The effect of hydroalcoholic extract of watercress on parameters of reproductive and sex hormones on the diabetic rats. J. Pharmaceut. Sci. Res. 9 (8), 1334−1338.

Narayani, M., Srivastava, S., 2017. Elicitation: a stimulation of stress in in vitro plant cell/tissue cultures for enhancement of secondary metabolite production. Phytochem. Rev. 16 (6), 1227−1252. https://doi.org/10.1007/s11101-017-9534-0.

Özyürek, M., Güçlü, K., Bektaşoğlu, B., Apak, R., 2007. Spectrophotometric determination of ascorbic acid by the modified CUPRAC method with extractive separation of flavonoids−La(III) complexes. Anal. Chim. Acta 588 (1), 88−95. https://doi.org/10.1016/j.aca.2007.01.078.

Quezada-Lázaro, R., Fernández-Zuñiga, E.A., García, A., Garza-González, E., Alvarez, L., Camacho-Corona, M.d.R., 2016. Antimycobacterial compounds from *Nasturtium officinale*. Afr. J. Trad. Complement. Alternat. Med. 13 (2), 31−34. https://doi.org/10.4314/ajtcam.v13i2.3.

Ramos, R., Bezerra, I., Ferreira, M., Soares, L., 2017. Spectrophotometric quantification of flavonoids in herbal material, crude extract, and fractions from leaves of *Eugenia uniflora* Linn. Pharmacogn. Res. 9 (3), 253−260. https://doi.org/10.4103/pr.pr_143_16.

Rao, U.M.V., Gaviraj, E.N., Veeresham, C., 2010. Effect of ethephon and ancymidol on the production of hypericins in shoot cultures of *Hypericum perforatum*. Indian Drugs 47 (6), 77−79.

Sadeghi, H., Mostafazadeh, M., Sadeghi, H., Naderian, M., Barmak, M.J., Talebianpoor, M.S., Mehraban, F., 2014. In vivo anti-inflammatory properties of aerial parts of *Nasturtium officinale*. Pharmaceut. Biol. 52 (2), 169−174. https://doi.org/10.3109/13880209.2013.821138.

Sajjalaguddam, R.R., Paladugu, A., 2015. Phenylalanine enhances quercetin content in in vitro cultures of abutilon indicum L. J. Appl. Pharmaceut. Sci. 5 (10), 80−84. https://doi.org/10.7324/JAPS.2015.501014.

Sánchez-Pujante, P.J., Borja-Martínez, M., Pedreño, M.Á., Almagro, L., 2017. Biosynthesis and bioactivity of glucosinolates and their production in plant in vitro cultures. Planta 246 (1), 19−32. https://doi.org/10.1007/s00425-017-2705-9.

Santamaria, A.R., Innocenti, M., Mulinacci, N., Melani, F., Valletta, A., Sciandra, I., Pasqua, G., 2012. Enhancement of viniferin production in *Vitis vinifera* L. cv. alphonse lavallée cell suspensions by low-energy ultrasound alone and in combination with methyl jasmonate. J. Agric. Food Chem. 60 (44), 11135−11142. https://doi.org/10.1021/jf301936u.

Schuchardt, J.P., Hahn, A., Greupner, T., Wasserfurth, P., Rosales-López, M., Hornbacher, J., Papenbrock, J., 2019. Watercress − cultivation methods and health effects. J. Appl. Bot. Food Qual. 92, 232−239. https://doi.org/10.5073/JABFQ.2019.092.032.

Shahani, S., Behzadfar, F., Jahani, D., Ghasemi, M., Shaki, F., 2017. Antioxidant and anti-inflammatory effects of *Nasturtium officinale* involved in attenuation of gentamicin-induced nephrotoxicity. Toxicol. Mech. Methods 27 (2), 107−114. https://doi.org/10.1080/15376516.2016.1258748.

Singleton, V.L., Orthofer, R., Lamuela-Raventós, R.M., 1999. Analysis of total phenols and other oxidation substrates and antioxidants by means of Folin-Ciocalteu reagent. Methods Enzymol. 299, 152–178. https://doi.org/10.1016/S0076-6879(99)99017-1.

Skrzypczak-Pietraszek, E., Piska, K., Pietraszek, J., 2018. Enhanced production of the pharmaceutically important polyphenolic compounds in *Vitex agnus castus* L. shoot cultures by precursor feeding strategy. Eng. Life Sci. 18 (5), 287–297. https://doi.org/10.1002/elsc.201800003.

Smetanska, I., 2008. Production of secondary metabolites using plant cell cultures. Adv. Biochem. Eng. Biotechnol. 111, 187–228. https://doi.org/10.1007/978-3-540-70536-9.

Sułkowska-Ziaja, K., Maślanka, A., Szewczyk, A., Muszyńska, B., 2017. Physiologically active compounds in four species of *Phellinus*. Nat. Prod. Commun. 12 (3), 363–366. https://doi.org/10.1177/1934578x1701200313.

Suroowan, S., Mahomoodally, M.F., 2016. A comparative ethnopharmacological analysis of traditional medicine used against respiratory tract diseases in Mauritius. J. Ethnopharmacol. 177, 61–80. https://doi.org/10.1016/j.jep.2015.11.029.

Szopa, A., Kokotkiewicz, A., Król, A., Luczkiewicz, M., Ekiert, H., 2018. Improved production of dibenzocyclooctadiene lignans in the elicited microshoot cultures of *Schisandra chinensis* (Chinese magnolia vine). Appl. Microbiol. Biotechnol. 102 (2), 945–959. https://doi.org/10.1007/s00253-017-8640-7.

Szopa, A., Kubica, P., Komsta, Ł., Walkowicz-Bożek, A., Ekiert, H., 2020. The effect of feeding culture media with biogenetic precursors on high production of depsides in agitated shoot cultures of black and red aronias. Plant Cell Tissue Organ Cult. 142 (2), 379–399. https://doi.org/10.1007/s11240-020-01869-4.

Teixidor-Toneu, I., Martin, G.J., Ouhammou, A., Puri, R.K., Hawkins, J.A., 2016. An ethnomedicinal survey of a Tashelhit-speaking community in the High Atlas, Morocco. J. Ethnopharmacol. 188, 96–110. https://doi.org/10.1016/j.jep.2016.05.009.

Thanh, N.T., Murthy, H.N., Yu, K.W., Hahn, E.J., Paek, K.Y., 2005. Methyl jasmonate elicitation enhanced synthesis of ginsenoside by cell suspension cultures of *Panax ginseng* in 5-l balloon type bubble bioreactors. Appl. Microbiol. Biotechnol. 67 (2), 197–201. https://doi.org/10.1007/s00253-004-1759-3.

Thibane, V.S., Ndhlala, A.R., Finnie, J.F., Van Staden, J., 2019. Modulation of the enzyme activity of secretory phospholipase A2, lipoxygenase and cyclooxygenase involved in inflammation and disease by extracts from some medicinal plants used for skincare and beauty. South Afr. J. Bot. 120, 198–203. https://doi.org/10.1016/j.sajb.2018.06.001.

Wielanek, M., Królicka, A., Bergier, K., Gajewska, E., Skłodowska, M., 2009. Transformation of *Nasturtium officinale*, *Barbarea verna* and *Arabis caucasica* for hairy roots and glucosinolate-myrosinase system production. Biotechnol. Lett. 31 (6), 917–921. https://doi.org/10.1007/s10529-009-9953-0.

World Flora Online, *Nasturtium officinale* R.Br. https://www.worldfloraonline.org/taxon/wfo-0000380323 (Accessed 15 January 2023).

Yaricsha, C.A., Katrin, Rissyelly, 2017. ACE Inhibitory activity, total phenolic and flavonoid content of watercress (*Nasturtium officinale* R. Br.) extract. Phcog. J. 9 (2), 249–251. https://doi.org/10.5530/pj.2017.2.42.

Yazdanparast, R., Bahramikia, S., Amin, A., 2008. *Nasturtium officinale* reduces oxidative stress and enhances antioxidant capacity in hypercholesterolaemic rats. Chem. Biol. Interact. 172 (3), 176–184. https://doi.org/10.1016/j.cbi.2008.01.006.

Yehuda, H., Soroka, Y., Zlotkin-Frušić, M., Gilhar, A., Milner, Y., Tamir, S., 2012. Isothiocyanates inhibit psoriasis-related proinflammatory factors in human skin. Inflamm. Res. 61 (7), 735–742. https://doi.org/10.1007/s00011-012-0465-3.

Yu, Z.Z., Fu, C.X., Han, Y.S., Li, Y.X., Zhao, D.X., 2006. Salicylic acid enhances jaceosidin and syringin production in cell cultures of *Saussurea medusa*. Biotechnol. Lett. 28 (13), 1027–1031. https://doi.org/10.1007/s10529-006-9035-5.

Yuan, J.M., Stepanov, I., Murphy, S.E., Wang, R., Allen, S., Jensen, J., Strayer, L., Adams-Haduch, J., Upadhyaya, P., Le, C., Kurzer, M.S., Nelson, H.H., Yu, M.C., Hatsukami, D., Hecht, S.S., 2016. Clinical trial of 2-phenethyl isothiocyanate as an inhibitor of metabolic activation of a tobacco-specific lung carcinogen in cigarette smokers. Cancer Prev. Res. 9 (5), 396–405. https://doi.org/10.1158/1940-6207.CAPR-15-0380.

Zafar, R., Zahoor, M., Shah, A.B., Majid, F., 2017. Determination of antioxidants and antibacterial activities, total phenolic, polyphenol and pigment contents in *Nasturtium officinale*. SILAE (Italo-Latin American Society of Ethnomedicine), Pakistan Pharmacologyonline 1, 11–18.

Zeb, A., 2015. Phenolic profile and antioxidant potential of wild watercress (*Nasturtium officinale* L.). SpringerOpen, Pakistan SpringerPlus 4 (1), 1–7. https://doi.org/10.1186/s40064-015-1514-5.

CHAPTER

10

Current knowledge on selected stilbenes and their promising applications

Marta Sharafan[1,2], Magdalena Anna Malinowska[2], Manon Ferrier[3], Nathalie Giglioli-Guivarc'h[3], Christophe Hano[4], Elżbieta Sikora[2], Arnaud Lanoue[3] and Agnieszka Szopa[1]

[1]*Department of Medicinal Plant and Mushroom Biotechnology, Medical College, Jagiellonian University, Kraków, Poland;* [2]*Department of Organic Chemistry and Technology, Faculty of Chemical Engineering and Technology, Cracow University of Technology, Kraków, Poland;* [3]*EA 2106 Biomolécules et Biotechnologies Végétales, UFR des Sciences Pharmaceutiques, Université de Tours, Tours, France;* [4]*Institut de Chimie Organique et Analytique, Université d'Orléans-CNRS, UMR 7311, Orléans, France*

Abbreviations

16HBE	bronchial epithelial line cells
4CL	4-coumarate-CoA ligase CosIng Cosmetic Ingredient
ABTS	2,2′-azino-bis-(3-ethylbenzothiazoline-6-sulfonic acid)
ACC	acetyl-CoA carboxylase
AEAC	L-ascorbic acid equivalent antioxidant capacity
ALP	alkaline phosphatase
B164A5	melanoma mouse cell line
B16F10	melanoma cell line
BALF	bronchoalveolar lavage fluid
C4H	cinnamate 4-hydroxylase
CaCo 2	colon adenocarcinoma cells
CAT	catalase
CCL3	C–C motif chemokine ligand 3
COX	cyclooxygenase
CPK	creatine phosphokinase
CRTC1	transcriptional coactivator 1
CUPRAC	cupric ion reducing antioxidant capacity
DMBA	7,12-dimethylbenz[a]anthracene
DMSO	dimethyl sulfoxide
DPPH	1,1-diphenyl-2-picrylhydrazyl
ECM	extracellular matrix
EFSA	European Food and Safety Authority
EMA	European Medicines Agency
ERK1/2	extracellular signal-regulated kinase 1/2
FDA	Food and Drug Administration
fMLP	N-formyl-L-methionyl-L-leucyl-L-phenylalanine

Phytochemicals for Health. https://doi.org/10.1016/B978-0-443-15366-2.00010-1

FRAP	ferric ion reducing antioxidant activity
FTH1	ferritin heavy chain 1
G6PDH	glucose-6P-dehydrogenase
GAC	gastric adenocarcinoma
GLUT4	glucose transporter type 4
GPDH	glycerol-3-phosphate dehydrogenase
GPX	glutathione peroxidase
GPX4	peroxidase 4
HaCaT	human keratinocyte cell line
HCC	hepatocellular carcinoma high-density lipoprotein
HDL	high-density lipoprotein
HFD	high-fat-diet
HHV	human herpesvirus
HOCl	hypochlorous acid scavenging activity
HP-β-CDs	hydroxypropyl-beta-cyclodextrins
HUVECs	human umbilical vein endothelial cells
IBAT	interscapular brown adipose tissue
IGF-1	insulin-like growth factor-1
IgE	immunoglobulin E
iNOS	inducible nitric oxide synthase
IGFBP-3	insulin-like growth factor-binding protein 3
IOLA	linoleic acid
IL	interleukin
JAK 1	janus kinase 1
L-DOPA	3,4-dihydroxyphenylalanine
LDH	lactate dehydrogenase
LDL	low-density lipoprotein
LPL	lipoprotein lipase
LPS	lipopolysaccharide
MAO-B	monoamine oxidase-B
MCP-1	monocyte/macrophage chemotactic protein-1
MDA	malondialdehyde
MDA-MB-231	human breast cancer cell line
MDC	macrophage-derived cytokine
MGO	methylglyoxal
MKP-1	mitogen-activated protein kinase phosphatase-1
MMOC	mouse mammary gland organ culture carcinogenesis model
MMP-9	matrix metalloproteinase-9
NF-κB	nuclear factor κ-light-chain-enhancer of activated beta cells
O2•−	superoxide anion radical scavenging activity
OGD/R	oxygen-glucose deprivation/reoxygenation
ORAC	oxygen radical scavenging capacity
ORAC-FL	ORAC-fluorescein
p38-MAPK	p38-mitogen-activated protein kinase
PAL	phenylalanine ammonia-lyase
PBMCs	peripheral blood mononuclear cells
PCNA	proliferating cell nuclear antigen
PGC-1γ	peroxisome proliferator—activated receptor-gamma coactivator 1 alpha

PGE2	prostaglandin E2
PlGF	placental growth factor
PTX3	pentraxin 3
QR	quinone reductase
QS	quorum-sensing
ROS	reactive oxygen species
SIRT	sirtuin 1
SLO	soybean lipoxygenase
SNP	sodium nitroprusside
SOD	superoxide dismutase
STAT	signal transducer and activator of transcription
STS	stilbene synthase
STZ	streptozotocin
T2DM	type 2 diabetic mice
TARC	thymus and activation-regulated cytokine
TEWL	transepidermal water loss
TFAM	mitochondrial-transcription-actor-A
TG	triglyceride
TGF-β1	transforming growth factor beta 1
TNF-α	tumor necrosis factor-alfa
TRAP	total reactive antioxidant potential
Trolox (Tx)	(±)-6-hydroxyl-2,5,7,8-tetramethlychromane-2-carboxylic acid
VCAM-1	vascular cell adhesion molecule 1
VEGF	vascular endothelial growth factor

1. Stilbenoids—General characteristic

Polyphenols are a large family of naturally occurring secondary metabolites presented in various plant species. They are involved in defense of plants against aggressions e.g., pathogens or UV radiation. Polyphenols can be divided into five main groups: phenolic acids, flavonoids, lignans, tannins, and stilbenoids (Bharadvaja et al., 2023; Pandey and Rizvi, 2009).

Stilbenoids are important phenolic compounds first isolated from the plants in 1899 (Xiao et al., 2008). The name "stilbene" is derived from the Greek word "stilbos," which means "shining" (Likhtenshtein, 2009). The term "stilbenoids" was proposed by Gorham in 1980 and reviewed by him in 1995 (Xiao et al., 2008). Stilbenoids are synthesized by at least 72 plants species especially by *Vitis vinifera* species (Teka et al., 2022; Valletta et al., 2021). The stilbenoids are regarded as phytoalexins, the primary role of which is to protect the plant against biotic (e.g., pathogenic microorganisms) and abiotic (e.g., UV radiation, heavy metals, air pollutants) stresses (Teka et al., 2022; Valletta et al., 2021).

Stilbenoids have been investigated for their health-promoting properties. One of the best-known stilbene, *trans*-resveratrol, is marked with antiinflammatory, anticancer, antidiabetic, and antiobesity activities (Banez et al., 2020; Vervandier-Fasseur and Latruffe, 2019; Zhu et al., 2017; Zu and Wang, 2020). It also possesses antioxidant, antiaging, and depigmentation properties, which determine the potential application of stilbenoids in cosmetics formulations (Akinwumi et al., 2018; Gülçin, 2010; Li et al., 2017). However, resveratrol derivatives also showed valuable biological properties, deserve attention, such as: piceatannol, polydatin, pterostilbene, ampelopsin A, oxyresveratrol, (+)-*trans*-ε-viniferin, *trans*-δ-viniferin, pallidol, scirpusin A, scirpusin B, α-viniferin, miyabenol C, vitisin A, vitisin B, and hopeaphenol. These compounds are also extremely interesting and are the subject of scientific research confirming their valuable biological properties, which is presented in the following sections of this chapter.

Due to various biological properties, there is an increasing interest directed to stilbenoids large-scale production. The major method of stilbene production is the extraction from plants. However, there are some other techniques, e.g., chemical synthesis and biotechnological methods, which also could be used in stilbenes production (Valletta et al., 2021).

2. Phytochemistry of stilbenoids

Stilbenoids (structurally compounds with 1,2-diphenylethylene core) belong to the class of phenolic compounds. They are relatively unreactive, colorless, and practically insoluble in water. Structurally, stilbenoids have 14 carbons (C6—C2—C6) skeleton, in which two aromatic rings are joined by an ethylene bridge. They exist in two isomeric forms: (*E*)-stilbene and (*Z*)-stilbene (Fig. 10.1). (*E*)-Stilbene (i.e., *trans*-stilbene) isomerizes to (*Z*)-stilbene (i.e., *cis*-stilbene) under the light influence. *trans*-Stilbene is not sterically hindered in contrast to *cis*-stilbene, which is therefore less stable (Fig. 10.2). The melting point of (*E*)-stilbene is 125 °C, while the melting point of (*Z*)-stilbene is 6°C (Likhtenshtein, 2009).

Due to the presence of phenol groups, stilbenoids can occur in hydroxylated, prenylated, geranylated forms, or combined with sugars to form glucosides (Shen et al., 2013; Teka et al., 2022). The stilbene derivatives are known as stilbenoids.

Stilbenoids can be classified to monomers and oligomers (dimers, trimers, and tetramers) (Fig. 10.2). For instance, among stilbene monomers, resveratrol, oxyresveratrol, pterostilbene, and piceatannol can be distinguished. They are characterized by the presence of two aromatic rings, which can bear prenyl, geranyl, or farnesyl chains (Mattio et al., 2020). Monomeric stilbenoids can be further divided into six types: stilbenoids, bibenzyls, bisbibenzyls, and phenanthrenoids. Stilbene oligomers are derived from the oxidative condensation of stilbene monomers (e.g., dimers, trimers, tetramers, and more) (Mattio et al., 2020; Waffo-Teguo et al., 2008).

Stilbenoids derivatives are easily synthesized and usually are thermally and chemically stable. They also possess absorption and fluorescence abilities (Likhtenshtein, 2009).

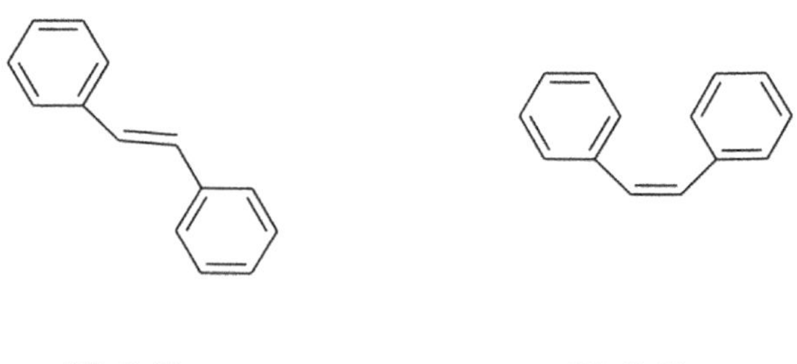

(*E*)-Stilbene (*Z*)-Stilbene

FIGURE 10.1

The chemical structure of (*E*)-stilbene and (*Z*)-stilbene.

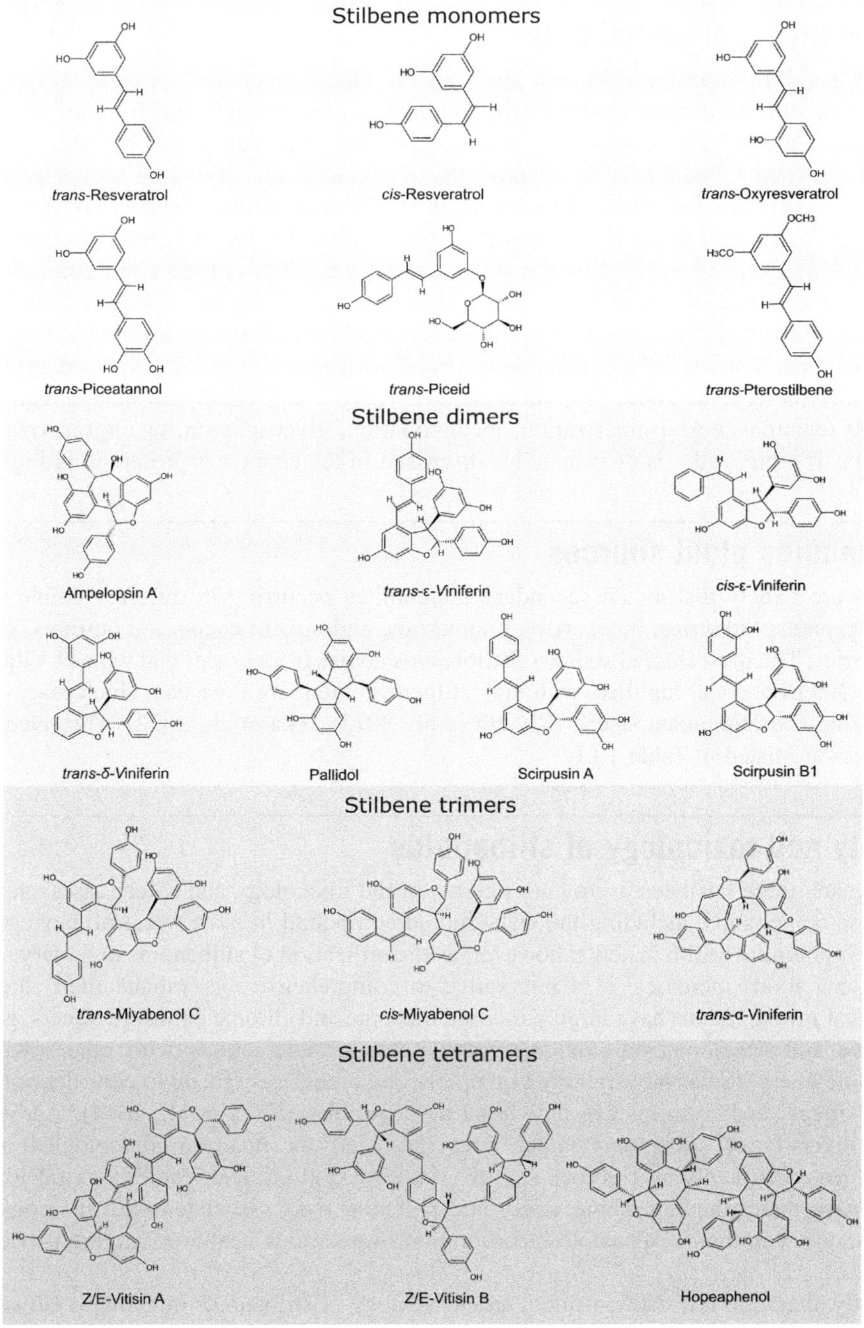

FIGURE 10.2

The selected stilbenoids and their classification basing on molecular weight (monomers, dimers, trimers, and tetramers).

3. Biosynthesis of stilbenoids

Stilbenoids are biosynthesized through the general phenylpropanoid pathway (Fig. 10.3). The biosynthesis of stilbenoids is generated by the shikimate pathway and begins from the aromatic amino acid L-phenylalanine deamination to *trans*-cinnamic acid catalyzed by the phenylalanine ammonia-lyase (PAL). Phenylalanine is converted to cinnamic acid, which in turn is hydroxylated to *p*-coumaric acid by the cinnamate 4-hydroxylase (C4H). *p*-Coumaric acid is converted by the 4-coumarate-CoA ligase (4CL) to *p*-coumaroyl-CoA followed by the condensation with three malonyl-CoA molecules catalyzed by the stilbene synthase (STS) (Billet et al., 2020; Pawlus et al., 2012; Teka et al., 2022; Tsopmo et al., 2013; Vermerris and Nicholson, 2006; Waffo-Teguo et al., 2008). STS is the primary enzyme of stilbenoid biosynthesis, which is responsible for the generation of stilbene backbone. Subsequently, tetraketide intermediate cyclization leads to the formation of stilbene ring structure (Chong et al., 2009; Tsopmo et al., 2013) (Fig. 10.3). Finally, stilbenoids can be modified by different reactions, e.g., isomerization, methoxylation, glycosylation, or oligomerization (Dávid et al., 2021). The biosynthesis of stilbenoids discussed in the chapter is presented in Fig. 10.3.

4. Stilbenoids plant sources

Stilbenoids are widely distributed secondary metabolites occurring in different edible plants. It is present in peanuts, bilberries, blueberries, cranberries, and also in cocoa, red currants, strawberries, and cowberries. The most studied source of stilbenoids comes from grapes and wine of Vitaceae family (*Vitis* sp.). Other botanical families, with high stilbene content are Pinaceae, Gnetaceae, Cyperaceae, Polygonaceae, and Dipterocarpaceae (Rivière et al., 2012; Teka et al., 2022). The selected stilbene plant sources are listed in Table 10.1.

5. Safety and toxicology of stilbenoids

In recent years, there has been a growing interest in the toxicology and safety assessment of stilbenoids. These compounds, including the well-studied compound *trans*-resveratrol, have gained attention for their potential health benefits; however, as the utilization of stilbenoids in dietary supplements and functional foods increases, it is imperative to comprehensively evaluate their safety profiles. Toxicological investigations have largely focused on acute and chronic exposure effects, genotoxicity, reproductive and developmental toxicities, and potential interactions with other pharmaceutical agents. While many studies have reported favorable outcomes, it is crucial to consider potential dose-dependent effects and variations in individual susceptibility (Silva et al., 2014). Additionally, the structural diversity of stilbenoids underscores the need for nuanced toxicological evaluations. Continued research involving rigorous in vitro, in vivo, and clinical studies is vital to establish a comprehensive understanding of the safety and potential risks associated with the consumption of stilbenoids, ensuring their responsible incorporation into various health-promoting products (Pavela et al., 2017).

Currently, there are few data on safety and toxicology of stilbenoids in humans. However, among all the stilbenoids, *trans*-resveratrol is the most studied in term of safety and toxicology. Resveratrol is believed to be safe, nontoxic, and well-tolerated for oral and dermal administration with a 50% lethal

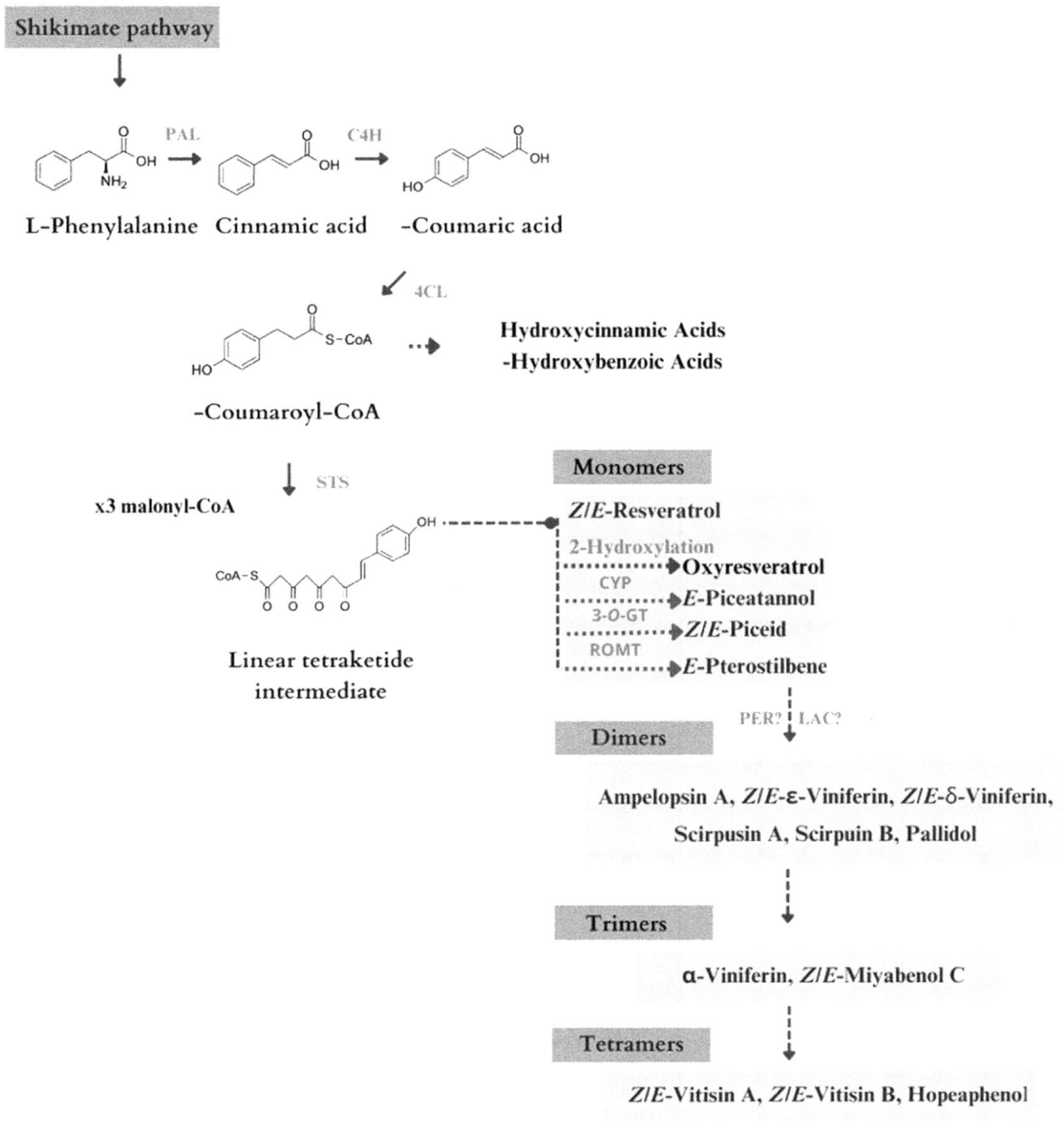

FIGURE 10.3

Stilbenoids general biosynthesis pathway in plants
Biosynthetic enzymes are presented in red: phenylalanine ammonia-lyase (PAL); cinnamate 4-hydroxylase (C4H); 4-coumarate:CoA (4CL); chalcone synthase (CHS); stilbene synthase (STS); 3-*O*-glucosyl-transferase (3-*O*-GT); (E-resveratrol)-*O*-methyltransferase ((R)-OMT); cytochrome P450 (CYP); peroxidase (PER); laccase (LAC) (Billet et al., 2020; Marant et al., 2024; Pawlus et al., 2012).

Table 10.1 Plant sources of the selected stilbenoids.

Compound group	Compound name	Family	References
Monomers	*trans*-Resveratrol, *cis*-Resveratrol	**Anacardiaceae** (*Pistacia vera*), **Bromeliaceae** (*Ananas comosus*), **Cannabaceae** (*Humulus lupulus*), **Cupressaceae** (*Juniperus macropoda*), **Cyperaceae** (*Carex baccans, Carex dimorpholepis, Cyperus eragrostis, Cyperus stoloniferus, Scirpus fluviatilis, Scirpus maritimus*), **Ericaceae** (*Vaccinium macrocarpon*), **Juglandaceae** (*Juglans regia*), **Leguminaceae** (*Arachis hypogea, Caesalpinia latisiliqua, Caragana jubata, Caragana sinica*), **Magnoliaceae** (*Magnolia officinalis*), **Melanthiaceae** (*Veratrum grandiflorum*), **Myrtaceae** (*Myrciaria dubia*), **Oleaceae** (*Olea europaea*), **Paeoniaceae** (*Paeonia lactiflora, Paeonia suffruticosa*), **Polygonaceae** (*Calligonum leucocladum, Polygonum cuspidatum, Polygonum multiflorum, Reynoutria japonica*), **Vitaceae** (*Vitis* sp)	(Dávid et al., 2021; Teka et al., 2022)
	Oxyresveratrol	**Caesalpiniaceae** (*Caesalpinia furfuracea*, Cassia garrettiana), **Gnetaceae** (*Gnetum africanum, Gnetum cuspidatum, Gnetum gnemonoides, Gnetum hainanse, Gnetum montanum, Gnetum pendulum*Fabaceae (*Spirotropis longifolia*), **Liliaceae** (*Schoenocaulon officinale*), **Melanthiaceae** (*Veratrum album, Veratrum dahuricum, Veratrum grandiflorum, Veratrum lobelianum, Veratrum maackii*), **Moraceae** (*Artocarpus lacucha, Artocarpus fulvicortex, Artocarpus gomezianus, Artocarpus heterophyllus, Artocarpus hirsutus, Artocarpus hypargyreus, Artocarpus lakoocha, Artocarpus nitidus, Artocarpus rigida, Artocarpus styracifolius, Artocarpus. thailandicus, Artocarpus xanthocarpus, Cudrania cochinchinensis, Cudrania tricuspidate, Maclura cochinchinensis, Maclura pomifera, Morus alba, Morus atropurpurea, Morus australis, Morus bombycis, Morus cathayana, Morus laevigata, Morus latifolia, Morus macroura, Morus multicaulis, Morus nigra, Morus rubra, Morus wittiorum, Morus yunnanensis*, **Myrtaceae** (*Melaleuca leucadendron*), **Poaceae** (*Chrysopogon aciculatis*), **Rutaceae** (*Glycosmis pentaphylla*), **Rosaceae** (*Prunus dulcis*), **Smilacaceae** (*Smilax china, Smilax microphylla*), **Vitaceae** (*Tetrastigma hemsleyanum*)	(Likhitwitayawuid, 2021; Teka et al., 2022)

Table 10.1 Plant sources of the selected stilbenoids.—cont'd

Compound group	Compound name	Family	References
	trans-Piceatannol *cis*-Piceatannol	**Cyperaceae** (*Carex appressa, Cyperus eragrostis, Cyperus stoloniferus, Scirpus californicus, Scirpus* fluviatilis), **Dioscoreaceae** (*Dioscorea antaly*), **Leguminaceae** (*Arachis hypogea, Caragana stenophylla, Caragana tibetica*), **Lophatherum** (*Lophatherum gracile*), **Myrtaceae** (*Rhodomyrtus tomentosa*), **Passifloraceae** (*Passiflora edulis*), **Vitaceae** (*Vitis amurensis, Vitis coignetiae, Vitis vinifera*)	(Dávid et al., 2021; Krambeck et al., 2023; Pawlus et al., 2012; Teka et al., 2022)
	trans-Piceid, *cis*-Piceid	**Cannabaceae** (*Humulus lupulus*), **Cupressaceae** (*Juniperus macropoda*), **Cyperaceae** (*Cyperus rotundus*), **Leguminaceae** (*Arachis hypogea*), **Polygonaceae** (*Polygonum cuspidatum*), **Vitaceae** (*Muscadinia rotundifolia, Vitis amurensis, Vitis coignetiae, Vitis labrusca, Vitis vinifera*)	(Karami et al., 2022; Pawlus et al., 2012; Teka et al., 2022)
	Pterostilbene	**Ericaceae** (*Vaccinium corymbosum, Vaccinium angustifolium*), **Leguminaceae** (*Arachis hypogea*), **Vitaceae** (*Vitis coignetiae, Vitis vinifera*)	(Pawlus et al., 2012; Teka et al., 2022)
Dimers	Ampelopsin A	**Dipterocarpaceae** (*Hopea dryobalanoides*), **Vitaceae** (*Vitis amurensis, Vitis betulifolia, Vitis coignetiae, Vitis heyneana, Vitis thumbergii, Vitis vinifera*)	(Pawlus et al., 2012; Teka et al., 2022)
	trans-ε-Viniferin	**Bombaceae** (*Bombax malabarica*), **Dipterocarpaceae** (*Dryobalanops lanceolata, Hopea parviflora, Vateria indica*), **Gnetaceae** (*Gnetum microcapum*), **Iridaceae** (*Iris lactea*), **Paeoniaceae** (*Paeonia lactiflora, Paeonia ostia, Paeonia suffruticosa*), **Polygonaceae** (*Rheum undulatum, Rheum lhasaense*) **Vitaceae** (*Cayratia trifolia, Cyphostemma crotalarioides, Parthenocissus quinquefolia, Vitis amurensis, Vitis berlandieri, Vitis betulifolia, Vitis chunganensis, Vitis cinerea, Vitis coignetiae, Vitis davidii, Vitis flexuosa, Vitis ficifolia, Vitis heyneana, Vitis labrusca, Vitis longii, Vitis riparia, Vitis rupestris, Vitis solonis longii, Vitis solonis richter, Vitis rotundifolia, Vitis thumbergii, Vitis vinifera, Vitis wilsoniae*)	(Fuloria et al., n.d.; Pawlus et al., 2012; Teka et al., 2022)
	trans-δ-Viniferin	**Gnetaceae** (*Gnetum hainanense*), **Polygonaceae** (*Rheum undulatum, Rheum maximowiczii*), **Vitaceae** (*Vitis labrusca, Vitis thumbergii, Vitis vinifera*)	(Fuloria et al., n.d.; Lotos: Natural Products Online, 2023; Teka et al., 2022)

Continued

Table 10.1 Plant sources of the selected stilbenoids.—cont'd

Compound group	Compound name	Family	References
	Pallidol	**Cyperaceae** (*Carex distachya*, *Carex folliculata*, Carex glauca, *Carex gynandra*, *Cyperus longus*), **Leguminaceae** (*Caragana sinica*), **Vitaceae** (*Cissus pallida*, *Vitis amurensis*, *Vitis vinifera*)	(Dávid et al., 2021; Lotos: Natural Products Online, 2023; Pawlus et al., 2012; Teka et al., 2022)
	Scirpusin A	**Cyperaceae** (*Cyperus longus*, *Cyperus rotundus*, *Scirpus californicus*, *Scirpus fluviatilis*, *Scirpus maritimus*), **Leguminaceae** (*Caragana rosea, Maackia amurensis*), **Vitaceae** (*Vitis vinifera*)	(Dávid et al., 2021; Teka et al., 2022)
	Scirpusin B	**Cyperaceae** (*Cyperus eragrostis*, *Cyperus longus*, *Cyperus rotundus*, *Scirpus californicus*, *Scirpus fluviatilis*, *Scirpus maritimus*), Dioscoreaceae (*Dioscorea antaly*), **Leguminaceae** (*Caragana rosea, Maackia amurensis*)	(Dávid et al., 2021; Teka et al., 2022)
Trimers	*trans*-Miyabenol C, *cis*-Miyabenol C	**Apiaceae** (*Foeniculum vulgare*), **Cyperaceae** (*Carex fedia, Carex glauca, Carex gynandra, Carex kobomugi, Carex pendula, Carex pumila*), Leguminaceae (*Caragana sinica, Caragana stenophylla*), **Vitaceae** (*Vitis vinifera*)	(Dávid et al., 2021; Teka et al., 2022)
	α-Viniferin	**Dipterocarpaceae** (*Dipterocarpus littoralis, Dryobalanops lanceolata, Hopea ponga, Shorea maxwelliana, Shorea ovalis, Shorea hemsleyana, Shorea seminis*), **Cyperaceae** (*Carex baccans, Carex humilis*), **Iridaceae** (*Iris clarkei*), **Leguminaceae** (*Caragana chamlagu, Caragana sinica, Caragana stenophylla*), **Saxifragaceae** (*Astilbe grandis*), **Vitaceae** (*Cayratia trifolia, Vitis quinquangularis, Vitis vinifera*)	(Dávid et al., 2021; Fuloria et al., n.d.; Teka et al., 2022)
Tetramers	Vitisin A (r2-Viniferin)	**Iridaceae** (*Iris lacteal*), **Vitaceae** (*Vitis amurensis, Vitis betulifolia, Vitis chunganensis, Vitis coignetiae, Vitis davidii, Vitis flexuosa, Vitis thumbergii, Vitis vinifera, Vitis wilsoniae*)	(Fuloria et al., n.d.; Pawlus et al., 2012; Teka et al., 2022)
	trans-Vitisin B (r-viniferin)	**Iridaceae** (*Iris lacteal*), **Vitaceae** (*Vitis amurensis, Vitis berlandieri, Vitis cinerea, Vitis coignetiae, Vitis heyneana, Vitis longii, Vitis riparia, Vitis rupestris, Vitis solonis longii, Vitis solonis richter, Vitis thunbergii* var. *taiwaniana*, *Vitis vinifera*)	(Fuloria et al., n.d.; Pawlus et al., 2012; Teka et al., 2022)
	Hopeaphenol	**Dipterocarpaceae** (*Dryobalanops lanceolata*), **Vitaceae** (*Vitis amurensis, Vitis betulifolia, Vitis chunganensis, Vitis flexuosa, Vitis vinifera*)	(Pawlus et al., 2012; Teka et al., 2022)

dose of 2 g/day. However, during in vivo studies, *trans*-resveratrol caused mild-to-moderate gastrointestinal symptoms, e.g., nausea or diarrhea in 2,5 and 5 g douses (Ratz-Łyko and Arct, 2019).

However, *trans*-resveratrol is no approved as a drug ingredient, but there is a list of dietary supplements, which contains *trans*-resveratrol as one of the active ingredients submitted by the Food and Drug Administration (FDA) starting from 2000 to 2010 (Table 10.2) (Food and Drug Administration (FDA), 2023). Moreover, in 2016, European Food and Safety Authority (EFSA) published a positive opinion about synthetic *trans*-resveratrol (≥99% (w/w)) concluding on its safety under proposed condition of use as a novel food (Safety of Synthetic Trans-resveratrol as a Novel Food Pursuant to Regulation (EC) No 258/97, 2016). In addition, in 2017, European Medicines Agency (EMA) assigned to *trans*-resveratrol an orphan designation status (EU/3/16/1827) granted for Luis Pereira de Almeida for the treatment of spinocerebellar ataxia (European Medicines Agency (EMA). Available Online, 2023).

Jadhav et al. (2016) also evaluated *trans*-resveratrol for normal toxicity by performing histopathological studies on the brain, heart, lung, liver, kidney, and stomach tissue. All the treated groups exhibited good arrangement of tissue. No edema, intracellular space, or edematous changes in exposed brain tissue were observed. No myocardial necrosis, infiltration of inflammatory cells, or myocardial degeneration was observed in the heart, as well as no infiltration of inflammatory cells or mononuclear cells of pulmonary parenchyma was found in the lung tissue. Furthermore, no necrosis was observed in the liver tissue, while no tubular degeneration, intratubular degeneration, or congestion was found in the kidney tissue. Thus, *trans*-resveratrol was confirmed to be safe in a wide range of human tissues (Jadhav et al., 2016).

6. Current applications of stilbenoids

6.1 Application in medicine

Due to the beneficial effect of stilbenoids on human health (Fig. 10.4), there is an increasing number of its application in medicine, food supplements, and cosmetics (Duarte et al., 2021).

Table 10.2 Examples of dietary supplements containing resveratrol with FDA approval.

Company	Active ingredient
Solgar Vitamin and Herb	**Resveratrol**
InterHealth Neutraceuticals, Inc.	***trans*-Resveratrol** (the root extract of *Polygonum cuspidatum* Siebold & Zucc.) (Protykin® RSV-5000 powder)
Ceres Living, Inc.	**Resveratrol** (AIO Premium Cellular Health)
Beverage Solutions & Bottling Company, Inc	Maqui Berry Organic Powder derived from whole berry with pulp, skin and seed of *Aristotelia chilensis* (Molina) Stuntz, Acai Berry Organic Powder derived from whole fruit with pulp, skin and seed of *Euterpe oleracea* Mart, and ***trans*-Resveratrol** 98% derived from the root of *Polygonum cuspidatum* Sieb Siebold (Milagro capsules with **Resveratrol**)

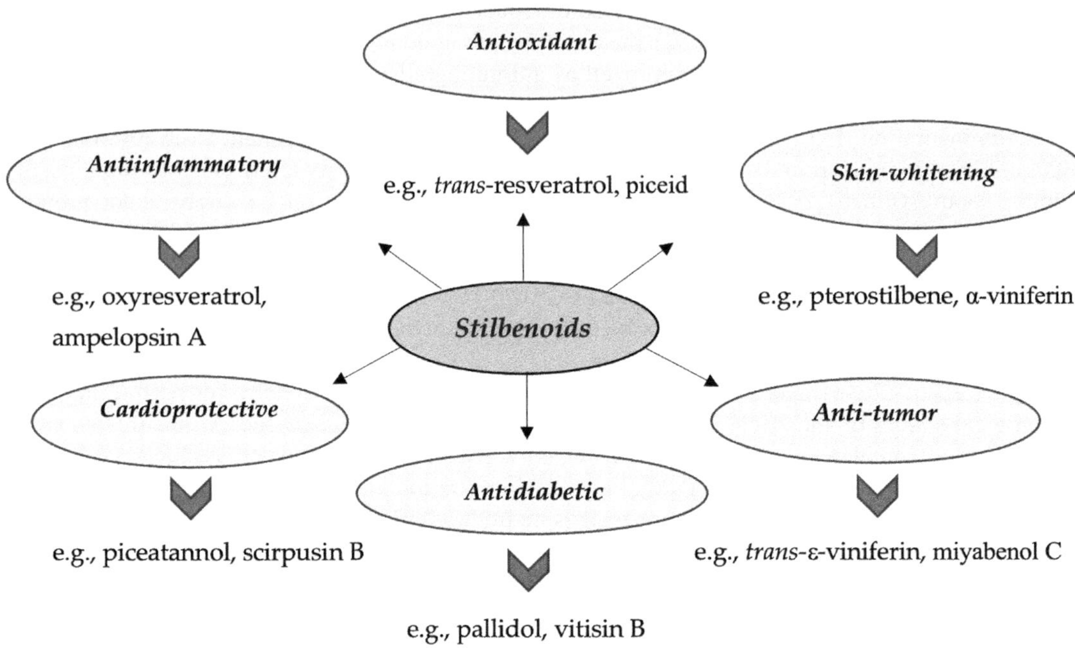

FIGURE 10.4

Stilbenoids' main directions of biological activity.

In industry, stilbenoids are used in nutraceuticals. Stilbenoids included in wine can be proposed to prevent the development of cardiovascular disease due to its association with the "*French paradox*" phenomenon (Renaud and de Lorgeril, 1992). Stilbenoids obtained from other sources, e.g., peanuts, pistachios, or chocolate, are also recommended mainly because of their health-beneficial effects (Javed et al., 2020).

In pharmaceutical industry, stilbenoids' biological properties are highly appreciated. Stilbenoids target interleukins, tumor necrosis factor α (TNF- α), nuclear factor kappa B (NF-κB), and cyclooxygenases (Javed et al., 2020).Therefore, they are considered as potential antiinflammatory drugs (Duarte et al., 2021). Along with antiinflammatory activity, stilbenoids can be proposed as anticancer or antidiabetic agents. Some of other stilbenoids, e.g., pterostilbene or piceatannol, could be potentially used for drug development for human disorders (Teka et al., 2022).

The most studied stilbenoid, *trans*-resveratrol, is proposed as an antioxidant, antiaging, and depigmenting agent (Akinwumi et al., 2018; Gülçin, 2010; Li et al., 2017). For instance, the well-known trademark Caudalie has a series of cosmetic products containing *trans*-resveratrol (Duarte et al., 2021). Other popular cosmetic companies, such as Sesderma or The Ordinary, also have resveratrol-containing cosmetics, which are focused on skin anti-aging (The Ordinary, 2023; Sesderma, 2023).

Apart from aforementioned applications, stilbenoids are used in material sciences and are involved in the manufacture of industrial dyes, dye lasers, optical brighteners, phosphors, and scintillators (Likhtenshtein, 2009).

The potential applications of individual stilbenoids in medicine are focused on in the following parts of this monograph. The applications of individual compounds have been characterized in detail and supported by scientific research.

6.2 Application in food industry

There is a wide range of dietary supplements, which contains mainly resveratrol and its derivative pterostilbene. However, regardless of their accessibility, their safety of use is not approved by FDA. Most resveratrol-containing supplements are produced in Europe and USA. The list of examples of dietary supplements, which contain resveratrol and/or pterostilbene, is presented in Table 10.3.

6.3 Application in cosmetology

There are five stilbenoid compounds described in the review listed in Cosmetic Ingredient Database (CosIng): resveratrol, hydroxyresveratrol, piceatannol, polydatin (i.e., piceid), and α-viniferin (*Available Online:* Www.Ec.Europa.Eu (Accessed on 14, 2023). There are also 10 resveratrol forms presented in CosIng: resveratrol tripentyl ether, resveratrol dimethyl ether maltotrioside, trisodium resveratrol triphosphate, resveratryl triacetate, glucosyl resveratrol, resveratrol dimethyl ether, resveratryl glucoside, resveratrol tributyl ether, resveratrol dimethyl ether glucoside, and resveratrol dimethyl ether oligosaccharides. There is also one extract containing resveratrol – INCI: Pichia/ Resveratrol Ferment Extract, which functions as skin conditioner (*Available Online:* Www.Ec.Europa.Eu (Accessed on 14, 2023) (Table 10.4).

Based on the analysis of the cosmetic industry, the dominant stilbene compound, which could be found in the cosmetics, is resveratrol. Other stilbenoids are not popular constituents in cosmetic formulations regardless of their valuable properties. The complex composition of these new natural ingredients containing stilbenoids can be explored using advance analytical tools such as metabolomics (Ferrier et al., 2024). As can be seen, most of cosmetics formulations containing resveratrol are characterized by their anti-aging properties (Ferrier et al., 2022) (Table 10.5).

7. Characteristics of the selected stilbenoids

7.1 Monomers

7.1.1 Trans-*resveratrol*

7.1.1.1 General information

Trans-resveratrol (3,4′,5-*trans*-trihydroxy-stilbene) (Fig. 10.5) is a stilbenoid occurring in many plants. Resveratrol consists of two hydroxylated phenol rings linked by a methylene bridge (Zhang et al., 2021). It is soluble in organic solvents including ethanol, methanol, acetone, and relatively insoluble in water (xPharm, 2010). Resveratrol exists in both *trans* and *cis* configurations. Due to ultraviolet and visible light exposure, *trans*-resveratrol is modified into *cis* isomer. *Trans*-resveratrol appears to have more prominent potential and is approved to be more stable form than *cis* isomer (Ahmadi & Ali Ebrahimzadeh, 2020; Galiniak et al., 2019; Kataria and Khatkar, 2019) (Fig. 10.5). The scientific studies concern *trans* isomer of resveratrol because it is naturally occurring form of resveratrol.

Table 10.3 Examples of dietary supplements containing resveratrol and/or pterostilbene.

Manufacturer/ Country	Trade name/Active compound	Function
Augment life (Germany) www.augmentlifeshop.com	*trans*-Resveratrol (99.5%)	Protection against oxidative stress, antiinflammatory, improvement of blood sugar level, antiaging
Warnke Vitalstoffe GmbH (Germany) www.warnke.de	Resveratrol (500 mg)	Not specified
Alliness (Poland) www.aliness.pl/	Natural resveratrol (250 mg) Veri-Te	Antioxidant, cardioprotective
Progress Labs (Poland) www.//progresslabs.pl	Resveratrol (50 mg)	Cardioprotective, antiaging, neuroprotective
Proton Labs (Poland) www.protonlabs.pl	*trans*-Resveratrol (98%)	Antiaging
Kenay AG (Poland) www.kenay.com.pl	NMN Uthever® with resveratrol (200 mg of *trans*-resveratrol) Pterostilbene 90% (Silbinol®) (90 mg pterostilbene)	Antiaging
Lab one (Poland) www. sklep.lab1.com	N°1 Heartvi Balance (100 mg)	Cardioprotective
OstroVit (Poland) www.ostrovit.com/pl	Resveratrol (150 mg of *trans*-resveratrol)	Antioxidant, cardioprotective, neuroprotective, regulation of glucose level
Yango (Poland) www.yango.pl	Resveratrol - extract 50%	Antioxidant
Doctors's best (USA) www.drbvitamins.com	Resveratrol French Red Wine Grape Extract (BioVin® - 5% *trans*-resveratrol) *trans*-Resveratrol (100 mg) with ResVinol *trans*-Resveratrol (200 mg) with Resvinol HighPotency *trans*-Resveratrol (600 mg)	Antioxidant, cardioprotective, antiaging, support for immune function
Jarrow formulas (USA) www.jarrow.com	Resveratrol synergy (resveratrol 200 mg; *trans*-pterostilbene (10 mg)) *trans*-pterostilbene (50 mg)	Antioxidant, cardioprotective, sirtuin activator, calorie-restriction mimetic

Table 10.3 Examples of dietary supplements containing resveratrol and/or pterostilbene.—cont'd

Manufacturer/ Country	Trade name/Active compound	Function
Life Extension (USA) www.lifeextension.com	Resveratrol (100 mg of *trans*-resveratrol) Optimized resveratrol (125 mg of *trans*-resveratrol) Optymized resveratrol Elite™ (30 mg of 18% *trans*-resveratrol)	Antioxidant, cardioprotective
NowFoods (USA) www. sklepnowfoods.pl	Resveratrol (200 mg)	Cardioprotective
Solgar (USA) www.solgar.com	*trans*-Resveratrol (100 mg) *trans*-Resveratrol (250 mg) with red wine extract *trans*-Resveratrol (500 mg)	Antioxidant
Swanson (USA) www.sklepswanson.pl	Resveratrol (250 mg) Resveratrol (30 mg) Complex Pterostilbene (50 mg)	Antioxidant, cardioprotective
Thorne research (USA) www.thorne.com	PolyResveratrol-SR (100 mg *trans*-resveratrol; 100 mg of pterostilbene)	Antiaging, cardioprotective, hepatoprotective

Resveratrol was first isolated from the roots of white hellebore (*Veratrum grandiflorum*) by Takaoka in 1939 (Galiniak et al., 2019; Tian and Liu, 2020). Nowadays, *trans*-resveratrol can be found in different nutritional foods, such as grapes, peanuts, cranberries, or blueberries as well as in processed products such as wine, particularly red one. It was first detected in grape skin in 1988 (Ifuku, 2017). Currently, resveratrol can be found in more than 30 plant families. Stems, roots, leaves, flowers, fruits, and seeds can be used as extracting parts for resveratrol (Bansal et al., 2018; Tian and Liu, 2020).

7.1.1.2 Biological activity

7.1.1.2.1 Antioxidant activity. Li et al. (2021) investigated the antioxidant activity of *trans*-resveratrol isolated from mulberry (*Morus alba*). The antioxidant activity was tested by in vitro single electron transfer-based methods (1,1-diphenyl-2-picrylhydrazyl (DPPH), 2,2′-azino-bis-(3-ethylbenzothiazoline-6-sulfonic acid (ABTS)) and hydrogen atom transfer-based method (oxygen radical scavenging capacity (ORAC)). In DPPH assay, *trans*-resveratrol exhibited IC_{50} of 15.54 μg/mL and showed a scavenging ability rate from 29.56% to 75.63% depending on the concentration. In

Table 10.4 Possible application of stilbenoid compounds according to CosIng database.

INCI	Functions
Resveratrol	Antioxidant, skin protecting
Resveratrol tripentyl ether	Skin conditioning
Resveratrol dimethyl ether maltotrioside	Antioxidant
Trisodium resveratrol triphosphate	Antioxidant, skin conditioning
Resveratryl triacetate	Skin conditioning, skin protecting
Glucosyl resveratrol	Antioxidant, antiplaque, humectant, skin conditioning
Resveratrol dimethyl ether	Antioxidant, skin conditioning
Resveratryl glucoside	Humectant
Resveratrol tributyl ether	Not reported
Resveratrol dimethyl ether glucoside	Antioxidant
Resveratrol dimethyl ether oligosaccharides	Antioxidant
Hydroxyresveratrol	Bleaching, skin conditioning, skin protecting
Piceatannol	Antioxidant
Polydatin	Antioxidant
α-Viniferin	UV filter

ABTS assay, *trans*-resveratrol showed higher antioxidant activity compared with L-ascorbic acid used as a reference with IC_{50} values of 2.86 and 5.18 μg/mL, respectively. In ORAC assay, *trans*-resveratrol also exhibited an antioxidant activity expressed by Trolox equivalent (23.12 μmol TE/g). In addition, *trans*-resveratrol pretreatment of HepG2 cells resulted in reactive oxygen species (ROS) production inhibition. Interestingly, *trans*-resveratrol treatment of HepG2 cells also reversed the 2,2-azobis(2-amidinopropane) dihydrochloride (AAPH)-induced decrease levels of catalase (CAT), superoxide dismutase (SOD), and glutathione (GSH) levels. It increased the levels of these antioxidant defenses depending on the concentration (Li et al., 2021).

Rosiak et al. (2022) evaluated the resistance of *trans*-resveratrol on electron beam irradiation by measuring its antioxidant properties. The antioxidant activity of nonirradiated and irradiated *trans*-resveratrol was carried out in vitro by using DPPH, ABTS, cupric ion reducing antioxidant capacity (CUPRAC), and ferric ion reducing antioxidant activity (FRAP) assays. It was confirmed that nonirradiated *trans*-resveratrol showed the best antioxidant activity in ABTS and FRAP of IC_{50} value of 2 μg/mL and 5.1 μg/mL, respectively. However, in case of irradiated *trans*-resveratrol, the antioxidant properties were increased in DPPH and CUPRAC ($IC_{50} = 19.7$ and 5.6 μg/mL respectively), which proposed to be associated with the enhancement of HAT (Hydrogen Atom Transfer) mechanism in aforementioned assays (Rosiak et al., 2022).

Table 10.5 Examples of resveratrol-containing cosmetics.

Manufacturer/ Country	Trade name/Form	Active ingredient	Indications
The Ordinary (Canada) www.theordinary.com/en-us	Resveratrol 3% + Ferulic Acid 3%	3% resveratrol	Antiaging, regenerating
Caudalie® www.pl.caudalie.com (France)	Resveratrol-Lift Serum	Resveratrol (likely from Palmitoyl grapevine shoot extract and *Paeonia lactiflora* root extract)	Lift firming, antiaging, moisturizing
Dr. Althea (Korea) www.doctoraltheaglobal.com	Resveratrol 345NA_Intensive repair cream	Resveratrol	Regenerating, skin brightening, rejunevating, skin elasticity improvement, moisturising, antiacne
Sesderma (Spain) www.sesderma.pl	Resveraderm Youth Gel Cream Resveraderm Liposomal Serum Resveraderm Antiox Nourishing Cream	Resveratrol dimethyl ether Resveratrol	Antiaging, skin elasticity improvement, skin-brightening Protecting against environmental factors, moisturizing, skin elasticity improvement Antiaging, skin-brightening, skin elasticity improvement
Revolution Beauty (UK) www.revolutionbeauty.com	Encapsulated Resveratrol Serum	Resveratrol	Calming, soothing, protective against environmental factors
Dermomedica (USA/ Poland) www.dermomedica.pl	Resveratrol Mineral Cream SPF 30	Pichia/Resveratrol Ferment Extract	Photoprotecting, antiaging, moisturizing, skin elasticity improvement
PCA skin (USA) www.pcaskin.com	Resveratrol Restorative Complex	1% Resveratrol	Antiaging, moisturizing
Skin Ceuticals (USA) www.skinceuticals.com	Resveratrol B E	1% Resveratrol	Skin firmness improvement, moisturizing, antiaging

Lucas-Abellán et al. (2011) tested the antioxidant activity of *trans*-resveratrol with the absence or presence of hydroxypropyl-β-cyclodextrins (HP-β-CDs), which allowed increasing the concentrations of this active molecule. The antioxidant potential was determined by three assays: DPPH, ABTS, and ORAC using Trolox as a standard. The best results were obtained using ORAC method (6 and 18 μM Tx) depending on *trans*-resveratrol concentration—0.5 and 1.5 μM Tx, respectively. The ABTS and DDPH assays appeared to be less sensitive (Lucas-Abellán et al., 2011).

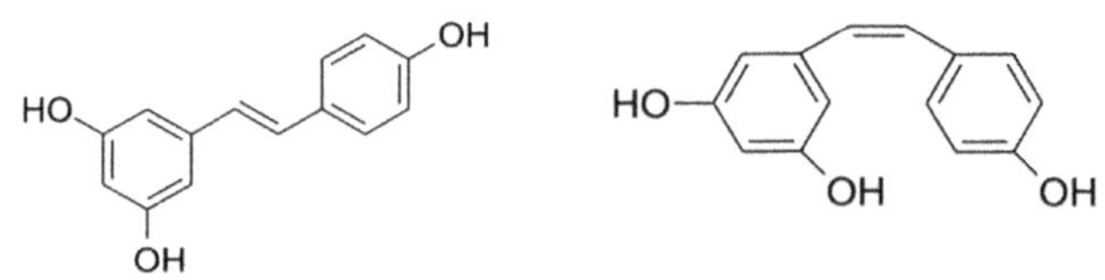

trans-Resveratrol *cis*-Resveratrol

FIGURE 10.5

The chemical structure of *trans*-resveratrol and *cis*-resveratrol.

Hangun-Balkir et al. (Hangun-Balkir and McKenney, 2012) evaluated the antioxidant activity of *trans*-resveratrol using DPPH assay. The results were measured spectrophotometrically and expressed as an inhibitory concentration IC_{50} ($IC_{50} = 4.70$ mg/mL) (Hangun-Balkir and McKenney, 2012).

Gülçin (2010) investigated the antioxidant activity of *trans*-resveratrol using different antioxidant assays: DPPH, ABTS, and N,N-dimethyl-p-phenylenediamine (DMPD). The chelating activity on ferrous ions (Fe^{2+}), hydrogen peroxide scavenging, and superoxide anion radical scavenging activities were also measured. In DPPH assay, *trans*-resveratrol exhibited higher antioxidant activity ($IC_{50} = 17.8\%$) in comparison to α-tocopherol ($IC_{50} = 28.3\%$) and Trolox ($IC_{50} = 25.5\%$) used as a reference. In ABTS test, *trans*-resveratrol showed effective radical cation scavenging activity ($EC_{50} = 6.96$ μg/mL). *Trans*-resveratrol was also found to be effective in DMPD assay with EC_{50} values of 9.5 μg/mL. The effective chelating effect of *trans*-resveratrol on ferrous ions was obtained at concentration 20 μg/mL, giving the values of 86.3%. The resveratrol superoxide anion radical scavenging activity was inhibited at the level of 71.8%, which was significantly higher in comparison to tested reference compounds. In addition, the hydrogen peroxide scavenging activity was found to be 19.5% at 30 μg/mL concentration (Gülçin, 2010).

Ferrier et al. (2022) identified *trans*-resveratrol as one of the key molecules that contributed the most to the antioxidant capacities within a complex mixture of polyphenols in numerous grapevine cane extracts. Among 42 metabolites identified in 44 grape varieties, *trans*-resveratrol, *trans*-piceatannol, *trans*-ε-Viniferin, hopeaphenol, isohopeaphenol, and *cis/trans*-vitisin B were found to be the main drivers of antioxidant capacity using two antioxidant tests, ABTS and DPPH, based on the mixed-mode (HAT/SET) antioxidant ability (Ferrier et al., 2022).

7.1.1.2.2 Anti-aging activity. Bastianetto et al. (2010) investigated protective effect of *trans*-resveratrol (1−30 μM) using human keratinocyte (HaCaT) cells. The cells were exposed to sodium nitroprusside (SNP), which provoked cell damages. Based on the results, *trans*-resveratrol exhibited a protective effect against SNP-induced toxicity ($EC_{50} = 14.7$ μM). *Trans*-resveratrol significantly reduced the increase of SNP-induced caspase-3 activity in a dose-dependent manner. The best effect was obtained at 10 μM concentration. In addition, the activity of caspase-9 was completely blocked by *trans*-resveratrol (Bastianetto et al., 2010).

Subedi et al. (2017) evaluated the efficacy of the *trans*-resveratrol-enriched rice (RR) in inhibiting the skin aging induced by ultraviolet B (UVB) radiation. It was confirmed that both *trans*-resveratrol (R) and *trans*-resveratrol-enriched rice significantly reduced the production of proapoptotic proteins—caspase 3, p53, Bax, and cytochrome C involved in cellular senescence. Moreover, both aforementioned *trans*-resveratrol forms downregulated the expression of proinflammatory cytokines—tumor necrosis factor α (TNF-α), IL-6 as well as inflammatory mediators—inducible nitric oxide synthase (iNOS) and cyclooxygenase-2 (COX-2), involved in extracellular matrix degradation and overall skin aging.

Baca et al. (2016) evaluated the potential of topical *trans*-resveratrol preparation on wound healing in pigs. There were used three concentrations of *trans*-resveratrol: 2, 10, and 50 mg/ml. As a control, carboxymethyl cellulose gel and silver sulfadiazine (SSD) were used. It was shown that the wound surface reduction (88.4%) was higher in comparison to control and SSD groups—86.9% and 77.2%, respectively (Baca et al., 2016).

7.1.1.2.3 Cardioprotective activity. Ting et al. (2022) studied the protective effect of resveratrol against myocardial ischemia–reperfusion (I/R) injury using oxygen-glucose deprivation/reoxygenation (OGD/R) model was in H9c2 cells. Based on the results, resveratrol treatment (10 μM/L) attenuated the oxidative stress induced by OGD/R through the malondialdehyde (MDA) level. Moreover, resveratrol suppressed ferroptosis through Fe^{2+} decrease and the increase expression of peroxidase 4 (GPX4) and ferritin heavy chain 1 (FTH1). In addition, resveratrol significantly reduced the protein expression level of autophagy markers (Beclin-1, NCOA4, LC3) (Ting et al., 2022).

Kazemirad and Kazerani (2020) investigated the cardioprotective properties of resveratrol on isolated hearts of Wistar rats. According to the results, infarct size was lower in resveratrol-treated group (17%). Moreover, the levels of catalase (CAT), superoxide dismutase (SOD), and glutathione peroxidase (GPX) were increased in resveratrol group (2.96, 1.64, 8.0 U/mg, respectively). In addition, resveratrol reduced the level of MDA (3.5 nM/mg) (Kazemirad and Kazerani, 2020).

Riba et al. (2017) evaluated the cardioprotective activity of resveratrol in postinfarction heart failure model after administration of Isoproterenol (ISO) to induce myocardial injury. According to the results, resveratrol improved the left ventricular function and decreased the myocardial hypertrophy and fibrosis, overall reducing the severity of heart failure. Resveratrol treatment reduced oxidative stress through iNOS and COX-2 decreased expression. Moreover, resveratrol reduced the plasma B-type natriuretic peptide (BNP) level (ca. 2.4 ng/mL), attenuated p38-mitogen-activated protein kinase (MAPK) and extracellular signal-regulated kinase 1/2 (ERK1/2) phosphorylation responsible for their inactivation and increased the mitogen-activated protein kinase phosphatase-1 (MKP-1), appearing to be cytoprotective consequently to heart failure (Riba et al., 2017). Sebai et al. (2011) examined the cardioprotective effect of resveratrol (RVT) on lipopolysaccharide (LPS)-induced oxidative stress in rat. According to the results, the activity of the SOD was significantly increased in LPS/RVT group (>20 U/mg). Moreover, the heart MDA levels were decreased by resveratrol (<0.6 nM/mg). In addition, tissue's NO levels were reduced (>3 μM/mg) in LPS/RVT-treated group, and the heart iron levels were increased (>100 μM/mg) (Sebai et al., 2011).

7.1.1.2.4 Anticancer activity. Chow et al. (2010) evaluated the ability of resveratrol to exhibit cancer preventive activity. CYP 1A2, 2D6, 2C9, and 3A4 enzyme activity was accessed by oral coadministration of CYP metabolic probe drugs: caffeine, dextromethorphan, losartan, and buspirone, respectively. It was confirmed that the administration of 1 mg of resveratrol for 4 weeks resulted in inhibition of phenotypic indices of aforementioned enzymes (Chow et al., 2010).

Boocock et al. (2007) examined the impact of 29 days of four different concentrations of *trans*-resveratrol (0.5, 0.1, 2.5, 5.0 g) oral administration on the activity of insulin-like growth factor-1 (IGF-1) and insulin-like growth factor-binding protein 3 (IGFBP-3) involved in tumor genesis in 40 healthy volunteers. It was confirmed that oral administration of resveratrol resulted in IGF-1 and IGFBP-3 decrease. ($P < .04$). The best results were obtained at the medium dose level (2.5 g) of resveratrol (The Ordinary, 2023).

Zhu et al. (2012) tested the influence of *trans*-resveratrol on the expression of methylation assessment of RAS-association domain family-1α (RASSF-1α) gene and prostaglandin-E_2 (PGE_2) associated with breast cancer progression. The expression of PGE_2 is related to cell proliferation, and its level is significantly higher in malignant breast tumors compared with normal breast tissue. The RASSF-1α methylation is directly related to the change in PGE2. *Trans*-resveratrol may correlate this change by the RASSF-1α methylation decrease. In the study, *trans*-resveratrol was administrated orally to 39 enrolled women for 12 weeks in two concentrations - 5 and 50 mg. It was investigated that higher concentration of *trans*-resveratrol was correlated with decrease in RASSF-1α methylation ($P = .047$) and as a result prostaglandin-E2 expression (Zhu et al., 2012).

Espinoza et al. (2017) evaluated the effect of *trans*-resveratrol on circulating immune cells in healthy individuals. It was proved that the repeated oral doses of resveratrol (1000 mg/day) for 28 days, significantly increased the number of circulating γδ T cells and regulatory T cells. Moreover, the plasma levels of the proinflammatory cytokines: tumor necrosis factor α (TNF-α) and monocyte chemoattractant protein−1 (MCP-1) were also significantly decreased, which could result in a lower cancer apparition/progression, as these biomarkers are related to cancer proliferation in many ways (Espinoza et al., 2017).

Nguyen et al. (Holcombe et al., 2009) studied the influence of *trans*-resveratrol contained in freeze-dried grape powder (GP) with others bioactive compounds, on wingless-related integration site (Wnt) signaling pathway in colonic mucosa of patients with colon cancer. It was approved that low concentration of *trans*-resveratrol/GP noticeably inhibits the Wnt pathway through target gene including myc, cyclinD1, in both normal and cancer type colonic mucosa, playing critical role in colon cancer development. The most significant reduction was obtained after 80 g/day of GP condition (Holcombe et al., 2009).

7.1.1.2.5 Antidiabetic activity. Szkudelska et al. (2021) studied the effect of *trans*-resveratrol on diabetes-associated dysregulation in Goto-Kakizaki (GK) rats. It was investigated that *trans*-resveratrol (20 mg/kg b.w. (body weight)) prevented the expression and phosphorylation levels of insulin receptor in the skeletal muscle of GK rats. In addition, the compound induced normalization of the protein's expression related to intracellular glucose transport (GLUT4 and TUG) in adipose tissue of GK rats, overall, allowing glucose tolerance and less insulin resistance in this diabetes model (Szkudelska et al., 2021).

Tomé-Carneiro et al. (2013) evaluated the changes in peripheral blood mononuclear cells (PBMCs) caused by *trans*-resveratrol-enriched grape extract in patients with type 2 diabetes mellitus, an inflammatory disease. The extract was administrated in two concentrations: 8 mg/day (first 6 months) and 16 mg/day (last 6 months). The results confirmed that resveratrol-enriched extract decreased the alkaline phosphatase (ALP) usually increase in type II diabetes. In addition, extract significantly reduced the expression of proinflammatory cytokines such as C−C motif chemokine ligand 3 (CCL3), interleukin 1β (IL-1β), interleukin 6 (IL-6), and TNF-α (Tomé-Carneiro et al., 2013).

7.1.1.2.6 Antiobesity activity. Gómez-Zorita et al. (2012) evaluated the effect of *trans*-resveratrol (15 and 45 mg/kg b.w./day) on liver fat accumulation and fatty acid oxidation in obese Zucker rats. Based on the obtained results, *trans*-resveratrol reduced nonesterified fatty acids (NEFA) and alkaline phosphatase levels (ALP) in both concentrations, which are related with obesity. Serum total cholesterol was significantly reduced in treated rats. Moreover, *trans*-resveratrol increased the activity of carnitine palmitoyltransferase-Ia (CPT-Ia) and acyl-coenzyme A oxidase (ACO) resulting in higher fatty acids oxidation and their break down (Gómez-Zorita et al., 2012).

In another study, Gómez-Zorita et al. (2013) tested the effect of *trans*-resveratrol given orally (15 mg/kg b.w./day) for 6 weeks to obese rats. The administration of this molecule resulted in a decrease of glucose-6P-dehydrogenase (G6PDH), acetyl-CoA carboxylase (ACC), and lipoprotein lipase (LPL) activity. The IL-6 and nuclear factor κ-light-chain-enhancer of activated β cells (NF-κB) expression was reduced following *trans*-resveratrol administration. Moreover, the serum levels of TNF-α, monocyte chemoattractant protein-1 (MCP1) and C-reactive protein (CRP) were decreased (Gómez-Zorita et al., 2013).

Alberdi et al. (2013) studied an impact of 6-week *trans*-resveratrol supplementation (30 mg/day) on thermogenesis in skeletal muscle and interscapular brown adipose tissue (IBAT) in Sprague–Dawley rats. The administration of *trans*-resveratrol resulted in mitochondrial-transcription-factor-A (TFAM), mitochondrial-protein-cytochrome-C-oxidase subunit-2 (COX2), sirtuin 1 (SIRT1), and peroxisome proliferator-activated receptor-gamma coactivator 1 alpha (PGC-1α) genes expression increase. Moreover, resveratrol increased the uncoupling proteins (UCP1, UPC3) expression in IBAT. Overall, these biomarkers are involved in body fat lowering observed in rats (Alberdi et al., 2013).

Timmers et al. (2011) examined the effect of 30-day oral supplementation of pure *trans*-resveratrol (150 mg/day) on metabolic changes in obese patients. The results showed that *trans*-resveratrol activated the AMP-activated protein kinase (AMPK) in muscle, which induced the SIRT1 and PGC-1α levels increase. The adipose tissue lipolysis was also decreased following *trans*-resveratrol application (Timmers et al., 2011).

7.1.1.2.7 Antifungal activity. Langcake et al. (Langcake and Pryce, 1977) evaluated the antifungal activity of *trans*-resveratrol against *Plasmopara viticola*. Based on the results, *trans*-resveratrol inhibited both zoospore release and mobility with IC_{50} values of >200 μg/mL. Moreover, *trans*-resveratrol inhibited spore germination *Cladosporium cucumerinum*, *Botrytis cinerea,* and *Pyricularia oryzae* (>200 μg/mL) (Langcake and Pryce, 1977).

In another study, Langcake (1981) also investigated the antifungal potential of *trans*-resveratrol against *P. viticola.* According to the results, *trans*-resveratrol inhibited the spore germination as well as zoospore release and mobility with IC_{50} values of >200 μg/mL (Langcake, 1981).

Pezet et al. (2004) studied the antifungal activity of *trans*-resveratrol against *P. viticola. Trans*-resveratrol inhibited the spores germination ($ED_{50} = 145$ μM) and zoospore mobility ($ED_{50} = 192$ μM) (Pezet et al., 2004).

Schnee et al. (2013) evaluated the antifungal activity of six compounds including *trans*-resveratrol against *P. viticola*. According to the results, *trans*-resveratrol inhibited zoospore mobility and sporulation of *P. viticola* with IC_{50} values of 122 and 121 μM, respectively (Schnee et al., 2013).

Adrian et al. (1997) investigated the antifungal activity of *trans*-resveratrol against *Botrytis cinerea*. Based on the results, *trans*-resveratrol (160 μg/mL) had a significant effect *B. cinerea* and resulted in complete inhibition of germination (Adrian et al., 1997).

7.1.2 Oxyresveratrol

7.1.2.1 General information

Oxyresveratrol (2,3′,4,5′-tetrahydroxystilbene) (Fig. 10.6) is a natural stilbenoid, which is believed to be derived from resveratrol. However, the statement still requires the experimental evidence (Likhitwitayawuid, 2021). Oxyresveratrol structure consists of *trans*-1,2-diphenylethylene nucleus

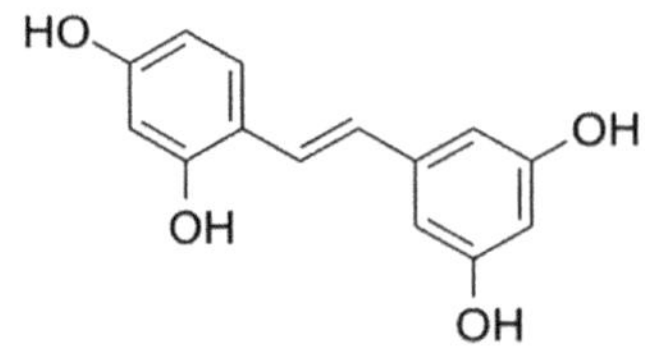

trans-Oxyresveratrol

FIGURE 10.6

The chemical structure of *trans*-oxyresveratrol.

with two OH groups on each aromatic ring. It occurs as free stilbene or as glycosidic form. Until recently, only mono- and di-O-glucosidic forms of oxyresveratrol have been reported in the literature i.e., 2-,4- or 3′-*O*-glucoside and 3′,4-diglucoside (i.e., mulberroside A) (Likhitwitayawuid, 2021).

Oxyresveratrol is also distributed in certain plant families i.e., *Caesalpiniaceae, Gnetaceae, Fabaceae, Liliaceae, Melanthiaceae, Moraceae, Myrtaceae, Poaceae, Rutaceae, Rosaceae, Smilacaceae, Vitaceae*. In food products, oxyresveratrol is naturally presented in grapes, peanuts, and mulberries (Galal et al., 2015; Likhitwitayawuid, 2021).

7.1.2.2 Biological activity

7.1.2.2.1 Antioxidant activity. Lorenz et al. (2003) evaluated the antioxidant activity of oxyresveratrol using DPPH assay. Based on the results, oxyresveratol has a significant radical scavenging activity ($IC_{50} = 28.9\ \mu M$) and appeared to be more potent than *trans*-resveratrol ($IC_{50} = 38.5\ \mu M$). Moreover, oxyresveratrol reduced the radical nitric oxide (NO) level (45.31 μM) in murine N9 microglial cells. It was also reported to be less cytotoxic toward microglial cells than *trans*-resveratrol ($IC_{50} = 243.8$ and 61.0 μM, respectively) (Lorenz et al., 2003).

Rodríguez-Bonilla et al. (2017) investigated the antioxidant activity of oxyresveratrol using three in vitro assays: ABTS, FRAP, and ORAC. The results were expressed as Trolox equivalent antioxidant capacity (TEAC). For the ABTS test, oxyresveratrol possessed an antioxidant activity higher in comparison to resveratrol (TEAC = 2.04 and 1.57 mM, respectively). Based on FRAP assay results, oxyresveratrol was also more effective than *trans*-resveratrol (TEAC = 0.52 and 0.47 mM). Due to ORAC method, oxyresveratrol was slightly less effective than *trans*-resveratrol (TEAC = 3.35 and 3.67 mM, respectively) (Rodríguez-Bonilla et al., 2017).

Jin et al. (2015) studied the antioxidant activity of 24 compounds derived from *Artocarpus xanthocarpus* including oxyresveratrol. The antioxidant activity was evaluated by DPPH, ABTS, and superoxide anion radical scavenging assays and introduced as a scavenging concentration (SC_{50}). In DPPH assay, oxyresveratrol showed stronger antioxidant potential than *trans*-resveratrol ($SC_{50} = 47.7$ and 86.9 μM, respectively). The same result was observed in ABTS assay ($SC_{50} = 28.8$ and 50.1 μM, respectively). The results of superoxide anion radical scavenging assay testify a better $O_2^{\bullet -}$ scavenging activity for oxyresveratrol in comparison to *trans*-resveratrol ($SC_{50} = 90.8$ and 102.1 μM, respectively) (Jin et al., 2015).

7.1.2.2.2 Anti-inflammatory activity. Kutil et al. (2015) investigated the inhibitory activity of oxyresveratrol on 5-lipoxygenase (5-LOX) and cyclooxygenase one and 2 (COX-1, COX-2) involved in proinflammatory activity. According to the results, oxyresveratrol possessed moderate 5-lipoxygenase inhibitory activity with IC_{50} values of 18.49 μM. Moreover, oxyresveratrol significantly inhibited the activity of COX-1 and COX-2 (IC_{50} = 2.79 and 14.71 μM, respectively) (Kutil et al., 2015).

Aziz et al. (2019) evaluated the anti-inflammatory activity of oxyresveratrol in ethanol-induced gastric ulcer. Oxyresveratrol was administrated in two concentrations: 30 and 50 mL/kg body weight. Oxyresveratrol treatment resulted in suppression of inflammatory infiltrate, erosion, edema, and ulcer. The gastric pH value was increased in both oxyresveratrol concentrations. Moreover, oxyresveratrol significantly decreased the mRNA expression levels of the inflammatory cytokines IL-6 both in low and high concentration (22.44 and 19.46, respectively). In addition, both low and high doses of oxyresveratrol reduced the expression of another inflammation marker, TNF-α (14.99 and 13.65, respectively). The treatment with both concentrations of oxyresveratrol also suppressed the NF-κB expression (18.83 and 14.67). In case of COXs, oxyresveratrol significantly inhibited COX-2 expression level in both doses (Aziz et al., 2019).

Ashraf et al. (2015) evaluated the anti-inflammatory activity of oxyresveratrol in mouse model of ovalbumin (OVA)-induced allergic airway inflammation. Oxyresveratrol was used in two concentrations: 7 and 15 mg/kg. Treatment with the low concentration oxyresveratrol resulted in the decrease of total leukocyte count both in blood and bronchoalveolar lavage fluid (BALF) (6.52 and 0.92, respectively). Moreover, the eosinophil count, implicated in the inflammatory response, was also significantly lowered in presence of low and high oxyresveratrol concentrations (0.33 and 0.1, respectively). In addition, oxyresveratrol reduced the expression of inflammatory cytokines IL-4, IL-5 and IL-13 (Ashraf et al., 2015), resulting in overall lowered OVA-induced allergic airway inflammation.

7.1.2.2.3 Tyrosinase inhibitory activity. Kim et al. (2002b) studied the inhibitory effect of oxyresveratrol on mushroom tyrosinase activity. Based on the results, oxyresveratrol showed significant inhibitory activity (IC_{50} = 1.2 μM) in comparison to kojic acid with IC_{50} values of 40.1 μM. Moreover, oxyresveratrol was potent against *L*-tyrosine oxidation in a dose-dependent manner, hence inhibiting its role in the melanogenesis. The highest result was obtained at a 100 μM concentration (97.3% of the enzyme inhibition) Kim et al. (2002b).

Kim et al. (2004) evaluated the oxyresveratrol inhibition activity on tyrosinase and 3,4-dihydroxyphenylalanine (L-DOPA) oxidation, which is important in the prevention of melanogenesis. It was confirmed that oxyresveratrol significantly inhibited the tyrosinase oxidation (IC_{50} = 1.2 μM). In addition, L-DOPA oxidation was also inhibited (IC_{50} = 2.3 μM). Both inhibition capacities were higher in comparison to *trans*-resveratrol (IC_{50} = 43.5 and 123.3 μM, respectively) (Kim et al., 2004).

Ortiz-Ruiz et al. (Kang et al., 2011) investigated the inhibitory activity of five constituents including oxyresveratrol isolated from 80% MeOH extract from *Morus bombycis* in mushroom tyrosinase inhibition assay. The results showed that oxyresveratrol was more potent in comparison to kojic acid used as a positive control (IC_{50} = 3.66 and 112.7 μM, respectively) (Kang et al., 2011).

7.1.2.2.4 Anticancer activity. Ferlinahayati et al. (2008) studied the cytotoxic effects of 10 phenolic constituents including oxyresveratrol and prenylated stilbene 4′-(2-methyl-2-buten-4-yl)oxyresveratrol from the wood of *Morus australis* against murine leukemia P-388 cells. Based on the results, 4′-(2-methyl-2-buten-4-yl) oxyresveratrol showed the strongest cytotoxic activity in comparison to all the tested constituents with IC_{50} values of 6.9 μM (Ferlinahayati et al., 2008).

Yang et al. (2019) investigated the anticancer activity of oxyresveratrol in human bladder cancer T24 cells. According to the results, oxyresveratrol showed a significant inhibition of 67.68% on T24 cells (IC_{50} = 47.46 μM) in comparison to *trans*-resveratrol (54.34%; IC_{50} = 69.33 μM) (Yang et al., 2019).

Abdel Bar et al. (2020) evaluated the antiproliferative activity of constituents including oxyresveratrol obtained from *Morus nigra* against human hepatocellular carcinoma cell line (Hep G2) and human breast cancer line (MCF-7). Oxyresveratrol showed high antiproliferative activity (IC_{50} = 37.83 μM) against HepG2 cell line. However, in case of MCF-7 cell line, oxyresveratrol was less potent with IC_{50} values of 219.6 μM (Abdel Bar et al., 2020).

7.1.2.2.5 Antiviral activity. Likhitwitayawuid et al. (2005) evaluated the antiviral activity of phenolic constituents including oxyresveratrol isolated from *Millettia erythrocalyx* and *Artocarpus lakoocha* against human herpesviruses 1 and 2 (HHV-1, HHV-2). In addition, the potential activity of oxyresveratrol against human immunodeficiency virus type 1 (HIV-1) was evaluated. According to the results, oxyresveratrol showed moderate antiviral activity against both HHV types (EC_{50} = 63.5 and 55.3 μM, respectively). Moreover, oxyresveratrol also possessed inhibitory activity against HIV-1 (EC_{50} = 28.2 μM) (Likhitwitayawuid et al., 2005).

Chuanasa et al. (2008) investigated the antiviral activity of oxyresveratrol against HHV-1, HHV-2, poliovirus (WPV) type 1 and measles virus in plague reduction assay in mice. Oxyresveratrol showed the best results antiviral activity for HSV-1 strain 7401 H and HSV two strain Baylor 186 with IC_{50} values of 19.8 and 18.7 μg/mL, respectively (Chuanasa et al., 2008).

7.1.2.2.6 Antibacterial activity. Sheng et al. (2015) studied the quorum-sensing (QS) inhibition activity of stilbenoids including oxyresveratrol and measured by spectrophotometric assay to quantify violacein produced by *Chromobacterium violaceum* CV026, in case of growth inhibition a white halo is visible. Based on the results, oxyresveratrol demonstrated potential QS inhibition activity in a dose-dependent manner. The best results were obtained at the highest concentration (150 μM) and caused the 86% reduction of violacein production. Moreover, the QS inhibition activity of oxyresveratrol against *Pseudomonas aeruginosa* PAO1 was investigated. Oxyresveratrol inhibited the pyocyanin production in dose-dependent manner. The significant inhibition was obtained at 400 μM concentration and resulted in 50.2% reduction of pyocyanin concentration (Sheng et al., 2015).

Phoolcharoen et al. (2013) evaluated the antibacterial activity of oxyresveratrol *Porphyromonas gingivalis* and *Aggregatibacter actinomycetemcomitans.* According to the results, oxyresveratrol showed antibacterial activity with MIC values of 0.07 and 0.08 mg/mL, respectively. The minimal bactericidal concentration (MBC) was also evaluated and values of 0.16 mg/mL for both pathogenic bacteria were measured (Phoolcharoen et al., 2013).

Mazimba et al. (2018) investigated the antibacterial activity of *Morus nigra* constituents including oxyresveratrol. Oxyresveratrol showed an antibacterial activity against *Bacillus subtilis*, *Micrococcus flavus*, *P. aeruginosa*, *Salmonella abony*, *Staphylococcus aureus* and *Streptococcus faecalis* with MIC values of 125 and 250 μg/mL depending on the bacteria strain (Mazimba et al., 2018).

7.1.2.2.7 Antifungal activity. Basset et al. (2012) evaluated the antifungal activity of the chemical constituents including oxyresveratrol isolated from the roots of *Spirotropis longifolia*. Oxyresveratrol showed antifungal activity against: *Candida tropicalis*, *Candida krusei*, *Microsporum canis*, *Microsporum gypseum*, *Trichophyton rubrum*, and *Trichophyton mentagrophytes*. The best results were obtained for *T. mentagrophytes* (MIC = 8 μg/mL) (Basset et al., 2012).

Kim et al. (Kim and Lee, 2018) studied the antifungal activity of oxyresveratrol using antifungal susceptibility assay. Based on the results, oxyresveratrol possessed antifungal activity against: *Candida albicans*, *Candida parapsilosis*, *Saccharomyces cerevisiae*, *Trichosporon beigelii* with MIC values of 5.0 μg/mlor all tested fungi (Kim and Lee, 2018).

7.1.3 Piceatannol

7.1.3.1 General information

Piceatannol (3,3′,4,5-*trans*-tetrahydroxystilbene) is *trans*-resveratrol 3-hydroxylated derivative (Fig. 10.7) (Pubchem, 2023). It is commonly found in *Vitis amurensis*, *Vitis coignetiae,* and *Vitis vinifera*, but it is also presented in *Lophatherum gracile* stems and leaves, *Passiflora edulis* seeds and fruits of *Rhodomyrtus tomentosa* (Krambeck et al., 2023; Pawlus et al., 2012; Sharafan et al., 2023).

7.1.3.2 Biological activity

7.1.3.2.1 Antioxidant activity. Santos et al. (n.d.) evaluated the antioxidant capacity of piceatannol (PIC) using DPPH method. Moreover, the superoxide anion radical scavenging activity ($O_2^{\bullet-}$) and hypochlorous acid (HOCl) scavenging activity were determined. Based on the results, piceatannol possessed an antioxidant activity in the DPPH assay with IC_{50} values of 6.3 μg/mL. Piceatannol showed the highest activity in HOCl scavenging activity and $O_2^{\bullet-}$ radical scavenging activity ($IC_{50} = 1.2$ and 7.3 μg/mL, respectively) compared with ethanolic extract of *Passiflora edulis* and quercetin, which were also analyzed in the study (Santos et al., n.d.).

Gerhäuser et al. (2003) also investigated the antioxidant activity of piceatannol using DPPH assay. According to the results, piceatannol showed much higher antioxidant activity compared with its analogue *trans*-resveratrol ($SC_{50} = 11.8$ and 82.0 μM, respectively). Moreover, piceatannol was tested for its inhibition of superoxide anion scavenging radical formation using xanthine/xanthine oxidase assay (X/XO). The results demonstrated that piceatannol inhibited the XO activity ($IC_{50} = 19.4$ μM) (Gerhäuser et al., 2003).

7.1.3.2.2 Anti-inflammatory activity. Zhu et al. (2020) investigated the anti-inflammatory activity of piceatannol in HaCaT cell line. Piceatannol showed the inhibitory activity against proinflammatory cytokines: IL-6, IL-8, and TNF-α. Moreover, piceatannol downregulated the p65 nuclear translocation, being a subunit of NF-κB, acting as a transcription factor implicated in inflammation process (Zhu et al., 2020).

Lee et al. (2022) studied the anti-inflammatory activity of piceatannol (500 nM) on *Dermatophagoides farina* extract (DFE)-induced atopic dermatitis (AD) in NC/Nga mice and HaCaT cell line.

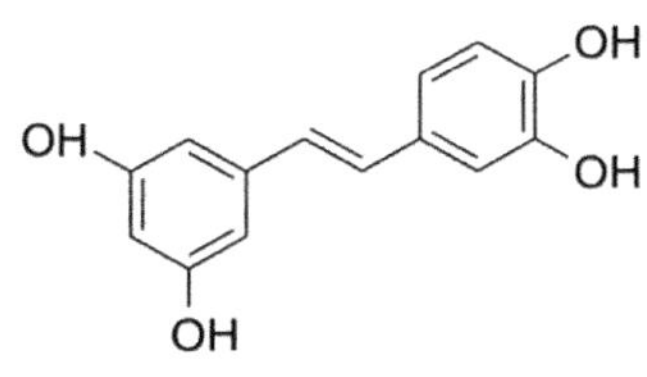

trans-Piceatannol

FIGURE 10.7

The chemical structure of *trans*-piceatannol.

Piceatannol significantly suppressed the symptoms associated with AD such as dryness, erythema, and scratching marks. Moreover, piceatannol reduced TEWL and improved skin hydration. In addition, the epidermal thickness was reduced by 46.69%. The number of eosinophiles and mast cells were reduced in piceatannol-treated group (40.49% and 59.49%, respectively). Piceatannol significantly decreased several cytokines and inflammatory biomarkers, such as the thymus and activation-regulated cytokine (TARC), macrophage-derived cytokine (MDC) and IgE by 75.09%, 76.91%, and 45.17%, respectively. In HaCaT cell line, piceatannol treatment (20 μM) inhibited TNF-α/IFN-γ-induced activation of janus kinase 1 (JAK 1), signal transducer, and activator of transcription 1 and 3 (STAT-1, STAT-3), belonging to the JAK/STAT pathway regulating cell life as well as immune and inflammatory response (Lee et al., 2022).

Yamamoto et al. (2017) tested the anti-inflammatory properties of piceatannol in RAW264.7 macrophages. Based on the results, piceatannol (30 μM/L) showed the significant decrease in expression levels of LPS-induced iNOS, TNF-α, and IL-6 (29.6%, 57.4%, and 72.7, respectively) (Yamamoto et al., 2017).

7.1.3.2.3 Anti-acne activity. Boutot et al. (n.d.) investigated the effect of gel containing *R. tomentosa* fruit extract and piceatannol (500 ppm) on voluntaries with acne. According to the results, the gel reduced imperfections associated with acne: blackheads, papules, inflammatory and noninflammatory lesions (Boutot et al., n.d.).

Zhu et al. (2020) evaluated the inhibitory activity of piceatannol on *Propionibacterium acnes*-induced proliferation and migration of keratinocytes in HaCaT cell line after a transfection method. The transfection experiment demonstrated that picetannol effectively curbed the proliferation and migration of HaCaT cells induced by *P. acnes*. This suppression was achieved by triggering the Nrf2 antioxidant pathway while concurrently inhibiting the NF-κB inflammatory pathway, without cytotoxicity for the HaCaT line (Zhu et al., 2020).

7.1.3.2.4 Anticancer activity. Gerhäuser et al. (2003) studied the chemopreventing activity of piceatannol on carcinogen-metabolizing enzymes. The induction of quinone reductase (QR) activity in Hepa1c1c7 murine hepatoma cell culture was evaluated. According to the results, piceatannol was potent against quinone reductase ($IC_{50} = > 21.1$ μM). The activity of piceatannol was higher compared with resveratrol ($IC_{50} = 29.4$ μM) (Gerhäuser et al., 2003).

Jayasooriya et al. (2012) evaluated the inhibiting activity of piceatannol on the matrix metalloproteinase-9 (MMP-9) expression in DU145 human prostate cancer cells, this endopeptidases are well-known mediators of tumor invasion and metastasis by damaging connective tissues. According to the results, piceatannol reduced the TNF-α-induced activity of MMP-9 in gelatin zymography method in dose-dependent manner. The best results were obtained for the 40 μM concentration. In addition, piceatannol reduced the MMP-9 luciferase reporter activity in luciferase assay. Moreover, inflammation markers were also affected, the NF-κB activity was downregulated in the presence of piceatannol through inhibiting nuclear translocation subunits p65 and p50. Furthermore, the TNF-α-induced Akt phosphorylation resulting in cell survival triggered in cancer was suppressed by piceatannol (Jayasooriya et al., 2012).

Kim et al. investigated (Kim and Ma, 2019) the anticancer activity of piceatannol in HT1080 malignant fibrosarcoma cells. Due to gelatin zymography, piceatannol suppressed the activity of phorbol 12-myristate 13-acetate (PMA)-induced gelatinolytic MMP-9. The potential of piceatannol against the MMPs activity was tested using culture supernatant of PG-treated and untreated HT1080 cells. According to the results, piceatannol reduced the level of MMPs. Based on wound migration assay, piceatannol inhibited the HT1080 migration. Moreover, the three-dimensional (3D) spheroid invasion assay showed the spheroid-like HT1080 cells invasion into extracellular matrix (ECM) was reduced by piceatannol compared with untreated group (75.42% and 81.91, respectively). Piceatannol also reduced the levels of proangiogenic factors including pentraxin 3 (PTX3), vascular endothelial growth factor (VEGF), and placental growth factor (PlGF). Moreover, piceatannol treatment resulted in reduced migration of human umbilical vein endothelial cells (HUVECs) in a dose-dependent manner. The best result was obtained for 20 μM of piceatannol resulting in 53.6% migration reduction. In addition, piceatannol showed no cytotoxicity against both HT1080 and HUVECs cells after 48 hours of incubation (Kim and Ma, 2019).

Aljabali et al. (2020) studied the anticancer potential of piceatannol-loaded nanoparticles (PIC-BSA) in human colon cancer cells (CaCo-2 and HT-29) compared with free piceatannol for reference. According to MTT assay, the nanoparticles exhibited increased cytotoxicity, inhibited cell migration, reduced colony formation, and effectively suppressed cancer cell invasion. Additionally, PIC-BSA significantly decreased p65 expression (Aljabali et al., 2020).

7.1.4 Piceid

7.1.4.1 General information

Piceid (3,4′,5-trihydroxystilbene-3-β-D-glucoside), also named polydatin, is stilbenoid naturally occurring phytoalexin present in the roots and rhizome of *Polygonum cuspidatum.* It can also be found in grapevine products, peanuts, hop pellets, chocolate, and cocoa-containing products (Du et al., 2013; Karami et al., 2022; Pubchem, 2023).

Piceid is a glucoside derivative of *trans*-resveratrol in which hydroxyl group is substituted by a glucoside group (Fig. 10.8). It exists in both *trans* and *cis* configuration. *Trans*-polydatin is more biologically active naturally occurring compared with *cis* isomer (Du et al., 2013; Karami et al., 2022).

7.1.4.2 Biological activity

7.1.4.2.1 Antioxidant activity. Li et al. (2021) evaluated the antioxidant activity of *trans*-piceid isolated from mulberry (*Morus alba*) using DPPH, ABTS and ORAC assays. The results showed that *trans*-piceid exhibited antioxidant activity with IC_{50} of 54.35 and 13.44 μg/mL using DPPH and ABTS methods, respectively. Moreover, *trans*-piceid exhibited IC_{50} values of 18.67 μmol TE/g in ORAC assays (Li et al., 2021).

Jayalakshmi et al. (Jayalakshmi and Devika, 2019) tested the antioxidant activity of *trans*-piceid using four in vitro methods: DPPH and ABTS assays as well as hydroxyl radical •OH and superoxide

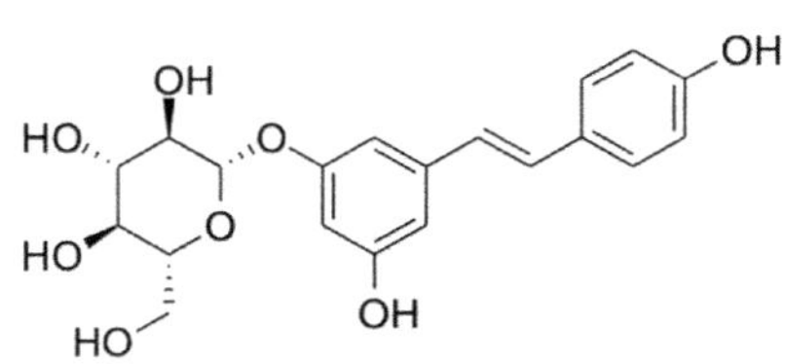

trans-Piceid

FIGURE 10.8

The chemical structure of *trans*-piceid.

anion radical ($O_2^{\bullet-}$) scavenging activities. In DPPH assay, piceid was potent in a dose-dependent manner with the percentage of DPPH radical scavenging activity values ranging from 53.53% to 88.63%. The percentage of ABTS radical cation inhibition values ranged from 33.72% to 83.72% depending on the concentration. In addition, hydroxyl and superoxide radical scavenging activity was also dose-dependent and ranging between of 22.26%–55.47% and 38.51%–69.96%, respectively (Jayalakshmi and Devika, 2019).

7.1.4.2.2 Anti-inflammatory activity. Matacchione et al. (2022) evaluated the anti-inflammatory activity of *trans*-piceid in young and senescent human endothelial cells (HUVECs). The results showed that *trans*-piceid significantly reduced the expression levels of inflammatory markers such as monocyte/macrophage chemotactic protein-1 (MCP-1) (58%) and IL-8 (57%) in young HUVEC and miR-126 (50%). In addition, the combination of *trans*-piceid, curcumin, and quercetin also effectively decreased the levels of the MCP-1, IL-8 and miR-126 both in young and senescent HUVECs (61%, 58%, 82% and 163%, 102%, 145%, respectively). The suppression of the aforementioned markers are important, because they are directly implicated in inflammatory process (Matacchione et al., 2022).

Lanzilli et al. (2012) tested the anti-inflammatory activity of different concentrations of *trans*-piceid on the production of IL-17 in peripheral blood mononuclear cells (PBMC). According to the results, *trans*-piceid almost completely inhibited by 98% IL-17 production after treatment at 20 μg/mL, which was more effective than *trans*-resveratrol at the same concentration (IC = 50%) (Lanzilli et al., 2012).

Chen et al. (2021) studied the anti-inflammatory activity of *trans*-piceid in LPS-induced macrophages. Based on the results, pretreatment with *trans*-piceid significantly decreased the levels of proinflammatory cytokines including TNF-α, IL-4, and IL-6 and increased the secretion of IL-10. In addition, *trans*-piceid reduced the expression levels of iNOS and COX-2. Moreover, the phosphorylation and nuclear translocation of NF-κB-p65 subunit signaling pathway implicated in the inflammatory response were inhibited by *trans*-piceid (Chen et al., 2021).

7.1.4.2.3 Cardioprotective activity. Tan et al. (2020) evaluated the cardioprotective effect of *trans*-piceid against hyperglycemia-induced myocardial injury in type 2 diabetic rats. *Trans*-polydatin treatment improved myocardial dysfunction in diabetic rats, reduced abnormal tissue changes by improving left ventricular ejection fraction and ventricular shortening fraction. It also reduced heart hypertrophy and fibrosis. In cultured H9c2 cells, pretreatment *trans*-piceid dose-dependently inhibited high glucose-induced cardiomyocyte injury. Further observation evidenced that *trans*-piceid suppressed the increase in the ROS levels, NADPH oxidase activity (125%), and inflammatory cytokines production, namely the myocardial NF-κB activity and nuclear translocation of p65, downregulated cytokines IL-1β, IL-6, TNF-α, and vascular cell adhesion molecule (VCAM-1) in diabetic myocardial tissues (Tan et al., 2020).

Miao et al. (2011) investigated the cardioprotective potential of *trans*-piceid against ischemia/reperfusion (I/R) injury. Based on the results, treatment with this compound resulted in infarct size reduction with value of 12.47% (the percentage of risk areas) in comparison to I/R group (23.89%). Moreover, the concentration of creatine phosphokinase (CPK) and lactate dehydrogenase (LDH) were limited in *trans*-Piceid (20 μg/kg) treatment group, whereas they are usually elevated in infarctus injuries. Furthermore, the SOD level was significantly higher in *trans*-Piceid group compared with I/R group (1057.3 and 749.7 U/g, respectively), allowing to counteract ROS generation after I/R injury (Miao et al., 2011).

Qi et al. (2015) studied the preventing potential of *trans*-piceid against angiotensin II-induced cardiac hypertrophy and myocardial superoxide generation on Male Sprague–Dawley rats. The results showed that 5 weeks of *trans*-piceid pretreatment (50 mg/kg/day) suppressed the perivascular and interstitial fibrosis by 82.6% and 84.7%, respectively. Moreover, *trans*-polydatin decreased ROS levels, NADPH oxidase activity and reduced the protein levels of Nox2 and Nox4, induced by angiotensin II infusion. TGF-β1 expression induced by angiotensin II was also reduced by *trans*-piceid (Qi et al., 2015).

7.1.4.2.4 Anticancer activity. Chen et al. (2020) evaluated the anticancer activity of *trans*-piceid against glioblastoma (GBM). According to the results, *trans*-polydatin suppressed the GBM cell viability according to increased concentrations. *Trans*-piceid treatment also caused the morphological changes in GBM cells leading toward cell apoptosis in MTT assay. Moreover, *trans*-piceid reduced in vivo the tumor growth U251 tumor-bearing nude mice. Furthermore, the tumor angiogenesis was impeded and the expression of Ki67, a cell proliferation factor, and proliferating cell nuclear antigen (PCNA) was suppressed by *trans*-piceid. *Trans*-piceid also decreased the expression of stemness markers (CD133, SOX2), G1 phase marker (Cyclin D1), antiapoptotic marker (Bcl-2), as well as mesenchymal markers (Vimentin and N-cadherin), leading to a reduction in cell proliferation activity. In addition, dose-dependent *trans*-piceid increase proapoptotic marker (Bax) and epithelial marker (E-cadherin). It was also suggested that the anticancer activity of *trans*-piceid could be related to the inhibition of EGFR-AKT/ERK1/2/STAT3-SOX2/snail signaling pathway involved in cell survival, proliferation, and also tumorigenesis (Chen et al., 2020).

Jiao et al. (2018) studied the studied the anticancer potential of *trans*-piceid in hepatocellular carcinoma (HCC). Based on the results, *trans*-piceid inhibited the cell proliferation, invasion, and migration of HCC cell line. The treatment increased cells apoptotic rates in HepG2 and SMMC-7721 cell dose-dependently (10, 30 and 100 μM Moreover, *trans*-piceid suppressed the expression of cell survival proliferation proteins namely caspase-3, β-catenin, c-myc, cyclin D1, and survivin at 30 μM concentration. It was also observed for the expression of Ki-67 in dose-dependent manner (Jiao et al., 2018).

Verma and Tiku (2022) studied the chemotherapeutic effect of *trans*-polydatin in lung cancer A549 lines, and no colonies were observed at 0.8 mM concentration. Moreover, *trans*-piceid inhibited the cell density in a dose-dependent manner. In addition, the expression of Ki67 was significantly decreased at 0.8–1 mM concentration of this stilbenoid. Furthermore, *trans*-piceid induced the apoptotic cell death (32%) at 1 mM concentration. The expression of proapoptotic molecules, procaspase-3 and procaspase-9, was significantly reduced by the higher concentrations of *trans*-piceid (0.8–1 mM) (Verma and Tiku, 2022).

7.1.4.2.5 Antiobesity activity. Wang et al. (2020) tested the potential of *trans*-piceid on lipid peroxidation inhibition ability. According to the results, piceid showed strong lipid peroxidation inhibition ability, which was higher compared with *trans*-resveratrol ($IC_{50} = 24.5$ and 38.7 nM, respectively) (Wang et al., 2020).

Zheng et al. (2019) evaluated the glucose and lipid metabolic regulation in high-fat-diet (HFD)-induced obese mice model. Results revealed that *trans*-piceid inhibited the adipogenesis of 3T3-L1 adipocytes through the decrease of lipid accumulation. Moreover, *trans*-polydatin treatment reduced the mRNA expression level of peroxisome proliferator-activated receptor gamma (PPARγ), positively regulating adipogenesis when activated. Furthermore, *trans*-piceid decreased the total cholesterol triglyceride (TG) and low-density lipoprotein (LDL) levels and increased the high-density lipoprotein (HDL) concentration in serum (Zheng et al., 2019).

Yousef et al. (2021) studied the protective effect of *trans*-piceid on pancreatic β-cells damage in rats. Based on the results, the *trans*-piceid treatment (50 mg/kg/b.w.) significantly ameliorated the blood glucose and insulin levels in the bloodstream of rats. Moreover, *trans*-piceid decreased the lipid peroxidation and reduced the triglycerides, total cholesterol, LDL, and free fatty acids (FFA) (Yousef et al., 2021).

7.1.5 Pterostilbene

7.1.5.1 General information

Pterostilbene (3,5-dimethoxy-4′-hydroxystilbene) is a dimethyl ether resveratrol analogy and the *trans* isomer occurred in plants as the more active form (Fig. 10.9). Structurally, pterostilbene consists of carbon skeleton with two phenyl rings, which are linked by the double-bonded ethylene bridge. It has one hydroxyl and two methoxy groups. The presence of two methoxy groups determines the stronger pharmacological potential of pterostilbene compared with resveratrol (Chan et al., 2019; Pubchem, 2023).

Similarly, to previously described stilbenoids, pterostilbene also behaves as phytoalexin, which was first isolated from *Pterocarpus marsupium* in 1940. In 1974, it was also found in *Vitis vinifera* leaves and berries as well as in *Vitis cognetiae* berries. The dietary sources of pterostilbene are rabbiteye blueberry, deerberries, and peanuts (Kim et al., 2020; Nagarajan et al., 2022; Yan et al., 2021).

7.1.5.2 Biological activity

7.1.5.2.1 Antioxidant activity. Rimando et al. (2002) evaluated the activity of *trans*-pterostilbene against peroxyl radicals ($ROO^{\bullet}$) scavenging activity using luminol-enhanced chemiluminescence. The total reactive antioxidant potential (TRAP) of *trans*-pterostilbene was 237 μM Trolox equivalent and appeared to be more effective in comparison to Trolox (144 μM) used as a reference (Rimando et al., 2002).

Acharya et al. (Acharya and Ghaskadbi, 2013) investigated the antioxidant capacity of different concentrations of *trans*-pterostilbene using DPPH and ABTS assays. Moreover, *trans*-pterostilbene activity against hydroxyl and hydrogen peroxide was evaluated. The results were expressed as an L-ascorbic acid equivalent antioxidant capacity (AEAC). The DPPH assay showed that *trans*-pterostilbene exhibited significant free radical scavenging activity according to the concentration. In ABTS assay, *trans*-pterostilbene also demonstrated dose-dependent inhibition with the best results at

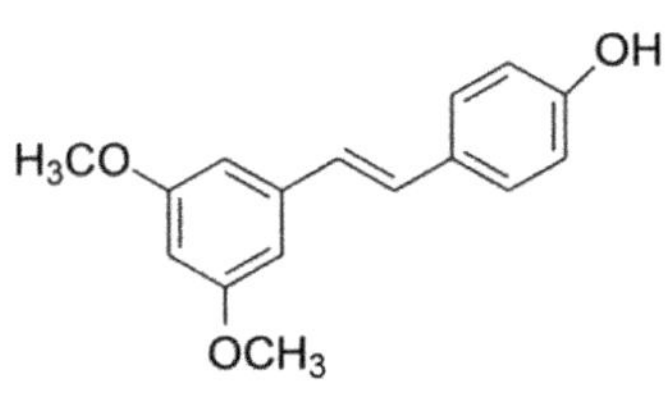

trans-Pterostilbene

FIGURE 10.9

The chemical structure of *trans*-pterostilbene.

0.20 mM concentration (1.42 mM AEAC). *Trans*-pterostilbene also showed the capacity to scavenge hydroxyl and hydrogen peroxide in dose-dependent manner with the highest results at 0.2 mM (0.09 mM AEAC). Moreover, pterostilbene inhibited superoxide scavenging activity by 55.6% (Acharya and Ghaskadbi, 2013).

Sun et al. (2021) tested the antioxidant activity of *trans*-pterostilbene encapsulated in a nanoemulsion system using DPPH method. The content of this molecule in the nanoemulsion was approximately 9.5%. According to DPPH result, pterostilbene showed significantly higher antioxidant activity, which was enhanced by the nanoemulsion form (Sun et al., 2021).

7.1.5.2.2 Anti-inflammatory activity. Rimando et al. (2002) investigated the activity of *trans*-pterostilbene against cyclooxygenases, synthesizing inflammation markers. The inhibition activity of *trans*-pterostilbene was evaluated by measuring the production of prostaglandin E_2 (PGE_2). As a result, *trans*-pterostilbene showed moderate inhibition activity against COX-1 with an IC_{50} of 19.8 μM (Rimando et al., 2002).

Xu et al. (2021) studied the anti-inflammatory activity of *trans*-pterostilbene in ovalbumin (OVA)-induced asthmatic mice cell cultures. They found that *trans*-pterostilbene reduces lung inflammation and airway hyperreactivity while promoting antioxidant activity. Cytokine levels IL-4, IL-5 and IL-13 were reduced in BALF as well as total and OVA-specific IgE. Mechanistically, *trans*-pterostilbene activates AMPK and suppresses ROS production through AMPK/Sirt1 and Nrf2/HO-1 pathways (Xu et al., 2021).

Liu et al. (2019) investigated the anti-inflammatory activity of *trans*-pterostilbene on astrocytic inflammation. The results showed that the stilbene attenuated astrocytic inflammation by the down-regulation of TNF-α, IL-1β and IL-6 and increased the activity of antioxidant enzymes, SOD and glutathione peroxide (GSH-Px). Moreover, *trans*-pterostilbene inhibited NF-κB phosphorylation involved into astrocytic inflammation (Liu et al., 2019).

7.1.5.2.3 Tyrosinase inhibition activity. An et al. (2022) evaluated the *trans*-pterostilbene capacity to inhibit melanogenesis in melanoma cell line (B16F10), in zebrafish embryos and in in vitro human skin. According to the results, *trans*-pterostilbene (3 μM) inhibited melanogenesis in B16F10 and tyrosinase activity. Moreover, *trans*-pterostilbene had inhibitory impact on number and length of dendrites. The compound also inhibited melanin synthesis and tyrosinase activity in zebrafish embryos. The results of Fontana–Masson staining showed that *trans*-pterostilbene (10 μM) inhibited melanogenesis in in vitro human skin. In addition, melanin content and tyrosinase activity were decreased by this metabolite (An et al., 2022).

Basri et al. (2021) studied the anti-melanogenic effect of *trans*-pterostilbene in UVB-irradiated melanoma mouse cell line (B164A5). According to the results, *trans*-pterostilbene (5 and 10 μM) inhibited tyrosinase activity in dose-dependent manner (37.14% and 58.36%, respectively). Moreover, melanin content was significantly decreased (27.34 μg/mL) by *trans*-pterostilbene (10 μM) (Basri et al., 2021).

Majeed et al. (2020) evaluated the skin-brightening capacity of 0.4% formulation of *trans*-pterostilbene in B16F1 mouse melanoma cells. The results showed that the compound significantly inhibited melanogenesis with IC_{50} of 0.55 μg/mL. Moreover, the tyrosinase activity was also down-regulated by *trans*-pterostilbene (IC_{50} = 7 μg/mL) (Majeed et al., 2020).

7.1.5.2.4 Anticancer activity. Rimando et al. (2002) studied the chemopreventing properties of *trans*-pterostilbene using mouse mammary gland organ culture (MMOC) carcinogenesis model induced by 7,12-dimethylbenz[a]anthracene (DMBA). According to the results, *trans*-pterostilbene significantly inhibited preneoplastic lesions with an ED_{50} of 4.8 μM (Rimando et al., 2002).

Zhao et al. (2020) investigated the potential sorafenib and *trans*-pterostilbene and its combination against gastric adenocarcinoma (GAC). Firstly, the results showed *trans*-pterostilbene inhibited the cell viability in human gastric cancer cell lines - N87 and MKN45 ($IC_{50} = 52.71$ and 65.63 μM, respectively). The combination of *trans*-pterostilbene and sorafenib resulted in cyclin D1, Cdk-2, Cdk-4, Cdk-6, and p62 decrease, while the level of apoptotic-related proteins e.g., LC3II, caspase-9, caspase-3, and PARP was increased. Moreover, the tumor volume and weight were significantly inhibited in the group treated with *trans*-pterostilbene/sorafenib combination to 132.27 mm^3 and 0.09 g, respectively in tumor xenografts model in N87 cells (Zhao et al., 2020).

Tam et al. (2018) tested the anticancer activity of *trans*-pterostilbene in combination with α-tocopheryl succinate (α-TOS) in breast cancer xenograft animal model. According to the results, the *trans*-pterostilbene/α-TOS combination exhibited strong inhibition effect on the signaling proteins (AKT and ERK) activation involved in cell proliferation. Moreover, the IC_{50} cell viability in human breast cancer cell line (MDA-MB-231) exposure to 5 and 10 μM of *trans*-pterostilbene and 10 μM of α-TOS was reduced ($IC_{50} = 25.1$ and 19.3 μM, respectively) (Tam et al., 2018).

7.1.5.2.5 Antiobesity activity. Hsu et al. (2012) evaluated the inhibitory effect of *trans*-pterostilbene on adipogenesis in 3T3-L1 adipocytes. The results showed that during adipose differentiation, *trans*-pterostilbene inhibited glycerol-3-phosphate dehydrogenase (GPDH), involved in TAG synthesis, by 61.5%. Moreover, the lipid accumulation was inhibited in 3T3-L1. Furthermore, the mRNA expression of leptin, resistin (or "resistance to insulin"), and fatty acid synthase (FAS) were reduced, leading toward less adipogenesis. Treatment with 10 μM of *trans*-pterostilbene upregulated the expression of adiponectin, implicated in the decrease of lipid and glucose content and the increase of insulin sensibility (Hsu et al., 2012).

Gómez-Zorita et al. (2015) studied the effect of 15 mg/kg/d and 30 mg/kg/d of *trans*-pterostilbene (P15 and P30) on glycemic control in rats. According to the results, the glucose level was lower in both concentrations (P15, P30) in comparison to control group (111, 108 and 128 mg/dL, respectively). Similarly, insulin level was also lower in *trans*-pterostilbene groups with values of 2.4 and 3.0 ng/mL compared with control group (3.2 ng/mL). In addition, the Homeostatic Model Assessment for Insulin Resistance (HOMA-IR) was reduced in both *trans*-pterostilbene groups (14.0 and 19.8, respectively). Furthermore, *trans*-pterostilbene-treated groups showed higher values of glucose transporter type 4 (GLUT4) protein expression (+58% and +52%) (Gómez-Zorita et al., 2015).

Seo et al. (2017) investigated the antiadipogenesis potential of *trans*-pterostilbene in 3T3-L1 cell line. Based on the results of Oil Red O (ORO) staining, *trans*-pterostilbene decreased the lipid accumulation. Moreover, the expression of C/EBPα, PPARγ, and AP2 was decreased (68%, 78%, and 84%, respectively). Furthermore, *trans*-pterostilbene decreased the triglyceride (TG) synthesis-associated enzymes e.g., Lipin1 (63.8%) and DGAT (37.5%) (Seo et al., 2017).

7.1.5.2.6 Antifungal activity. Houillé et al. (2014) evaluated the activity of pterostilbene against 11 *Candida* species: *C. albicans*, *C. dubliniensis*, *C. famata*, *C. glabrata*, *C. guilliermondii*, *C. inconspicua*, *C. kefyr*, *C. krusei*, *C. lusitaniae*, *C. norvegensis*, *C. parapsilopsis*, and *C. tropicalis*. Based on the results, *trans*-pterostilbene showed higher antifungal activity against *Candida* species compared with *trans*-resveratrol with minimum inhibitory concentration (MIC_{95}) value ranging from 7 to 28 μg/mL (Houillé et al., 2014).

Pezet et al. (2004) studied the antifungal activity of *trans*-pterostilbene against *P. viticola. Trans*-pterostilbene inhibited the spores germination ($ED_{50} = 14.70\ \mu M$) and zoospore mobility ($ED_{50} = 28.30\ \mu M$) (Pezet et al., 2004).

Adrian et al. (1997) investigated the antifungal activity of *trans*-pterostilbene against *B. cinerea.* Based on the results, *trans*-pterostilbene (20 µg/mL) had a significant effect on germination of *B. cinerea* and resulted in 42% inhibition (Adrian et al., 1997).

7.2 Dimers

7.2.1 Ampelopsin A

7.2.1.1 General information

Ampelopsin A (Fig. 10.10) is resveratrol dimer reported in *Hopea dryobalanoides*, *V. amurensis, V. thunbergiiV. thunbergii, V. betulifolia, V. coignetiae, V. heyneana,* and *V. vinifera* (Kim et al., 2012; Pubchem, 2023; Teka et al., 2022; Wang et al., 2011).

7.2.1.2 Biological activity

7.2.1.2.1 Antioxidant activity. Sáez et al. (2018) evaluated the antioxidant activity of 12 stilbenoids including ampelopsin A using three different assays: ABTS, CUPRAC, and ORAC-fluorescein (FL). The results were expressed as Trolox equivalent (TE). Based on ABTS and CUPRAC assay, ampelopsin A showed moderate antioxidant activity with IC_{50} values of 11.97 and 4.13 mmol TE/g, respectively. Moreover, ampelopsin A showed higher activity compared with *trans*-resveratrol in ORAC-FL assay (30.37 and 21.85 mmol TE/g, respectively) (Sáez et al., 2018).

7.2.1.2.2 Anti-inflammatory activity. Wang et al. (2011) investigated the anti-inflammatory activity of six compounds including ampelopsin A isolated from *V. thunbergii* in lipopolysaccharide-induced arthritis. According to the results, ampelopsin A significantly inhibited PGE2 compared with other tested compounds with IC_{50} values of 15.52 µM (Wang et al., 2011).

7.2.1.2.3 Anticancer activity. Saroyobudiono et al. (2008) investigated the cytotoxic effect of six stilbenoids including ampelopsin A against murine leukemia P-388 cells. According to the results, ampelopsin A showed one of the best results among tested compounds with IC_{50} of 17.0 µM (Saroyobudiono et al., 2008).

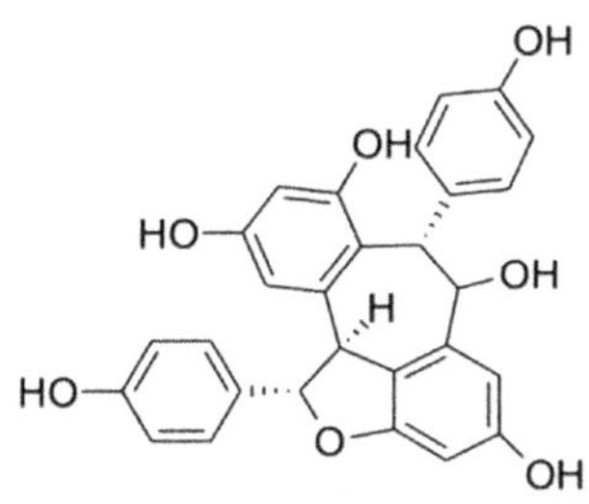

Ampelopsin A

FIGURE 10.10

The chemical structure of ampelopsin A.

Ha et al. (2009) studied the anticancer potential of 10 stilbenoids including ampelopsin A against murine leukemia L1210 and human K562 cancer cell lines. Ampelopsin A exhibited significant anticancer activity with IC_{50} values of 30.6 and 38.6 μM, respectively (Ha et al., 2009).

Lim et al. (2012) evaluated the anticancer activity of three compounds including ampleopsin A against survival of MCF-7 breast cancer cells. Based on the results, ampelopsin A downregulated the expression of sphingosine kinase 1 (SK1) activity involved in cancer development (Lim et al., 2012).

7.2.1.2.4 Antiobesity activity. Kim et al. (2014) studied the inhibitory activity of seven stilbenoids including ampelopsin A isolated from the *V. vinifera* roots against pancreatic lipase. The results showed that mpelopsin A exhibited stronger pancreatic lipase inhibitory activity compared with *trans*-resveratrol ($IC_{50} = 143.6$ and >200 μM, respectively) (Kim et al., 2014).

7.2.1.2.5 Antifungal activity. Schnee et al. (2013) evaluated the antifungal activity of six compounds including ampelopsin A against *Plasmopara viticola*. According to the results, ampelospin A inhibited zoospore mobility and sporulation of *P. viticola* with IC_{50} values of 124 and 282 μM, respectively (Schnee et al., 2013).

Lambert et al. (2012) studied the content of eight stilbenoids including ampleopsin Ain plants inoculated by *Diplodia seriata* and *Neofusicoccum parvum.* According to the results, the ampelopsin A content in *D. seriata* and *N. parvum* group was significantly higher compared with the control group (151, 89 and 64 μg/g FW, respectively) (Lambert et al., 2012).

7.2.1.2.6 Neuroprotective activity. Kim et al. (2012) evaluated the neuroprotective activity of ampelopsin A isolated from *V. amurensis* against ischemic brain damage in rats. Based on the results, ampelopsin A pretreatment of cortical neurons resulted in glutamate-induced neuronal death reduction 92.1%. Moreover, antiapoptotic protein BcL-2 was also inhibited by ampelopsin. Furthermore, ampelopsin increased the expression of proapoptotic protein BAX (Kim et al., 2012).

7.2.2 Trans-*ε-viniferin*

7.2.2.1 General information

Viniferins are the resveratrol derivatives, existing in six different forms: α-Viniferin (Fig. 10.11), ε-Viniferin, *δ*-Viniferin, ω-Viniferin (Fig. 10.12), R-Viniferin (vitisin A), and R2-Viniferin (vitisin B)

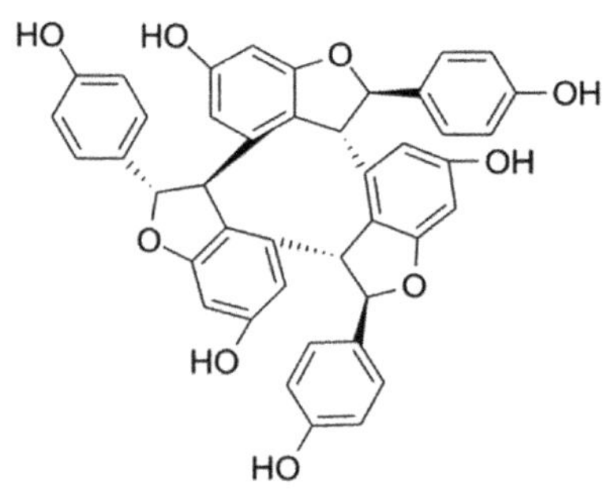

α-Viniferin

FIGURE 10.11

The chemical structure of α-viniferin.

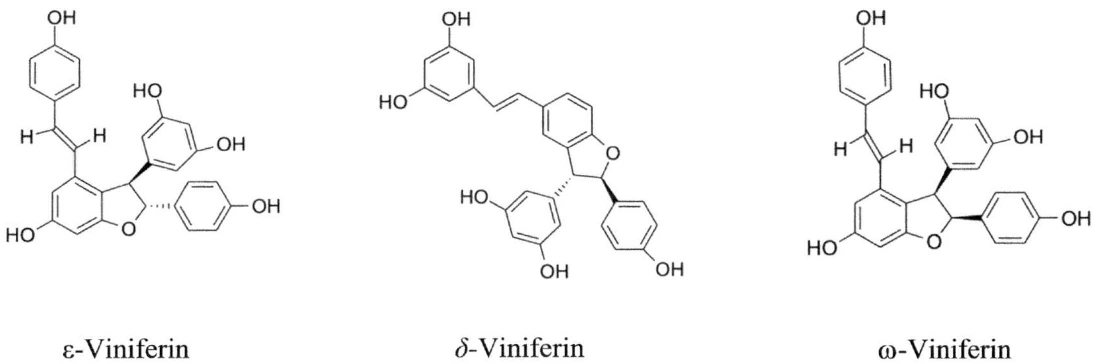

FIGURE 10.12

The chemical structure of ε-viniferin, δ-viniferin, and ω-viniferin.

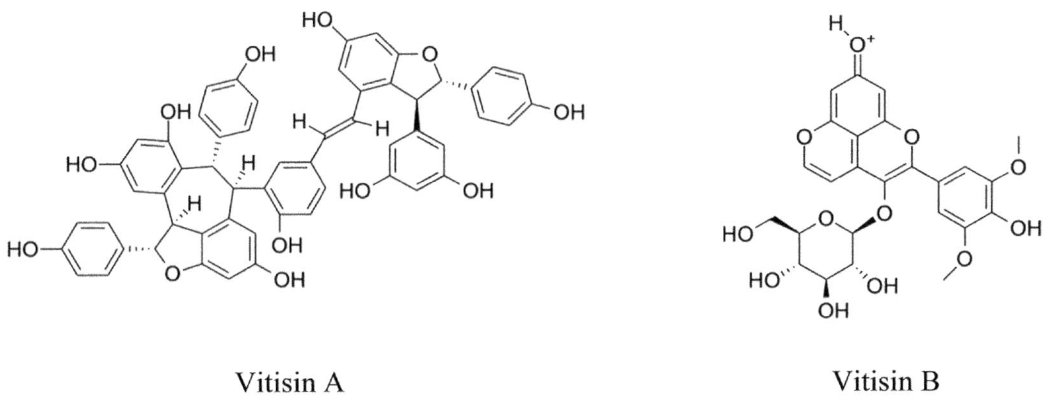

FIGURE 10.13

The chemical structure of vitisin A and vitisin B.

(Fig. 10.13). ε-Viniferin has two isomeric forms: *trans* and *cis* (Fig. 10.14) (Fuloria et al., n.d.; Pubchem, 2023).

Trans-ε-Viniferin ((2S,3R)-*trans*-epsilon-viniferin) is a resveratrol dimer mainly presented in woody plants. *Trans*-ε-viniferin (Fig. 10.12) is assumed to be the most studied isomer of ε-Viniferin. The UV radiation induces the isomerization of *trans*-ε-Viniferin to *cis*-ε-Viniferin (Fig. 10.14). It was firstly detected in grapevine leaves in 1977 (Beaumont et al., 2022; Gómez-Zorita et al., 2023). It has been reported in *Cyphostemma crotalarioides, Paeonia lactiflora, Paeonia ostii, Paeonia suffruticosa, Vitis amurensis, Vitis vinifera,* and in other species (Table 10.1) (Beaumont et al., 2022; Fuloria et al., n.d.; Gómez-Zorita et al., 2023).

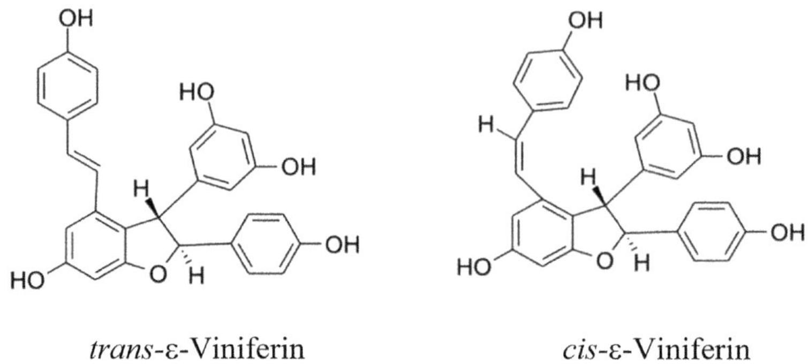

FIGURE 10.14

The chemical structure of *trans*-ε-viniferin and *cis*-ε-viniferin.

7.2.2.2 Biological activity

7.2.2.2.1 Antioxidant activity. Privat et al. (2002) evaluated the antioxidant activity of *trans*-ε-viniferin against superoxide radicals in dimethyl sulfoxide (DMSO) medium. Based on the results, *trans*-ε-viniferin exhibited the stronger antioxidant capacity ($IC_{50} = 0.14$ mM) compared with *trans*-resveratrol and synthetic stilbenic derivatives (e.g., 4-hydroxystilbene, 4,4′-dihydroxystilbene, 3,5-dihydroxystilbene, and trimethylresveratrol) with IC_{50} value ranging from 0.82–1.68 mM, respectively (Privat et al., 2002).

Biais et al. (2017) studied the antioxidant capacity of main constituents obtained from grapevine shoot extract including *trans*-ε-viniferin using ORAC assay. According to the results, *trans*-ε-viniferin exhibited the strongest antioxidant activity (98.7 mmol TE/g) compared with *trans*-resveratrol (27.9 mmol TE/g) using Trolox as a standard (Biais et al., 2017).

7.2.2.2.2 Anti-inflammatory activity. Liao et al. (2021) studied the anti-inflammatory activity of *trans*-ε-viniferin obtained from the roots of *Vitis thunbergii* against N-formyl-L-methionyl-L-leucyl-L-phenylalanine (fMLP)-induced superoxide anion production in human neutrophils. Based on the results, *trans*-ε-viniferin inhibited superoxide anion production depending on the concentration ($IC_{50} = 2.30$ and 9.80 μM). Moreover, ε-viniferin (3–30 μM) inhibited the cathepsin release, also contributing to inflammation (Liao et al., 2021).

Ha et al. (2018) evaluated the anti-inflammatory activity of oligostilbenoids including *trans*-ε-viniferin isolated from *Vitis heyneana* in LPS-induced RAW 264.7 macrophages. The results showed that *trans*-ε-viniferin decreased inflammation markers NO, PGE2, and COX-2 production in RAW 264.7 macrophages in dose-dependent manner. The best results were obtained for the concentration of 30 μM with 50% decrease. Moreover, *trans*-ε-viniferin inhibited the NF-κB activity with the strongest activity at 30 μM concentration (Ha et al., 2018).

7.2.2.2.3 Tyrosinase inhibition activity. Kim et al. (2002a) investigated the tyrosinase inhibition activity of seven stilbenoids including *trans*-ε-viniferin and *cis*-ε-viniferin isolated from the seeds of *Paeonia lactiflora*. According to the results, *trans*-ε-viniferin was stronger tyrosinase inhibitor in comparison to *cis*-ε-viniferin with IC_{50} values of 51.5 and 56.3 μM, respectively (Kim et al., 2002a).

Morel-Salmi et al. (2014) studied the anti-tyrosinase activity of *trans*-ε-Viniferin. The results demonstrated that *trans*-ε-viniferin possesses strong tyrosinase inhibition activity with IC_{50} values of 4.1 μM, compared with *trans*-resveratrol (52.8 μM) (Morel-Salmi et al., 2014).

Malinowska et al. (2020) investigated the tyrosinase inhibition for *trans*-ε-viniferin to compare with *trans*-resveratrol and kojic acid at the concentration of 10 μM. The activity of both stilbenoids was relatively high; however, *trans*-ε-viniferin was a more potent brightening agent than *trans*-resveratrol (76% and 75%, respectively). Kojic acid exhibited the strongest whitening activity with IC_{50} value 16.9 μM, while IC_{50} value of 52.93 μM was obtained for *trans*-ε-viniferin and 60.75 μM for *trans*-resveratrol. Importantly, the results of the docking data indicate that *trans*-resveratrol and *trans*-ε-viniferin exhibited tyrosinase inhibition potential. Nonetheless, the binding affinity of trans-ε-viniferin exhibits a minor increase, calculated at −7.73 kcal/mol in contrast to −5.95 kcal/mol, which was calculated for *trans*-resveratrol. This is attributed to the specific involvement of interactions with histidine residues located in the protein's amino acid sequence (His85 and His244) through hydrogen bonding and π-π stacking. In contrast, trans-resveratrol demonstrates only singular hydrogen bond interaction with methionine residue (Met280). The two stilbenoids bind to tyrosinase at the same site, thus confirming their competitive inhibition mechanism (Malinowska et al., 2020).

7.2.2.2.4 Anticancer activity. Özdemir et al. (2018) evaluated the apoptotic effect of *trans*-ε-viniferin against glioblastoma (GBM). Based on the results, the most significant decrease in C6 glioma cell line was obtained when the combination of *trans*-ε-viniferin and cisplastin (CDDP) was used. Moreover, the apoptotic index was found to be 50% higher after 48 hours of incubation (Özdemir et al., 2018).

Colin et al. (2008) compared the anti-proliferative activity of *trans*-resveratrol related compounds including *trans*-ε-viniferin in human hepatoma cell line (HepG2). *Trans*-ε-viniferin showed anti-proliferative activity with IC_{50} value of 58.4 μM. Moreover, *trans*-ε-viniferin treatment (60 μM) resulted in complete inhibition of cell proliferation and causing cells death (Colin et al., 2008).

7.2.2.2.5 Antifungal activity. Langcake et al. (Langcake and Pryce, 1977) evaluated the antifungal activity of *trans*-ε-viniferin against *P. viticola*. Based on the results, *trans*-ε-viniferin inhibited both zoospore release and motity with IC_{50} values of 19 and 12.5 μg/mL, respectively (Langcake and Pryce, 1977).

Pezet et al. (2004) studied the antifungal activity of *trans*-ε-viniferin against *P. viticola*. *Trans*-ε-viniferin inhibited the spores germination ($ED_{50} = 71.20$ μM) and zoospore mobility ($ED_{50} = 73$ μM) (Pezet et al., 2004).

Schnee et al. (2013) evaluated the antifungal activity of six compounds including *trans*-ε-viniferin against *P. viticola*. According to the results, *trans*-resveratrol inhibited zoospore mobility and sporulation of *P. viticola* with IC_{50} values of 66 and 63 μM, respectively (Schnee et al., 2013).

Lambert et al. (2012) studied the content of eight stilbenoids including *trans*-ε-viniferin in plants inoculated by *Diplodia seriata* and *Neofusicoccum parvum*. According to the results, the *trans*-ε-viniferin content in *D. seriata* and *N. parvum* group was significantly higher compared with the control group (1208, 772 and 649 μg/g FW, respectively) (Lambert et al., 2012).

7.2.2.2.6 Neuroprotective activity. Freyssin et al. (2022) investigated the neuroprotective activity of *trans*-ε-viniferin in an Alzheimer's mouse model. Based on the results, *trans*-ε-viniferin reduced (79.5%) the density of hippocampal amyloid deposits. Moreover, the expression of GFAP, IBA1, and IL-1β, key markers of Alzheimer's disease and neuro inflammation, was decreased by *trans*-ε-viniferin (46.5%, 15.5%, and 75% respectively) after 7 months of treatment (Freyssin et al., 2022).

Zhang et al. (2020) studied the neuroprotective effect of *trans*-ε-viniferin in rotenone-induced cell model of Parkinsons's disease. Human neuroblastoma SH-SY5Y cells were incubated at different concentrations of *trans*-ε-viniferin (50—10 μM) and different time intervals (0—48 hours). According to MTT assay, the cell viability was significantly improved after 24 hours treatment with *trans*-ε-viniferin by 40.0%. Moreover, the pretransfection of cells with SIRT3 or fork headbox O3 (FOXO3) shRNA in combination with ε-viniferin resulted in attenuation of neuroprotective effect (Zhang et al., 2020).

7.2.3 Trans-δ-v*iniferin*

7.2.3.1 General information

*Trans-δ-*viniferin ((2S,3S)-*trans*-delta-viniferin) (Fig. 10.15) is a resveratrol dehydrodimer, which is a *trans*-ε-Viniferin isomer. It has been found in *Gnetum hainanense*, *Rheum maximowiczii,* and *Vitis vinifera* (*Lotos: Natural Products Online*, 2023; Pubchem, 2023).

7.2.3.2 Biological activity

7.2.3.2.1 Antioxidant activity. Ficarra et al. (2016) evaluated the antioxidant effect of *trans-δ*-viniferin using DPPH and hydroxyl radical scavenging models. According to DPPH results, *trans-δ*-viniferin showed antioxidant activity with IC_{50} values of 50% at concentration higher than 100 μM. Moreover, *trans-δ*-viniferin inhibited hydroxyl radicals with inhibition value of 58% at a concentration of 10 μM (Ficarra et al., 2016).

Sáez et al. (2018) studied the antioxidant activity of oligostilbenoids including *trans-δ*-viniferin using ABTS, CUPRAC, and ORAC-fluorescein (FL) in vitro assays. The results were expressed as Trolox equivalent. Based on ABTS and CUPRAC assay, *trans-δ*-viniferin showed moderate antioxidant activity compared with all tested stilbenoids (13.40 and 4.38 mmol TE/g, respectively). Based on the ORAC-FL results, *trans-δ*-viniferin demonstrated higher antioxidant activity compared with *trans*-resveratrol with a value of 30.89 mmol TE/g (Sáez et al., 2018).

7.2.3.2.2 Antiangiogenic activity. Wu et al. (2020) tested the protective effect of resveratrol and its dimers including *trans*-δ-viniferin in vascular endothelial cells (VECs). Pretreatment with resveratrol (10 μM), ε-viniferin (10 μM), and *trans-* δ-viniferin (5 μM) resulted in suppression of H_2O_2-

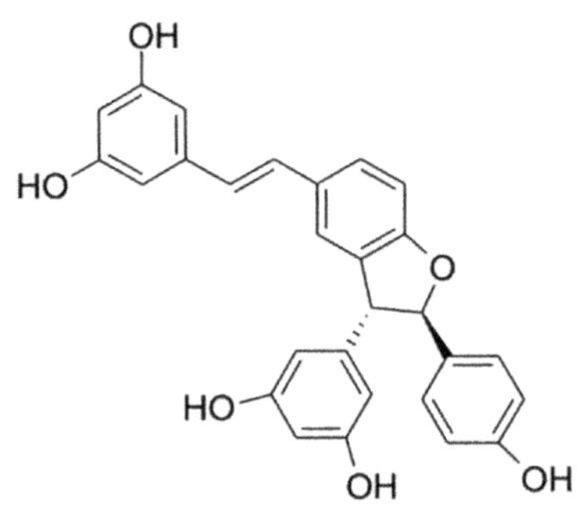

trans-δ-Viniferin

FIGURE 10.15

The chemical structure of *trans-δ*-viniferin.

dependent reduction cell viability, which indicates the resistance of the tested compounds against oxidative stress. Moreover, both tested compounds stimulated the wound repair of VECs decreasing the expression of SIRT1 and HO-1 expression involved in wound repair process (Wu et al., 2020).

7.2.3.2.3 Antidiabetic activity. Mattio et al. (2019) investigated the inhibitory activity of eight stilbenoid compounds including *trans*-δ-viniferin against pancreatic α-amylase, involved in polysaccharide break down and postprandial hyperglycaemia, using acarbose as a reference. Based on the results, among all tested compounds, *trans*-δ-viniferin demonstrated the highest α-amylase inhibitory activity (~90%) comparable with acarbose (86%) at the same concentrations (10 mM) (Mattio et al., 2019).

7.2.3.2.4 Antifungal activity. Pezet et al. (2004) studied the antifungal activity of *trans*-δ-viniferin against *P. viticola*. *Trans*-δ-resveratrol inhibited the spores germination ($ED_{50} = 14.70\ \mu M$) and zoospore mobility ($ED_{50} = 14.60\ \mu M$) (Pezet et al., 2004).

Timperio et al. (2012) in their research also found that *trans*-δ-viniferin formed by the bioconversion of *trans*-resveratrol actively inhibited the growth of *B. cinerea*. Although infection by *B. cinerea* typically occurs during the bloom stage, the remarkable sensitivity of the HPLC-MS analysis- enabled the detection of measurable levels of viniferins even in the early prebloom stages of Cabernet Sauvignon flowers. Delta-viniferin is thus one of the phytoalexins in control of antifungal response potential (Timperio et al., 2012).

7.2.4 Pallidol

7.2.4.1 General information

Pallidol (syn. pallidol A) (Fig. 10.16) is a naturally occurring tetracyclic homodimer obtained by cyclodimerization of resveratrol. It has been firstly isolated from *Cissus pallida* in 1986. In 2001, pallidol was discovered in red wine. Later, pallidol was found in 19 red wines, where its level was comparable with resveratrol (0,5–4.8 mg/L). It is also reported in *Caragana sinica* (He et al., 2009; Pubchem, 2023; Teka et al., 2022).

7.2.4.2 Biological activity

7.2.4.2.1 Antioxidant activity. Biais et al. (2017) evaluated the antioxidant activity of pallidol obtained from grapevine shoot extract using ORAC assay. Based on the results, pallidol showed moderate

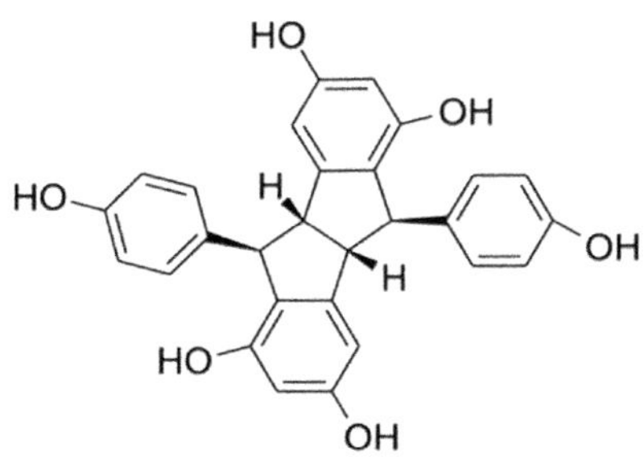

Pallidol

FIGURE 10.16

The chemical structure of pallidol.

antioxidant capacity (28.8 mmolTE/g. Its activity was higher compared with vitisin A and B (18.6 and 4.7 mmol TE/g, respectively) (Biais et al., 2017).

Sáez et al. (2018) evaluated the antioxidant activity of oligostilbenoids including pallidol using ORAC-FL assay. Based on the results, pallidol showed the highest antioxidant activity compared with tested stilbenoids with a value of 56.24 mmol TE/g (Sáez et al., 2018).

7.2.4.2.2 Anticancer activity. Silva et al. (2012) studied the antiproliferative activity of pallidol obtained from *Cenchrus echinatus* against colon adenocarcinoma cells ($CaCo_2$). According to the results, pallidol showed antiproliferative activity of IC_{50} equal 306.0 μg/mL (Silva et al., 2012).

7.2.4.2.3 Antiangiogenic activity. Fernandez-Cruz et al. (2019) determined the antiangiogenic potential of stilbenoids including pallidol against vascular endothelial growth factor (VEGF). Based on the results, pallidol showed anti-VEGF potential with IC_{50} values of 4.42 μM (Fernandez-Cruz et al., 2019).

7.2.4.2.4 Antidiabetic activity. Mattio et al. (2019) investigated the inhibitory activity of stilbenoid compounds including pallidol against pancreatic α-amylase using acarbose as a reference. According to the results, all tested compounds showed modest inhibitory activity (21%–45%) compared with acarbose (86%) where pallidol was one of the most potent ones (Mattio et al., 2019).

7.2.5 Scirpusins

7.2.5.1 General information

Scirpusin A and B (Fig. 10.17) are the dimers of *trans*-resveratrol. Both scirpusin A and B can appear in *trans* and *cis* form (Fig. 10.17). Naturally, scirpusins are found in Cyperaceae species: *Cyperus eragrostis*, *Cyperus longus*, *Scirpus californicus*, *Scirpus fluviatilis*, *Scirpus maritimus,* and in *Vitis vinifera* (Dávid et al., 2021; Pubchem, 2023).

7.2.5.2 Biological activity

7.2.5.2.1 Antioxidant activity. Sáez et al. (2018) studied the antioxidant activity of oligostilbenoids including scirpusin A using ORAC-FL in vitro assay. According to the results, scirpusin A demonstrated moderate antioxidant activity compared with tested compounds with value of 34.99 mmol TE/g (Sáez et al., 2018).

Utkina and Kulesh (2012) evaluated the antioxidant activity of scirpusin A and B using DPPH and ABTS assays. According to DPPH assay, scirpusin A and scirpusin B showed similar scavenging activity with IC_{50} values of 16.6 and 14.4 μM, respectively. Similar values were also obtained in ABTS method (1.75 and 1.77 mM, respectively). Moreover, both compounds demonstrated strong inhibitory activity in inhibition of initiated oxidation of linoleic acid (IOLA) method with inhibitory values of 91.6% and 93.0%, respectively (Utkina and Kulesh, 2012).

Dávid et al. (Schmeda-Hirschmann et al., 1996) compared the inhibitory activity of scirpusin A and B isolated from *S. californicus* against xanthine oxidase. The scirpusin A showed higher inhibitory potential compared with scirpusin B with IC_{50} values of 3.6 and 6.0 μM, respectively (Schmeda-Hirschmann et al., 1996).

Morikawa et al. (2010) investigated the antioxidant activity of 14 stilbenoids including *trans*-scirpusin A and B isolated from *Cyperus longus* using DPPH assay. According to the results, *trans*-scirpusin B showed stronger radical scavenging activity compared with *trans*-scirpusin A with SC_{50} values of 2.8 and 8.2 μM, respectively. Moreover, the aforementioned compounds were more potent compared with *trans*-resveratrol and pallidol ($SC_{50} = 24$ and 29 μM, respectively) (Morikawa et al., 2010).

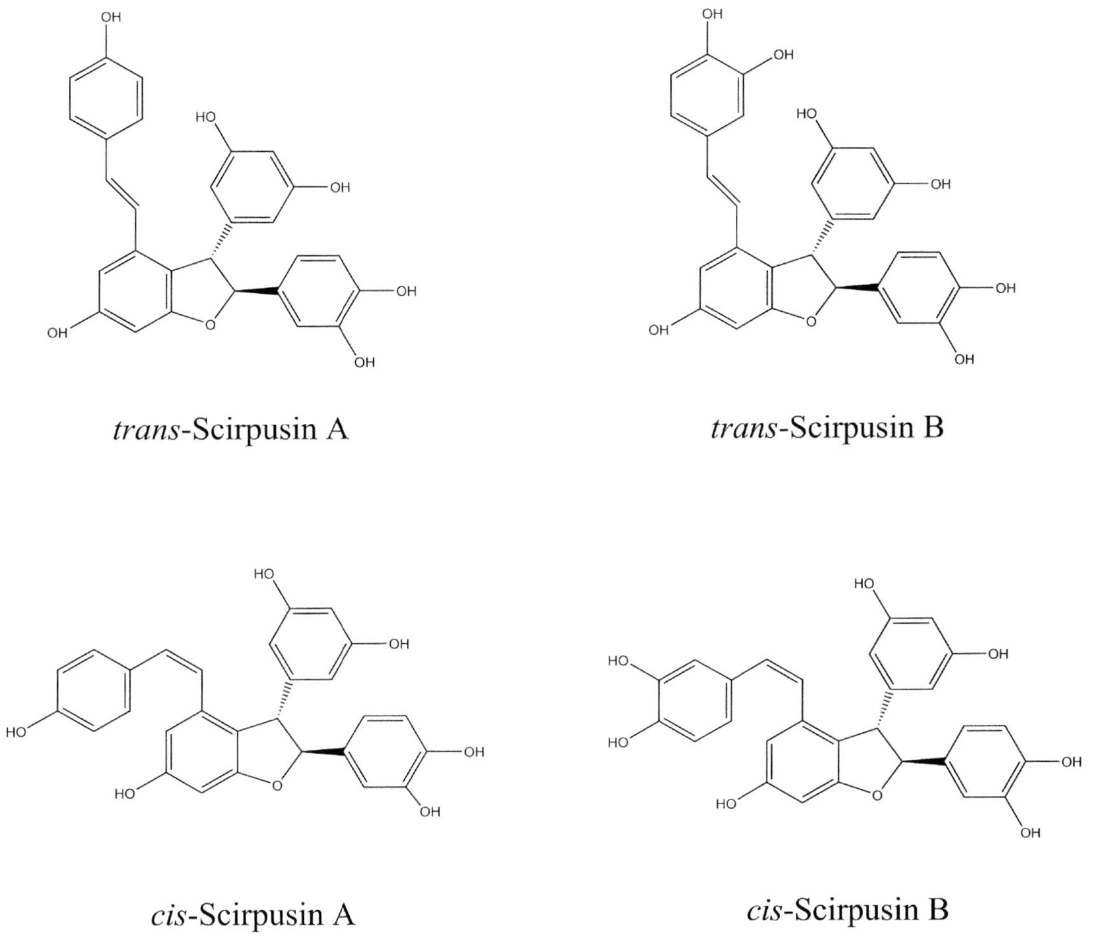

FIGURE 10.17

The chemical structure of *trans*-and *cis*-scirpusin A and *trans*-and *cis*-scirpusinB.

7.2.5.2.2 Cardioprotective activity. Arraki et al. (2017) investigated the vasorelaxant properties of six stilbenoid compounds including scirpusin B by its potential to inhibit arginase competing with NO synthase leading to reduced NO production involved in endothelial dysfunction and cardiovascular diseases. Based on the results, scirpusin B showed arginase inhibitory activity with IC_{50} value of 22.6 μM, which was higher compared with piceatannol (12.6 μM) (Arraki et al., 2017).

7.2.5.2.3 Anticancer activity. Hong et al. (2017) studied the anticancer activity of *trans*-Scirpusin A against colorectal cancer Her2/CT26 cells in mice. Based on the results, *trans*-scirpusin A inhibited the tumor growth and decreased its weight. Besides, in a combination with docetaxel, *trans*-scirpusin A exaggerated its effect and inhibited the cancer growth to a greater extent. Moreover, *trans*-scirpusin A induced autophagy in tumor cells, which is suggested to be associated with trans-Scirpusin A-stimulated AMPK activation and mTORC1 pathway inhibition involved in cell survival (Hong et al., 2017).

7.3 Trimers

7.3.1 Miyabenol C

7.3.1.1 General information

Miyabenol C is a resveratrol trimer and can occur in the form of *trans* and *cis* (Fig. 10.18), or in combination of *cis* and *trans* resveratrol monomer. It occurs naturally as *trans*-isomer in *Caragana sinica*, *Caragana stenophylla*, *Foeniculum vulgare* and *Vitis vinifera* (Fuloria et al., n.d.; Pubchem, 2023; Teka et al., 2022; Xue et al., 2014).

7.3.1.2 Biological activity

7.3.1.2.1 Antioxidant activity. Ono et al. (1997) investigated the antioxidant activity of six stilbenoids trimers including *trans*-miyabenol C and *cis*-miyabenol C. Both tested compounds showed stronger antioxidant activity compared with *t*-butylhydroxyanisole (BHA) used as a reference. Moreover, *trans*-miyabenol C and *cis*-miyabenol C significantly inhibited hyaluronidase activation with inhibition ratio of 100% and 98%, respectively at a final concentration of 400 μM (Ono et al., 1997).

Sáez et al. (2018) studied the antioxidant activity of oligostilbenoids including (*E*)-*trans-cis*-miyabenol C using ORAC-FL in vitro assay. The results were expressed as Trolox equivalent (TE). Based on the results, (*E*)-*trans-cis*-miyabenol C possessed strong antioxidant activity with values of 40.66 mmol TE/g. In addition, the antioxidant potential of (*E*)-*trans-cis*-miyabenol C was higher

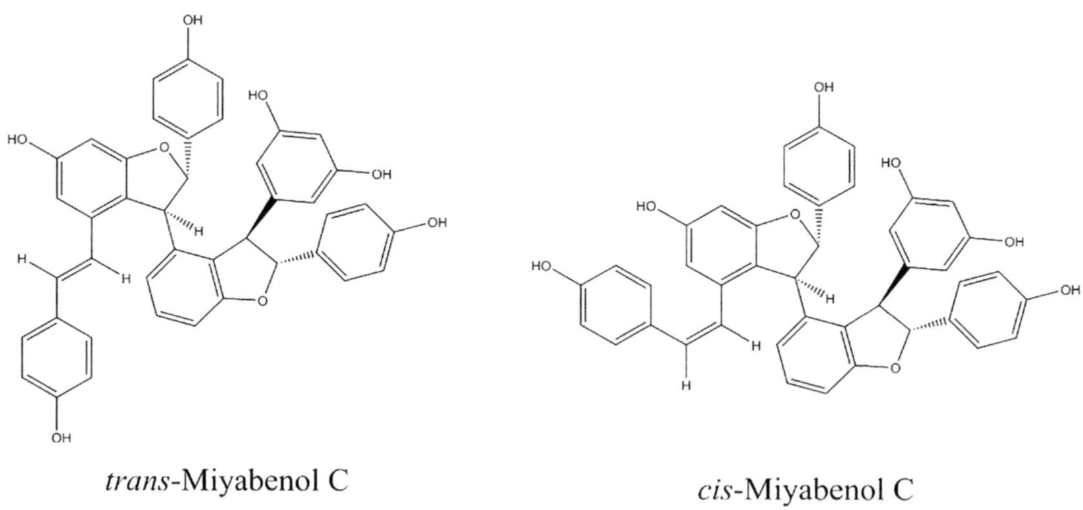

FIGURE 10.18

The chemical structure of *trans*-miyabenol C and *cis*-miyabenol C.

compared with *trans*-resveratrol, *trans*-piceatannol, *trans*-ε-viniferin, and *trans*-δ-viniferin (25.59, 27.15, 21.85, and 30.89 mmol TE/g, respectively).

Marino de et al. (De Marino et al., 2007) evaluated the antioxidant activity of 12 compounds including *trans* and *cis*-miyabenol C. According to the results, *cis*-miyabenol C showed stronger antioxidant potential compared with *trans*-miyabenol C with EC_{50} values of $2.56 \cdot 10^{-5}$ M and $7.08 \cdot 10^{-5}$ M, respectively (De Marino et al., 2007).

7.3.1.2.2 Anticancer activity. Barjot et al. (2007) investigated the anticancer activity of three stilbenoids including *trans*-miyabenol C obtained from *Vitis vinifera* against myeloid and lymphoid cell lines. Based on the results, *trans*-miyabenol C showed the highest antiproliferative activity in all tested cell lines compared with resveratrol and *trans*-ε-viniferin (U266, RPMI8226, Jurkat, K562, U937) with IC_{50} values of 12.1, 20.8, 29.4, 18.7, 10.8 μM, respectively (Barjot et al., 2007).

González-Sarrías et al. (De Marino et al., 2007) tested the anticancer activity of stilbenoids including *trans*-miyabenol C against human colon cancer (HCT-116, HT-29, Caco-2) and normal human colon (CCD-18Co) cell lines. Based on the results, *trans*-miyabenol C demonstrated moderate antiproliferative activity with IC_{50} values of 59.1, 84.2, 72.8, 115.0 μM after 48 hours treatment. In addition, *trans*-miyabenol C was significantly more effective compared with *trans*-resveratrol and pallidol (IC_{50} = 159.5, 164.7, 130.6, 189.7 μM and 167.2, 170.5, 159.7, 165.2 μM a, respectively) (De Marino et al., 2007).

7.3.1.2.3 Antifungal activity. Lambert et al. (2012) studied the content of eight stilbenoids including *trans*-miyabenol C in plants inoculated by *Diplodia seriata* and *Neofusicoccum parvum.* According to the results, the *trans*-miyabenol C content in *D. seriata* and *N. parvum* group was significantly higher compared with the control group (487, 276 and 202 μg/g FW, respectively) (Lambert et al., 2012).

7.3.1.2.4 Neuroprotective activity. Hu et al. (2015) studied the neuroprotective effect of *trans*-miyabenol C in N2a695 cells against pathogenesis of Alzheimer's disease. The N2a695 cells were treated with increasing doses of miyabenol C (0.1−100 μM) for 10 hours. According to the results, *trans*-miyabenol C decreased dose-dependently, the secretion of damaging Aβ40 and Aβ42 (amyloid β) when in amyloid plaque found in Alzheimers's patients' brains. The strongest effect was observed at 20 μM of *trans*-miyabenol C. Moreover, *trans*-miyabenol C inhibited sAPP levels in both the cortex and hippocampus, which are usually increased in Alzheimer's patients. Furthermore, the levels of Aβ42 and Aβ40 in TBSX-soluble fractions were also reduced by *trans*-miyabenol C treatment (Hu et al., 2015).

7.3.2 α-Viniferin

7.3.2.1 General information

α-Viniferin (alpha-Viniferin) (Fig. 10.11) is a trimer obtained from the condensation of three resveratrol monomers. This compound is found in *Astilbe grandis, Caragana chamlagu, Caragana sinica, Carex baccans, Carex humilis, Cayratia trifolia, Dipterocarpus littoralis, Shorea roxburghii, Hopea exalata, Hopea ponga, Iris clarkei, Shorea maxwelliana, Shorea ovalis, Shorea seminis, Vitis heyneana, Vitis quinquangularis,* and *Vitis vinifera* (Dávid et al., 2021; Fuloria et al., n.d.; Pubchem, 2023).

7.3.2.2 Biological activity

7.3.2.2.1 Antioxidant activity. Jin et al. (2012) evaluated the antioxidant activity of α-viniferin using DPPH assay. Based on the results, α-viniferin possessed antioxidant activity with IC_{50} values of >200 μM (Jin et al., 2012).

Lulan et al. (2020) studied the antioxidant activity of α-viniferin using DPPH and ABTS assays. Based on the DPPH results, α-viniferin showed significant free radical scavenging activity with a 76.11% of inhibition. In ABTS assay, α-viniferin also demonstrated strong inhibitory activity (97.64%) (Lulan et al., 2020).

7.3.2.2.2 Anti-inflammatory activity. Chung et al. (2003) investigated the anti-inflammatory activity of α-viniferin in animal model of carrageenin-induced paw edema. α-Viniferin was given by oral administration and in form of intravenous injection. Oral administration of 30 mg/kg of α-viniferin resulted in significant antiinflammatory activity dependent on time. The best results were obtained after 4 hours for both α-viniferin oral administration and intravenous injection (42.4% and 32.5%, respectively). Moreover, α-viniferin demonstrated inhibitory activity against COX-2 with IC_{50} value of 4.9 μM. Furthermore, α-viniferin inhibited NO production with IC_{50} values of 2.7 μM in LPS-activated murine macrophages Raw264.7. In addition, the synthesis of COX-2 transcript was also inhibited by α-viniferin in Raw 264.7 macrophages (Chung et al., 2003).

Dilshara et al. (2014) studied the activity of α-viniferin isolated from *Clematis mandshurica* on the release of proinflammatory mediators in BV2 microglial cells. It was proved that α-viniferin suppressed the expression of proinflammatory genes, e.g., iNOS and COX-2 through the inactivation of NF-κB Moreover, treatment with α-viniferin downregulated the NO and PGE2 production (6.2 μM and 181 pg/ml, respectively) (Dilshara et al., 2014).

7.3.2.2.3 Skin-brightening activity. Yun et al. (2018) evaluated the anti-melanogenic potential of *Caragana sinica* and its active compound α-viniferin in improving the facial hyperpigmentation. C. sinica and its compound α-viniferin suppressed melanin production in melanocytes activated by α-melanocyte-stimulating hormone (α-MSH), histamine, or cAMP. Topical application of C. sinica containing α-viniferin reduced melanin levels in patients with facial melasma and freckles. At the molecular level, α-viniferin accelerated the inactivation of protein kinase A (PKA) by promoting the reassociation of its catalytic and regulatory subunits in melanocytes exposed to elevated cAMP. This disrupted a melanogenic feedback loop, leading to reduced melanin pigmentation through the inhibition of CREB phosphorylation and CRTC1 dephosphorylation, which downregulated MITF-M and Tyro gene expression involved in the activation of melanogenesis (Yun et al., 2018).

7.3.2.2.4 Anticancer activity. Chiou et al. (2022) examined the anticancer activity of α-viniferin. According to the results, α-viniferin inhibited both the invasion and migration of TGF-β1 (transforming growth factor beta 1)-induced in A549 cells. Moreover, α-viniferin suppressed the expression of vimentin a filament protein usually overexpressed in various cancers, known as a marker of EMT (epithelial—mesenchymal transition) correlated with tumor growth, metastasis, being induced by vimentin and other transcription factors of the SMAD signaling pathway. Thus,α-viniferin was found to inhibit the SMAD signaling pathway through MMP2, Zeb1, Snail, p-SMAD2, p-SMAD3, and ABCG2 expression in A549 cells (Chiou et al., 2022).

Huang et al. (2023) investigated the anticancer potential of α-viniferin in non-small-cell lung cancer cells. The results showed that α-viniferin demonstrated antiproliferative potential in NCI—H460 in a dose- and time-dependent manner. The best results were obtained for 50 μM concentration of α-viniferin. Furthermore, α-viniferin induced apoptosis in NCI—H460 cells in dose-dependent manner. Treatment with the highest concentration of α-viniferin (30 μM) resulted in presence of 35.0% of apoptotic cells (Huang et al., 2023).

7.3.2.2.5 Antidiabetic activity. Lulan et al. (2020) studied the antidiabetic potential of α-viniferin extracted from *Dipterocarpus littoralis*. Based on the results, α-viniferin inhibited enzymes hydrolyzing polysaccharides into glucose, thereby increasing blood glucose levels, alpha-glucosidase, and alpha-amylase, with IC_{50} values of 256.17 and 212.79 μg/mL, respectively (Lulan et al., 2020).

Sasikumar et al. (2019) evaluated the antidiabetic activity of six resveratrol oligomers including α-viniferin obtained from *Hopea ponga*. According to the results, α-viniferin was the most potent in inhibition of α-glucosidase with IC_{50} values of 7.17 μM. Moreover, α-viniferin significantly inhibited α-amylase and glycation with IC_{50} values of 4.85 and 27.10 μM, respectively (Sasikumar et al., 2019).

7.3.2.2.6 Antifungal activity. Langcake and Pryce (1977) evaluated the antifungal activity of α-viniferin against *P. viticola*. Based on the results, α-viniferin inhibited both zoospore release and motity with IC_{50} values of 35 and 11 μg/mL, respectively (Langcake and Pryce, 1977).

7.4 Tetramers

7.4.1 Vitisins

7.4.1.1 General information

Vitisin A (r2-viniferin) and vitisin B (r-viniferin) (Fig. 10.13) are the resveratrol tetramers, in *trans* or *cis* isomers, usually *trans* isomers are more studied. Vitisin A has been found in *Iris lacteal*, *Vitis thumbergii* and *Vitis vinifera*. Vitisin B occurs in *Iris lacteal, Vitis vinifera,* and additionally in *Vitis* species (Table 10.1) (Fuloria et al., n.d.; Pubchem, 2023; Teka et al., 2022).

7.4.1.2 Biological activity

7.4.1.2.1 Anti-inflammatory activity. Sung et al. (2009) evaluated the anti-inflammatory activity of vitisin A in RAW 264.7 macrophage cells. Based on the results, vitisin A inhibited NO production and iNOS expression preventing the NF-κB activation in LPS-induced macrophage cells. The best results were obtained at 10 μM concentration ($P < .05$ vs. LPS). Moreover, vitisin A inhibited the extracellular signal-regulated kinase 1/2 (ERK1/2) and p38 phosphorylation, involved in the production of inflammatory mediators (Sung et al., 2009).

Chang et al. (2017) examined the anti-inflammatory potential of five stilbenoids including vitisin A. Based on the results, vitisin A demonstrated moderate inhibitory activity with the best results obtained at 10 μM concentration (Chang et al., 2017).

7.4.1.2.2 Anticancer activity. Wang et al. (2019) studied the anticancer activity of vitisin B isolated from the seeds of *Iris lactea* against human breast cancer MDA-MB-231 cell line. Treatment with vitisin B (0.681 μM) resulted in 50% inhibition of FAS overall reaction activity. Moreover, the viability of MDA-MB-231 cells was reduced by vitisin B with IC_{50} values of 9.32 μM. Furthermore, the MDA-MB-231 cell apoptosis reached 54.96% after vitisin B (8 μg/mL) treatment (Wang et al., 2019).

Wu et al. (2013) investigated the anticancer potential of vitisin B isolated from *Vitis thunbergii* against human leukemia (HL-60) cells. Based on the results, vitisin B significantly inhibited the HL-60 cells proliferation by 52.4% and 73.2% at higher concentrations (12.5 and 25 μM, respectively). Moreover, vitisin B stimulated HL-60 cell apoptosis by the induction of sub-G1 population by 21.3% and 35.9% at 12.5 and 25 μM concentration, respectively. Furthermore, the caspase-3, -8, and -9 activity was increased by vitisin B (Wu et al., 2013).

7.4.1.2.3 Antidiabetic activity. Liu et al. (2023) studied the antidiabetic activity of vitisin A and B obtained from *Vitis thunbergii* in nicotinamide/streptozotocin (STZ)-induced type 2 diabetic mice (T2DM). Daily vitisin B treatment showed significant hypoglycemic activity associated with fasting blood glucose (BG) levels reduction from 263.16 mg/dL to 215.34 mg/dL after 21 days in T2DM mice (Liu et al., 2023).

7.4.1.2.4 Antiobesity activity. Yuan et al. (2023) evaluated the hypocholesterolemic activity of vitisin A in HepG2 cells. According to the results, vitisin A decreased intracellular cholesterol in HepG2 cells in dose-dependent manner. The best results were obtained with higher vitisin A concentration (200 μM) and resulted in 30% cholesterol decrease. Moreover, vitisin A downregulated the mRNA expression of HMGCR, an enzyme involved in the biosynthesis of cholesterol, compared with the high-cholesterol model group. Furthermore, the mRNA expression of SREBP2, PPARγ, and ABCA1, involved in cholesterol, lipid, and glucose metabolism, was increased with the best results at 200 μM vitisin A concentration (Yuan et al., 2023).

Kim et al. (2008) tested the antiobesity potential of vitisin A and B through the inhibiting the adipocyte differentiation. According to the results, vitisin A and B showed better adipocyte differentiation inhibitory activity compared with resveratrol with IC_{50} values of 5.0, 21.7, and 38.4 μM, respectively (Kim et al., 2008).

7.4.1.2.5 Neuroprotective activity. Chen et al. (2022) examined the neuroprotective activity of vitisin A and B in methylglyoxal (MGO)-induced cell deaths in SH-SY5Y cell model. Based on the results, both vitisin A and B showed significantly increased the MGO-induced cell viability. Moreover, vitisin A and B showed inhibitory activity against acetylcholinesterase (AChE) with IC_{50} values of 1.29 and 1.33 μM, respectively, allowing the increased availability of acetylcholine by improving cognitive function. Furthermore, the monoamine oxidase-B (MAO-B) activity was inhibited by both compounds (IC_{50} = 4.94 and 1.08 μM, respectively) (Chen et al., 2022).

7.4.1.2.6 Antifungal activity. Schnee et al. (2013) evaluated the antifungal activity of six compounds including *trans*-vitisin B against *P. viticola*. According to the results, *trans*-vitisin B inhibited zoospore mobility and sporulation of *P. viticola* with IC_{50} values of 13 and 12 μM, respectively (Schnee et al., 2013).

7.4.2 Hopeaphenol

7.4.2.1 General information

Hopeaphenol (Fig. 10.19) is a resveratrol tetramer, naturally presented in *Ampelopsis brevipedunculata, Dryobalanops lanceolata, Hopea hainanensis, Shorea uliginosa,* and *Vitis* species (Table 10.1) (Pubchem, 2023; Teka et al., 2022).

FIGURE 10.19

The chemical structure of hopeaphenol.

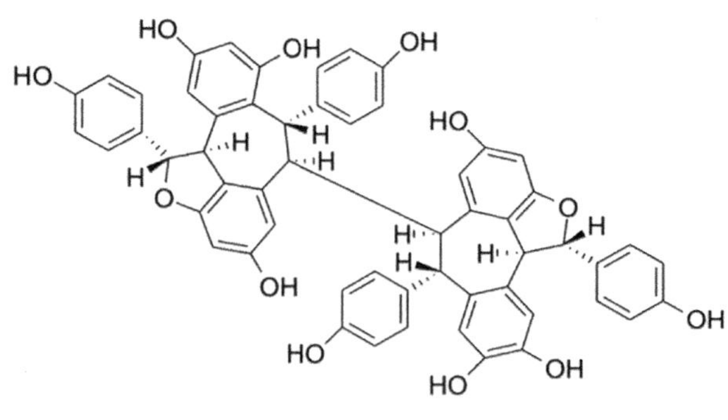

Hopeaphenol

7.4.2.2 Biological activity

7.4.2.2.1 Antioxidant activity. Biais et al. (2017) studied the antioxidant capacity of main constituents obtained from grapevine shoot extract including hopeaphenol using ORAC assay. The results showed that hopeaphenol demonstrated stronger antioxidant activity compared with isohopeaphenol and *trans*-resveratrol with IC_{50} values of 22.1, 24.8, 27.9 mmolTE/g, respectively.

Subramanian et al. (2015) evaluated the antioxidant activity of hopeaphenol using DPPH, ferric reducing power, hydrogen peroxide, and superoxide anion radical scavenging activity assays. Based on DPPH results, hopeaphenol showed antioxidant activity with 62.66% inhibition. In ferric reducing power, hopeaphenol showed significant reducing power activity ($IC_{50} = 47.55$ μg/mL). In addition, hopeaphenol (100 μg/mL) was also potent in hydrogen peroxide and superoxide anion radical scavenging activity with values of 54.64% and 64.48%, respectively (Subramanian et al., 2015).

7.4.2.2.2 Anti-inflammatory activity. Perez (2020) investigated the inhibitory activity of 20 stilbenoids including hopeaphenol on NO production and ROS formation in LPS-stimulated murine macrophages. Based on the results, hopeaphenol significantly suppressed the NO production with 91.8% inhibition at 10 μM concentration. Moreover, hopeaphenol decreased ROS generation by 60% with 10 μM concentration (Perez, 2020).

7.4.2.2.3 Cardioprotective activity. Su et al. (2015) studied the angiotensin I converting enzyme (ACE) inhibitory activity of 10 resveratrol derivatives including hopeaphenol from *Ampelopsis brevipedunculata*. Hopeaphenol (8 μM) showed strong inhibitory activity (98.32%) against ACE among all tested compounds, enabling the management of hypertension and heart failure (Su et al., 2015).

7.4.2.2.4 Anticancer activity. Moriyama et al. (2016) evaluated the antimelanoma activity of some resveratrol oligostilbenoids including hopeaphenol isolated from the bark of *Shorea roxburghii* against SK-MEL-28 melanoma cells. Based on the results, hopeaphenol showed significant antimelanoma activity compared with *trans*-resveratrol with IC_{50} values of 3.6 and 21.0 μM, respectively. Moreover, hopeaphenol reduced the cell population in S phase (during which the DNA is replicated before cell division/proliferation) (11.0%) (Moriyama et al., 2016).

7.4.2.2.5 Antifungal activity. Lambert et al. (2012) studied the content of eight stilbenoids including hopeaphenol in plants inoculated by *Diplodia seriata* and *Neofusicoccum parvum*. According to the results, the hopeaphenol content in *D. seriata* and *N. parvum* group was significantly higher compared with the control group (1887, 1252 and 1147 μg/g FW, respectively) (Lambert et al., 2012).

7.4.2.2.6 Neuroprotective activity. Chen et al. (2022) examined the inhibitory activity of hopeaphenol against AChE in MGO-induced cell death in SH-SY5Y cell model. Based on the results, hopeaphenol significantly inhibited AChE activity with IC_{50} values of 8.06 μM (Chen et al., 2022).

8. Conclusions

Stilbenoids are one of the most interesting and currently quite explored groups of polyphenol compounds. They occurred in more than 30 plant families and linked with a wide range of health-promoting activities. Most of them have proven strong antioxidant potential, but also cardioprotective, anticancer, and antidiabetic properties, which are highly valued in medicine (Table 10.6). Thus, they are the subject of increasing scientific interest and are being suggested as potential candidates for therapeutical drugs development.

Table 10.6 The summary of biological activity of stilbenoids.

Compound name	Biological activity	References
trans-Resveratrol	Antioxidant	(Ferrier et al., 2022; Gülçin, 2010; Hangun-Balkir and McKenney, 2012; Li et al., 2021; Lucas-Abellán et al., 2011; Rosiak et al., 2022)
	Antiaging	(Baca et al., 2016; Bastianetto et al., 2010; Subedi et al., 2017)
	Cardioprotective	(Kazemirad and Kazerani, 2020; Ting et al., 2022; Riba et al., 2017; Sebai et al., 2011)
	Anticancer	(Boocock et al., 2007; Chow et al., 2010; Espinoza et al., 2017; Holcombe et al., 2009; Zhu et al., 2012)
	Antidiabetic	(Szkudelska et al., 2021; Tomé-Carneiro et al., 2013)
	Antiobesity	(Alberdi et al., 2013; Gómez-Zorita et al., 2012, 2013; Timmers et al., 2011)
	Antifungal	(Adrian et al., 1997; Langcake, 1981; Langcake and Pryce, 1977; Pezet et al., 2004; Schnee et al., 2013)
Oxyresveratrol	Antioxidant	(Ashraf et al., 2015; Aziz et al., 2019; Jin et al., 2015; Kutil et al., 2015; Lorenz et al., 2003; Rodríguez-Bonilla et al., 2017)
	Tyrosinase inhibition	(Kang et al., 2011; Kim et al., 2002b, 2004)
	Anticancer	(Abdel Bar et al., 2020; Ferlinahayati et al., 2008; Yang et al., 2019)
	Antiviral	(Chuanasa et al., 2008; Likhitwitayawuid et al., 2005)
	Antibacterial	(Mazimba et al., 2018; Phoolcharoen et al., 2013; Sheng et al., 2015)
	Antifungal	(Basset et al., 2012; Kim and Lee, 2018)
trans-Piceatannol	Antioxidant	(Gerhäuser et al., 2003; Santos et al., n.d.)
	Anti-inflammatory	(Lee et al., 2022.; Yamamoto et al., 2017; Zhu et al., 2020)
	Antiacne	(Boutot et al., n.d.; Zhu et al., 2020)
	Anticancer	(Aljabali et al., 2020; Gerhäuser et al., 2003; Jayasooriya et al., 2012; Kim and Ma, 2019)

Table 10.6 The summary of biological activity of stilbenoids.—cont'd

Compound name	Biological activity	References
trans-Piceid	Antioxidant	(Jayalakshmi and Devika, 2019; Li et al., 2021)
	Anti-inflammatory	(Chen et al., 2021; Lanzilli et al., 2012; Matacchione et al., 2022)
	Cardioprotective	(Miao et al., 2011; Tan et al., 2020; Qi et al., 2015)
	Anticancer	(Chen et al., 2020; Jiao et al., 2018; Verma and Tiku, 2022)
	Antiobesity	(Wang et al., 2020; Yousef et al., 2021; Zheng et al., 2019)
trans-Pterostilbene	Antioxidant	(Acharya and Ghaskadbi, 2013; Rimando et al., 2002; Sun et al., 2021)
	Anti-inflammatory	(Liu et al., 2019; Rimando et al., 2002; Xu et al., 2021)
	Tyrosinase inhibition	(An et al., 2022; Basri et al., 2021; Majeed et al., 2020)
	Anticancer	(Rimando et al., 2002; Tam et al., 2018; Zhao et al., 2020)
	Antiobesity	(Gómez-Zorita et al., 2015; Hsu et al., 2012; Seo et al., 2017)
	Antifungal activity	(Adrian et al., 1997; Houillé et al., 2014; Pezet et al., 2004)
Ampelopsin A	Antioxidant	(Sáez et al., 2018)
	Anti-inflammatory	(Wang et al., 2011)
	Anticancer	(Ha et al., 2009; Lim et al., 2012; Saroyobudiono et al., 2008)
	Antiobesity	(Kim et al., 2014)
	Antifungal	(Lambert et al., 2012; Schnee et al., 2013)
	Neuroprotective	(Kim et al., 2012)
trans-ε-Viniferin	Antioxidant	(Biais et al., 2017; Privat et al., 2002)
	Anti-inflammatory	(Ha et al., 2018; Liao et al., 2021)
	Tyrosinase inhibition	(Kim et al., 2002a; Malinowska et al., 2020; Morel-Salmi et al., 2014)
	Anticancer	(Colin et al., 2008; Özdemir et al., 2018)
	Antifungal	(Lambert et al., 2012; Langcake and Pryce, 1977; Pezet et al., 2004; Schnee et al., 2013)
	Neuroprotective	(Freyssin et al., 2022; Zhang et al., 2020)

Continued

Table 10.6 The summary of biological activity of stilbenoids.—cont'd

Compound name	Biological activity	References
*trans-δ-*Viniferin	Antioxidant	(Ficarra et al., 2016; Sáez et al., 2018)
	Antiangiogenic	(Wu et al., 2020)
	Antidiabetic	(Mattio et al., 2019)
	Antifungal	(Pezet et al., 2004; Timperio et al., 2012)
Pallidol	Antioxidant	(Biais et al., 2017; Sáez et al., 2018)
	Anticancer	(Silva et al., 2012)
	Antiangiogenic	(Fernandez-Cruz et al., 2019)
	Antidiabetic	(Mattio et al., 2019)
trans-Scirpusin A	Antioxidant	(Morikawa et al., 2010; Sáez et al., 2018; Schmeda-Hirschmann et al., 1996; Utkina and Kulesh, 2012)
	Anticancer	(Hong et al., 2017)
trans-Scirpusin B	Antioxidant	(Morikawa et al., 2010; Schmeda-Hirschmann et al., 1996; Utkina and Kulesh, 2012)
	Cardioprotective	(Arraki et al., 2017)
trans-Miyabenol	Antioxidant	(De Marino et al., 2007; Ono et al., 1997)
	Anticancer	(Barjot et al., 2007; De Marino et al., 2007)
	Antifungal	(Lambert et al., 2012)
	Neuroprotective	(Hu et al., 2015)
cis-Miyabenol (*E*)-*trans*-*cis*-miyabenol C	Antioxidant	(De Marino et al., 2007; Ono et al., 1997; Sáez et al. 2018)
α-Viniferin	Antioxidant	(Jin et al., 2012; Lulan et al., 2020)
	Anti-inflammatory	(Chung et al., 2003; Dilshara et al., 2014)
	Skin-brightening	(Yun et al., 2018)
	Anticancer	(Chiou et al., 2022; Huang et al., 2023)
	Antidiabetic	(Lulan et al., 2020; Sasikumar et al., 2019)
	Antifungal	(Langcake and Pryce, 1977)
Vitisin A	Anti-inflammatory	(Chang et al., 2017; Sung et al., 2009)
	Antidiabetic	(Liu et al., 2023)
	Antiobesity	(Kim et al., 2008; Yuan et al., 2023)
	Neuroprotective	(Chen et al., 2022)
Vitisin B	Anticancer	(Wang et al., 2019; Wu et al., 2013)
	Antidiabetic	(Liu et al., 2023)
	Antiobesity	(Kim et al., 2008)
	Neuroprotective	(Chen et al., 2022)
	Antifungal	(Schnee et al., 2013)

Table 10.6 The summary of biological activity of stilbenoids.—cont'd

Compound name	Biological activity	References
Hopeaphenol	Antioxidant	(Biais et al., 2017; Subramanian et al., 2015)
	Anti-inflammatory	(Perez, 2020)
	Cardioprotective	(Su et al., 2015)
	Antifungal	(Lambert et al., 2012)

Some of them, e.g., oxyresveratrol or pterostilbene, are also acknowledged with skin-whitening potential and could be widely applied in the cosmetic industry. *Trans*-resveratrol is the dominant stilbenoid included in cosmetics formulations. There are 11 forms of resveratrol listed in CosIng database (Table 10.4). Most of them possess antioxidant and skin-conditioning activity highly appreciated in cosmetics. Moreover, *trans*-resveratrol is present in the wide range of dietary supplements (solely or in a combination with pterostilbene), where it exhibits antioxidant and cardioprotective activities (Table 10.3).

The safety of stilbenoids is not much examined; however, there are some reports that approve the safe application of *trans*-resveratrol. For instance, FDA approved some of the supplements, which contain *trans*-resveratrol (Table 10.2). In addition, in 2016, EFSA approved the synthetic *trans*-resveratrol as a "novel food". However, the safety of other stilbenoids is still needed further examination.

To sum up, stilbenoids are potent phenolic compounds, which are highly appreciated in nutraceutical, pharmaceutical, and cosmetic industries. Due to their valuable, multidirectional potential, confirmed by scientific research, it is important to expand knowledge about their safety and versatility of applications.

Acknowledgments

The research was funded by the project PHC Polonium 2022 (Project N° 49223SG), joint research projects between France and Poland, entitled "Cosmetic potential of viticulture byproducts as novel functional ingredients for skin barrier recovery." Authors gratefully acknowledge the Région Center Val de Loire for funding the project INNOCOSM from ARD Cosmetosciences.

References

Abdel Bar, F.M., Abbas, G.M., Gohar, A.A., Lahloub, M.F.I., 2020. Antiproliferative activity of stilbene derivatives and other constituents from the stem bark of Morus nigra L. Nat. Prod. Res. 34 (24), 3506–3513. https://doi.org/10.1080/14786419.2019.1573236.

Acharya, J.D., Ghaskadbi, S.S., 2013. Protective effect of Pterostilbene against free radical mediated oxidative damage. BMC Compl. Alternative Med. 13. https://doi.org/10.1186/1472-6882-13-238.

Adrian, M., Jeandet, P., Veneau, J., Weston, L.A., Bessis, R., 1997. Biological activity of resveratrol, a stilbenic compound from grapevines, against Botrytis cinerea, the causal agent for gray mold. J. Chem. Ecol. 23 (7), 1689–1702. https://doi.org/10.1023/B:JOEC.0000006444.79951.75.

Ahmadi, R., Ali Ebrahimzadeh, M., 2020. Resveratrol – a comprehensive review of recent advances in anticancer drug design and development. Eur. J. Med. Chem. 200, 112356. https://doi.org/10.1016/j.ejmech.2020.112356.

Akinwumi, B.C., Bordun, K.A.M., Anderson, H.D., 2018. Biological activities of stilbenoids. Int. J. Mol. Sci. 19 (3). https://doi.org/10.3390/ijms19030792.

Alberdi, G., Rodríguez, V.M., Miranda, J., Teresa Macarulla, M., Churruca, I., Portillo, M.P., 2013. Thermogenesis is involved in the body-fat lowering effects of resveratrol in rats. Food Chem. 141 (2), 1530–1535. https://doi.org/10.1016/j.foodchem.2013.03.085.

Aljabali, A.A.A., Bakshi, H.A., Hakkim, F.L., Haggag, Y.A., Al-Batanyeh, K.M., Al Zoubi, M.S., Al-Trad, B., Nasef, M.M., Satija, S., Mehta, M., Pabreja, K., Mishra, V., Khan, M., Abobaker, S., Azzouz, I.M., Dureja, H., Pabari, R.M., Dardouri, A.A.K., Kesharwani, P., Gupta, G., Shukla, S.D., Prasher, P., Charbe, N.B., Negi, P., Kapoor, D.N., Kumar Chellappan, D., Silva, M.W.da, Thompson, P., Kamal, D., Paul, M.C., Tambuwala, M.M., Aljabali, A.A.A., et al., Correction, 2020. Albumin nano-encapsulation of piceatannol enhances its anticancer potential in colon cancer via down regulation of nuclear p65 and HIF-1α. Cancers 2020, 12, 113. Cancers 12 (12), 3587. https://doi.org/10.3390/cancers12123587.

An, X., Lv, J., Wang, F., 2022. Pterostilbene inhibits melanogenesis, melanocyte dendricity and melanosome transport through cAMP/PKA/CREB pathway. Eur. J. Pharmacol. 932, 175231. https://doi.org/10.1016/j.ejphar.2022.175231.

Arraki, K., Totoson, P., Decendit, A., Badoc, A., Zedet, A., Jolibois, J., Pudlo, M., Demougeot, C., Girard-Thernier, C., 2017. Cyperaceae species are potential sources of natural mammalian arginase inhibitors with positive effects on vascular function. J. Nat. Prod. (Lloydia) 80 (9), 2432–2438. https://doi.org/10.1021/acs.jnatprod.7b00197.

Ashraf, M.I., Shahzad, M., Shabbir, A., 2015. Oxyresveratrol ameliorates allergic airway inflammation via attenuation of IL-4, IL-5, and IL-13 expression levels. Cytokine 76 (2), 375–381. https://doi.org/10.1016/j.cyto.2015.09.013.

Aziz, R.S., Siddiqua, A., Shahzad, M., Shabbir, A., Naseem, N., 2019. Oxyresveratrol ameliorates ethanol-induced gastric ulcer via downregulation of IL-6, TNF-α, NF-κB, and COX-2 levels, and upregulation of TFF-2 levels. Biomed. Pharmacother. 110, 554–560. https://doi.org/10.1016/j.biopha.2018.12.002.

Baca, M.E., Rapp, D.A., Gm Knoll, R.D., 2016. Ford, abstract:evaluation of the potential for improved wound healing through the usage of a topical resveratrol preparation. Plast Reconstr. Surg. Glob. Open. https://doi.org/10.1097/01.GOX.00005030522016.

Banez, M.J., Geluz, M.I., Chandra, A., Hamdan, T., Biswas, O.S., Bryan, N.S., Von Schwarz, E.R., 2020. A systemic review on the antioxidant and anti-inflammatory effects of resveratrol, curcumin, and dietary nitric oxide supplementation on human cardiovascular health. Nutr. Res. 78, 11–26. https://doi.org/10.1016/j.nutres.2020.03.002.

Bansal, M., Singh, N., Pal, S., Dev, I., Ansari, K.M., 2018. Chemopreventive role of dietary phytochemicals in colorectal cancer. Adv. Mol. Toxicol. 2, 69–121. https://doi.org/10.1016/B978-0-444-64199-1.00004-X.

Barjot, C., Tournaire, M., Castagnino, C., Vigor, C., Vercauteren, J., Rossi, J.F., 2007. Evaluation of antitumor effects of two vine stalk oligomers of resveratrol on a panel of lymphoid and myeloid cell lines: comparison with resveratrol. Life Sci. 81 (23–24), 1565–1574. https://doi.org/10.1016/j.lfs.2007.08.047.

Basri, D.F., Lew, L.C., Muralitharan, R.V., Nagapan, T.S., Ghazali, A.R., 2021. Pterostilbene inhibits the melanogenesis activity in UVB-irradiated B164A5 cells. Dose Res. 19 (4). https://doi.org/10.1177/15593258211047651.

Basset, C., Rodrigues, A.M.S., Eparvier, V., Silva, M.R.R., Lopes, N.P., Sabatier, D., Fonty, E., Espindola, L.S., Stien, D., 2012. Secondary metabolites from Spirotropis longifolia (DC) Baill and their antifungal activity against human pathogenic fungi. Phytochemistry 74, 166–172. https://doi.org/10.1016/j.phytochem.2011.10.011.

Bastianetto, S., Dumont, Y., Duranton, A., Vercauteren, F., Breton, L., Quirion, R., Deli, M.A., 2010. Protective action of resveratrol in human skin: possible involvement of specific receptor binding sites. PLoS One 5 (9), e12935. https://doi.org/10.1371/journal.pone.0012935.

Beaumont, P., Courtois, A., Atgié, C., Richard, T., Krisa, S., 2022. In the shadow of resveratrol: biological activities of epsilon-viniferin. J. Physiol. Biochem. 78 (2), 465–484. https://doi.org/10.1007/s13105-022-00880-x.

Bharadvaja, N., Gautam, S., Singh, H., 2023. Natural polyphenols: a promising bioactive compounds for skin care and cosmetics. Mol. Biol. Rep. 50 (2), 1817–1828. https://doi.org/10.1007/s11033-022-08156-9.

Biais, B., Krisa, S., Cluzet, S., Da Costa, G., Waffo-Teguo, P., Mérillon, J.-M., Richard, T., 2017. Antioxidant and cytoprotective activities of grapevine stilbenes. J. Agric. Food Chem. 65 (24), 4952–4960. https://doi.org/10.1021/acs.jafc.7b01254.

Billet, K., Malinowska, M.A., Munsch, T., Unlubayir, M., Dugé De Bernonville, T., Besseau, S., Courdavault, V., Oudin, A., Pichon, O., Clastre, M., Giglioli-Guivarc'h, N., Lanoue, A., 2020. Stilbenoid-Enriched Grape Cane Extracts for the Biocontrol of Grapevine Diseases. Springer Science and Business Media LLC, pp. 215–239.

Boocock, D.J., Faust, G.E.S., Patel, K.R., Schinas, A.M., Brown, V.A., Ducharme, M.P., Booth, T.D., Crowell, J.A., Perloff, M., Gescher, A.J., Steward, W.P., Brenner, D.E., 2007. Phase I dose escalation pharmacokinetic study in healthy volunteers of resveratrol, a potential cancer chemopreventive agent. Cancer Epidemiol. Biomarkers Prev. 16 (6), 1246–1252. https://doi.org/10.1158/1055-9965.EPI-07-0022.

Boutot, C., Gervason, S., Dreux-Zigha, A., Cabannes, M., Berthon, J.Y., Metton, F.E. Natural Anti-acne Extract Aids Healthy Skin Microbiota.

Chan, E.W.C., Wong, C.W., Tan, Y.H., Foo, J.P.Y., Wong, S.K., Chan, H.T., 2019. Resveratrol and pterostilbene: a comparative overview of their chemistry, biosynthesis, plant sources and pharmacological properties. J. Appl. Pharmaceut. Sci. 9 (7), 124–129. https://doi.org/10.7324/JAPS.2019.90717.

Chang, C.-I., Chien, W.-C., Huang, K.-X., Hsu, J.-L., 2017. Anti-inflammatory effects of vitisinol A and four other oligostilbenes from Ampelopsis brevipedunculata var. Hancei. Molecules 22 (7), 1195. https://doi.org/10.3390/molecules22071195.

Chen, Y., Niu, J., Li, L., Li, Z., Jiang, J., Zhu, M., Dong, T., Zhang, J., Shi, C., Xu, P., Lu, Y., Jiang, Y., Liu, P., Chen, W., 2020. Polydatin executes anticancer effects against glioblastoma multiforme by inhibiting the EGFR-AKT/ERK1/2/STAT3-SOX2/Snail signaling pathway. Life Sci. 258, 118158. https://doi.org/10.1016/j.lfs.2020.118158.

Chen, L.-G., Wang, C.-C., Lee, Y.-S., Sie, Y.-Y., Chang, C.-I., Hou, W.-C., Vitisin, a, 2022. Resveratrol tetramer, improves scopolamine-induced impaired learning and memory functions in amnesiac ICR mice. Biomedicines 2.

Chen, G., Yang, Z., Wen, D., Guo, J., Xiong, Q., Li, P., Zhao, L., Wang, J., Wu, C., Dong, L., 2021. Polydatin has anti inflammatory and antioxidant effects in LPS induced macrophages and improves DSS induced mice colitis. Immun. Inflamm. Dis. 9 (3), 959–970. https://doi.org/10.1002/iid3.455.

Chiou, W.C., Huang, C., Lin, Z.J., Hong, L.S., Lai, Y.H., Chen, J.C., Huang, H.C., 2022. α-Viniferin and ε-viniferin inhibited TGF-β1-induced epithelial-mesenchymal transition, migration and invasion in lung cancer cells through downregulation of vimentin expression. MDPI, Taiwan Nutrients 14 (11). https://doi.org/10.3390/nu14112294.

Chong, J., Poutaraud, A., Hugueney, P., 2009. Metabolism and roles of stilbenes in plants. Plant Sci. 177 (3), 143–155. https://doi.org/10.1016/j.plantsci.2009.05.012.

Chow, H.H.S., Garland, L.L., Hsu, C.H., Vining, D.R., Chew, W.M., Miller, J.A., Perloff, M., Crowell, J.A., Alberts, D.S., 2010. Resveratrol modulates drug- and carcinogen-metabolizing enzymes in a healthy volunteer study. Cancer Prev. Res. 3 (9), 1168–1175. https://doi.org/10.1158/1940-6207.CAPR-09-0155.

Chuanasa, T., Phromjai, J., Lipipun, V., Likhitwitayawuid, K., Suzuki, M., Pramyothin, P., Hattori, M., Shiraki, K., 2008. Anti-herpes simplex virus (HSV-1) activity of oxyresveratrol derived from Thai medicinal plant: mechanism of action and therapeutic efficacy on cutaneous HSV-1 infection in mice. Antivir. Res. 80 (1), 62–70. https://doi.org/10.1016/j.antiviral.2008.05.002.

Chung, E.Y., Kim, B.H., Lee, M.K., Yun, Y.P., Lee, S.H., Min, K.R., Kim, Y., 2003. Anti-inflammatory effect of the oligomeric stilbene α-viniferin and its mode of the action through inhibition of cyclooxygenase-2 and inducible nitric oxide synthase. Planta Med. 69 (8), 710–714. https://doi.org/10.1055/s-2003-42787.

Colin, D., Lancon, A., Delmas, D., Lizard, G., Abrossinow, J., Kahn, E., Jannin, B., Latruffe, N., 2008. Antiproliferative activities of resveratrol and related compounds in human hepatocyte derived HepG2 cells are associated with biochemical cell disturbance revealed by fluorescence analyses. Biochimie 90 (11–12), 1674–1684. https://doi.org/10.1016/j.biochi.2008.06.006.

Dávid, C.Z., Hohmann, J., Vasas, A., 2021. Chemistry and pharmacology of Cyperaceae stilbenoids: a review. Molecules 9. https://doi.org/10.3390/molecules26092794.

De Marino, S., Gala, F., Borbone, N., Zollo, F., Vitalini, S., Visioli, F., Iorizzi, M., 2007. Phenolic glycosides from Foeniculum vulgare fruit and evaluation of antioxidative activity. Phytochemistry 68 (13), 1805–1812. https://doi.org/10.1016/j.phytochem.2007.03.029.

Dilshara, M.G., Lee, K.T., Kim, H.J., Lee, H.J., Choi, Y.H., Lee, C.M., Kim, L.K., Kim, G.Y., 2014. Anti-inflammatory mechanism of α-viniferin regulates lipopolysaccharide-induced release of proinflammatory mediators in BV2 microglial cells. Cell. Immunol. 290 (1), 21–29. https://doi.org/10.1016/j.cellimm.2014.04.009.

Du, Q.-H., Peng, C., Zhang, H., 2013. Polydatin: a review of pharmacology and pharmacokinetics. Pharmaceut. Biol. 51 (11), 1347–1354. https://doi.org/10.3109/13880209.2013.792849.

Duarte, Ramalhete, C., Rijo, Reis, M.A., Ferreira, M.-J.U., 2021. Handbook of Dietary Phytochemicals. Springer, pp. 1005–1029.

Espinoza, J.L., Trung, L.Q., Inaoka, P.T., Yamada, K., An, D.T., Mizuno, S., Nakao, S., Takami, A., 2017. The repeated administration of resveratrol has measurable effects on circulating T-cell subsets in humans. Oxid. Med. Cell. Longev. 2017. https://doi.org/10.1155/2017/6781872.

www.ec.europa.eu. (2023). (Accessed 14 2023).

European Medicines Agency (EMA), 2023. Available online.

Ferrier, M., Billet, K., Drouet, S., Tungmunnithum, D., Malinowska, M.A., Marchal, C., Dedet, S., Giglioli-Guivarc'h, N., Hano, C., Lanoue, A., 2022. Identifying major drivers of antioxidant activities in complex polyphenol mixtures from grape canes. Molecules 13, 1–13. https://doi.org/10.3390/molecules27134029.

Ferlinahayati, Syah, Y.M., Juliawaty, L.D., Achmad, S.A., Hakim, E.H., Takayama, H., Said, I.M., Latip, J., 2008. Phenolic constituents from the wood of Morus australis with cytotoxic activity . Verlag der Zeitschrift fur Naturforschung. Indonesia Zeitschrift fur Naturforschung - Section C J. Biosci. 63 (1–2), 35–39. https://doi.org/10.1515/znc-2008-1-207.

Fernandez-Cruz, E., Cerezo, A.B., Cantos-Villar, E., Richard, T., Troncoso, A.M., Garcia-Parrilla, M.C., 2019. Inhibition of VEGFR-2 phosphorylation and effects on downstream signaling pathways in cultivated human endothelial cells by stilbenes from vitis spp. J. Agric. Food Chem. 67 (14), 3909–3918. https://doi.org/10.1021/acs.jafc.9b00282.

Ferrier, M., Gémin, M.P., Malinowska, M.A., Abdallah, C., Magot, F., Birer-Williams, C., Giglioli-Guivarc'h, N., Hano, C., Lanoue, A., 2024. Metabolomics applications in natural cosmetics: addressing the new challenges of bio-sourced ingredients. Phytochem. Rev. https://doi.org/10.1007/s11101-024-10048-8.

Ficarra, S., Tellone, E., Pirolli, D., Russo, A., Barreca, D., Galtieri, A., Giardina, B., Gavezzotti, P., Riva, S., De Rosa, M.C., 2016. Insights into the properties of the two enantiomers of trans-δ-viniferin, a resveratrol derivative: antioxidant activity, biochemical and molecular modeling studies of its interactions with hemoglobin. Mol. Biosyst. 12 (4), 1276–1286. https://doi.org/10.1039/c5mb00897b.

Food and Drug Administration (FDA), 2023.

Freyssin, A., Bilan, A.R., Fauconneau, B., Galineau, L., Serrière, S., Tauber, C., Perrin, F., Guillard, J., Chalon, S., Page, G., 2022. Trans ε-viniferin decreases amyloid deposits with greater efficiency than resveratrol in an Alzheimer's mouse model. Front. Neurosci. 15. https://doi.org/10.3389/fnins.2021.803927.

Fuloria, S., Khattulanuar, N.S., Rani, N.N.I.M., Ravi, S., Subramaniyan, S J., Begum, M.P., Chidambaram, K.V.S., Safi, S.Z., Wu, Y.S., Nordin, M.N.H.M., Kumarasamy, V., Fuloria, N.K. Biosynthesis and pharmacology of viniferin: potential resveratrol-derived molecules for new drug discovery.

Galal, A.M., Avula, B., Khan, I.A., 2015. Utility of Thin-Layer Chromatography in the Assessment of the Quality of Botanicals Instrumental Thin-Layer Chromatography. Elsevier Inc, United States, pp. 479–504. https://doi.org/10.1016/B978-0-12-417223-4.00018-2. Available from: http://www.sciencedirect.com/science/book/9780124172234.

Galiniak, S., Aebisher, D., Bartusik-Aebisher, D., 2019. Health benefits of resveratrol administration. Acta Biochimica Polonica, Poland Acta Biochimica Polonica 66 (1), 13–21. https://doi.org/10.18388/abp.2018_2749.

Gerhäuser, C., Klimo, K., Heiss, E., Neumann, I., Gamal-Eldeen, A., Knauft, J., Liu, G.Y., Sitthimonchai, S., Frank, N., 2003. 1 2003/01 Mutation Research - Fundamental and Molecular Mechanisms of Mutagenesis 10.1016/S0027-5107(02)00332-9 00275107 163 172 Elsevier Germany Mechanism-based in vitro screening of potential cancer chemopreventive agents. http://www.journals.elsevier.com/mutation-research-fundamental-and-molecular-mechanisms-of-mutagenesis/, 523-524.

Gómez-Zorita, S., Fernández-Quintela, A., MacArulla, M.T., Aguirre, L., Hijona, E., Bujanda, L., Milagro, F., Martínez, J.A., Portillo, M.P., 2012. Resveratrol attenuates steatosis in obese Zucker rats by decreasing fatty acid availability and reducing oxidative stress. Br. J. Nutr. 107 (2), 202–210. https://doi.org/10.1017/S0007114511002753.

Gómez-Zorita, S., Fernández-Quintela, A., Lasa, A., Hijona, E., Bujanda, L., Portillo, M.P., 2013. Effects of resveratrol on obesity-related inflammation markers in adipose tissue of genetically obese rats. Nutrition 29 (11–12), 1374–1380. https://doi.org/10.1016/j.nut.2013.04.014.

Gómez-Zorita, S., Fernández-Quintela, A., Aguirre, L., MacArulla, M.T., Rimando, A.M., Portillo, M.P., 2015. Pterostilbene improves glycaemic control in rats fed an obesogenic diet: involvement of skeletal muscle and liver. Food Funct. 6 (6), 1968–1976. https://doi.org/10.1039/c5fo00151j.

Gómez-Zorita, S., Milton-Laskibar, I., Eseberri, I., Beaumont, P., Courtois, A., Krisa, S., Portillo, M.P., 2023. Beneficial effects of ε-viniferin on obesity and related health alterations. Nutrients 15 (4), 928. https://doi.org/10.3390/nu15040928.

Gülçin, İ., 2010. Antioxidant properties of resveratrol: a structure–activity insight. Innovat. Food Sci. Emerg. Technol. 11 (1), 210–218. https://doi.org/10.1016/j.ifset.2009.07.002.

Ha, D.T., Chen, Q.C., Hung, T.M., Youn, U.J., Ngoc, T.M., Thuong, P.T., Kim, H.J., Seong, Y.H., Min, B.S., Bae, K., 2009. Stilbenes and oligostilbenes from leaf and stem of Vitis amurensis and their cytotoxic activity. Arch Pharm. Res. (Seoul) 32 (2), 177–183. https://doi.org/10.1007/s12272-009-1132-2.

Ha, D.T., Long, P.T., Hien, T.T., Tuan, D.T., An, N.T.T., Khoi, N.M., Oanh, H., Hung, T.M., 2018. Anti-inflammatory effect of oligostilbenoids from Vitis heyneana in LPS-stimulated RAW 264.7 macrophages via suppressing the NF-KB activation. Chem. Cent. J. 12 (1). https://doi.org/10.1186/s13065-018-0386-5.

Hangun-Balkir, Y., McKenney, M.L., 2012. Determination of antioxidant activities of berries and resveratrol. Green Chem. Lett. Rev. 5 (2), 147–153. https://doi.org/10.1080/17518253.2011.603756.

He, S., Jiang, L., Wu, B., Pan, Y., Sun, C., 2009. Pallidol, a resveratrol dimer from red wine, is a selective singlet oxygen quencher. Biochem. Biophys. Res. Commun. 379 (2), 283–287. https://doi.org/10.1016/j.bbrc.2008.12.039.

Holcombe, R.F., Nguyen, A., Martinez, Stamos, M.J., Pat Moyer, M., Hope, K.P., Holcombe, R.F., 2009. Results of a phase I pilot clinical trial examining the effect of plant-derived resveratrol and grape powder on Wnt pathway target gene expression in colonic mucosa and colon cancer. Cancer Manag. Res. 1, 25–37. https://doi.org/10.2147/cmar.s4544.

Hong, E.H., Heo, E.Y., Song, J.H., Kwon, B.E., Lee, J.Y., Park, Y., Kim, J., Chang, S.Y., Chin, Y.W., Jeon, S.M., Ko, H.J., 2017. Trans-scirpusin A showed antitumor effects via autophagy activation and apoptosis induction of colorectal cancer cells. Oncotarget 8 (25), 41401–41411. https://doi.org/10.18632/oncotarget.17388.

Houillé, B., Papon, N., Boudesocque, L., Bourdeaud, E., Besseau, S., Courdavault, V., Enguehard-Gueiffier, C., Delanoue, G., Guérin, L., Bouchara, J.P., Clastre, M., Giglioli-Guivarc'H, N., Guillard, J., Lanoue, A., 2014. Antifungal activity of resveratrol derivatives against Candida Species. J. Nat. Prod. (Lloydia) 77 (7), 1658–1662. https://doi.org/10.1021/np5002576.

Hsu, C.-L., Lin, Y.-J., Ho, C.-T., Yen, G.-C., 2012. Inhibitory effects of garcinol and pterostilbene on cell proliferation and adipogenesis in 3T3-L1 cells. Food Funct. 3 (1), 49–57. https://doi.org/10.1039/c1fo10209e.

Hu, J., Lin, T., Gao, Y., Xu, J., Jiang, C., Wang, G., Bu, G., Xu, H., Chen, H., Zhang, Y.-wu, Ohno, M., 2015. The resveratrol trimer miyabenol C inhibits β-secretase activity and β-amyloid generation. PLoS One 10 (1), e0115973. https://doi.org/10.1371/journal.pone.0115973.

Huang, C., Lin, Z.J., Chen, J.C., Zheng, H.J., Lai, Y.H., Huang, H.C., 2023. α-Viniferin-Induced apoptosis through downregulation of SIRT1 in non-small cell lung cancer cells. Pharmaceuticals 16 (5). https://doi.org/10.3390/ph16050727.

Ifuku, O., 2017. Botanical Ingredients Cosmetic Science and Technology: Theoretical Principles and Applications. Elsevier Inc, Japan, pp. 305–320. https://doi.org/10.1016/B978-0-12-802005-0.00020-3.

Javed, A., Umar, H.M.I., Khan, F., Niaz, K., 2020. Recent Advances in Natural Products Analysis. Elsevier, pp. 119–129.

Jayalakshmi, P., Devika, P.T., 2019. Assessment of in vitro antioxidant activity study of polydatin. J. Pharmacogn. Phytochem. 8 (4), 55–58.

Jayasooriya, R.G.P.T., Lee, Y.G., Kang, C.H., Lee, K.T., Choi, Y.H., Park, S.Y., Hwang, J.K., Kim, G.Y., 2012. Piceatannol inhibits MMP-9-dependent invasion of tumor necrosis factor-α-stimulated DU145 cells by suppressing the Akt-mediated nuclear factor-κB pathway. Oncol. Lett. 5 (1), 341–347. https://doi.org/10.3892/ol.2012.968.

Jadhav, P., Bothiraja, C., Pawar, A., 2016. Resveratrol-piperine loaded mixed micelles: formulation, characterization, bioavailability, safety and in vitro anticancer activity. RSC Adv. 6 (114), 112795–112805. https://doi.org/10.1039/C6RA24595A.

Jiao, Y., Wu, Y., Du, D., 2018. Polydatin inhibits cell proliferation, invasion and migration, and induces cell apoptosis in hepatocellular carcinoma. Braz. J. Med. Biol. Res. 51 (4). https://doi.org/10.1590/1414-431x20176867.

Jin, Q., Han, X.H., Hong, S.S., Lee, C., Choe, S., Lee, D., Kim, Y., Hong, J.T., Lee, M.K., Hwang, B.Y., 2012. Antioxidative oligostilbenes from Caragana sinica. Bioorg. Med. Chem. Lett 22 (2), 973–976. https://doi.org/10.1016/j.bmcl.2011.12.012.

Jin, Y.J., Lin, C.C., Lu, T.M., Li, J.H., Chen, I.S., Kuo, Y.H., Ko, H.H., 2015. Chemical constituents derived from Artocarpus xanthocarpus as inhibitors of melanin biosynthesis. Phytochemistry 117, 424–435. https://doi.org/10.1016/j.phytochem.2015.07.003.

Kang, K.B., Du Kim, S., Kim, T.B., Jeong, E.J., Kim, Y.C., Sung, J.H., Sung, S.H., 2011. Tyrosinase inhibitory constituents of Morus bombycis cortex. Nat. Prod. Sci. 17 (3), 198–201.

Karami, A., Fakhri, S., Kooshki, L., Khan, H., 2022. Polydatin: pharmacological mechanisms, therapeutic targets, biological activities, and health benefits. Molecules 27 (19), 6474. https://doi.org/10.3390/molecules27196474.

Kataria, R., Khatkar, A., 2019. Resveratrol in various pockets: a review. Curr. Top. Med. Chem. 19 (2), 116–122. https://doi.org/10.2174/1568026619666190301173958.

Kazemirad, H., Kazerani, H.R., 2020. Cardioprotective effects of resveratrol following myocardial ischemia and reperfusion. Mol. Biol. Rep. 47 (8), 5843–5850. https://doi.org/10.1007/s11033-020-05653-7.

Kim, S., Lee, D.G., 2018. Oxyresveratrol-induced DNA cleavage triggers apoptotic response in Candida albicans. Microbiology (United Kingdom) 164 (9), 1112–1121. https://doi.org/10.1099/mic.0.000696.

Kim, A., Ma, J.Y., 2019. Piceatannol-3-O-β-D-glucopyranoside (PG) exhibits in vitro anti-metastatic and anti-angiogenic activities in HT1080 malignant fibrosarcoma cells. Phytomedicine 57, 95–104. https://doi.org/10.1016/j.phymed.2018.12.017.

Kim, H.-J., Ha, S.-C., Park, S.-W., 2002a. Inhibition of tyrosinase and lipoxygenase activities by resveratrol and its derivatives from seeds of Paeonia lactiflora. Prevent. Nutrit. Food Sci. 7 (4), 447–450. https://doi.org/10.3746/jfn.2002.7.4.447.

Kim, Y.M., Yun, J., Lee, C.K., Lee, H., Min, K.R., Kim, Y., 2002b. Oxyresveratrol and hydroxystilbene compounds. Inhibitory effect on tyrosinase and mechanism of action. J. Biol. Chem. 277 (18), 16340–16344. https://doi.org/10.1074/jbc.M200678200.

Kim, D.H., Kim, J.H., Baek, S.H., Seo, J.H., Kho, Y.H., Oh, T.K., Lee, C.H., 2004. Enhancement of tyrosinase inhibition of the extract of Veratrum patulum using cellulase. Biotechnol. Bioeng. 87 (7), 849–854. https://doi.org/10.1002/bit.20189.

Kim, S.h., Park, H.S., Lee, M.s., Cho, Y.J., Kim, Y.S., Hwang, J.T., Sung, M.J., Kim, M.S., Kwon, D.Y., 2008. Vitisin A inhibits adipocyte differentiation through cell cycle arrest in 3T3-L1 cells. Biochem. Biophys. Res. Commun. 372 (1), 108–113. https://doi.org/10.1016/j.bbrc.2008.04.188.

Kim, J.Y., Jeong, H.Y., Lee, H.K., Kim, S., Hwang, B.Y., Bae, K., Seong, Y.H., 2012. Neuroprotection of the leaf and stem of Vitis amurensis and their active compounds against ischemic brain damage in rats and excitotoxicity in cultured neurons. Phytomedicine 19 (2), 150–159. https://doi.org/10.1016/j.phymed.2011.06.015.

Kim, Y.M., Lee, E.W., Eom, S.H., Kim, T.H., 2014. Pancreatic lipase inhibitory stilbenoids from the roots of Vitis vinifera. Int. J. Food Sci. Nutr. 65 (1), 97–100. https://doi.org/10.3109/09637486.2013.832172.

Kim, H., Seo, K.-H., Yokoyama, W., 2020. Chemistry of pterostilbene and its metabolic effects. J. Agric. Food Chem. 68 (46), 12836–12841. https://doi.org/10.1021/acs.jafc.0c00070.

Krambeck, K., Santos, D., Sousa Lobo, J.M., Amaral, M.H., 2023. Benefits of skin application of piceatannol—a minireview. Australas. J. Dermatol. 64 (1), e21. https://doi.org/10.1111/ajd.13937.

Kutil, Z., Kvasnicova, M., Temml, V., Schuster, D., Marsik, P., Cusimamani, E.F., Lou, J.D., Vanek, T., Landa, P., 2015. Effect of dietary stilbenes on 5-lipoxygenase and cyclooxygenases activities in vitro. Int. J. Food Prop. 18 (7), 1471–1477. https://doi.org/10.1080/10942912.2014.903416.

Lambert, C., Bisson, J., Waffo-Téguo, P., Papastamoulis, Y., Richard, T., France Corio-Costet, M.-, Mérillon, J.-M., Cluzet, S., 2012. Phenolics and their antifungal role in grapevine wood decay: focus on the botryosphaeriaceae family. J. Agric. Food Chem. 60 (48), 11859–11868. https://doi.org/10.1021/jf303290g.

Langcake, P., 1981. Disease resistance of Vitis spp. and the production of the stress metabolites resveratrol, ε-viniferin, α-viniferin and pterostilbene. Physiol. Plant Pathol. 18 (2), 213–226. https://doi.org/10.1016/s0048-4059(81)80043-4.

Langcake, P., Pryce, R.J., 1977. A new class of phytoalexins from grapevines. Experientia 33 (2), 151–152. https://doi.org/10.1007/bf02124034.

Lanzilli, G., Cottarelli, A., Nicotera, G., Guida, S., Ravagnan, G., Fuggetta, M.P., 2012. Anti-infl ammatory effect of resveratrol and polydatin by in vitro IL-17 modulation. Inflammation (1), 240–248.

Lee, C.H., Yang, H., Park, J.H.Y., Kim, J.-E., Lee, K.W., 2022. Piceatannol, a metabolite of resveratrol, attenuates atopic dermatitis by targeting Janus kinase 1. Phytomedicine. https://doi.org/10.1016/j.phymed.2022.153981.

Li, J., Zhang, C.X., Liu, Y.M., Chen, K.L., Chen, G., 2017. A comparative study of anti-aging properties and mechanism: resveratrol and caloric restriction. Oncotarget 8 (39), 65717–65729. https://doi.org/10.18632/oncotarget.20084.

Li, Z., Chen, X., Liu, G., Li, J., Zhang, J., Cao, Y., Miao, J., 2021. Antioxidant activity and mechanism of resveratrol and polydatin isolated from mulberry (Morus alba L.). Molecules 26 (24), 7574. https://doi.org/10.3390/molecules26247574.

Liao, H.-R., Lin, C.-H., Chen, J.-J., Liu, F.-C., Tseng, C.-P., 2021. The anti-inflammatory effect of ε-viniferin by specifically targeting formyl peptide receptor 1 on human neutrophils. Chem. Biol. Interact. 259. https://doi.org/10.1016/j.cbi.2021.109490.

Likhitwitayawuid, K., 2021. Oxyresveratrol: sources, productions, biological activities, pharmacokinetics, and delivery systems. Molecules 26 (14), 4212. https://doi.org/10.3390/molecules26144212.

Likhitwitayawuid, K., Sritularak, B., Benchanak, K., Lipipun, V., Mathew, J., Schinazi, R.F., 2005. Phenolics with antiviral activity from Millettia erythrocalyx and Artocarpus lakoocha. Nat. Prod. Res. 19 (2), 177–182. https://doi.org/10.1080/14786410410001704813.

Likhtenshtein, G., 2009. Stilbenes: applications in chemistry. Life Sci. Mater. Sci. 1–2.

Lim, K.G., Gray, A.I., Pyne, S., Pyne, N.J., 2012. Resveratrol dimers are novel sphingosine kinase 1 inhibitors and affect sphingosine kinase 1 expression and cancer cell growth and survival. Br. J. Pharmacol. 166 (5), 1605–1616. https://doi.org/10.1111/j.1476-5381.2012.01862.x.

Liu, H., Wu, X., Luo, J., Wang, X., Guo, H., Feng, D., Zhao, L., Bai, H., Song, M., Liu, X., Guo, W., Li, X., Yue, L., Wang, B., Qu, Y., 2019. Pterostilbene attenuates astrocytic inflammation and neuronal oxidative injury after ischemia-reperfusion by inhibiting NF-κB phosphorylation. Front. Immunol. 10. https://doi.org/10.3389/fimmu.2019.02408.

Liu, Y.H., Lin, Y.S., Sie, Y.Y., Wang, C.C., Chang, C.I., Hou, W.C., Vitisin, B., 2023. A resveratrol tetramer from Vitis thunbergii var. taiwaniana, ameliorates impaired glucose regulations in nicotinamide/streptozotocin-induced type 2 diabetic mice. J. Tradition. Complement. Med. 13 (5), 479–488. https://doi.org/10.1016/j.jtcme.2023.05.003.

Lorenz, P., Roychowdhury, S., Engelmann, M., Wolf, G., Horn, T.F.W., 2003. Oxyresveratrol and resveratrol are potent antioxidants and free radical scavengers: effect on nitrosative and oxidative stress derived from microglial cells. Nitric Oxide 9 (2), 64–76. https://doi.org/10.1016/j.niox.2003.09.005.

Lotos: Natural Products Online, 2023, p. 2023.

Lucas-Abellán, C., Mercader-Ros, M.T., Zafrilla, M.P., Gabaldón, J.A., Núñez-Delicado, E., 2011. Comparative study of different methods to measure antioxidant activity of resveratrol in the presence of cyclodextrins. Food Chem. Toxicol. 49 (6), 1255–1260. https://doi.org/10.1016/j.fct.2011.03.004.

Lulan, T.Y.K., Fatmawati, S., Santoso, M., 2020. Taslim Ersam, α-VINIFERIN as a potential antidiabetic and antiplasmodial extracted from Dipterocarpus littoralis. Heliyon 6 (5), e04102. https://doi.org/10.1016/j.heliyon.2020.e04102.

Majeed, M., Majeed, S., Jain, R., Mundkur, L., Rajalakshmi, H.R., Lad, P.S., Neupane, P., 2020. An open-label single-arm, monocentric study assessing the efficacy and safety of natural pterostilbene (Pterocarpus marsupium) for skin brightening and antiaging effects. Clin. Cosmet. Invest. Dermatol. 13, 105–116. https://doi.org/10.2147/CCID.S238358.

Malinowska, M.A., Billet, K., Drouet, S., Munsch, T., Unlubayir, M., Tungmunnithum, D., Giglioli-Guivarc'H, N., Hano, C., Lanoue, A., 2020. Grape cane extracts as multifunctional rejuvenating cosmetic ingredient: evaluation of sirtuin activity, tyrosinase inhibition and bioavailability potential. Molecules 25 (9). https://doi.org/10.3390/molecules25092203.

Marant, B., Crouzet, J., Flourat, A.L., Jeandet, P., Aziz, A., Courot, E., 2024. Key-enzymes involved in the biosynthesis of resveratrol-based stilbenes in *Vitis* spp.: a review. Phytochem Rev. https://doi.org/10.1007/s11101-024-09958-4.

Matacchione, G., Valli, D., Silvestrini, A., Giuliani, A., Sabbatinelli, J., Giordani, C., Coppari, S., Rippo, M.R., Cristina Albertini, M., Olivieri, F., Curcumin, 2022. Polydatin and quercetin synergistic activity protects from high-glucose-induced inflammation and oxidative stress. Antioxidants 11 (6), 1037. https://doi.org/10.3390/antiox11061037.

Mattio, L.M., Marengo, M., Parravicini, C., Eberini, I., Dallavalle, S., Bonomi, F., Iametti, S., Pinto, A., 2019. Inhibition of pancreatic α-amylase by resveratrol derivatives: biological activity and molecular modelling evidence for cooperativity between viniferin enantiomers. Molecules 24 (18), 3225. https://doi.org/10.3390/molecules24183225.

Mattio, L.M., Catinella, G., Pinto, A., Dallavalle, S., 2020. Natural and nature-inspired stilbenoids as antiviral agents. Eur. J. Med. Chem. https://doi.org/10.1016/j.ejmech.2020.112541.

Mazimba, O., Majinda, R.R.T., Motlhanka, D., 2018. Antioxidant and Antibacterial Constituents from Morus Nigra, vol. 7, pp. 1–4.

Miao, Q., Wang, S., Miao, S., Wang, J., Xie, Y., Yang, Q., 2011. Cardioprotective effect of polydatin against ischemia/reperfusion injury: roles of protein kinase C and mito KATP activation. Phytomedicine 19 (1), 8–12. https://doi.org/10.1016/j.phymed.2011.06.023.

Morel-Salmi, C., Julia, A., Vigor, C., Vercauteren, J., 2014. A huge PVDF adsorption difference between resveratrol and ε-viniferin allows to quantitatively purify them and to assess their anti-tyrosinase property. Chromatographia 77 (13–14), 957–961. https://doi.org/10.1007/s10337-014-2707-8.

Morikawa, T., Xu, F., Matsuda, H., Yoshikawa, M., 2010. Structures of novel norstilbene dimer, longusone A, and three new stilbene dimers, longusols A, B, and C, with antiallergic and radical scavenging activities from Egyptian natural medicine Cyperus longus. Chem. Pharm. Bull. 58 (10), 1379–1382. https://doi.org/10.1248/cpb.58.1379Japan.

Moriyama, H., Moriyama, M., Ninomiya, K., Morikawa, T., Hayakawa, T., 2016. Inhibitory effects of oligostilbenoids from the bark of *Shorea roxburghii* on malignant melanoma cell growth: implications for novel topical anticancer candidates. Biol. Pharm. Bull. 39 (10), 1675–1682. https://doi.org/10.1248/bpb.b16-00420.

Nagarajan, S., Mohandas, S., Ganesan, K., Xu, B., Mohanram Ramkumar, K., 2022. New insights into dietary pterostilbene: sources, metabolism, and health promotion effects. Molecules 27 (19), 6316. https://doi.org/10.3390/molecules27196316.

Ono, M., Masuoka, C., Ito, Y., Yujiro, N., Kinjo, J., Nohara, T., 1997. Antioxidative and antihyaluronidase activities of some constituents from foeniculi fructus(fruit of Foeniculum vulgare MILLER). Food Sci. Technol. Int. Tokyo 3 (1), 53–55. https://doi.org/10.3136/fsti9596t9798.3.53.

Özdemir, F., Apaydın, E., Önder, N.İ., Şen, M., Ayrım, A., Öğünç, Y., İncesu, Z., 2018. Apoptotic effects of ε-viniferin in combination with cis-platin in C6 cells. Turkey Cytotechnol. 70 (3), 1061–1073. https://doi.org/10.1007/s10616-018-0197-5.

Pandey, K.B., Rizvi, S.I., 2009. Plant polyphenols as dietary antioxidants in human health and disease. Oxid. Med. Cell. Longev. 2 (5), 270–278. https://doi.org/10.4161/oxim.2.5.9498India.

Pavela, R., Waffo-Teguo, P., Biais, B., Richard, T., Mérillon, J.M., 2017. Vitis vinifera canes, a source of stilbenoids against Spodoptera littoralis larvae. J. Pest. Sci. 90 (3), 961–970. https://doi.org/10.1007/s10340-017-0836-1.

Pawlus, A.D., Waffo-Téguo, P., Shaver, J., Mérillon, J.M., 2012. Stilbenoid chemistry from wine and the genus vitis, a review. J. Int. Sci. Vigne Vin 46 (2), 57–111.

Perez, A.I., 2020. Antitumoral actions of natural stilbenes derived from Vitis vinifera. Cell Biol.

Pezet, R., Gindro, K., Viret, O., Richter, H., 2004. Effects of resveratrol, viniferins and pterostilbene on Plasmopara viticola zoospore mobility and disease development. Vitis - J. Grapevine Res. 43 (3), 145–148.

Phoolcharoen, W., Sooampon, S., Sritularak, B., Likhitwitayawuid, K., Kuvatanasuchati, J., Pavasant, P., 2013. Anti-periodontal pathogen and anti-inflammatory activities of oxyresveratrol . Natural Product Incorporation. Thailand Natural Prod. Commun. 8 (5), 613–616. https://doi.org/10.1177/1934578x1300800518.

Privat, C., Telo, J.P., Bernardes-Genisson, V., Vieira, A., Souchard, J.P., Nepveu, F., 2002. Antioxidant properties of trans-ε-Viniferin as compared to stilbene derivatives in aqueous and nonaqueous media. J. Agric. Food Chem. 50 (5), 1213–1217. https://doi.org/10.1021/jf010676t.

Pubchem, 2023. National Center for Biotechnology Information (NCBI). Retrieved from https://pubchem.ncbi.nlm.nih.gov

Qi, Z., Tan, Y., Zhang, N., Yao, F., 2015. Polydatin prevents angiotensin II-induced cardiac hypertrophy and myocardial superoxide generation. Exp. Biol. Med. 240 (10), 1352–1361. https://doi.org/10.1177/1535370214561958.

Ratz-Łyko, A., Arct, J., 2019. Resveratrol as an active ingredient for cosmetic and dermatological applications: a review. J. Cosmet. Laser Ther. 21 (2), 84–90. https://doi.org/10.1080/14764172.2018.1469767.

Renaud, S., de Lorgeril, M., 1992. Wine, alcohol, platelets, and the French paradox for coronary heart disease. Lancet 339 (8808), 1523–1526. https://doi.org/10.1016/0140-6736(92)91277-f.

Riba, A., Deres, L., Sumegi, B., Toth, K., Szabados, E., Halmosi, R., 2017. Cardioprotective effect of resveratrol in a postinfarction heart failure model. Oxid. Med. Cell. Longev. 2017, 1–10. https://doi.org/10.1155/2017/6819281.

Rimando, A.M., Cuendet, M., Desmarchelier, C., Mehta, R.G., Pezzuto, J.M., Duke, S.O., 2002. Cancer chemopreventive and antioxidant activities of pterostilbene, a naturally occurring analogue of resveratrol. J. Agric. Food Chem. 50 (12), 3453–3457. https://doi.org/10.1021/jf0116855.

Rivière, C., Pawlus, A.D., Mérillon, J.M., 2012. Natural stilbenoids: distribution in the plant kingdom and chemotaxonomic interest in Vitaceae. Nat. Prod. Rep. 29 (11), 1317–1333. https://doi.org/10.1039/c2np20049j.

Rodríguez-Bonilla, P., Gandía-Herrero, F., Matencio, A., García-Carmona, F., López-Nicolás, J.M., 2017. Comparative study of the antioxidant capacity of four stilbenes using ORAC, ABTS+, and FRAP techniques. Food Anal. Methods 10 (9), 2994–3000. https://doi.org/10.1007/s12161-017-0871-9.

Rosiak, N., Cielecka-Piontek, J., Skibiński, R., Lewandowska, K., Bednarski, W., Zalewski, P., 2022. Antioxidant potential of resveratrol as the result of radiation exposition. Antioxidants 11.

Sáez, V., Pastene, E., Vergara, C., Mardones, C., Hermosín-Gutiérrez, I., Gómez-Alonso, S., Victoria Gómez, M., Theoduloz, C., Riquelme, S., von Baer, D., 2018. Oligostilbenoids in Vitis vinifera L. Pinot Noir grape cane extract: isolation, characterization, in vitro antioxidant capacity and anti-proliferative effect on cancer cells. Food Chem. 265, 101–110. https://doi.org/10.1016/j.foodchem.2018.05.050.

Safety of synthetic trans-resveratrol as a novel, 2016. Food pursuant to Regulation (EC) No 258/97. EFSA J. 14 (1), 1–30.

Santos, D., Xavier, F.A.R., Silva, J.A, Merlin, JPJ Goulart, J.P.J. MOF rupasinghe, HPV, antidiabetic, antiglycation, and antioxidant activities of ethanolic seed extract of Passiflora edulis and piceatannol.

Saroyobudiono, H., Juliawaty, L.D., Syah, Y.M., Achmad, S.A., Hakim, E.H., Latip, J., Said, I.M., 2008. Oligostilbenoids from Shorea gibbosa and their cytotoxic properties against P-388 cells. Indonesia J. Natural Med. 62 (2), 195–198. https://doi.org/10.1007/s11418-007-0205-0.

Sasikumar, P., Lekshmy, K., Sini, S., Prabha, B., Kumar, N.A., Sivan, V.V., Jithin, M.M., Jayamurthy, P., Shibi, I.G., Radhakrishnan, K.V., 2019. Isolation and characterization of resveratrol oligomers from the stem bark of Hopea ponga (Dennst.) Mabb. And their antidiabetic effect by modulation of digestive enzymes, protein glycation and glucose uptake in L6 myocytes. J. Ethnopharmacol. 236, 196–204. https://doi.org/10.1016/j.jep.2019.01.046.

Schmeda-Hirschmann, G., Gutiérrez, M.I., Loyola, J.I., Zúñiga, J., 1996. Biological activity and xanthine oxidase inhibitors fromScirpus californicus (C. A. Mey.) steud. Phytother Res. 10 (8), 683–685. https://doi.org/10.1002/(sici)1099-1573(199612)10:8<683::aid-ptr916>3.0.co;2-t.

Schnee, S., Queiroz, E.F., Voinesco, F., Marcourt, L., Dubuis, P.H., Wolfender, J.L., Gindro, K., 2013. Vitis vinifera canes, a new source of antifungal compounds against Plasmopara viticola, Erysiphe necator, and Botrytis cinerea. J. Agric. Food Chem. 61 (23), 5459–5467. https://doi.org/10.1021/jf4010252.

Sebai, H., Sani, M., Aouani, E., Ghanem-Boughanmi, N., 2011. Cardioprotective effect of resveratrol on lipopolysaccharide-induced oxidative stress in rat. Drug Chem. Toxicol. 34 (2), 146–150. https://doi.org/10.3109/01480545.2010.494666.

Seo, Y.J., Kim, K.J., Koh, E.J., Choi, J., Lee, B.Y., 2017. Anti-adipogenesis mechanism of pterostilbene through the activation of heme oxygenase-1 in 3T3-L1 cells. Phytomedicine 33, 7–13. https://doi.org/10.1016/j.phymed.2017.05.009.

Sesderma. (2023). www.sesderma.com/es_es/prehome.

Sharafan, M., Malinowska, M.A., Ekiert, H., Kwaśniak, B., Sikora, E., Szopa, A., 2023. Vitis vinifera (vine grape) as a valuable cosmetic Raw material. Pharmaceutics 15 (5), 1372. https://doi.org/10.3390/pharmaceutics15051372.

Shen, T., Xie, C.F., Wang, X.N., Lou, H.X., 2013. Stilbenoids Natural Products: Phytochemistry, Botany and Metabolism of Alkaloids, Phenolics and Terpenes. Springer Berlin Heidelberg, China, pp. 1901–1949. https://doi.org/10.1007/978-3-642-22144-6_63.

Sheng, J.-Y., Chen, T.-T., Tan, X.-J., Chen, T., Jia, A.-Q., 2015. The quorum-sensing inhibiting effects of stilbenoids and their potential structure-activity relationship. Bioorg. Med. Chem. Lett 22, 5217–5220. https://doi.org/10.1016/j.bmcl.2015.09.064.

Silva, A.A., Haraguchi, S.K., Cellet, T.S.P., Schuquel, I.T.A., Sarragiotto, M.H., Vidotti, G.J., De Melo, J.O., Bersani-Amado, C.A., Zanoli, K., Nakamura, C.V., 2012. Resveratrol-derived stilbenoids and biological activity evaluation of seed extracts of Cenchrus echinatus L. Nat. Prod. Res. 26 (9), 865–868. https://doi.org/10.1080/14786419.2011.561538.

Silva, F., Figueiras, A., Gallardo, E., Nerín, C., Domingues, F.C., 2014. Strategies to improve the solubility and stability of stilbene antioxidants: a comparative study between cyclodextrins and bile acids. Food Chem. 145, 115–125. https://doi.org/10.1016/j.foodchem.2013.08.034.

Su, P.S., Doerksen, R.J., Chen, S.H., Sung, W.C., Juan, C.C., Rawendra, R.D.S., Chen, C.R., Li, J.W., Aisha, Huang, T.C., Liao, M.H., Chang, C.I., Hsu, J.L., 2015. Screening and profiling stilbene-type natural products with angiotensin-converting enzyme inhibitory activity from Ampelopsis brevipedunculata var. hancei (Planch.) Rehder. J. Pharmaceut. Biomed. Anal. 108, 70–77. https://doi.org/10.1016/j.jpba.2015.01.053.

Subedi, L., Lee, T.H., Wahedi, H.M., Baek, S.H., Kim, S.Y., 2017. Resveratrol-enriched rice attenuates UVB-ROS-Induced skin aging via downregulation of inflammatory cascades. Oxid. Med. Cell. Longev. 2017. https://doi.org/10.1155/2017/8379539.

Subramanian, R., Raj, V., Manigandan, K., Elangovan, N., 2015. Antioxidant activity of hopeaphenol isolated from Shorea roxburghii stem bark extract. J. Taibah Univ. Sci. 9 (2), 237–244. https://doi.org/10.1016/j.jtusci.2014.11.004.

Sun, F.M., Chou, Y.J., Huang, Q., Hu, J.Y., Ting, Y., 2021. Making concentrated pterostilbene highly bioavailable in pressure processed phospholipid nanoemulsion. Processes 9 (2), 1–15. https://doi.org/10.3390/pr9020294.

Sung, M.J., Davaatseren, M., Kim, W., Park, S.K., Kim, S.H., Jeon Hur, H., Kim, M.S., Kim, Y.S., Kwon, D.Y., 2009. Vitisin A suppresses LPS-induced NO production by inhibiting ERK, p38, and NF-κB activation in RAW 264.7 cells. Int. Immunopharm. 9 (3), 319–323. https://doi.org/10.1016/j.intimp.2008.12.005.

Szkudelska, K., Deniziak, M., Sassek, M., Szkudelski, I., Noskowiak, W., Szkudelski, T., 2021. Resveratrol affects insulin signaling in type 2 diabetic goto-kakizaki rats. Int. J. Mol. Sci. 22 (5), 2469. https://doi.org/10.3390/ijms22052469.

Tam, K.W., Ho, C.T., Tu, S.H., Lee, W.J., Huang, C.S., Chen, C.S., Wu, C.H., Lee, C.H., Ho, Y.S., 2018. α-Tocopherol succinate enhances pterostilbene anti-tumor activity in human breast cancer cells in vivo and in vitro. Oncotarget 9 (4), 4593–4606. https://doi.org/10.18632/oncotarget.23390.

Tan, Y.Y., Chen, L.X., Fang, L., Zhang, Q., 2020. Cardioprotective effects of polydatin against myocardial injury in diabetic rats via inhibition of NADPH oxidase and NF-κB activities. BMC Complement. Med. Therap. 20 (1). https://doi.org/10.1186/s12906-020-03177-y.

Teka, T., Zhang, L., Ge, X., Li, Y., Han, L., Yan, X., 2022. Stilbenes: source plants, chemistry, biosynthesis, pharmacology, application and problems related to their clinical application-a comprehensive review. Phytochemistry 197. https://doi.org/10.1016/j.phytochem.2022.113128.

The Ordinary, 2023. www.theordinary.com/en-us.

Tian, B., Liu, J., 2020. Resveratrol: a review of plant sources, synthesis, stability, modification and food application. J. Sci. Food Agric. 100 (4), 1392–1404. https://doi.org/10.1002/jsfa.10152.

Timmers, S., Konings, E., Bilet, L., Houtkooper, R.H., Van De Weijer, T., Goossens, G.H., Hoeks, J., Van Der Krieken, S., Ryu, D., Kersten, S., Moonen-Kornips, E., Hesselink, M.K.C., Kunz, I., Schrauwen-Hinderling, V.B., Blaak, E.E., Auwerx, J., Schrauwen, P., 2011. Calorie restriction-like effects of 30 days of resveratrol supplementation on energy metabolism and metabolic profile in obese humans. Cell Metabol. 14 (5), 612–622. https://doi.org/10.1016/j.cmet.2011.10.002.

Timperio, A.M., D'Alessandro, A., Fagioni, M., Magro, P., Zolla, L., 2012. Production of the phytoalexins *trans*-resveratrol and delta-viniferin in two economy-relevant grape cultivars upon infection with *Botrytis cinerea* in field conditions. Plant Physiol. Biochem. 50, 65–71. https://doi.org/10.1016/j.plaphy.2011.07.008.

Ting, L., Tan, Y., Ouyang, S., He, J., Liu, L., 2022. Resveratrol protects against myocardial ischemia-reperfusion injury via attenuating ferroptosis. Gene 808, 145968. https://doi.org/10.1016/j.gene.2021.145968.

Tomé-Carneiro, J., Larrosa, M., Yáñez-Gascón, M.J., Dávalos, A., Gil-Zamorano, J., Gonzálvez, M., García-Almagro, F.J., Ruiz Ros, J.A., Tomás-Barberán, F.A., Espín, J.C., García-Conesa, M.T., 2013. One-year supplementation with a grape extract containing resveratrol modulates inflammatory-related microRNAs and cytokines expression in peripheral blood mononuclear cells of type 2 diabetes and hypertensive patients with coronary artery disease. Pharmacol. Res. 72, 69–82. https://doi.org/10.1016/j.phrs.2013.03.011.

Tsopmo, A., Awah, F.M., Kuete, V., 2013. Lignans and Stilbenes from African Medicinal Plants Medicinal Plant Research in Africa: Pharmacology and Chemistry. Elsevier Inc., Canada, pp. 435–478. https://doi.org/10.1016/B978-0-12-405927-6.00012-6.

Utkina, N.K., Kulesh, N.I., 2012. Antioxidant activity of polyphenols and polyphenol complex from the far-eastern tree Maackia amurensis. Pharmaceut. Chem. J. 46 (8), 488–491. https://doi.org/10.1007/s11094-012-0831-z.

Valletta, A., Maria Iozia, L., Leonelli, F., 2021. Impact of environmental factors on stilbene biosynthesis. Plants 10 (1), 90. https://doi.org/10.3390/plants10010090.

Verma, N., Tiku, A.B., 2022. Polydatin-induced direct and bystander effects in A549 lung cancer cell line. Routl. India Nutrit. Cancer 74 (1), 237–249. https://doi.org/10.1080/01635581.2020.1870705.

Vermerris, W., Nicholson, R., 2006. Phenolic Compound Biochemistry Phenolic Compound Biochemistry. Springer, Netherlands, United States, pp. 1–276. https://doi.org/10.1007/978-1-4020-5164-7.

Vervandier-Fasseur, D., Latruffe, N., 2019. The potential use of resveratrol for cancer prevention. Molecules 24 (24), 4506. https://doi.org/10.3390/molecules24244506.

Waffo-Teguo, P., Krisa, S., Richard, T., Mérillon, J.-M., 2008. Grapevine Stilbenes and Their Biological Effects. Springer Science and Business Media LLC, pp. 25–54. https://doi.org/10.1007/978-3-540-74603-4_2.

Wang, K.T., Chen, L.G., Tseng, S.H., Huang, J.S., Hsieh, M.S., Wang, C.C., 2011. Anti-inflammatory effects of resveratrol and oligostilbenes from vitis thunbergii var. taiwaniana against lipopolysaccharide-induced arthritis. J. Agric. Food Chem. 59 (8), 3649–3656. https://doi.org/10.1021/jf104718g.

Wang, X., Jiang, B., Lv, H., Liang, Y., Ma, 2019. Vitisin B as a novel fatty acid synthase inhibitor induces human breast cancer cells apoptosis. Am J Transl Res 11 (8).

Wang, W.-D., Wu, Z.-C., Wang, Y.-F., Wang, J.-Z., Wang, Z.-P., 2020. Evaluation of free radical scavenging and antioxidation capacity of resveratrol and polydatin. Adv. Eng. Res. 63−67. https://doi.org/10.2991/aer.k.201203.013.

Wilson, C., 2010. xPharm, the Comprehensive Pharmacology Reference, pp. 1−9.

Wu, S.S., Chen, L.G., Lin, R.J., Lin, S.Y., Lo, Y.E., Liang, Y.C., 2013. Cytotoxicity of (-)-vitisin B in human leukemia cells. Drug Chem. Toxicol. 36 (3), 313−319. https://doi.org/10.3109/01480545.2012.720990.

Wu, C.W., Nakamoto, Y., Hisatome, T., Yoshida, S., Miyazaki, H., 2020. Resveratrol and its dimers ε viniferin and δ-viniferin in red wine protect vascular endothelial cells by a similar mechanism with different potency and efficacy. Kaohsiung J. Med. Sci. 36 (7), 535−542. https://doi.org/10.1002/kjm2.12199.

Xiao, K., Zhang, H.J., Xuan, L.J., Zhang, J., Xu, Y.M., Bai, D.L., 2008. Stilbenoids: chemistry and bioactivities. Stud. Nat. Prod. Chem. 34 (C), 453−646. https://doi.org/10.1016/S1572-5995(08)80032-4.

Xu, C., Song, Y., Wang, Z., Jiang, J., Piao, Y., Li, L., Jin, S., Li, L., Zhu, L., Yan, G., 2021. Pterostilbene suppresses oxidative stress and allergic airway inflammation through AMPK/Sirt1 and Nrf2/HO-1 pathways. Immun. Inflamm. Dis. 9 (4), 1406−1417. https://doi.org/10.1002/iid3.490.

Xue, Y.Q., Di, J.M., Luo, Y., Cheng, K.J., Wei, X., Shi, Z., 2014. Resveratrol oligomers for the prevention and treatment of cancers. Oxid. Med. Cell. Longev. 2014. https://doi.org/10.1155/2014/765832.

Yamamoto, T., Li, Y., Hanafusa, Y., Yeh, Y.S., Maruki-Uchida, H., Kawakami, S., Sai, M., Goto, T., Ito, T., Kawada, T., 2017. Piceatannol exhibits anti-inflammatory effects on macrophages interacting with adipocytes. Food Sci. Nutr. 5 (1), 76−85. https://doi.org/10.1002/fsn3.366.

Yan, Z.-B., Liang, J.-L., Niu, F.-X., Shen, Y.-P., Liu, J.-Z., 2021. Enhanced production of pterostilbene in *Escherichia coli* through directed evolution and host strain engineering. Front. Microbiol. 12. https://doi.org/10.3389/fmicb.2021.710405.

Yang, Y., Zhang, G., Li, C., Wang, S., Zhu, M., Wang, J., Yue, H., Ma, X., Zhen, Y., Shu, X., 2019. Metabolic profile and structure-activity relationship of resveratrol and its analogs in human bladder cancer cells. Cancer Manag. Res. 11, 4631−4642. https://doi.org/10.2147/CMAR.S206748.

Yousef, A.I., Shawki, H.H., El-Shahawy, A.A., El-Twab, S.M.A., Moneim, A.-, Oishi, A., 2021. Polydatin mitigates pancreatic β-cell damage through its antioxidant activity. Biomed. Pharmacother. https://doi.org/10.1016/j.biopha.2020.111027.

Yuan, Y., Zhu, Y., Li, Y., Li, X., Jiao, R., Bai, W., 2023. Cholesterol-lowering activity of vitisin A is mediated by inhibiting cholesterol biosynthesis and enhancing LDL uptake in HepG2 cells. Int. J. Mol. Sci. 4. https://doi.org/10.3390/ijms24043301.

Yun, C.Y., Ko, S.M., Choi, Y.P., Kim, B.J., Lee, J., Kim, J.M., Kim, J.Y., Song, J.Y., Kim, S.H., Hwang, B.Y., Hong, J.T., Han, S.B., Kim, Y., 2018. α-Viniferin improves facial hyperpigmentation via accelerating feedback termination of cAMP/PKA- signaled phosphorylation circuit in facultative melanogenesis. Theranostics 8 (7), 2031−2043. https://doi.org/10.7150/thno.24385.

Zhang, S., Ma, Y., Feng, J., 2020. Neuroprotective mechanisms of ϵ-viniferin in a rotenone-induced cell model of Parkinson's disease: significance of SIRT3-mediated FOXO3 deacetylation. Neural Regener. Res. 15 (11), 2143−2153. https://doi.org/10.4103/1673-5374.282264.

Zhang, L.X., Li, C.X., Kakar, M.U., Khan, M.S., Wu, P.F., Amir, R.M., Dai, D.F., Naveed, M., Li, Q.Y., Saeed, M., Shen, J.Q., Rajput, S.A., Li, J.H., 2021. Resveratrol (RV): a pharmacological review and call for further research. Biomed. Pharmacother. 143. https://doi.org/10.1016/j.biopha.2021.112164.

Zhao, T., Wang, C., Huo, X., He, M.L., Chen, J., 2020. Pterostilbene enhances sorafenib's anticancer effects on gastric adenocarcinoma. J. Cell Mol. Med. 24 (21), 12525−12536. https://doi.org/10.1111/jcmm.15795.

Zheng, L., Wu, J., Mo, J., Guo, L., Wu, X., Bao, Y., 2019. Polydatin inhibits adipose tissue inflammation and ameliorates lipid metabolism in high-fat-fed mice. BioMed Res. Int. 1–10. https://doi.org/10.1155/2019/7196535.

Zhu, W., Qin, W., Zhang, K., Rottinghaus, G.E., Chen, Y.C., Kliethermes, B., Sauter, E.R., 2012. Trans -resveratrol alters mammary promoter hypermethylation in women at increased risk for breast cancer. Nutr. Cancer 64 (3), 393–400. https://doi.org/10.1080/01635581.2012.654926.

Zhu, X., Wu, C., Qiu, S., Yuan, X., Li, L., 2017. Effects of resveratrol on glucose control and insulin sensitivity in subjects with type 2 diabetes: systematic review and meta-analysis. Nutr. Metabol. 14 (1). https://doi.org/10.1186/s12986-017-0217-z.

Zhu, T., Fang, F., Sun, D., Yang, S., Zhang, X., Yu, X., Yang, L., Piceatannol Inhibits, P., 2020. Acnes–induced keratinocyte proliferation and migration by downregulating oxidative stress and the inflammatory response. Inflammation 43 (1), 347–357. https://doi.org/10.1007/s10753-019-01125-8.

Zu, Y., Wang, S., 2020. Resveratrol-loaded liposomes: browning subcutaneous white adipose tissue for combating obesity in C57BL/6 J mice. Curr. Dev. Nutr. 4. https://doi.org/10.1093/cdn/nzaa063_107.

CHAPTER

Role of abiotic stress factors from natural and anthropogenic sources in plant-environment interaction 11

Sabina Licen, S. Fornasaro, M. Crosera, G. Adami and P. Barbieri
University of Trieste, Department of Chemical and Pharmaceutical Sciences, Trieste, Italy

1. Introduction

Plants that grow in an open environment can be subjected to abiotic stressors of different origins, produced both by natural and anthropogenic sources. Natural stressors typically include temperature, solar radiation, drought, and soil salinity. Anthropogenic stressors are mainly pollutants that can be absorbed by the plant from air, water, soil, heavy metals (Manwani et al., 2022), emerging concern organic substances (Pullagurala et al., 2018), particulate matter (PM) (Mandal et al., 2023), inorganic gases (e.g., NO_x, SO_2, O_3), and micro and nanoplastics (Le et al., 2023). Except for the latter, the other pollutants can also have a natural origin, but usually, they don't have a high or long-term impact. Moreover, abiotic stressor sources can be interconnected because some pollutants are the main cause of climate change and, therefore, of a variation in natural stressors such as temperature and drought (Duan et al., 2024) (Fig. 11.1).

The stressors induce a variety of physiological and biochemical responses in the plants. They can trigger the production of Reactive Oxygen Species (ROS) that cause damage to the cell apparatus (Hlihor et al., 2022; Pullagurala et al., 2018). The increase in the concentration of ROS induces the plant to activate defense mechanisms such as the production of reactive oxygen detoxification enzymes and antioxidant molecules such as glutathione, ascorbic acid, tocopherols, phenolic compounds, terpenes, and flavonoids (Mandal et al., 2023). Typical physiological responses are changes in the foliar pH and relative water content, as well as root amino acid and sugar content. Morphological responses are also generated, such as changes in the leaf and root anatomy and micromorphology (Viana et al., 2022; Ghosh et al., 2023). The highest impact of stressors can inhibit plant growth, cause important damage to the tissues, and, finally, lead to plant death. On the other hand, these responses allow us to use plants as bioindicators (Ghosh et al., 2023; Zsigmond et al., 2021), and to select resilient species for phytoremediation purposes (Nedjimi, 2021; Lee et al., 2020).

As a result, when considering the use of plants for their medicinal attributes or micronutrients content, it has to take into account both the possible contamination by environmental pollutants and the alteration of the content and variety of bioactive compounds due to stressors.

Phytochemicals for Health. https://doi.org/10.1016/B978-0-443-15366-2.00011-3

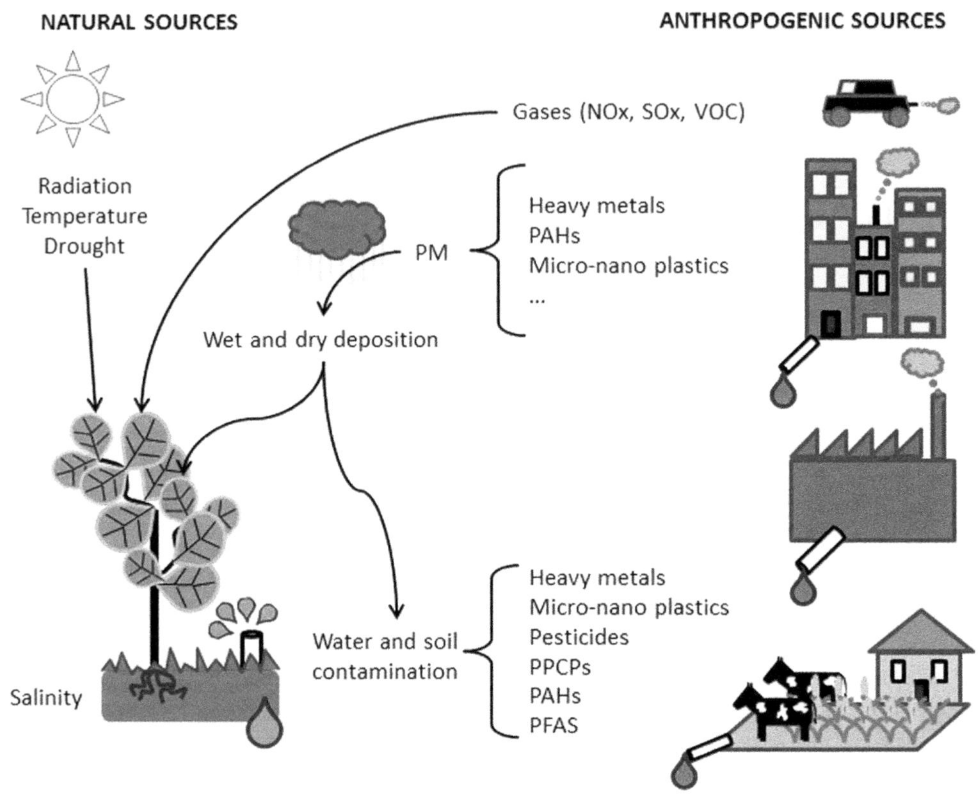

FIGURE 11.1

Natural and anthropogenic sources of plant abiotic stressors.

PAHs, polycyclic aromatic hydrocarbons; *PFAS*, per- and polyfluoroalkyl substances; *PM*, particulate matter; *PPCPs*, pharmaceutical and personal care products; *VOC*, volatile organic compounds.

2. Natural stressors

Natural stressors cause a variety of responses by the plant to counteract the stressor effects and enhance the resilience and survival probability. Light is necessary for photosynthetic activity, but high levels of radiation and, in particular, high levels of UV light can cause damage to the leaf surface thus inducing the plant to produce antioxidant substances to mitigate the light effect (Aqeel et al., 2023; Zhang et al., 2014). Enzymatic reactions can be altered when the plant is subjected to high temperatures that are far from the optimal temperature for enzyme activity. Temperature can also affect cell-membrane integrity, photosynthesis, and respiration regulation (Jamloki et al., 2021). Drought affects the stomata regulation and thus plant transpiration. It also affects respiration, and chlorophyll production and can lead to an osmotic unbalance that induces the plant to produce soluble substances for equilibrating the osmotic pressure (Li et al., 2020). A similar effect is induced by high salinity that furthermore

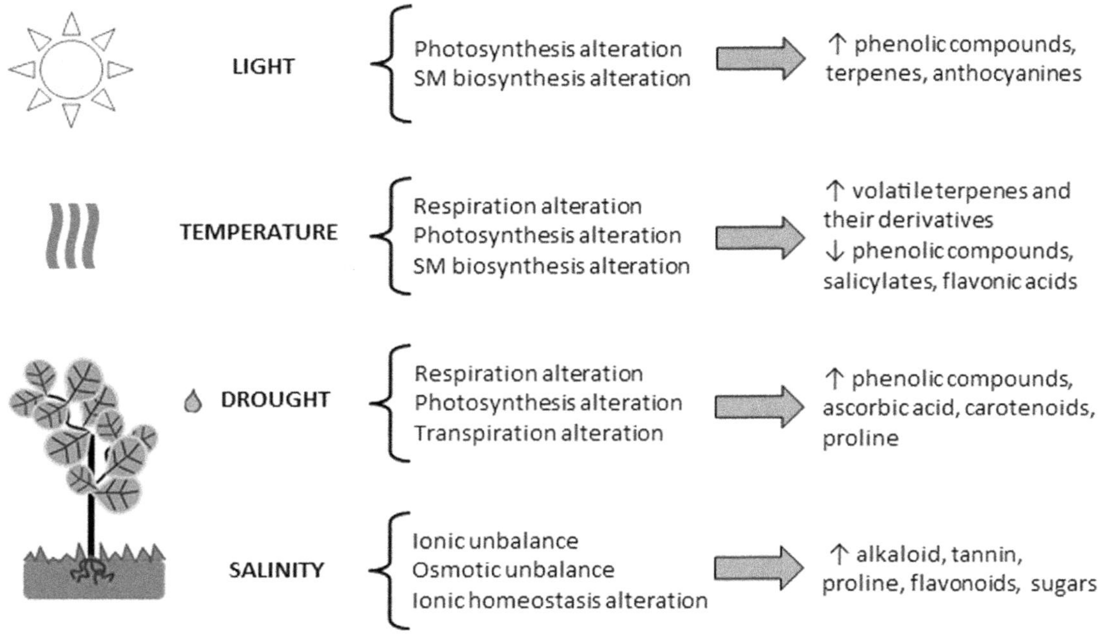

FIGURE 11.2

Effect of natural stressors on the plant bioactive compounds.
SM, secondary metabolites.

affects redox and ionic homeostasis regulation (Xiao and Zhou, 2023). The main effects of the above-mentioned factors on plant bioactive compounds are reported in Fig. 11.2.

The stressors' influence can be also concurrent and can affect the secondary metabolites (SMs) production in a synergic or antinergic way (Zandalinas et al., 2018). Often the factors depend on the altitude, seasonality, soil composition, and the geographical origin (Zidorn, 2010; Al-Qahtani et al., 2020). This aspect allows us to trace the geographical origin of plants and food by analyzing their composition (Dantas et al., 2023; Kaseke et al., 2020; Masetti et al., 2021). As a consequence, the stressors can dramatically affect the resulting chemical composition and yield of essential oils (Aqeel et al., 2023; Gonçalves et al., 2024). Some examples of the effect of natural stressors on the chemical content of aromatic/medicinal plants are reported in Table 11.1.

3. Anthropogenic stressors

Pollutants generated by anthropogenic activities can be a threat to plant health and can affect the plant's chemical composition, both directly penetrating into the plant tissues and inducing metabolic alterations. The above-mentioned compounds can be heavy metals and a variety of organic compounds, engineered nanomaterials (ENMs), and micro and nanoplastics (MNPs) that are usually referred to as contaminants of emerging concern (CECs) (Pullagurala et al., 2018). These pollutants

Table 11.1 Examples of stressors and their effect on some plant species: high levels of stressors can induce high production (↑) or low production (↓) of bioactive compounds.

Stressor	Species	Effect	References
Light	Tea (*Camellia sinensis*)	↑ flavonoids; ↓ β-carotene, free amino acids, caffeine, and benzoic acid derivatives	(Zhang et al., 2014)
Light	Ginger (*Zingiber officinale*)	↑ total phenolic compounds; ↓flavonid compounds	(Ghasemzadeh et al., 2010)
Light	Clary sage (*Salvia sclarea* L.)	↑ linalool, α-terpineol; ↓shading linalyl acetate, sclareol, and germacrene D	(Kumar et al., 2013)
Light	Wild marigold (*Tagetes minuta* L.)	↑ ocimene and dihydrotagetones; ↓ tagetone, and ocimenone	(Kumar et al., 2014)
Light	Thyme (*Thymus vulgaris* L.)	↑ ρ-Cymene and γ-terpinene; ↓ chlorophyll *a*, chlorophyll *b*, and thymol	(da Cunha Honorato et al., 2023)
Temperature	Clary sage (*Salvia sclarea* L.)	↑ linalyl acetate; ↓ ocimene (Z) beta, linalool, and manoyl oxide	(Kumar et al., 2017)
Temperature	St John's wort (*Hypericum perforatum*)	↑ hypericin, pseudohypericin, and hyperforin	(Zobayed et al., 2005)
Temperature	Rose periwinkle (*Catharanthus roseus*)	↑ vindoline, catharanthine, vinblastine, and vincristine	(Guo et al., 2007)
Temperature	Lemon thyme (*Thymus pulegioides* L.)	↑ geraniol	(Ložienė et al., 2021)
Temperature	Peppermint (*Mentha x piperita L.*)	↑ pulegone, menthyl acetate, and p-3-menthene; ↓ menthol	(Heydari et al., 2018)
Salinity	Rapeseed (*Brassica napus* L.)	↓ protein content, Fe, Mn and Zn	(Chakraborty et al., 2016)
Salinity	Peppermint (*Mentha x piperita L.*)	↑ menthone; ↓ -(−)menthol, isomenthone, isomenthyl acetate and linalool	(Aziz et al., 2008)
Salinity	Lemongrass (*Cymbopogon flexuosus*)	↑ proline, soluble sugars, and polyamines; ↓ chlorophyll	(Mukarram et al., 2022)
Salinity	Sweet basil (*Ocimum basilicum* cv. Keshkeni luvelu)	↑ 1,8-cineole and *tau*-muurolol; ↓ linalool and α-cadinol	(Farsaraei et al., 2020)
Salinity	Rosemary (*Rosmarinus officinalis* L.)	↑ phellandrene; ↓ dill ether, dihydrocarvone and dillapiole	(Dehghani Bidgoli et al., 2019)

Drought	Sage plant (*Salvia officinalis*)	↑ 1,8-cineole and camphor; ↓ α– and β-thujone, borneol	(Ramezani et al., 2020)
Drought	German thyme (*Thymus carmanicus* L.)	↑ thymol and borneol; ↓ carvacrol and γ-terpinene	(Bahreininejad et al., 2014)
Drought	Fennel (*Foeniculum vulgare Mill*)	↑ fenchone and limonene; ↓ estragole	(Coban et al., 2018)
Drought	Greek oregano (*Origanum vulgare* ssp. *hirtum)*	↓ p-cymene and thymol	(Giannoulis et al., 2020)
Drought	Sweet basil (*Ocimum basilicum* L.)	↑ methylchavicol and methyleugenol; ↓ eugenol	(Abdollahi Mandoulakani et al., 2017)
Heavy metals	Wild mint (*Mentha longifolia* L.)	↑ menthone; ↓ carotenoids, pulegone	(Gharib et al., 2021)
Cu (combined with salinity)	Spearmint (*Mentha spicata* L.)	↑ limonene; ↓ carvone, 1,8-cineole	(Chrysargyris et al., 2019)
Cd	Thornapple (*Datura stramonium* L.)	↑ lubimin, 3-hydroxylubimin	(Furze et al., 1991)
As	Creat (*Andrographis paniculata*)	↑ andrographolide	(Kumar et al., 2017)
Cd	Summer savory (*Satureja hortensis* L.)	↑ carbohydrates, anthocyanins; ↓ chlorophyll	(Azizollahi et al., 2019)
Cr	Holy basil (*Ocimum tenuiflorum* L.)	↑ eugenol, proline; ↓ chlorophyll, carotenoids, ascorbic acid	(Rai et al., 2004)

are monitored because of their transfer ability along the trophic chain and thus for their possible harmfulness to human health (Gupta et al., 2019; Hlihor et al., 2022; Maddela et al., 2022). The pollution source can sometimes be tough to identify because of the possible long-range transport and chemical transformation of the pollutants from the source to the impacted site (Luo et al., 2023; Kanakidou et al., 2018). Pollutants can penetrate the plant by different pathways according to their chemical nature and environmental compartment source. Moreover, they can generate oxidative stress that induces the plant to activate a response and, in particular, the production of antioxidant substances such as polyphenols and terpenes.

3.1 Heavy metals

Some heavy metals (Fe, Cu, Zn, Mn, Se) are indispensable for plant health when present in small concentrations, but if their concentration is high, they can become toxic for the plant (Nedjimi, 2021). The sources of heavy metals can be related both to agricultural management activities and uncontrolled long-range sources. The former are associated with soil treatment such as irrigation with wastewater or sewage water and the application of fertilizers, composts, and manure (Gupta et al., 2019). The latter is associated with possible contamination from a variety of industrial activities such as landfills, ore extraction, oil refineries, cement production, power plants, smelting plants, and degradation of asphalt surfaces, vehicle brakes, and vehicle tyres (Hlihor et al., 2022; Nnaji et al., 2023). Contaminants' uptake mainly occurs from the roots and the leaves. The foliar uptake takes place via stomata, or cuticular cracks. It depends on the leaf surface (Shahid et al., 2017). The root uptake strongly depends on soil properties such as pH, humic substances content, cation-exchange capacity and soil microbiome. The contaminants can then be translocated to other parts of the plant through the xylem stream. The distribution and accumulation of heavy metals in the different parts of the plant depend on the metal speciation and plant species peculiarities (Alsafran et al., 2022). These stressors make the plant start a tolerance strategy as it starts to emit fluids that can bind metals and keep them in the rhizosphere, extracellular spaces, or vacuoles. This mechanism maintains the metals out of the cytosol, where they can hinder some important functions (Gupta et al., 2019; Nedjimi, 2021; Asare et al., 2023). Moreover, the translocation from the roots to the leaves via the xylem can be useful for eliminating some volatile metals such as As, Hg, and Se (Hlihor et al., 2022). Some heavy metals can enter the plant in the form of ENMs, such as nanoparticles (Pullagurala et al., 2018). Some examples of the effects on the chemical content of aromatic/medicinal plants are reported in Table 11.1.

3.2 Organic pollutants

A variety of organic compounds can impact plant health causing metabolism alterations and tissue damage. Moreover, they can bioaccumulate and pose a threat to human health when the plants are assumed as food or used for aromatic/medicinal properties. The attention is especially focused on the CECs, such as polycyclic aromatic hydrocarbons (PAHs), pharmaceutical and personal care products (PPCPs), pesticides, per- and polyfluoroalkyl substances (PFASs). The plant uptake varies according to the CECs physical-chemical characteristics: the main properties that affect the uptake are the hydrophobicity, the ionizability at different pH, and the molecular size (Wei et al., 2023). The organic pollutants can be metabolized for reducing their toxicity and also volatilized from the leaves. Some species show degradation capabilities for specific organic compounds (Doucette et al., 2018).

Pesticides are used for controlling the biotic stressors of plants: insects, fungi, rodents, worms, and weeds. Their large use worldwide has led to a ubiquitous spread, also at long distances from the source. They can be absorbed by the roots, but are more likely to be absorbed by the leaves during irrigation (Mikes et al., 2009). In general, they negatively affect plant growth. It has been shown that plants cultivated in an indoor environment are potentially exposed to higher relative concentrations of pesticides than those grown in an outdoor environment, in which the pesticide concentration is diluted by weathering agents (wind, water, light) (Hwang et al., 2015).

PAHs are mainly produced by incomplete combustion of fossil fuels and biomasses. They are hydrophobic substances and, usually, they are classified as LMW (low molecular weight) and HMW (high molecular weight). The former is easier to be taken up from the plant both from the roots and the leaves, the latter is more likely to be retained by the roots. It was shown that the PAHs bioaccumulation is directly proportional to the tissue lipid content of the plant (Lin et al., 2007). Moreover, a difference in site bioaccumulation has been observed for monocotyledons and dicotyledons, the former showed the presence both in the roots and the shoots, the latter in the roots only (Gworek et al., 2016).

PFASs are artificial compounds containing highly stable C–F bonds that are used in a variety of consumer products (Giesy and Kannan, 2002). PFASs mainly access the plant tissues via root uptake, but they can also be absorbed by the leaves, especially the more volatile ones (Wang et al., 2020a,b). They can be transported both by apoplastic and symplastic routes, thus they are more likely to be bioaccumulated in cell organelles with respect to other CECs (Li et al., 2022). Short-chain PFASs can be rather easily translocated from the roots to the aerial parts of the plants, whereas the long-chain ones are highly retained by the cell membrane and macromolecules inside it (Zhou et al., 2021). It was shown that low concentrations of PFASs can stimulate plant growth, but high concentrations produce DNA oxidative injury and metabolism downregulation, thus hampering the plant development (Lan et al., 2018).

PPCPs describe a large variety of products such as pharmaceutical compounds, antibiotics, hormones, disinfectants, cosmetics, detergents, and pest repellents. Considering that in the plant there is a lack of a proper excretory system, the plants try to modify the contaminants to reduce their toxicity with a series of processes known as the "green liver", as the overall behavior can be considered similar to that of the animal liver (Sandermann, 1994). The processes usually involve three phases: (i) chemical modification of the pollutant; (ii) conjugation; and (iii) isolation or separation of the conjugates (Wei et al., 2023). The intended detoxification process may produce metabolites that are more harmful than their parent compounds (Ben Mordechay et al., 2018). The uptake potential strongly depends on the plant species with a higher accumulation capacity for leafy vegetables than fruit vegetables (Christou et al., 2019). Soil amendments such as biochar can reduce the bioavailability of these substances. The PPCPs uptake negatively affects plant growth and development causing direct damage to specific plant organs or metabolic alterations (Wei et al., 2023). Antibiotics can be easily taken up by the plants via apoplastic transport and they are usually accumulated in the intercellular spaces where they can undergo biochemical reactions that can increase their toxicity (Tian et al., 2019), the transformation of nitrofuran into genotoxic hydrazine-containing metabolites (Di Marco et al., 2014). The estrogens mainly accumulate in the roots that can retain lipophilic contaminants (Gao et al., 2021). An increase of estrogen contamination can induce a translocation in the shoots but with a lower concentration than in the roots. The photosynthetic metabolism tends to decrease and thus the chlorophyll and carotenoids content, whereas the shoot biomass tends to increase (Adeel et al., 2023).

3.3 Particulate matter and inorganic gases

PM and inorganic gases (e.g., NO_x, SO_2, O_3) can be mainly absorbed by aerial parts of the plants, nevertheless, PM can be present in water and soil due to mechanisms of dry and wet deposition (see) and thus can be taken up by roots (Anand et al., 2022). O_3 was found to be the most detrimental agent followed by PM (Mahmood et al., 2020). O_3 induces peroxidation in membrane lipids, chlorophyll reduction, visible foliar injury, and a reduced capacity to produce antioxidant metabolites (Sinha et al., 2015). Long-term exposure was shown to reduce carbon and nitrogen content and plant development, comprising a reduction of total biomass and leaf area (Wang et al., 2020a,b). NO_x are taken up by the plant during respiration simultaneously with O_2. They affect physiological stomata processes and reduce the photosynthetic rate (Hu et al., 2015).

SO_2 at low concentrations can be used as a nutrient due to its sulfur content, whereas at high concentrations it is harmful to the plant (Agrawal, 2003). The main effects are a reduction of ascorbic acid and chlorophyll content, an increase in proline concentration, and reduction and browning effect on the leaf surface (Vwioko et al., 2018).

PM can be emitted by a variety of anthropogenic sources (see) and it can vary in size and composition according to the source (Ghosh et al., 2023). It can contain several types of pollutants, such as those described in par. 3.1 and par. 3.2. The PM uptake by leaves mainly depends on their characteristics, such as leaf size, shape, roughness, thrichomes presence, and wax content. Broader leaves with the presence of hair, high roughness, and high wax content can capture high concentrations of PM (Mandal et al., 2023). It has shown that the access stomata also depends on PM shape, spherical and rounded particles can enter easier than other shapes (Fatima et al., 2021). PM exposure can hamper the light to access the chloroplasts, thus reducing the photosynthesis rate (Popek et al., 2018). Moreover, it can block stomata and their gas exchange mechanism necessary for respiration and photosynthesis (Qadir et al., 2019). Other effects on physiology and metabolism strongly depend on PM composition (Mandal et al., 2023). PM can also act as a micro and nanoplastic carrier (see par. 3.4).

3.4 Micro and nanoplastics

MNPs are polymeric materials contained in products prepared using this specific size range or that derive from the degradation of macro-plastics by physical, chemical, or biological agents (Saeedi, 2023). The most represented polymers present in the MNPs are polyvinyl chloride (PVC), polypropylene (PP), and low-density polyethylene (LDPE) (Liu et al., 2023). They are ubiquitous and can mainly come into contact with the plants through soil and atmosphere (Kumar et al., 2020; Le et al., 2023). The MNPs penetration pathway depends on their size, shape, surface charge, and composition (Hua et al., 2023). MNPs of large size can be adsorbed by roots and block their stomata, but are scarcely translocated in the plant parts (Giorgetti et al., 2020). Smaller MNPs can be absorbed by the root system, where they can aggregate, causing damage to the cell wall and cytoplasmic membrane (Gao et al., 2021). Nano-plastics can be absorbed by cracks in leaves or roots, by endocytosis, or by aquaporins (Liu et al., 2023). MNPs can show various shapes, such as spheres, fibers, and fragments. Microspheres are more likely to be absorbed, whereas fibers are entrapped by the trichome, especially on the leaf surface (Bi et al., 2020). Lipophilic MNPs can be accumulated on the leaf surface with a high content of waxes, whereas positively charged MNPs can easily be absorbed by the plant because the cell walls are negatively charged (Zhu et al., 2021). MNPs can migrate along the plant both by symplastic and apoplastic transport (Huang et al., 2022). MNPs can produce a variety of effects such as oxidative stress, root cell clogging, photosynthesis rate reduction, alteration of amino acid and sugar content, reduced development (Chang et al., 2022; Lian et al., 2021). The toxic effect can be enhanced by organic pollutants and heavy metals adsorbed on the MNPs (Maguire and Gardner, 2023; Naziri et al., 2023).

4. Biomonitoring and phytoremediation

Plant species can be used for pollution biomonitoring, both quantifying the stress and the pollutant content (Nadgórska−Socha et al., 2017; Picó et al., 2023; da Costa Filho et al., 2022). The Air Pollution Tolerance Index (APTI), which comprises the evaluation of leaf parameters such as ascorbic acid, chlorophyll, water content, and pH, is widely used (Ghosh et al., 2023). The long-term variation of pollutant content can be evaluated with species that are tolerant to pollution stress (Doucette et al., 2018). Some examples are: camphorwood *Cinnamomum camphora* (Yin et al., 2020), bay laurel *Laurus nobilis* (Fellet et al., 2016), red pine *Pinus sylvestris* (Kalugina et al., 2017), wild rosemary *Rhododendron tomentosum* (Metrak et al., 2016), yarrow *Achillea millefolium* L. (Nujkić et al., 2020), common ivy *Hedera helix* L. (Przybysz et al., 2014).

The concurrent tolerance and bioaccumulation ability of some species towards specific pollutants can be used for environmental phytoremediation, which is a sustainable technique for removing pollutants from environmental compartments, especially soil, by using plants (Mishra and Chandra, 2022). Phytoremediation takes advantage of the defensive mechanisms used by the plant for mitigating the stress due to pollutants, such as phytostabilization, phytovolatilization, phytoextraction, phytofiltration, rhizodegradation, and phytodegradation (Nedjimi, 2021; Lee et al., 2020) (Fig. 11.3).

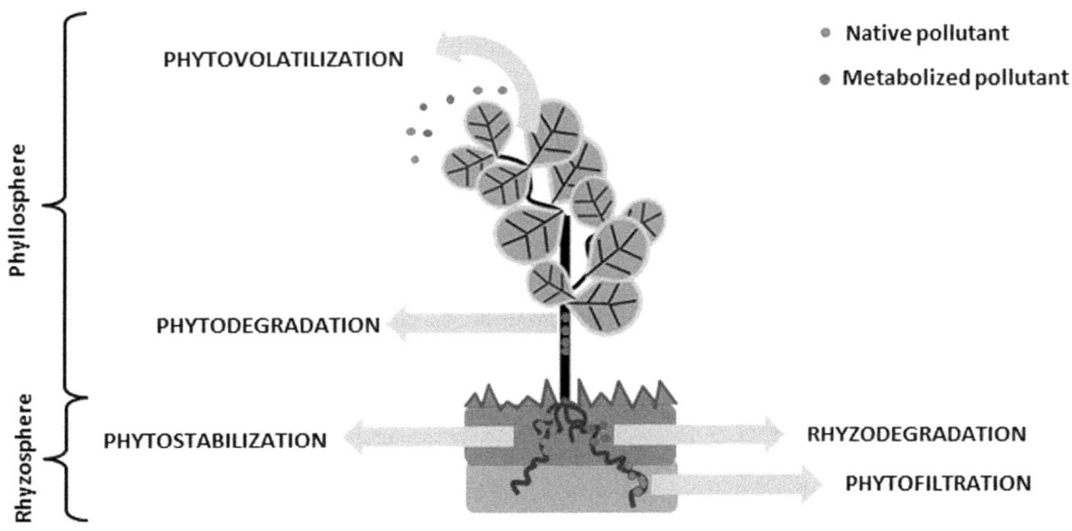

FIGURE 11.3

Phytoremediation processes.

Table 11.2 Examples of plant species used for phytoremediation.

Species	Pollutant/s	References
Lemongrass (*Cymbopogon citratus*)	Cd	(Sulastri et al., 2022)
Lavender (*Lavandula angustifolia* L.)	Ni, Pb	(Rasouli et al., 2023)
Soapwort (*Saponaria officinalis* L.)	Pb, As, Cd	(Nujkić et al., 2020)
Sweet wormwood (*Artemisia annua* L.)	As	(Kumari and Pandey-Rai 2018)
Onion (*Allium cepa* L.)	Se	(Karasinski et al., 2017)
Sunflower (*Helianthus annuus*)	Persistent organic contaminants	(Weyens et al., 2009)
Mosquito fern (*Azolla* sp.)	antibiotics	(Maldonado et al., 2022)
Common ivy (*Hedera helix* L.)	Toluene	(Dela Cruz et al., 2014)
Alfalfa (*Medicago sativa*)	PAHs	(Xiao et al., 2015)
Vetiver (*Vetiveria zizanioides* L. Nash)	Hydrocarbons	(Brandt et al., 2006)
Sweet flag (*Acorus calamus*)	Diclofenac, sulfamethoxazole	(Sochacki et al., 2018)
Indian shot (*Canna indica*)	Ibuprofen, naproxen, ketoprofen	(Zhang et al., 2018)

Phytostabilization occurs when pollutants are stored in the plant roots or sequestrated by the root exudates to avoid their migration. This mechanism typically stabilizes heavy metals by metal-chelating proteins (Peer et al., 2006; Skuza et al., 2022). Phytovolatilization is a mechanism with which the plant converts pollutants into more volatile forms for excretion by the stomata; usually it involves organic compounds and some volatile heavy metals such as Hg and As. Some volatile organic compounds are released without the need for biotransformation processes but only through transpiration (Tollsten and Müller, 1996). Phytoextraction takes advantage of the ability of some species to bioaccumulate high amounts of contaminants (Jacobs et al., 2018). Phytofiltration involves the same mechanism as phytostabilization but is mainly used to detoxify the water compartment (Moreno et al., 2008). Rhizodegradation involves the microbiome activity of symbiotic species in the rhizosphere (Zahoor et al., 2017). Phytodegradation occurs when pollutants are translocated in the inner parts of the plant and undergo biotransformations that reduce their toxicity (Nnaji et al., 2023). Some examples of aromatic/medicinal plants used for phytoremediation are reported in Table 11.2.

References

Abdollahi Mandoulakani, B., Eyvazpour, E., Ghadimzadeh, M., 2017. The effect of drought stress on the expression of key genes involved in the biosynthesis of phenylpropanoids and essential oil components in basil (*Ocimum basilicum* L.). Phytochemistry 139, 1–7. https://doi.org/10.1016/J.PHYTOCHEM.2017.03.006.

Adeel, M., Zain, M., Shakoor, N., Azeem, I., Hussain, M., Ahmad, M.A., Chaudhary, S., Zaheer, U., Aziz, M.A., Ahmar, S., Yukui, R., Xu, M., 2023. Estrogens in plants and emerging risks to human health. Environ. Int. 178. https://doi.org/10.1016/j.envint.2023.107985.

Agrawal, M., 2003. Plant Responses to Atmospheric Sulphur. Springer Science and Business Media LLC, pp. 279–293. https://doi.org/10.1007/978-94-017-0289-8_15.

Al-Qahtani, H., Alfarhan, A.H., Al-Othman, Z.M., 2020. Changes in chemical composition of *Zilla spinosa* Forssk. medicinal plants grown in Saudi Arabia in response to spatial and seasonal variations. Saudi J. Biol. Sci. 27, 2756–2769. https://doi.org/10.1016/j.sjbs.2020.06.035.

Alsafran, M., Usman, K., Ahmed, B., Rizwan, M., Saleem, M.H., Al Jabri, H., 2022. Understanding the phytoremediation mechanisms of potentially toxic elements: a proteomic overview of recent advances. Front. Plant Sci. 13. https://doi.org/10.3389/fpls.2022.881242.

Anand, P., Mina, U., Khare, M., Kumar, P., Kota, S.H., 2022. Air pollution and plant health response-current status and future directions. Atmos. Pollut. Res. 13 (8). https://doi.org/10.1016/j.apr.2022.101508.

Aqeel, U., Aftab, T., Khan, M.M.A., Naeem, M., 2023. Regulation of essential oil in aromatic plants under changing environment. J. Appl. Res. Med. Aromat. Plants 32, 100441. https://doi.org/10.1016/j.jarmap.2022.100441.

Asare, M.O., Száková, J., Tlustoš, P., 2023. The fate of secondary metabolites in plants growing on Cd-, As-, and Pb-contaminated soils—a comprehensive review. Environ. Sci. Pollut. Res. 30, 11378–11398. https://doi.org/10.1007/s11356-022-24776-x.

Aziz, E.E., Al-Amier, H., Craker, L.E., 2008. Influence of salt stress on growth and essential oil production in peppermint, pennyroyal, and apple mint. J. Herbs. Spices Med. Plants 14, 77–87. https://doi.org/10.1080/10496470802341375.

Azizollahi, Z., Ghaderian, S.M., Ghotbi-Ravandi, A.A., 2019. Cadmium accumulation and its effects on physiological and biochemical characters of summer savory (*Satureja hortensis* L.). Int. J. Phytoremediation 21, 1241–1253. https://doi.org/10.1080/15226514.2019.1619163.

Bahreininejad, B., Razmjoo, J., Mirza, M., 2014. Effect of water stress on productivity and essential oil content and composition of Thymus carmanicus. J. Essent. Oil Bear. Plants 17, 717–725. https://doi.org/10.1080/0972060X.2014.901605.

Ben Mordechay, E., Tarchitzky, J., Chen, Y., Shenker, M., Chefetz, B., 2018. Composted biosolids and treated wastewater as sources of pharmaceuticals and personal care products for plant uptake: a case study with carbamazepine. Environ. Pollut. 232, 164–172. https://doi.org/10.1016/j.envpol.2017.09.029.

Bi, M., He, Q., Chen, Y., 2020. What roles are terrestrial plants playing in global microplastic cycling? Environ. Sci. Technol. 54, 5325–5327. https://doi.org/10.1021/ACS.EST.0C01009.

Brandt, R., Merkl, N., Schultze-Kraft, R., Infante, C., Broll, G., 2006. Potential of vetiver (Vetiveria zizanioides (L.) Nash) for phytoremediation of petroleum hydrocarbon-contaminated soils in Venezuela. Int. J. Phytoremediation 8 (4), 273–284. https://doi.org/10.1080/15226510600992808.

Chakraborty, K., Sairam, R.K., Bhaduri, D., 2016. Effects of different levels of soil salinity on yield attributes, accumulation of nitrogen, and micronutrients in *Brassica* spp. J. Plant Nutr. 39, 1026–1037. https://doi.org/10.1080/01904167.2015.1109105.

Chang, X., Fang, Y., Wang, Y., Wang, F., Shang, L., Zhong, R., 2022. Microplastic pollution in soils, plants, and animals: a review of distributions, effects and potential mechanisms. Sci. Total Environ. 850, 157857. https://doi.org/10.1016/j.scitotenv.2022.157857.

Christou, A., Papadavid, G., Dalias, P., Fotopoulos, V., Michael, C., Bayona, J.M., Piña, B., Fatta-Kassinos, D., 2019. Ranking of crop plants according to their potential to uptake and accumulate contaminants of emerging concern. Environ. Res. 170, 422–432. https://doi.org/10.1016/j.envres.2018.12.048.

Chrysargyris, A., Papakyriakou, E., Petropoulos, S.A., Tzortzakis, N., 2019. The combined and single effect of salinity and copper stress on growth and quality of Mentha spicata plants. J. Hazard Mater. 368, 584–593. https://doi.org/10.1016/j.jhazmat.2019.01.058.

Coban, F., Ozer, H., Ors, S., Sahin, U., Yildiz, G., Cakmakci, T., 2018. Effects of deficit irrigation on essential oil composition and yield of fennel (Foeniculum vulgare Mill) in a high-altitude environment. J. Essent. Oil Res. 30 (6), 457–463. https://doi.org/10.1080/10412905.2018.1496156.

da Costa Filho, B.M., Duarte, A.C., Santos, T.A.P.R., 2022. Environmental monitoring approaches for the detection of organic contaminants in marine environments: a critical review. Trends Environ. Anal. Chem. 33, e00154. https://doi.org/10.1016/J.TEAC.2022.E00154.

da Cunha Honorato, A., Nohara, G.A., de Assis, R.M.A., Maciel, J.F.A., de Carvalho, A.A., Pinto, J.E.B.P., Bertolucci, S.K.V., 2023. Colored shade nets and different harvest times alter the growth, antioxidant status, and quantitative attributes of glandular trichomes and essential oil of Thymus vulgaris L. J. Appl. Res. Med. Aromat. Plants 35. https://doi.org/10.1016/j.jarmap.2023.100474.

Dantas, D.L.L., Pereira, G.E., de Souza, A.L., dos Santos Lima, M., 2023. Chemometric analysis for authentication of 'Syrah' and 'Tempranillo' red wines of San Francisco Valley-Brazil compared to wines from other world regions by the molecular profile in HPLC. J. Food Sci. Technol. 60 (7), 2050–2062. https://doi.org/10.1007/s13197-023-05739-7.

Dehghani Bidgoli, R., Azarnezhad, N., Akhbari, M., Ghorbani, M., 2019. Salinity stress and PGPR effects on essential oil changes in Rosmarinus officinalis L. Agric. Food Secur. 8 (1). https://doi.org/10.1186/s40066-018-0246-5.

Dela Cruz, M., Müller, R., Svensmark, B., Pedersen, J.S., Christensen, J.H., 2014. Assessment of volatile organic compound removal by indoor plants-a novel experimental setup. Environ. Sci. Pollut. Control Ser. 21 (13), 7838–7846. https://doi.org/10.1007/s11356-014-2695-0.

Di Marco, G., Gismondi, A., Canuti, L., Scimeca, M., Volpe, A., Canini, A., Pichersky, E., 2014. Tetracycline accumulates in I beris sempervirens L . through apoplastic transport inducing oxidative stress and growth inhibition. Plant Biol. 16 (4), 792–800. https://doi.org/10.1111/plb.12102.

Doucette, W.J., Shunthirasingham, C., Dettenmaier, E.M., Zaleski, R.T., Fantke, P., Arnot, J.A., 2018. A review of measured bioaccumulation data on terrestrial plants for organic chemicals: metrics, variability, and the need for standardized measurement protocols. Environ. Toxicol. Chem. 37 (1), 21–33. https://doi.org/10.1002/etc.3992.

Duan, X., Gu, H., Lam, S.S., Sonne, C., Lu, W., Li, H., Chen, X., Peng, W., 2024. Recent progress on phytoremediation of urban air pollution. Chemosphere 349, 140821. https://doi.org/10.1016/j.chemosphere.2023.140821.

Farsaraei, S., Moghaddam, M., Pirbalouti, A.G., 2020. Changes in growth and essential oil composition of sweet basil in response of salinity stress and superabsorbents application. Sci. Hortic. (Amsterdam) 271, 109465. https://doi.org/10.1016/J.SCIENTA.2020.109465.

Fatima, S., Sehgal, A., Mishra, S.K., Mina, U., Goel, V., Vijayan, N., Tawale, J.S., Kothari, R., Ahlawat, A., Sharma, C., 2021. Particle composition and morphology over urban environment (New Delhi): plausible effects on wheat leaves. Environ. Res. 202, 111552. https://doi.org/10.1016/j.envres.2021.111552.

Fellet, G., Pošćić, F., Licen, S., Marchiol, L., Musetti, R., Tolloi, A., Barbieri, P., Zerbi, G., 2016. PAHs accumulation on leaves of six evergreen urban shrubs: a field experiment. Atmos. Pollut. Res. 7 (5), 915–924. https://doi.org/10.1016/j.apr.2016.05.007.

Furze, J.M., Rhodes, M.J.C., Parr, A.J., Robins, R.J., Withehead, I.M., Threlfall, D.R., 1991. Abiotic factors elicit sesquiterpenoid phytoalexin production but not alkaloid production in transformed root cultures of Datura stramonium. Plant Cell Rep. 10 (3), 111–114. https://doi.org/10.1007/BF00232039.

Gao, F., Shen, Y., Sallach, J.B., Li, H., Liu, C., Li, Y., 2021. Direct prediction of bioaccumulation of organic contaminants in plant roots from soils with machine learning models based on molecular structures. Environ. Sci. Technol. 55 (24), 16358–16368. https://doi.org/10.1021/acs.est.1c02376.

Gharib, F.A., Mansour, K.H., Ahmed, E.Z., Galal, T.M., 2021. Heavy metals concentration, and antioxidant activity of the essential oil of the wild mint (Mentha longifolia L.) in the Egyptian watercourses. Int. J. Phytoremediation 23 (6), 641–651. https://doi.org/10.1080/15226514.2020.1847035.

Ghasemzadeh, A., Jaafar, H.Z.E., Rahmat, A., Wahab, P.E.M., Halim, M.R.A., 2010. Effect of different light intensities on total phenolics and flavonoids synthesis and anti-oxidant activities in young ginger varieties (Zingiber officinale Roscoe). Int. J. Mol. Sci. 11 (10), 3885–3897. https://doi.org/10.3390/ijms11103885.

Ghosh, S., Dutta, R., Mukhopadhyay, S., 2023. A review on seasonal changes in particulate matter accumulation by plant bioindicators: effects on leaf traits. Water. Air. Soil Pollut. 234, 1–18. https://doi.org/10.1007/S11270-023-06549-5/FIGURES/2.

Giannoulis, K.D., Kamvoukou, C.A., Gougoulias, N., Wogiatzi, E., 2020. Irrigation and nitrogen application affect Greek oregano (Origanum vulgare ssp. hirtum) dry biomass, essential oil yield and composition. Ind. Crop. Prod. 150. https://doi.org/10.1016/j.indcrop.2020.112392.

Giesy, J.P., Kannan, K., 2002. Peer reviewed: perfluorochemical surfactants in the environment. Environ. Sci. Technol. 36 (7), 146A. https://doi.org/10.1021/es022253t.

Giorgetti, L., Spanò, C., Muccifora, S., Bottega, S., Barbieri, F., Bellani, L., Castiglione, M.R., 2020. Exploring the interaction between polystyrene nanoplastics and Allium cepa during germination: internalization in root cells, induction of toxicity and oxidative stress. Plant Physiol. Biochem. 149, 170–177. https://doi.org/10.1016/j.plaphy.2020.02.014.

Gonçalves, S., Mansinhos, I., Romano, A., 2024. Influence of biotic and abiotic factors on the production and composition of essential oils. Essent. Oils Extr. Methods Appl. 69–97. https://doi.org/10.1002/9781119829614.CH4.

Guo, X.R., Yang, L., Yu, J.H., Tang, Z.H., Zu, Y.G., 2007. Alkaloid variations in Catharanthus roseus seedlings treated by different temperatures in short term and long term. J. For. Res. 18 (4), 313–315. https://doi.org/10.1007/s11676-007-0063-3.

Gupta, N., Yadav, K.K., Kumar, V., Kumar, S., Chadd, R.P., Kumar, A., 2019. Trace elements in soil-vegetables interface: translocation, bioaccumulation, toxicity and amelioration - a review. Sci. Total Environ. 651, 2927–2942. https://doi.org/10.1016/j.scitotenv.2018.10.047.

Gworek, B., Klimczak, K., Kijeńska, M., Gozdowski, D., 2016. Comparison of PAHs uptake by selected Monocotyledones and Dicotyledones from municipal and industrial sewage sludge. Environ. Sci. Pollut. Control Ser. 23 (19), 19461–19470. https://doi.org/10.1007/s11356-016-7130-2.

Heydari, M., Zanfardino, A., Taleei, A., Bushehri, A.A.S., Hadian, J., Maresca, V., Sorbo, S., Napoli, M.D., Varcamonti, M., Basile, A., Rigano, D., 2018. Effect of heat stress on yield, monoterpene content and antibacterial activity of essential oils of *Mentha* x *piperita* var. Mitcham and *Mentha arvensis* var. piperascens. Molecules 23, 1903. https://doi.org/10.3390/MOLECULES23081903.

Hlihor, R.M., Roșca, M., Hagiu-Zaleschi, L., Simion, I.M., Daraban, G.M., Stoleru, V., 2022. Medicinal plant growth in heavy metals contaminated soils: responses to metal stress and induced risks to human health. Toxics 10 (9). https://doi.org/10.3390/toxics10090499.

Hu, Y., Bellaloui, N., Tigabu, M., Wang, J., Diao, J., Wang, K., Yang, R., Sun, G., 2015. Gaseous NO2 effects on stomatal behavior, photosynthesis and respiration of hybrid poplar leaves. Acta Physiol. Plant. 37 (2). https://doi.org/10.1007/s11738-014-1749-8.

Hua, Z., Ma, S., Ouyang, Z., Liu, P., Qiang, H., Guo, X., 2023. The review of nanoplastics in plants: detection, analysis, uptake, migration and risk. TrAC, Trends Anal. Chem. 158, 116889. https://doi.org/10.1016/j.trac.2022.116889.

Huang, D., Chen, H., Shen, M., Tao, J., Chen, S., Yin, L., Zhou, W., Wang, X., Xiao, R., Li, R., 2022. Recent advances on the transport of microplastics/nanoplastics in abiotic and biotic compartments. J. Hazard Mater. 438, 129515. https://doi.org/10.1016/j.jhazmat.2022.129515.

Hwang, J.I., Lee, S.E., Kim, J.E., 2015. Plant uptake and distribution of endosulfan and its sulfate metabolite persisted in soil. PLoS One 10, e0141728. https://doi.org/10.1371/JOURNAL.PONE.0141728.

Jacobs, A., Drouet, T., Noret, N., 2018. Field evaluation of cultural cycles for improved cadmium and zinc phytoextraction with *Noccaea caerulescens*. Plant Soil 430, 381–394. https://doi.org/10.1007/S11104-018-3734-2/TABLES/5.

Jamloki, A., Bhattacharyya, M., Nautiyal, M.C., Patni, B., 2021. Elucidating the relevance of high temperature and elevated CO2 in plant secondary metabolites (PSMs) production. Heliyon 7 (8), e07709. https://doi.org/10.1016/j.heliyon.2021.e07709.

Kalugina, O.V., Mikhailova, T.A., Shergina, O.V., 2017. *Pinus sylvestris* as a bio-indicator of territory pollution from aluminum smelter emissions. Environ. Sci. Pollut. Res. 24, 10279–10291. https://doi.org/10.1007/S11356-017-8674-5/FIGURES/8.

Kanakidou, M., Myriokefalitakis, S., Tsigaridis, K., 2018. Aerosols in atmospheric chemistry and biogeochemical cycles of nutrients. Environ. Res. Lett. https://doi.org/10.1088/1748-9326/aabcdb.

Karasinski, J., Wrobel, K., Escobosa, A.R.C., Konopka, A., Bulska, E., Wrobel, K., Allium, C.L., 2017. Response to sodium selenite (Se(IV)) studied in plant roots by a LC-MS-based proteomic approach. J. Agric. Food Chem. 65 (19), 3995–4004. https://doi.org/10.1021/acs.jafc.7b01085.

Kumar, M., Xiong, X., He, M., Tsang, D.C.W., Gupta, J., Khan, E., Harrad, S., Hou, D., Ok, Y.S., Bolan, N.S., 2020. Microplastics as pollutants in agricultural soils. Environ. Pollut. 265. https://doi.org/10.1016/j.envpol.2020.114980.

Kaseke, T., Opara, U.L., Fawole, O.A., 2020. Fatty acid composition, bioactive phytochemicals, antioxidant properties and oxidative stability of edible fruit seed oil: effect of preharvest and processing factors. Heliyon 6. https://doi.org/10.1016/j.heliyon.2020.e04962.

Kumar, R., Kaundal, M., Sharma, S., Thakur, M., Kumar, N., Kaur, T., Vyas, D., Kumar, S., 2017. Effect of elevated [CO 2] and temperature on growth, physiology and essential oil composition of Salvia sclarea L. in the western Himalayas. J. Appl. Res. Med. Aromat. Plants 6, 22–30. https://doi.org/10.1016/j.jarmap.2017.01.001.

Kumar, R., Sharma, S., Pathania, V., 2013. Effect of shading and plant density on growth, yield and oil composition of clary sage (*Salvia sclarea* L.) in north western Himalaya. J. Essent. Oil Res. 25, 23–32. https://doi.org/10.1080/10412905.2012.742467.

Kumar, R., Sharma, S., Ramesh, K., Pathania, V., Prasad, R., 2014. Irradiance stress and plant spacing effect on growth, biomass and quality of wild marigold (Tagetes minuta L.) – an industrial crop in western Himalaya. J. Essent. Oil Res. 26 (5), 348–358. https://doi.org/10.1080/10412905.2014.935497.

Kumari, A., Pandey-Rai, S., 2018. Enhanced arsenic tolerance and secondary metabolism by modulation of gene expression and proteome profile in Artemisia annua L. after application of exogenous salicylic acid. Plant Physiol. Biochem. 132, 590–602. https://doi.org/10.1016/j.plaphy.2018.10.010.

Lan, Z., Zhou, M., Yao, Y., Sun, H., 2018. Plant uptake and translocation of perfluoroalkyl acids in a wheat–soil system. Environ. Sci. Pollut. Control Ser. 25 (31), 30907–30916. https://doi.org/10.1007/s11356-018-3070-3.

Le, V.G., Nguyen, M.K., Nguyen, H.L., Lin, C., Hadi, M., Hung, N.T.Q., Hoang, H.G., Nguyen, K.N., Tran, H.T., Hou, D., Zhang, T., Bolan, N.S., 2023. A comprehensive review of micro- and nano-plastics in the atmosphere: occurrence, fate, toxicity, and strategies for risk reduction. Nam Sci. Total Environ. 904. https://doi.org/10.1016/j.scitotenv.2023.166649.

Lee, B.X.Y., Hadibarata, T., Yuniarto, A., 2020. Phytoremediation Mechanisms in Air Pollution Control: a Review. Water, Air, Soil Pollut. 231, 437. https://doi.org/10.1007/s11270-020-04813-6.

Li, Y., Kong, D., Fu, Y., Sussman, M.R., Wu, H., 2020. The effect of developmental and environmental factors on secondary metabolites in medicinal plants. Plant Physiol. Biochem. 148, 80–89. https://doi.org/10.1016/j.plaphy.2020.01.006.

Li, J., Sun, J., Li, P., 2022. Exposure routes, bioaccumulation and toxic effects of per- and polyfluoroalkyl substances (PFASs) on plants: a critical review. Environ. Int. https://doi.org/10.1016/j.envint.2021.106891.

Lian, J., Liu, W., Meng, L., Wu, J., Chao, L., Zeb, A., Sun, Y., 2021. Foliar-applied polystyrene nanoplastics (PSNPs) reduce the growth and nutritional quality of lettuce (Lactuca sativa L.). Environ. Pollut. 280, 116978. https://doi.org/10.1016/j.envpol.2021.116978.

Lin, H., Tao, S., Zuo, Q., Coveney, R.M., 2007. Uptake of polycyclic aromatic hydrocarbons by maize plants. Environ. Pollut. 148 (2), 614–619. https://doi.org/10.1016/j.envpol.2006.11.026.

Liu, Y., Ben, Y., Che, R., Peng, C., Li, J., Wang, F., 2023. Uptake, transport and accumulation of micro- and nanoplastics in terrestrial plants and health risk associated with their transfer to food chain - a mini review. Sci. Total Environ. 902, 166045. https://doi.org/10.1016/j.scitotenv.2023.166045.

Ložienė, K., Vaičiulytė, V., Maždžierienė, R., 2021. Influence of meteorological conditions on essential oil composition in geraniol-bearing *Thymus pulegioides* and *Thymus* hybrid. Acta Physiol. Plant. 43, 1–9. https://doi.org/10.1007/S11738-021-03201-9/TABLES/4.

Luo, M., Liu, X., Legesse, N., Liu, Y., Wu, S., Han, F.X., Ma, Y., 2023. Evaluation of agricultural non-point source pollution: a review. Water Air Soil Pollut. 234 (10). https://doi.org/10.1007/s11270-023-06686-x.

Maddela, N.R., Ramakrishnan, B., Dueñas-Rivadeneira, A.A., Venkateswarlu, K., Megharaj, M., 2022. Chemicals/materials of emerging concern in farmlands: sources, crop uptake and potential human health risks. Environ. Sci.: Process. Impacts 24 (12), 2217–2236. https://doi.org/10.1039/d2em00322h.

Maguire, L.W., Gardner, C.M., 2023. Fate and transport of biological microcontaminants bound to microplastics in the soil environment. Sci. Total Environ. 892. https://doi.org/10.1016/j.scitotenv.2023.164439.

Mahmood, F., Khokhar, M.F., Mahmood, Z., 2020. Examining the relationship of tropospheric ozone and climate change on crop productivity using the multivariate panel data techniques. J. Environ. Manage. 272, 111024. https://doi.org/10.1016/J.JENVMAN.2020.111024.

Maldonado, I., Moreno Terrazas, E.G., Vilca, F.Z., 2022. Application of duckweed (*Lemna* sp.) and water fern (*Azolla* sp.) in the removal of pharmaceutical residues in water: state of art focus on antibiotics. Sci. Total Environ. 838, 156565. https://doi.org/10.1016/j.scitotenv.2022.156565.

Mandal, M., Das, S., Roy, A., Rakwal, R., Jones, O.A.H., Popek, R., Agrawal, G.K., Sarkar, A., 2023. Interactive relations between plants, the phyllosphere microbial community, and particulate matter pollution. Sci. Total Environ. 890. https://doi.org/10.1016/j.scitotenv.2023.164352.

Manwani, S., Devi, P., Singh, T., Yadav, C.S., Awasthi, K.K., Bhoot, N., Awasthi, G., 2022. Heavy metals in vegetables: a review of status, human health concerns, and management options. Environ. Sci. Pollut. Res. 30 (28), 71940–71956. https://doi.org/10.1007/S11356-022-22210-W.

Masetti, O., Sorbo, A., Nisini, L., 2021. Nmr tracing of food geographical origin: the impact of seasonality, cultivar and production year on data analysis. Separations 8. https://doi.org/10.3390/separations8120230.

Metrak, M., Aneta, E., Wiłkomirski, B., Staszewski, T., Suska-Malawska, M., 2016. Interspecific differences in foliar 1 PAHs load between Scots pine, birch, and wild rosemary from three polish peat bogs. Environ. Monit. Assess. 188 (8). https://doi.org/10.1007/s10661-016-5465-2.

Mikes, O., Cupr, P., Trapp, S., Klanova, J., 2009. Uptake of polychlorinated biphenyls and organochlorine pesticides from soil and air into radishes (Raphanus sativus). Environ. Pollut. 157 (2), 488–496. https://doi.org/10.1016/j.envpol.2008.09.007.

Mishra, B., Chandra, M., 2022. Evaluation of phytoremediation potential of aromatic plants: a systematic review. J. Appl. Res. Med. Aromat. Plants 31, 100405. https://doi.org/10.1016/j.jarmap.2022.100405.

Moreno, F.N., Anderson, C.W.N., Stewart, R.B., Robinson, B.H., 2008. Phytofiltration of mercury-contaminated water: volatilisation and plant-accumulation aspects. Environ. Exp. Bot. 62 (1), 78–85. https://doi.org/10.1016/j.envexpbot.2007.07.007.

Mukarram, M., Khan, M.M.A., Zehra, A., Petrik, P., Kurjak, D., 2022. Suffer or survive: decoding salt-sensitivity of lemongrass and its implication on essential oil productivity. Front. Plant Sci. 13. https://doi.org/10.3389/fpls.2022.903954.

Nadgórska–Socha, A., Kandziora-Ciupa, M., Trzesicki, M., Barczyk, G., 2017. Air pollution tolerance index and heavy metal bioaccumulation in selected plant species from urban biotopes. Chemosphere 183, 471–482. https://doi.org/10.1016/j.chemosphere.2017.05.128.

Naziri, A., Mina, T., Manoli, K., Beretsou, V.G., Christou, A., Michael, C., Agathokleous, E., Fatta-Kassinos, D., 2023. Looking into the effects of co-contamination by micro(nano)plastics in the presence of other pollutants on irrigated edible plants. Sci. Total Environ. 892. https://doi.org/10.1016/j.scitotenv.2023.164618.

Nedjimi, B., 2021. Phytoremediation: a sustainable environmental technology for heavy metals decontamination. SN Appl. Sci. 3 (3). https://doi.org/10.1007/s42452-021-04301-4.

Nnaji, N.D., Onyeaka, H., Miri, T., Ugwa, C., 2023. Bioaccumulation for heavy metal removal: a review. SN Appl. Sci. 5 (5). https://doi.org/10.1007/s42452-023-05351-6.

Nujkić, M., Milić, S., Spalović, B., Dardas, A., Alagić, S., Ljubić, D., Papludis, A., Saponaria, O.L., Achillea, M.L., 2020. As possible indicators of trace elements pollution caused by mining and metallurgical activities in Bor, Serbia. Environ. Sci. Pollut. Control Ser. 27 (36), 44969–44982. https://doi.org/10.1007/s11356-020-10371-5.

Peer, W.A., Baxter, I.R., Richards, E.L., Freeman, J.L., Murphy, A.S., 2006. Phytoremediation and hyper-accumulator plants. Top. Curr. Genet. 14, 299–340. https://doi.org/10.1007/4735_100.

Picó, Y., Campo, J., Alfarhan, A.H., El-Sheikh, M.A., Barceló, D., 2023. Wild and ruderal plants as bioindicators of global urban pollution by air, water and soil in Riyadh and Abha, Saudi Arabia. Sci. Total Environ. 888. https://doi.org/10.1016/j.scitotenv.2023.164166.

Popek, R., Łukowski, A., Grabowski, M., 2018. Influence of particulate matter accumulation on photosynthetic apparatus of *Physocarpus opulifolius* and *Sorbaria sorbifolia*. Polish J. Environ. Stud. 27, 2391–2396. https://doi.org/10.15244/PJOES/78626.

Przybysz, A., Sæbø, A., Hanslin, H.M., Gawroński, S.W., 2014. Accumulation of particulate matter and trace elements on vegetation as affected by pollution level, rainfall and the passage of time. Sci. Total Environ. 481 (1), 360–369. https://doi.org/10.1016/j.scitotenv.2014.02.072.

Pullagurala, V.L.R., Rawat, S., Adisa, I.O., Hernandez-Viezcas, J.A., Peralta-Videa, J.R., Gardea-Torresdey, J.L., 2018. Plant uptake and translocation of contaminants of emerging concern in soil. Sci. Total Environ. 636, 1585–1596. https://doi.org/10.1016/j.scitotenv.2018.04.375.

Qadir, S.U., Raja, V., Siddiqui, W.A., Zzafar, M., Abd_allah, E.F., Hashem, A., Alam, P., Ahmad, P., 2019. Fly-ash pollution modulates growth, biochemical attributes, antioxidant activity and gene expression in pith-ecellobium dulce (Roxb) benth. Plants 8 (12). https://doi.org/10.3390/plants8120528.

Rai, V., Vajpayee, P., Singh, S.N., Mehrotra, S., 2004. Effect of chromium accumulation on photosynthetic pigments, oxidative stress defense system, nitrate reduction, proline level and eugenol content of Ocimum tenuiflorum L. Plant Sci. 167 (5), 1159–1169. https://doi.org/10.1016/j.plantsci.2004.06.016.

Ramezani, S., Abbasi, A., Sobhanverdi, S., Shojaeiyan, A., Ahmadi, N., 2020. The effects of water deficit on the expression of monoterpene synthases and essential oils composition in Salvia ecotypes. Physiol. Mol. Biol. Plants 26 (11), 2199–2207. https://doi.org/10.1007/s12298-020-00892-1.

Rasouli, F., Hassanpouraghdam, M.B., Pirsarandib, Y., Aazami, M.A., Asadi, M., Ercisli, S., Mehrabani, L.V., Puglisi, I., Baglieri, A., 2023. Improvements in the biochemical responses and Pb and Ni phytoremediation of lavender (*Lavandula angustifolia* L.) plants through *Funneliformis mosseae* inoculation. BMC Plant Biol. 23. https://doi.org/10.1186/s12870-023-04265-0.

Saeedi, M., 2023. How microplastics interact with food chain: a short overview of fate and impacts. J. Food Sci. Technol. https://doi.org/10.1007/s13197-023-05720-4.

Sandermann, H., 1994. Higher plant metabolism of xenobiotics: the 'green liver' concept. Pharmacogenetics 4 (5), 225–241. https://doi.org/10.1097/00008571-199410000-00001.

Shahid, M., Dumat, C., Khalid, S., Schreck, E., Xiong, T., Niazi, N.K., 2017. Foliar heavy metal uptake, toxicity and detoxification in plants: a comparison of foliar and root metal uptake. J. Hazard Mater. 325, 36–58. https://doi.org/10.1016/j.jhazmat.2016.11.063.

Sinha, B., Singh Sangwan, K., Maurya, Y., Kumar, V., Sarkar, C., Chandra, B.P., Sinha, V., 2015. Assessment of crop yield losses in Punjab and Haryana using 2 years of continuous in situ ozone measurements. Atmos. Chem. Phys. 15 (16), 9555–9576. https://doi.org/10.5194/acp-15-9555-2015.

Skuza, L., Szućko-Kociuba, I., Filip, E., Bożek, I., 2022. Natural molecular mechanisms of plant hyperaccumulation and hypertolerance towards heavy metals. Int. J. Mol. Sci. 23 (16), 9335. https://doi.org/10.3390/ijms23169335.

Sochacki, A., Felis, E., Bajkacz, S., Nowrotek, M., Miksch, K., 2018. Removal and transformations of diclofenac and sulfamethoxazole in a two-stage constructed wetland system. Ecol. Eng. 122, 159–168. https://doi.org/10.1016/j.ecoleng.2018.07.039.

Sulastri, Y.S., Mukhlis, Damanik, R.I., Sabrina, T., 2022. Study of aromatic plants on cadmium exposure through anatomical structure, shoot root ratio, essential oil and heavy metals. Asian J. Plant Sci. 21 (4), 565–573. https://doi.org/10.3923/ajps.2022.565.573.

Tian, R., Zhang, R., Uddin, M., Qiao, X., Chen, J., Gu, G., 2019. Uptake and metabolism of clarithromycin and sulfadiazine in lettuce. Environ. Pollut. 247, 1134–1142. https://doi.org/10.1016/j.envpol.2019.02.009.

Tollsten, L., Müller, P.M., 1996. Volatile organic compounds emitted from beech leaves. Phytochemistry 43 (4), 759–762. https://doi.org/10.1016/0031-9422(96)00272-5.

Viana, V.E., Aranha, B.C., Busanello, C., Maltzahn, L.E., Panozzo, L.E., de Oliveira, A.C., Rombaldi, C.V., Pegoraro, C., 2022. Metabolic profile of canola (Brassica napus L.) seedlings under hydric, osmotic and temperature stresses. Plant Stress 3. https://doi.org/10.1016/j.stress.2022.100059.

Vwioko, D.E., Okoekhian, I., Ogwu, M.C., 2018. Stress analysis of *Amaranthus hybridus* L. and lycopersicon esculentum Mill. Exposed to sulphur and nitrogen dioxide. Pertanika J. Trop. Agric. Sci. 41, 1169–1191.

Wang, H., Xing, H., Wang, Y.F., Zhai, H., Mei-Ling, T., Du, Y.P., 2020a. Ozone risk assessment of grapevine ‘Cabernet Sauvignon’ using open-top chambers. Sci. Hortic. 260. https://doi.org/10.1016/j.scienta.2019.108874.

Wang, T.T., Ying, G.G., Shi, W.J., Zhao, J.L., Liu, Y.S., Chen, J., Ma, D.D., Xiong, Q., 2020b. Uptake and translocation of Perfluorooctanoic acid (PFOA) and Perfluorooctanesulfonic acid (PFOS) by wetland plants: tissue- A nd cell-level distribution visualization with desorption electrospray ionization mass spectrometry (DESI-MS) and transmission electron microscopy equipped with energy-dispersive spectroscopy (TEM-EDS). Environ. Sci. Technol. 54 (10), 6009–6020. https://doi.org/10.1021/acs.est.9b05160.

Wei, H., Tang, M., Xu, X., 2023. Mechanism of uptake, accumulation, transport, metabolism and phytotoxic effects of pharmaceuticals and personal care products within plants: a review. Sci. Total Environ. 892. https://doi.org/10.1016/j.scitotenv.2023.164413.

Weyens, N., van der Lelie, D., Taghavi, S., Vangronsveld, J., 2009. Phytoremediation: plant–endophyte partnerships take the challenge. Curr. Opin. Biotechnol. 20 (2), 248–254. https://doi.org/10.1016/j.copbio.2009.02.012.

Xiao, F., Zhou, H., 2023. Plant salt response: perception, signaling, and tolerance. Front. Plant Sci. 13. https://doi.org/10.3389/fpls.2022.1053699.

Xiao, N., Liu, R., Jin, C., Dai, Y., 2015. Efficiency of five ornamental plant species in the phytoremediation of polycyclic aromatic hydrocarbon (PAH)-contaminated soil. Ecol. Eng. 75, 384–391. https://doi.org/10.1016/j.ecoleng.2014.12.008.

Yin, S., Tan, H., Hui, N., Ma, Y., Lu, T., Sun, N., Liu, C., 2020. Polycyclic aromatic hydrocarbons in leaves of Cinnamomum camphora along the urban–rural gradient of a megacity: distribution varies in concentration and potential toxicity. Sci. Total Environ. 732, 139328. https://doi.org/10.1016/j.scitotenv.2020.139328.

Zahoor, M., Irshad, M., Rahman, H., Qasim, M., Afridi, S.G., Qadir, M., Hussain, A., 2017. Alleviation of heavy metal toxicity and phytostimulation of Brassica campestris L. by endophytic Mucor sp. MHR-7. Ecotoxicol. Environ. Saf. 142, 139–149. https://doi.org/10.1016/j.ecoenv.2017.04.005.

Zandalinas, S.I., Mittler, R., Balfagón, D., Arbona, V., Gómez-Cadenas, A., 2018. Plant adaptations to the combination of drought and high temperatures. Physiol. Plantarum 162 (1), 2–12. https://doi.org/10.1111/ppl.12540.

Zhang, Q., Shi, Y., Ma, L., Yi, X., Ruan, J., Heil, M., 2014. Metabolomic analysis using ultra-performance liquid chromatography-quadrupole-time of flight mass spectrometry (UPLC-Q-TOF MS) uncovers the effects of light intensity and temperature under shading treatments on the metabolites in tea. PLoS One 9 (11), e112572. https://doi.org/10.1371/journal.pone.0112572.

Zhang, X., Jing, R., Feng, X., Dai, Y., Tao, R., Vymazal, J., Cai, N., Yang, Y., 2018. Removal of acidic pharmaceuticals by small-scale constructed wetlands using different design configurations. Sci. Total Environ. 639, 640–647. https://doi.org/10.1016/j.scitotenv.2018.05.198.

Zhou, Y., Zhou, Z., Lian, Y., Sun, X., Wu, Y., Qiao, L., Wang, M., 2021. Source, transportation, bioaccumulation, distribution and food risk assessment of perfluorinated alkyl substances in vegetables: a review. Food Chem. 349, 129137. https://doi.org/10.1016/j.foodchem.2021.129137.

Zhu, J., Wang, J., Chen, R., Feng, Q., Zhan, X., 2021. Cellular Process of Polystyrene Nanoparticles Entry into Wheat Roots. Environ. Sci. Technol. 2022, 6436–6444. https://doi.org/10.1021/ACS.EST.1C08503.

Zidorn, C., 2010. Altitudinal variation of secondary metabolites in flowering heads of the Asteraceae: trends and causes. Phytochemistry Rev. 9 (2), 197–203. https://doi.org/10.1007/s11101-009-9143-7.

Zobayed, S.M.A., Afreen, F., Kozai, T., 2005. Temperature stress can alter the photosynthetic efficiency and secondary metabolite concentrations in St. John's wort. Plant Physiol. Biochem. 43, 977–984. https://doi.org/10.1016/J.PLAPHY.2005.07.013.

Zsigmond, A.R., Száraz, A., Urák, I., 2021. Macro and trace elements in the black pine needles as inorganic indicators of urban traffic emissions. Environ. Pollut. 291, 118228. https://doi.org/10.1016/J.ENVPOL.2021.118228.

CHAPTER

12 New active principle ingredient (API) (considering new formulations and new functionalized ingredients)

Maria Chiara Cristiano[1], Antonia Mancuso[2], Rosy Cavaliere[2] and Donatella Paolino[2]

[1]*Department of Medical and Surgical Sciences, University Magna Graecia of Catanzaro, Catanzaro, Italy;* [2]*Department of Experimental and Clinical Medicine, University Magna Graecia of Catanzaro, Catanzaro, Italy*

1. Introduction

The physiological functions of the skin are well known, as well as its barrier function to the percutaneous penetration of exogenous substances, both accidentally encountered and deliberately applied, mainly carried out by the stratum corneum (Madison, 2003).

In detail, the *stratum corneum* acts as a two-way barrier, controlling the loss of water, electrolytes, and other body constituents, while preventing the entry of harmful or unwanted molecules from the external environment and limiting the passage of molecules with pharmacological activity. The percutaneous permeation of active compounds is strictly related to the skin structure; the physicochemical characteristics of the penetrant and/or the vehicle in which the penetrant is delivered; the dosage and conditions of administration. Due to the abundant presence of ceramides, cholesterol, and free fatty acids, the *stratum corneum* appears particularly lipophilic, and this composition influences the penetration of active compounds (Menon et al., 2012). For this reason, it is very difficult to think of a passive diffusion through the *stratum corneum* of a molecule characterized by highly hydrophilic nature because it cannot overcome the hydrophobic *stratum corneum* layer. At the same time, a hydrophobic substance is able to enter the *stratum corneum,* but it remains stored inside it due to the hydrophilicity of the next living layer (Bolzinger et al., 2012).

Many natural compounds possess physicochemical properties not suitable for direct passage through the skin, and consequently, their pharmacological action can be exploited only for local conditions, following a topical application. On the contrary, some natural active ingredients could show their pharmacological action on systemic levels or on deeper skin layers. The topical route remains a convenient strategy of administration, since it meets the patient's favor, reduces the risk of widespread side effects and the hepatic first pass effect typical of the oral route of administration. For these reasons, various approaches have been designed and implemented by the scientific community, in such a way as to increase the percutaneous permeation of active molecules of natural origin, using conventional and unconventional, minimally- or no-invasive, specific or no-specific methods.

Phytochemicals for Health. https://doi.org/10.1016/B978-0-443-15366-2.00012-5

These approaches can be divided into two big classes (Patel and Shah, 2018): physical and chemical strategies. The physical methods include stripping, iontophoresis, electroporation, ultrasound (thermal and citational), thermal ablation, mechanical abrasion, and microneedles. Many of these strategies require the use of devices and some of these promises to significantly expand the spectrum of drugs that can be administered independent of their physicochemical characteristics. Unfortunately, the use of this type of approach is not without drawbacks, related to irreversible disruption of stratum corneum, a potential skin discomfort and pain, and above all, an unspecific percutaneous passage of several molecules (Dragicevic and Maibach, 2018).

The chemical strategies to enhance trans(dermal) drug delivery include compounds that interact with the lipid structures of the *stratum corneum* increasing permeability (Williams and Barry, 2012). The most common chemical enhancers are represented by solvent such as ethanol or surfactants such as sodium dodecyl (lauryl) sulfate. The main advantages of their use are related to their low cost, the possibility to include them in conventional topical formulation, and the lack of need for complex device. On the contrary, these chemical enhancers can induce skin irritation or toxicity when present and administered at high concentrations or for a long time of exposure (Dragicevic et al., 2015).

Among the chemical strategies, smart drug delivery systems are much more efficient, selective, and safe. In this chapter, we explain the state of the art of the most common drug delivery systems used to improve the percutaneous permeation of natural compounds.

2. Smart delivery systems

Drug delivery systems are a valid tool to deliver therapeutic agents into the body, with the aim to reach specifically the action site of the entrapped drug. Transdermal drug delivery systems are innovative systems, mostly nanosized, able to improve the passage of the delivered drug through the *stratum corneum*, reaching the deeper skin layers or, in some cases, the systemic circulation.

The main exponents of this category of chemical enhancers are the lipid-based nanocarriers, such as liposomes, ethosomes, transfersomes, niosomes, cubosomes, SLN, NLC, micro- and nanoemulsions. The success of this class of devices is strictly related to their peculiar composition, often made up of ingredients "generally recognized as safe" (GRAS), biodegradable, and highly biocompatible because they often self-organize to form structures that mimic cell membranes.

The use of nanotechnology is gaining ground in various fields of application, such as pharmaceutical, cosmetics, and food fields. In recent years, the use of nanosystems for the delivery of compounds derived from natural sources is receiving particular attention from the scientific world. The inclusion of natural products in lipid-based nanocarriers has many advantages, since the nanosystems are able to improve the stability, solubility, and bioavailability as well as to protect the active compound from external agents such as light and to improve its pharmacological effects (Puglia et al., 2017).

2.1 Liposomes

Liposome is the most famous vesicular systems used in different fields of application and defined as self-assembling and cell-resembling lipid-based colloidal delivery systems (Shishir et al., 2018). They are spherical vesicles made up of phospholipids that form a bilayer surrounding an aqueous core

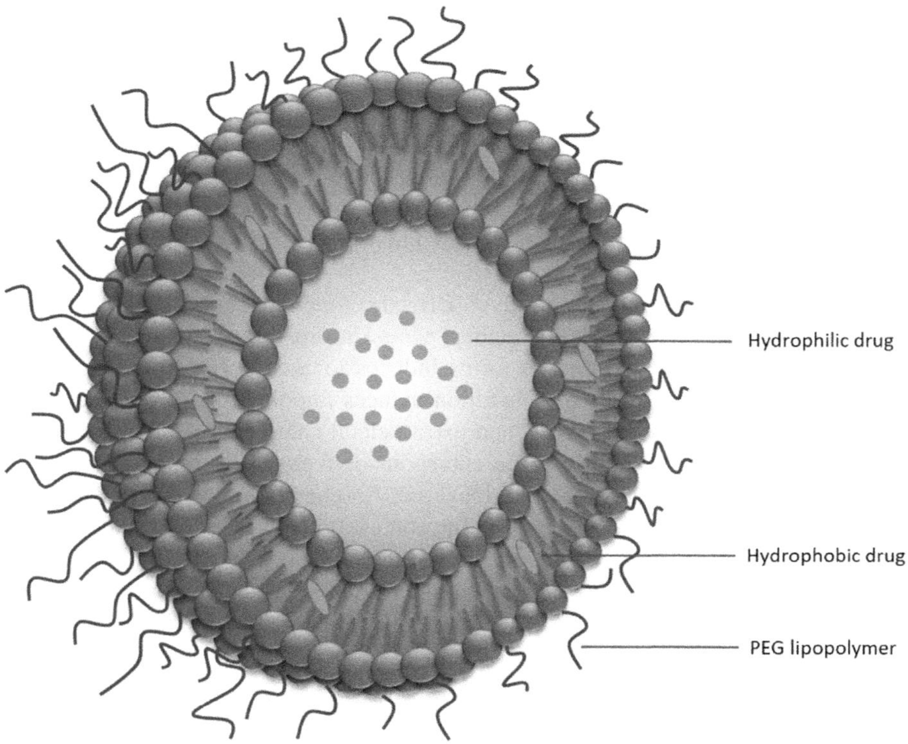

FIGURE 12.1

Schematic representation of a liposomal structure. Liposomes are able to accommodate both hydrophobic and hydrophilic compounds in the phospholipid bilayer or in the internal aqueous core, respectively. The presence of polyethylene glycol (PEG) affects the bilayer packing and liposomal properties.

(Fig. 12.1) (Mancuso et al., 2020). Thanks to this organization and amphiphilic compositions, liposomes are able to include one or more drugs characterized by different physicochemical features. In detail, lipophilic and hydrophilic drugs can enter the phospholipid bilayer and the aqueous core, respectively, while amphiphilic drugs can localize in both compartments.

Liposomes were not designed as topical drug delivery systems, but their use for the treatment of local skin diseases was subsequently investigated. Unfortunately, their rigid structure does not permit to pass stratum corneum and to reach the deeper skin layers as they remained confined to the most superficial layers of skin (Van Tran et al., 2019). Nevertheless, several research groups have investigated the ability of liposomes to guarantee an accumulation of the entrapped drug at the administered site on the skin, since their constituents can interact with the lipid structures of the skin, also leading to an increased skin hydration (Patel et al., 2000; Van Tran et al., 2019). This peculiarity is defined as "penetration enhancement effect" and promotes the local release of the entrapped active compound on the skin, as demonstrated by Masini et al. (1993) (Esposito et al., 2019). The liposomal system, prepared by the authors, was able to improve the skin deposition of the entrapped retinoic acid, probably due to the presence of dipalmitoyl phosphatidylcholine in its composition, which enhanced the interaction with skin lipids. The effects of the topical administration of retinoic acid–loaded liposomes were compared with that recorded by a conventional alcoholic gel containing the same concentration of retinoic acid. Results from in vitro studies highlighted a greater presence of the active into epidermis and dermis (41% and 13%, respectively) when delivered by liposomes compared with the results recorded by the gel form (18% and 8%, respectively). These results were corroborated by in vivo experiments, confirming the potentiality of dipalmitoyl phosphatidylcholine-liposomes to reduce the typical systemic effects of retinoic acid and to improve its local effects against acne disease (Masini et al., 1993).

Retinoic acid is an example of natural product-derived compound that has demonstrated its efficacy against some skin diseases and in particular against acne. Unfortunately, the topical administration of retinoic acid by using conventional dosage forms induces local skin irritation that leads to patient discontinuing therapy, while the oral administration induces important systemic side effects, such as increased serum triglycerides, hyperostosis, and extraskeletal calcification (Beckenbach et al., 2015; Milosheska and Roškar, 2022). Therefore, the cutaneous delivery of retinoic acid represents the solution for overcoming the limitation of its traditional use. Another natural compound able to reduce the signs of facial acne, tea tree oil, was loaded into liposomes. In detail, the liposomes designed by Biju et al. showed a double action, since, on the one hand, they increased the stability and the solubility of encapsulated essential oil, and on the other hand, they were able to improve the specific follicular delivery of the antimicrobial agent, if compared with microemulsion, multiple emulsion and colloidal vesicles, thus permitting to exploit the pharmacological effects at low concentrations (Biju et al., 2005).

Liposomes can be made up of different lipid components, and the composition is the key for the effectiveness of these vesicular systems. The modulation of the lipid composition permits to obtain systems characterized by different features in terms of superficial charge, interaction with skin lipid structure, drug release, and affinity. The importance of the liposomal composition in their efficacy was highlighted by Zhao et al. who compared propylene glycol liposomes systems with traditional liposomes, made up of hydrogenated phosphatidylcholine, cholesterol, and Tween 80, and with ethosomes, which are vesicular systems containing ethanol that will be described in the next section (Zhao et al., 2013). The different nanosystems were prepared as carriers for curcumin, a natural polyphenol having multiple health benefits (Hewlings and Kalman, 2017). In the study of Zhao and coworkers, the results of entrapment efficacy (EE) confirmed the importance of the nanosystems composition, since propylene glycol liposomes showed the greater EE (92.74% $\pm$ 3.44%) with respect to traditional liposomes and ethosomes (70.43% $\pm$ 4.68% and 85.50% $\pm$ 3.87%, respectively) (Zhao et al., 2013). Moreover, in vivo studies showed that topical treatment to carrageenan-induced paw edema in rat, using propylene glycol liposomes, exerted maximum antiinflammatory activity up to 8 hours, and inhibition was maintained up to 24 hours compared with ethosomes and traditional liposomes. These results confirmed that propylene glycol was able to induce an increase in the flexibility of lipid bilayer of liposomes, with respect to ethanol present in ethosomes, and consequently, a greater skin interaction and topical delivery of curcumin.

Another research group that confirmed the importance of the modulation of liposomal composition for their topical administration is represented by Caddeo et al. (2014). In this work, liposomal systems made of polyethylene glycol 400 (PEG400) were designed for quercetin topical delivery. Quercetin is a natural flavonoid characterized by several beneficial effects, such as scavenging of oxygen radicals, prevention of lipid peroxidation, and antiinflammatory activity, and by a safe profile. Its multiple pharmacological activities could be exploited for the treatment of skin pathologies and imperfections, but its poor water solubility and low skin permeability severely limit this use of the active compound (Bose et al., 2013). In this study, PEG was chosen as penetration enhancer, characterized by a double activity when incorporated into liposomes for quercetin delivery. In detail, PEG was able to reduce the interfacial tension between water and a hydrophobic drug, acting as solubilizing agent, and it created a less-polar environment that resulted in more lipophilic active molecules going into solution. In fact, quercetin was entrapped with greater concentration in PEG-liposomes compared with traditional liposomal formulation. On the other hand, PEG was able to show its penetration enhancement ability, reducing the *stratum corneum* barrier function, thus creating an easier pathway for PEG-liposomes, which resulted in themselves being more fluidized thanks to PEG action on the vesicular bilayer. All this has led to good in vivo results on mice topically treated with 12-O-Tetradecanoylphorbol 13-acetate-inducer inflammatory edema, with a significant amelioration of the edema and infiltration after treatments with quercetin-loaded PEG-liposomes compared with traditional liposomes or conventional dosage form of quercetin.

There are scientific evidences for the topical efficacy of liposomes, and in particular, the applicability of liposomes as topical carrier of active compounds is still not ascertained, due to the conflicting scientific results described in the literature. To date, there are two schools of thought: some researchers affirm that liposomes are suitable candidates for transcutaneous delivery (Choi and Maibach, 2005), other ones affirm that conventional liposomes can only carry out a superficial action, not being able to cross the *stratum corneum* (Gupta et al., 2012).

The discordant results may be due to differences in the composition of the vesicles, in the preparation methods, in the transition temperatures of the lipids used, etc.; therefore, they may be due to all the variables in the experimental conditions. In any case, it seems to be a common idea that conventional liposomes are unable to penetrate deep layers of the skin, but they can result in a suitable system for the treatment of "surface" skin diseases (Lai et al., 2020).

As already affirmed, the composition of conventional liposomes (made of phospholipids with/without cholesterol) makes their bilayer particularly rigid, limiting and hindering their passage through the intact *stratum corneum*. Based on the effect of lipid composition on nanocarriers rigidity, new vesicular systems have been designed by modifying the composition and by introducing other additives to the formulation. Cevc and Blume in 1992 demonstrated the efficacy of edge activators, such as sodium cholate, in conferring high deformability to liposomes. Starting from their idea, transfersomes have been obtained. Another modification in liposomes composition permitted the realization of ethosomes, elastic nanosystems characterized by lipid bilayers and amply recognized as topical delivery systems (Cevc and Blume, 1992). The elasticity of ethosomes, which are nanosystems born by an idea of Touitou et al., was a consequence of the addition of high ethanol concentrations in the liposome composition (Touitou et al., 2000).

2.2 Deformable vesicles

2.2.1 Ethosomes

Ethosomes were introduced for the first time in the 1990s, and they were obtained from a modification in liposomal composition. In detail, ethosomes owe their name to the high concentration of ethanol in their composition (20%–50%) (Dayan and Touitou, 2000; Touitou et al., 2000). The presence of

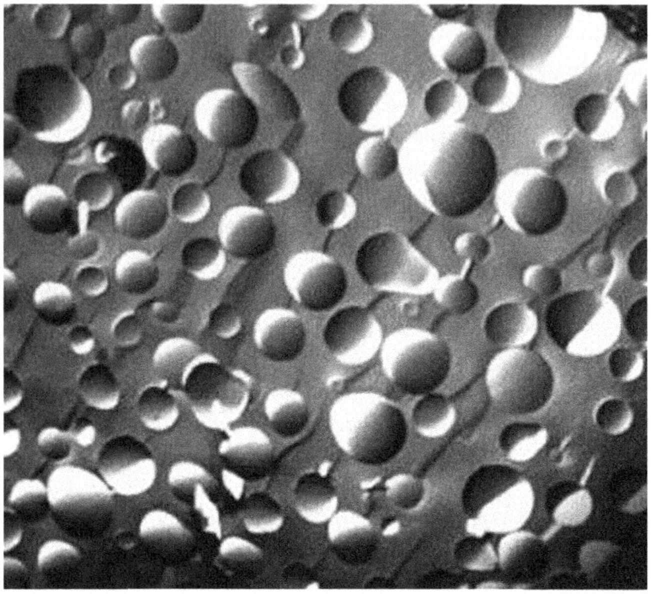

FIGURE 12.2

Freeze-fracture electron micrograph of a generic ethosomal formulation.

From Paolino, D., Lucania, G., Mardente, D., Alhaique, F., Fresta, M., 2005. Ethosomes for skin delivery of ammonium glycyrrhizinate: in vitro percutaneous permeation through human skin and in vivo antiinflammatory activity on human volunteers. J. Contr. Release 106(1–2), 99–110. https://doi.org/10.1016/j.jconrel.2005.04.007.

ethanol confers to ethosomes a softer and malleable nature with respect to conventional liposomes (Esposito et al., 2019). The classical composition of ethosomes represents a variation in the classical composition of liposomes, and it corresponds to a mixture of phospholipids, ethanol, and water (Fig. 12.2). This particular composition and preparation methods lead to obtaining multilamellar vesicles, as previously shown by transmission electron micrograph experiments (Touitou et al., 2000). The ethanol present in the structure acts directly on ethosomes structure, conferring malleability, but it is also able to fluidize skin lipid structure acting as penetration enhancer (Garg et al., 2017; Touitou and Godin, 2007).

Even if ethosomes were shown to be able to reach the deeper skin layers, the exact mechanism of permeation through skin is still unclear. Yang et al. have proposed that the penetration and the permeation of these nanosystems are expected to occur through fusion with cell membranes, thanks to a synergic action of phospholipids and ethanol, as also demonstrated by Niu et al. (2019) (Touitou et al., 2000; Yang et al., 2017). In fact, the activity of a penetration enhancer as a structural component of ethosomes resulted to be greater with respect to the action of the free ethanol.

The active interaction between ethosomes and skin structures was investigated and confirmed by Cristiano et al. analyzing the transepidermal water loss (TEWL) both on skin section and human volunteers after the topical administration of ethosomes, transfersomes, and niosomes (Cristiano et al., 2020). TEWL corresponds to the volume of water lost through the skin, and it is an indication of the integrity of skin barriers, namely TEWL values increase by decreasing the barrier effect of the skin (Gardien et al., 2016). This research work demonstrated two important aspects: first of all, the tested nanocarriers interact with the skin in a temporary and reversible way; and secondly, the ethosomes induce a less marked increase in TEWL than the same concentration of free ethanol, and a more rapid recovery of the basal skin values of TEWL. Based on these results, it was possible to affirm that ethosomes, such as transfersomes and niosomes, can be used as suitable colloidal nanocarriers to increase the percutaneous permeation of drugs after topical application without damaging the native structure of the skin (Fig. 12.3) (Cristiano et al., 2020).

The advantage of the use of ethosomes as topical drug delivery systems is not only their high deformability and elasticity but also their ability to provide enhanced transport of both hydrophilic and lipophilic drugs with preference of the latter, in comparison with free drug form, conventional

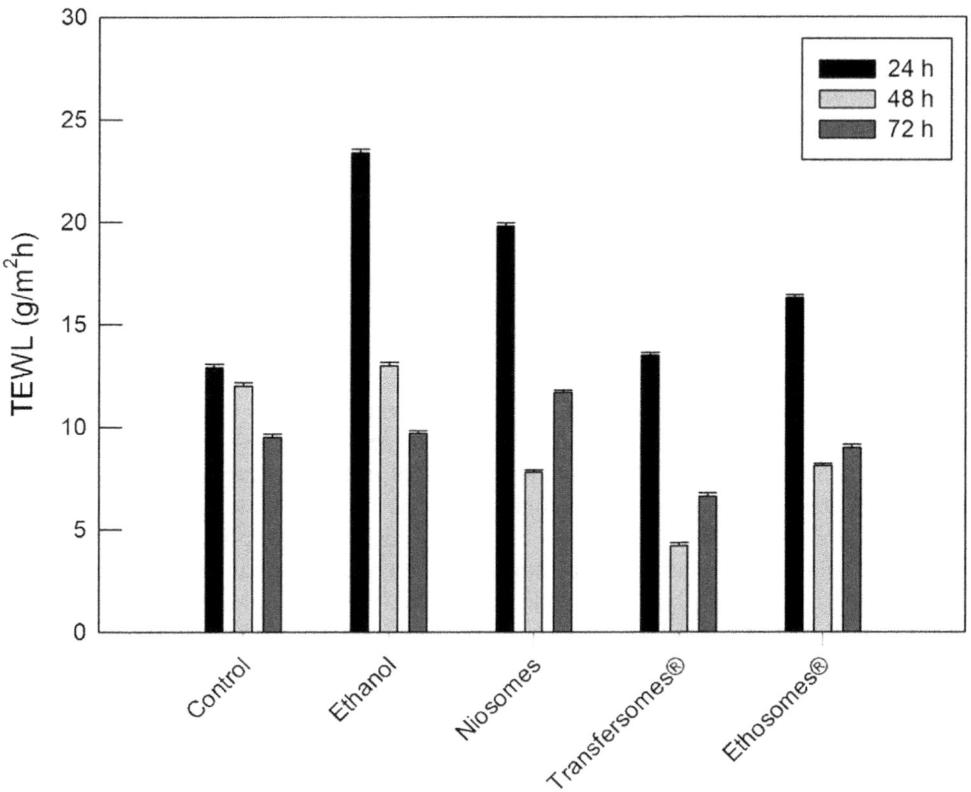

FIGURE 12.3

Example of in vivo study carried out on healthy human volunteers, measuring TEWL values of skin after topical application of niosomes, ethosomes and transfersomes.

From Cristiano, M.C., Froiio, F., Mancuso, A., Iannone, M., Fresta, M., Fiorito, S., Celia, C., Paolino, D., 2020. In vitro and in vivo trans-epidermal water loss evaluation following topical drug delivery systems application for pharmaceutical analysis. J. Pharm. Biomed.

liposomes and alcoholic phospho solution (Jaiswal et al., 2016; Rattanapak et al., 2012). Ethosomes are characterized by high formulation versatility, since by balancing the ethanol/lecithin ratio, it is possible to obtain specific and variable dimensions of the nanosystems. In detail, it has been demonstrated that mean size of ethosomes decreases with increase of ethanol concentration used during preparation; on the contrary, low percentage of lecithin led to a decrease of ethosomal mean sizes (Paolino et al., 2005). About the delivery of hydrophobic compounds using ethosomes, recently, Lin et al. have demonstrated the ability of ethosomes to contain at the same time two natural hydrophobic molecules, berberine (BBR) and evodiamine (EVO) (Lin et al., 2020). These compounds are two main active substances of a classical Chinese herbal medicine, Zuo Jin Wan, and they boast a synergistic inhibitory effect on various cell lines. The authors proposed ethosomes, as a suitable topical drug delivery system for BBR and EVO application in the melanoma treatment. The necessary choice to use an innovative delivery system is due to the poor aqueous solubility and the low skin penetration of active compounds. During in vitro skin deposition studies, BBR and EVO were quantified in higher concentrations in dermis and epidermis when delivered by ethosomes in comparison with the control (same concentrations of BBR and EVO solution). This result probably was due to the interaction between ethanol and lipid molecules, which improved lipid fluidity and cell membrane permeability. The detection of BBR and EVO in the epidermis after the administration of ethosomes confirmed the ability of nanosystems to squeeze through the *stratum corneum* and to deposit drugs in the epidermis. Also the anticancer activity of BBR and EVO was improved thanks to their delivery in ethosomes. In fact, in vitro studies carried out on B16 melanoma cells showed a good anticancer activity of loaded ethosomes already at 4.65 mg/mL and 0.56 mg/mL of BBR and EVO, respectively. In detail, at these low concentrations, BBR- and EVO-loaded ethosomes induced a cell mortality equal to 71.64% compared with 47.53% of control solution. These results demonstrated the enhanced transmembrane transport of ethosomes, which led to more BBR and EVO entering into B16 melanoma cells (Yu et al., 2015).

About the delivery of natural compounds, several research groups focused their studies on the possible use of ethosomes for the cutaneous delivery. Recently, our research group proposed ethosomes for skin delivery of sulforaphane, a natural dietary isothiocyanate obtained from cruciferous vegetables, characterized by high pharmacological activity, above all anticancer activity, but also poor percutaneous permeation ability (Cristiano et al., 2019). This research work demonstrated the effect of phospholipid/ethanol ratio on the ability of ethosomes to contain the natural compound. Among the different formulations tested, ethosomes made up of 40% (w/v) ethanol and 2% (w/v) phospholipon 90G showed increased percutaneous permeation of sulforaphane during in vitro experiments performed on human stratum corneum and epidermis membranes; on the other hand, experiments carried out on melanoma cell line, SK-MEL 28, showed an improved anticancer activity of sulforaphane-loaded ethosomes in comparison with the free active compound.

The influence of ethosomal composition on physicochemical characteristics and on linoleic acid delivery was also evaluated by Celia et al. (2012). The authors reported that at the same concentration of linoleic acid (0.05% w/v), the mean size of prepared ethosomes decreased until twofold when the ethanol percentage was increased from 30% to 45% (v/v). Maintaining constant the ethanol concentration (45%), a similar trend in mean size variation was observed when lecithin and linoleic acid were increased up to 3% and 0.1% (w/v), respectively, obtaining average size suitable for topical application of nanosystems. The selected ethosomal formulation made up of 40% and 45% of ethanol, 3% of lecithin, and 0.1% linoleic acid showed the best percutaneous permeation profiles when compared with linoleic acid hydro-alcoholic solution and a formulation of transfersomes. In fact, the ethosomes significantly ($P < .05$) improved the percutaneous permeation of linoleic acid through stratum corneum and epidermis (SCE) membranes, leading to a permeated amount of linoleic acid over 80%. This improved permeation of ethosomes seemed to be related to the dynamic interaction between vesicles and SCE membranes, which contributes to ethosomes delivery properties with respect to the free form of linoleic acid and transfersomes.

Like sulforaphane, also paeonol is characterized by not suitable physicochemical characteristics for topical administration, such as low water solubility and poor stability (Xu et al., 2017). Paeonol is a natural compound derived from peonies, characterized by several pharmacological activities, i.e., antiinflammatory, antiallergic, anticancer activities (Kim et al., 2004; Wu et al., 2016; Zong et al., 2018). The study performed by Ma et al. aimed to evaluate the possibility of use ethosomes for transdermal delivery of paeonol (Ma et al., 2018). The authors prepared several ethosomes formulations using different percentages of lecithin and ethanol. The best physicochemical characteristics were obtained for ethosomes composed of 25% of ethanol and 2.5% of lecithin, obtaining the higher OD (overall desirability) index. The in vitro permeation study and skin deposition of paeonol delivered from ethosomes resulted 1.6-fold and 2.5-fold higher, respectively, compared with that recorded from a hydroalcoholic solution composed of the same percentage (25%) of ethanol. Ethosomes characterized by that specific composition were able to form a reservoir on the skin, retarding paeonol release and enhancing skin retention. Facing a medium-high percentage of ethanol in the formulation of tested ethosomes, the authors also evaluated their skin tolerability. The studies were carried out on two groups of rats; group 1 was exposed to ethosomes, whereas rats of group 2 were taken as a control. Thanks to the similarity between ethosomal components and skin structure, the in vivo tests on nanosystems showed their great safety as a carrier for transcutaneous delivery with no associated irritation.

Another natural compound delivered using ethosomes was ammonium glycyrrhizinate, an amphiphilic antiinflammatory compound obtained from *Glycyrrhiza Glabra*. In the work of Paolino et al., ammonium glycyrrhizinate at the tested concentration (0.3% w/v) did not influence the mean size of ethosomes, highlighting the high physicochemical stability of these elastic carriers (Paolino et al., 2005). Moreover, two important results were obtained in this experimental work. First of all, ethosomes demonstrated their ability to improve the skin permeation of ammonium glycyrrhizinate in a significant way ($P < .001$) with respect to drug-hydroalcoholic solution, acting as drug delivery system. The same experiments were carried out evaluating the effects of empty ethosomes incorporated in a hydroalcoholic solution of ammonium glycyrrhizinate. The empty ethosomes have demonstrated their ability as penetration enhancers, inducing a modification of lipid structure of skin and consequently improving passage through *stratum corneum* compared with free ammonium glycyrrhizinate. The two preparations were evaluated and compared also by in vivo experiments on human volunteers. A noninvasive reflectance spectrophotometry was used following the extent of an induced erythema derived from the topical treatment with nicotinate solution. The posttreatment with ammonium glycyrrhizinate-loaded ethosomes showed a reduction of induced erythema more rapidly with respect to the ethosomal-free solution of ammonium glycyrrhizinate.

Ethosomes have been also used to enhance the dermal and transdermal delivery of cannabidiol (Lodzki et al., 2003). Cannabidiol was proposed in 1997 as drug for symptomatic ameliorative of collagen-induced rheumatoid arthritis in mice after intraperitoneal and oral administration (Malfait et al., 2000). Lodzki et al. have demonstrated that the topical administration of cannabidiol-loaded ethosomes on nude mice skin induced a significant accumulation of drug (Lodzki et al., 2003). Moreover, in this study the pharmacological effects of cannabidiol-loaded ethosomes were evaluated in mice after induction of carrageenan-induced aseptic paw inflammation. The experimental data collected by authors showed that topical pretreatment with ethosomes results in a systematically efficient delivery to the inflamed organ and in therapeutical antiinflammatory levels at the action site. All this demonstrated that ethosomes are effective not only as reservoir systems, but they are also able to permeate through deeper layers of skin and release the loaded drug at systemic level.

2.2.2 Transfersomes

In the early 1990s, Cevc and Blume were the first to describe a new vesicular carrier, similar to liposomes and characterized by greater deformability and elasticity, i.e., transfersomes or ultradeformable vesicles. In detail, transfersomes derived from liposomes are vesicular carriers, characterized by at least one inner aqueous compartment surrounded by a lipid bilayer (Rai et al., 2017). As for ethosomes, also in the case of Transfersomes, the chemical components used during the preparation have a key role in determining their features. In this case, thanks to the presence of one or more edge activators, the phospholipid-based vesicles could be able to deform themselves and squeeze as intact vesicles through narrow pores of the skin, this overcoming the main drawback of conventional liposomes (Fernández-García et al., 2020). An edge activator is generally a single chain surfactant with a high radius of curvature, and it is able to destabilize phospholipid bilayer of vesicles, increasing their deformability (Lai et al., 2020; Bhasin and Londhe, 2018). Another positive effect of edge activator included in transfersomes formulation is its action as solubilizing agent for hydrophobic drugs; in this way the drug entrapment efficiency of the formulation results improved. The main components used as edge activator for preparation of transferosomes are sodium cholate, sodium deoxycholate, some Spans, and Tweens as Tween 20, 60, and 80.

The choice of edge activator and its concentration is very important, because they influence the final physicochemical characteristics of transfersomes. For example, Paolino et al. prepared transfersomes for asiaticoside topical delivery using Phospholipon 100G and different sodium cholate molar fractions (0.00–0.35) (Paolino et al., 2012). Asiaticoside is an active component of *Centella asiatica* that promotes fibroblast proliferation and collagen biosynthesis (He et al., 2023). The Dynamic Light Scattering analysis showed that the increase in sodium cholate molar fraction induced a reduction of the mean sizes of vesicles and an increase of size distribution. As previously demonstrated also by Hildebrand et al. an excessive amount of edge activator could lead to the formation of various phospholipid/sodium cholate aggregates, such as mixed vesicles, micelles, opened vesicles, and others (Hildebrand et al., 2004). These mixed vesicles are normally characterized by large number and very small sizes. Also net surface charge of vesicles was influenced by sodium cholate molar fraction; as it increased, there was an increase in the negative charge obtaining a value of -33.7 ± 3.9 mV for a molar fraction of 0.35. Thanks to the DSC model of sodium cholate in lecithin bilayers, authors confirmed that this decreasing value was due to the increase in carboxylate group belonging to edge activator along the surface of the transfersomes bilayer. About asiaticoside entrapment efficiency, no significant differences were observed using a molar fraction of edge activator <0.2; on the contrary, a progressive reduction was recorded for entrapment efficiency of transfersomes when molar fraction was increased beyond 0.2, due to the presence of mixed vesicles and aggregates that were characterized by poor ability to contain the drug. In view of the results, authors identified transferosomes made of Phospholipon 100G and sodium cholate in molar fraction of 0.2, the best nanosystems for topical delivery of asiaticoside, as also confirmed by their in vivo studies.

To date, the net surface charge is a sensitive issue to be considered during the design of transfersomes. Some research groups argue that a negative surface charge is essential for the transcutaneous efficacy of transfersomes, while others affirm that a positive charge is necessary for a good transfersomes–skin interaction (Kitagawa and Kasamaki, 2006; Ogiso et al., 2001; Song and Kim, 2006). This last point of view is based on the evidence that lipid structures of *stratum corneum* are rich in negatively charged lipids, conferring a generalized negative charge to skin surface. Cationic drug delivery systems showed a higher skin permeation of entrapped drugs compared with negatively charged liposomes, probably because the negatively charged *stratum corneum* favors the electrostatic attraction of positively charged nanosystems, promoting skin drug permeation (Kitagawa and Kasamaki, 2006). Based on these opposing points of view, Lin et al. wanted to test different types of surface-charged transfersomes for topical administration of imperatorin (IMP) (Lin et al., 2018a). IMP is an active compound obtained from *Angelica dahurica*, a Chinese herbal medicine plant, characterized by several pharmacological effects, including antiinflammatory, antibacterial, and antitumor activity (Deng et al., 2020; Li et al., 2019; Mi et al., 2017; Nasser et al., 2019). The oral administration of this natural compound is limited by its low bioavailability, its poor absorption, and its extensive metabolism (Lin et al., 2018a); on the contrary, a conventional topical administration might fail due to high nonpolarity of IMP. For this reason, the use of suitable topical delivery system is required to improve the cutaneous delivery of the active compound. The authors prepared four formulation: conventional liposomes; conventional transfersomes made up of phosphatidylcholine, cholesterol and Tween 20; cationic and anionic transfersomes characterized by the same composition of conventional transfersomes but with the addition of stearylamine and dicetyl phosphate, respectively. Thanks to the presence of stearylamine, cationic transfersomes were characterized by a net positively surface charge equal to $+20.22$ mV $\pm$ 0.96; while anionic transfersomes containing dicetyl phosphate showed a greater negative charge (-42.63 mV $\pm$ 0.80) compared with conventional liposomes and transfersomes, which respectively showed a charge of -34.60 mV $\pm$ 0.81 and -24.87 mV $\pm$ 1.87. Among all the realized samples, cationic transfersomes reported the highest entrapment efficiency (60.32% $\pm$ 2.82), due to an increased hydrophobic area within the phosphatidylcholine bilayer for imperatorin by the positively charged combination. Another important effect consequent to the presence of stearylamine is the increased deformability of transfersomes. With respect to other used components in nanocarriers preparation, stearylamine is characterized by a high radius of curvature, which could destabilize and increase the deformability of vesicles (Elsayed et al., 2007). In any case, as already demonstrated, all types of transferosomes were more deformable compared with liposomes (cation transferosomes > transferosomes > anionic transfersomes > conventional liposomes). The same trend was obtained for in vitro skin permeation profiles. In fact, cationic liposomes permitted a good permeation of IMP within 24 hours equal to 330.50 μg $\pm$ 9.68, which was the greatest recorded permeated amount of this work and ~3.5 times greater than IMP solution.

The exact skin penetration mechanism of transfersomes is not yet completely defined, but of course, their ultradeformability is essential for the delivery of active compounds through skin. In any case a fundamental requirement is to administer transfersomes in nonocclusive conditions in order to keep unchanged the transepidermal osmotic gradient across the skin (Opatha et al., 2020). In fact, transferosomes seem to be greedy for moisture, and the difference in water content between skin surface (~15%) and the epidermis (~75%) is the driving force that pushes transfersomes to move through *stratum corneum* and toward deeper skin layers (Lai et al., 2020). In detail, Schatzlein and Cevc proposed that the topical application of transfersomes induces a natural water loss from vesicles and to avoid dehydration, transfersomes are forced to modify their shape, to enter the skin, and to reach more hydrated skin layers (Schätzlein and Cevc, 1998). This process was defined as "xerophobia" by transfersomes creators. The shape modification does not alter *stratum corneum* integrity, and it is important for successfully delivering hydrophilic and/or lipophilic entrapped drug(s) (Cristiano et al., 2020).

In subsequent years, other researchers confirmed the need for a nonocclusive application condition to allow transfersomes to pass through skin; but many of them stated that after crossing the *stratum corneum*, the release of the trapped drug instantly occurred so transferosomes were not able to reach the deep layers (Honeywell-Nguyen et al., 2002; Honeywell-Nguyen and Bouwstra, 2003). Some years later, Celia et al. proposed another mechanism of action for transfersomes (Celia et al., 2012). Authors wanted to compare the ability of ethosomes and trasferosomes to delivery linoleic acid through skin. They found that the amount of linoleic acid, which permeated through stratum corneum and epidermidis membranes, was over 80% when delivered by ethosomes, while the percutaneous permeation of linoleic acid—loaded transfersomes was very low. In this scenario, authors demonstrated that ultradeformable vesicles were able to across *stratum corneum*, but the high affinity of their components (phospholipids, sodium cholate, and the same linoleic acid) for outer skin structures caused the vesicular carriers to remain entrapped and accumulated in the *stratum corneum*. These results highlighted the possibility to obtain with transfersomes a "skin depot system," as previously demonstrated (Honeywell-Nguyen et al., 2002; Honeywell-Nguyen and Bouwstra, 2003).

The skin depot effect proposed for transferosomes can be used for the treatment of skin diseases localized in specific layers of the skin. For example, Cosco et al. demonstrated that the multidrug transfersomes induced a gradual release of resveratrol and 5-fluorouracil after their accumulation in skin layers (Cosco et al., 2015). This depot system was permitted to improve the anticancer activity of the two entrapped drugs on skin cancer. In the same article, the authors investigated and proved the interaction and accumulation of rhodaminated transfersomes within skin cancer cell lines, proving their potential efficacy for cutaneous administration (Fig. 12.4).

Another mechanism of penetration was proposed for transfersomes by Trotta et al. (2002). The research group prepared novel transfersomes using dipotassium glycyrrhizinate as edge activator. The natural compound, extracted from liquorice root, had a double role: it was able to modify the integrity and the fluidity of transfersomal membrane, acting as functional components, and at the same time, it showed a specific antiinflammatory activity, very useful in the dermatitis treatments. Based on their results, the authors affirmed that deformable vesicles penetrated across the virtual channels of the *stratum corneum* and under transcutaneous hydration force, as proposed by Cevc; but they argued that the improved drug skin deposition recorded was due to the fusion of vesicles to the skin, following the increased fluidity of lipid bilayer containing dipotassium glycyrrhizinate.

Many research groups believe in the effective applicability of transfersomes for topical delivery of natural compounds. Recently, Caddeo et al. proposed transfersomes as topical delivery systems for tocopherol acetate, a natural phenolic compound with a well-known antioxidant activity (Caddeo et al., 2014). In this work, effects of composition on physicochemical characteristics of nanosystems were evaluated. In detail, the transfersomes were prepared using different edge activators (Tween 20, 40, 60, and 80) and were characterized. The ability to contain effectiveness tocopherol concentrations was related to the Tween used during the preparation; entrapment efficiency increased as a function of the

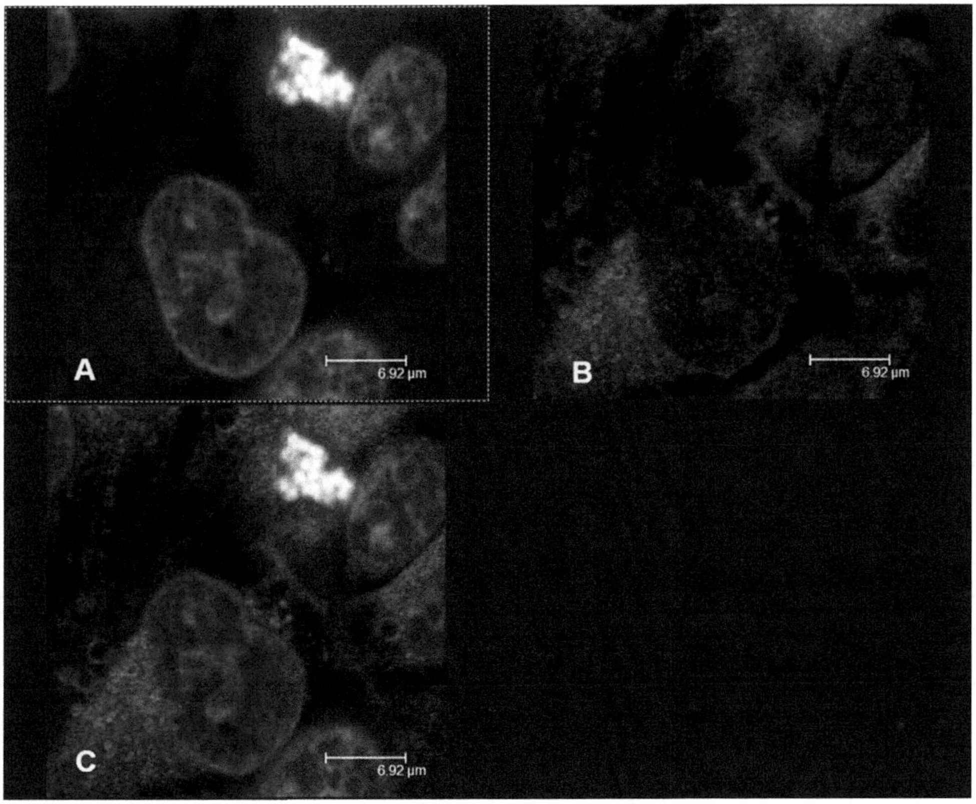

FIGURE 12.4

Confocal laser scanning micrographs showing the interaction between transfersomes and skin cell lines. Interaction and accumulation of rhodamine-labeled transfersomes within Colo-38 cells using: (A) Hoechst filter and (B) TRITC filter; and (C) superimposition of figures.

From Cosco, D., Paolino, D., Maiuolo, J., Marzio, L.D., Carafa, M., Ventura, C.A., Fresta, M., 2015. Ultradeformable liposomes as multidrug carrier of resveratrol and 5-fluorouracil for their topical delivery. Int. J. Pharm. 489(1–2), 1–10. https://doi.org/10.1016/j.ijpharm.2015.04.056.

length of the Tween fatty acid chain. In detail, transfersomes made up by Tween 80 reported the best results in terms of mean size (76 nm) and tocopherol entrapment efficiency (90%); moreover, the chosen transfersomes were able to improve the deposition of tocopherol in the different skin strata and its antioxidant activity, ensuring the in vitro repair of skin wound.

Another recent study focused the attention on the topical delivery of a natural compound. Researchers demonstrated the effectiveness of transferosomes in ammonium glycyrrhizinate delivery through human skin (Barone et al., 2020). They evaluated if the presence of ammonium glycyrrhizinate (0.3% w/v) in transfersomes influenced the traditional deformability of vesicles. Performing a deformability test based on extrusion processes, the authors reported a not statistically significant difference between the deformability index of empty transfersomes (12.9) and ammonium glycyrrhizinate-loaded transfersomes (12.8). In this way they demonstrated both the high deformability of transfersomes, because of their ability to cross intact through 50 nm pore size membrane filters used for analysis and the ability of ammonium glycyrrhizinate to not affect the supramolecular structure of nanocarriers. Further in vivo studies were carried out on healthy human volunteers, evaluating the antiinflammatory activity of ammonium glycyrrhizinate-loaded transfersomes, compared with hydroalcoholic solution of the active compound. Since the study included an efficacy analysis, the volunteers' skin was previously treated with 0.2% w/v methyl nicotinate solution, in order to chemically induce the erythema. Subsequently, the erythema was treated with saline solution (used as control), ammonium glycyrrhizinate-loaded transfersomes, or a hydroalcoholic solution of ammonium glycyrrhizinate. The results, recorded as erythema index variation (ΔE.I.), probed the increased antiinflammatory efficacy of ammonium glycyrrhizinate when it was loaded into transfersomes; in fact, transfersomes have demonstrated to significantly decrease the ΔE.I. already after 30 minutes of application on the forearm of human volunteers, compared with control and ammonium glycyrrhizinate solution. Probably, these promising results were consequent to the ability of tested nanocarriers to allow the transport of ammonium glycyrrhizinate through the skin. Moreover, phospholipids and edge activator, sodium cholate, induced a formation of ammonium glycyrrhizinate depot systems in the skin tissue, thus proving a continuous release of drug (Jyoti Prakash and Pramod, 2019).

Also epigallocatechin-3-gallate (EGCG), an antioxidant compound abundantly available in green tea, was encapsulated within transfersomes (Avadhani et al., 2017; Brand and Jendrzejewski, 2008). The aim of Avadhani et al. was to combine in a unique nanocarrier the natural compound EGCG and hyaluronic acid (HA), synergizing their antioxidant and antiaging effects. Thanks to the optimization of formulations, EGCG and HA-loaded transfersomes showed a better antioxidant activity in terms of free radical-scavenging effect compared to free drugs. Moreover, in vitro experiments carried out on a human keratinocytes cell model reported a greater suppression of MDA (malondialdehyde) and ROS levels after UV irradiation when transfersomes were used. The UV irradiation, without any treatment, induced high level of MDA (9.84 $\pm$ 0.87 nmol/mg protein), while the treatment with EGCG- and HA-loaded transfersomes decreased the MDA concentration until 1.08 $\pm$ 0.17 nmol/mg protein, demonstrating a more defensive mechanism in human keratinocytes against UV damage by EGCG-loaded transferosomes than free EGCG. Moreover, in vitro skin permeation studies have demonstrated that transfersomes were able to improve ~2.8 times EGCG skin permeation, compared with the free form of active compound. Based on these in vitro results, the authors confirmed the ability of transfersomes to penetrate intercellular lipids and to increase the lipid fluidity, decreasing the density of skin lipid layers.

Topical and painless application of resveratrol is an attractive alternative method to other administration routes, avoiding gastrointestinal problems, destabilization of natural compound, and first-pass metabolism effects (Alonso et al., 2017; Intagliata et al., 2019). Generally, topical application and skin permeation from conventional dosage of resveratrol are limited by its low water solubility and its low stability due to photoisomerization that reduces resveratrol activity. For this reason, Tosato et al. wanted to evaluate the effect of nanoencapsulation of resveratrol in different drug delivery systems, to find the best carrier for resveratrol topical application (Tosato et al., 2018). Through physicochemical characterization of nanosystems, the authors demonstrated that transfersomes were the best carriers for resveratrol because they showed the best entrapment efficiency. In detail, transfersomes led to encapsulation efficiency equal to 91.8%, while ethosomes and conventional liposomes showed values equal to 51.0% and 23.0%, respectively. Despite ethosomes and transferosomes both being elastic and deformable vesicles, the different result was probably due to the presence of Tween 80, used as edge activator, and the low ethanol concentration. Tween 80 increased the fluidity of the bilayer, thus facilitating the accommodation of resveratrol in the lipid structure without altering the typical mobility of carriers (Tosato et al., 2018).

Particular results were obtained by Wu et al. for resveratrol loaded transferosomes (Wu et al., 2019). Unlike many other research groups, the authors of this work showed that the encapsulation of resveratrol in the transferosomes did not improve its antioxidant activity, but rather preserved it over time. In detail, the authors performed four different antioxidant activity experiments, and all these reported that the antioxidant activity of resveratrol loaded transferosomes was comparable with that of free resveratrol, indicating that the resveratrol activity was not affected after coating. At the same time, results confirmed that resveratrol did not lose its activity during the encapsulation process, further confirming that transfersomes preparation process did not alter the pharmacological characteristics of resveratrol. Based on these results, it is possible to affirm that transfersomes could be also used to improve the stability of natural compounds (Wu et al., 2019; Banu et al., 2018; Strugała et al., 2017).

2.3 Other topical drug delivery systems

As transfersomes and ethosomes, also niosomes were obtained from modification of liposomal composition (Mancuso et al., 2020). Niosomes are so-called for the presence of nonionic surfactants in their composition (cetyl alcohol, decyl glucoside, Brij, polysorbates, spans, and poloxamers) (Kaur and Kumar, 2018). Niosomes are characterized, as liposomes, by a highly ordered structure, with an external bilayer and an aqueous core; but unlike their precursors, niosomes are characterized by higher vesicle stability, greater resistance to lipid peroxidation thanks to nonionic surfactants. Moreover, they have good biocompatibility, low toxicity, and low preparation cost (Moghassemi and Hadjizadeh, 2014). As well as for physicochemical stability, nonionic surfactants are important because they can act as permeation enhancers, improving the skin delivery of the encapsulated drug into niosomes. The exact contribution to cutaneous permeation of nonionic surfactants included in the niosomes is not jet clear, but probably they fuse with the skin surface, inducing permeation of the delivered compound through the *stratum corneum*, modifying the barrier property of skin (Cristiano et al., 2020).

Despite the potentiality of niosomes, the research works for skin diseases treatments with natural compound-loaded niosomes are few. For example, Manconi et al. (2006) studied the ability of niosomes to improve skin permeation of retinoic acid for acne treatment. For this aim, the authors performed in vitro studies using Franz diffusion cells and newborn pig skin interposed between receptor and donor compartments of the permeation system. Authors demonstrated that the skin permeation of retinoic acid—loaded niosomes, prepared using alkyl polyglucosides as nonionic surfactants, was strongly influenced by the vesicles structure, mean size, and surface charge; in particular, the best results were obtained with small negatively charged niosomes, saturated with retinoic acid, showing a greater cutaneous retention of drug compared with positively charged vesicles. On the contrary, the positively charged niosomes were highly able to permeate through the skin.

Negi et al. (2017) developed and evaluated resveratrol loaded niosomes, prepared using two different process methods: thin film hydration (TFH) and ether injection method (EJM). The EJM method leads to obtaining a smaller particle size of niosomes, because the components were dispersed homogeneously at molecular level in comparison with the TFH method; on the contrary, TFH permitted to obtain high entrapment efficiency due to the biggest mean size. Moreover, the authors included niosomes in a three-dimensional network based on Carbopol 934 hydrogel, making niosomal suspension more acceptable for topical application. Ex vivo drug permeation studies, employing Wistar rat skin, showed a better drug transport profile of optimized niosomal gel (88.4%) than drug suspension (62.7%), as well as the better results in terms of resveratrol skin retention (4.76 ± 0.21 μg/cm^2 with niosomal gel and 0.98 ± 0.01 μg/cm^2 with drug suspension). The reasons for this better profile could be the absorption and fusion of niosomes on the surface skin or the reduction of the barrier properties of *stratum corneum* resulting from penetration enhancement induced by nonionic surfactants (Friedrich et al., 2015). Finally, in vivo antiinflammatory activity of niosomal gel was compared with that recorded from standard sodium diclofenac gel, evaluating the reduction of carrageenan-induced paw edema. Results from in vivo studies showed that niosomal gel sufficiently reduced the induced edema in rats and permitted to obtain prolonged therapeutic effect compared with the conventional topical dosage form.

Ammonium glycyrrhizinate was entrapped also in niosomal formulation by Paolino et al. (2007). In this study, niosomes were prepared by using thin layer evaporation method and using a particular nonionic surfactant, α,ω-hexadecyl-bis-(1-aza-18-crown-6) (Bola-surfactant)-Span 80-cholesterol (2:3:1 M ratio). Experimental data showed that bola-niosomes were characterized by suitable mean size for topical administration, ~400 nm, and an entrapment efficiency equal to 40% with respect to the ammonium glycyrrhizinate amount used (3 mg/mL) for preparation. During in vitro human skin percutaneous permeation studies, the authors compared the ammonium glycyrrhizinate solution, ammonium glycyrrhizinate-loaded niosomes, and a mixture of free ammonium glycyrrhizinate and niosomes. The permeation of drug from the solution through the *stratum corneum* was lower than that obtained from bola-niosomes, i.e., 258.23 ± 41.7 μg/cm^2 hour^{-1} and 989.28 ± 77.8 μg/cm^2 hour^{-1}, respectively. Moreover, the physical mixture (free ammonium glycyrrhizinate and empty bola-niosomes) improved the percutaneous permeation of drug with respect to its solution. Thanks to these double results, the authors demonstrated that the in vitro percutaneous permeation effects of niosomes were due to both penetration enhancement effect and edge activator action.

These in vitro results were confirmed by in vivo studies, evaluating the antiinflammatory activity of ammonium glycyrrhizinate-loaded niosomes on healthy human volunteers. The nanocarriers were able to reduce an induced erythema much more rapidly and efficiently than the aqueous solution of ammonium glycyrrhizinate.

Cubosomes are another important carrier to be taken into account as smart delivery system, and they represent an evolution of the bicontinuous cubic phase (Esposito et al., 2019). Cubic phases are considered appropriate drug delivery systems for the skin and mucosae due to some of their features, such as bioadhesion, biocompatibility, and ability to solubilize hydrophilic and hydrophobic compounds (Shah et al., 2001). The main limit of cubic phase is their high gel like viscosity, which makes it difficult to handle and to administer the formulation. To solve this problem, starting from cubic phase composition, cubosomes, considered as nanoparticles characterized by a cubic-ordered inner structures (Siekmann et al., 2002), were obtained by the fragmentation of lipid base (glyceryl monooleate) in the presence of surfactant (Poloxamer P407) using excess water (Garg et al., 2017). Several methods of preparation were proposed, for example, "top-down approach" (Peng et al., 2015), "bottom-up approach", spray drying method (Akhlaghi et al., 2016; Sharma et al., 2020). Cubosomes are able to contain both hydrophilic and lipophilic molecules, respectively, in the aqueous channel of the cubic reticulum and in the lipid domain (Fresta et al., 2020). Since the viscosity of cubosomes is very low, their topical application can be benefited by the inclusion of cubosomes within hydrogel, or adding a colloidal polymer such as xanthan gum into the dispersion (Liu et al., 2015; Mancuso et al., 2023).

Cubosomes seem to be able to penetrate skin and mucosa, thanks to the similarity between its inner structure and epithelium cell, enhancing drug bioavailability (Sguizzato et al., 2021). Recently, Nasr et al. evaluated the applicability of cubosomes for the topical delivery of colchicine (Nasr et al., 2020). Colchicine (COL) is a natural extracted alkaloid obtained from *Cochicum autumnale* L. plant, and it is normally used for gouty arthritis treatment, but also for psoriasis and actinic keratosis (Angelidis et al., 2018). The main limits of treatments with colchicine are its poor bioavailability and the potential side effects associated with its oral and parenteral administration. Cubosomes seemed a good solution to permit transdermal delivery of colchicine, reducing its limitation. Among eight samples prepared by Nasr et al. (2020) the formulations prepared with 0.625 gr of GMO and 31.25 P407 (F2) and with 1.25 gr of GMO and 125 mg of P407 (F8) were chosen for ex vivo permeation studies, based on physicochemical characteristics and ability to contain colchicine. For ex vivo studies, rabbit skin has been utilized as an alternative for human skin. After 24°hours of experiments, the amounts of colchicine deposited in the skin were 239.12 ± 19.80, 164.48 ± 12.27, and 154.99 ± 9.57 $\mu g/cm^2$ for F2, F8, and COL aqueous solution, respectively. This deposition induced by cubosomes could be useful in certain dermatological conditions such as atopic dermatitis, actinic keratosis, and psoriasis. The amount of colchicine deposited in the skin after F8 application was relatively lower than that of F2, probably due to the greater zeta potential of F8 ($+27.90$ mV $\pm$ 0.45) with respect to F2 ($+21.20$ mV $\pm$ 0.31). Before Nasr et al., other research groups had demonstrated that positively charged systems containing phytosphingosine were more effective in skin diffusion of entrapped drugs than the negatively charged ones (Hoeller et al., 2009).

The potential use of cubosomes as effective topical drug delivery system for natural active ingredient has been recently demonstrated by Mancuso et al. (2023). In this case cubosomes with two different compositions were used for improving the percutaneous permeation and the antioxidant activity of resveratrol. Resveratrol (RSV) is a natural phenolic compound, obtained from grapes, berries, peanuts, and red wine (Intagliata et al., 2019; Nastiti et al., 2020). This natural compound is famous for its potential health benefits as anticancer, antiinflammatory, and its antioxidant properties as a function of its concentration and of the duration of the treatment (acute or chronic use) but also for its physicochemical properties not suitable to guarantee its passage through the stratum corneum after its skin application (Carrizzo et al., 2013; Csiszar, 2011; Yousef et al., 2017). In the scientific work of Mancuso et al. the presence of PEG derivatives combined with GMO and P407 in the formulation seems to influence on antioxidant activity of resveratrol loaded cubosomes and the ability of nanosystems to retain suitable amounts of drug, with respect to the formulation made up of only GMO and P407 (Mancuso et al., 2023). Thanks to the use of cubosomes, the loaded resveratrol showed a higher protective effect on NCTC2544 cells: when cells were treated with free Resveratrol and resveratrol loaded cubosomes and then exposed to prooxidant agent, the loaded resveratrol induced a reduced release of LDH from oxidized cells. This result demonstrated that, thanks to the cubosomes, resveratrol led to a reduction in the release of LDH ($31.56\% \pm 0.85$) with respect to oxidized but not treated cells ($56.45\% \pm 1.65$), as symptom of effective antioxidant protection on human skin cells.

Another natural compound useful the treatment of psoriasis and other diseases, such as apocrine chromhidrosis and contact allergy, is capsaicin, a natural alkaloid extracted from fruit of the capsicum plant family (Peng et al., 2015). Also in this case, the topical application of capsaicin could be advantageous in terms of reduction of systemic side effects, first-pass metabolism in the gastrointestinal tract, short half-life after intravenous administration. Several topical drug delivery systems were tested for capsaicin delivery, as transfersomes, ethosomes, liposomes for the treatment of pain-related disorders (Kumar Sarwa et al., 2015; Sarwa et al., 2015; Zhu et al., 2015). The main aim of topical systems is the optimization of skin-targeted delivery of capsaicin, ensuring the main absorption in the skin and reducing systemic absorption as compared with the conventional topical application, as cream, lotion, ointment. For this reason, Peng et al. developed cubosomes delivering capsaicin (Peng et al., 2010). The authors analyzed two different cubosomal formulations, characterized in their composition by phytantriol (PYT) (F1) and glycerol monooleate (GMO) (F2), in association with poloxamer 407 and capsaicin (250 μg/mL). The formulations, independent of their composition, were able to contain high percentage of capsaicin, recording an entrapment efficiency equal to 97.14% ± 0.54% and 97.58% ± 0.53%, respectively, for F1 and F2. This good result was attributable to the unique structure of cubosomes and the hydrophobic capsaicin that was dissolved in GMO o PYT and distributed in the lipid layers (Peng et al., 2010). In vitro skin retention tests showed that both cubosomal formulations were able to enhance the retention of capsaicin in stratum corneum and epidermis and derma with respect to a cream used as control. This result is important because it confirmed the cubosomes' ability to retard the diffusion of the drug from the skin to the systemic blood stream, enhancing the topical effects of capsaicin.

Solid Lipid Nanoparticles (SLNs) are particular drug delivery systems made of solid macromolecular matrices able to solubilize different molecules within their structures (Alsaad et al., 2020; Ganesan and Narayanasamy, 2017; Mendoza-Muñoz et al., 2021). The constituent lipids remain solid at both temperature of human body and room temperature. These solid lipid structures permit to obtain a controlled drug release. SLNs are characterized by wide size range (40−1000 nm), higher drug stability, entrapment of both hydrophilic and lipophilic drugs, atoxicity, easy large-scale production, and the possibility of lyophilization. The typical solid consistency of SLN core can be modulated using mixtures of solid and liquid lipids, obtaining the so-called nanostructured lipid carriers (NLC), which are characterized by imperfect and disordered lipid matrix, increasing drug loading (Alsaad et al., 2020; Ganesan and Narayanasamy, 2017; Mendoza-Muñoz et al., 2021). The potential of lipid nanoparticles in the cosmetic field is well known and is mainly related to the similar composition between their constituent lipid matrix and the skin lipids and to the ability of carriers to improve skin hydration and elasticity (Borges et al., 2020). Lipid nanoparticles have been used to load several phenolic compounds having different beneficial properties. Recent investigations considered phenolic-loaded lipid nanoparticles with the aim to produce sunscreen formulations less harmful compared with that containing synthetic compounds. Among phenolic compounds, quercetin has already shown interesting properties in preventing UVB-induced oxidative skin damage, and it has been considered from several research groups as constituent for sunscreen formulations (Tripathi et al., 2022). Bose and coworkers prepared different quercetin-loaded SLNs and performed in vitro studies of permeation using full thickness human skin, and they found that greater amount of the active compound was localized within the skin when loaded into lipid nanoparticles compared with a control sample with particles in micrometric range (Bose et al., 2013). The detection of the natural flavonoid into the skin was highly desirable because epidermis represented the main target for the effectiveness of quercetin as scavenger of free radicals. Another research group loaded phenolic compounds, known as curcuminoids, in SLNs. Curcuminoids recently gained considerable attention due to their antiaging properties (Plianbangchang et al., 2007). For this reason, the authors wanted to test the efficacy as antiaging agent of the natural compound when it was loaded into SLNs and the safety of the facial cream produced. Curcuminoids-loaded SLN cream was applied on 30 volunteers and different parameters were monitored within 8°weeks to check the good state of skin. Measurements of hydration (Corneometer CM 825), skin wrinkles (%SEw, Visiometer SV 600 USB), elasticity, and viscoelasticity (Cutometer 580 MPA) showed the greater ability of the sample to improve skin healthy if compared with the administration of a cream base. Interesting values have been recorded confirming the maintenance of the efficacy of the natural compound loaded into these carriers. Skin irritation was also monitored recording values of Transepidermal water loss, pH variation, and physician's observation and no sign of skin irritation occurred during the analysis, thus showing that curcuminoids-loaded SLN cream was a promising and safe antiaging preparation.

The combination of more phenolic compounds has been recently proposed by Friedrich and coworkers as an intelligent and immediate way to treat cutaneous disease (Friedrich et al., 2015). Curcumin and resveratrol are two known phenolic compounds having antimicrobial, antiinflammatory, and antioxidant properties. A part of their individual efficacy, increased therapeutic effects have been reported for the coloading of the two natural compounds. However, their low water solubility limited their use in clinical practice. For this reason, authors decided to codeliver curcumin and resveratrol into NLCs. Studies of permeation through the skin proved that resveratrol loaded NLCs reached deeper skin layers when it was coloaded with curcumin, with respect to the encapsulation of only resveratrol, and this was probably due to interactions occurred between curcumin and the *stratum corneum,* which enhanced the skin absorption of the codelivered active compound. Further in vitro studies have been performed by authors on primary human skin cells showing the uptake into skin cells of polyphenols within 24 hours, leading to specific intracellular effects.

These studies probed the efficacy of NLCs and SLNs in the cosmetic field making lipid nanoparticles good candidates for topical treatment of skin diseases.

2.4 Use of natural ingredients for nanosystems design

Natural active ingredients can be used not only as delivered drugs within nanosystems for several applications but also as components of their structures. The design of naturally based nanosystems is increasingly pursued since the use of structural natural compounds could be beneficial in reducing potential no tolerability of nanosystems. Generally, the main exponents of topical drug delivery systems are made up of natural lipids presenting a high dermophilicity and hydration (Esposito et al., 2007). An example of less known topical nanosystems made up of natural ingredients is represented by UFAsomes. UFAsomes are Unsaturated Fatty Acid vesicles characterized by lipid bilayer and a fatty acid composite (Bhattacharya, 2021). Among the unsaturated fatty acids, oleic acid is the one most used for nanosystems preparation, and several research groups have exploited its benefits for effective topical activity of the resulting nanosystems (Bruno et al., 2022; Kumar et al., 2018). For example, it is well known the ability of oleic acid to act as permeation enhancer by causing disordering and fluidization of skin lipids and, as consequence, by reducing skin barrier function (Atef et al., 2022; Da Silva et al., 2011). Moreover, oleic acid is known for its wound healing, antioxidant, and antiinflammatory properties that it can explain on skin diseases (Lin et al., 2018b). Often oleic acid was associated with others unsaturated fatty acid in the construction of naturally based nanosystems for topical use such as linoleic acid. In some cases, linoleic acid was used for its ability to induce a lightening effect on the skin (Ando et al., 1998); this effect is a consequence of the inhibition of melanin production in melanocytes, and this inhibition seems to be related to the number of unsaturated bonds and their position in linoleic acid structure. Moreover, linoleic acid is known for its skin barrier repairing properties, due to the effect on keratinocytes proliferation and lipid synthesis (Danby et al., 2013). In a scientific work, Cristiano et al. wanted to combine the topical benefits of linoleic and oleic acids for the topical delivery of another natural ingredient, ammonium glycyrrhizinate (Cristiano et al., 2021). The obtained UFAsomes showed to be able to increase the percutaneous permeation of ammonium glycyrrhizinate overcoming the issue related to physicochemical characteristics of the natural compound in free form (15.0 ± 4.9 μg/cm^2 hour^{-1} and 16.6 ± 5.2 μg/cm^2 hour^{-1} for two UFAsomes formulations respect to 8.1 ± 1.1 μg/cm^2 hour^{-1}of free form of ammonium glycyrrhizinate). Moreover, thanks to the probable interaction between nanovesicles and cell membranes, the drug-loaded UFAsomes have demonstrated the improved antioxidant activity of ammonium, when tested on NCTC 2544 cells.

3. Conclusions

The importance and the effectiveness of smart drug delivery systems in the treatment of cutaneous diseases were the object of many research works, above all in natural active compounds fields. As is well known, many substances of natural origin have remarkable pharmacological activities, useful for the treatment of skin diseases on an inflammatory, oxidative, and/or tumor basis. Unfortunately, the physicochemical characteristics of these substances often do not allow a conventional topical administration. In this scenario, liposomes, deformable vesicles and other vesicles modified with different hydrophilic/lipophilic additives, play a key role in facilitating the permeation through the skin of the active ingredients, improving their repartition coefficient (logP), physicochemical stability and reducing their degradation. For example, natural active compounds such as flavonoids, obtained a better absorption profile when topically delivered through nanovesicles. Each of these smart drug delivery systems is characterized by good delivery capabilities but also by limitations that have been discussed in this book chapter. Some limitations have already been overcome making changes to their composition and preparation methods and creating new nanosystems. Despite this, and despite the most recent technologies represented above all by transferosomes and ethosomes, the number of products that have been achieved in clinical practice and the world market is still limited. This aspect can be partly explained by some problems of smart drug delivery systems not yet completely solved, such as the difficulty of guaranteeing reproducibility from batch to batch in large productions compared with laboratory practice; or the costs of starting materials, procedures, and controls.

Acknowledgment

None Declare.

Conflict of interest

None Declare.

References

Akhlaghi, S.P., Ribeiro, I.R., Boyd, B.J., Loh, W., 2016. Impact of preparation method and variables on the internal structure, morphology, and presence of liposomes in phytantriol-Pluronic® F127 cubosomes. Colloids Surf. B Biointerfaces 145, 845–853. https://doi.org/10.1016/j.colsurfb.2016.05.091.

Alonso, C., Martí, M., Barba, C., Carrer, V., Rubio, L., Coderch, L., 2017. Skin permeation and antioxidant efficacy of topically applied resveratrol. Arch. Dermatol. Res. 309 (6), 423–431. https://doi.org/10.1007/s00403-017-1740-5.

Alsaad, A.A.A., Hussien, A.A., Gareeb, M.M., 2020. Solid lipid nanoparticles (SLN) as a novel drug delivery system: a theoretical review. Sys. Rev. Pharm. 11 (5), 259–273. https://doi.org/10.31838/srp.2020.5.39.

Ando, H., Ryu, A., Hashimoto, A., Oka, M., Ichihashi, M., 1998. Linoleic acid and α-linolenic acid lightens ultraviolet-induced hyperpigmentation of the skin. Arch. Dermatol. Res. 290 (7), 375–381. https://doi.org/10.1007/s004030050320.

Angelidis, C., Kotsialou, Z., Kossyvakis, C., Vrettou, A.R., Zacharoulis, A., Kolokathis, F., Kekeris, V., Giannopoulos, G., 2018. Colchicine pharmacokinetics and mechanism of action. Curr. Pharm. Des. 24 (6), 659–663. https://doi.org/10.2174/1381612824666180123110042.

Atef, B., Ishak, R.A.H., Badawy, S.S., Osman, R., 2022. Exploring the potential of oleic acid in nanotechnology-mediated dermal drug delivery: an up-to-date review. J. Drug Deliv. Sci. Technol. 67, 103032. https://doi.org/10.1016/j.jddst.2021.103032.

Avadhani, K.S., Manikkath, J., Tiwari, M., Chandrasekhar, M., Godavarthi, A., Vidya, S.M., Hariharapura, R.C., Kalthur, G., Udupa, N., Mutalik, S., 2017. Skin delivery of epigallocatechin-3-gallate (EGCG) and hyaluronic acid loaded nano-transfersomes for antioxidant and anti-aging effects in UV radiation induced skin damage. Drug Deliv. 24 (1), 61–74. https://doi.org/10.1080/10717544.2016.1228718.

Banu, R., Gerding, J., Franklin, C., Sikazwe, D., Horton, W., Török, M., Davis, J., Cheng, K.H., Nakazwe, M., Mochona, B., 2018. 5-Dimethoxy-2-nitrobenzohydrazides and 1-(1-benzylpiperidin-4-yl) ethan-1-ones as potential antioxidant/cholinergic endowed small molecule leads. Sci. Pharm. 4 (1).

Barone, A., Cristiano, M.C., Cilurzo, F., Locatelli, M., Iannotta, D., Di Marzio, L., Celia, C., Paolino, D., 2020. Ammonium glycyrrhizate skin delivery from ultradeformable liposomes: a novel use as an anti-inflammatory agent in topical drug delivery. Colloids Surf. B Biointerfaces 193, 111152. https://doi.org/10.1016/j.colsurfb.2020.111152.

Beckenbach, L., Baron, J.M., Merk, H.F., Löffler, H., Amann, P.M., 2015. Retinoid treatment of skin diseases. Eur. J. Dermatol. 25 (5), 384–391. https://doi.org/10.1684/ejd.2015.2544.

Bhasin, B., Londhe, V.Y., 2018. An overview of transfersomal drug delivery. Int. J. Pharm. Sci. Res. 9 (6), 2175–2184. https://doi.org/10.13040/IJPSR.0975-8232.9(6).2175-84.

Bhattacharya, S., 2021. Preparation and characterizations of glyceryl oleate ufasomes of terbinafine hydrochloride: a novel approach to trigger Candida albicans fungal infection. Future J. Pharm. Sci. 7 (1). https://doi.org/10.1186/s43094-020-00143-w.

Biju, S.S., Ahuja, A., Khar, R.K., 2005. Tea tree oil concentration in follicular casts after topical delivery: determination by high-performance thin layer chromatography using a perfused bovine udder model. J. Pharm. Sci. 94 (2), 240–245. https://doi.org/10.1002/jps.20250.

Bolzinger, M.A., Briançon, S., Pelletier, J., Chevalier, Y., 2012. Penetration of drugs through skin, a complex rate-controlling membrane. Curr. Opin. Colloid Interface Sci. 17 (3), 156–165. https://doi.org/10.1016/j.cocis.2012.02.001.

Borges, A., Freitas, V., Mateus, N., Fernandes, I., Oliveira, J., 2020. Solid lipid nanoparticles as carriers of natural phenolic compounds. Antioxidants 10. https://doi.org/10.3390/antiox9100998.

Bose, S., Du, Y., Takhistov, P., Michniak-Kohn, B., 2013. Formulation optimization and topical delivery of quercetin from solid lipid based nanosystems. Int. J. Pharm. 441 (1–2), 56–66. https://doi.org/10.1016/j.ijpharm.2012.12.013.

Brand, R.M., Jendrzejewski, J.L., 2008. Topical treatment with (-)-epigallocatechin-3-gallate and genistein after a single UV exposure can reduce skin damage. J. Dermatol. Sci. 50 (1), 69–72. https://doi.org/10.1016/j.jdermsci.2007.11.008.

Bruno, M.C., Gagliardi, A., Mancuso, A., Barone, A., Tarsitano, M., Cosco, D., Cristiano, M.C., Fresta, M., Paolino, D., 2022. Oleic acid-based vesicular nanocarriers for topical delivery of the natural drug thymoquinone: improvement of anti-inflammatory activity. J. Contr. Release 352, 74–86. https://doi.org/10.1016/j.jconrel.2022.10.011.

Caddeo, C., Díez-Sales, O., Pons, R., Fernàndez-Busquets, X., Fadda, A.M., Manconi, M., 2014. Topical anti-inflammatory potential of quercetin in lipid-based nanosystems: in vivo and in vitro evaluation. Pharm. Res. 31 (4), 959–968. https://doi.org/10.1007/s11095-013-1215-0.

Carrizzo, A., Forte, M., Damato, A., Trimarco, V., Salzano, F., Bartolo, M., Maciag, A., Puca, A.A., Vecchione, C., 2013. Antioxidant effects of resveratrol in cardiovascular, cerebral and metabolic diseases. Food Chem. Toxicol. 61, 215–226. https://doi.org/10.1016/j.fct.2013.07.021.

Celia, C., Cilurzo, F., Trapasso, E., Cosco, D., Fresta, M., Paolino, D., 2012. Ethosomes® and transfersomes® containing linoleic acid: physicochemical and technological features of topical drug delivery carriers for the potential treatment of melasma disorders. Biomed. Microdevices 14 (1), 119–130. https://doi.org/10.1007/s10544-011-9590-y.

Cevc, G., Blume, G., 1992. Lipid vesicles penetrate into intact skin owing to the transdermal osmotic gradients and hydration force. Biochim. Biophys. Acta Biomembr. 1104 (1), 226–232. https://doi.org/10.1016/0005-2736(92)90154-e.

Choi, M.J., Maibach, H.I., 2005. Liposomes and niosomes as topical drug delivery systems. Skin Pharmacol. Physiol. 18 (5), 209–219. https://doi.org/10.1159/000086666.

Cosco, D., Paolino, D., Maiuolo, J., Marzio, L.D., Carafa, M., Ventura, C.A., Fresta, M., 2015. Ultradeformable liposomes as multidrug carrier of resveratrol and 5-fluorouracil for their topical delivery. Int. J. Pharm. 489 (1–2), 1–10. https://doi.org/10.1016/j.ijpharm.2015.04.056.

Cristiano, M.C., Froiio, F., Mancuso, A., Iannone, M., Fresta, M., Fiorito, S., Celia, C., Paolino, D., 2020. In vitro and in vivo trans-epidermal water loss evaluation following topical drug delivery systems application for pharmaceutical analysis. J. Pharm. Biomed. Anal. 186.

Cristiano, M.C., Froiio, F., Spaccapelo, R., Mancuso, A., Nisticò, S.P., Udongo, B.P., Fresta, M., Paolino, D., 2019. Sulforaphane-loaded ultradeformable vesicles as a potential natural nanomedicine for the treatment of skin cancer diseases. Pharmaceutics 1. https://doi.org/10.3390/pharmaceutics12010006.

Cristiano, M.C., Mancuso, A., Fresta, M., Torella, D., De Gaetano, F., Ventura, C.A., Paolino, D., 2021. Topical unsaturated fatty acid vesicles improve antioxidant activity of ammonium glycyrrhizinate. Pharmaceutics 4. https://doi.org/10.3390/pharmaceutics13040548.

Csiszar, A., 2011. Anti-inflammatory effects of resveratrol: possible role in prevention of age-related cardiovascular disease. Ann. N. Y. Acad. Sci. 1215 (1), 117–122. https://doi.org/10.1111/j.1749-6632.2010.05848.x.

Da Silva, E.R., Faria De Freitas, Z.M., Brito Gitirana, L.D., Ricci-Júnior, E., 2011. Improving the topical delivery of zinc phthalocyanine using oleic acid as a penetration enhancer: in vitro permeation and retention. Drug Dev. Ind. Pharm. 37 (5), 569–575. https://doi.org/10.3109/03639045.2010.529144.

Danby, S.G., Alenezi, T., Sultan, A., Lavender, T., Chittock, J., Brown, K., Cork, M.J., 2013. Effect of olive and sunflower seed oil on the adult skin barrier: implications for neonatal skin care. Pediatr. Dermatol. 30 (1), 42–50. https://doi.org/10.1111/j.1525-1470.2012.01865.x.

Dayan, N., Touitou, E., 2000. Carriers for skin delivery of trihexyphenidyl HCl: ethosomes vs. liposomes. Biomaterials 21 (18), 1879–1885. https://doi.org/10.1016/S0142-9612(00)00063-6.

Deng, M., Xie, L., Zhong, L., Liao, Y., Liu, L., Li, X., 2020. Imperatorin: a review of its pharmacology, toxicity and pharmacokinetics. Eur. J. 879. https://doi.org/10.1016/j.ejphar.2020.173124.

Dragicevic, N., Maibach, H., 2018. Combined use of nanocarriers and physical methods for percutaneous penetration enhancement. Adv. Drug Deliv. Rev. 127, 58–84. https://doi.org/10.1016/j.addr.2018.02.003.

Dragicevic, N., Atkinson, J.P., Maibach, H.I., 2015. Chemical Penetration Enhancers: Classification and Mode of Action Percutaneous Penetration Enhancers Chemical Methods in Penetration Enhancement: Modification of the Stratum Corneum. Springer Berlin Heidelberg, Serbia, pp. 11–27. https://doi.org/10.1007/978-3-662-47039-8_2.

Elsayed, M.M.A., Abdallah, O.Y., Naggar, V.F., Khalafallah, N.M., 2007. Lipid vesicles for skin delivery of drugs: reviewing three decades of research. Int. J. Pharm. 332 (1–2), 1–16. https://doi.org/10.1016/j.ijpharm.2006.12.005.

Esposito, E., Drechsler, M., Mariani, P., Sivieri, E., Bozzini, R., Montesi, L., Menegatti, E., Cortesi, R., 2007. Nanosystems for skin hydration: a comparative study. Int. J. Cosmet. Sci. 29 (1), 39–47. https://doi.org/10.1111/j.1467-2494.2007.00362.x.

Esposito, E., Nastruzzi, C., Sguizzato, M., Cortesi, R., 2019. Nanomedicines to treat skin pathologies with natural molecules. Curr. Pharm. Des. 25 (21), 2323–2337. https://doi.org/10.2174/1381612825666190709210703.

Fernández-García, R., Lalatsa, A., Statts, L., Bolás-Fernández, F., Ballesteros, M.P., Serrano, D.R., 2020. Transferosomes as nanocarriers for drugs across the skin: quality by design from lab to industrial scale. Int. J. Pharm. 573. https://doi.org/10.1016/j.ijpharm.2019.118817.

Fresta, M., Mancuso, A., Cristiano, M.C., Urbanek, K., Cilurzo, F., Cosco, D., Iannone, M., Paolino, D., 2020. Targeting of the pilosebaceous follicle by liquid crystal nanocarriers: in vitro and in vivo effects of the entrapped minoxidil. Pharmaceutics 11. https://doi.org/10.3390/pharmaceutics12111127.

Friedrich, R.B., Kann, B., Coradini, K., Offerhaus, H.L., Beck, R.C.R., Windbergs, M., 2015. Skin penetration behavior of lipid-core nanocapsules for simultaneous delivery of resveratrol and curcumin. Eur. J. Pharm. Sci. 78, 204–213. https://doi.org/10.1016/j.ejps.2015.07.018.

Ganesan, P., Narayanasamy, D., 2017. Lipid nanoparticles: different preparation techniques, characterization, hurdles, and strategies for the production of solid lipid nanoparticles and nanostructured lipid carriers for oral drug delivery. Sustain. Chem. Pharm. 6, 37–56. https://doi.org/10.1016/j.scp.2017.07.002.

Gardien, K.L.M., Baas, D.C., de Vet, H.C.W., Middelkoop, E., 2016. Transepidermal water loss measured with the Tewameter TM300 in burn scars. Burns 42 (7), 1455–1462. https://doi.org/10.1016/j.burns.2016.04.018.

Garg, V., Singh, H., Bimbrawh, S., Singh, S.K., Gulati, M., Vaidya, Y., Kaur, P., 2017. Ethosomes and transfersomes: principles, perspectives and practices. Curr. Drug Deliv. 14 (5). https://doi.org/10.2174/1567201813666160520114436.

Gupta, M., Agrawal, U., Vyas, S.P., 2012. Nanocarrier-based topical drug delivery for the treatment of skin diseases. Expert Opin. Drug Deliv. 9 (7), 783–804. https://doi.org/10.1517/17425247.2012.686490.

He, Z., Hu, Y., Niu, Z., Zhong, K., Liu, T., Yang, M., Ji, L., Hu, W., 2023. A review of pharmacokinetic and pharmacological properties of asiaticoside, a major active constituent of Centella asiatica (L.) Urb. J. Ethnopharmacol. 302, 115865. https://doi.org/10.1016/j.jep.2022.115865.

Hewlings, S.J., Kalman, D.S., 2017. Curcumin: a review of its effects on human health. Foods 6 (10). https://doi.org/10.3390/foods6100092.

Hildebrand, A., Beyer, K., Neubert, R., Garidel, P., Blume, A., 2004. Solubilization of negatively charged DPPC/DPPG liposomes by bile salts. J. Colloid Interface Sci. 279 (2), 559–571. https://doi.org/10.1016/j.jcis.2004.06.085.

Hoeller, S., Sperger, A., Valenta, C., 2009. Lecithin based nanoemulsions: a comparative study of the influence of non-ionic surfactants and the cationic phytosphingosine on physicochemical behaviour and skin permeation. Int. J. Pharm. 370 (1–2), 181–186. https://doi.org/10.1016/j.ijpharm.2008.11.014.

Honeywell-Nguyen, P.L., Bouwstra, J.A., 2003. The in vitro transport of pergolide from surfactant-based elastic vesicles through human skin: a suggested mechanism of action. J. Contr. Release 86 (1), 145–156. https://doi.org/10.1016/S0168-3659(02)00415-7.

Honeywell-Nguyen, P.L., Frederik, P.M., Bomans, P.H.H., Junginger, H.E., Bouwstra, J.A., 2002. Transdermal delivery of pergolide from surfactant-based elastic and rigid vesicles: characterization and in vitro transport studies. Pharm. Res. 19 (7), 991–997. https://doi.org/10.1023/A:1016466406176.

Intagliata, S., Modica, M.N., Santagati, L.M., Montenegro, L., 2019. Strategies to improve resveratrol systemic and topical bioavailability: an update. Antioxidants 8. https://doi.org/10.3390/antiox8080244.

Jaiswal, P.K., Kesharwani, S., Kesharwani, R., Dilip, K.P., 2016. Ethosome: a new technology used as topical & transdermal delivery system. J. Drug Deliv. Ther. 6 (3). https://doi.org/10.22270/jddt.v6i3.1245.

Katahira, N., Murakami, T., Kugai, S., Yata, N., Takano, M., 1999. Enhancement of topical delivery of a lipophilic drug from charged multilamellar liposomes. J. Drug Target. 6 (6), 405–414. https://doi.org/10.3109/10611869908996847.

Kaur, D., Kumar, S., 2018. Niosomes: present scenario and future aspects. J. Drug Deliv. Therap. 8 (5), 35–43. https://doi.org/10.22270/jddt.v8i5.1886.

Kim, S.H., Kim, S.A., Park, M.K., Kim, S.H., Park, Y.D., Na, H.J., Kim, H.M., Shin, M.K., Ahn, K.S., 2004. Paeonol inhibits anaphylactic reaction by regulating histamine and TNF-α. Int. Immunopharmacol. 4 (2), 279–287. https://doi.org/10.1016/j.intimp.2003.12.013.

Kumar, P., Singh, S., Handa, V., Kathuria, H., 2018. Oleic acid nanovesicles of minoxidil for enhanced follicular delivery. Medicines 5 (3), 103. https://doi.org/10.3390/medicines5030103.

Kitagawa, S., Kasamaki, M., 2006. Enhanced delivery of retinoic acid to skin by cationic liposomes. Chem. Pharmaceut. Bull. 54 (2), 242–244. https://doi.org/10.1248/cpb.54.242.

Kumar Sarwa, K., Rudrapal, M., Mazumder, B., 2015. Topical ethosomal capsaicin attenuates edema and nociception in arthritic rats. Drug Deliv. 22 (8), 1043–1052. https://doi.org/10.3109/10717544.2013.861041.

Lai, F., Caddeo, C., Manca, M.L., Manconi, M., Sinico, C., Fadda, A.M., 2020. What's new in the field of phospholipid vesicular nanocarriers for skin drug delivery. Int. J. Pharm. 583. https://doi.org/10.1016/j.ijpharm.2020.119398.

Li, Y.Z., Chen, J.H., Tsai, C.F., Yeh, W.L., 2019. Anti-inflammatory property of imperatorin on alveolar macrophages and inflammatory lung injury. J. Nat. Prod. 82 (4), 1002–1008. https://doi.org/10.1021/acs.jnatprod.9b00145.

Lin, H., Lin, L., Choi, Y., Michniak-Kohn, B., 2020. Development and in-vitro evaluation of co-loaded berberine chloride and evodiamine ethosomes for treatment of melanoma. Int. J. Pharm. (1192). https://doi.org/10.1016/j.ijpharm.2020.119278.

Lin, H.W., Xie, Q.C., Huang, X., Ban, J.F., Wang, B., Wei, X., Chen, Y.Z., Lu, Z.F., 2018a. Increased skin permeation efficiency of imperatorin via charged ultradeformable lipid vesicles for transdermal delivery. Int. J. Nanomed. 13, 831–842. https://doi.org/10.2147/IJN.S150086.

Lin, T.K., Zhong, L., Santiago, J.L., 2018b. Anti-inflammatory and skin barrier repair effects of topical application of some plant oils. Int. J. Mol. Sci. 19 (1). https://doi.org/10.3390/ijms19010070.

Liu, M., Chen, M., Xu, P., Yang, Z., 2015. Nanostructured cubosomes as a platform for oral drug delivery. Curr. Pharm. Biotechnol. 16 (4), 313–321. https://doi.org/10.2174/1389201016666150118134805.

Lodzki, M., Godin, B., Rakou, L., Mechoulam, R., Gallily, R., Touitou, E., 2003. Cannabidiol - transdermal delivery and anti-inflammatory effect in a murine model. J. Contr. Release 93 (3), 377–387. https://doi.org/10.1016/j.jconrel.2003.09.001.

Ma, H., Guo, D., Fan, Y., Wang, J., Cheng, J., Zhang, X., 2018. Paeonol-loaded ethosomes as transdermal delivery carriers: design, preparation and evaluation. Molecules 23 (7), 1756. https://doi.org/10.3390/molecules23071756.

Madison, K.C., 2003. Barrier function of the skin: "la raison d'être" of the epidermis. J. Invest. Dermatol. 121 (2), 231–241. https://doi.org/10.1046/j.1523-1747.2003.12359.x.

Malfait, A.M., Gallily, R., Sumariwalla, P.F., Malik, A.S., Andreakos, E., Mechoulam, R., Feldmann, M., 2000. The nonpsychoactive cannabis constituent cannabidiol is an oral anti-arthritic therapeutic in murine collagen-induced arthritis. Proc. Natl. Acad. Sci. U S A 97 (17), 9561–9566. https://doi.org/10.1073/pnas.160105897.

Manconi, M., Sinico, C., Valenti, D., Lai, F., Fadda, A.M., 2006. Niosomes as carriers for tretinoin: III. A study into the in vitro cutaneous delivery of vesicle-incorporated tretinoin. Int. J. Pharm. 311 (1–2), 11–19. https://doi.org/10.1016/j.ijpharm.2005.11.045.

Mancuso, A., Cristiano, M.C., Fresta, M., Paolino, D., 2020. The challenge of nanovesicles for selective topical delivery for acne treatment: enhancing absorption whilst avoiding toxicity. Int. J. Nanomed. 15, 9197–9210. https://doi.org/10.2147/IJN.S237508.

Mancuso, A., Tarsitano, M., Cavaliere, R., Fresta, M., Cristiano, M.C., Paolino, D., 2023. Gelled liquid crystal nanocarriers for improved antioxidant activity of resveratrol. Gels 11. https://doi.org/10.3390/gels9110872.

Masini, V., Bonte, F., Meybeck, A., Wepierre, J., 1993. Cutaneous bioavailability in hairless rats of tretinoin in liposomes or gel. J. Pharm. Sci. 82 (1), 17–21. https://doi.org/10.1002/jps.2600820104.

Mendoza-Muñoz, N., Urbán-Morlán, Z., Leyva-Gómez, G., De La Luz Zambrano-Zaragoza, M., Piñón-Segundo, E., Quintanar-Guerrero, D., 2021. Solid lipid nanoparticles: an approach to improve oral drug delivery. J. Pharm. Pharm. Sci. 24, 509–532. https://doi.org/10.18433/jpps31788.

Menon, G.K., Cleary, G.W., Lane, M.E., 2012. The structure and function of the stratum corneum. Int. J. Pharm. 435 (1), 3–9. https://doi.org/10.1016/j.ijpharm.2012.06.005.

Mi, C., Ma, J., Wang, K.S., Zuo, H.X., Wang, Z., Li, M.Y., Piao, L.X., Xu, G.H., Li, X., Quan, Z.S., Jin, X., 2017. Imperatorin suppresses proliferation and angiogenesis of human colon cancer cell by targeting HIF-1α via the mTOR/p70S6K/4E-BP1 and MAPK pathways. J. Ethnopharmacol. 203, 27–38. https://doi.org/10.1016/j.jep.2017.03.033.

Milosheska, D., Roškar, R., 2022. Use of retinoids in topical antiaging treatments: a focused review of clinical evidence for conventional and nanoformulations. Adis, Slovenia Adv. Ther. 39 (12), 5351–5375. https://doi.org/10.1007/s12325-022-02319-7.

Moghassemi, S., Hadjizadeh, A., 2014. Nano-niosomes as nanoscale drug delivery systems: an illustrated review. J. Contr. Release 185 (1), 22–36. https://doi.org/10.1016/j.jconrel.2014.04.015.

Nasr, M., Younes, H., Abdel-Rashid, R.S., 2020. Formulation and evaluation of cubosomes containing colchicine for transdermal delivery. Drug Deliv. Transl. Res. 10 (5), 1302–1313. https://doi.org/10.1007/s13346-020-00785-6.

Nasser, M.I., Zhu, S., Hu, H., Huang, H., Guo, M., Zhu, P., 2019. Effects of imperatorin in the cardiovascular system and cancer. Biomed. Pharmacother. 120, 109401. https://doi.org/10.1016/j.biopha.2019.109401.

Nastiti, C.M., Ponto, T., Mohammed, Y., Roberts, M.S., Benson, H.A., 2020. Novel nanocarriers for targeted topical skin delivery of the antioxidant resveratrol. Pharmaceutics 12 (2). https://doi.org/10.3390/pharmaceutics12020108.

Negi, P., Aggarwal, M., Sharma, G., Rathore, C., Sharma, G., Singh, B., Katare, O.P., 2017. Niosome-based hydrogel of resveratrol for topical applications: an effective therapy for pain related disorder(s). Biomed. Pharmacother. 88, 480–487. https://doi.org/10.1016/j.biopha.2017.01.083.

Niu, X.Q., Zhang, D.P., Bian, Q., Feng, X.F., Li, H., Rao, Y.F., Shen, Y.M., Geng, F.N., Yuan, A.R., Ying, X.Y., Gao, J.Q., 2019. Mechanism investigation of ethosomes transdermal permeation. Int. J. Pharm. X 1. https://doi.org/10.1016/j.ijpx.2019.100027.

Ogiso, T., Yamaguchi, T., Iwaki, M., Tanino, T., Miyake, Y., 2001. Effect of positively and negatively charged liposomes on skin permeation of drugs. J. Drug Target. 9 (1), 49–59. https://doi.org/10.3109/10611860108995632.

Opatha, S.A.T., Titapiwatanakun, V., Chutoprapat, R., 2020. Transfersomes: a promising nanoencapsulation technique for transdermal drug delivery. Pharmaceutics 12 (9). https://doi.org/10.3390/pharmaceutics12090855.

Paolino, D., Lucania, G., Mardente, D., Alhaique, F., Fresta, M., 2005. Ethosomes for skin delivery of ammonium glycyrrhizinate: in vitro percutaneous permeation through human skin and in vivo anti-inflammatory activity on human volunteers. J. Contr. Release 106 (1–2), 99–110. https://doi.org/10.1016/j.jconrel.2005.04.007.

Paolino, D., Muzzalupo, R., Ricciardi, A., Celia, C., Picci, N., Fresta, M., 2007. In vitro and in vivo evaluation of Bola-surfactant containing niosomes for transdermal delivery. Biomed. Microdevices 9 (4), 421–433. https://doi.org/10.1007/s10544-007-9046-6.

Paolino, D., Cosco, D., Cilurzo, F., Trapasso, E., Morittu, V.M., Celia, C., Fresta, M., 2012. Improved in vitro and in vivo collagen biosynthesis by asiaticoside-loaded ultradeformable vesicles. J. Contr. Release 162 (1), 143–151. https://doi.org/10.1016/j.jconrel.2012.05.050.

Patel, V.B., Misra, A., Marfatia, Y.S., 2000. Topical liposomal gel of tretinoin for the treatment of acne: research and clinical implications. Pharm. Dev. Technol. 5 (4), 455–464. https://doi.org/10.1081/PDT-100102029.

Patel, A.V., Shah, B.N., 2018. Transdermal drug delivery system: a review. Pharma Sci. Monit. (1).

Peng, X., Wen, X., Pan, X., Wang, R., Chen, B., Wu, C., 2010. Design and in vitro evaluation of capsaicin transdermal controlled release cubic phase gels. AAPS PharmSciTech 11 (3), 1405–1410. https://doi.org/10.1208/s12249-010-9481-1.

Peng, X., Zhou, Y., Han, K., Qin, L., Dian, L., Li, G., Pan, X., Wu, C., 2015. Characterization of cubosomes as a targeted and sustained transdermal delivery system for capsaicin. Drug Des. Dev. Ther. 9, 4209–4218. https://doi.org/10.2147/DDDT.S86370.

Plianbangchang, P., Tungpradit, W., Tiyaboonchai, W., 2007. Efficacy and safety of curcuminoids loaded solid lipid nanoparticles facial cream as an anti-aging agent. Naresuan Univ. J. 15 (2), 73–81.

Prakash, A.J., Pramod, K., 2019. The effect of Transferosomes on skin permeation of Ketorolac tromethamine. Adv. Pharm. Res. 01–13. https://doi.org/10.36218/apr/1949694.

Puglia, C., Lauro, M.R., Tirendi, G.G., Fassari, G.E., Carbone, C., Bonina, F., Puglisi, G., 2017. Modern drug delivery strategies applied to natural active compounds. Expert Opin. Drug Deliv. 14 (6), 755–768. https://doi.org/10.1080/17425247.2017.1234452.

Rai, S., Pandey, V., Rai, G., 2017. Transfersomes as versatile and flexible nano-vesicular carriers in skin cancer therapy: the state of the art. Nano Rev. Exp. 8 (1). https://doi.org/10.1080/20022727.2017.1325708.

Rattanapak, T., Young, K., Rades, T., Hook, S., 2012. Comparative study of liposomes, transfersomes, ethosomes and cubosomes for transcutaneous immunisation: characterisation and in vitro skin penetration. J. Pharm. Pharmacol. 64 (11), 1560–1569. https://doi.org/10.1111/j.2042-7158.2012.01535.x.

Sarwa, K.K., Mazumder, B., Rudrapal, M., Verma, V.K., 2015. Potential of capsaicin-loaded transfersomes in arthritic rats. Drug Deliv. 22 (5), 638–646. https://doi.org/10.3109/10717544.2013.871601.

Schätzlein, A., Cevc, G., 1998. Non-uniform cellular packing of the stratum corneum and permeability barrier function of intact skin: a high-resolution confocal laser scanning microscopy study using highly deformable vesicles (Transfersomes). Br. J. Dermatol. 138 (4), 583–592. https://doi.org/10.1046/j.1365-2133.1998.02166.x.

Sguizzato, M., Esposito, E., Cortesi, R., 2021. Lipid-based nanosystems as a tool to overcome skin barrier. Int. J. Mol. Sci. 15. https://doi.org/10.3390/ijms22158319.

Shah, J.C., Sadhale, Y., Chilukuri, D.M., 2001. Cubic phase gels as drug delivery systems. Adv. Drug Deliv. Rev. 47 (2–3), 229–250. https://doi.org/10.1016/S0169-409X(01)00108-9.

Sharma, P., Dhawan, S., Nanda, S., 2020. Cubosome: a potential liquid crystalline carrier system. Curr. Pharm. Des. 26 (27), 3300–3316. https://doi.org/10.2174/1381612826666200617162424.

Shishir, M.R.I., Xie, L., Sun, C., Zheng, X., Chen, W., 2018. Advances in micro and nano-encapsulation of bioactive compounds using biopolymer and lipid-based transporters. Trends Food Sci. Technol. 78, 34–60. https://doi.org/10.1016/j.tifs.2018.05.018.

Siekmann, B., Bunjes, H., Koch, M.H.J., Westesen, K., 2002. Preparation and structural investigations of colloidal dispersions prepared from cubic monoglyceride-water phases. Int. J. Pharm. 244 (1–2), 33–43. https://doi.org/10.1016/S0378-5173(02)00298-3.

Song, Y.K., Kim, C.K., 2006. Topical delivery of low-molecular-weight heparin with surface-charged flexible liposomes. Biomaterials 27 (2), 271–280. https://doi.org/10.1016/j.biomaterials.2005.05.097.

Strugała, P., Tronina, T., Huszcza, E., Gabrielska, J., 2017. Bioactivity in vitro of quercetin glycoside obtained in *Beauveria bassiana* culture and its interaction with liposome membranes. Molecules 9. https://doi.org/10.3390/molecules22091520.

Tosato, M.G., Maya Girón, J.V., Martin, A.A., Krishna Tippavajhala, V., Fernández Lorenzo de Mele, M., Dicelio, L., 2018. Comparative study of transdermal drug delivery systems of resveratrol: high efficiency of deformable liposomes. Mater. Sci. Eng. C 90, 356–364. https://doi.org/10.1016/j.msec.2018.04.073.

Touitou, E., Godin, B., 2007. Dermal drug delivery with ethosomes: therapeutic potential. Therapy 4 (4), 465–472. https://doi.org/10.2217/14750708.4.4.465Israel.

Touitou, E., Dayan, N., Bergelson, L., Godin, B., Eliaz, M., 2000. Ethosomes - novel vesicular carriers for enhanced delivery: characterization and skin penetration properties. J. Contr. Release 65 (3), 403–418. https://doi.org/10.1016/S0168-3659(99)00222-9.

Tripathi, N., Verma, S., Vyas, M., Yadav, N.S., Gain, S., Khatik, G.L., 2022. Nanoformulations of quercetin: a potential phytochemical for the treatment of UV radiation induced skin damages. Braz. J. Pharm. Sci. 58. https://doi.org/10.1590/s2175-97902020000118744.

Trotta, M., Peira, E., Debernardi, F., Gallarate, M., 2002. Elastic liposomes for skin delivery of dipotassium glycyrrhizinate. Int. J. Pharm. 241 (2), 319–327. https://doi.org/10.1016/S0378-5173(02)00266-1.

Van Tran, V., Moon, J.Y., Lee, Y.C., 2019. Liposomes for delivery of antioxidants in cosmeceuticals: challenges and development strategies. J. Contr. Release 300, 114–140. https://doi.org/10.1016/j.jconrel.2019.03.003.

Williams, A.C., Barry, B.W., 2012. Penetration enhancers. Adv. Drug Deliv. Rev. 64, 128–137. https://doi.org/10.1016/j.addr.2012.09.032.

Wu, P.S., Li, Y.S., Kuo, Y.C., Tsai, S.J.J., Lin, C.C., 2019. Preparation and evaluation of novel transfersomes combined with the natural antioxidant resveratrol. Molecules 3. https://doi.org/10.3390/molecules24030600.

Wu, J., Xue, X., Zhang, B., Cao, H., Kong, F., Jiang, W., Li, J., Sun, D., Guo, R., 2016. Enhanced antitumor activity and attenuated cardiotoxicity of Epirubicin combined with Paeonol against breast cancer. Tumor Biol. 37 (9), 12301–12313. https://doi.org/10.1007/s13277-016-5088-9.

Xu, Y., Zhu, J.Y., Lei, Z.M., Wan, L.J., Zhu, X.W., Ye, F., Tong, Y.Y., 2017. Anti-proliferative effects of paeonol on human prostate cancer cell lines DU145 and PC-3. J. Physiol. Biochem. 73 (2), 157–165. https://doi.org/10.1007/s13105-016-0537-x.

Yang, L., Wu, L., Wu, D., Shi, D., Wang, T., Zhu, X., 2017. Mechanism of transdermal permeation promotion of lipophilic drugs by ethosomes. Int. J. Nanomed. 12, 3357–3364. https://doi.org/10.2147/IJN.S134708.

Yousef, M., Vlachogiannis, I.A., Tsiani, E., 2017. Effects of resveratrol against lung cancer: in vitro and in vivo studies. Nutrients 11. https://doi.org/10.3390/nu9111231.

Yu, X., Du, L., Li, Y., Fu, G., Jin, Y., 2015. Improved anti-melanoma effect of a transdermal mitoxantrone ethosome gel. Biomed. Pharmacother. 73, 6–11. https://doi.org/10.1016/j.biopha.2015.05.002.

Zhao, Y.Z., Lu, C.T., Zhang, Y., Xiao, J., Zhao, Y.P., Tian, J.L., Xu, Y.Y., Feng, Z.G., Xu, C.Y., 2013. Selection of high efficient transdermal lipid vesicle for curcumin skin delivery. Int. J. Pharm. 454 (1), 302–309. https://doi.org/10.1016/j.ijpharm.2013.06.052.

Zhu, Y., Wang, M., Zhang, J., Peng, W., Firempong, C.K., Deng, W., Wang, Q., Wang, S., Shi, F., Yu, J., Xu, X., Zhang, W., 2015. Improved oral bioavailability of capsaicin via liposomal nanoformulation: preparation, in vitro drug release and pharmacokinetics in rats. Arch. Pharm. Res. 38 (4), 512–521. https://doi.org/10.1007/s12272-014-0481-7.

Zong, S., Pu, Y., Li, S., Xu, B., Zhang, Y., Zhang, T., Wang, B., 2018. Beneficial anti-inflammatory effect of paeonol self-microemulsion-loaded colon-specific capsules on experimental ulcerative colitis rats. Artif. Cell Nanomed. Biotechnol. 46 (Suppl. 1), 324–335. https://doi.org/10.1080/21691401.2017.1423497.

PART 3

Organic phytochemicals state-of-the-art

CHAPTER

Alkaloids' wide world: Biological activity, extraction, quantification 13

Adriana Arigò and Genny Grasselli
University of Urbino Carlo Bo, Department of Pure and Applied Sciences, Urbino, Italy

1. Alkaloids' role and classification

Plants produce various low molecular weight compounds that assume different tasks. Mainly, those compounds can be divided into three groups: primary metabolites, which are directly involved in plants' growth and development; secondary metabolites, which include phenolics, terpenes, and nitrogen-containing compounds with the role to defend and guide the plants to biotic/abiotic environmental interactions; hormones, which mediate the organismal processes and metabolism (Erb and Kliebenstein, 2020).

In this chapter, we are going to focus on alkaloids, a class of secondary metabolites chemically identified as nitrogen-containing compounds that usually possess one or more nitrogen atoms incorporated in a heterocyclic ring (Debnath et al., 2018). Different studies demonstrated that those compounds show different effects on human and animal health due to their alkaline properties given by nitrogen atoms (Debnath et al., 2018; Gomes et al., 2022; Tittikpina et al., 2023; Faisal et al., 2023). Because they originate from different precursors, alkaloids are characterized by different chemical skeletons, which confer different biological activities (Gañán et al., 2023); some of them captured the interest of pharmaceutical industries because of their therapeutic potentials, while others are well known for their acute dosage-dependent toxicity, where their excessive ingestion causes harmful side effects, such as tremors, gastrointestinal problems, convulsion, till death (Bhambhani et al., 2021; Beaudoin and Facchini, 2014). The European Food Safety Authority (EFSA) established limits for dietary exposure for some alkaloid classes, which can represent a risk to human and animal health. For example, the acute reference dose (ARfD) for tropane alkaloids was set at 16 ng/kg of body weight (Human Acute Exposure Assessment to Tropane Alkaloids, 2018), while a lower confidence limit on benchmark dose associated with a 10% response ($BMDL_{10}$) of 273 μg/kg of body weight was established for pyrrolizidine alkaloids due to the carcinogenic risks (Knutsen et al., 2017). The ARfDs established for ergot alkaloids and morphine opium alkaloids are 1 μg/kg (Scientific Opinion on Ergot Alkaloids in Food and Feed, 2012) and 10 μg/kg of body weight (Scientific Opinion on the Risks for Public Health Related to the Presence of Opium Alkaloids in Poppy Seeds, 2011; Update of the Scientific Opinion on Opium Alkaloids in Poppy Seeds, 2018), respectively; for quinolizidine alkaloids, the lowest observed adverse effect level (LOAEL) was 2.5 mg/kg of body weight per day regarding the animals' health (Scientific Opinion on the Risks for Animal and Human Health Related

Phytochemicals for Health. https://doi.org/10.1016/B978-0-443-15366-2.00013-7

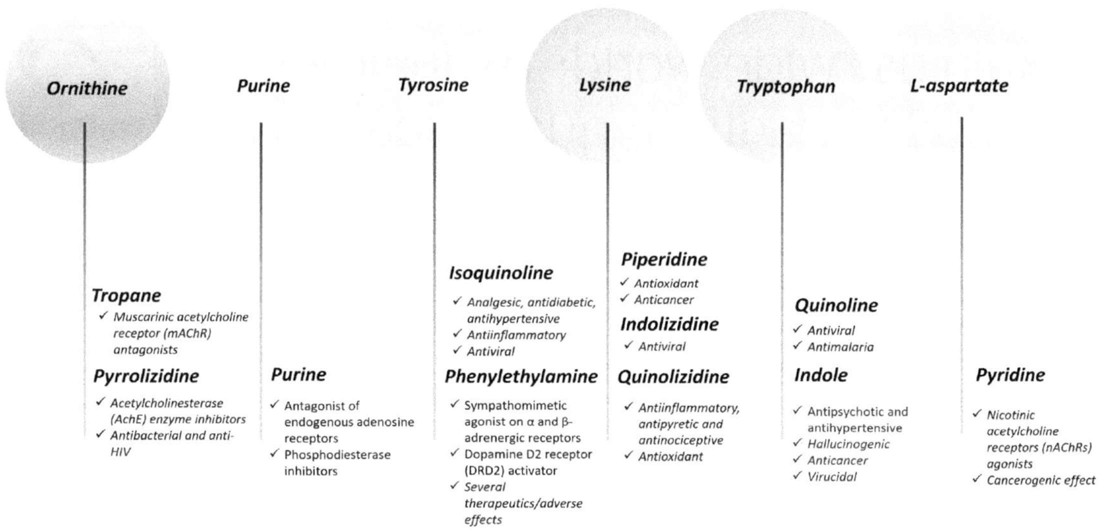

FIGURE 13.1

Classification of the main classes of alkaloids based on their precursor molecules and biological activities.

to the Presence of Quinolizidine Alkaloids in Feed and Food, in Particular in Lupins and Lupin-Derived Products, 2019). Hence, in recent years, the development of new sensitive and specific methods to determine the quantity of some alkaloids in food is acquiring a wide interest.

Alkaloids classification can be based on botanical origin, structural similarities, starting molecules, or biological/pharmacological activities (Dey et al., 2020; Casado et al., 2022). Because of the enormous number of alkaloids classes present in nature and their innumerable biological activities, in this chapter, the main alkaloids classes are classified based on the starting molecules (Gañán et al., 2023; Bhambhani et al., 2021; Koleva et al., 2012) and discussed on the basis of the most interesting and recent biological evidence (Fig. 13.1):

- derived from ornithine: tropane and pyrrolizidine alkaloids;
- derived from purine (adenine/guanine): purine alkaloids;
- derived from tyrosine: isoquinoline and phenylethylamine alkaloids;
- derived from lysine: piperidine, indolizidine, quinolizidine alkaloids;
- derived from tryptophan: indole and quinoline alkaloids;
- derived from L-aspartate: pyridine alkaloids.

2. Alkaloids' classes: Chemical structures, plant sources, biological activities, and traditional uses

2.1 Derived from ornithine

Tropane alkaloids (TAs). TAs are a class of alkaloids that includes more than 200 compounds, characterized by a two-ringed structure with pyrrolidine and a piperidine ring sharing a single nitrogen atom and two carbon atoms (tropane ring), arising from ornithine (Gañán et al., 2023; Kohnen-Johannsen and Kayser, 2019). The most well-known source of this alkaloids class is the Solanaceae family (Kohnen-Johannsen and Kayser, 2019; Scientific Opinion of the Panel on Contaminants in the Food Chain on a Request from the European Commission on Tropane Alkaloids (from Datura Sp.) as Undesirable Substances in Animal Feed, 2008; González-Gómez et al., 2022b), but their occurrence is well documented also in other plant families, such as Brassicaceae (Cruciferae) (Kohnen-Johannsen and Kayser, 2019), Convolvulaceae (Scientific Opinion of the Panel on Contaminants in the Food Chain on a Request from the European Commission on Tropane Alkaloids (from Datura Sp.) as Undesirable Substances in Animal Feed, 2008), Erythroxylaceae (Kohnen-Johannsen and Kayser, 2019), Euphorbiaceae (Adamse and Van Egmond, 2010), Oleaceae, Proteaceae Rhizophoraceae (Adamse et al., 2014), and Meliaceae (Debnath et al., 2018). The main compounds belonging to this class are hyoscyamine, atropine, scopolamine, cocaine, and catuabine (Debnath et al., 2018; Human Acute Exposure Assessment to Tropane Alkaloids, 2018; Kohnen-Johannsen and Kayser, 2019; Perharič et al., 2013; Roque Bravo et al., 2022). Scopolamine and hyoscyamine have been used for a long time against a variety of diseases, due to their anticholinergic characteristic; in fact, they act as competitive and nonselective muscarinic acetylcholine receptor (mAChR) antagonists, preventing the binding of the physiological neurotransmitter acetylcholine. Diseases such as depression, Parkinson's, and Alzheimer's show a muscarinic cholinergic system dysfunction; hence, antagonists of the muscarinic system represent a possibility for drug development (Kohnen-Johannsen and Kayser, 2019). Some studies reported intoxications due to the consumption of TA-containing plants or foods contaminated with these plants (Human Acute Exposure Assessment to Tropane Alkaloids, 2018; Adamse and Van Egmond, 2010; Adamse et al., 2014). These compounds may avoid the binding of acetylcholine with the muscarine receptor (González-Gómez et al., 2022b) causing side effects such as tachycardia, muscle spasms, hallucinogenic, and even lethal episodes (Kohnen-Johannsen and Kayser, 2019). Atropine and scopolamine were found in some foods, such as tea, herbs for infusions, spices, hempseed, and cereals; for this reason, TAs captured the attention of the EFSA Panel on Contaminants in the Food Chain (CONTAM Panel) in a Scientific Opinion due to their toxic effects at high dose, establishing an ARfD, based on the results for decreased heart rate in human volunteer study, of 16 ng/kg body weight per day as the sum of hyoscyamine and scopolamine (Human Acute Exposure Assessment to Tropane Alkaloids, 2018).

Pyrrolizidine alkaloids (PAs). PA is a group of compounds that originated from ornithine and are mainly found in the genera *Senecio* and *Eupatorium* (Asteraceae, tribes Senecioneae and Eupatorieae), many genera of the Boraginaceae (e.g., *Heliotropium*), rather isolated within the Fabaceae, the large subtropical genus *Crotalaria*, and certain genera of the Orchidaceae (e.g., Phalaenopsis) (Gañán et al., 2023; Hartmann, 1999; Moreira et al., 2018). PAs can be divided into four groups, retronecine-, heliotridine-, otonecine- and platynecine-types; the compounds, except the platynecine-type, are characterized by double bound in 1,2 position, which can be correlated with their toxicity (Hartmann, 1999). In 1992, Taylor and coworkers (Taylor et al., 1992), in an in vitro study, demonstrated for the first time the anti-HIV activity of a PA isolated from *Alexa leiopetala*. This polyhydroxylated PA, named 7,7a-diepialexine, showed activity against the HIV-1 virus infection due to its capacity to contrast the HIV growth by reducing the cleavage of the precursor HIV-1 glycoprotein 160 (gp160) (Taylor et al., 1992; Mohammadi Ziarani et al., 2022). 7-O-angeloylechinatine-N-oxide, together with 3′-O-acetylheliosupine-N-oxide, heliosupine-N-oxide, and heliosupine, extracted from *Solenanthus lanatus*, act by inhibiting the acetylcholinesterase (AchE) enzyme, which has the function of catalyzing the hydrolysis of acetylcholine (Ach) and other neurotransmitters. Substances that can inhibit the AchE represent an interesting possibility to treat Alzheimer's disease, myasthenia gravis, and diabetic urinary bladder (Wei et al., 1970). In addition, other PAs showed some activities for the treatment of ulcerogenic disease and stomach pain in mice and rats (Toma et al., 2004). PAs are toxicologically inactive compounds that became dangerous after bio-activation processes. Bio-activation routes include hydrolysis to form necines and necic acids, N-oxidation to form pyrrolizidine alkaloids N-oxides (PANO), and oxidation to form pyrrolic esters or dehydropyrrolizidine alkaloids (DHPAs). Their bio-activation occurs in the liver, where they show the highest toxicity; a human in vivo study demonstrated that the metabolites toxic activity can be ascribed to their capacity to bind DNA and cellular proteins, damaging the hepatocytes and sinusoidal cells and causing veno-occlusive disease (VOD) (Zhu et al., 2017; Dusemund et al., 2018). In addition, metabolites formed after the activation can reach other target organs, such as the lungs and kidneys, traveling through the circulatory system, causing pulmonary hypertension, cardiac hypertrophy, kidney degenerative injuries, and death (Casado et al., 2022; Moreira et al., 2018; Dusemund et al., 2018; Field et al., 2015; Prakash et al., 1999; Ma et al., 2018). A large number of in vitro and in vivo studies demonstrated also that PAs show genotoxicity (Allemang et al., 2018; Hadi et al., 2021; Li et al., 2020; Buchmueller et al., 2022) and tumorigenicity activity (Chou et al., 2003; Luckert et al., 2015; He et al., 2023) due to their capacity to form adducts with the DNA, sister chromatid exchange, and chromosomal aberration (Wang et al., 2005). Despite the toxicity, recently some studies demonstrated that PAs show many interesting biological properties (Taylor et al., 1992; da Silva Negreiros Neto et al., 2016). Usaramine, monocrotaline, and azido-retronecine demonstrated to be powerful compounds with antimicrobial activities; in particular, Themístocle's research group demonstrated that usuramine shows antibiofilm activity against *S. epidermidis*, reducing more than 50% of biofilm formation, and mococrotaline can inhibit the 80% of parasite growth of T. vaginalis, which is involved in many health complications in women (da Silva Negreiros Neto et al., 2016). Recently, Cantero and coworkers showed that pyrrolizidine alkaloids found in propolis from *S. postica* have an antibacterial effect against Gram-negative bacterial pathogens, including *Escherichia coli* D31-streptomycin resistant.

2.2 Derived from lysine

Piperidine alkaloids. Piperidine alkaloids are an alkaloids class derived from lysine and their scaffold is the most found in U.S. FDA-approved pharmaceuticals (Wu et al., 2022). Several family plants are sources of piperidine alkaloids; among these, we can find *Piper nigrum* L. (Piperaceae), *Piper longum* L., *C. maculatum*, *Nicotiana* spp. (Solanaceae), *N. glauca*, *N. tabacum*, *Lobelia* spp. (Campanulaceae), and *L. inflata* L (Bhambhani et al., 2021; Xu et al., 2019; Qasem et al., 2022). The most bioactive piperidine alkaloid in *Piper nigrum* L. is piperine, which shows different therapeutic properties; an in vivo study in mice and rats demonstrated the antiinflammatory effect of piperine at doses of 10 and 15 mg/kg after 30 minutes, and analgesic effect at doses of 5 mg/kg after 120 minutes (Tasleem et al., 2014). Piperine also showed an antioxidant effect, being able to inhibit monoamine oxidase (MAO). MAO catalyzes the oxidative deamination of serotonin, dopamine, and norepinephrine and is involved in several neurological disorders. In the same study, piperine was demonstrated to also possess antidepressant-like activity (Lee et al., 2005). Other reported therapeutic effects are cognitive-enhancing effects and antiulcer (Li et al., 2020). Another piperidine alkaloid component is coniine, which is produced by *C. maculatum*, well-known for its toxic properties. It acts causing a neuromuscular blockage, leading to death. Poisonings were reported among animals that feed on vegetative parts or seeds. Instead, its chronic toxicity effect is explicated in pregnant animals, causing multiple congenial contractures in offspring. Human toxicity occurs only for accidental ingestion of plant parts; studies reported acute renal failure and neurological effects (Vetter, 2004; Rizzi et al., 1991; Bowman and Sanghvi, 1963). Recently, some anticancer activities were attributed to piperidine alkaloids. Piperine shows antibreast cancer activity, acting on different signaling pathways; one of them involves the inhibition of the PI3K/Akt/mTOR signaling pathway. Cancer cell lines treated with piperine displayed the inhibition of the Akt signaling pathway and, consequently, apoptosis (Zhang et al., 2013; Greenshields et al., 2015; Mitra et al., 2022). Piperine demonstrated anticancer activities also against other types of cancer cells (Mitra et al., 2022).

Indolizidine alkaloids. Indolizidine alkaloids include compounds that present an indolizidine ring system in their structures. Indolizidine compounds were isolated from different plant families, such as Asclepiadaceae, Convolvulaceae, Moraceae, and Orchidaceae, from the fungus *Rhizoctonia* and animal sources (Gellert, 1982). This class of compounds includes different groups, such as hydroxylated indolizidines, phenanthroindolizidines, and simpler indolizidines (Gellert, 1982; Michael, 2007). Castanospermine and swainsonine are two indolizine alkaloids that act as alpha-mannosidase and beta-glucosidase inhibitors; this confers them several therapeutic activities (Elbein and Molyneux, 1999); in particular, this activity gives castanospermine the antiviral activity against human immunodeficiency viruses (HIVs) (Taylor et al., 1992; Pan et al., 1983). Castanospermine showed in vitro activity also against other viruses, such as influenza virus, and bovine diarrhea virus, and in vitro and in vivo activity against dengue virus (DEN) (Whitby et al., 2005).

Quinolizidine alkaloids (QAs). QAs are found in different plant species, such as *Sophora flavescens Ait.* and *S. alopecuroides* L. The most well-known compounds belonging to this class are sparteine, lupanine, a-isolupanine, 13a-hydroxylupanine, anagyrine, matrine, sophoridine, sophocarpine, cytisine, and aloperine (Koleva et al., 2012; Reinhard et al., 2006; Liu et al., 2008). QA alkaloids have several biological and pharmacological activities, such as antiinflammatory, antipyretic, and antinociceptive (Liu et al., 2008). Matrine, sophoridine, sophocarpine, and cytisine have structural properties that allow them to bind acetylcholine receptors (nAChRs) in *P. americana*; the nitrogen atom and the carbonyl oxygen atom positions confer them the ability to bind, despite low affinity, this receptor. This activity laid the foundation for novel insecticidal nAChR ligands (Liu et al., 2008). Oxymatrine (OMT) demonstrated to improve the diabetes-associated cognitive deficit in rats, acting in different ways. In humans, people affected by diabetes can show progressive dementia; OMT can decrease blood glucose levels, leading to the reduction of oxidative stress, the inhibition of the tumor necrosis factor (TNF) alpha-induced nuclear factor kappa (NF-kB) signaling pathways, and the improvement of learning and memory functions in diabetic rats (Wang and Jia, 2014). More recently, Zhang et al.(Zhang et al., 2018) isolated five novel quinolizidine-based alkaloids, named sophalines E, F, G, H, and I in *S. alopecuroides*; the second compound was the first example of matrine-indolizine alkaloids and showed antihepatitis B virus activity.

2.3 Derived from L-aspartate

Pyridine alkaloids. Pyridine alkaloids are a class of alkaloids derived from L-aspartate. The most representative plant sources are *Nicotiana tabacum* (Solanaceae), *Laburnum anagyroides* (Leguminosae), *Areca catechu* (Arecaceae), *Conium maculatum* (Umbelliferae), *Lobelia inflata* (Lobeliaceae),

Cystisus scoparium, and *Lupinus luteus* (Fabaceae). The main molecules belonging to this class are nicotine, cytisine, arecoline, arecaidine, guvacoline, guvacine, ricinine, coniine, lobeline, pelletierine, and sparteine. Some compounds of this class showed effects on the central nervous system (CNS) (Lin et al., 2020), capturing the attention of the scientific community because the N atom in their structure plays a central role in drug development (Vitaku et al., 2014; Pennington and Moustakas, 2017). In particular, nicotine and its structural related compound, cytisine and epibadine (an amphibian-derived alkaloid), which are nicotinic acetylcholine receptors (nAChRs), agonists, leading to the development of smoking cessation drugs and dementia drugs. Nicotine and cytisine are the only pyridine-marketed alkaloids. Nicotine is the most well-known compound belonging to pyridine alkaloids; it occurs in tobacco products and in products used to break the smoking habit. Nicotine can bind nicotinic cholinergic receptors mainly expressed in neurons and neuromuscular junctions, helping the release of neurotransmitters, which lead to a modification of behavior in human beings. Being the major addictive compound of tobacco smoke, nicotine effect on human health was largely studied. Cytisine instead is a selective nicotinic cholinergic agonist and, such as nicotine, is sold as a commercial product for smoking cessation due to its action as a partial agonist of α4-β2 nicotinic acetylcholine receptors, causing a reduction in the nicotine effect on dopamine release (Walker et al., 2014). At a low level in the bloodstream, nicotine does not represent a carcinogenic effect; however, it seems to show carcinogenic properties at high doses despite the mechanism being already unclear (Chellappan, 2022). In vivo and in vitro studies (Singh et al., 2011; Wu et al., 2011) demonstrate that the nicotinic acetylcholine receptors (nAChRs), which are activated by nicotine, may activate some signaling pathways involved in tumorigenic effects. However, different studies demonstrated that nicotine can have a cancerogenic effect using a different mechanism, which involved or not the nAChRs. In addition, metabolites produced starting from nicotine also showed carcinogenic properties; nicotine is metabolized in the liver and 70%–80% is converted into cotinine. As demonstrated by Tomohisa et al. (Nakada et al., 2012), in an in vivo study on mice, this metabolite showed tumor-promoting effects, such as other two tobacco-specific carcinogens, generated by nicotine during smoking, nicotine-derived nitrosamine ketone (NNK), and N-nitrosonornicotine (NNN), which can contribute to oncogenesis in lung cancers (Arredondo et al., 2006). Despite the negative effect, more recently, Grizzell et al. proved that cotinine can lead to some beneficial for stress effects, memory, and synapse, inhibiting the glycogen synthase kinase 3 β in the hippocampus and prefrontal cortex in mice (Grizzell et al., 2014).

2.4 Derived from tryptophan

Quinoline alkaloids. Quinoline alkaloids are a class of compounds derived from tryptophan, including more than 200 molecules, which acquired notoriety due to their properties against multiresistant malaria. Their structures, based on N-heterocyclic aromatic rings, play important roles against different diseases. The main plant sources of quinoline alkaloids are *Cinchona officinalis* (Rubiaceae), *Strychnos Nux-Vomica* (Loganiaceae), *Veratrum sabadibla* (Melanthiaceae), and *Strychnos nuxvomica* (Loganiaceae). The molecules belonging to this class can be divided into marketed and nonmarketed quinoline alkaloids; the main marketed quinoline compounds act as antimalaria drugs and among them, there are quinidine, dihydroquinine (a natural impurity found in quinine pharmaceutical formulation), hydroxyquinine (a principal metabolite of quinine in human) (Nontprasert et al., 1996). In 1820, for the first time, quinine was isolated from the Cinchona tree and used for malaria treatment.

However, in 2006, the World Health Organization (WHO) stopped the use of this compound, due to its toxicity and the development of *Plasmodium* sp. resistance (Shang et al., 2018). Later, different antimalaria drugs were developed, based on quinoline's structures; the most famous is chloroquinine. Its mechanism of action is efficacious only in the blood stages of the malaria parasite in which the parasite degrades hemoglobin (Foley and Tilley, 1997; Kaur et al., 2010). Quinoline-based drugs, such as amodiaquinire, piperaquinine, primaquine, quinine, and mefloquine, were developed against malaria, mostly acting at the blood stages of the parasite's life cycle (Kaur et al., 2010). Recently, an in vitro study demonstrated also that antimalarial inhibitors alkaloids and alkaloid-based drugs, such as chloroquine, hydroxy-chloroquine, and pyronaridine, showed antiviral activities against SARS-CoV-2; the mechanism of action lays decreasing the viral replication (Gendrot et al., 2020). In addition, the increase of cytokines interleukin IL-6 and IL-10 in some patients acts as an immune response to the SARS-CoV-2. Hydroxychloroquine and chloroquine can also decrease the immune factor levels, due to their immunomodulatory effects, preventing the progression of the disease (Yao et al., 2020).

Indole alkaloids. Indole alkaloids are a class of alkaloids having as main plant sources Rauwolfia serpentina, (Apocynaceae), Claviceps purpurea (Clavicipitaceae), Pausinystalia yohimbe (Rubiaceae), Catharanthus roseus (Apocynaceae), Catharanthus roseus (Apocynaceae), and Argyreia nervosa (Convolvulaceae) (Debnath et al., 2018). The peculiarity of this class is that they are produced also by some psychoactive mushrooms, mainly Claviceps and Cordyceps. The most well-known arising from mushrooms are ergot alkaloids-indole type (Plazas and Faraone, 2023; Mitra et al., 2021). Because of the enormous number of molecules belonging to this class, they were classified into different subgroups yohimbans, strychnos alkaloids, heteroyohimbans, vinca alkaloids, kratom alkaloids, ergolines/clavinet alkaloids, beta-carbolines, tryptamines, and Tabernanthe iboga alkaloids (Plazas and Faraone, 2023). Indole alkaloids are formed by a six-membered benzene ring fused to a five-membered pyrrole ring, containing an N atom in the pyrrole ring, which confers basic properties to these compounds, responsible for their therapeutic activities (Plazas and Faraone, 2023). Indole alkaloids are largely used as drugs; for example, reserpine showed antipsychotic and antihypertensive effects (Torchiana et al., 1970), while ergotamine and dihydroergotamine, which is a hydrogenated ergot alkaloid synthesized by unsaturated bound reduced in ergotamine, are involved in migraine treatment due to their vasoconstrictor action on pain-producing intracranial extracerebral blood vessels at the antimigraine receptors (Silberstein and McCrory, 2003). Despite their therapeutic effects, as mentioned earlier, being well-known also for their hallucinogenic effects; mushroom-derived indole alkaloids, such as psilocybin and psilocin, sharing similar structural properties with serotonin, can interact with serotonin (5-HT_{2A}) receptor, which is responsible for hallucinogens effects in humans and animals (Araújo et al., 2015). D-lysergic acid diethylamide (LSD) is probably the most famous ergot derivative with potent psychoactive properties (Plazas and Faraone, 2023). More recently, a wider variety of therapeutic properties were attributed to indole alkaloids. Jerantinine A, combined with δ-tocotrienol, showed cytotoxic activities against human lung adenocarcinoma cells (A549); they can prevent the G2/M cell cycle progression, causing cellular apoptosis, due to the disruption of the microtubule (Mitra et al., 2021; Abubakar and Loh, 2019). Some β-carboline (βC) indole alkaloids showed virucidal effects; harmane was demonstrated to have antiinfluenza activities on avian influenza A/H5N1 virus and seasonal human A/H1N1 virus, while harmalol acts against A/H1N1 and A/H5N1 viruses, both affecting the viral replication (Hegazy et al., 2023).

2.5 Derived from tyrosine

Isoquinoline alkaloids. Isoquinoline alkaloids are natural compounds, derived from tyrosine, which includes about 4000 molecules making it the second-largest class of alkaloids. The main plant sources are *Papaver somniferum* (Papaveraceae), *Berberis aristate*, and *Berberis lyceum* (Berberidaceae). Morphine was the first isoquinoline alkaloid isolated from *Papaver somniferum* (Sharma et al., 2023); other well-known molecules belonging to this class of compounds are papaverine, barberine, and codeine (Debnath et al., 2018). Isoquinoline alkaloids are characterized by no structural homogeneity; they are divided into 13 classes, which include benzyltetrahydro-isoquinoline, benzylisoquinolines, aporphines, protoberberine, benzophenanthridine, phthalideisoquinoline, morphinans, pavines, cularines, bisbenzylisoquinolines (BBI), naphythylisoquinoline, promorphinans, and ipecac alkaloids. Several biological activities are associated with those compounds; papaverine shows vasodilation and relaxation of smooth muscle tone actions; instead, barberine is well-known for its antidiabetic effect, suppression of cell growth, antagonist action of bacterial endotoxin and antihypertensive, antiinflammatory, antioxidant, antidepressant, and hepatoprotective action (Debnath et al., 2018). Morphine and codeine are known for their notable analgesics properties (Yang and Stöckigt, 2010). In 2017, Guarino et al. (Guarino et al., 2017) conducted a study in a placebo and a control group for 52 weeks, to evaluate the effects of berberine combined with silymarin on abdominal fat in obese patients with type 2 diabetes mellitus (T2DM). The result of the study confirmed that berberine leads to an improvement of all the parameters in the berberine-treated group (Guarino et al., 2017). In detail, it acts as cholesterol-lowering, reducing cardiovascular risk factors (Guarino et al., 2017; Kong et al., 2005), protecting against atherosclerosis (Guarino et al., 2015), due to the antiinflammatory, antioxidant, and smooth muscle cell antiproliferative properties (Guarino et al., 2017; Zhang et al., 2008); improving endothelial dysfunction; lowering uric acid, which high levels in the blood are involved in metabolism diseases, such as diabetes mellitus, dyslipidemia, arterial hypertension, and the metabolic syndrome and especially as a major reversible risk factor for metabolic and cardiovascular diseases (Kanbay et al., 2016); increasing adipose and muscle cell glucose utilization (Guarino et al., 2017); decreasing gut glucose adsorption and, consequently, the body weight, hypothesizing that it may act as an alpha-glucosidase inhibitor, inhibiting the disaccharidases activities and decreasing the glucose transportation across the intestinal epithelium (Guarino et al., 2017; Yin et al., 2008). However, several in vitro and in vivo studies demonstrated that berberine is a beneficial compound with hypoglycemic and lipid-lowering effects, becoming a candidate for type 2 diabetes (Guarino et al., 2017; Kong et al., 2005; Guarino et al., 2015; Yin et al., 2008; Lee et al., 2006). New studies were performed regarding the biological activities of isoquinoline (Sharma et al., 2023; Xia et al., 2021; Warowicka et al., 2020). Berberine showed antiviral activities against different viruses, such as HIV, HSV, HCMV, and human papillomavirus, interfering with some pathways such as nuclear factor-kappa B (NF-κB), mitogen-activated protein kinase/extracellular-signal-regulated kinase (MEK/ERK), and mammalian target of rapamycin and the adenosine monophosphate-activated protein kinase (AMPK/mTOR) to contrast the viral replication (Warowicka et al., 2020). Another isoquinoline alkaloid, litcubanine, demonstrated, in an in vitro study, to reduce the activation of inflammatory macrophages by lipopolysaccharide (LPS) through the Nf-kB pathway, reducing the inflammatory mediators' levels (Xia et al., 2021). Noscapine, instead, showed an inhibitory effect against cytokine storm in COVID-19 (Ebrahimi, 2020).

Phenylethylamine alkaloids. Phenylethylamine alkaloids are nonheterocyclic alkaloids, and the main plants sources are *Ephedra sinica*, *E. viridis*, *E. fragilis*, *E. distachya*, *E. ciliate*, *Capsicum frutescens*, *C. annuum*, *C. chinense*, *C. baccatum*. Ephedrine and hordenine are the main compounds belonging to marketed phenylethylamine alkaloid compounds (Debnath et al., 2018; Bhambhani et al., 2021). Ephedrine is present in many pharmaceutical formulations, and its effects are mediated by the interactions with adrenergic receptors; it acts as a sympathomimetic agonist on alpha and beta-adrenergic receptors, and also as an indirect sympathetic activation, enhancing the norepinephrine release from sympathetic neurons (Debnath et al., 2018; Abourashed et al., 2003). Those pharmacological mechanisms seem to lead to therapeutic and also adverse effects. Ephedrine is used for a wide variety of therapeutic uses, such as during spinal anesthesia to avoid low blood pressure, due to its capacity to increase blood pressure, or for bronchial asthma, due to its bronchodilation effects. Two ephedrine isomers 1R,2S-ephedrine, and 1R,2R-pseudoephedrine have been used as nasal decongestants, bronchodilators, and CSN stimulants (Liu et al., 2008; Wang and Jia, 2014; Vansal and Feller, 1999). However, many adverse effects were associated with ephedrine administration (Frank et al., 1991). Most recently, hordenine has been reported to have several therapeutical properties (Li et al., 2023; Su et al., 2018; Zhou et al., 2018). Hongi et al. (Li et al., 2023) in an in vivo study reported that hordenine can activate the dopamine D2 receptor (DRD2), one subtype of dopamine receptor, involved in the regulation of locomotor activity, drug addiction, cognition, and memory. With Parkinson's disease being characterized by a loss of dopaminergic neurons, hordenine acts against Parkinson's disease such as motor deficits in mice and nematodes (Li et al., 2023). In another study, conducted by Zhou et al., it was demonstrated that hordenine may act as a competitive inhibitor for signaling molecules and quorum sensing (QS)-based agent against *Pseudomonas aeruginosa*, inhibiting the production of virulence factors involved in quorum sensing (Zhou et al., 2018). For the first time, Su et al. conducted an in vivo study aimed to evaluate the antidiabetics, antiinflammatory, and antifibrotic role of hordenine in combination with insulin. Despite the molecular mechanism involved in the hordenine-insulin synergistic effect, which was not completely elucidated, hordenine demonstrated to be a candidate to prevent diabetic nephropathy (Su et al., 2018). About ephedrine, studies demonstrated that cotreatment with methanolic extract of *Ephedra alata* in mice protects the liver and kidney from injuries caused by oxidative stress, and DNA damage, due to the treatment with a high dose of chemotherapeutic cisplatin (Sioud et al., 2020).

2.6 Derived from purine

Purine alkaloids. Purine alkaloids are derived from purine nucleotides, and the most famous compound belonging to this class is caffeine (1,3,7-trimethylxanthine) (Zulak et al., 2006; Ashihara et al., 2008). The main caffeine sources are coffee (*Coffea arabica*), tea (*Camellia sinensis*), and matè (*Ilex paraguariensis*). Other compounds are included in this class; theobromine, produced by cacao, coca tea(*Camellia ptilophylla*) and *Camellia irrawadiensis*, and methyluric acid produced by *Coffea dewevrei*, *Coffea liberica*, *C. excelsa* and kucha tea (*Camellia assamica*) (Zulak et al., 2006). The effects on the human body induced by caffeine exposure are caused by different biological activities. Caffeine acts as an antagonist of endogenous adenosine receptors as they show a similar structure; this interaction induces vasoconstriction and increases blood pressure (Zulak et al., 2006; Martínez-Pinilla et al., 2015). Other activities include the inhibition of phosphodiesterase and the regulation of intracellular calcium (Martínez-Pinilla et al., 2015; Tazzeo et al., 2012). Even theobromine is considered an antagonist of adenosine receptors, carrying out the function of neuroregulation in the CNS (Martínez-Pinilla et al., 2015).

3. Extraction techniques

The development of novel, sensitive, and reliable analytical procedures for alkaloids detection and quantification in different matrices is acquiring a lot of interest. Because of the variety of alkaloid biological activities, which include toxic effects, but are especially therapeutic, analytical methods for trace levels detection, metabolites identification, and isomers separation are necessary (Gañán et al., 2023; Tajabadi et al., 2022; Ahmadi et al., 2023; Luger et al., 2022). In addition, the development of green analytical strategies is nowadays mandatory, without losing analytical performance (Voelker et al., 2021; Tobiszewski et al., 2009; Płotka-Wasylka, 2018).

The extraction method represents the most important step in sample preparation and can directly influence the analytical methods' performance (Poole et al., 2016). Optimizing the extraction step allows not only to remove interfering compounds, avoiding detection and quantification errors but also to reach low LODs and LOQs (Poole et al., 2016). Alkaloids are usually analyzed in a wide variety of complex matrices, such as food, plant extracts, and biological fluids and tissues, and the choice of the extraction method depends on the sample matrices and the properties of the analyte (Chen et al., 2019; de Araújo et al., 2022; Casado et al., 2022; Kaczyński and Łozowicka, 2020; Wu et al., 2019). Further, some of the most recent techniques are reported, based on the type of matrix.

Extraction in plant parts and other matrices. The characterization of alkaloid profiles in plants and processed products is important due to the multiple benefits that plant extracts showed on human health (Lai et al., 2023; Santos M et al., 2022). As for foods and biological matrices, different extraction techniques are used for alkaloids extraction from roots, leaves, bulbs, natural and processed samples of tobacco, stem, flowers, dry fruits, tea extracts, and others. SPE and its variants, Quick, Easy, Cheap, Effective, Rugged, and Safe (QuEChERS), solid-phase microextraction (SPME), ultrasound-assisted extraction (UAE), microextraction by packed sorbent (MEPS), supported liquid extraction (SLE). and LLE represent some of the most recently used extraction techniques. SPE technique, also for plant parts and extracts, was performed using different kinds of sorbents. For the extraction of tetrahydropalmatine (THP), tetrahydroberberine (THB), and tetrahydrocoptisine (THC) in Rhizoma Corydalis, concentrated ammonia solution-methanol, 1:20, v/v was used as extraction solvent; after that, the methanolic extract was diluted in water and purified with a C18 SPE cartridge. The samples were eluted for gravity, and vacuum was applied in the end to collect the solution excess. The recovery values were between 98.9% and 101.3% for all compounds at three concentrations (Sun et al., 2022). Another work compared four purification processes for the analysis of five poppy alkaloids, and other compounds, in acetaminophen-caffeine-aspirin tablets in hot pot broth, were compared. For the enhanced matrix removal-lipid (EMR-lipid), the sample was added by an internal standard solution, 5 mL of hydrochloric acid (0.1 mol), and the pH was adjusted to neutral, then acetonitrile was added to extract the analytes. After centrifugation, supernatants were transferred to an already activated EMR-lipid tube and centrifuged again. 3 g of anhydrous magnesium sulfate was added to the collected supernatants and centrifuged again inside polypropylene tubes. Another purification material, C_{18}/PSA mixed powders, was used; the procedure was the same as the previous material, less than for the last step, where the centrifuge tubes contained C_{18}/PSA, instead of polypropylene, and 3 mL petroleum ether was added to remove fat. For the last two methods of purification, SPE was performed; the column used was Waters Oasis MCX and HLB. EMR-lipid provides the highest recovery values between 82% and 113% (Song et al., 2022a). A method was developed for alkaloids determination in surface water using an online SPE coupled to hydrophilic interaction liquid chromatography high-resolution mass spectrometry (online SPE–HILIC-HRMS); the samples, previously filtrated and spiked with the internal standard, were loaded on the Isolute ENV+on-line SPE column, and the elution was performed using the HILIC-HRMS mobile phases. The recovery was calculated between 87.4% and 116% (Vo Duy et al., 2022). For the extraction of toxic and nontoxic alkaloids from Corydalis Rhizoma, a nano-SiO_2-assisted miniaturized matrix solid-phase dispersion (MMSPD) was developed; this technique consists of a dispersion of ground sample in a solid sorbent. The mixture is then loaded in an SPE column (Hoff and Pizzolato, 2018). In detail, the sample powders were spiked with standard solutions and added with 5–50 mg of nano-SiO_2 and were transferred to an SPE cartridge in which a 3-mL polyethylene (PE) frit is placed at its bottom and another covered the mixture with slight pressure with a syringe plunger. Then, the mixture was eluted whit 500 μL of eluent at room temperature and collected in a centrifugal tube. The recovery values were between 80.5% and 104.6% (Yu et al., 2022). In another study, the MSPD microextraction procedure was used with a polypropylene cartridge; in this case, the samples were blended with citric acid for the extraction of alkaloid compounds in *Fritillariae thunbergii* Bulbus. Also, in this case, good recovery values were obtained (Wang et al., 2019). Santos et al. developed, instead, a miniaturized matrix solid-phase dispersion (μMSPD) for β-carboline alkaloids in tobacco samples; the lyophilized tobacco sample and 100 mg of Discovery DPA-6S were blended to obtain a homogeneous mixture loaded in SPE column and eluted with an aqueous solution containing 1% of formic acid. Good recovery values, between 87% and 93%, were obtained (Santos M et al., 2022). TAs in leafy vegetable samples were extracted with an optimized methodology of a miniaturized QuEChERS (μ-QuEChERS); lyophilized samples were mixed with water and vortexed, then 1 mL of acetonitrile was added to the mixtures. After shaking, the partitioning salts mixtures ($MgSO_4$, NaCl, $C_6H_5Na_3O_7$ 2 H_2O, and $C_6H_6Na_2O_7$ 1.5 H_2O, in proportions of 4:1:1:0.5) were added. After vortex, ultrasound agitation, and centrifugation, the supernatants were collected in Eppendorf containing $MgSO_4$ and primary-secondary amine (PSA) for the cleanup step. The recoveries were between 90% and 100% for atropine, and scopolamine, at three spiked concentration levels (González-Gómez et al., 2022a). A modified QuEChERS procedure was used also for the extraction of 15 toxic alkaloids in forage grass; the analytes were extracted using acetonitrile/ammonium carbonate (200 mg/L), 84:16 (v/v, pH 8.5), vortexed for 15 minutes, and extracted under ultrasound. The mixture was centrifuged and the supernatant collected and evaporated. The residue was dissolved with 500 μL acetonitrile and 500 μL of 5 mmol/L ammonium acetate solution fortified with formic acid 0.1% (v/v) and added with 50 mg of PSA, the mixture was centrifuged. In the end, the supernatant was filtered with a nylon membrane and analyzed (Qie et al., 2021). Another technique, microwave-assisted extraction (MAE), resulted in the best choice for the extraction of vasicine and vasicinone from different plant parts of *Justicia adhatoda* L. The dried samples were dissolved in different solvents and placed in the microwave oven for 6 minutes at 1200 W. The best solvent for the extraction resulted in methanol and the recovery, calculated for three different concentrations of standard solutions, were 13.92%, 53.53%, and 64.13% for vasicine, and 46.15%, 88.55%, and 91.99% for vasicinone (Padhiari et al., 2020). As for biological matrices, also for the extraction of alkaloids from tobacco, molecularly imprinted polymers (MIPs) based on Fe_3O_4@SiO_2 and polyamidoamine (PAMAM) dendrimers were used; adding PAMAM dendrimers, the technique showed a higher number of functional groups, improving the efficiency and reaching recovery values between 84.2% and 93.5% (Dai et al., 2022). In 2022, a novel green extraction technique, in-situ cloud point-reinforced ionic liquid-assisted mechanochemical extraction (IS-CPR-IL-MCE), was developed for the extraction of five alkaloids from lotus leaves. In brief, the powdered samples were added by water, and the ultrasonic extraction was performed; then, after centrifugation, 0.3–0.6 g Triton X-100 and 1.0–3.0 g ammonium sulfate were added to 7-mL lotus leaf extracts, and the pH was adjusted by 1.0 mol/L hydrochloric acid solution or 1.0 mol/L sodium hydroxide solution after ultrasound for 10 minutes. After another centrifugation and hot water bath at 30°–70 degrees, the enrichment of the supernatant was reached. The recovery values were between 80.4% and 93.7% for all compounds at three concentration levels (Zhu et al., 2022). In 2021, for the first time, the in-situ formation of ion pair-assisted liquid-liquid microextraction (LLME) technique has been applied for alkaloids extraction from Rhizoma Coptidis (RC). The technique is a miniaturized liquid–liquid extraction process (LLE), in which the analytes are extracted due to their solubilities in two immiscible liquids; the miniaturized version was applied in different applications due to the low solvent consumption. The use of ion pair, a duplex of charged particles (one positive and one negative), the reagent is chosen to protonate alkaloids and deprotonate organic dyes, in an acidic environment, to improve the concentration and extraction by bonding reagent (dichloromethane, chloroform, and heptane). The average recovery values were between 87.20% and 102.50% (Zhu et al., 2021).

Extraction in food. Many kinds of alkaloids are detected in food and beverage because of their toxic effects in high doses. Mainly, they are studied in honey, spices, cereal, derived products, tea, and milk samples. The most recently used extraction techniques include solid phase extraction (SPE), dispersive solid-phase extraction (DSPE), ultrasound-assisted extraction (UAE), assisted liquid–liquid extraction (SALLE), and QuEChERs. In a work, Hungerford and coworkers developed a method for 30 Pyrrolizidine alkaloids (PAs) detection in honey samples, extracting the analytes with the SPE technique. In detail, the supernatant, obtained from the honey samples dissolved in H_2SO_4 (0.05 M, 10 mL) and centrifugated, was applied to an Agilent SPE Bond Elut 100 mg LRC-SCX columns, and the analytes were eluted with 3% ammonia in methanol showing recoveries between 53% and 108% for most compounds, except for erucifoline, erucifoline N-oxide, jacobine N-oxide, and seneciphylline N-oxide showing low recoveries, between 16% and 36% (Hungerford et al., 2019). In another work, Lattanzio and his research group monitored 12 ergot Alkaloids (ergometrine, ergometrinine, ergosine, ergosinine, ergotamine, ergotaminine, ergocornine, ergocorninine, α isomers of ergocryptine, α isomers of ergocryptinine, ergocristine, and ergocristinine) in Italian food commodities between 2017 and 2020. The sample analyzed included 18 cereal grains, 16 flours, and 31 other cereals derivatives. The extraction of the analytes was performed using an SPE column MycoSep 150 Ergot; the acetonitrile: ammonium carbonate extraction solution was previously added to the sample and, after shaking and centrifugation, the extract was collected and loaded on the SPE column. In this work, the recoveries, obtained at three spiking levels (2.5, 5, 10 μg/kg), were between 87% and 114% (Lattanzio et al., 2021). Two different extraction procedures, SPE and QuEChERS, were compared in terms of extraction efficiency for PAs extraction in honey, tea, herbal tinctures, and milk samples. The SPE procedure, using a *Strata-X sorbent* cartridge, was performed using 6 mL of methanol as an eluent; the samples of honey and tea were previously extracted with 40 mL of 0.2% of formic acid in deionized water shaking for 30 minutes and centrifuged, then the supernatants were collected. About the QuEChERS procedure, after the extraction of the samples with a solution of 50% acetonitrile with 1% of formic acid, a salts mixture was added; the samples were then shacked and centrifuged. Next, the supernatants were stored in the refrigerator at −80°C for 30 minutes, centrifuged for 18 minutes, evaporated at 50°C, and dissolved in deionized water containing 1% of formic acid. In both extraction procedures, the extracts were analyzed after filtration with polyvinylidene difluoride (PVDF) filters, 0.22 μm pore size. The extraction efficiency for all matrices resulted in 76 ± 30% and 73 ± 15% for the QuEChERS and the SPE procedure, respectively (Martins et al., 2022). A modified QuEChERS procedure was used to determine ergot alkaloids in cereal samples. Herein, 4 mL of extraction mixture (ACN:5 mM ammonium carbonate; 85:15, v/v) was added to the samples. The mixture was agitated and centrifuged, and the supernatant was placed in a falcon tube containing a C18:Z-Sep+ (1:1, w/w) powder as a cleanup sorbent. The residues of evaporated supernatants, obtained by shaking and centrifugation, were reconstituted with MeOH: water (50:50, v/v). The recovery for all 12 compounds was between 90% and 109% at two fortifying concentration levels (Carbonell-Rozas et al., 2021). PAs and PANOs were also extracted from honey samples, pollen and solid forms of dietary supplements, and black and green tea (herbal infusions) using Salting-out Assisted Liquid–Liquid Extraction (SALLE). The aqueous extracts were obtained by each matrix using the same SALLE procedure; the aqueous solution was then brought to a concentration of 1 M of $MgSO_4 \cdot 7H_2O$, 1.5 M Na_2SO_4 and, after centrifugation, 2 mL of the solutions was extracted with ACN by the vortex. After centrifugation, the organic phases were collected and evaporated. The extraction efficiency was between 69% and 109% for all compounds in all matrices (Rizzo et al., 2023). The pulsed ultrasound-assisted extraction (UAE) followed by SBA-15 purification was proposed for the first time by Casado-Hidalgo et al. for the extraction of opium alkaloids in biscuits and sponge cakes. The procedures were optimized to reduce solvent volume and time. In detail, after samples pretreatment, where the samples (biscuits and sponge cakes) were manually grounded and passed through an 18-mesh sieve, 10 mL of methanol with 1% of formic acid was added to 2.5 g of sample, and the solutions were mixed using vortex; then they were exposed to 75% amplitude for 5 minutes in pulse mode falcon tube immersed in a glass with crushed ice to prevent heating. The mixtures were then centrifuged, and the supernatants were loaded in an SBA-15 cartridge for the SPE procedure, which resulted in recoveries between 79% and 101% (Casado et al., 2022).

Extraction in biological fluids and tissues. The accurate detection of some kinds of alkaloids in biological tissues and fluids and in cells is acquiring interest for multiple reasons. For example, the detection of toxic alkaloids and their metabolites may help to understand when a fatal intoxication occurs (Natori et al., 2022), or the detection of different nicotine metabolites in biological fluids can be used to determine the phenotype individuals for cytochrome P450 2A6 (CYP2A6) activity, useful to assess a pharmacotherapy for tobacco addiction (Jacob et al., 2011). The study of alkaloids metabolites is useful also to determine their pharmacological effects and benefits (Li et al., 2020), or if they can be used as disease markers. Recently, it has been reported that reduced levels of caffeine and its metabolites in the blood, in comparison with a control group, may be a marker of early-stage Parkinson's disease (Han et al., 2020). For the extraction of alkaloids from biological fluids and tissues different extraction techniques have been used, such as QuEChERS, in-syringe assisted vortex-induced salt-enhanced liquid–liquid microextraction technique (ISA-VI-SE-LLME), SPE, microwave-assisted extraction (MAE), microextraction by packed sorbent (MEPS), supported liquid extraction (SLE), magnetic solid-phase extraction (MSPE), liquid–liquid extraction coupled with SPE and ultrasound-enhanced and surfactant-assisted dispersive liquid–liquid microextraction (USA-DLLME). As for extraction from foods, the most used technique is SPE. A study reported that infants born from mothers whose during pregnancy were kratom user, showed abstinence syndrome; for this reason, SPE technique was used for the extraction of mitragynine (MG), and speciociliatine (SC) from the umbilical cord (Hughs et al., 2023). After a preextraction procedure of umbilical cord homogenates, the supernatants containing the analytes were loaded on the SPE column, which was the Biotage Evolute Express CX 60 (cation exchange column), and the eluent mixture was 2:20:78 ammonium hydroxide/isopropanol/dichloromethane (Hughs et al., 2023). SPE was recently used also for caffeine and its metabolites extraction from blood samples, allowing preconcentration and sample purification. Different SPE cartridges were tested, including Agela Cleanert PEP-2, Agela CleanertiM ODS C18N, Welchrom C18E, PIKMA ProElut C8, and Waters HLB (500 mg, 6 mL); however, the Welchrom C18E gives the highest recovery. A phosphoric acid—water solution (pH 3) was used to remove the plasma-interfering compounds, and the recovery was between 88% and 105.9% for all compounds (Han et al., 2020). A novel SPE adsorbent carbonized π-conjugated polymer-coated porous silica (SiO2@C-π-CP) was developed by Jin et al. for the extraction of berberine from urine samples (Jin et al., 2022). The recovery was between 98% and 103%, with an RSD of 5%, in line with the recoveries obtained with another adsorbent for the extraction of compounds in real samples; however, the synthesis of SiO2@C-π-CP resulted in cheaper and more reliable using HPLC coupled to UV, differentially from some of the other adsorbents, which require HPLC coupled to MS/MS or TOF/MS (Peng et al., 2017; Hu et al., 2020). Another kind of adsorbent for SPE used for alkaloid extraction from complex matrices is SupelMIP TSNA. Molecularly imprinted polymers (MIPs) are polymers selective for specific analytes (Pichon et al., 2020); tobacco-specific nitrosamines (TSNAs) MIPs are available in trade and are specific for four TSNAs (Mulder et al., 2022). SupelMIP TSNA selectivity was studied also for nicotine, which is a TSNA precursor, and its metabolite, cotinine. Cotinine showed a recovery higher than 80%; instead, nicotine showed high variability between recovery values, probably due to the structural/chemical differences with its products (Mulder et al., 2022). A variant of SPE was magnetic solid-phase extraction (MSPE). The magnetic adsorbent in this study was a covalent organic polymer (COP), magnetic sulfonated COP ($COP{-}SO_3H@Fe_3O_4$), which was dissolved in the samples and stirred for 10 minutes. After this step, the analyte particles are bound to the nanoparticles and separated from the matrix using a magnet. The analytes were then desorbed by sonication using 400 μL of methanol-acetic acid (9/1, v/v). Good recovery values, between 91% and 110%, were obtained (Ma et al., 2023). QuEChERS has been used also for biological matrices. Aconities and their metabolites, which show high toxicity, have been recently analyzed in postmortem blood using a modified QuEChERS procedure. QuECheRS salt (80 mg $MgSO_4$ and 20 mg CH_3COONa) was added to a tube with acetonitrile, distilled water, and whole blood samples; the tube was then mixed, incubated at room temperature, and centrifuged. The supernatant was evaporated, and then the residues were reconstituted. The recovery was between 52.3% and 76%; the low recovery of aconine may be associated with its high hydrophilicity and the consequent affinity for the aqueous layer. Despite the recovery, the matrix effects (about 100%) suggested that this QuEChERS method can remove almost totally the interfering compounds from the matrix (Natori et al., 2022). In-syringe assisted vortex-induced salt-enhanced liquid—liquid microextraction technique (ISA-VI-SE-LLME) is another extraction technique recently developed by Pasupuleti et al. for the extraction of five toxic areca nut alkaloids from saliva and urine samples. In detail, the sample and 0.05% of formic acid are loaded in a plastic syringe with a tip cap and vortexed; then the extraction solvent (isopropyl alcohol), 1.0 mL of 0.1 M (NH_4OH), and 4 g of ammonium sulfate $(NH_4)_2SO_4$ salt were added, vortexed, and centrifuged. In the end, the extract was collected from the tip and pushed by the syringe piston. The recovery for saliva samples was between 87.9% and 110.1%, while for urine samples between 81.5% and 115.1%

(Pasupuleti et al., 2022). Areca nut and tobacco biomarkers were also extracted from hair samples using microwave-assisted extraction (MAE). Analytes extraction started by adding methanol/TFA (8.5:1.5, v/v) mixture and internal standard to the samples, and then MAE was performed for 5 minutes. The samples were then evaporated and reconstituted. The extraction efficiency was more than 100% for all the analytes (Chen et al., 2022). Opiate alkaloids from blood were extracted using microextraction by packed sorbent (MEPS), which was applied for routine analyses of hospital samples. Blood samples were previously pretreated for protein precipitation. MEPS procedure consisted of a sorbent (80% C8 and 20% SCX, cation exchange) conditioned, before the sample load passages; formic acid was added to maintain positively charged analytes, favoring their retention on the sorbent. Then, the methanol was used as an eluent solvent. The absolute recovery for all compounds corresponded to 6% and 22% (Prata et al., 2019). The authors specified that the low recovery values depend on different factors, such as the precipitation step added to remove components from the matrix that can clog the sorbent, and the small amount of sorbet used, being a miniaturized procedure; however, the method demonstrated good accuracy and precision (Prata et al., 2019). Another study focused on the extraction of alkaloids from blood, reported the use of supported liquid extraction (SLE) for the extraction of 16 alkaloids from blood samples using ethyl acetate with recovery between 79.2% and 95.8% (Chen et al., 2019). Always considering body fluids, another rapid technique, based on ultrasound-enhanced and surfactant-assisted dispersive liquid–liquid microextraction (USA-DLLME), was developed for mirtazapine and its metabolites extraction. The procedure consists of adding analytes, contained in a phosphate buffer solution, and Brij35 (0.05 mM), used as an emulsifier, in a vial. Then, the extraction solvent was rapidly added to the sample solution, shaken manually, and sonicated at ambient temperature to form a cloudy solution. The solution was then centrifuged, dispersed droplets, and collected in a conical tube. The recovery values ranged from 82.7% to 116.5% for urine and from 88.5% to 111.4% for serum (Yan et al., 2022).

4. Analytical methods

Traditional analytical methods. The analytical methods developed for alkaloids characterization are mainly based on high-performance liquid chromatography (HPLC) (Velichkova et al., 2022; González-Gómez et al., 2022b; Lekhak et al., 2022) and gas chromatography (GC) (Prata et al., 2019; Pereira et al., 1999; GOLDBAUM et al., 1963) coupled to tandem mass spectrometry (MS/MS) (González-Gómez et al., 2022a; Cui et al., 2018). Some studies reported also the use of different detectors, such as a diode array detector (DAD), but most recently, it is coupled with MS/MS to provide more reliable results for targeted and nontargeted analysis (Gackowski et al., 2021). Different methods were developed for the simultaneous detection of different kinds of alkaloids from different matrices reaching very low limits of detection (LODs) and limits of quantification (LOQs) values. Regarding HPLC, recently, Qie et al. validated a method for the detection of 15 toxic alkaloids belonging to different classes (EAs, Pas, and TAs) in forage grass, which present a risk to human health; the method was applied to 134 forage grass samples, founding different contamination between sample belonging to different herbage species and different regions. Reverse-phase chromatography (RP) was performed using a C18 column, and a mass spectrometer equipped with an electrospray ionization (ESI) source, used in positive mode (Qie et al., 2021). Twenty-one PAs were also analyzed in teas by using a different HPLC-MS/MS method. Teas may be contaminated with PAs during growing, or simply those compounds are naturally produced by tea plants; hence, careful detection is required. As in the study previously reported, the separation was carried out using a C18 column and ESI in positive mode was used as the ionization source. LOD and LOQ values ranged respectively between 0.1 and 2.3 μg/kg and 0.3 and 6.9 μg/kg (Kwon et al., 2021). Other methods were validated for the detection and quantification of alkaloids: norephedrine, norpseudoephedrine, ephedrine, pseudoephedrine, and methylephedrine were analyzed by LC-PDA in *Ephedrae herba* (Song et al., 2022b); the same technique was applied to determine N-nitrosonornicotine (NNN), nicotine, cotinine, and morphine in human urine samples (Mulder et al., 2022). Atropine and scopolamine extracted from thyme, basil, and coriander samples were analyzed by LC-MS/MS (González-Gómez et al., 2022b); as nut alkaloids (ASAs), arecoline, arecaidine, N-methylnipecotic acid (NMPA), guvacoline and guvacine, and tobacco-specific alkaloids (TSAs), nicotine, cotinine, and nornicotine, which were detected and quantified from saliva samples, studying the transformation and distribution of alkaloids in the oral cavity (Chang et al., 2022). Another separation method, supercritical fluid chromatography (SFC), was applied by Huang et al. for the separation of eight isoquinoline alkaloids: corypalmine, berbamine, magnoflorine, columbamine, jatrorrhizine, coptisine, plamatine, and berberine. The study was conducted by extracting the eight alkaloids from four different parts of *M. bealei*; the anti-acetylcholinesterase (AChE) activity was also evaluated for the four extracts. The SFC-DAD performance was compared with LC performance, demonstrating to provide a faster separation; LODs and LOQs ranged respectively from 0.4 to 2.3 μg/mL and from 1.5 to 7.5 μg/mL (Huang et al., 2006). Other SFC-DAD methods were used for alkaloids separation and detection, such as tetrahydrocoptisine, corydaline, tetrahydropalmatine, protopine, allocryptopine, columbamine, and dehydrocavidine Rhizoma corydalis (Yue et al., 2022); an SFC-PDA method was developed for pyrrolizidine alkaloids from different *Senecio* species (Zwerger et al., 2023).

Other separation techniques used to analyze these natural compounds include capillary electrophoresis (CE) (Yan et al., 2022; Pérez-Alcaraz et al., 2019; Chu et al., 2022) and high-performance thin-layer chromatography (HPTLC). CE can separate compounds based on their mass-to-charge ratios (*m/z*) (Monton and Soga, 2007; Issaq, 1999). CE is used in different modes, which differ from the separation principle, such as capillary zone electrophoresis (CZE) and Micellar electrokinetic chromatography (MEKC), which allows the separation of neutrals compounds due to their partitioning on charged micelles but cannot be coupled with MS (Gackowski et al., 2021; Monton and Soga, 2007; Hurtado-Fernández et al., 2010; Hirayama et al., 2014). MEKC technique was used for the detection of caffeine and its metabolites in human plasma after SPE; the running buffer was composed of 35 mM phosphate, pH of 10.5, and 25 mM SDS. The separation voltage was 15 kV. The analyses were carried out in less than 12 minutes and the LOD and LOQ were between 4.0 and 7.5 ng/mL and 0.3 and 0.5 μg/mL respectively (Han et al., 2020). In 2019, Yu et al. developed a method for the detection of sesquiterpene alkaloids of *Tripterygium wilfordii*. A new CE mode, microemulsion electrokinetic chromatography (MEEKC), coupled with a DAD detector was used for compound separations (Yu et al., 2019). The microemulsion buffer was prepared using 0.5% (w/v, 17 mmol/L) SDS, 3.0% (v/v) n-butanol, 0.5% (v/v) n-hexane, 3.0% (v/v) ethanol, 7.0% (v/v) acetonitrile and 86.5% (v/v) 5 mmol/L $Na_2B_4O_7$−10 mmol/L NaH_2PO_4 buffer (pH 8.5), and the sensitivity of the method was improved with an on-line signal amplification technique, the large volume sample stacking (LVSS). The LVSS allows the injection of larger volumes of sample in a capillary stacked preventing changing of the peak shape. The LOD and LOQ values ranged between 0.003 and 0.009 (μg/mL) and between 0.01 and 0.03 (μg/mL), respectively (Yu et al., 2019). Another online pre-concentration technique, electrokinetic supercharging (EKS), was coupled to CZE, for alkaloid separation in rat fecal samples. LOD values for the two compounds, lycorine and galanthamine, were 6.0 ng/mL and 8.0 ng/mL, respectively (Chu et al., 2022). Regarding thin-layer chromatography (TLC), the principle is that the mobile phase is transported through the layer by capillary forces due to a decrement in the free energy of the solvent when it goes into the porous structure of the layer (Poole, 2003). It allows a low-cost analysis of complex matrices with the least possible sample cleanup procedures, and it was widely used for screening steps, due to the possibility of stationary-phase single use; parallel samples separation (high-throughput analyses); the possibility of postchromatography derivatization; the possibility of performing simultaneously different analytes protocols (Poole, 2003). HPTLC coupled to a densitometric scanning under a mercury lamp at a wavelength 351 nm in absorbance mode was used for quantification of berberine in rhizomes of *Coptis teeta* wall. The LOD and LOQ values were respectively, 30 ng/band and 70 ng/band (Goswami et al., 2019). HPTLC is reported as a method for vasicine and vasicinone quantification in different parts of *Justicia adhatoda* using. In this case, different extraction techniques were tested, and the best sensitivity was obtained using UAE as an extraction technique. The LOD and LOQ for vaccine and vasicinone were respectively, 6 and 20 ng and 10 and 32 ng (Padhiari et al., 2020). A method for simultaneous detection and quantification of more alkaloids, berberine, berbamine, palmatine, magnoflorine, and jatrorrhizine, from *Berberis aristate* was validated by Basera et al. using HPTLC; the densiometric scanning of the plates was performed at 254 and 366 nm. The LODs ranged from 55.68 to 323.24 ng for all the compounds, instead, LOQs resulted between 168.74 and 979.51 ng (Basera et al., 2021).

Other studies reported also the use of laser desorption ionization (LDI) MS and matrix-assisted laser desorption/ionization mass spectrometry (MALDI-MSI) for the analysis of natural products in crude plant extracts (Le Pogam et al., 2020; Wu et al., 2022). The advantages of LDI-MS and MALDI-MS use, concerning traditional techniques, regard mainly the possibility to avoid sample preparation steps; in particular, MALDI-MS can reduce the matrix effects and shows high salt tolerance (Le Pogam et al., 2020). A matrix-assisted laser desorption/ionization mass spectrometry imaging (MALDI-MS) was used as an imagining technique for the in situ detection of 10 alkaloids in different *A. catechu* fruit tissues; the spatial distribution of secondary metabolites is useful for understanding plant physiological activities (Wu et al., 2022). Other recent studies involved MALDI-MS (Yamagaki and Osawa, 2024) and LDI coupled with Fourier-transform ion cyclotron mass spectrometry (FT-ICR MS) (Islam et al., 2022) for alkaloid analyses.

To improve the separation, ion mobility spectrometry (IMS) was used coupled with different ionization techniques, such as thermionic sodium ion source (Asadi et al., 2019), electrospray ionization (ESI) (Carbonell-Rozas et al., 2022), and corona discharge ionization source (Asadi et al., 2019). IMS separation is based on the drift time of ions, and it can separate compounds coeluted from the LC column (Carbonell-Rozas et al., 2022). IMS was used for isoquinoline alkaloids detection (morphine, codeine, and thebaine) in seeds, stems, leaves, and capsules of seven Papaver species, and corona discharge ionization was used as the ionization technique. The LODs values were between 0.2 and 0.05 mg/L, while LOQ values were between 0.5 and 0.5 mg/L (Tajabadi et al., 2022). In another study, HPLC was coupled with ESI-IMS for detecting arecoline, guvacine, and arecaidine in areca nut shells, areca nut seeds, green areca nut, smoked areca nut, and simotang (decoction of four powered drugs used in traditional medicine). The LODs and LOQs were between 0.026 and 0.110 μg/mL, and between 0.010 and 0.370 μg/mL (Lu et al., 2022). Other methods based on IMS were established for alkaloids detection, among them: Zhen Shi et al.(Shi et al., 2022) used IMS coupled to Q-TOF for the detection of alkaloids in lotus plumule; Carbonell-Rozas et al.(Carbonell-Rozas et al., 2022) used LC coupled to IMS for EAs detection in cereal samples.

One of the most complete and recent methods was developed in 2022 for separation and detection of 30 pyrrolizidine alkaloids, 24 pyrrolizidine alkaloid N-oxides, and 2 tropane alkaloids in milk samples (Klein et al., 2022) (Fig. 13.2). Recently, given the large interest to speed up the analyses and reduce the sampling time, operating in situ/online sampling steps, many portable devices have been evaluated, such as Voelker and his group compared four portable devices for rapid detection of mitragynine in suspected kratom products. The four devices were: (1) direct Analysis in real time with thermal desorption mass spectrometry (DART-TD-MS); (2) handheld mass spectrometer; (3) corona-based IMS; (4) FT-IR spectrometer (Voelker et al., 2021). Among them, DART-TD-MS and corona-based IMS allow direct analysis without extraction techniques, increasing the greenness of the

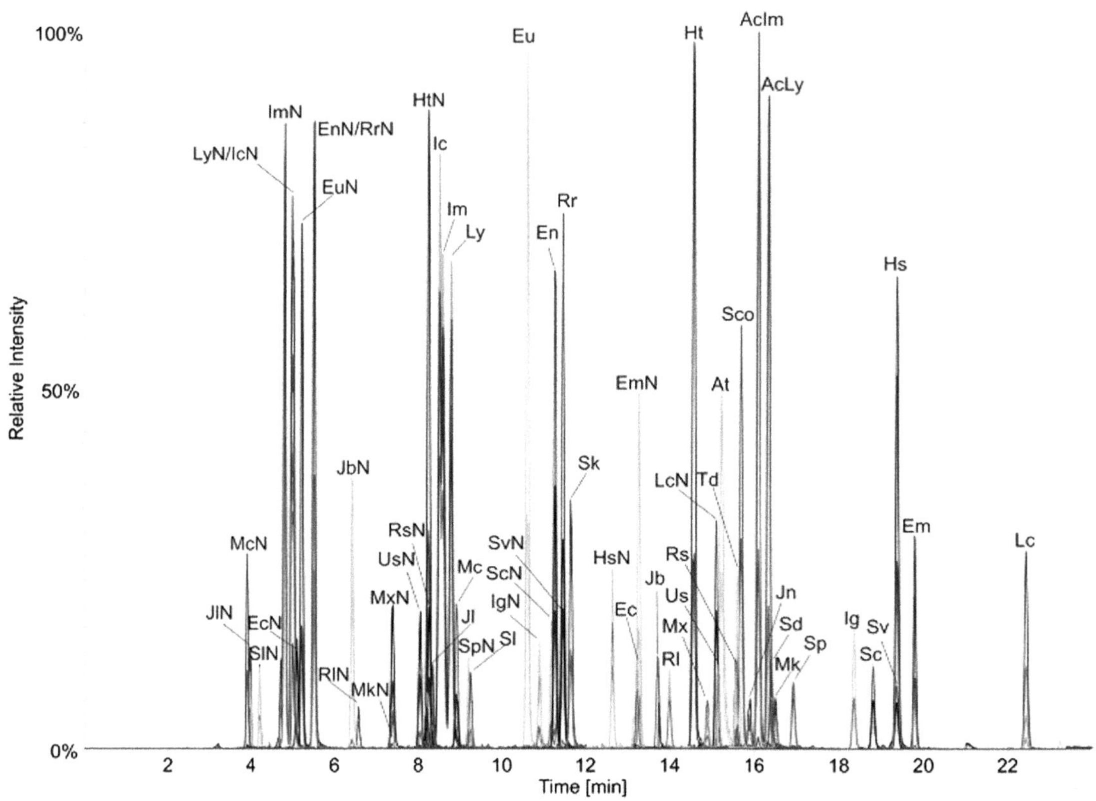

FIGURE 13.2

LC-MS/MS methods for separation and detection of 30 pyrrolizidine alkaloids, 24 pyrrolizidine alkaloid N-oxides, and 2 tropane alkaloids in milk samples.

Adapted from Klein, L.M., Gabler, A.M., Rychlik, M., Gottschalk, C., Kaltner, F. A sensitive LC-MS/MS method for isomer separation and quantitative determination of 51 pyrrolizidine alkaloids and two tropane alkaloids in cow's milk. Anal. Bioanal. Chem. 2022, 8107–8124.

methods, instead the other devices needed samples extraction with chloroform. Based on the results, the authors observed that DART-TD-MS and IMS can be used for the initial screening, eliminating negative samples (Voelker et al., 2021). Other low-cost methods were proposed, such as paper spray ionization (PS) coupled to MS/MS, used for methylxanthines in cocoa products and drugs (Bartella et al., 2019). PS is an ambient ionization source that allows in situ detection. Theobromine, caffeine, and theophylline's LODs were respectively 137.6 mg/100 g, 44.7 mg/100 g 35.2 mg/cpr, while LOQs were 68.7 mg/100 g, 14.7 mg/100 g, and 16.3 mg/cpr (Bartella et al., 2019).

Innovative techniques. As mentioned before, in the past years, the development of methods that minimize sample preparation/extraction steps and, sometimes, allow in situ analysis using handheld devices are acquiring growing interest, due to their versatility, low costs, speed, and low solvents consumption (Baranwal et al., 2022). Herein some of the novel methods are mentioned, including electrochemical methods, which include also ion-selective electrodes (ISEs) (Luger et al., 2022), nanosensors (Sargazi et al., 2022; Ambrin et al., 2019), and others (Ahmadi et al., 2023; João et al., 2021; Ognjanović et al., 2022; Dushna et al., 2022; Sheibani et al., 2019; Xiao et al., 2022), surface-enhanced Raman scattering (SERS) (Qin et al., 2021), time-resolved fluorescent microsphere immunochromatographic test strip (TRFM-ICTS) (Zhou et al., 2022), and electrochemiluminescence (ECL) (Xiao et al., 2022; Brown et al., 2019). Electrochemical sensors find large use in different fields, such as forensics, illicit drugs, healthcare, sweat monitoring, and others (Ferreira et al., 2019) and are formed by two units: receptor and physicochemical transducer. The receptor, after analytes interaction, generates an output signal, hence is important that it shows high specificity for the analytes. The transducer, instead, has the function of generating a value from the analyte–receptor interaction (Baranwal et al., 2022). About alkaloids, recently a sensor with a surface electrode modification was developed for thebaine detection in urine, reaching LOD 2.4 nM (Sheibani et al., 2019). The technique involved two protons and two electrons, producing a secondary amine (Sheibani et al., 2019). As thebaine, also morphine was detected in urine samples using a modified electrode with $FeWO_4$, reaching an LOD of 0.59 μM and LOQ of 1.94 μM (Ognjanović et al., 2022). ISEs, instead, use a polymer-based membrane. The membrane material plays a fundamental role in the selectivity of this kind of sensor. The method was developed by Lunger et al. for tropane, atropine, and scopolamine detection as a proof-of-concept, demonstrating that the portable device is not influenced by compounds that show similarity with the analytes of interest (Luger et al., 2022). Another sensor type, ECL, was developed to study PAs; ECL involves the generation of species at electrode surfaces that, due to electron-transfer reactions, form excited states and emit light (Richter, 2004). Tested molecules were monocrotaline, senecionine N-oxide, retrorsine, and senkirkine, reaching LODs between 0.02 and 0.07 μM (Xiao et al., 2022). A method for TAs detection and quantification based on SERS detection, instead, was proposed by Qin et al. In particular, they developed a new substrate material (Ag@Au for silver nanowires surface) to improve the detection of five compounds; LODs were between 10^{-9} M and 10^{-10} M (Qin et al., 2021). Regarding another technique, TRFM-ICTS, a novel method was developed to give a faster and more sensitive method for higenamine (HIG) detection, which is prohibited by Anti-Doping Agency (WASA). TRFM-ICTS works using an indirect competitive immune reaction between HIG-BSA (antigen) in the sample and Anti-HIG-BSA monoclonal antibodies (mAb) for the detection of HIG in functional food was performed using TRFM-ICTS (Zhou et al., 2022).

References

Abourashed, E.A., El-Alfy, A.T., Khan, I.A., Walker, L., 2003. Ephedra in perspective—a current review. Phytother Res. 17 (7), 703−712. https://doi.org/10.1002/ptr.1337.

Abubakar, I.B., Loh, H.S., 2019. Potentiation of in vitro apoptotic effects of δ-tocotrienol and jerantinine A on human lung adenocarcinoma cells. J. HerbMed Pharmacol. 8 (4), 333−338. https://doi.org/10.15171/jhp.2019.49.

Adamse, P., Van Egmond, H.P., 2010. Tropane Alkaloids in Food, RIKILT—Institute of Food Safety.

Adamse, P., van Egmond, H.P., Noordam, M.Y., Mulder, P.P.J., De Nijs, M., 2014. Tropane alkaloids in food: poisoning incidents. Qual. Assur. Saf. Crop Foods 6 (1), 15−24. https://doi.org/10.3920/QAS2013.0314.

Ahmadi, E., Zarei, E., Asghari, A., 2023. Electrochemical sensor based on graphene and tungsten disulfide nanoparticles for determination of noscapine and papaverine. Ionics 29 (4), 1579−1591. https://doi.org/10.1007/s11581-023-04895-2.

Allemang, A., Mahony, C., Lester, C., Pfuhler, S., 2018. Relative potency of fifteen pyrrolizidine alkaloids to induce DNA damage as measured by micronucleus induction in HepaRG human liver cells. Food Chem. Toxicol. 121, 72−81. https://doi.org/10.1016/j.fct.2018.08.003.

Ambrin, G., Ahmad, M., Alqarawi, A.A., Hashem, A., Abd_Allah, E.F., Ahmad, A., 2019. Conversion of cytochrome P450 2D6 of human into a FRET-based tool for real-time monitoring of ajmalicine in living cells. Front. Bioeng. Biotechnol. 7. https://doi.org/10.3389/fbioe.2019.00375. Available from: http://journal.frontiersin.org/journal/bioengineering-and-biotechnology#archive.

Araújo, A.M., Carvalho, F., Bastos, M.d.L., Guedes de Pinho, P., Carvalho, M., 2015. The hallucinogenic world of tryptamines: an updated review. Arch. Toxicol. 89 (8), 1151−1173. https://doi.org/10.1007/s00204-015-1513-x.

Arredondo, J., Chernyavsky, A.I., Grando, S.A., 2006. The nicotinic receptor antagonists abolish pathobiologic effects of tobacco-derived nitrosamines on BEP2D cells. J. Cancer Res. Clin. Oncol. 132 (10), 653−663. https://doi.org/10.1007/s00432-006-0113-9.

Asadi, M., Valadbeigi, Y., Tabrizchi, M., 2019. Thermionic sodium ion source versus corona discharge in detection of alkaloids using ion mobility spectrometry. Int. J. Ion Mobil. Spectrometr. 22 (2), 51−58. https://doi.org/10.1007/s12127-019-00249-9.

Ashihara, H., Sano, H., Crozier, A., 2008. Caffeine and related purine alkaloids: biosynthesis, catabolism, function and genetic engineering. Phytochemistry 69 (4), 841−856. https://doi.org/10.1016/j.phytochem.2007.10.029.

Baranwal, J., Barse, B., Gatto, G., Broncova, G., Kumar, A., 2022. Electrochemical sensors and their applications: a review. Chemosensors 10 (9), 363. https://doi.org/10.3390/chemosensors10090363.

Bartella, L., Di Donna, L., Napoli, A., Siciliano, C., Sindona, G., Mazzotti, F., 2019. A rapid method for the assay of methylxanthines alkaloids: theobromine, theophylline and caffeine, in cocoa products and drugs by paper spray tandem mass spectrometry. Food Chem. 278, 261−266. https://doi.org/10.1016/j.foodchem.2018.11.072.

Basera, I.A., Girme, A., Bhatt, V.P., Shah, M.B., 2021. A validated high-performance thin-layer chromatography method for the simultaneous estimation of berberine, berbamine, palmatine, magnoflorine and jatrorrhizine from Berberis aristata. J. Planar Chromatogr. Mod. TLC 34 (2), 147−155. https://doi.org/10.1007/s00764-021-00094-9.

Beaudoin, G.A.W., Facchini, P.J., 2014. Benzylisoquinoline alkaloid biosynthesis in opium poppy. Planta 240 (1), 19−32. https://doi.org/10.1007/s00425-014-2056-8.

Bhambhani, S., Kondhare, K.R., Giri, A.P., 2021. Diversity in chemical structures and biological properties of plant alkaloids. Molecules 26 (11). https://doi.org/10.3390/molecules26113374. Available from: https://www.mdpi.com/1420-3049/26/11/3374/pdf.

Bowman, W.C., Sanghvi, I.S., 1963. Pharmacological actions of hemlock (conium maculatum) alkaloids. J. Pharm. Pharmacol. 15 (1), 1–25. https://doi.org/10.1111/j.2042-7158.1963.tb12738.x.

Brown, K., McMenemy, M., Palmer, M., Baker, M.J., Robinson, D.W., Allan, P., Dennany, L., 2019. Utilization of an electrochemiluminescence sensor for atropine determination in complex matrices. Anal. Chem. 91 (19), 12369–12376. https://doi.org/10.1021/acs.analchem.9b02905.

Buchmueller, J., Enge, A.M., Peters, A., Ebmeyer, J., Küpper, J.H., Schäfer, B., Braeuning, A., Hessel-Pras, S., 2022. The chemical structure impairs the intensity of genotoxic effects promoted by 1,2-unsaturated pyrrolizidine alkaloids in vitro. Food Chem. Toxicol. 164. https://doi.org/10.1016/j.fct.2022.113049. Available from: http://www.elsevier.com/locate/foodchemtox.

Carbonell-Rozas, L., Gámiz-Gracia, L., Lara, F.J., García-Campaña, A.M., 2021. Determination of the main ergot alkaloids and their epimers in oat-based functional foods by ultra-high performance liquid chromatography tandem mass spectrometry. Molecules 26 (12). https://doi.org/10.3390/molecules26123717. Available from. https://www.mdpi.com/1420-3049/26/12/3717/pdf.

Carbonell-Rozas, L., Hernández-Mesa, M., Righetti, L., Monteau, F., Lara, F.J., Gámiz-Gracia, L., Bizec, B.L., Dall'Asta, C., García-Campaña, A.M., Dervilly, G., 2022. Ion mobility-mass spectrometry to extend analytical performance in the determination of ergot alkaloids in cereal samples. J. Chromatogr. A 1682. https://doi.org/10.1016/j.chroma.2022.463502. Available from: http://www.elsevier.com/locate/chroma.

Casado, N., Morante-Zarcero, S., Sierra, I., 2022. The concerning food safety issue of pyrrolizidine alkaloids: an overview. Trends Food Sci. Technol. 120, 123–139. https://doi.org/10.1016/j.tifs.2022.01.007.

Chang, Y.J., Muthukumaran, R.B., Chen, J.L., Chang, H.Y., Hung, Y.C., Hu, C.W., Chao, M.R., 2022. Simultaneous determination of areca nut- and tobacco-specific alkaloids in saliva by LC-MS/MS: distribution and transformation of alkaloids in oral cavity. J. Hazard Mater. 426. https://doi.org/10.1016/j.jhazmat.2021.128116. Available from: http://www.elsevier.com/locate/jhazmat.

Chellappan, S., 2022. Smoking cessation after cancer diagnosis and enhanced therapy response: mechanisms and significance. Curr. Oncol. 29 (12), 9956–9969. https://doi.org/10.3390/curroncol29120782.

Chen, R., Ning, Z., Zheng, C., Yang, Y., Zhang, C., Ou, X., Chen, K., Yu, H., Wei, X., Zhao, Q., He, J., 2019. Simultaneous determination of 16 alkaloids in blood by ultrahigh-performance liquid chromatography-tandem mass spectrometry coupled with supported liquid extraction. J. Chromatogr., B: Anal. Technol. Biomed. Life Sci. 1128. https://doi.org/10.1016/j.jchromb.2019.121789. Available from: http://www.elsevier.com/inca/publications/store/5/0/2/6/8/9.

Chen, H.C., Chen, Y.Y., Chao, M.R., Chang, Y.Z., 2022. Validation of a high-throughput method for simultaneous determination of areca nut and tobacco biomarkers in hair using microwave-assisted extraction and isotope dilution liquid chromatography tandem mass spectrometry. J. Pharmaceut. Biomed. Anal. 216. https://doi.org/10.1016/j.jpba.2022.114775. Available from: http://www.elsevier.com/locate/jpba.

Chou, M.W., Yan, J., Nichols, J., Xia, Q., Beland, F.A., Chan, P.C., Fu, P.P., 2003. Correlation of DNA adduct formation and riddelliine-induced liver tumorigenesis in F344 rats and B6C3F1 mice. Cancer Lett. 193 (2), 119–125. https://doi.org/10.1016/S0304-3835(03)00045-4.

Chu, C., Lian, L., Gao, W., Zang, Y., Zou, Y., Lv, Y., Li, C., Tong, S., Yan, J., Chen, S., 2022. An efficient matrix solid-phase dispersion extraction combined with online preconcentration by electrokinetic supercharging in capillary electrophoresis for the sensitive determination of two alkaloids in complicated samples matrix. Microchem. J. 178. https://doi.org/10.1016/j.microc.2022.107314. Available from: http://www.elsevier.com/inca/publications/store/6/2/0/3/9/1.

Cui, M.C., Chen, S.J., Wang, H.H., Li, Z.H., Chen, H.J., Chen, Y., Zhou, H.B., Li, X., Chen, J.W., 2018. Metabolic profiling investigation of Fritillaria thunbergii Miq. by gas chromatography–mass spectrometry. J. Food Drug Anal. 26 (1), 337–347. https://doi.org/10.1016/j.jfda.2016.10.003.

da Silva Negreiros Neto, T., Gardner, D., Hallwass, F., Leite, A.J.M., Guimarães de Almeida, C., Nunes Silva, L., Roque, A.de A., Gobbi de Bitencourt, F., Guimarães Barbosa, E., Tasca, T., Macedo, A.J., Vieira de

Almeida, M., Brandt Giordani, R., 2016. Activity of pyrrolizidine alkaloids against biofilm formation and Trichomonas vaginalis. Biomed. Pharmacother. 83, 323–329. https://doi.org/10.1016/j.biopha.2016.06.033.

Dai, Y., Jiang, W., Zheng, Y., He, P., Zhu, R., Fei, J., Xu, W., Liu, C., Hong, J., 2022. Recyclable molecularly imprinted polymers based on Fe3O4@SiO2 and PAMAM dendrimers for the determination of myosmine in cigarettes. Microchim. Acta 189 (12). https://doi.org/10.1007/s00604-022-05529-6. Available from. https://www.springer.com/journal/604.

de Araújo, R.L., de Pinho, C.L.C., Farias, F.O., Igarashi-Mafra, L., Mafra, M.R., Crinum, L., 2022. Species as a potential source of alkaloids: extraction methods and relevance on medicinal and pharmacological. South Afr. J. Bot. 151, 720–734.

Debnath, B., Singh, W.S., Das, M., Goswami, S., Singh, M.K., Maiti, D., Manna, K., 2018. Role of plant alkaloids on human health: a review of biological activities. Mater. Today Chem. 9, 56–72. https://doi.org/10.1016/j.mtchem.2018.05.001.

Dey, P., Kundu, A., Kumar, A., Gupta, M., Lee, B.M., Bhakta, T., Dash, S., Kim, H.S., 2020. Analysis of alkaloids (indole alkaloids, isoquinoline alkaloids, tropane alkaloids). Recent Advances in Natural Products Analysis. Elsevier, South Korea, pp. 505–567. https://doi.org/10.1016/B978-0-12-816455-6.00015-9. Available from: https://www.elsevier.com/books/recent-advances-in-natural-products-analysis/nabavi/978-0-12-816455-6.

Dusemund, B., Nowak, N., Sommerfeld, C., Lindtner, O., Schäfer, B., Lampen, A., 2018. Risk assessment of pyrrolizidine alkaloids in food of plant and animal origin. Food Chem. Toxicol. 115, 63–72. https://doi.org/10.1016/j.fct.2018.03.005.

Dushna, O., Dubenska, L., Vojs, M., Marton, M., Patsay, I., Ivakh, S., Plotycya, S., 2022. Highly sensitive determination of atropine in pharmaceuticals, biological fluids and beverage on planar electrochemical cell with working boron-doped diamond electrode. Electrochim. Acta 432, 141182. https://doi.org/10.1016/j.electacta.2022.141182.

Ebrahimi, S.A., 2020. Noscapine, a possible drug candidate for attenuation of cytokine release associated with SARS-CoV-2. Drug Dev. Res. 81 (7), 765–767. https://doi.org/10.1002/ddr.21676. Available from: http://onlinelibrary.wiley.com/journal/10.1002/(ISSN)1098-2299.

Elbein, A.D., Molyneux, R.J., 1999. Alkaloid Glycosidase Inhibitors. Elsevier BV, pp. 129–160. https://doi.org/10.1016/b978-0-08-091283-7.00098-9.

Erb, M., Kliebenstein, D.J., 2020. Plant secondary metabolites as defenses, regulators, and primary metabolites: the blurred functional trichotomy. Plant Physiol. 184 (1), 39–52. https://doi.org/10.1104/pp.20.00433.

Faisal, S., Badshah, S.L., Kubra, B., Emwas, A.H., Jaremko, M., 2023. Alkaloids as potential antivirals. A comprehensive review. Nat. Product. Bioprospect. 13 (1). https://doi.org/10.1007/s13659-022-00366-9. Available from: https://www.springer.com/journal/13659.

Ferreira, P.C., Ataíde, V.N., Silva Chagas, C.L., Angnes, L., Tomazelli Coltro, W.K., Longo Cesar Paixão, T.R., Reis de Araujo, W., 2019. Wearable electrochemical sensors for forensic and clinical applications. Trends in Anal. Chem. 119. https://doi.org/10.1016/j.trac.2019.115622.

Field, R.A., Stegelmeier, B.L., Colegate, S.M., Brown, A.W., Green, B.T., 2015. An in vitro comparison of the cytotoxic potential of selected dehydropyrrolizidine alkaloids and some N-oxides. Toxicon 97, 36–45. https://doi.org/10.1016/j.toxicon.2015.02.001.

Foley, M., Tilley, L., 1997. 2 1997/02 International Journal for Parasitology 10.1016/S0020-7519(96)00152-X 00207519 2 231 240 Australia Quinoline Antimalarials: Mechanisms of Action and Resistance 27.

Frank, M., Weckman, T.J., Wood, T., Woods, W.E., Tai, C.L., Chang, S.L., Ewing, A., Blake, J.W., Tobln, T., 1991. Hordenine : pharmacology, pharmacokinetics and behavioural effects in the horse, 22, 437–441.

Gackowski, M., Przybylska, A., Kruszewski, S., Koba, M., Mądra-Gackowska, K., Bogacz, A., 2021. Recent applications of capillary electrophoresis in the determination of active compounds in medicinal plants and pharmaceutical formulations. Molecules 26 (14), 4141. https://doi.org/10.3390/molecules26144141.

Gañán, J., Martínez-García, G., Morante-Zarcero, S., Pérez-Quintanilla, D., Sierra, I., 2023. Nanomaterials-modified electrochemical sensors for sensitive determination of alkaloids: recent trends in the application to biological, pharmaceutical and agri-food samples. Microchem. J. 184, 108136. https://doi.org/10.1016/j.microc.2022.108136.

Gellert, E., 1982. The indolizidine alkaloids. J. Nat. Prod. 45 (1), 50−73. https://doi.org/10.1021/np50019a005.

Gendrot, M., Andreani, J., Boxberger, M., Jardot, P., Fonta, I., Le Bideau, M., Duflot, I., Mosnier, J., Rolland, C., Bogreau, H., Hutter, S., Scola, B.La, Pradines, B., 2020. Antimalarial drugs inhibit the replication of SARS-CoV-2: an in vitro evaluation. Trav. Med. Infect. Dis. 37, 101873. https://doi.org/10.1016/j.tmaid.2020.101873.

Goldbaum, L.R., Schloegel, E.L., Dominguez, A.M., 1963. Application of gas chromatography to toxicology. Prog. Chem. Toxicol. 1, 11−52. https://doi.org/10.1016/b978-0-12-536501-7.50006-0.

Gomes, J.V.D., Tosta, C.L., Cunha Neto, Á.d., Fagg, C.W., Silva, C.A.G., Gomes-Copeland, K.K.P., Magalhães, P.O., Fonseca-Bazzo, Y.M., Jamal, C.M., Silveira, D., 2022. Chemical profile and biological activity of Crinum americanum L. (Amaryllidaceae). South Afr. J. Bot. 146, 25−35. https://doi.org/10.1016/j.sajb.2021.09.034.

González-Gómez, L., Gañán, J., Morante-Zarcero, S., Pérez-Quintanilla, D., Sierra, I., 2022a. Mesostructured silicas as cation-exchange sorbents in packed or dispersive solid phase extraction for the determination of tropane alkaloids in culinary aromatics herbs by HPLC-MS/MS. Toxins 14 (3), 218. https://doi.org/10.3390/toxins14030218.

González-Gómez, L., Morante-Zarcero, S., Pereira, J.A.M., Câmara, J.S., Sierra, I., 2022b. Improved analytical approach for determination of tropane alkaloids in leafy vegetables based on μ-QuEChERS combined with HPLC-MS/MS. Toxins 14 (10). https://doi.org/10.3390/toxins14100650. Available from: http://www.mdpi.com/journal/toxins/.

Goswami, A.K., Gogoi, N., Shakya, A., Sharma, H.K., 2019. Development and validation of high-performance thin-layer chromatographic method for quantification of berberine in rhizomes of coptis teeta wall, an endangered species collected from Arunachal Pradesh, India. J. Chromatogr. Sci. 57 (5), 411−417. https://doi.org/10.1093/chromsci/bmz009.

Greenshields, A.L., Doucette, C.D., Sutton, K.M., Madera, L., Annan, H., Yaffe, P.B., Knickle, A.F., Dong, Z., Hoskin, D.W., 2015. Piperine inhibits the growth and motility of triple-negative breast cancer cells. Cancer Lett. 357 (1), 129−140. https://doi.org/10.1016/j.canlet.2014.11.017.

Grizzell, J.A., Iarkov, A., Holmes, R., Mori, T., Echeverria, V., 2014. Cotinine reduces depressive-like behavior, working memory deficits, and synaptic loss associated with chronic stress in mice. Behav. Brain Res. 268, 55−65. https://doi.org/10.1016/j.bbr.2014.03.047.

Guarino, G., Della Corte, T., Sofia, M., Carbone, L., Marino, G., Martedì, E., Gentile, S., 2015. Metabolic effects of the association Berberis aristata/Silybum marianum: a preliminary double-blind, placebo-controlled study in obese patients with type 2 diabetes. Nutrafoods 14 (4), 181−188. https://doi.org/10.1007/s13749-015-0052-7.

Guarino, G., Strollo, F., Carbone, L., Della Corte, T., Letizia, M., Marino, G., Gentile, S., 2017. Bioimpedance analysis, metabolic effects and safety of the association berberis aristataisilybum marianum: a 52-week double-blind, placebo-controlled study in obese patients with type 2 diabetes. J. Biol. Regul. Homeost. Agents 31 (2), 495−502.

Hadi, N.S.A., Bankoglu, E.E., Schott, L., Leopoldsberger, E., Ramge, V., Kelber, O., Sievers, H., Stopper, H., 2021. Genotoxicity of selected pyrrolizidine alkaloids in human hepatoma cell lines HepG2 and Huh6. Mutat. Res. Genet. Toxicol. Environ. Mutagen 861−862. https://doi.org/10.1016/j.mrgentox.2020.503305. Available from: http://www.elsevier.com/wps/find/journaldescription.cws_home/522820/description#description.

Han, Y., Xun, L., Wang, X., Yang, S., Sun, Z., Shi, H., Xu, X., Hu, S., 2020. Detection of caffeine and its main metabolites for early diagnosis of Parkinson's disease using micellar electrokinetic capillary chromatography. Electrophoresis 41 (16−17), 1392−1399. https://doi.org/10.1002/elps.202000059. Available from: http://onlinelibrary.wiley.com/journal/10.1002/(ISSN)1522-2683.

Hartmann, T., 1999. Chemical ecology of pyrrolizidine alkaloids. Planta 207 (4), 483–495. https://doi.org/10.1007/s004250050508.

He, X., Xia, Q., Zhu, L., He, Y., Bryant, M.S., Lin, G., Fu, P.P., 2023. Formation of DHP-DNA adducts from rat liver microsomal metabolism of 1,2-unsaturated pyrrolizidine alkaloid-containing plant extracts and dietary supplements. Chem. Res. Toxicol. 36 (2), 243–250. https://doi.org/10.1021/acs.chemrestox.2c00321.

Hegazy, A., Mahmoud, S.H., Elshaier, Y.A.M.M., Shama, N.M.A., Nasr, N.F., Ali, M.A., El-Shazly, A.M., Mostafa, I., Mostafa, A., 2023. Antiviral activities of plant-derived indole and β-carboline alkaloids against human and avian influenza viruses. Sci. Rep. 13 (1). https://doi.org/10.1038/s41598-023-27954-0. Available from: https://www.nature.com/srep/.

Hirayama, A., Wakayama, M., Soga, T., 2014. Metabolome analysis based on capillary electrophoresis-mass spectrometry. TrAC, Trends Anal. Chem. 61, 215–222. https://doi.org/10.1016/j.trac.2014.05.005.

Hoff, R.B., Pizzolato, T.M., 2018. Combining extraction and purification steps in sample preparation for environmental matrices: a review of matrix solid phase dispersion (MSPD) and pressurized liquid extraction (PLE) applications. TrAC, Trends Anal. Chem. 109, 83–96. https://doi.org/10.1016/j.trac.2018.10.002.

Hu, K., Pang, T., Shi, Y., Cheng, J., Huang, Y., 2020. Facile preparation of a magnetic porous organic frameworks for highly sensitive determination of eight alkaloids in urine samples based UHPLC-MS/MS. Microchem. J. 157, 105048. https://doi.org/10.1016/j.microc.2020.105048.

Huang, Y., Wang, T., Jiang, Z., 2006. Fast analysis of alkaloids from different parts of Mahonia bealei (Fort.) Carr. Studied for their anti-Alzheimer's activity using supercritical fluid chromatography. J. Separ. Sci. 44.

Hughs, M., Kish-Trier, E., O'Brien, A., McMillin, G.A., 2023. Analysis of mitragynine and speciociliatine in umbilical cord by LC-MS-MS for detecting prenatal exposure to kratom. J. Anal. Toxicol. 46 (9), 957–964. https://doi.org/10.1093/jat/bkac064.

Human acute exposure assessment to tropane alkaloids. EFSA J. 16 (2), 2018. https://doi.org/10.2903/j.efsa.2018.5160. Available from. https://efsa.onlinelibrary.wiley.com/doi/epdf/10.2903/j.efsa.2018.5160.

Hungerford, N.L., Carter, S.J., Anuj, S.R., Tan, B.L.L., Hnatko, D., Martin, C.L., Sharma, E., Yin, M., Nguyen, T.T.P., Melksham, K.J., Fletcher, M.T., 2019. Analysis of pyrrolizidine alkaloids in Queensland honey: using low temperature chromatography to resolve stereoisomers and identify botanical sources by uhplc-ms/ms. Toxins 11 (12). https://doi.org/10.3390/toxins11120726. Available from: https://www.mdpi.com/2072-6651/11/12.

Hurtado-Fernández, E., Gómez-Romero, M., Carrasco-Pancorbo, A., Fernández-Gutiérrez Alberto, A., 2010. Application and potential of capillary electroseparation methods to determine antioxidant phenolic compounds from plant food material. J. Pharmaceut. Biomed. Anal. 53 (5), 1130–1160. https://doi.org/10.1016/j.jpba.2010.07.028.

Islam, S., Alam, R., Kim, S., 2022. Improved coverage of plant metabolites using powder laser desorption/ionization coupled with Fourier-transform ion cyclotron mass spectrometry. Food Chem. 373, 131541. https://doi.org/10.1016/j.foodchem.2021.131541.

Issaq, H.J., 1999. Capillary electrophoresis of natural products-II. Electrophoresis 20 (15–16), 3190–3202. https://doi.org/10.1002/(SICI)1522-2683(19991001)20:15/16<3190::AID-ELPS3190>3.0.CO;2-K. Available from: http://onlinelibrary.wiley.com/doi/10.1002/elps.v34.9-10/issuetoc.

Jacob, P., Yu, L., Duan, M., Ramos, L., Yturralde, O., Benowitz, N.L., 2011. Determination of the nicotine metabolites cotinine and trans-3'-hydroxycotinine in biologic fluids of smokers and non-smokers using liquid chromatography-tandem mass spectrometry: biomarkers for tobacco smoke exposure and for phenotyping cytochrome P450 2A6 activity. J. Chromatogr., B: Anal. Technol. Biomed. Life Sci. 879 (3–4), 267–276. https://doi.org/10.1016/j.jchromb.2010.12.012.

Jin, P., Zhu, F., Jiang, M., Takafuji, M., Ihara, H., Nie, L., Liu, H., 2022. Carbonized π-conjugated polymer-coated porous silica: preparation and evaluating its extraction ability for berberine. Microchim. Acta 189 (11). https://doi.org/10.1007/s00604-022-05496-y. Available from. http://www.springer/at/mca.

João, A.F., Rocha, R.G., Matias, T.A., Richter, E.M., Petruci, J.F.S., Muñoz, R.A.A., 2021. 3D-printing in forensic electrochemistry: atropine determination in beverages using an additively manufactured graphene-polylactic acid electrode. Microchem. J. 167. https://doi.org/10.1016/j.microc.2021.106324. Available from: http://www.elsevier.com/inca/publications/store/6/2/0/3/9/1.

Kaczyński, P., Łozowicka, B., 2020. A novel approach for fast and simple determination pyrrolizidine alkaloids in herbs by ultrasound-assisted dispersive solid phase extraction method coupled to liquid chromatography—tandem mass spectrometry. J. Pharmaceut. Biomed. Anal. 187, 113351. https://doi.org/10.1016/j.jpba.2020.113351.

Kanbay, M., Jensen, T., Solak, Y., Le, M., Roncal-Jimenez, C., Rivard, C., Lanaspa, M.A., Nakagawa, T., Johnson, R.J., 2016. Uric acid in metabolic syndrome: from an innocent bystander to a central player. Eur. J. Intern. Med. 29, 3—8. https://doi.org/10.1016/j.ejim.2015.11.026.

Kaur, K., Jain, M., Reddy, R.P., Jain, R., 2010. Quinolines and structurally related heterocycles as antimalarials. Eur. J. Med. Chem. 45 (8), 3245—3264. https://doi.org/10.1016/j.ejmech.2010.04.011.

Klein, L.M., Gabler, A.M., Rychlik, M., Gottschalk, C., Kaltner, F., 2022. A sensitive LC-MS/MS method for isomer separation and quantitative determination of 51 pyrrolizidine alkaloids and two tropane alkaloids in cow's milk. Anal. Bioanal. Chem. 414, 8107—8124. https://doi.org/10.1007/s00216-022-04344-5.

Knutsen, H.K., Alexander, J., Barregård, L., Bignami, M., Brüschweiler, B., Ceccatelli, S., Cottrill, B., Dinovi, M., Edler, L., Grasl-Kraupp, B., Hogstrand, C., Hoogenboom, L., Nebbia, C.S., Oswald, I.P., Petersen, A., Rose, M., Roudot, A.C., Schwerdtle, T., Vleminckx, C., Vollmer, G., Wallace, H., Ruiz Gomes, J.A., Binaglia, M., 2017. Risks for human health related to the presence of pyrrolizidine alkaloids in honey, tea, herbal infusions and food supplements. EFSA J. 15 (7). https://doi.org/10.2903/j.efsa.2017.4908. Available from: http://efsa.onlinelibrary.wiley.com/journal/18314732/.

Kohnen-Johannsen, K.L., Kayser, O., 2019. Tropane alkaloids: chemistry, pharmacology, biosynthesis and production. Molecules 24 (4). https://doi.org/10.3390/molecules24040796. Available from: https://www.mdpi.com/1420-3049/24/4/796/pdf.

Koleva, I.I., van Beek, T.A., Soffers, A.E.M.F., Dusemund, B., Rietjens, I.M.C.M., 2012. Alkaloids in the human food chain—natural occurrence and possible adverse effects. Mol. Nutr. Food Res. 56 (1), 30—52. https://doi.org/10.1002/mnfr.201100165.

Kong, W., Wei, J., Abidi, P., Lin, M., Inaba, S., Li, C., Wang, Y., Wang, Z., Si, S., Pan, H., Wang, S., Wu, J., Wang, Y., Li, Z., Liu, J., Jiang, J.D., 2005. Berberine is a novel cholesterol-lowering drug working through a unique mechanism distinct from statins. Nat. Rev. Drug Discov. 4.

Kwon, Y., Koo, Y., Jeong, Y., 2021. Determination of pyrrolizidine alkaloids in teas using liquid chromatography—tandem mass spectrometry combined with rapid-easy extraction. Foods 10 (10), 2250. https://doi.org/10.3390/foods10102250.

Lai, Q., Yang, C.J., zhang, Q., Zhuang, M., Ma, Y.H., Lin, C.Y., Zeng, G.Z., Yin, J.L., 2023. Alkaloid from Alstonia yunnanensis diels root against gastrointestinal cancer: acetoxytabernosine inhibits apoptosis in hepatocellular carcinoma cells. Front. Pharmacol. 13. https://doi.org/10.3389/fphar.2022.1085309. Available from: http://www.frontiersin.org/Pharmacology.

Lattanzio, V.M.T., Verdini, E., Sdogati, S., Caporali, A., Ciasca, B., Pecorelli, I., 2021. Undertaking a new regulatory challenge: monitoring of ergot alkaloids in Italian food commodities. Toxins 13 (12). https://doi.org/10.3390/toxins13120871. Available from. https://www.mdpi.com/2072-6651/13/12/871.

Le Pogam, P., Richomme, P., Beniddir, M.A., Duong, T.H., Bernadat, G., Schinkovitz, A., 2020. A thorough evaluation of matrix-free laser desorption ionization on structurally diverse alkaloids and their direct detection in plant extracts. Anal. Bioanal. Chem. 412 (27), 7405—7416. https://doi.org/10.1007/s00216-020-02872-6.

Lee, S.A., Hong, S.S., Han, X.H., Hwang, J.S., Oh, G.J., Lee, K.S., Lee, M.K., Hwang, B.Y., Ro, J.S., 2005. Piperine from the fruits of Piper longum with inhibitory effect on monoamine oxidase and antidepressant-like activity. Chem. Pharm. Bull. 53 (7), 832—835. https://doi.org/10.1248/cpb.53.832 (South Korea).

Lee, Y.S., Kim, W.S., Kim, K.H., Yoon, M.J., Cho, H.J., Shen, Y., Ye, J.M., Lee, C.H., Oh, W.K., Kim, C.T., Hohnen-Behrens, C., Gosby, A., Kraegen, E.W., James, D.E., Kim, J.B., 2006. Berberine, a natural plant product, activates AMP-activated protein kinase with beneficial metabolic effects in diabetic and insulin-resistant states. Diabetes 55 (8), 2256–2264. https://doi.org/10.2337/db06-0006 (South Korea).

Lekhak, M.M., Patel, S.B., Otari, S.S., Lekhak, U.M., Ghane, S.G., 2022. Bioactive potential and RP-HPLC detection of phenolics and alkaloids (lycorine and galanthamine) from ultrasonic-assisted extracts of Crinum roots. South Afr. J. Bot. 149, 923–936. https://doi.org/10.1016/j.sajb.2021.07.024.

Li, H., Gao, M., Chen, Z., Zhou, Z., Li, W., Zhang, X., Jiang, X., Luo, L., Li, F., Wang, G., Zhang, Y., Huang, X., Zhu, J., Fan, S., Wu, X., Huang, C., 2023. Drug Hordenine improves Parkinsonian-like motor deficits in mice and nematodes by activating dopamine D2 receptor-mediated signaling. Phytother Res. 37, 3296–3308. https://doi.org/10.1002/ptr.7790.

Li, X., He, X., Chen, S., Guo, X., Bryant, M.S., Guo, L., Manjanatha, M.G., Zhou, T., Witt, K.L., Mei, N., 2020. Evaluation of pyrrolizidine alkaloid-induced genotoxicity using metabolically competent TK6 cell lines. Food Chem. Toxicol. 145. https://doi.org/10.1016/j.fct.2020.111662. Available from: http://www.elsevier.com/locate/foodchemtox.

Li, Y., Li, M., Wang, Z., Wen, M., Tang, J., 2020. Identification of the metabolites of piperine via hepatocyte incubation and liquid chromatography combined with diode-array detection and high-resolution mass spectrometry. Rapid Commun. Mass Spectrom. 34 (23). https://doi.org/10.1002/rcm.8947.

Lin, S.X., Curtis, M.A., Sperry, J., 2020. Pyridine alkaloids with activity in the central nervous system. Bioorg. Med. Chem. 28 (24). https://doi.org/10.1016/j.bmc.2020.115820. Available from: http://www.journals.elsevier.com/bioorganic-and-medicinal-chemistry/.

Liu, L., Alam, M.S., Hirata, K., Matsuda, K., Ozoe, Y., 2008. Actions of quinolizidine alkaloids on *Periplaneta americana* nicotinic acetylcholine receptors. Pest Manag. Sci. 64 (12), 1222–1228. https://doi.org/10.1002/ps.1622.

Lu, Y., Yu, J., Liu, W., Jing, G., Li, W., Liu, W., 2022. Quantitative determination of major alkaloids in areca nut products by high-performance liquid chromatography coupled with ion mobility spectrometry. J. Separ. Sci. 45 (24), 4469–4477. https://doi.org/10.1002/jssc.202200361. Available from: http://onlinelibrary.wiley.com/journal/10.1002/(ISSN)1615-9314.

Luckert, C., Hessel, S., Lenze, D., Lampen, A., 2015. Disturbance of gene expression in primary human hepatocytes by hepatotoxic pyrrolizidine alkaloids: a whole genome transcriptome analysis. Toxicol. Vitro 29 (7), 1669–1682. https://doi.org/10.1016/j.tiv.2015.06.021.

Luger, S., Mayerhuber, L., Weigelhofer, G., Hein, T., Holzer, B., Hametner, C., Fruhmann, P., 2022. Development of ion-selective electrodes for tropane, atropine, and scopolamine – a concept for the analysis of tropane alkaloids. Electroanalysis 34 (10), 1579–1586. https://doi.org/10.1002/elan.202100594. Available from: http://onlinelibrary.wiley.com/journal/10.1002/(ISSN)1521-4109.

Ma, C., Liu, Y., Zhu, L., Ji, H., Song, X., Guo, H., Yi, T., 2018. Determination and regulation of hepatotoxic pyrrolizidine alkaloids in food: a critical review of recent research. Food Chem. Toxicol. 119, 50–60. https://doi.org/10.1016/j.fct.2018.05.037.

Ma, S., He, Z., Teng, Q., Wang, R., 2023. One-pot synthesis of magnetic sulfonated covalent organic polymer for extraction of protoberberine alkaloids in herbs and human plasma. J. Separ. Sci. 46 (2). https://doi.org/10.1002/jssc.202200613. Available from: http://onlinelibrary.wiley.com/journal/10.1002/(ISSN)1615-9314.

Martínez-Pinilla, E., Oñatibia-Astibia, A., Franco, R., 2015. The relevance of theobromine for the beneficial effects of cocoa consumption. Front. Pharmacol. 6. https://doi.org/10.3389/fphar.2015.00030. Available from: http://journal.frontiersin.org/article/10.3389/fphar.2015.00030/full.

Martins, J., Fedorenko, D., Pavlenko, R., Berzina, Z., Bartkevics, V., 2022. Nanoflow liquid chromatography mass spectrometry method for quantitative analysis and target ion screening of pyrrolizidine alkaloids in honey, tea, herbal tinctures, and milk. J. Chromatogr. A 1676, 463269. https://doi.org/10.1016/j.chroma.2022.463269.

Michael, J.P., 2007. Indolizidine and quinolizidine alkaloids. Nat. Prod. Rep. 24 (1), 191–222. https://doi.org/10.1039/b509525p.

Mitra, S., Prova, S.R., Sultana, S.A., Das, R., Nainu, F., Emran, T.B., Tareq, A.M., Uddin, M.S., Alqahtani, A.M., Dhama, K., Simal-Gandara, J., 2021. Therapeutic potential of indole alkaloids in respiratory diseases: a comprehensive review. Phytomedicine 90. https://doi.org/10.1016/j.phymed.2021.153649. Available from: http://www.urbanfischer.de/journals/phytomed.

Mitra, S., Anand, U., Jha, N.K., Shekhawat, M.S., Saha, S.C., Nongdam, P., Rengasamy, K.R.R., Proćków, J., Dey, A., 2022. Anticancer applications and pharmacological properties of piperidine and piperine: a comprehensive review on molecular mechanisms and therapeutic perspectives. Front. Pharmacol. 12. https://doi.org/10.3389/fphar.2021.772418. Available from. http://www.frontiersin.org/Pharmacology.

Mohammadi Ziarani, G., Jamasbi, N., Mohajer, F., 2022. Recent advances on the synthesis of natural pyrrolizidine alkaloids: alexine, and its stereoisomers. Nat. Product. Bioprospect. 12 (1). https://doi.org/10.1007/s13659-022-00324-5. Available from. https://link.springer.com/journal/13659.

Monton, M.R.N., Soga, T., 2007. Metabolome analysis by capillary electrophoresis-mass spectrometry. J. Chromatogr. A 1168 (1–2), 237–246. https://doi.org/10.1016/j.chroma.2007.02.065.

Moreira, R., Pereira, D.M., Valentão, P., Andrade, P.B., 2018. Pyrrolizidine alkaloids: chemistry, pharmacology, toxicology and food safety. Int. J. Mol. Sci. 19 (6). https://doi.org/10.3390/ijms19061668. Available from: http://www.mdpi.com/1422-0067/19/6/1668/pdf.

Mulder, H.A., Pearcy, A.C., Halquist, M.S., 2022. Characterization of molecularly imprinted polymers for the extraction of tobacco alkaloids and their metabolites in human urine. Biomed. Chromatogr. 36 (6). https://doi.org/10.1002/bmc.5361. Available from: http://onlinelibrary.wiley.com/journal/10.1002/(ISSN)1099-0801.

Nakada, T., Kiyotani, K., Iwano, S., Uno, T., Yokohira, M., Yamakawa, K., Fujieda, M., Saito, T., Yamazak, H., Imaida, K., Kamataki, T., 2012. Lung tumorigenesis promoted by anti-apoptotic effects of cotinine, a nicotine metabolite through activation of PI3K/Akt pathway. J. Toxicol. Sci. 37 (3), 555–563. https://doi.org/10.2131/jts.37.555.

Natori, Y., Kamioka, S., Yoshimoto, T., Ishii, A., 2022. A simple and rapid method for quantifying aconitines and their metabolites in whole blood by modified QuEChERS and liquid chromatography/tandem mass spectrometry (LC/MS/MS). Forensic Sci. Int. 341, 111475. https://doi.org/10.1016/j.forsciint.2022.111475.

Nontprasert, A., Pukrittayakamee, S., Kyle, D.E., Vanijanonta, S., White, N.J., 1996. Antimalarial activity and interactions between quinine, dihydroquinine and 3-hydroxyquinine against Plasmodium falciparum in vitro. Trans. R. Soc. Trop. Med. Hyg. 90 (5), 553–555. https://doi.org/10.1016/S0035-9203(96)90320-X.

Ognjanović, M., Nikolić, K., Bošković, M., Pastor, F., Popov, N., Marciuš, M., Krehula, S., Antić, B., Stanković, D.M., 2022. Electrochemical determination of morphine in urine samples by tailoring FeWO4/CPE sensor. Biosensors 12 (11). https://doi.org/10.3390/bios12110932. Available from: http://www.mdpi.com/journal/biosensors/.

Padhiari, B.M., Ray, A., Jena, S., Champati, B.B., Sahoo, A., Mohanty, S., Nayak, S., 2020. Simultaneous quantification of vasicine and vasicinone in different parts of Justicia adhatoda using high-performance thin-layer chromatography–densitometry: comparison of different extraction techniques and solvent systems. J. Planar Chromatogr. Mod. TLC 33 (6), 599–607. https://doi.org/10.1007/s00764-020-00070-9.

Pan, Y.T., Hori, H., Saul, R., Elbein, A.D., Sanford, B.A., Molyneux, R.J., 1983. Castanospermine inhibits the processing of the oligosaccharide portion of the influenza viral hemagglutinin. Biochemistry 22 (16), 3975–3984. https://doi.org/10.1021/bi00285a038.

Pasupuleti, R.R., Lee, C.H., Osborne, P.G., Wu, M.T., Ponnusamy, V.K., 2022. Rapid green analytical methodology for simultaneous biomonitoring of five toxic areca nut alkaloids using UHPLC-MS/MS for predicting health hazardous risks. J. Hazard Mater. 422. https://doi.org/10.1016/j.jhazmat.2021.126923. Available from: http://www.elsevier.com/locate/jhazmat.

Peng, L.Q., Ye, L.H., Cao, J., Chang, Y.x., Li, Q., An, M., Tan, Z., Xu, J.J., 2017. Cyclodextrin-based miniaturized solid phase extraction for biopesticides analysis in water and vegetable juices samples analyzed by ultra-high-performance liquid chromatography coupled with quadrupole time-of-flight mass spectrometry. Food Chem. 226, 141–148. https://doi.org/10.1016/j.foodchem.2017.01.006.

Pennington, L.D., Moustakas, D.T., 2017. The necessary nitrogen atom: a versatile high-impact design element for multiparameter optimization. J. Med. Chem. 60 (9), 3552–3579. https://doi.org/10.1021/acs.jmedchem.6b01807.

Pereira, A.S., Amaral, A.C.F., Barnes, R.A., Cardoso, J.N., Aquino Neto, F.R., 1999. Identification of isoquinoline alkaloids in crude extracts by high temperature gas chromatography-mass spectrometry. Phytochem. Anal. 10 (5), 254–258. https://doi.org/10.1002/(SICI)1099-1565(199909/10)10:5<254::AID-PCA462>3.0.CO;2-Q.

Pérez-Alcaraz, A., Borrull, F., Aguilar, C., Calull, M., 2019. Enantioselective determination of cathinones in urine by high pressure in-line SPE–CE. Electrophoresis 40 (14), 1762–1770. https://doi.org/10.1002/elps.201900072. Available from: http://onlinelibrary.wiley.com/journal/10.1002/(ISSN)1522-2683.

Perharič, L., Juvan, K.A., Stanovnik, L., 2013. Acute effects of a low-dose atropine/scopolamine mixture as a food co'ntaminant in human volunteers. J. Appl. Toxicol. 33 (9), 980–990. https://doi.org/10.1002/jat.2797.

Pichon, V., Delaunay, N., Combès, A., 2020. Sample preparation using molecularly imprinted polymers. Anal. Chem. 92 (1), 16–33. https://doi.org/10.1021/acs.analchem.9b04816.

Plazas, E., Faraone, N., 2023. Indole alkaloids from psychoactive mushrooms: chemical and pharmacological potential as psychotherapeutic agents. Biomedicines 11 (2), 461. https://doi.org/10.3390/biomedicines11020461.

Płotka-Wasylka, J., 2018. A new tool for the evaluation of the analytical procedure: green Analytical Procedure Index. Talanta 181, 204–209. https://doi.org/10.1016/j.talanta.2018.01.013.

Poole, C.F., 2003. Thin-layer chromatography: challenges and opportunities. J. Chromatogr. A 1000 (1–2), 963–984. https://doi.org/10.1016/S0021-9673(03)00435-7.

Poole, C., Mester, Z., Miró, M., Pedersen-Bjergaard, S., Pawliszyn, J., 2016. Extraction for analytical scale sample preparation (IUPAC Technical Report). Pure Appl. Chem. 88 (7), 649–687. https://doi.org/10.1515/pac-2015-0705.

Prakash, A.S., Pereira, T.N., Reilly, P.E.B., Seawright, A.A., 1999. Pyrrolizidine alkaloids in human diet. Mutat. Res. Genet. Toxicol. Environ. Mutagen 443 (1–2), 53–67. https://doi.org/10.1016/S1383-5742(99)00010-1.

Prata, M., Ribeiro, A., Figueirinha, D., Rosado, T., Oppolzer, D., Restolho, J., Araújo, A.R.T.S., Costa, S., Barroso, M., Gallardo, E., 2019. Determination of opiates in whole blood using microextraction by packed sorbent and gas chromatography-tandem mass spectrometry. J. Chromatogr. A 1602, 1–10. https://doi.org/10.1016/j.chroma.2019.05.021.

Qasem, A.M.A., Rowan, M.G., Blagbrough, I.S., 2022. Poisonous piperidine plants and the biodiversity of nor-diterpenoid alkaloids for leads in drug Discovery: experimental aspects. Int. J. Mol. Sci. 23 (20). https://doi.org/10.3390/ijms232012128. Available from: http://www.mdpi.com/journal/ijms.

Qie, M., Li, S., Guo, C., Yang, S., Zhao, Y., 2021. Study of the occurrence of toxic alkaloids in forage grass by liquid chromatography tandem mass spectrometry. J. Chromatogr. A 1654, 462463. https://doi.org/10.1016/j.chroma.2021.462463.

Qin, Y., Wu, Y., Wang, B., Wang, J., Yao, W., 2021. Facile synthesis of Ag@Au core-satellite nanowires for highly sensitive SERS detection for tropane alkaloids. J. Alloys Compd. 884, 161053. https://doi.org/10.1016/j.jallcom.2021.161053.

Reinhard, H., Rupp, H., Sager, F., Streule, M., Zoller, O., 2006. Quinolizidine alkaloids and phomopsins in lupin seeds and lupin containing food. J. Chromatogr. A 1112 (1–2), 353–360. https://doi.org/10.1016/j.chroma.2005.11.079.

Richter, M.M., 2004. Electrochemiluminescence (ECL). Chem. Rev. 104 (106), 3003–3036. https://doi.org/10.1021/cr020373d.

Rizzi, D., Basile, C., Di Maggio, A., Sebastio, A., Introna Jnr, F., Rizzi, R., Scatizzi, A., De Marco, S., Smialek, J.E., 1991. Clinical spectrum of accidental hemlock poisoning: neurotoxic manifestations, rhabdomyolysis and acute tubular necrosis. Nephrol. Dial. Transplant. 6 (12), 939–943. https://doi.org/10.1093/ndt/6.12.939.

Rizzo, S., Celano, R., Lisa Piccinelli, A., Serio, S., Russo, M., Rastrelli, L., 2023. An analytical platform for the screening and identification of pyrrolizidine alkaloids in food matrices with high risk of contamination. Food Chem. 406, 135058. https://doi.org/10.1016/j.foodchem.2022.135058.

Roque Bravo, R., Faria, A.C., Brito-da-Costa, A.M., Carmo, H., Mladenka, P., Dias da Silva, D., 2022. Remião and on behalf of the OEMONOM researchers, cocaine: an updated overview on chemistry, detection, biokinetics, and pharmacotoxicological aspects including abuse pattern. Toxins 14.

Santos M, L.F.S., Silva, A., Gaujac, S.A., 2022. Navickiene, miniaturized matrix solid-phase dispersion methodology (MMSPD) for determination of β-carboline alkaloids in tobacco samples by UPLC-ESI-Q-TOF/MSE. J. Pharmaceut. Biomed. Anal. 218.

Sargazi, S., Fatima, I., Hassan Kiani, M., Mohammadzadeh, V., Arshad, R., Bilal, M., Rahdar, A., Díez-Pascual, A.M., Behzadmehr, R., 2022. Fluorescent-based nanosensors for selective detection of a wide range of biological macromolecules: a comprehensive review. Int. J. Biol. Macromol. 206, 115–147. https://doi.org/10.1016/j.ijbiomac.2022.02.137.

Scientific Opinion on the risks for public health related to the presence of opium alkaloids in poppy seeds. EFSA Panel on Contaminants in the Food Chain (CONTAM) 9, 2011.

Scientific Opinion on Ergot alkaloids in food and feed. EFSA Panel on Contaminants in the Food Chain (CONTAM) 10, 2012.

Scientific opinion on the risks for animal and human health related to the presence of quinolizidine alkaloids in feed and food, in particular in lupins and lupin-derived products. EFSA Panel on Contaminants in the Food Chain (CONTAM) 17, 2019.

Scientific opinion of the Panel on Contaminants in the food chain on a request from the European commission on tropane alkaloids (from Datura sp.) as undesirable substances in animal feed. EFSA J. 691, 2008, 1–55.

Shang, X.F., Morris-Natschke, S.L., Yang, G.Z., Liu, Y.Q., Guo, X., Xu, X.S., Goto, M., Li, J.C., Zhang, J.Y., Lee, K.H., 2018. Biologically active quinoline and quinazoline alkaloids part II. Med. Res. Rev. 38 (5), 1614–1660. https://doi.org/10.1002/med.21492. Available from: http://onlinelibrary.wiley.com/journal/10.1002/(ISSN)1098-1128.

Sharma, D., Sharma, N., Manchanda, N., Prasad, S.K., Sharma, P.C., Thakur, V.K., Rahman, M.M., Dhobi, M., 2023. Bioactivity and in silico studies of isoquinoline and related alkaloids as promising antiviral agents: an insight. Biomolecules 13 (1). https://doi.org/10.3390/biom13010017. Available from. http://www.mdpi.com/journal/biomolecules.

Sheibani, N., Kazemipour, M., Jahani, S., Foroughi, M.M., 2019. A novel highly sensitive thebaine sensor based on MWCNT and dandelion-like Co3O4 nanoflowers fabricated via solvothermal synthesis. Microchem. J. 149. https://doi.org/10.1016/j.microc.2019.103980. Available from: http://www.elsevier.com/inca/publications/store/6/2/0/3/9/1.

Shi, M.Z., Yu, Y.L., Zhu, S.C., Gu, Y.X., Yue, Z.X., Yan, T.C., Zheng, H., Cao, J., 2022. Boron nitride nanosheet-assisted matrix solid-phase dispersion microextraction of alkaloids from lotus plumule by high-performance liquid chromatography coupled with ultraviolet detection and ion mobility quadrupole time-of-flight mass spectrometry. Electrophoresis 43 (4), 581–589. https://doi.org/10.1002/elps.202100286. Available from: http://onlinelibrary.wiley.com/journal/10.1002/(ISSN)1522-2683.

Silberstein, S.D., McCrory, D.C., 2003. Ergotamine and dihydroergotamine: history, pharmacology, and efficacy. Headache 43 (2), 144–166. https://doi.org/10.1046/j.1526-4610.2003.03034.x.

Singh, S., Pillai, S., Chellappan, S., 2011. Nicotinic acetylcholine receptor signaling in tumor growth and metastasis. J. Oncol. 2011, 1–11. https://doi.org/10.1155/2011/456743.

Sioud, F., Ben Toumia, I., Lahmer, A., Khlifi, R., Dhaouefi, Z., Maatouk, M., Ghedira, K., Chekir-Ghedira, L., 2020. Methanolic extract of Ephedra alata ameliorates cisplatin-induced nephrotoxicity and hepatotoxicity through reducing oxidative stress and genotoxicity. Environ. Sci. Pollut. Control Ser. 27 (11), 12792–12801. https://doi.org/10.1007/s11356-020-07904-3.

Song, J., Guo, T., Wu, W., Liao, M., Xiao, Q., Qiao, B., Wan, Y., Dai, Q., Zhou, Y., 2022a. Simultaneous determination of five poppy husk alkaloids and three main components of acetaminophen-caffeine-aspirin tablets in hot pot broth by ultra-high-performance liquid chromatography–tandem mass spectrometry. Separ. Sci. 5 (11), 616–625. https://doi.org/10.1002/sscp.202200088.

Song, C., Yu, D., Jin, G., Ding, J., Zhou, H., Guo, Z., Liang, X., 2022b. High-performance liquid chromatography quantitative analysis of ephedrine alkaloids in Ephedrae Herba on a perfluorooctyl stationary phase. J. Separ. Sci. 45 (5), 1051–1058. https://doi.org/10.1002/jssc.202100645. Available from: http://onlinelibrary.wiley.com/journal/10.1002/(ISSN)1615-9314.

Su, S., Cao, M., Wu, G., Long, Z., Cheng, X., Fan, J., Xu, Z., Su, H., Hao, Y., Li, G., Peng, J., Li, S., Wang, X., 2018. Hordenine protects against hyperglycemia-associated renal complications in streptozotocin-induced diabetic mice. Biomed. Pharmacother. 104, 315–324. https://doi.org/10.1016/j.biopha.2018.05.036.

Sun, J., Dai, J., Zhou, Y., Shang, H., 2022. Enantioseparation and quantitative analysis of three alkaloids in Rhizoma Corydalis by chiral LC-MS/MS. J. Pharmaceut. Biomed. Anal. 220, 115008. https://doi.org/10.1016/j.jpba.2022.115008.

Tajabadi, F., Khalighi-Sigaroodi, F., Nahooji, M.G., Ghiaci-Yekta, M., Ghasemi, S.V., 2022. Selective and simple determination of isoquinoline alkaloids in papaver species by ion mobility spectrometry. Iran. J. Pharm. Res. (IJPR) 21 (1). https://doi.org/10.5812/ijpr-127037. Available from: https://brieflands.com/articles/ijpr-127037.pdf.

Tasleem, F., Azhar, I., Ali, S.N., Perveen, S., Mahmood, Z.A., 2014. Analgesic and anti-inflammatory activities of Piper nigrum L. Asian Pac. J. Tropical Med. 7 (1), S461. https://doi.org/10.1016/S1995-7645(14)60275-3.

Taylor, D.L., Nash, R., Fellows, L.E., Kang, M.S., Tyms, A.S., 1992. Naturally occurring pyrrolizidines: inhibition of α-glucosidase 1 and anti-HIV activity of one steroisomer. Antiviral Chem. Chemother. 3 (5), 273–277. https://doi.org/10.1177/095632029200300504.

Tazzeo, T., Bates, G., Roman, H.N., Lauzon, A.M., Khasnis, M.D., Eto, M., Janssen, L.J., 2012. Caffeine relaxes smooth muscle through actin depolymerization. Am. J. Physiol. Lung Cell Mol. Physiol. 303 (4), L334. https://doi.org/10.1152/ajplung.00103.2012Canada.

Tittikpina, N.K., Katragunta, K., Avula, B., Ali, Z., Khan, I.A., 2023. Strategy for the quality control of herbal preparations made of Sarcocephalus latifolius: development and validation of a UHPLC-PDA method for quantification of angustoline and strictosamide and chemical profiling using LC-QToF. Phytochem. Anal. 34 (1), 105–126. https://doi.org/10.1002/pca.3183. Available from: http://onlinelibrary.wiley.com/journal/10.1002/(ISSN)1099-1565.

Tobiszewski, M., Mechlińska, A., Zygmunt, B., Namieśnik, J., 2009. Green analytical chemistry in sample preparation for determination of trace organic pollutants. TrAC, Trends Anal. Chem. 28 (8), 943–951. https://doi.org/10.1016/j.trac.2009.06.001.

Toma, W., Trigo, J.R., De Paula, A.C.B., Brito, A.R.M.S., 2004. Preventive activity of pyrrolizidine alkaloids from Senecio brasiliensis (Asteraceae) on gastric and duodenal induced ulcer on mice and rats. J. Ethnopharmacol. 95 (2–3), 345–351. https://doi.org/10.1016/j.jep.2004.08.017.

Torchiana, M.L., Porter, C.C., Stone, C.A., Hanson, H.M., 1970. Some biochemical and pharmacological actions of α-methylphenylalanine. Biochem. Pharmacol. 19 (5), 1601–1614. https://doi.org/10.1016/0006-2952(70)90148-6.

Update of the Scientific Opinion on opium alkaloids in poppy seeds. EFSA Panel on Contaminants in the Food Chain (CONTAM) 16, 2018.

Vansal, S.S., Feller, D.R., 1999. Direct effects of ephedrine isomers on human β-adrenergic receptor subtypes. Biochem. Pharmacol. 58 (5), 807–810. https://doi.org/10.1016/S0006-2952(99)00152-5.

Velichkova, S., Theunis, M., Naessens, T., Pieters, L., Foubert, K., 2022. Development and validation of an HPLC-DAD method for Rapid quantification of vasicine in Adhatoda vasica leaves and commercial products. Heliyon 8 (8). https://doi.org/10.1016/j.heliyon.2022.e10226. Available from: http://www.journals.elsevier.com/heliyon/.

Vetter, J., 2004. Poison hemlock (Conium maculatum L.). Food Chem. Toxicol. 42 (9), 1373–1382. https://doi.org/10.1016/j.fct.2004.04.009.

Vitaku, E., Smith, D.T., Njardarson, J.T., 2014. Analysis of the structural diversity, substitution patterns, and frequency of nitrogen heterocycles among U.S. FDA approved pharmaceuticals. J. Med. Chem. 57 (24), 10257–10274. https://doi.org/10.1021/jm501100b.

Vo Duy, S., Munoz, G., Dinh, Q.T., Zhang, Y., Simon, D.F., Sauvé, S., 2022. Fast screening of saxitoxin, neosaxitoxin, and decarbamoyl analogues in fresh and brackish surface waters by on-line enrichment coupled to HILIC-HRMS. Talanta 241. https://doi.org/10.1016/j.talanta.2022.123267. Available from: https://www.journals.elsevier.com/talanta.

Voelker, S.E., Kern, S.E., Falconer, T.M., Thatcher, M.D., Skelton, D.M., Gaston, K.W., Lanzarotta, A.C., Lorenz, L.M., Batson, J.C.S., Litzau, J.J., 2021. Evaluation of four field portable devices for the rapid detection of mitragynine in suspected kratom products. J. Pharmaceut. Biomed. Anal. 201. https://doi.org/10.1016/j.jpba.2021.114104. Available from: http://www.elsevier.com/locate/jpba.

Walker, N., Howe, C., Glover, M., McRobbie, H., Barnes, J., Nosa, V., Parag, V., Bassett, B., Bullen, C., 2014. Cytisine versus nicotine for smoking cessation. N. Engl. J. Med. 371 (25), 2353–2362. https://doi.org/10.1056/NEJMoa1407764. Available from: http://www.nejm.org/doi/pdf/10.1056/NEJMoa1407764.

Wang, S.B., Jia, J.P., 2014. Oxymatrine attenuates diabetes-associated cognitive deficits in rats. Acta Pharmacol. Sin. 35 (3), 331–338. https://doi.org/10.1038/aps.2013.158.

Wang, Y.P., Fu, P.P., Chou, M.W., 2005. 1 2005/01 International Journal of Environmental Research and Public Health 10.3390/ijerph2005010074 16604601 1 74 79 MDPI United States Metabolic activation of the tumorigenic pyrrolizidine alkaloid, retrorsine, leading to DNA adduct formation in vivo, 2. http://www.mdpi.org/ijerph/papers/ijerph2005010074.pdf.

Wang, Q.Y., Dong, X., Yang, J., Zhen, X.T., Ye, L.H., Chu, C., Wang, B., Hu, Y.H., Zheng, H., Cao, J., 2019. Solid acids assisted matrix solid-phase dispersion microextraction of alkaloids by capillary electrophoresis coupled with quadrupole time-of-flight mass spectrometry. J. Separ. Sci. 42 (23), 3579–3588. https://doi.org/10.1002/jssc.201900647. Available from: http://onlinelibrary.wiley.com/journal/10.1002/(ISSN)1615-9314.

Warowicka, A., Nawrot, R., Goździcka-Józefiak, A., 2020. Antiviral activity of berberine. Arch. Virol. 165 (9), 1935–1945. https://doi.org/10.1007/s00705-020-04706-3.

Wei, X., Ruan, W., Vrieling, K., 1970. Current knowledge and perspectives of pyrrolizidine alkaloids in pharmacological applications. Molecules 26.

Whitby, K., Pierson, T.C., Geiss, B., Lane, K., Engle, M., Zhou, Y., Doms, R.W., Diamond, M.S., 2005. Castanospermine, a potent inhibitor of dengue virus infection in vitro and in vivo. J. Virol. 79 (14), 8698–8706. https://doi.org/10.1128/JVI.79.14.8698-8706.2005.

Wu, C.H., Lee, C.H., Ho, Y.S., 2011. Nicotinic acetylcholine receptor-based blockade: applications of molecular targets for cancer therapy. Clin. Cancer Res. 17 (11), 3533–3541. https://doi.org/10.1158/1078-0432.CCR-10-2434Taiwan.

Wu, F., Yang, S., Wang, L., Wei, W., Ding, C.F., 2019. Simultaneous enrichment and analysis of tobacco alkaloids by microextraction coupled with mass spectrometry using a poly (N-isopropyl-acrylamide-co-divinylbenzene-co-N, N'-methylene diacrylamide) monolithic column. Talanta 198, 118–127. https://doi.org/10.1016/j.talanta.2019.01.073.

Wu, J., Cui, C., Zhao, H., Zhou, G., Qin, L., Li, X., Chen, L., Wang, X., Wan, Y., 2022. In-situ detection and imaging of Areca catechu fruit alkaloids by MALDI-MSI. Ind. Crop. Prod. 188, 115533. https://doi.org/10.1016/j.indcrop.2022.115533.

Wu, Z.L., Zhang, W.Y., Zhong, J.C., Huang, X.J., Xu, W., Chen, M.F., Weng, S.Q., Zhang, D.M., Che, C.T., Ye, W.C., Wang, Y., 2022. Angiogenesis-inhibitory piperidine alkaloids from the leaves of microcos paniculata. J. Nat. Prod. 85 (2), 375–383. https://doi.org/10.1021/acs.jnatprod.1c00960.

Xia, H., Liu, Y., Xia, G., Liu, Y., Lin, S., Guo, L., 2021. Novel isoquinoline alkaloid litcubanine A—a potential anti-inflammatory candidate. Front. Immunol. 12. https://doi.org/10.3389/fimmu.2021.685556. Available from: https://www.frontiersin.org/journals/immunology#.

Xiao, Y., Yi, H., Wang, G., Chen, S., Xiang, L., Wu, Q., Zhang, S., Deng, K., He, Y., Yang, X., 2022. Electrochemiluminescence sensor for point-of-care detection of pyrrolizidine alkaloids. Talanta 249, 123645. https://doi.org/10.1016/j.talanta.2022.123645.

Xu, R., Chen, X., Wang, X., Yu, L., Zhao, W., Ba, Y., Wu, X., 2019. Development and validation of an ultra-high performance supercritical fluid chromatography-photodiode array detection-mass spectrometry method for the simultaneous determination of 12 compounds in Piper longum L. Food Chem. 298, 125067. https://doi.org/10.1016/j.foodchem.2019.125067.

Yamagaki, T., Osawa, T., 2024. Diaminoanthraquinone Enhances Alkaloid Ionization in MALDI-MS.

Yan, Q.T., Chen, M.C., Hsieh, M.M., 2022. Ultrasensitive analysis of mirtazapine and its metabolites enantiomers in body fluids using ultrasound-enhanced and surfactant-assisted dispersive liquid–liquid microextraction followed by polymer-mediated stacking in capillary electrophoresis. J. Chromatogr. A 1678. https://doi.org/10.1016/j.chroma.2022.463328. Available from: http://www.elsevier.com/locate/chroma.

Yang, L., Stöckigt, J., 2010. Trends for diverse production strategies of plant medicinal alkaloids. Nat. Prod. Rep. 27 (10), 1469–1479. https://doi.org/10.1039/c005378c.

Yao, X., Ye, F., Zhang, M., Cui, C., Huang, B., Niu, P., Liu, X., Zhao, L., Dong, E., Song, C., Zhan, S., Lu, R., Li, H., Tan, W., Liu, D., 2020. In vitro antiviral activity and projection of optimized dosing design of hydroxychloroquine for the treatment of severe acute respiratory syndrome coronavirus 2 (SARS-CoV-2). Clin. Infect. Dis. 71 (15), 732–739. https://doi.org/10.1093/cid/ciaa237.

Yin, J., Xing, H., Ye, J., 2008. Efficacy of berberine in patients with type 2 diabetes mellitus. Metab. Clin. Exp. 57 (5), 712–717. https://doi.org/10.1016/j.metabol.2008.01.013.

Yu, L., Cong, R., Zhou, X., Xu, H., Chen, L., Lai, X., Li, Q., Chu, K., Xu, W., 2019. Simultaneous detection of Tripterygium wilfordii sesquiterpene alkaloids by microemulsion electrokinetic chromatography coupled with large volume sample stacking. Microchem. J. 148, 449–455. https://doi.org/10.1016/j.microc.2019.04.073.

Yu, Y.L., Jin, H.F., Shi, Y., Cao, J., 2022. Synchronous microextraction of active and toxic compounds from medicinal plant using nano-SiO2 assisted miniaturized matrix solid-phase dispersion. Microchem. J. 183. https://doi.org/10.1016/j.microc.2022.108099. Available from: http://www.elsevier.com/inca/publications/store/6/2/0/3/9/1.

Yue, Z.X., Gu, Y.X., Yan, T.C., Li, M.H., Zheng, H., Cao, J., 2022. Ion pair-based mobile phase additives to improve the separation of alkaloids in supercritical fluid chromatography. J. Pharmaceut. Biomed. Anal. 208. https://doi.org/10.1016/j.jpba.2021.114467. Available from: http://www.elsevier.com/locate/jpba.

Zhang, B.J., Xu, D., Guo, Y., Ping, J., Chen, L.B., Wang, H., 2008. Protection by and anti-oxidant mechanism of berberine against rat liver fibrosis induced by multiple hepatotoxic factors. Clin. Exp. Pharmacol. Physiol. 35 (3), 303–309. https://doi.org/10.1111/j.1440-1681.2007.04819.x.

Zhang, X., Li, X.r., Zhang, J., 2013. Current status and future perspectives of PI3K and mTOR inhibitor as anticancer drugs in breast cancer. Curr. Cancer Drug Targets 13 (2), 175–187. https://doi.org/10.2174/1568009611313020007.

Zhang, Y.B., Yang, L., Luo, D., Chen, N.H., Wu, Z.N., Ye, W.C., Li, Y.L., Wang, G.C., Sophalines, E.-I., 2018. Five quinolizidine-based alkaloids with antiviral activities against the hepatitis B virus from the seeds of Sophora alopecuroides. Org. Lett. 20 (18), 5942–5946. https://doi.org/10.1021/acs.orglett.8b02637.

Zhou, J.W., Luo, H.Z., Jiang, H., Jian, T.K., Chen, Z.Q., Jia, A.Q., 2018. Hordenine: a novel quorum sensing inhibitor and antibiofilm agent against *Pseudomonas aeruginosa*. J. Agric. Food Chem. 66 (7), 1620–1628. https://doi.org/10.1021/acs.jafc.7b05035.

Zhou, S., Xu, X., Wang, L., Liu, L., Hua, K., Xu, C., 2022. Rapid, on-site quantitative determination of higenamine in functional food using a time-resolved fluorescence microsphere test strip. Food Chem. 387, 132859. https://doi.org/10.1016/j.foodchem.2022.132859.

Zhu, L., Xue, J., Xia, Q., Fu, P.P., Lin, G., 2017. The long persistence of pyrrolizidine alkaloid-derived DNA adducts in vivo: kinetic study following single and multiple exposures in male ICR mice. Arch. Toxicol. 91 (2), 949–965. https://doi.org/10.1007/s00204-016-1713-z.

Zhu, S.C., Shi, M.Z., Yu, Y.L., Jiao, Y.H., Zheng, H., Liu, F.M., Cao, J., 2021. Situ Formation of ion pair assisted liquid-liquid microextraction of natural alkaloids by response surface methodology. Microchem. J. 171 (5).

Zhu, S.C., Shi, M.Z., Yu, Y.L., Cao, J., 2022. Simultaneous extraction and enrichment of alkaloids from lotus leaf by in-situ cloud point-reinforced ionic liquid assisted mechanochemical extraction technology. Ind. Crop. Prod. 183. https://doi.org/10.1016/j.indcrop.2022.114968. Available from: http://www.elsevier.com/inca/publications/store/5/2/2/8/2/5.

Zulak, K.G., Liscombe, D.K., Ashihara, H., Facchini, P.J., 2006. Alkaloids. Plant Secondary Metabolites. Occurrence, Structure and Role in the Human Diet. Blackwell Publishing Ltd., Oxford, pp. 102–131.

Zwerger, M., Zelger, J., Ganzera, M., 2023. Separation of pyrrolizidine alkaloids in different Senecio species using ultra-high performance supercritical fluid chromatography. J. Pharmaceut. Biomed. Anal. 228, 115310. https://doi.org/10.1016/j.jpba.2023.115310.

CHAPTER

14 Terpenoids (extraction, quantification, and biological activities)

Cecilia Cagliero, Barbara Sgorbini, Arianna Marengo and Patrizia Rubiolo

University of Turin, Department of Drug Science and Technology, Turin, Italy

1. Introduction

Terpenes belong to the largest class of secondary (specialized) metabolites, considering that about 30,000 terpenes are currently known in the literature. They comprise a wide range of compounds with different physicochemical properties, such as molecular weight, polarity, volatility, etc. Terpenes consist of isoprene units (namely five carbons) linked together. The term "terpenes," originating from turpentine (lat. *Balsamum terebinthinae*), refers to the hydrocarbons, while "terpenoids" are modified terpenes with various functional groups (alcohols and their glycosides, esters, ethers, aldehydes, ketones, carboxylic acids). Terpenes are classified into different groups, i.e., monoterpenes, sesquiterpenes, diterpenes, sesterpenes, triterpenes and tetraterpenes (carotenoids), depending on the number of isoprene units present in the structure. Fig. 14.1 shows the classification of terpenes.

Terpene compounds are all formed from the C5 substrates dimethylallyl diphosphate (DMAPP) and isopentenyl diphosphate (IPP), usually by initial condensation of DMAPP with one or more IPP molecules in a 1-4 or "head-to-tail" fashion to form geranyl diphosphate (GPP, C10 structure precursor), farnesyl diphosphate (FPP, C15 precursor), or geranylgeranyl diphosphate (GGPP, C20 precursor). FPP and GGPP can then condense "head-to-head" to form squalene (precursor of C30, sterols such as cholesterol) or phytoene (precursor of C40, the so-called carotenoids such as β-carotene).

The precursors of DMAPP and IPP originate from two independent pathways localized in different cellular compartments: the mevalonate (MVA) and the methylerythritol phosphate (MEP) pathways, both characterized by several enzymatic steps. The MVA pathway is used by most eukaryotes (including humans) as well as archaea, while the MEP pathway is found in most eubacteria. In plants, both pathways occur, with the MEP pathway typically operating in the plastids, while the MVA pathway is found in the cytosol: sterols (triterpenes) are produced via mevalonate, while hemi-(C5), mono- and di-terpenes, and carotenoids (tetraterpenes) are produced via MEP (Dewick, 2009).

In the MVA pathway, three molecules of acetyl-coenzyme A are used to form mevalonic acid. Two molecules first combine in a Claisen condensation to form acetoacetyl-CoA, and a third is incorporated via a stereospecific aldol addition, forming the branched-chain ester β-hydroxy-β-methyl glutaryl-CoA(HMG-CoA). The conversion of HMG-CoA to (*3R*)-mevalonic acid involves a two-step reduction of the thioester group to a primary alcohol. The six-carbon (*3R*)-mevalonic acid is converted to the

Phytochemicals for Health. https://doi.org/10.1016/B978-0-443-15366-2.00014-9

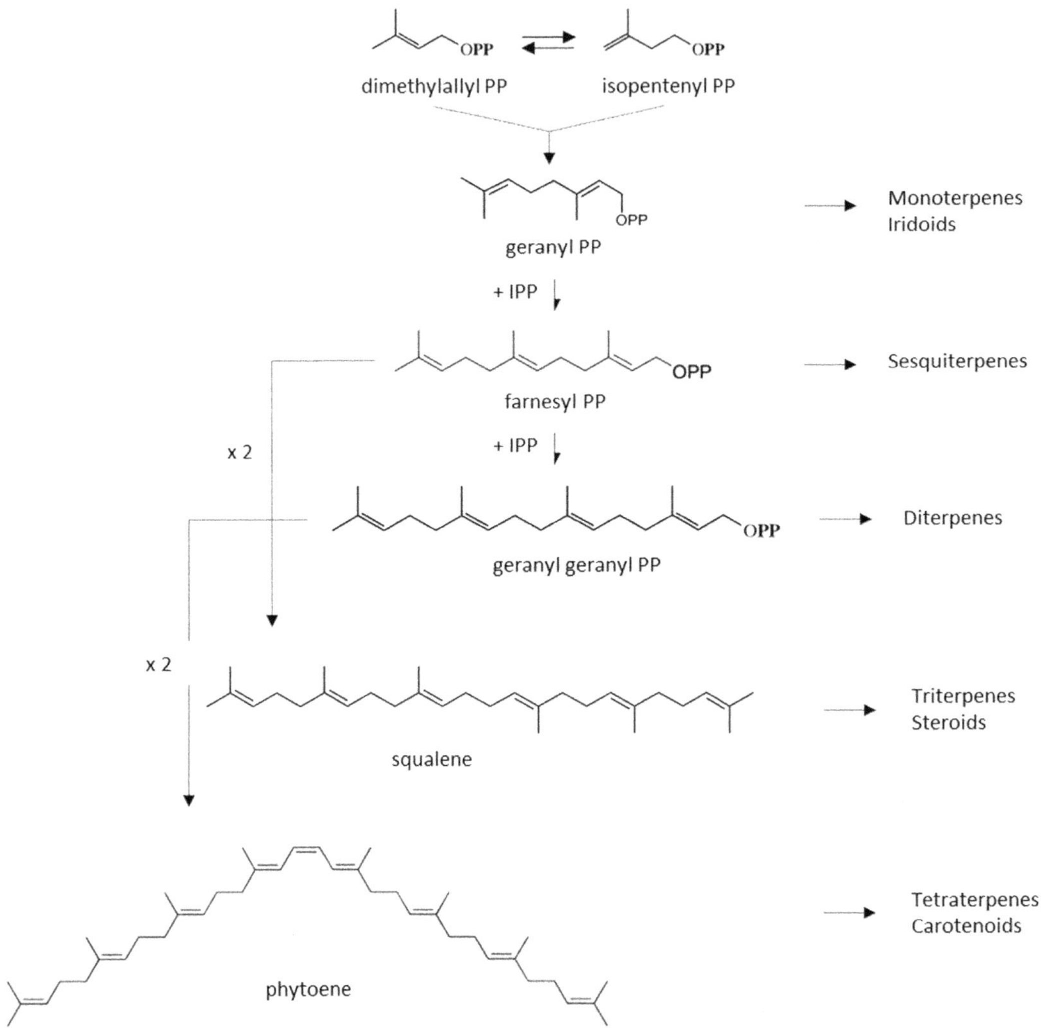

FIGURE 14.1

Classification of terpenes.

five-carbon phosphorylated isoprene units in a series of reactions, beginning with phosphorylation of the primary alcohol group, followed by decarboxylation and dehydration, eventually forming IPP. IPP is then isomerized into the other isoprene unit, DMAPP, by an isomerase enzyme that stereospecifically removes the pro-R proton (HR) from C-2 and incorporates a proton from water into C-4. Although the isomerization is reversible, the equilibrium is strongly on the side of DMAPP.

The MEP pathway, originally known as the "non-mevalonate pathway," begins with the biosynthesis of 1-deoxy-D-xylulose-5-phosphate (DXP) from the intermediates of the glycolytic pathway, pyruvic acid and glyceraldehyde-3-phosphate, with the loss of the pyruvate carboxyl. In the next step, DXP is converted to MEP. MEP is then converted to the cyclic intermediate methylerythritol-2,4-cyclodiphosphate (MEcDP) by three reactions, namely, cytidylation, phosphorylation, and cyclization. MEcDP is converted to hydroxymethylbutenyl diphosphate (HMBDP). Finally, HMDBP is reduced to IPP and DMAPP.

Monoterpenes (molecular formula $C_{10}H_{16}$) consist of 10 carbon atoms with two isoprene units. They are structurally divided into acyclic, monocyclic and bicyclic compounds. Some “irregular” monoterpenes arise from the nonclassic coupling of IPP and DMAPP. The compounds belonging to this class usually have a strong aroma and odor and are used in various fields. Mixtures of different monoterpene-based essential oils are used as fragrances to make perfumes and in other cosmetics. Most monoterpenes are biologically active, in particular, they show strong antibacterial activities.

Sesquiterpenes ($C_{15}H_{24}$) consist of three isoprene units and occur in linear, cyclic, bicyclic, and tricyclic forms. Sesquiterpenes also occur in the form of lactone rings. Artemisinin, a sesquiterpene lactone, is one of the most active compounds and can be found in the aerial parts of *Artemisia annua* L.

Diterpenoids belong to a versatile group of chemical constituents found in various natural sources and have the molecular formula $C_{20}H_{32}$ and four isoprene units. They have been found in some insects and various animal organisms but they are especially widespread in plants. Some of them are ubiquitous (e.g., the phytohormones gibberellins) whereas others have a more limited distribution.

Sesterpenes are composed of 25 carbon atoms with five isoprene units and the molecular formula $C_{25}H_{40}$. They occur naturally in fungi, marine organisms, insects, sponges, lichens, and insect protective waxes, whereas they are rarely biosynthesized in vascular plants.

Triterpenes usually contain 30 carbon atoms consisting of six isoprene units ($C_{30}H_{48}$). They are obtained via the squalene biosynthetic pathway. Triterpenes have many methyl groups and can be oxidized into alcohols, aldehydes, and carboxylic acids, making them complex and biologically differentiated. Triterpenes have many active sites for glycosylation, which converts them into another large class of compounds, saponins (triterpene glycoside).

Finally, the carotenoid group includes several hundreds tetraterpenoid compounds consisting of a sequence of eight isoprene units. Their multiple double bonds provide them with a characteristic chromophore that explains their yellow or orange color and tendency to get oxidized. The hydrocarbons are commonly called carotenes while the hydroxylated derivatives xanthophylls. They are either acyclic or comprise one or two penta- or hexacyclic rings at one or both ends. Fig. 14.2 reports the structures of some representative terpenes.

2. Extraction

As mentioned in the introduction, terpenes comprise a wide range of compounds with different physicochemical properties, such as molecular weight, polarity, volatility, etc. Therefore, the isolation of these compounds from plant material requires the use of specific techniques depending on the properties of the target compounds. In this section, the main methods for the isolation of terpenes and their subsequent analytical determination are described, with a particular focus on the most innovative and environmentally friendly methods. Fig. 14.3 illustrates the main techniques used for the extraction of the different classes of terpenes.

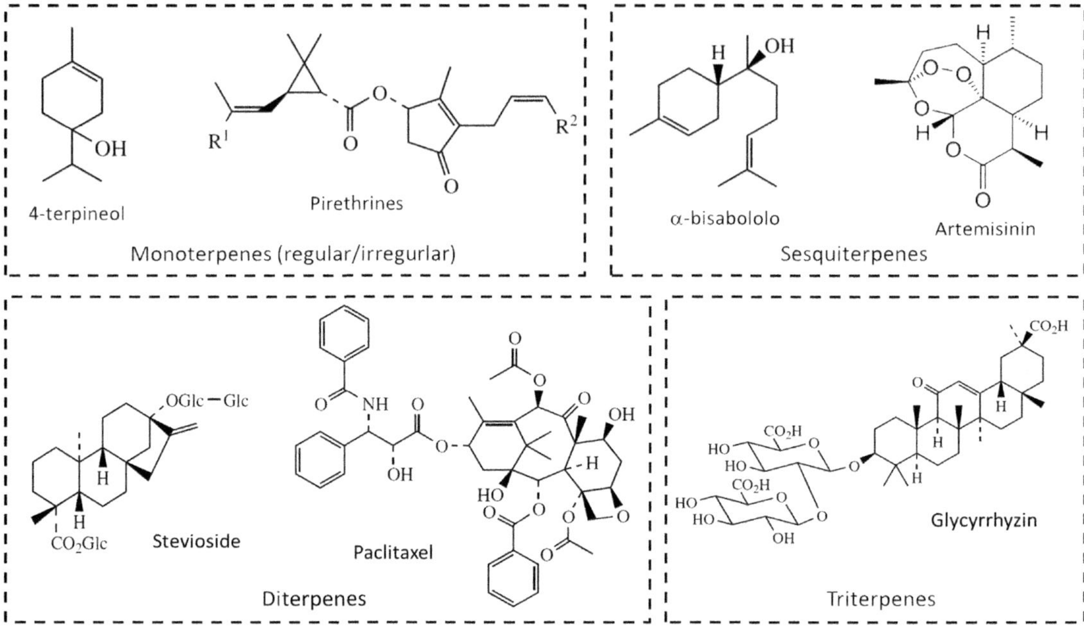

FIGURE 14.2

Representative structures of the main classes of terpenes.

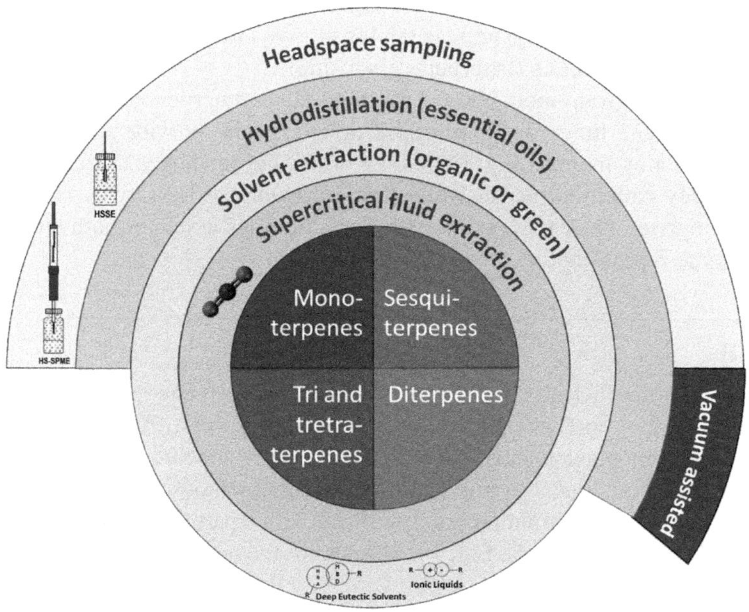

FIGURE 14.3

Scheme of the main techniques used for the extraction of the different classes of terpenes.

Conventional methods mainly include distillation for volatile and semivolatile terpenoids and solvent extraction (Bruneton, 2016). In solvent extraction, the choice of solvent is mainly based on the polarity of the target compounds. In addition to more traditional techniques such as Soxhlet extraction or maceration, technologies that increase extraction kinetics, such as ultrasound-assisted extraction (UAE) and microwave-assisted extraction (MAE), are increasingly being used.

2.1 Distillation

Regarding the isolation of volatile components, special attention should be paid to essential oils. According to ISO 9235:2021, essential oils (EOs) are "products obtained from a natural raw material of plant origin, by steam distillation, by mechanical processes from the epicarp of citrus fruits or by dry distillation after separation of the aqueous phase – if any – by physical processes", in particular, essential oils obtained by steam distillation are "essential oils which are obtained by steam distillation with the addition of water to the still (hydrodistillation) or without addition of water to the still (directly by steam)" while dry distilled essential oils are obtained by distillation of wood, bark, roots or gums without the addition of water or steam (Aromatic Natural Raw Materials —Vocabulary, 2013). A similar definition is provided by the European Pharmacopeia (European Pharmacopeia 11.0, 2023). Therefore, all products obtained by different methods and techniques (i.e., solvent extraction, head-space sampling, etc.) cannot be called EOs.

EOs are thus very complex mixtures of volatile (and semivolatile) constituents and, from a chemical point of view, include mainly mono-, sesqui- (and in some cases di-) terpenes and/or mono-, sesqui- and diterpenoids, together with some phenolic compounds (e.g., phenylpropanoids) derived from the shikimic acid pathway, or other minor components derived from the degradation of fatty acids or terpenes (Dewick, 2009).

EOs can therefore be placed within the more general framework of a plant's volatile fraction, which encompasses a range of other extraction approaches and techniques that yield samples that are representative (though almost always not comparable) to the volatiles that characterize the plant matrix under study, such as headspace, flavors, fragrances, aromas and extracts. The term "volatile fraction" generally defines a mixture of volatile compounds in a matrix of plant origin that can be sampled based on their ability to evaporate spontaneously and/or under appropriate conditions or techniques (Bicchi et al., 2011; Bicchi and Maffei, 2012). Thus, mono- and sesquiterpenes and terpenoids can be found in a volatile plant fraction.

2.2 Headspace sampling

Headspace (HS) extraction is nowadays one of the most commonly used techniques to characterize a plant's volatile fraction. It can be described as "green" because it is a solvent-free technique that aims to sample the gaseous or vapor phase in equilibrium (or not) with a solid or liquid matrix in order to characterize its composition (Sgorbini et al., 2014). The HS technique can be performed either in dynamic (D- HS) or static (S– HS) mode, depending on the instruments and procedures used.

D-HS is a nonequilibrium gas extraction process that typically consists of the following steps: (i) the continuous removal of volatiles released from a matrix into the empty volume of the sealed enclosure system in which they reside by an inert gas flowing either through or over the system; (ii) the concentration of analytes in a collection system; and (iii) the recovery of collected analytes by desorption using intensive treatments such as solvent back-extraction or thermal desorption. The use of D-HS for the analysis of plant volatiles dates back to the 1960s, when it was introduced by Wahlroos (Wahlroos and Nikkilä, 1966) and immediately used by Herout for the sampling of volatiles from flowers (Herout, 1967). It is currently one of the most applied approaches for the in vivo collection of biogenic volatile compounds (BVOCs), which, as mentioned earlier, include a wide range of mono- and sesquiterpenes (Cagliero et al., 2021a,b). The success of the technique is mainly due to its versatility, as different trapping systems and materials can be chosen. Although it is an environmentally friendly technique, the correct setup of D-HS requires relatively complex equipment and the standardization of several parameters if reliable sampling is to be achieved, while the desorption step involves either the use of solvents or high energy consumption (Liberto et al., 2017).

In S-HS, a liquid or solid sample reaches equilibrium with its vapor phase and the target analytes are transferred to the headspace according to their partition coefficients (Sgorbini et al., 2014). The term "static" means that there is no air flow in the sample chamber, so the headspace is representative of the sample emissions. The main advantages of this "one-step gas extraction technique" for the study of a volatile plant fraction are its simplicity, versatility, and ease of automation. Moreover, the S-HS technique makes a "picture" of the metabolites released from the sample. On the other hand, its major limitation is the lack of analyte enrichment or accumulation, resulting in low sensitivity. To overcome this limit, high-concentration capacity (HCC-HS) techniques have been developed since the 1990s. They act as a "bridge" between traditional static (S-HS) and dynamic head-space techniques (D-HS), as analytes are enriched either by sorption or adsorption from a vapor in static equilibrium in a solid or liquid phase. The main advantages of HCC-HS techniques include the fact that i) they are as simple, fast, easy to automate and reliable as S-HS, but at the same time have an analyte concentration capability often comparable to that of D-HS; ii) sampling times are almost always compatible with analysis times without affecting the final results; iii) they usually avoid the use of environmentally harmful solvents; and iv) they can often be combined with the analytical step online.

Although several HCC-HS techniques have been developed and applied for the extraction of volatile terpenes, HS-solid phase microextraction (HS-SPME) is undoubtedly the most successful (Zhang and Pawliszyn, 1993). This technique has become widely used in the analysis of volatile plant compounds and is now routinely employed. In HS-SPME, target analytes are transferred from matrix to a polymeric fiber coating in two sequential steps (matrix/headspace and headspace/fiber), and recovery is maximized when equilibrium is reached throughout the process. The time required to reach equilibrium depends on several factors, including the properties of analytes, matrix, and fiber coating. HS-SPME is highly sustainable because it is a solvent-free technique that can be automated. However, conventional HS-SPME shows its limits when it comes to reaching equilibrium with semivolatiles (such as diterpenoids). In this case, long extraction times and high extraction temperatures are required for the recovery of these analytes. These limitations can be overcome by vacuum-assisted HS-SPME (Vac−HS−SPME), a technique in which low pressure is applied during HS-SPME sampling. Vacuum facilitates the volatilization of semivolatiles by reducing the resistance in the thin gas film layer at the sample/headspace interface. Analytes are transferred to the headspace more rapidly and the time required to reach equilibrium is reduced. Sampling under vacuum greatly improves the extraction kinetics (velocity) of less volatile compounds, resulting in high extraction efficiency and sensitivity over time (Psillakis, 2017). For example, Vac−HS−SPME has been successfully used to distinguish the resins of *Boswellia* spp. on the basis of their volatile and semivolatile fractions (Capetti et al., 2020). The frankincense resins are indeed an unusual example of a natural raw material in which each class of terpenoids is present in similar proportions. The semivolatile diterpenoids (serratol, incensole, incensole acetate) are commonly used to distinguish samples of different species and origins. However, conventional HS-SPME requires long sampling times for medium to low volatile markers, while EO hydrodistillation is time-consuming and requires significant sample amounts. An environmentally friendly method combining Vac–HS–SPME and fast GC-MS with ionic liquid-based stationary phases has therefore been developed to distinguish incense resins. By sampling only 100 mg of frankincense resin for 30 minutes by Vac–HS–SPME it was possible to extract all three classes of terpenoids effectively, with a 3-fold improvement in the response for the semi-volatile diterpenes. Optimization of a fast GC method enabled baseline separation of all markers in 20 minutes. In addition, the sampling step, which is fully automated, was combined online with GC analysis, resulting in a total analysis time of 50 minutes. The method was successfully applied to commercial frankincense samples and showed a significant improvement in sustainability compared to previously reported methods. Recently, high-capacity (arrow) fibers have also been developed. They consist of a steel rod coated with a volume of sorbent material up to 10 times that of the corresponding conventional SPME fiber, while their dimensions remain compatible with a standard GC liner for thermal desorption of analytes by adjusting the injection port with a special kit. SPME Arrows have been used in the extraction of monoterpenes and sesquiterpenes from kava (*Piper methysticum* G. Forst) (Cheung et al., 2022).

Stir bar sorptive extraction (SBSE) is another HCC technique that can be used for the recovery of volatile terpenoids. It was introduced in 1999 by Sandra and Cramers and their groups (Baltussen et al., 1999) to improve the recovery of trace compounds from liquid samples. In 2000, the use of SBSE was extended to vapor phase sampling and named headspace sorptive extraction (Bicchi et al., 2000). In this technique, analytes are recovered with a thick polydimethylsiloxane (PDMS) film applied to a glass-coated magnetic stir bar suspended in the headspace of the matrix (the devices are known by the trade name Twister). After sampling, the PDMS stir bar is placed in a thermal desorption system, from where the sampled analytes are transferred on-line to a GC or GC-MS system. HS-SBSE has been used, for example, for the extraction of several monoterpenoids from the headspace of various aromatic plants (ex vivo and in vivo) for the determination of their enantiomeric composition (Ruberto et al., 2002). Another approach to improve the efficiency and sensitivity of sampling, especially in S-HS, is to increase the active contact surface of the sampling device through a more favorable surface-to-volume ratio of the extraction phase by using flexible, thin PDMS tapes in a technique called sorptive tape extraction (STE) or thin film microextraction (TFME). This technique can be applied in HS sampling as well as in direct contact and has been used, for instance, for the extraction of terpenes from aromatic plants (Bicchi et al., 2007) and for in vivo studies (Boggia et al., 2015).

2.3 Supercritical fluid extraction

Another very interesting and versatile technique for the extraction of terpenes (both volatile and nonvolatile) is supercritical fluid extraction (SFE). Supercritical fluids have been used to isolate natural products since the late 1970s, but for a long time, the applications were related to only a few products. SFE can be considered a green extraction technique because it uses a green and renewable solvent: carbon dioxide. Although other fluids can be used for plant extraction (e.g., propane), CO_2 has great advantages, such as nontoxicity, thermodynamic parameters that facilitate its use in a supercritical state (above 31.1°C and 7.4 MPa), and polarity that can be adjusted physically by changing the extraction temperature and pressure. It is also cheap and recognized as safe by the FDA and EFSA. Another advantage is that CO_2 is gaseous at room temperature and pressure, which makes (not volatile) analyte recovery very easy and provides solvent-free analytes. Also important is the ability of SFE with CO_2 to work at low temperatures with a nonoxidizing medium, allowing extraction of thermally labile or easily oxidized compounds. The main disadvantage of SC-CO_2 is its low polarity. This problem can be overcome by using polar modifiers (co-solvents) to change the polarity of the supercritical fluid and increase its solubilizing power toward the analyte of interest (Herrero et al., 2010). Extraction of terpenoids is one of the major targets driving SFE research (de Melo et al., 2014). Numerous articles address the SFE of volatile terpenes as an alternative to the distillation of essential oils, although SFE extract and essential oils should not be confused (see the detailed definition of essential oil) (de Melo et al., 2014). Other studies focused on SFE extraction of nonvolatile bioactive sesquiterpenoids such as artemisinin (Quispe-Condori et al., 2005) or parthenolide (Čretnik et al., 2005) and the coffee diterpenoids cafestol and kahweol (Barbosa et al., 2014; Araújo and Sandi, 2007). Triterpenic acids such as ursolic acid, oleanolic acid, betulinic acid, and betulonic acid have also been successfully extracted, especially from *Eucalyptus globulus* Labill (Domingues et al., 2012). Within the group of triterpenoids, also phytosterols (such as β-sitosterol, campesterol, and stigmasterol) have been the subject of several publications dealing with SFE. These compounds have been reported in SFE extracts of *Anoectochilus roxburghii* (Wall.) Lindl., amaranth (*Amaranthus* spp.), corn (*Zea mays* L.), eucalyptus, grape (*Vitis vinifera* L.), Kalahari melon (*Citrullus lanatus* (Thunb.) Matsum. & Nakai), mulberry (*Morus alba* L.), orange (*Citrus sinensis* (L.) Osbeck, syn *Citrus* × *aurantium* L.), pumpkin (*Cucurbita* spp.), sea buckthorn (*Hippophae rhamnoides* L.), soybean (*Glycine* spp.), and sunflower (*Helianthus annuus* L.) (de Melo et al., 2014). However, the largest class of terpenes extracted by SFE are probably tetraterpenoids, with a very large number of studies focusing on the extraction of β-carotene or lycopene (de Melo et al., 2014).

2.4 Unconventional solvents

Besides SFE, nonvolatile terpenes are mainly isolated by conventional extraction methods which may require toxic solvents and/or high consumption of energy and plant material. However, in the last decades, great efforts have been made to improve the sustainability of extraction for analytical purposes according to the Green Analytical Chemistry (GAC) guidelines (Gałuszka et al., 2013). In this sense, there are three main strategies to improve the characteristics of conventional sample preparation methods that can be used for the extraction of terpenes from plant matrices: (i) miniaturization of the analytical method, which can be achieved by the development of microextraction techniques that preconcentrate the analytes and thus reduce the consumption of reagents and samples, and (ii) the development of environmentally friendly extraction phases (solvents and sorbents) that either reduce or, if possible, eliminate the use of toxic organic solvents. In particular, research into new environmentally friendly solvents and sorbents has become one of the most important topics in the field of sample preparation, and several examples related to the extraction of terpenes (Cagliero et al., 2022; Mastellone et al., 2021).

Ionic liquids (ILs) are a class of molten salts with melting points at or below 100°C. In most cases, they consist of an organic cation and an organic or inorganic anion and are characterized by high thermal stability, low vapor pressure, and varying viscosity, conductivity, and miscibility in different solvents. Unlike conventional organic solvents, ILs are characterized by negligible vapor pressure at room temperature, which prevents solvent loss to the atmosphere and reduces their environmental impact. ILs have been successfully used as solvents in extraction and separation processes for bioactive compounds (including terpenoids), in simple solid-liquid extractions (SLE), and in microwave-assisted (MA) and ultrasound-assisted (UA) extractions. The high polarity and high thermal stability of ILs are crucial for their fast and uniform heating under microwave irradiation, which is why they can be considered excellent microwave absorbers. They can also be used in UAE, especially to extract thermolabile compounds at room temperature. The tunability of their structure and, consequently, of their properties mean that ILs can be used in metabolomics studies as well as for selective extraction of specific compounds. For example, ILs can be used in metabolomics studies for the simultaneous determination of volatile and nonvolatile components (Li et al., 2018; Liu et al., 2019), which drastically reduces the time and reagent consumption. An IL-mediated microwave-assisted hydrodistillation with a liquid-liquid extraction was used for the distillation of EO and the simultaneous extraction of nonvolatile flavonoids from *Helichrysum arenarium* (L.) Moench (Liu et al., 2019). Similarly, ultrasound- and microwave-assisted simultaneous distillation extraction with ionic liquids can be applied to increase the extraction of volatile and nonvolatile components. This approach was successfully applied to the determination of flavonoids and volatiles in sea buckthorn leaves (*Hippophae rhamnoides* L.), dramatically reducing extraction times and reagents consumed, and achieving higher analyte recovery and similar volatile composition compared to conventional hydrodistillation (Li et al., 2018). Thanks to their versatility, ILs can be used in liquid phase extraction of natural products in various (micro)extraction modes. For example, they can form aqueous biphasic systems (ABSs) when an aqueous IL solution is mixed with inorganic salts that promote phase separation and increase analyte enrichment. This approach has been used, for example, for the extraction of ginsenosides (triterpenoid saponins) from the flower buds of *Panax ginseng* C.A. Mey (Liang et al., 2019). Alternatively, dispersive liquid-liquid microextraction (DLLME) can be used if the samples are in liquid form. In this case, hydrophobic ILs are dispersed in the water solution using a dispersing solvent and then recovered and analyzed. Again, ultrasonic energy can be used to accelerate the mass transfer process and improve the extraction efficiency. Recently, irregular monoterpene pyrethrin residues were detected in herbal teas by an ultrasound-assisted, temperature-controlled approach IL-DLLME coupled to HPLC-UV analysis (Zhang et al., 2020).

Deep eutectic solvents (DESs) are another interesting class of green solvents. This family of phases consists of systems formed from a hydrogen bond acceptor (HBA) and a hydrogen bond donor (HBD). As the term "eutectic" implies, they are characterized by lower melting points than their individual components. They are usually associated with ILs because they share similar physicochemical properties with them, including low vapor pressure, low flammability, and high thermal stability. However, unlike ILs, DESs are characterized by ease of production, low cost of their synthetic raw materials, and, more importantly, lower toxicity and higher biodegradability because they can be prepared by incorporating ionic or neutral species (Farooq et al., 2020). In particular, HBAs and HBDs of natural origin, such as amines, sugars, carboxylic acids, and polyalcohols, are attractive source for the preparation of biodegradable and biocompatible DES. These natural deep eutectic solvents, namely NADES, have found promising applications in the extraction of natural products, including terpenes (Farooq et al., 2020). Some monoterpenes can also be selected as HBDs or HBAs for the preparation of hydrophobic NADES and used for the extraction of other terpenes. For example, menthol was used in combination with 1-propanol for the extraction of taxanes from the needles of *Podocarpus chinensis* Wall. ex Benn. (syn. *Taxus chinensis* Roxb.), with higher efficiency compared to conventional solvents (e.g., MeOH or EtOH) (Fan et al., 2020). Interestingly, the use of DES with a water content of 80% resulted in a higher extraction efficiency than when a conventional hydro-alcoholic solvent was used, despite the apolar nature of the taxanes. The same HBA in combination with different carboxylic acids was successfully used in the extraction of four cannabinoids from the inflorescences of *Cannabis sativa* L. (Křížek et al., 2018). In addition, the ability of DES to extract volatile compounds has been documented. For example, a rich profile of volatile compounds was detected by headspace SPME of DES extracts of peppermint leaves with even better results than conventional methods (Jeong et al., 2018).

3. Quantification

As discussed in detail in previous chapters, an analytical technique (or method) is fully acceptable as a reference standard only if it can provide reliable quantitative data throughout the analytical process (i.e., sample preparation, analysis, and data processing). Unfortunately, however, relative or absolute quantification is still considered ambiguous, mainly because, if performed correctly, it is sometimes quite complex and time-consuming, especially when the volatile fraction is involved (Bicchi et al., 2008a,b; Cicchetti et al., 2008). This section addresses the basic approaches to the quantification of volatile terpenes.

Regardless of the method used to collect the volatile fraction (steam distillation, hydrodistillation, mechanical methods, or HCC-HS), some basic guidelines must be followed to obtain reliable quantitative results: (a) the variability of a plant matrix requires careful standardization, both in collection and sample preparation; (b) an adequate number of samples must be analyzed to obtain a representative composition of the volatile fraction of the species under study (i.e., the averaged composition of at least three samples from distinctly different populations).

3.1 Quantification approaches in essential oil analysis

In the analysis of EO, quantification is a subject that is not easy to handle, not only because the identification of components is generally prioritized to the detriment of quantification, but also because the approach to it is often ambiguous. This issue has been critically discussed by Bicchi et al. (2008a,b). Due to the complexity of EOs and the various methods used to quantify their components, there can be no single absolute approach to quantification. The most widely used approaches to EOs are: (a) relative percentage abundance; (b) percentage abundance normalized by internal standards; (c) "absolute" or true quantification of one or more constituents; and (d) quantification by a validated method.

Relative percentage abundance is the most commonly used approach, although it is often misused, especially when comparing compositions within a set of EO samples. Relative percentage abundance results are only accurate when used to evaluate the relative ratios of components within the same sample. If a group of EOs is to be compared, the raw data should first be normalized against an internal standard and the percentage abundance should be calculated against a fixed number of selected components that serve as markers and are usually common to all samples studied. In addition, normalized percentage data can also be used for profile analysis. This is a statistical method that allows us to discriminate or classify samples within a set based on the abundance of a specific number of markers characteristic of the EO under study. The most widely used approach is multivariate analysis, particularly principal component analysis (PCA), a method that can explain differences within a set of samples characterized by an appropriate number of components (variables) by linearly combining those that explain most of the variability. Normalized percentage abundance is not sufficient to solve all quantification problems with respect to an EO. In some cases, an EO must be characterized by determining the concentration or absolute amount of one (or more) of its markers. An "absolute" quantification is therefore required. The analyte(s) under investigation must be quantified by an internal (or external) standard after standardization of the chromatographic results, using a calibration curve generated in the operative concentration range with the pure standard of the analyte under investigation. Since pure standards are not always commercially available or difficult to isolate, an accepted compromise is to use compounds belonging to the same class (hydrocarbons, aldehydes, alcohols, esters, etc.) of the analyte under study and having a structure as similar as possible to the analyte. A more accurate alternative, when using a Flame Ionization Detector (FID), is to use predicted FID Relative Response Factors (RRFs), based on combustion enthalpies and molecular structure. This approach allows quantification without external standards. It was introduced by De Saint Laumer et al. (2010) and applied to various quantification problems, e.g., by Tissot et al. (2012) for quantification of markers of bergamot EO, purity assessment of unstable or reactive analytes, and profiling of dynamic mixtures stored at different temperature and pH conditions. The predicted FID RRFs have also been validated by Filippi et al. (2013) for GC×GC-FID applications, namely for the characterization of vetiver EOs of different geographical origins by GC×GC-MS and GC×GC-FID and for the study of the qualitative, quantitative and enantiomeric determination of *Mentha* and *Lavandula* species by parallel dual secondary column dual detection GC×GC (Sgorbini et al., 2015a,b).

3.2 Quantification approaches for headspace sampling

The issue of quantification is even more complex when it comes to headspace sampling (Bicchi et al., 2008a,b; Sgorbini et al., 2019). One of the most common practices associated with the characterization of the volatile fraction is cross-sample comparison. Analytes and/or informative markers are identified in each headspace profile of a sample and compared using quantitative indicators generally based on chromatographic peak areas (raw areas, percentage area) or internal standard (IS) normalization (normalized area). These approaches are based on relative/normalized responses of the analyte and, although accepted by the scientific community for various applications, may be inaccurate or misleading when considered as concentration indicators for the target compound(s). These indicators are suitable for cross-sample analysis of solid or liquid samples, provided that the effect of the condensed phase (matrix effect) on the release of the analyte into the HS is comparable or known. In many cases, the matrix effect can have a dramatic effect on analyte recovery and lead to erroneous conclusions. For solid samples, such as plants, all parameters that cause the volatile release phenomenon known as the matrix effect must be clearly characterized or at least verified in order to obtain reproducible, accurate and, most importantly, meaningful results. Appropriate techniques capable of compensating the matrix effect are referred to as quantification approaches and are generally based on external or internal calibration with authentic standards or stable isotopologues of the target analytes. The most widely used techniques are: i) standard addition (SA) by spiking the sample with known incremental amounts of the analyte(s), ii) stable isotope dilution assay (SIDA), which is a specific application of SA, iii) multiple headspace extraction (MHE). Whichever approach is used, HS linearity conditions must be established during sampling to ensure accurate quantification and avoid errors. This condition is verified when the amount of plant material under the sampling conditions applied is sufficient to release the minimum amount of analyte appropriate to achieve the sensitivity of the method without saturating the HS. Linearity is easy to achieve for trace and subtrace target analytes, or at least in those studies where the focus is on a few compounds, but it becomes difficult when multiple analytes are quantified.

The SA method is one of the most popular approaches for quantification of headspace constituents (Souza-Silva et al., 2015; Uribe and Cosio, 2021). It consists of a series of measurements in which the original sample and sample aliquots spiked with increasing and known amounts of reference compounds are subjected to analytical determination. There are two calibration approaches for SA: "single-point" calibration and "multi-point" calibration. Single-point calibration involves analysis of the original sample and the sample spiked with a known amount of the analyte standard. The standard must be added in appropriate amounts to obtain a signal of the analyte under investigation with an intensity increased by at least 20% compared to the original sample. The limitations of this approach are possible discrepancies in analyte recovery or linearity. Multipoint calibration overcomes these drawbacks thanks to a two-step procedure: (a) analysis of the original sample and spiked with several increasing amounts of the investigated analyte(s) to establish a linear calibration equation, (b) extrapolation of the amount of analyte originally present in the sample from the calibration equation. Although the single-point approach is very useful for routine work, when analyzing multiple samples, the addition of multiple standard amounts (up to six levels) is recommended to improve accuracy. The main limitations of SA are (a) the large number of measurements required for a specific calibration for each analyte, (b) the difficulty of obtaining analyte standard(s), and (c) the inhomogeneous distribution of the added standard.

SIDA is a method in which the internal standard (IS) is a labeled stable isotope of the target analyte. This ensures that the standard and analyte have the same chemical and physical properties, but different molecular weights, which can be detected using MS (or better GC-MS). It was introduced by the groups of Schieberle and Grosch (Schieberle and Grosch, 1987) for the reliable quantification of analytes (especially at trace levels) in liquid samples in the food sector and extended to HS applications by Steinhaus et al., in 2003 for the accurate quantification of linalool enantiomers in beer after HS-SPME sampling (Steinhaus et al., 2003). An analyte stably labeled with an appropriate isotope, preferably in a diagnostic position, is by definition an ideal IS. Any fragmentation of the target analyte is mirrored in the isotopologue. This symmetrical behavior overcomes a drawback of the conventional IS, when the target analyte/IS ratio may change during sample clean-up, especially when using a multistep approach, due to their different chemical structures. With SIDA, the target analyte can be easily quantified by spiking the sample with a known amount of a labeled standard after determining a correction factor between the ratio of isotopologues and the intensities of their selected ions by (SIM)-MS. The ability to select ions that are diagnostic for the analyte(s) under investigation makes SIDA less time consuming than SA and suitable for highly specific quantification. The main critical issues are: (a) the limited commercial availability of labeled standards, (b) the stability of labeling, which must be maintained throughout the analytical process to keep the isotope ratio constant, and (c) the same ratio of native analyte to isotope analog must be maintained in all sample compartments.

The multiple headspace extraction (MHE) methods was introduced by Suzuki et al. (1970) and McAuliffe (1971) and developed and made available for routine analysis thanks to the studies of Kolb (Kolb and Ettre, 2006). It is a dynamic gas extraction performed stepwise at regular time intervals. Kolb and Ettre referred to it as "a repeated liquid extraction of a sample in a separatory funnel: in each step, part of the analyte present is removed until no analyte remains in the original sample (Kolb and Ettre, 2006)." MHE also removes a portion of the headspace after the initial sampling, so a new equilibrium must be established, with a lower amount of analyte yielding a less intense peak in the subsequent HS-GC analysis. The total amount of a volatile(s) can therefore be extracted if a suitable number of samplings are performed. Thus, by processing an appropriate amount of matrix, it is possible to create an exponential decay in analyte amount and peak areas and determine the total area in the sample by summing the areas of the target peaks at each "extraction". Thus, the total area of the analyte measured in this way is not affected by the sample matrix, i.e., MHE avoids the matrix effect. MHE can be performed as an external standard calibration and consists of three experimental steps: (i) exhaustive extraction of analytes from calibration solutions within a concentration/quantity range corresponding to the concentrations of the real sample, (ii) exhaustive extraction of analytes from representative samples (solid, liquid or heterogeneous) to define HS linearity limits, (iii) application of the MHE procedure to the samples of interest. In the first two steps, a cumulative instrumental response function is established after repeated successive extractions of HS from the same aliquot of calibration solutions or representative samples. In routine work, it is not necessary to continue with extractions until the analyte has been completely removed from the matrix because the MHE follows a logarithmic path. However, four to six consecutive extractions are recommended to verify the completeness of the extraction process for all analytes of interest.

The total area of the analyte(s) under investigation is determined with the following Eq. (14.1):

$$A_T = \sum_{(i=1)}^{\infty} A_i = \frac{A_1}{(1 - e^{-q})} = \frac{A_1}{(1 - Q)} \tag{14.1}$$

where A_1 is the analyte area after the first analysis; A_T is the total area of the investigated analyte, A_i is the peak area obtained in the *i*th extraction, -q ($Q = $ e-q) is a constant representing the exponential decline during the stepwise MHE procedure.

The term q can be obtained by plotting the natural logarithm of the chromatographic peak areas as a function of the number of extractions. From this, a linear regression Eq. (14.2) can be calculated as :

$$\ln A_i = -q(i-1) + \ln A_1 \tag{14.2}$$

where i is the number of extraction steps, b is the intercept on the y-axis, and a is the slope.

Q (e^{-q}) is analyte- and matrix-dependent and is a constant for those samples exerting comparable matrix effects. The Q-values can be used to evaluate the actual release of the target analyte under certain conditions.

When applied to calibration solutions, MHE provides data for external calibration. The curves can be used to estimate the amount of analyte in the sample using a simplified procedure in which the analyte response after the first extraction step (A_1) is sufficient for accurate quantitation. Fig. 14.4 illustrates the MHE procedure and operative steps.

In this sense, MHE has been used to quantify the volatile fraction of herbs and spices, for quality control, and to quantify aromatic markers (in many cases terpenes) (Sgorbini et al., 2015a,b). The results highlight the reliability of MHE when used for direct quantification of volatile markers in solid matrices by showing that an average Q value can be used to quantify one or more analytes with an automated extraction (experiment) for each sample. This is particularly important when analyzing a large number of samples of the same homogeneous matrix. In the study, MHE-SPME was also

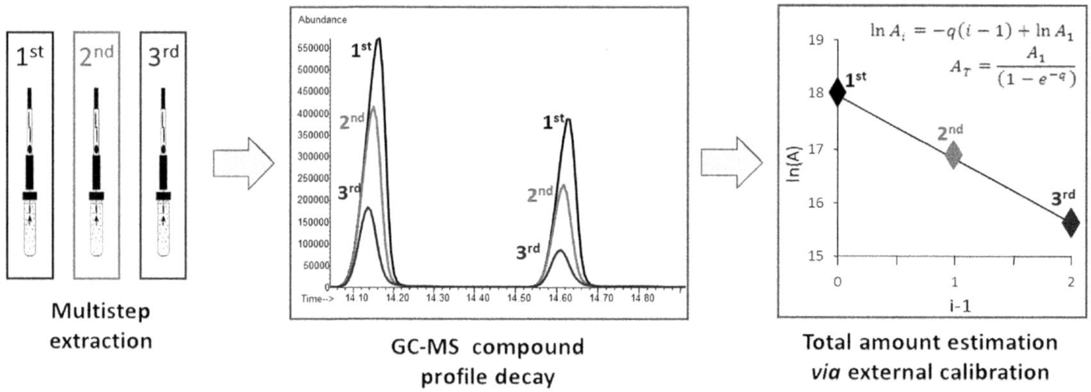

FIGURE 14.4

Multiple headspace extraction (MHE) by solid phase microextraction (SPME) operative steps.

successfully combined with nonseparative methods to accelerate control analysis when one or more markers need to be quantified from solid matrices, provided that they have specific diagnostic ion(s) throughout the MS fingerprint.

3.3 Determination of enantiomeric distribution

Another aspect that must be considered for accurate characterization of a plant matrix is the enantiomeric configuration of its chiral components (including terpenes), because it is well-known that different enantiomers can have different biological properties. While for the more complex terpenoids (sesqui- and diterpenoids) their bioguided synthesis usually results in a single enantiomer, for monoterpenes the enantiomer distribution can vary depending on plant species and environmental conditions. For this reason, this evaluation is essential. Chiral recognition can only be performed when the two enantiomers are separated. In this respect, derivatized cyclodextrins (CDs) are the most popular chiral selectors for enantioselective GC analysis (ES-GC) of monoterpenes.

Enantiomeric distribution is usually expressed as enantiomeric excess (ee), enantiomeric composition (ec) or enantiomeric ratio (er) (Cagliero et al., 2021a,b).

Enantiomeric excess, also known as enantiomeric purity, expresses the extra abundance of one enantiomer over the other and is defined as:

$$ee = \frac{E_1 - E_2}{E_1 + E_2} \tag{14.3}$$

where E_1 and E_2 are the areas of the enantiomers, E_1 being the major enantiomer; *ee* ranges from 0 for racemic mixture to one for pure E_1. In routine practice, *ee* is often expressed as a percentage.

$$ee\% = \frac{E_1 - E_2}{E_1 + E_2} \times 100 = \%E_1 - \%E_2 \tag{14.4}$$

Enantiomeric composition (*ec*) is defined as the fraction of one enantiomer in a mixture:

$$ec = x_E = \frac{E_{1(2)}}{E_1 + E_2} \tag{14.5}$$

In this case too, in routine, *ec* is in general expressed as a percentage:

$$ec\% = x_E = \frac{E_{1(2)}}{E_1 + E_2} \times 100 \tag{14.6}$$

Lastly, the enantiomeric ratio, *er*, is defined as:

$$er = \frac{E_1}{E_2} \tag{14.7}$$

where E_1 is the area of the major enantiomer; *er* extends from $er = 1$, for a racemic mixture, to $er = \infty$, when only E_1 is detected.

In routine analysis, a correct determination of these parameters preferably requires the baseline separation (i.e., resolution ≥ 1.5) of the two enantiomers.

4. Biological activity

In this section, the main biological activities of terpenes are described, with particular attention to the traditional uses of plants in Europe that mainly contain this class of compounds.

4.1 Biological activity of essential oils and exudates containing terpenes

As mentioned above, essential oils are mainly characterized by volatile monoterpenes and sesquiterpenes. Their activity is sometimes confused with that of the plant from which they were distilled. In addition, it is important to remember that essential oils are mixtures of different compounds, and therefore it is very difficult to study and describe their biological and pharmacological effects. In general, they show antiseptic activity, spasmolytic and sedative properties, and irritant properties, but it is important to consider their potential (acute) toxicity, especially when ingested orally (Bruneton, 2016).

Table 14.1 reports the main activities ascribed to the most used essential oils for which the Committee on Herbal Medicinal Products (HMPC) has identified traditional use in Europe (European Union (EU) Herbal Monograph, 2023) and that contains terpenes and terpenoids as main markers.

Terpenes are also found in plant exudates and especially in oleoresins, as produced by the Pinaceae family, which consist mainly of nonvolatile (resinous) and volatile terpenoids, as well as in gum resins, which consist mainly of resinous compounds and gums (such as gutta gum), and in gum oleoresins, which consist mainly of resinous compounds, gums, and certain amounts of volatile compounds (myrrh, olibanum, galbanum). As mentioned before, frankincense resin contains each terpene class

Table 14.1 Representative essential oils with a traditional or well-established used in Europe.

Botanical name of the plant from which the EO is obtained	Family	Main part of the plant distilled	Main terpene/ terpenoids markers	Traditional or well-established uses (indicated by the HMPC)
Carum carvi L.	Apiaceae	Fruit	Monoterpene and monoterpenoids (limonene carvone)	Relief of problems of the digestive system such as bloating and flatulence.
Eucalyptus globulus Labill., *Eucalyptus polybractea* R.T. Baker, *Eucalyptus smithii* R.T. Baker	Myrtaceae	Leaf	Monoterpenoids (1,8-cineole)	Relief of cough associated with the common cold and symptoms of localized muscle pain
Juniperus communis L.	Cupressaceae	Galbuli	Monoterpenes (a-pinene, b-pinene, myrcene, sabinene)	Increase the amount of urine to achieve flushing of the urinary tract as an adjuvant in minor urinary tract complaints. Symptomatic relief of digestive disorders such as dyspepsia and flatulence. Adjuvant in the relief of minor muscular and articular pain.
Lavandula angustifolia Mill.	Lamiaceae	Aerial part	Monoterpenoids (linalool, linalyl acetate)	Relief of mild symptoms of mental stress and exhaustion, and to aid sleep.
Matricaria recutita L.	Compositae	Flower	Sesquiterpenoids (a-bisabolol, bisabolol oxide A and B)	Treatment of irritation of the skin and areas around the anus and genitals, once serious conditions have been excluded by a medical doctor.
Melaleuca alternifolia (Maiden & Betche) Cheel	Myrtaceae	Leaf	Monoterpenoids (4-terpineol, 1,8-cineole)	Treatment of small superficial wounds, insect bites and small boils (furuncles and mild acne), for the relief of itching and irritation in mild athlete's foot, and for the treatment of minor inflammation of the lining of the mouth.

Mentha x piperita L.	Lamiaceae	Leaf	Monoterpenoids (menthol, menthone, 1,8-cineole)	Preparations that are taken by mouth relieve minor belly spasms, flatulence and belly pain, especially in patients with irritable bowel syndrome. Preparations that are rubbed on the skin of the forehead and temples can be used for relief of mild headaches. Digestive function. Regular gastrointestinal motility and elimination of gases. Functioning of the first respiratory tract.
Rosmarinus officinalis L.	Lamiaceae	Leaf	Monoterpenes and monoterpenoids (α-pinene, 1,8-cineole, linalool, camphor)	Relief of minor muscular and articular pain and in minor peripheral circulatory disorders.
Thymus vulgaris L., *Thymus zygis* L.	Lamiaceae	Aerial part	Monoterpenes and monoterpenoids (γ-terpinene, *p*-cymene, thymol)	Treatment of chesty (phlegm-producing) cough associated with colds, and as an application to the skin or as a bath additive for the relief of symptoms of cough and colds.
Valeriana officinalis L.	Caprifoliaceae	Root	Monoterpenoids (bornyl acetate, myrtenyl isovaleriate and acetate; sesquiterpenoids (kessane derivatives)	Relief of mild symptoms of mental stress and to aid sleep.

(mono-, sesqui-, di- and triterpenoids) in fairly similar proportions (Baldovini and Filippi, 2017). The boswellic acids, which are pentacyclic triterpenoidic acids, are nonvolatile markers and have been shown to possess antiinflammatory properties, while the essential oil from frankincense resins is a common raw material in perfumery.

4.2 Biological activity of different classes of terpenes and plants containing terpenes

With respect to the other classes of terpenes, some general considerations may be made (Bruneton, 2016). Iridoids usually have a defensive function for the plant and are involved in plant-animal interaction. Their use from a pharmacological point of view is limited; some of them have an anti-inflammatory effect, especially when applied topically. They can be components of allopathic medicines (valerian) or phytotherapeutic products (devil's claw or olive). Pyrethrins are the only irregular monoterpenes that are used. They are insecticides that are lox toxic to humans and other mammals. The sesquiterpenoid lactones (which are very common in the Compositae family) are rarely used therapeutically, except for the artemisinin derivatives (which are antimalarials). They are often responsible for the bitterness of some plants and have been associated with possible contact allergies. From a therapeutic point of view, the most important diterpenes are the tricyclic terpenes of yew. These compounds, known as taxanes, have been known for their toxicity since ancient times, but two of them (paclitaxel and docetaxel) are used in chemotherapy to treat advanced ovarian cancer and other various cancers such as breast and lung cancer. Other health properties of diterpenes include the antioxidant properties of phenolic diterpenes from some species of the Lamiaceae family such as rosemary and sage, and the sweetening properties of stevioside from *Stevia rebaudiana* Bertoni. Triterpenes and steroids form a group of secondary metabolites of major importance for therapeutic and industrial applications. Saponins usually have hemolytic properties due to their interaction with erythrocyte membrane sterols. Several saponin-containing drugs are used for their antiinflammatory and anti-edematous effects, others are used for their antitussive and expectorant properties, as well as for their analgesic properties, modulation of cholesterol uptake, immune regulation, and cell-protective effects. Apart from their pharmacological potential, saponins are also of interest because some of them (such as glycyrrhizin) have sweetening properties. Cardioactive glycosides have a steroidal genin of the cardenolide or bufadienolide type attached to a short sugar chain (usually an oligosaccharide). They have a narrow margin of safety but are useful in the long-term management of cardiac insufficiency since they act on the force and rate of cardiac contraction and electrophysiological properties. Few other steroids and triterpenes have found therapeutic applications. Interest in carotenoids is diverse: β-carotene and closely related compounds are precursors of vitamin A, they are thought to have preventive effects against degenerative diseases, they may be used in the treatment of photosensitization because they interfere with photooxidation, and they are used in the pharmaceutical and food industries as natural, effective, and nontoxic colorants.

Table 14.2 lists representative plants that contain terpenes and terpenoids as major markers and for which the Committee on Herbal Medicinal Products (HMPC) has identified traditional use in Europe (European Union (EU) Herbal Monograph, 2023).

Table 14.2 Representative plants containing terpenes and terpenoids as major compounds with a traditional or well-established used in Europe.

Plant	Main part of the plant used	Main terpenes/terpenoids markers	Traditional or well-established uses (indicated by the HMPC)
Aesculus hippocastanum L. (horse chestnut)	Seed	Triterpenoid saponins based on two aglycones–protoaescigenin and barringtogenol C	Relief of symptoms of discomfort and heaviness of legs related to minor venous circulatory disturbances, relief of signs of bruises, such as local edema and hematoma
Calendula officinalis L. (marigold)	Flower	Triterpenoid saponins Triterpene alcohols Carotenoids (predominately lutein and zeaxanthin)	Symptomatic treatment of minor inflammations of the skin (such as sunburn) and as an aid in healing of minor wounds, symptomatic treatment of minor inflammations in the mouth or the throat
Centella asiatica (L.) Urb. (gotu kola)	Aerial part	Pentacyclic triterpenoids (asiaticoside and madecassoside)	Aid in healing of minor wounds
Commiphora myrrha Engler. (myrrh)	Gum-resin	Sesquiterpenoids (furano-sesquiterpene)	Minor ulcers and inflammation in the mouth, treatment of minor wounds and small boils
Eleutherococcus senticosus Maxim. (Eleutherococcus)	Root	Triterpenoid saponins	Symptoms of asthenia such as fatigue and weakness
Gentiana lutea L. (yellow gentian)	Root	Secoiridoids (gentiopicroside, amarogentin)	Temporary loss of appetite and for mild stomach and gut complaints such as indigestion

Continued

Table 14.2 Representative plants containing terpenes and terpenoids as major compounds with a traditional or well-established used in Europe.—cont'd

Plant	Main part of the plant used	Main terpenes/terpenoids markers	Traditional or well-established uses (indicated by the HMPC)
Glycyrrhiza glabra L.; *Glycyrrhiza inflata* Batalin; *Glycyrrhiza uralensis* Fisch. (Liquorice)	Root	Triterpenoid saponins (mostly glycyrrhizin)	Relief of digestive symptoms including burning sensation and dyspepsia, expectorant in cough associated with cold
Grindelia spp. (gum plant)	Aerial part	Diterpenic acids (grindelic acid and derivatives) Triterpenoid saponines	Relief of cough associated with cold
Harpagophytum procumbens DC ex Meiss. (Devil's claw)	Root	Harpagoside (iridoid)	Relief of minor joint pain
Hedera helix L. (ivy)	Leaf	Triterpenoid saponins (hederagenin, oleanolic acid and bayogenin as main aglycones)	Expectorant in case of productive cough
Olea europaea L. (olive)	Leaf	Secoiridoids (oleuropein)	Water elimination through the kidneys in mild cases of water retention
Panax ginseng C.A. Mey. (ginseng)	Root	Triterpenoid saponins (ginsenosides)	Symptoms of asthenia such as fatigue and weakness
Rosmarinus officinalis L. (rosemary)	Leaf	Monoterpenes Diterpenoids (carnosol, carnosic acid, rosmariquinone) Triterpenoids (ursolic and oleanolic acid)	Relief of minor muscular and articular pain and in minor peripheral circulatory disorders

Ruscus aculeatus L. (butcher's broom)	Rhizome	Steroidal saponins (derivatives of ruscogenin)	Relief of symptoms of discomfort and heaviness of legs related to minor venous circulatory disturbances, relief of itching and burning associated with hemorrhoids
Salvia officinalis L. (sage)	Leaf	Mono and sesquiterpenes Phenolic diterpenes (e.g., rosmanol) Triterpenoids (ursolic acid, oleanolic acid, α-amyrin and β-amyrin	Relief of mild dyspeptic complaints, excessive sweating, inflammations in the mouth or the throat, minor skin inflammations
Tanacetum parthenium Sch. Bip. (feverfew)	Aerial part	Parthenolide (sesquiterpenoid lactone)	Prevention of migraine
Urtica dioica L. (nettle)	Root	Sterols (β-sitosterol)	Relief of lower urinary tract symptoms related to benign prostatic hyperplasia
Valeriana officinalis L. (valerian)	Root	Iridoids (valtrate, isovaltrate, acevaltrate) Sesquiterpenoids (valerenic acid and derivatives, valeranone	Relief of mild nervous tension and sleep disorders

References

Araújo, J.M.A., Sandi, D., 2007. Extraction of coffee diterpenes and coffee oil using supercritical carbon dioxide. Food Chem. 101 (3), 1087–1094. https://doi.org/10.1016/j.foodchem.2006.03.008.

Aromatic Natural Raw Materials —Vocabulary, vol. 9215. ISO, 2013.

Baldovini, N., Filippi, J.J., 2017. Natural Fragrant Raw Materials Springer Handbooks. Springer, France, pp. 11–12. https://doi.org/10.1007/978-3-319-26932-0_3.

Baltussen, E., Sandra, P., David, F., Cramers, C., 1999. Stir bar sorptive extraction (SBSE), a novel extraction technique for aqueous samples: theory and principles. J. Microcolumn Sep. 11 (10), 737–747. https://doi.org/10.1002/(SICI)1520-667X(1999)11:10<737::AID-MCS7>3.0.CO;2-4.

Barbosa, H.M.A., De Melo, M.M.R., Coimbra, M.A., Passos, C.P., Silva, C.M., 2014. Optimization of the supercritical fluid coextraction of oil and diterpenes from spent coffee grounds using experimental design and response surface methodology. J. Supercrit. Fluids 85, 165–172. https://doi.org/10.1016/j.supflu.2013.11.011.

Bicchi, C., Maffei, M., 2012. The plant volatilome: methods of analysis. Methods Mol. Biol. 918, 289–310. https://doi.org/10.1007/978-1-61779-995-2_15.

Bicchi, C., Cordero, C., Iori, C., Rubiolo, P., Sandra, P., 2000. Headspace Sorptive Extraction (HSSE) in the headspace analysis of aromatic and medicinal plants. HRC J. High Resolut. Chromatogr. 23 (9), 539–546. https://doi.org/10.1002/1521-4168(20000901)23:9<539::AID-JHRC539>3.0.CO;2-3.

Bicchi, C., Cordero, C., Liberto, E., Rubiolo, P., Sgorbini, B., Sandra, P., 2007. Sorptive tape extraction in the analysis of the volatile fraction emitted from biological solid matrices. J. Chromatogr. A 1148 (2), 137–144. https://doi.org/10.1016/j.chroma.2007.03.007.

Bicchi, C., Cordero, C., Liberto, E., Sgorbini, B., Rubiolo, P., 2008a. Headspace sampling of the volatile fraction of vegetable matrices. J. Chromatogr. A 1184 (1–2), 220–233. https://doi.org/10.1016/j.chroma.2007.06.019.

Bicchi, C., Liberto, E., Matteodo, M., Sgorbini, B., Mondello, L., Zellner, B.d.A., Costa, R., Rubiolo, P., 2008b. Quantitative analysis of essential oils: a complex task. Flav. Frag. J. 23 (6), 382–391. https://doi.org/10.1002/ffj.1905.

Bicchi, C., Cagliero, C., Rubiolo, P., 2011. New trends in the analysis of the volatile fraction of matrices of vegetable origin: a short overview. A review. Flav. Frag. J. 26 (5), 321–325. https://doi.org/10.1002/ffj.2059.

Boggia, L., Sgorbini, B., Bertea, C.M., Cagliero, C., Bicchi, C., Maffei, M.E., Rubiolo, P., 2015. Direct contact - sorptive tape extraction coupled with gas chromatography - mass spectrometry to reveal volatile topographical dynamics of lima bean (Phaseolus lunatus L.) upon herbivory by Spodoptera littoralis Boisd. BMC Plant Biol. 15 (1). https://doi.org/10.1186/s12870-015-0487-4.

Bruneton, J., 2016. Pharmacognosie, Phytochimie, Plantes Médicinales. Cinquième Edition, Lavoisier Tec & Doc.

Cagliero, C., Mastellone, G., Marengo, A., Bicchi, C., Sgorbini, B., Rubiolo, P., 2021a. Analytical strategies for in-vivo evaluation of plant volatile emissions - a review. Anal. Chim. Acta 1147, 240–258. https://doi.org/10.1016/j.aca.2020.11.029.

Cagliero, C., Sgorbini, B., Cordero, C., Liberto, E., Rubiolo, P., Bicchi, C., 2021b. Separation of Stereoisomers by Gas Chromatography Gas Chromatography. Elsevier, Italy, pp. 581–614. https://doi.org/10.1016/B978-0-12-820675-1.00015-0.

Cagliero, C., Marengo, A., Sgorbini, B., Rubiolo, P., 2022. Green Strategies for Analysis of Natural Products Green Approaches for Chemical Analysis. Elsevier, Italy, pp. 85–125. https://doi.org/10.1016/B978-0-12-822234-8.00005-6.

Capetti, F., Rubiolo, P., Bicchi, C., Marengo, A., Sgorbini, B., Cagliero, C., 2020. Exploiting the versatility of vacuum-assisted headspace solid-phase microextraction in combination with the selectivity of ionic liquid-based GC stationary phases to discriminate Boswellia spp. resins through their volatile and semivolatile fractions. J. Separ. Sci. 43 (9–10), 1879–1889. https://doi.org/10.1002/jssc.202000084.

Cheung, C., Baker, J.D., Byrne, J.M., Perrault, K.A., 2022. Investigating volatiles as the secondary metabolome of Piper methysticum from root powder and water extracts using comprehensive two-dimensional gas chromatography. J. Ethnopharmacol. 294. https://doi.org/10.1016/j.jep.2022.115346.

Cicchetti, E., Merle, P., Chaintreau, A., 2008. Quantitation in gas chromatography: Usual practices and performances of a response factor database. Flav. Frag. J. 23 (6), 450–459. https://doi.org/10.1002/ffj.1906.

Čretnik, L., Škerget, M., Knez, Z., 2005. Separation of parthenolide from feverfew: performance of conventional and high-pressure extraction techniques. Separ. Purif. Technol. 41 (1), 13–20. https://doi.org/10.1016/j.seppur.2004.03.011.

de Melo, M.M.R., Silvestre, A.J.D., Silva, C.M., 2014. Supercritical fluid extraction of vegetable matrices: applications, trends and future perspectives of a convincing green technology. J. Supercrit. Fluids 92, 115–176. https://doi.org/10.1016/j.supflu.2014.04.007.

De Saint Laumer, J.Y., Cicchetti, E., Merle, P., Egger, J., Chaintreau, A., 2010. 8 1 2010/08/01 Analytical Chemistry 10.1021/ac1006574 00032700 15 6457 6462 Switzerland Quantification in Gas Chromatography: Prediction of Flame Ionization Detector Response Factors from Combustion Enthalpies and Molecular Structures 82.

Dewick, P.M., 2009. Medicinal Natural Products: A Biosynthetic Approach: Third Edition Medicinal Natural Products: A Biosynthetic Approach. undefined John Wiley and Sons, undefined, third ed. John Wiley and Sons, pp. 1–539. https://doi.org/10.1002/9780470742761.

Domingues, R.M.A., Oliveira, E.L.G., Freire, C.S.R., Couto, R.M., Simões, P.C., Neto, C.P., Silvestre, A.J.D., Silva, C.M., 2012. Supercritical fluid extraction of Eucalyptus globulus bark-A promising approach for triterpenoid production. Int. J. Mol. Sci. 13 (6), 7648–7662. https://doi.org/10.3390/ijms13067648Portugal.

2023 7 European Pharmacopeia 11.0, 2023. https://pheur.edqm.eu/app/11-0/search/.

2023 6 European Union (EU) herbal monograph, 2023. https://www.ema.europa.eu/en/medicines/field_ema_web_categories%253Aname_field/Herbal/field_ema_herb_outcome/european-union-herbal-monograph-254.

Fan, X.H., Chang, Y.H., Wang, L.T., Zhu, Y.W., Dong, M.Z., Lv, M.J., An, J.Y., Yang, Q., Jiao, J., Meng, D., Fu, Y.J., 2020. A simple and efficient sample preparation for taxanes in Taxus chinensis needles with natural menthol-based aqueous deep eutectic solvent. J. Separ. Sci. 43 (7), 1339–1347. https://doi.org/10.1002/jssc.201900754.

Farooq, M.Q., Abbasi, N.M., Anderson, J.L., 2020. Deep eutectic solvents in separations: methods of preparation, polarity, and applications in extractions and capillary electrochromatography. J. Chromatogr. A 1633. https://doi.org/10.1016/j.chroma.2020.461613.

Filippi, J.J., Belhassen, E., Baldovini, N., Brevard, H., Meierhenrich, U., 2013. Qualitative and quantitative analysis of vetiver essential oils by comprehensive two-dimensional gas chromatography and comprehensive two-dimensional gas chromatography/mass spectrometry. J. Chromatogr. A. 1288, 127–148. https://doi.org/10.1016/j.chroma.2013.03.002.

Gałuszka, A., Migaszewski, Z., Namieśnik, J., 2013. The 12 principles of green analytical chemistry and the significance mnemonic of green analytical practices. TrAC, Trends Anal. Chem. 50, 78–84. https://doi.org/10.1016/j.trac.2013.04.010.

Herout, V., 1967. Gas chromatography of terpendoids. Planta Med. 68–78. https://doi.org/10.1055/s-0028-1100015.

Herrero, M., Mendiola, J.A., Cifuentes, A., Ibáñez, E., 2010. Supercritical fluid extraction: recent advances and applications. J. Chromatogr. A 1217 (16), 2495–2511. https://doi.org/10.1016/j.chroma.2009.12.019.

Jeong, K.M., Jin, Y., Yoo, D.E., Han, S.Y., Kim, E.M., Lee, J., 2018. One-step sample preparation for convenient examination of volatile monoterpenes and phenolic compounds in peppermint leaves using deep eutectic solvents. Food Chem. 251, 69–76. https://doi.org/10.1016/j.foodchem.2018.01.079.

Kolb, B., Ettre, L.S., 2006. Static Headspace-Gas Chromatography: Theory and Practice. John Wiley & Sons.

Křížek, T., Bursová, M., Horsley, R., Kuchař, M., Tůma, P., Čabala, R., Hložek, T., 2018. Menthol-based hydrophobic deep eutectic solvents: towards greener and efficient extraction of phytocannabinoids. J. Clean. Prod. 193, 391–396. https://doi.org/10.1016/j.jclepro.2018.05.080.

Li, C., Zhang, J., Zhao, C., Yang, L., Zhao, W., Jiang, H., Ren, X., Su, W., Li, Y., Guan, J., 2018. Separation of the main flavonoids and essential oil from seabuckthorn leaves by ultrasonic/microwave-assisted simultaneous distillation extraction. R. Soc. Open Sci. 5 (7), 180133. https://doi.org/10.1098/rsos.180133.

Liang, Q., Zhang, J., Su, X., Meng, Q., Dou, J., 2019. Extraction and separation of eight ginsenosides from flower buds of Panax ginseng using aqueous ionic liquid-based ultrasonic-assisted extraction coupled with an aqueous biphasic system. Molecules 4. https://doi.org/10.3390/molecules24040778.

Liberto, E., Cagliero, C., Cordero, C., Rubiolo, P., Bicchi, C., Sgorbini, B., 2017. Fractionated dynamic headspace sampling in the analysis of matrices of vegetable origin in the food field. J. Chromatogr. A 1489, 18–28. https://doi.org/10.1016/j.chroma.2017.02.009.

Liu, X., Jing, X., Li, G., 2019. A process to acquire essential oil by distillation concatenated liquid-liquid extraction and flavonoids by solid-liquid extraction simultaneously from Helichrysum arenarium (L.) Moench inflorescences under ionic liquid-microwave mediated. Separ. Purif. Technol. 209, 164–174. https://doi.org/10.1016/j.seppur.2018.07.028.

Mastellone, G., Marengo, A., Sgorbini, B., Rubiolo, P., Cagliero, C., 2021. New phases for analytical scale extraction from plants: current and future trends. TrAC, Trends Anal. Chem. 141, 116288. https://doi.org/10.1016/j.trac.2021.116288.

McAuliffe, C., 1971. Gas chromatographic determination of solutes by multiple phase equilibrium. Chem Technol. 1, 46–51.

Psillakis, E., 2017. Vacuum-assisted headspace solid-phase microextraction: a tutorial review. Anal. Chim. Acta 986, 12–24. https://doi.org/10.1016/j.aca.2017.06.033.

Quispe-Condori, S., Sánchez, D., Foglio, M.A., Rosa, P.T.V., Zetzl, C., Brunner, G., Meireles, M.A.A., 2005. Global yield isotherms and kinetic of artemisinin extraction from Artemisia annua L leaves using supercritical carbon dioxide. J. Supercrit. Fluids 36 (1), 40–48. https://doi.org/10.1016/j.supflu.2005.03.003.

Ruberto, G., Biondi, D.M., Barbagallo, C., Meli, R., Savoca, F., 2002. Enantioselective analysis of monoterpene compounds in essential oils by stir bar sorptive extraction (SBSE)-enantio-MDGC-MS. Flav. Frag. J. 17 (1), 32–40. https://doi.org/10.1002/ffj.1033.

Schieberle, P., Grosch, W., 1987. Quantitative analysis of aroma compounds in wheat and rye bread crusts using a stable isotope dilution assay. J. Agric. Food Chem. 35 (2), 252–257. https://doi.org/10.1021/jf00074a021.

Sgorbini, B., Cagliero, C., Cordero, C., Liberto, E., Rubiolo, P., Bicchi, C., 2014. 2014 Encyclopedia of Analytical Chemistry 1 10 John Wiley & Sons Headspace Sampling and Gas Chromatography of Plants: A Successful Combination to Study the Composition of a Plant Volatile Fraction.

Sgorbini, B., Bicchi, C., Cagliero, C., Cordero, C., Liberto, E., Rubiolo, P., 2015a. Herbs and spices: characterization and quantitation of biologically-active markers for routine quality control by multiple headspace solid-phase microextraction combined with separative or non-separative analysis. J. Chromatogr. A 1376, 9–17. https://doi.org/10.1016/j.chroma.2014.12.007.

Sgorbini, B., Cagliero, C., Boggia, L., Liberto, E., Reichenbach, S.E., Rubiolo, P., Cordero, C., Bicchi, C., 2015b. Parallel dual secondary-column-dual detection comprehensive two-dimensional gas chromatography: a flexible and reliable analytical tool for essential oils quantitative profiling. Flav. Frag. J. 30 (5), 366–380. https://doi.org/10.1002/ffj.3255.

Sgorbini, B., Cagliero, C., Liberto, E., Rubiolo, P., Bicchi, C., Cordero, C., 2019. Strategies for accurate quantitation of volatiles from foods and plant-origin materials: a challenging task. J. Agric. Food Chem. https://doi.org/10.1021/acs.jafc.8b06601.

Souza-Silva, É.A., Gionfriddo, E., Pawliszyn, J., 2015. A critical review of the state of the art of solid-phase microextraction of complex matrices II. Food analysis. TrAC, Trends Anal. Chem. 71, 236–248. https://doi.org/10.1016/j.trac.2015.04.018.

Steinhaus, M., Fritsch, H.T., Schieberle, P., 2003. Quantitation of (R)- and (S)-Linalool in beer using solid phase microextraction (SPME) in combination with a stable isotope dilution assay (SIDA). J. Agric. Food Chem. 51 (24), 7100–7105. https://doi.org/10.1021/jf0347057.

Suzuki, M., Tsuge, S., Takeuchi, T., 1970. Gas chromatographic estimation of occluded solvents in adhesive tape by periodic introduction method. Anal. Chem. 42 (14), 1705–1708. https://doi.org/10.1021/ac50160a035.

Tissot, E., Rochat, S., Debonneville, C., Chaintreau, A., 2012. Flavour Fragr. J . 27.

Uribe, C., Cosio, E., 2021. Combination of single-point standard addition calibration and natural internal standardization for quantification of terpenes in Pisco samples. LWT 147. https://doi.org/10.1016/j.lwt.2021.111551.

Wahlroos, O., Nikkilä, O.E., 1966. Quantitative determination of volatile compounds from food and evaluation of the gas flushing method for their concentration. Acta chemica Scandinavica. 20 (1), 197–205.

Zhang, Z., Pawliszyn, J., 1993. Headspace solid-phase microextraction. Anal. Chem. 65 (14), 1843–1852. https://doi.org/10.1021/ac00062a008.

Zhang, R., Tan, Z., Zhao, J., Wen, Y., Fan, S., Liu, C., 2020. Determination of pyrethroid residues in herbal tea using temperature-controlled ionic liquid dispersive liquid-liquid microextraction by high performance liquid chromatography. Sci. Rep. 10. https://doi.org/10.1038/s41598-020-61755-z.

CHAPTER

Phenolic compounds (extraction, quantification, biological activities) 15

Marina Russo[1], Ludovica Spagnuolo[2], Giovanna Cafeo[1] and Laura Dugo[2]

[1]*Messina Institute of Technology c/o Department of Chemical, Biological, Pharmaceutical and Environmental Sciences, Former Veterinary School, University of Messina, Messina, Italy;* [2]*Department of Science and Technologies for Sustainable Development and One Health, University Campus Bio-Medico di Roma, Rome, Italy*

1. Introduction

Phenolic compounds represent the major class of phytochemicals in our diet, typically abundant in plat-derived food. Fruits, vegetables, herbs, cereals, tea, coffee, nuts, seeds, and some alcoholic and nonalcoholic beverages are rich sources of these molecules (Arranz et al., 2010), also known as secondary plant metabolites (González-Vallinas et al., 2013).

The classes of phenolic compounds include phenolic acids, flavonoids, stilbenes, and lignans (Manach et al., 2004). Phenolic acids are nonflavonoid phenolic compounds which can be further divided into two main types: derivatives of benzoic acid and derivatives of cinnamic acid. The hydroxybenzoic acid content of edible plants is generally low, except certain red fruits, black radishes and onions. Examples of benzoic acid derivatives include compounds derived from gallic acid, while cinnamic acid derivatives are caffeic acid, chlorogenic acid, ferulic acid, and rosmarinic acid (Tsao, 2010).

Flavonoids have a common basic structure consisting of two aromatic rings bound together by three carbon atoms that form an oxygenated heterocycle and based on the variation in the type of heterocycle involved, can be categorized as flavonols (e.g., quercetin, kaempferol), flavones (e.g., apigenin, luteolin), flavanones (e.g., naringenin and hesperetin), isoflavones (e.g., genistein and daidzein), flavanols (e.g., catechin, epigallocatechin gallate), anthocyanins (e.g., cyanidin, malvidin), and flavonolignans (Gasmi et al., 2022; Menezes et al., 2017; Pandey and Rizvi, 2009).

Stilbenes are a group of naturally phenolic compounds found in various plants such as grapes and berries are synthesized by plants in response to various stresses. The most widely studied is resveratrol (3,5,4′-trihydroxy-trans stilbene), which is found in *Vitis* species (grapes), red wine, and other plant species (Rivière et al., 2012).

Lignans are a class of plant secondary metabolites belonging to the group of diphenolic compounds, which can be classified into several subgroups according to the way the oxygen is incorporated into the skeleton and the cyclization pattern. These compounds are found in several species in the plant kingdom including members of pteridophytes, gymnosperms, and angiosperms (Zálešák et al., 2019) (Fig. 15.1).

Phytochemicals for Health. https://doi.org/10.1016/B978-0-443-15366-2.00015-0

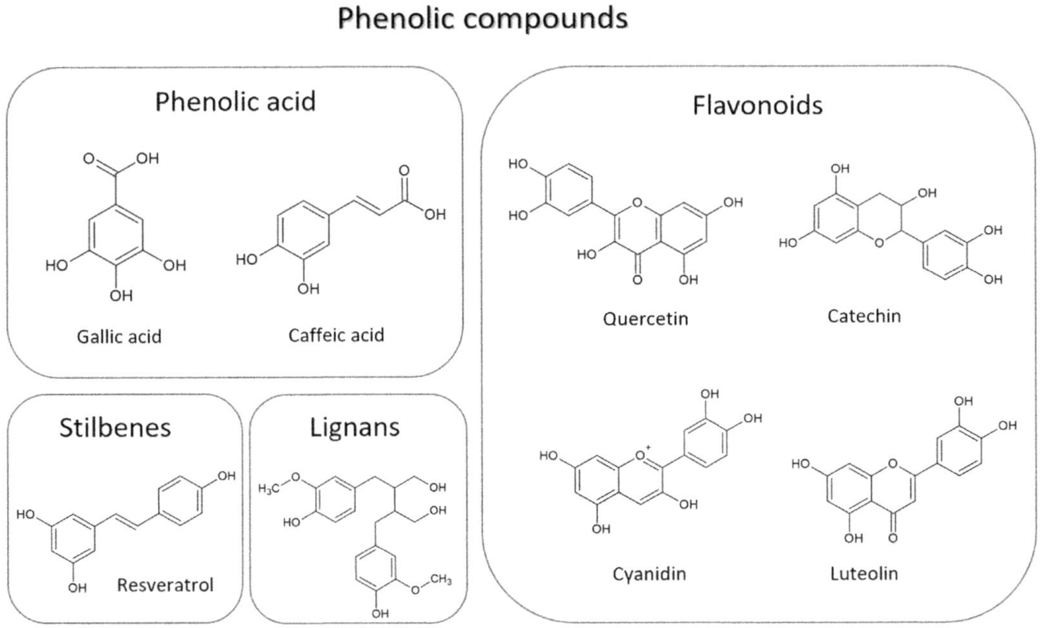

FIGURE 15.1 Main phenolic compounds and their chemical structures.

In plants, phenolic compounds exert protection against different types of stress, reactive oxygen and nitrogen species, UV light, pathogens, parasites and plant predators (Asensi et al., 2011). In humans, they possess biological activities with positive effects on health, such as antioxidative and antiinflammatory, and have shown prevention of various pathologies including cardiovascular diseases and cancer (Fresco et al., 2010; De Pace and Colombo, 2019) (Fig. 15.2). For example, phenolic compounds can protect from cardiovascular diseases through antiplatelet, antiinflammatory, and antioxidant effects in addition to their potential to boost high-density lipoprotein levels (HDL) and enhance endothelial functions, improve lipid profiles, blood pressure, insulin resistance, and systemic inflammation (Sanches-Silva et al., 2020).

Dietary phenolic compounds are used in the management and treatment of several illnesses and act on a variety of cellular and molecular processes throughout the human body. For example, it has been reported that the long-term use of polyphenol-rich diets could protect against some unhealthy conditions, tumors, osteoporosis, type 2 diabetes, stomach complications, pancreatitis, lung damage and neurodegenerative diseases (Cory et al., 2018).

2. Extraction techniques

The extraction techniques involve the separation of phytochemicals from inactive plant portions, based on their solubilities in various solvents. Phenolic compound extraction can be carried out with conventional and nonconventional techniques, as reported in Table 15.1. When deciding to undertake an extraction of phenolic compounds, it must be taken into consideration that these are a very complex

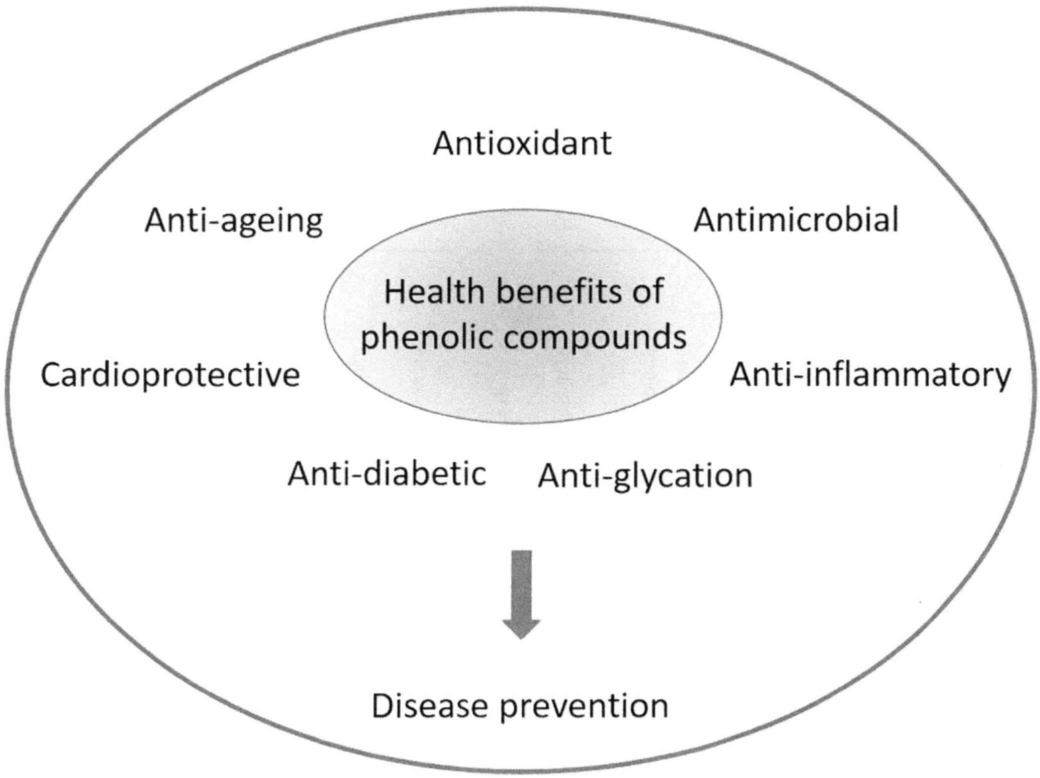

FIGURE 15.2 Schematic representation of beneficial effects and biological properties of phenolic compounds in human health.

family of molecules. Therefore, considering the complexity of this type of matrices, the choice of the extraction method as well as the type of solvent is crucial to recovering and purifying the components prior to their analysis.

Another very important aspect to take into consideration, before carrying out the extraction, is the storage and pretreatment of the material. Normally, in order to preserve the phenolic compounds, plant materials were stored in refrigerators or freezers (4 or −20°C) in the dark. Sample pretreatment consisted of three consecutive steps: (a) washing with distilled water in order to remove coarse material from the surface (dirt, dust, soil); (b) drying the material in the air (room temperature), in an oven or stove (at 50−60°C) or a freeze dryer; (c) homogenize the sample in a mortar or grinder to increase the exposure of the particulate matter to the extraction process (Sharma and Kaushik, 2021; Banu and Cathrine, 2015).

Homogenized samples can be subjected to solvent or Soxhlet extractions if you are interested in doing conventional extraction techniques. On the other hand, ultrasound-assisted (UAE), microwave-assisted (MAE), assisted solvent (ASE), supercritical fluid (SFE), and solid phase (SPE) extractions are valid extraction methods that belong to the category of nonconventional techniques Table 15.1.

Table 15.1 Extraction methods.

Extraction methods		Solvents	Temperature	Pressure	Time of extraction
Conventional methods	LLE	Water, aqueous and nonaqueous solvents	Room temperature or cold (4−20°C)	Atmospheric	1−2 hours
	Maceration	Water, aqueous and nonaqueous solvents	Room temperature or cold (4−20°C)	Atmospheric	3−7 days (even months)
	Percolation	Water, aqueous and nonaqueous solvents	Room temperature or under heat (35−70°C)	Atmospheric	2 hours−1 day
	Decoction	Water	Room temperature − 50°C	Atmospheric	1−2 hours
	Infusion	Water	80−100°C	Atmospheric	5−30 minute
	Soxhlet	Organic solvents (petroleum ether, chloroform, acetone, ethyl acetate, methanol)	39.6−64.7°C	Atmospheric	6−24 hours
Nonconventional methods	UAE	Water, acidified water, alcohols, acetone	20−80°C	Atmospheric	10 minutes−1 hour
	MAE	Polar solvents (water, alcohols) medium-polar solvents (acetone, ethyl acetate)	40−120°C	Atmospheric	30 seconds−20 minute
	UMAE	Water, alcohols, acetone, ethyl acetate	20−120°C	Atmospheric	30 seconds−30 minute
	ASE	Water, aqueous and nonaqueous solvents	40−200°C	50−100 bar	5−20 minute
	HHPE	Polar and nonpolar solvents	Room temperature	1000−8000 bar	5 minutes−1 hours
	SFE	Supercritical fluids S−CO_2, S−H_2O	30−80°C	350−700 bar	10 minutes−2 hours
	SCWE	Water	80−300°C	20−120 bar	5 minutes−2 hours
	EAE	Carbohydrases; proteases; endopeptidases; cellulases (+buffer)	33−67°C	Atmospheric	10 minutes−4 hours
	ILs	Ionic liquids	25−80°C	Atmospheric	30 minutes−1 hours
	DES/NADES	Deep eutectic solvents (glycerine, ethaline, reline, etc.)	25−105°C	Atmospheric	30 minutes−1 hour
	PEF	Water, aqueous and nonaqueous solvent	20−50°C	1.32−1.64 bar or atmospheric pressure	5 minutes−2 days

2.1 Conventional extraction methods

These are the most traditional extraction techniques used for the extraction of phenolic compounds. Conventional extraction methods have the advantage of being easy to perform conventional laboratory equipment is sufficient to achieve a good result. However, they have some disadvantages, among which: are time-consuming, risk of degradation of thermolabile phenolic compounds due to heat and require the use of large quantities of solvents (often not eco-friendly).

2.1.1 Solvent extraction

The chemical principle on which solvent extractions are based concerns the polarity of the analytes to be extracted. The most suitable solvent to be able to carry out an extraction will be the one with polarity closest to the molecules of our interest, and which is immiscible with the other molecules that characterize the plant matrix. Water, methanol, ethanol, ethyl acetate and acetone are used in the extraction of polar compounds while solvents such as hexane and dichloromethane are used in the extraction of nonpolar compounds. For toxicological and environmental reasons, ethanol is preferred (Mir-Cerdà et al., 2023). If the plant material is in liquid form, a liquid-liquid extraction (LLE) will be carried out. On the other hand, if the sample is solid, the technique consists of a typical solid-liquid extraction (SLE).

Traditional LLE extraction technique involves the use of a separating funnel and is carried out by mechanical shaking. Typically, phenolic compounds extraction from vegetable oils was carried out using LLE (Dugo et al., 2020). A polar solvent mixture was selected (methanol:water) because it is immiscible with the matrix and affine to polar compounds. On the other hand, phenolic compounds extraction from water-liquid samples (such as juices, waste waters, beverages) was carried out with nonpolar organic solvents such as ethyl acetate (Dugo et al., 2009; Russo et al., 2021).

SLE can be done in four different ways: maceration, percolation, decoction and infusion. In the maceration process, the powdered material is agitated in a container with the appropriate solvent at room temperature for a minimum of 3 days until the soluble matter is dissolved. Percolation can be defined as a short maceration (24 hours) and is carried out in a percolation tank. Decoction extraction is the most suitable technique to extract water-soluble and heat-stable components. This kind of extraction involves the use of water as solvent (often at 50°C) and is carried out under constant stirring. Finally, infusion follows the same procedure as the other SLE techniques (soaking of powered plant material in boiling water) but with a shorter extraction time (5–30 minutes). Maceration is doubtless the most time-consuming process (Salah and Allouche, 2019).

2.1.2 Soxhlet extraction

This technique is a conventional extraction procedure frequently used to recover phenolic compounds from solid samples. Soxhlet extraction, also known as hot continuous extraction, uses circulating hot and cold steam. The extraction solvent from the round base flask is heated and reaches boiling point, vaporizes, condenses in the coolant, and falls into the cellulose thimble containing the plant/natural material. The sample container is filled until it reaches the siphon, and the overflowed liquid falls back into the base flask to start a new extraction cycle. The solvent is recovered, and the number of extracted compounds becomes more concentrated with each cycle (Abubakar and Haque, 2020) .

Methanol is normally used to extract phenolic compounds from plant material (Kumar et al., 2023), and sequential Soxhlet extraction is common in the determination of secondary metabolites. Selection of various solvents in order of increasing polarity, such as ethyl acetate, methanol and ethanol, was largely applied in the extraction of phenolic compounds from plants. However, toxic gaseous emissions and subsequent exposure to numerous dangerous compounds are a serious downside of this technique. In other words, this extraction procedure is not environmental-friendly and could have negative effects on the operator's health.

2.2 Non-conventional extraction methods

Due to limitations associated with conventional extraction procedures, the current trend is to develop new fast, selective and environmentally friendly extraction methods. In fact, nonconventional extraction techniques present several advantages with respect to the conventional ones: they offer lower solvent consumption, higher speed of methodology, higher extraction rates and they have the possibility of automation.

2.2.1 Ultrasound assisted extraction (UAE)

Ultrasound-assisted extraction, also known as sonication extraction, is based on the application of high-frequency sounds (20–2000 kHz). Ultrasounds break the plant cell wall, increase the solvent penetration and promote the release of bioactive compounds from the sample into the solvent. The high intensity of the waves generates high pressure and temperature through bubble cavitation and produces this disruptive effect in the matrix. Several parameters influence the extraction yields: such as the nature of the solvent, cycle of extractions, frequency ultrasounds (kHz), and extraction time and temperature (Kumar et al., 2021). Most of the extractions reported in the literature were conducted at temperatures that do not exceed 50°C in the process. This is due to the possible degradation of the phenolic compounds at high temperatures. The time ranges from 10 to 90 minutes depending on the type of compound and the matrix. Water, alcohols, acetone and their mixtures are usually used as solvents. Acidified water is used for pectin extraction (Maran et al., 2017).

This kind of procedure presents numerous advantages. In particular, the extraction time, solvent and energy consumption are reduced, so it is a low-cost procedure. Thanks to the involvement of low-temperature extraction, this type of technique is suitable for thermolabile compounds such as anthocyanins and other phenolic compounds. The main disadvantage lies in the possible formation of free radicals at very high ultrasound energies, which can alter the phytochemical compounds of interest (Salah and Allouche, 2019).

2.2.2 Microwave-assisted extraction (MAE)

It is an extractive technique based on microwave radiation and conduction heat transfer. Microwaves are made of electric and magnetic fields that oscillate in frequencies from 300 to 300 GHz and wavelengths from 1 cm to 1 m (Altemimi et al., 2017). The common frequency of microwave equipment is 2450 MHz, which is equivalent to 600–700 W power, and the extraction temperature is in the range from 50 to 100°C. The solid sample is placed in closed vessels with the solvent and heated by microwave energy, which encourages the breakdown of the plant material's cell walls and the interaction of the matrix with the extraction solvent.

As with other extraction techniques, the main parameters to take into account are the amount of extraction solvent, the temperature and the time. Solvents with a strong ability to absorb microwaves (with high dielectric constants) are the best choices for Microwave assisted extraction (MAE), such as water, methanol, ethanol, and their mixtures with water, as they heat easily (Kumar et al., 2023). Medium polarity solvents such as diethyl ether, acetone and ethyl acetate have also been used. Solvents such as *n*-hexane, and petroleum ether are normally not able to absorb heat, so could be not appropriate to MAE. However, if they were used as a mixture with a more polar solvent (acetone and ethyl acetate) could be employed as an extraction medium for nonpolar phytochemicals (Akhtar et al., 2019).

The basic challenge of this extraction method is to maximize extraction yield without influencing the chemical composition of the natural substances present in plant tissue. According to this purpose, the main advantages of this technique are that it considerably reduces the extraction time (usually less than 30 minutes) and decreases solvent consumption in comparison to other conventional techniques. However, the temperatures applied are usually higher and the possible degradation of heat-sensitive compounds must also be taken into consideration. Therefore, this procedure produces the greatest results with smaller phenolic compounds resistant to thermal degradation like gallic acid, quercetin, isoflavin, and trans-resveratrol (Sharma and Kaushik, 2021). It is rarely used when it concerns the extraction of polymeric polyphenols, such as anthocyanins and tannins (Akhtar et al., 2019).

2.2.3 Ultrasound/microwave-assisted extraction (UMAE)

This extraction technique is based on coupling two radiation procedures. As previously described, UAE improves the solvent penetration into the plant sample using cavitation, while MAE takes advantage of heat to extract plant bioactive compounds. Combining the two techniques, ultrasound/microwave-assisted extraction allows obtaining higher extraction yields. UMAE has been successfully used to extract caffeic acid from *Arctium lappa*, flavonoids from *Spatholobus suberectus*, and lycopene from tomatoes (Khoddami et al., 2013).

2.2.4 Accelerated solvent extraction (ASE)

This is a solid–liquid extraction methodology, also known as pressurized liquid extraction (PLE), that is performed in an automated system at high temperatures and pressures. Because of the high pressure, the solvents stay liquid even at high temperatures. The high temperatures and pressures force the solvent to penetrate the matrix accelerating the kinetics of the extraction process. The extraction time varies from 5 to 20 minutes, temperature varies from 50°C to a maximum of 200°C and pressures from 5 to 10 MPa (Da Silva et al., 2023; Kumar et al., 2023). It has been used by many scientists to extract natural products such as anthocyanins and saponins (Zhang et al., 2018).

The extraction takes place in the absence of oxygen and light, so the occurrence of oxidation reactions is limited. ASE also reduces the solvent and time requirements and has better repeatability compared to conventional techniques. The main drawback is that at these extreme conditions, it can cause heat degradation and generate unwanted compounds, such as 5-(hydroxymethyl) furfural formed during the thermal decomposition of carbohydrates (Shi et al., 2022). The disadvantages were, however, a lower rate of recovery of thermosensitive polyphenols at high temperatures and incomplete extraction due to a low solvent volume (Chiriac et al., 2021).

2.2.5 High hydrostatic pressure extraction (HHPE)

It is an advanced method used to extract phenolic compounds from plant materials. HHPE uses super-high hydraulic pressure in the range between 1000 and 8000 bar. The extraction is generally conducted at ambient temperature. Several types of solvents (from polar to nonpolar) can be used based on the phytochemicals to be extracted. The very high pressure used causes the breaking of cell constituents increasing solvent penetration and extraction yields without heat use. Excellent results have been obtained from the use of HHPE to extract polyphenols from *Maclura pomifera*, grape by-products, propolis, green tea leaves, and juice of cashew apples (Khoddami et al., 2013). However, a main disadvantage of this technique is the expensive instrumentation required.

2.2.6 Supercritical fluid extraction (SFE)

It is an eco-friendly extraction technique that uses a supercritical fluid as an extraction solvent. A supercritical fluid is any substance that, at its critical point, has chemical-physical properties intermediate between a gas and a liquid. Supercritical carbon dioxide ($SC-CO_2$) is frequently referred to as an ideal solvent for the extraction of nonpolar components such as lipids, due to its low polarity (Shi et al., 2022). On the other hand, phenolic compounds have a low solubility value in $SC\text{-}CO_2$, resulting in low extraction yields. To overcome this problem, the solubility properties can be increased using cosolvent as modifiers. As cosolvents water, ethanol, methanol, acetone, acetonitrile and mixtures of them can be used (Radzali et al., 2020).

In this extraction method, the solid plant material (dried and ground) is placed in columns and a pressurized solvent, sometimes with a cosolvent, at its critical point flows through the column allowing the extraction of compounds from the solid matrix. Supercritical fluids can be used for the extraction of thermally labile molecules. Different classes of biologically active compounds (volatile compounds, lycopene, carotene, chlorophylls) have been extracted from different bio-resources (Geeta et al., 2020). SFE is considered an eco-friendly and sustainable extraction methodology because the extracts are solvent-free. The main drawback of this method includes the need for large capital investments in equipment and very high pressures. Moreover, low extraction yields achieved are a disadvantage compared to extraction using liquid solvents.

2.2.7 Subcritical water extraction (SCWE)

Subcritical water extraction (SCWE) is another eco-friendly extraction method that can be effectively used to isolate phenolic compounds. Water becomes subcritical when the temperature is between 100 and 347°C and the pressure is below 220 bar to preserve its liquid form. So, this method is also known as superheated water, pressurized water or hot liquid water extraction (Khoddami et al., 2013). Because water is employed as the solvent, SCWE has several advantages over conventional procedures such as environmental friendliness but also excellent extract quality and short extraction times. Moreover, this extraction procedure is useful to extract a wide range of polar to low-polar phenolic compounds.

2.2.8 Enzyme-assisted extraction (EAE)

Enzymatic treatment of plant samples is a nontoxic method suitable for the extraction of phenolic compounds from plants. It is based on the use of enzymes that hydrolyze the components of cell membranes, favoring the re-reading of compounds and increasing the extraction rate. Enzymes normally employed for this kind of extraction method are: cellulases, hemicellulases, carbohydrases, pectinases and endopeptidases. Extraction methodology can be carried out by adding a mixture of these enzymes to the sample with a buffer in order to work at specific pH conditions. Limitations in the use of enzymes are pH, temperature and, the knowledge of phytochemicals composition in the sample of interest. For this reason, optimal conditions for enzyme activity are: temperatures below 60°C and pH of four−6.5 (Chiriac et al., 2021; Shi et al., 2022; Nadar et al., 2018).

2.2.9 Extraction with ionic liquids (ILs)

ILs are a class of organic salts that occur in liquid form for temperatures below 100°C. These organic salts are composed of an organic cation (such as imidazolium, pyrrolidinium, pyridinium tetraalkyl ammonium and, tetraalkyl phosphonium) and an anion of organic or inorganic nature (like, tetrafluoroborate, hexafluorophosphate and, bromide) (Chiriac et al., 2021). Due to a large number of possible combinations of cations and anions, ionic liquids have a wide range of physical and chemical properties: high or low viscosity, hydrophobic or hydrophilic nature, miscibility with water or with other different organic phases, and specific electrochemical properties. Ionic liquids have been successfully used to extract isoflavones by UAE from soy (Chiriac et al., 2021).

2.2.10 Extraction with natural deep eutectic solvents (NADES)

In the last years, research has focused on the use of alternative solvents to extract bioactive compounds from plant sources. In this context, deep eutectic solvents have gained much popularity because of their low cost, easy preparation, biodegradable properties, and low toxicity. They are formed by a mixture of a variety of hydrogen bond donors (sugars, alcohols, organic acids or carbohydrates) and hydrogen bond acceptors (choline chloride or betaine derivatives) (Kalyniukova et al., 2021). Bond donors and acceptors are solids, but when mixing at high temperatures they become a liquid. They are considered a green alternative to the commonly used toxic organic solvents (Dheyab et al., 2021). Various DESs have been used in several advanced techniques such as UAE, MAE and ASE to extract bioactive compounds from diverse matrices (Cui et al., 2018; Kalyniukova et al., 2021; Hsieh et al., 2019). An exemplary application was carried out by Hsieh et al. (2019) for the extraction of gingerols in ginger; they tested different types of DESs in UAE and observed that the 1:4 L-carnitine:1,3-butanediol combination (at 40°C for 30 minutes) exhibited superior extraction capability than ethanol and water. Concluding, polyphenols, saponins, alkaloids, carotenoids and other pigments, can be extracted using NADES (Herrero, 2023).

2.2.11 Pulsed electric field extraction (PEF)

It is a nonthermal extraction method that applies high and short voltage pulses (nanoseconds to microseconds) to destroy the cell wall and membrane structures, with consequent release of metabolites and increased extraction yield. The fact that this extraction method is carried out at room temperature and in a few seconds plays a crucial role in minimizing the deterioration of bioactive molecules present in the matrices (Shi et al., 2022).

3. Quantitative techniques

Despite the countless scientific articles published over the years, simultaneous quantification of all phenolic compounds belonging to different structural groups is still a challenge. Many analytical methods can be used for the quantification of phytochemicals and their selection depends on the physical and chemical properties of the compounds of interest. Spectrophotometric methods are very easy techniques for quantification of total phenolic content in plants. In the last few decades, new approaches have been used, such as gas chromatography (GC) and high-pressure liquid chromatography (HPLC) coupled with spectroscopic techniques. The main quantitative methods are described below.

3.1 Spectrophotometric techniques

Spectrophotometric methods are useful for the quantification of the total content of selected groups of phytochemicals. They are the most extensive techniques used for this purpose because of their simplicity and low cost.

Folin-Ciocalteau and Folin-Denis methods are widely used to determine the total content of phenolic compounds in plant samples. Both methods are based on the ability of phenols to react with reagents containing tungsten and molybdenum at basic pH. This reaction is a chemical reduction that gives rise to an intense blue color and light absorption spectrum around 760 nm (Folin and Ciocalteau, 1927; Munteanu and Apetrei, 2021).

The vanillin and dimethylaminocinnamaldehyde assays have been used to determine the total content of proanthocyanidins (condensed tannins). These techniques are based on the binding of the aldehydic group of vanillin or dimethylaminocinnamaldehyde to C-6 of the catechin A cycle, forming a red chromophore complex (Schofield et al., 2001).

The butanol-HCl and bovine serum albumin methods are two other colorimetric methods used to determine proanthocyanidins. The basis of butanol-HCl method is the break of interflavonoid linkages in proanthocyanidin using hot acid, followed by an auto-oxidation that converts flavan-3-ols to anthocyanidin. These reactions form a red solution with a maximum absorbance of around 550 nm. In the second method, the sample was treated with bovine serum albumin forming an insoluble tannin-protein complex that precipitates. Then, the complex is dissolved in an alkaline sodium dodecyl sulfate-triethanolamine solution and reacted with a ferric chloride solution. Finally, a violet complex with a maximum absorbance at around 510 nm is generated (Khoddami et al., 2013).

Complexation with aluminum trichloride ($AlCl_3$) has been used to determine total flavonoid content (Shraim et al., 2021). The addition of $AlCl_3$ forms an acid complex with the oxo and hydroxyl groups of the flavonoids forming a yellow-colored Al(III)-flavonoid chelate complex that can be measured spectrophotometrically in the range 410–423 nm.

Acidic 2,4-dinitrophenylhydrazine was used in another spectrophotometric method to determine flavonones and dihydroflavonols. These classes of molecules interact with 2,4-dinitrophenylhydrazine and the absorbance is measured at 486 nm (Khoddami et al., 2013).

There are a lot of other spectrophotometric methods used to determine phenolics. These techniques provide useful qualitative as well as quantitative information. However, the main disadvantage is that they provide an estimate of the total content, and do not allow quantification of individual compounds (Ignat et al., 2011).

3.2 Chromatographic techniques

They are based on the physical separation of the components of a mixture by distribution between two phases, a stationary phase, and a mobile phase, based on the different interaction capacities of each component. Among the most widely used in the determination of the different classes of phytochemicals are gas chromatography (GC) and liquid chromatography (LC) coupled to various detectors.

3.2.1 Gas chromatography

GC coupled to mass spectrometry (MS) is suitable for the analysis of terpenoids and other volatile compounds. It is used for the quantification of components of the oily fraction (nonpolar or medium polarity compounds) or highly volatile compounds (including alkanes, esters, alcohols, aromatic hydrocarbons, terpene derivatives, and steroids). The use of the GC-MS technique for the analysis of nonvolatile or semivolatile compounds, such as polyphenols, is uncommon due to the prior derivatization step required in order to increase the volatility and stability of the compounds (Chiriac et al., 2021). Nonpolar capillary columns coated with 5% phenyl methylpolysiloxane or equivalent, of dimensions generally 30 m × 0.25 mm × 0.25 μm film thickness are the most widely used in the separation, using high purity helium as carrier gas usually at a flow rate of 1 mL/minute. The injector temperature is around 250–270°C, and temperature ramps are employed usually starting in the range 40–100°C to a final temperature between 200–350°C. In GC-MS the mass spectrometer operates in electron impact (EI) mode at 70 eV (Rao et al., 2023; Sagbo et al., 2020). Tebbi et al. (2022) used GC-MS to evaluate the characterization and quantification of volatile compounds in the leaves of *Clematis flammula* L. They identified a total of 10 compounds, and protoanemonin was the predominant component accounting for almost 64%.

3.2.2 Liquid chromatography

LC, especially in combination with UV/Vis spectrometry diode array detection (DAD) and tandem mass spectrometry (MS/MS), is typically used for the analysis of nonvolatile compounds, such as phenolic acids, anthocyanins, flavonoids, tocopherols and carotenoids. Due to the high solubility of these classes of compounds, reversed-phase (RP) LC is used, where a nonpolar packed column and mobile phases of a polar nature are used as stationary phases. The most commonly employed analytical columns are C_{18} (silica bonded with alkyl chains), and to a lesser extent C_8, for the separation of these classes of compounds (Chiriac et al., 2021; Zhang et al., 2019) of variable length, diameter and particle size; usually water (phase A) with organic solvents such as methanol or acetonitrile (B) is used as mobile phase composition; although C_{30} columns are suitable for separating lipid-soluble vitamins and phytochemicals, such as carotenoids, chlorophylls, and tocopherols, using MeOH and methyl tert-butyl ether, (MTBE) as mobile phases (Cortés-Herrera et al., 2019; Cervantes-Paz et al., 2023). In order to improve the separation, phosphoric acid, acetic acid, trifluoroacetic acid or formic acid are usually added to the mobile phase as additives, at varying percentages. Typically, the separation takes place using gradient elution, as it achieves a better separation of the analytes, although the isocratic mode has provided good resolution in some studies. Normal phase (NP) LC is also used by some authors, although to a lesser extent than RP, and has been applied in the analysis of flavan-3-ols and procyanidins according to their degree of polymerization. In this type of technique, analytical columns with more polar packing such as dihydroxypropyl group bonded to silica gel and organic solvent mixtures are used as mobile phases with almost no water (Goto et al., 2021). Finally, an alternative increasingly used and applied in the analysis of phytochemicals is hydrophilic interaction liquid chromatography (HILIC) separation, which is used in multidimensional techniques in the first dimension combined with RP in the second dimension has shown excellent results in the separation of polyphenols in a variety of matrices with a high-resolution capacity (Arena et al., 2023).

3.3 Spectroscopic techniques

These are instrumental methods based on the interaction of electromagnetic radiation with molecules to identify them or determine their concentration. In the quantitative analysis of phytochemical compounds, the most widely used are described below.

3.3.1 Ultra violet/vis spectroscopy

The region from 190 to 600 nm is of special interest in the determination of phytochemicals since most of them show their maximum absorption in this wavelength range. They can be used for identification and quantification by knowing the absorption maximum corresponding to their structural characteristics (e.g., total phenolic extract (280 nm), flavones (320 nm), phenolic acids (360 nm), etc.) (Lama-Muñoz and Contreras, 2022). UV/vis detectors and DAD are widely used together with LC because of their simplicity and low cost. The main disadvantage of UV/vis detection in the identification of natural compounds is its low sensitivity, resolution, and specificity since when two compounds co-elute it is not possible to distinguish them. Therefore, it is common to use MS detection in parallel to obtain complementary information (Sharma and Kaushik, 2021).

3.3.2 Infrared spectroscopy

Infrared spectroscopy (comprising the infrared region (IR) from 700 to 1 nm) has also been applied allowing rapid and nondestructive analysis. It relates to vibrational changes when molecules are excited with infrared radiation (Salah and Allouche, 2019). It is divided into three types according to the wavelength range: near-IR (0.78−2.5 μm), mid-IR (2.5−50 μm) and far-IR (50−1000 μm) (Johnson et al., 2023). IR spectrometry is not only of great utility for structural elucidation of new molecules but also shows great potential for quantification purposes and relative prediction of phytochemical levels in natural matrices (Johnson et al., 2020). In particular, near-IR and mid-IR have been applied to perform compositional predictions in a variety of matrices, including foods such as grapes, tomatoes, olive oil, cocoa bean, red wine, and tea, among others (Johnson et al., 2023; Lama-Muñoz and Contreras, 2022). Polyphenols, anthocyanins, carotenoids, and ascorbic acid have been the most studied bioactive compounds. Hernánzez-Hernández et al. (Hernández-Hernández et al., 2021) demonstrated the feasibility of NIR in the analysis of bioactive compounds in intact cocoa bean husk; by means of regression models, they could relate the constituents with the Visible + NIR reflectance spectrum and determined the total content of sugars and phenols, and the content of individual components, such as catechin, theobromine and epigallocatechin, obtaining very good correlations. A possible strategy is to couple GC or LC to Fourier Transform Infrared (FTIR) spectrometers (third generation of infrared spectrometers). This combination allows for the separation of the components through the capillary or analytical column, before collecting the FTIR spectrum of each eluted compound, which provides detailed information on the functional groups of the compounds (Johnson et al., 2023). An interesting application is that of Lucarini et al. (2020) who employed FTIR in the mid-IR region to determine qualitatively and quantitatively the profile of phenolic compounds along with that of fatty acids in grape seeds by analyzing the main functional groups. The main advantage over other analytical methods is that it is nondestructive, and the major drawback is the complexity of the spectrum which makes it difficult to interpret (Johnson et al., 2023).

3.3.3 Mass spectrometry

MS is a powerful and advanced detection technique that has high sensitivity and selectivity in quantitative analysis. It is broadly used as a detection method for LC and GC. There are two types of spectrometers: low-resolution and high-resolution spectrometers. Low-resolution spectrometry employs quadrupole (Q), ion traps (IT) and triple quadrupole (QqQ) and is used in selective screening methods, identification, and quantification of phytochemicals. For data acquisition, two or three transitions (product) resulting from the fragmentation of the parent precursor are generally monitored using multiple reaction monitoring (MRM) mode. Low-resolution spectrometry employs time-of-flight (TOF) and Orbitrap and is popular in the characterization of natural compounds as it provides complete information on the exact mass, elemental composition, and chemical structure of the compounds. Hybrids such as quadrupole-TOF (QTOF) and ion-trap Orbitrap (IT-Orbitrap) offer even greater confirmatory capability, as they allow the molecular formula and fragmentation pattern of the compounds to be generated (Lama-Muñoz and Contreras, 2022). High-resolution spectrometry is especially advantageous in the identification of new compounds (comprehensive screenings) while low-resolution spectrometry is especially advantageous in the identification and quantification of target compounds (selective screenings), although the former can also be used for quantitative purposes. The ionization source generally applied to MS hyphenation with LC is electrospray ionization (ESI), while for GC-MS, EI is the source of choice (at 70 eV). However, other ionization sources such as atmospheric pressure chemical ionization (APCI) have been applied providing even more complementary information to the above (Chiriac et al., 2021; Pereira-Coelho et al., 2023; Zhang et al., 2019; Ignat et al., 2011; Sharma and Kaushik, 2021).

3.4 Other quantitative techniques

Capillary electrophoresis (CE) is a relatively new separation technique compared to LC and has been used as an alternative in the analysis of various classes of phytochemicals. It is an analytical technique that applies an electric field to separate sample components, and these are separated by their electrophoretic mobility. Only charged molecules migrate through the capillary. The velocity depends on the charge, size, shape, and resistance of the particle (Gackowski et al., 2021; Shi et al., 2022). The most commonly used detectors in phytochemical analysis are UV/Vis DAD, fluorescence or amperometry detectors. It has been applied to the analysis of polyphenols, coumarins alkaloids and amino acids in herbal raw materials (Gackowski et al., 2021). Capillary electrophoresis capillaries have very narrow diameters which results in good and improved resolution compared to LC. The limitation is that pH control is critical, as small variations affect the charge of the molecules and have a greater impact than with LC (Przybylska et al., 2021).

Electrochemical techniques are those in which the potential, current or charge in the electrochemical cell serves as the analytical signal. In this context, biosensors with electrochemical detection have been proposed as alternatives to chromatographic and spectrometric techniques, in order to avoid the often-tedious sample preparation step. They have proven to be reliable on-site and specific techniques and have been applied for the quantification of various phytochemicals, such as phenolic compounds (Verma et al., 2023). This type of sensor is composed of two components, a biological part (analyte and receptor) being the enzyme-based sensor one of the most explored, and a transducer. They electrochemically convert nonactive analytes into products by catalysis according to the enzyme that performs the oxidation and reduction on the surface of the working electrode. The advantages include high sensitivity, precision, reliability, and short response time (Verma et al., 2023; Bounegru and Apetrei, 2021). Although the main limitation is the leaching of the enzyme from the substrate and its reduced activity when the enzymes are immobilized on the support. Bounegru and Apetrei (2021) developed a new electrochemical biosensor based on carbon nanofibers modified with cobalt phthalocyanine and laccase, to determine p-coumaric acid. The biosensor showed optimal linearity characteristics 0.4–6.4 μM and low LOD and LQ 4.83×10^{-7} M and 1.61×10^{-6} M, respectively, and was applied to different phytoproducts, obtaining p-coumaric acid concentrations comparable to those obtained with the standard FTIR method.

Fig. 15.3 outlines the workflow of the entire process described in this chapter, from sample pretreatment, and extraction, to quantification of principal phytocompounds. Many studies combine the implementation of two or more different analytical methodologies since they allow obtaining complementary and more complete compositional information, as well as more accurate qualitative and quantitative data. An interesting combined application was carried out by Cervantes-Paz et al. (2023) who used GC-MS, HPC-MS and FTIR for the identification and quantification of fatty acids and lipid-soluble phytochemicals in avocado during its ripening process. In this work, GC was used for the determination of fatty acids, HPLC-MS for the determination of carotenes, chlorophylls, and tocopherols, and FTIR to determine the degree of degradation of unsaturated fatty acids at different ripening stages.

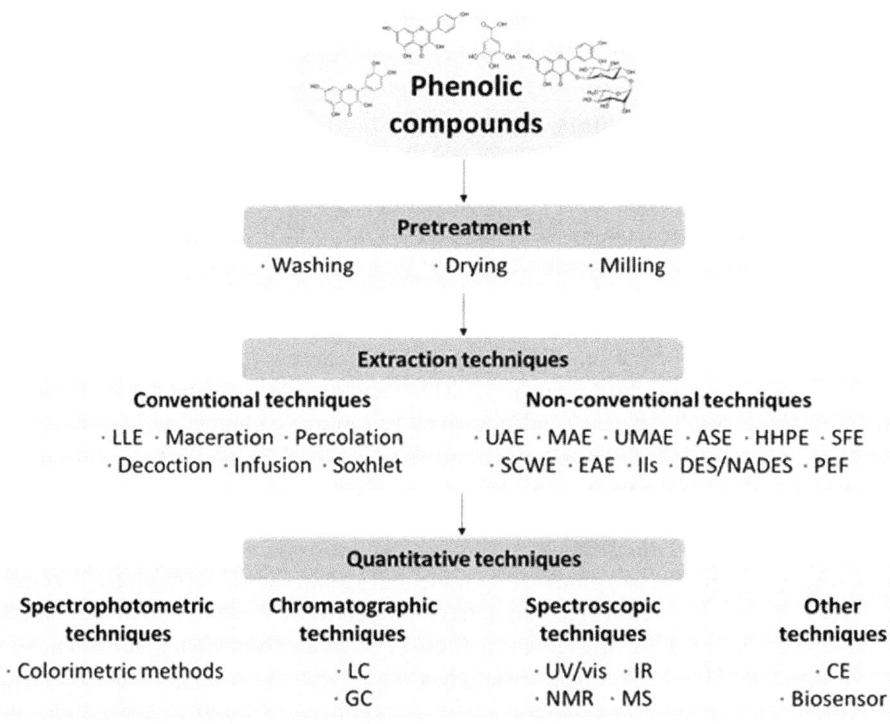

FIGURE 15.3 Representation of the possible workflow of the extraction and quantitative analysis of phenolic compounds from plants.

ASE, Assisted Extraction Solvent; *CE*, capillary electrophoresis; *DES*, Deep Eutectic Solvent; *EAE*, Enzymatic Assisted Extraction; *GC*, gas chromatography; *HHPE*, High Hydrostatic Pressure Extraction; *ILs*, Ionic Liquids; *IR*, infrared; *LC*, liquid chromatography; *LLE*, liquid–liquid extraction; *MAE*, Microwave Assisted Extraction; *MS*, mass spectrometry; *NADES*, Natural Deep Eutectic Solvents; *NMR*, Nuclear magnetic resonance spectroscopy; *PEF*, Pulse electric field; *SCWE*, Subcritical Water Extraction; *SFE*, Supercritical Fluid Extraction; *UAE*, ultrasound assisted extraction; *UMAE*, Ultrasound/Microwave Assisted Extraction.

4. Biological activities

4.1 Antioxidant and anti-inflammatory activities

Inflammation and oxidative stress are potentially major risk factors for the development of chronic diseases, including cancer, neurodegenerative, cardiovascular, and metabolic pathologies (Charles et al., 2021).

The increase in the production of reactive oxygen species (ROS) is associated with oxidative stress and protein oxidation; this activates an inflammatory response with the production of molecules and different inflammatory signals are triggered by protein oxidations (Malireddy et al., 2012).

Phenolic compounds have substantial health-promoting effects. Then, once they reach the human body, they protect cells from oxidative stress and inflammation. Indeed, a high dietary intake of fruits and vegetables rich in phenolic compounds is correlated with preventive effects against oxidative stress-mediated diseases (Rasouli et al., 2017). It was proposed that the antioxidant activity of phenolic compounds in biological systems was mainly due to their redox properties, chelating ability and capacity to transfer electrons due to the multiple hydroxyl groups contained in the chemical structure (Zhang and Tsao, 2016).

Several phenolic compounds have shown antioxidant effects: epigallocatechin-3-gallate, gallic acid, oleuropein, resveratrol and curcumin appear to be good antioxidants, since they are able to reduce ROS production and enhance antioxidant defense (Guo et al., 2020); grape seeds and olive oil polyphenols' rich extracts can be used as food additives for their antioxidant properties; p-coumaric and caffeic acids can be used as food preservatives for their antimicrobial activity (Martillanes et al., 2017). Ellagic acid (EA) possesses multiple effects: direct scavenging of free radicals, sequestration of potential oxidants, regulation of enzyme activity, modulation of cell signaling, and regulation of gene expression, with a good protective effect against oxidative stress (Marković et al., 2013).

Many biological and epidemiological investigations have indicated that dietary phenolic compounds, when regularly consumed with diet, are beneficial for the prevention of chronic inflammatory disorders (Hsiao et al., 2020). Several research studies have shown that phenolic compounds were able to modulate different cell signaling pathways associated with inflammation, both in vitro and in vivo, highlighting their beneficial role as therapeutic tools in multiple acute and chronic disorders (Yahfoufi et al., 2018).

The main pathway modulated by phenolic compounds is represented by nuclear factor-kappa β (NF-κB) activation. It was proposed that phenolic compounds blocked proinflammatory gene expression due to an effect on transcriptional activity suppression in response to inflammatory stimuli (Jantan et al., 2021). Among phenolic compounds, epigallocatechin gallate (EGCG) blocks NF-κB activation in human epithelial cells and downregulates the expression of iNOS (inducible nitric oxide synthase) and NO (nitric oxide) production in macrophages resulting in its immunomodulation (Leiherer et al., 2013). In addition, EGCG showed neuroprotective effect by inhibition of toll-like receptor 4 (TLR4)/NF-κB pathway (inflammasome activation) in microglia both in vitro and in an in vivo murine model of Alzheimer (Zhong et al., 2019).

Considering studies on inflammation, both in cells and animal models, curcumin decreased levels of proinflammatory mediators such as Interleukin-1 (IL-1), IL-1β, IL-6, IL-8, IL-17, IL-27, Tumor necrosis factor-α (TNF-α), iNOS and NO (Sadeghi et al., 2018).

Plants rich in flavonoids such as rutin, apigenin and quercetin have shown antiinflammatory properties due to the inhibition of NF-κB, activator protein-1 (AP-1), IL-1β, TNF-α, IL-6, IL-8, and cyclooxygenase 2 (COX2) (Al-Khayri et al., 2022). In addition, flavonoids can regulate many pathways such as mitogen-activated protein kinase (MAPK), extracellular signal-regulated kinase (ERK), Phosphoinositide three kinase (PI3K)/protein kinase B (Akt), and protein kinase-related pathways to reduce oxidative stress and inflammation (Ullah et al., 2020).

4.2 Antiglycation effects

The advanced glycation end products (AGEs) are derived by a nonenzymatic reaction of sugar with protein side chains. Enhanced production of AGEs plays an important role in the pathogenesis of diabetic complications as well as in natural aging, renal failure, oxidative stress, and chronic inflammation (Chaudhuri et al., 2018).

The synthetic inhibitor of AGEs, such as aminoguanidine (AMG), already used in type 1 diabetic conditions can cause side effects, like weakened liver, anemia, vomiting, gastrointestinal disorder, therefore the search for natural AGEs inhibitors could thus represent a valid alternative approach to synthetic inhibitors (Thornalley, 2003).

Phytochemicals exhibit several antiglycation mechanisms, including effects on glucose metabolism, amelioration of oxidative stress, scavenging of dicarbonyl species, and up/down-regulation of gene expression (Khangholi and Majid, 2016). In the last 3 decades, there has been a significant increase in the search for antiAGEs agents of natural origin.

Considering isolated natural compounds, phloretin (a dihydrochalcone) was able to capture active dicarbonyl compounds such as methylglyoxal (MGO) as well as curcumin which can directly capture MGO by forming curcumin-MGO adducts, leading to a reduction in AGEs formation (Zhong et al., 2019; Alizadeh et al., 2019). Kaempferol, the well-known antioxidant flavonol aglycone, was detected to inhibit the early stages of AGEs formation by scavenging MGO in physiological conditions, forming mono-MGO and di-MGO adduct (Yang et al., 2017). In another study, it was shown that phloretin and EGCG exhibited antiglycative properties in high glucose-stimulated human retinal pigment epithelial (HRPE) cells (Sampath et al., 2016). Caffeic acid (CA) significantly inhibits AGEs-induced intracellular ROS generation with respect to the control, in human umbilical vein endothelial cells (HUVECs) (Cao et al., 2019).

Considering plant extract, rosmarinus officinalis L. (Lamiaceae), which mainly contains rosmarinic acid, carnosic acid, and carnosol, possesses antioxidant activity and antiglycation properties comparable to AG; camellia sinensis L. (Theaceae), which is a rich source of EGCG and (−)-epicatechin 3-O-gallate (ECG), has strong antioxidant properties and inhibits the accumulation of AGEs, such as N-ε-carboxymethyl-L-lysine (CML) and N-ε-carboxyethyl-L-lysine (CEL) and the activation of receptor for AGEs (RAGE) (Wu et al., 2011). In the glucose-glycated BSA models, the addition of green tea extract inhibits final AGEs formation similarly to AMG (Ramlagan et al., 2017). Hazelnut skin polyphenol-rich extract has been shown to protect macrophage cell viability from damage induced by AGEs and reduce levels of pro-inflammatory cytokines and ROS in vitro (Spagnuolo et al., 2023).

4.3 Antidiabetic properties

Diabetes is a heterogeneous disease, a metabolic disorder characterized by the dysregulation of glucose metabolism resulting from anomalies in insulin secretion or action, divided into two categories: Type 1 Diabetes Mellitus (T1DM) and Type 2 Diabetes Mellitus (T2DM) (Tuomi et al., 2014). Several studies showed that phenolic compounds isolated from plants could be used for the prevention and treatment of diabetes, due to their antioxidant and antiinflammation abilities (Pannucci et al., 2023).

Among phenolic compounds, resveratrol decreases blood glucose in animals with hyperglycemia, by stimulating glucose uptake due to increased action of glucose transporter in the plasma membrane (Huang et al., 2020). Curcumin shows hypoglycemic action due to activation of Peroxisome Proliferator Activated Receptor Gamma (PPARγ), the site of action of the thiazolidinedione class of antidiabetic drugs and reduction of AGEs-induced complications in diabetes mellitus (Pivari et al., 2019). The possible mechanism of EA on glucose lowering may be through potentiation of insulin secretion from β-cells or due to enhanced transport of blood glucose to the peripheral tissue and reduction of the formation of glycosylated hemoglobin as shown in a rat model of diabetes (Malini and Kanchana, 2011). Other studies reported that quercetin shows several positive effects in treating T2DM by reducing hyperglycemia, enzyme levels, liver glucose content and improving glucose tolerance and enhancing the subsequent release of insulin and pancreatic cell regeneration (Chen et al., 2016; Haddad et al., 2015).

The regulation of carbohydrate metabolism and hyperglycemia by the inhibition of α-amylase and α-glucosidase represents a promising strategy for the treatment of diabetes. Some phenolic compounds slow down the absorption of glucose through the inhibition of these two enzymes, which are critical in the digestion of carbohydrates (Sun et al., 2019).

4.4 Antiaging activities

Human aging is often accompanied by the deterioration of health. The increase of oxidative processes in cells and tissues is one of the most important factors influencing age-related diseases. Aging is often accompanied by the development of metabolic disorders (the most common of which is T2DM) as well as neurodegenerative diseases, the development of inflammatory processes, all changes that have a significant impact on maintaining the quality of life (Liguori et al., 2018).

It has been shown that natural compounds have the potential to reduce the risk of aging-related diseases. Indeed, quercetin, a flavonoid, shows antioxidant, antiapoptotic and antiinflammatory properties, taking an important role in the treatment of aging-related diseases (Chondrogianni et al., 2010). In addition, curcumin, resveratrol and quercetin show a potential role in limiting aging processes by modulating multiple pathways implicated in different aspects of senescence and aging, which contribute to aging and aging-related diseases (Spagnuolo et al., 2023).

Several studies have demonstrated that some phenolic compounds ameliorated the harmful effects of UVA on aged dermal fibroblasts. Damaged collagen fibrils in the UV-affected dermis are caused by matrix-degrading metalloproteinases (MMPs) synthesis that take part in inflammatory processes by modulating chemokine activity. Apigenin, curcumin and myricetin decreased the expression of MMP-1 in UVA-irradiated human dermal fibroblasts (HDFs) (Choi et al., 2016).

Resveratrol is known for its potential to slow the progression of many age-related human diseases, such as some types of cancer, neurological disorders (e.g., Alzheimer's disease), cardiovascular diseases and diabetes (Berman et al., 2017). Catechins, a class of polyphenols abundant in *Camelia sinensis*, are used in the cosmetic industry because they have demonstrated skin antiaging effects (Hong et al., 2013). In addition, catechins and their derivatives were able to extend the lifespan of *Caenorhabditis elegans*, *Drosophila melanogaster* and *Mus musculus* (Pallauf et al., 2017).

4.5 Bioavailability

The bioaccessibility and bioavailability of many dietary phenolic compounds are not well defined. Phenolic compounds are known to have health-promoting effects and they are also used as functional foods, nutraceuticals or cosmeceuticals. However, their application is limited due to the problems related to their stability and water solubility as well as their low bioaccessibility and bioavailability which depends on metabolic processes mediated by the liver, intestine and microbiota (Ozkan et al., 2020).

Phenolic compounds may be first hydrolyzed by gastric fluids in the stomach and later metabolized by the enzymes of intestinal cells or catabolized by the colonic microflora, which may drastically affect the absorption of these molecules through the gut barrier (Velderrain-Rodríguez et al., 2014). Different experimental in vitro as well as in vivo models have been used to explore the bioavailability, pharmacokinetics and metabolism of bioactive compounds in order to better understand the role of metabolized derivatives and to explain how compounds can be held responsible for the observed health effects (Fernández-Ochoa et al., 2022).

To date, several technologies have been developed to improve drug delivery systems such as the encapsulation of compounds using nanoparticles or various emulsions, to protect them against enzymatic degradation during digestion and increase intestinal uptake, resulting in higher gut concentration as well as increased plasma levels (Jafari and McClements, 2017).

References

Abubakar, A.R., Haque, M., 2020. Preparation of medicinal plants: basic extraction and fractionation procedures for experimental purposes. J. Pharm. Bioallied. Sci. 12, 1–10. https://doi.org/10.4103/jpbs.JPBS_175_19.

Akhtar, I., Javad, S., Yousaf, Z., Iqbal, S., Jabeen, K., 2019. Microwave assisted extraction of phytochemicals an efficient and modern approach for botanicals and pharmaceuticals. Pak. J. Pharm. Sci. 32 (1), 223–230.

Alizadeh, M., Kheirouri, S., 2019. Curcumin against advanced glycation end products (AGEs) and AGEs-induced detrimental agents. Crit. Rev. Food. Sci. Nutr. 59 (7), 1169–1177. https://doi.org/10.1080/10408398.2017.1396200.

Al-Khayri, J.M., Sahana, G.R., Nagella, P., Joseph, B.V., Alessa, F.M., Al-Mssallem, M.Q., 2022. Flavonoids as potential anti-inflammatory molecules: a review. Molecules 27 (9). https://doi.org/10.3390/molecules27092901.

Altemimi, A., Lakhssassi, N., Baharlouei, A., Watson, D.G., Lightfoot, D.A., 2017. Phytochemicals: extraction, isolation, and identification of bioactive compounds from plant extracts. Plants 6 (4). https://doi.org/10.3390/plants6040042.

Arena, K., Cacciola, F., Miceli, N., Taviano, M.F., Cavò, E., Murphy, R.E., Dugo, P., Mondello, L., 2023. Determination of the polyphenolic content of berry juices using focusing-modulated comprehensive two-dimensional liquid chromatography coupled to mass spectrometry detection. Anal. Bioanal. Chem. 415 (13), 2371–2382. https://doi.org/10.1007/s00216-022-04216-y.

Arranz, S., Silván, J.M., et al., 2010. Nonextractable polyphenols, usually ignored, are the major part of dietary polyphenols: a study on the Spanish diet. Mol. Nutr. Food. Res. 54 (11), 1646–1658. https://doi.org/10.1002/mnfr.200900580. PMID: 20540148.

Asensi, M., Ortega, A., Mena, S., Feddi, F., Estrela, J.M., 2011. Natural polyphenols in cancer therapy. Crit. Rev. Clin. Lab Sci. 48 (5–6), 197–216. https://doi.org/10.3109/10408363.2011.631268.

Banu, K.S., Cathrine, L., 2015. General techniques involved in phytochemical analysis. Int. J. Adv. Res. Chem. Sci. 2, 25–32.

Berman, A.Y., Motechin, R.A., Wiesenfeld, M.Y., Holz, M.K., 2017. The therapeutic potential of resveratrol: a review of clinical trials. Npj Precis. Oncol. 1 (1). https://doi.org/10.1038/s41698-017-0038-6.

Bounegru, A.V., Apetrei, C., 2021. Development of a novel electrochemical biosensor based on carbon nanofibers—cobalt phthalocyanine—laccase for the detection of p-coumaric acid in phytoproducts. Int. J. Mol. Sci. 22 (17), 9302. https://doi.org/10.3390/ijms22179302.

Cao, X., Xia, Y., Zeng, M., Wang, W., He, Y., Liu, J., 2019. Caffeic acid inhibits the formation of advanced glycation end products (AGEs) and mitigates the AGEs-induced oxidative stress and inflammation reaction in human umbilical vein endothelial cells (HUVECs). Chem. Biodivers. 16 (10). https://doi.org/10.1002/cbdv.201900174.

Cervantes-Paz, B., Yahia, E.M., Nuñez-Vilchis, A., 2023. Identification and quantification of fatty acids and lipid-soluble phytochemicals using GC-MS, HPLC-MS, and FTIR and their association with quality parameters during avocado ripening. J. Food Sci. 88, 119—132. https://doi.org/10.1111/1750-3841.16390.

Charles, D., Gethings, L.A., Potts, J.F., Burney, P.G.J., Garcia-Larsen, V., 2021. Mass spectrometry-based metabolomics for the discovery of candidate markers of flavonoid and polyphenolic intake in adults. Sci. Rep. 11 (1). https://doi.org/10.1038/s41598-021-85190-w.

Chaudhuri, J., Bains, Y., Guha, S., Kahn, A., Hall, D., Bose, N., Gugliucci, A., Kapahi, P., 2018. The role of advanced glycation end products in aging and metabolic diseases: bridging association and causality. Cell Metabol. 28 (3), 337—352. https://doi.org/10.1016/j.cmet.2018.08.014.

Chen, S., Jiang, H., Wu, X., Fang, J., 2016. Therapeutic effects of quercetin on inflammation, obesity, and type 2 diabetes. Mediat. Inflamm. 2016, 1—5. https://doi.org/10.1155/2016/9340637.

Chiriac, E.R., Chitescu, C.L., Geană, E.I., Gird, C.E., Socoteanu, R.P., Boscencu, R., 2021. Advanced analytical approaches for the analysis of polyphenols in plants matrices—a review. Separations 8 (5). https://doi.org/10.3390/separations8050065.

Choi, S., Youn, J., Kim, K., Joo, D.H., Shin, S., Lee, J., Lee, H.K., An, I.S., Kwon, S., Youn, H.J., Ahn, K.J., An, S., Cha, H.J., 2016. WaveApigenin inhibits UVA-induced cytotoxicity in vitro and prevents signs of skin aging in vivo. Int. J. Mol. Med. 38 (2), 627—634. https://doi.org/10.3892/ijmm.2016.2626.

Chondrogianni, N., Kapeta, S., Chinou, I., Vassilatou, K., Papassideri, I., Gonos, E.S., 2010. Anti-ageing and rejuvenating effects of quercetin. Exp. Gerontol. 45 (10), 763—771. https://doi.org/10.1016/j.exger.2010.07.001.

Cortés-Herrera, C., Chacón, A., Artavia, G., Granados-Chinchilla, F., 2019. Simultaneous LC/MS analysis of carotenoids and fat-soluble vitamins in Costa Rican avocados (Persea americana mill.). Molecules 24 (24), 4517. https://doi.org/10.3390/molecules24244517.

Cory, H., Passarelli, S., Szeto, J., Tamez, M., Mattei, J., 2018. The role of polyphenols in human health and food systems: a mini-review. Front. Nutr. 5. https://doi.org/10.3389/fnut.2018.00087.

Cui, Q., Liu, J.Z., Wang, L.T., Kang, Y.F., Meng, Y., Jiao, J., Fu, Y.J., 2018. Sustainable deep eutectic solvents preparation and their efficiency in extraction and enrichment of main bioactive flavonoids from sea buckthorn leaves. J. Clean. Prod. 184, 826—835. https://doi.org/10.1016/j.jclepro.2018.02.295.

Da Silva, L.C., Viganó, J., Sanches, V.L., De Souza Mesquita, L.M., Pizani, R., Rostagno, M.A., 2023. Simultaneous extraction and analysis of apple pomace by gradient pressurized liquid extraction coupled in-line with solid-phase extraction and on-line with HPLC. Food Chem. 407. https://doi.org/10.1016/j.foodchem.2022.135117.

De Pace, N.L., Colombo, J., 2019. Diagnosis, Prevention, and Treatment for Mind-Body Wellness. Springer International Publishing, Cham (Switzerland). https://doi.org/10.1007/978-3-030-17016-5.

Dheyab, A.S., Bakar, M.F.A., Alomar, M., Sabran, S.F., Hanafi, A.F.M., Mohamad, A., 2021. Deep eutectic solvents (DESs) as green extraction media of beneficial bioactive phytochemicals. Separations 8 (10). https://doi.org/10.3390/SEPARATIONS8100176.

Dugo, L., Russo, M., Cacciola, F., Mandolfino, F., Salafia, F., Vilmercati, A., Fanali, C., Casale, M., De Gara, L., Dugo, P., Mondello, L., Rigano, F., 2020. Determination of the phenol and tocopherol content in Italian high-quality extra-virgin olive oils by using LC-MS and multivariate data analysis. Food Anal. Methods 13 (5), 1027–1041. https://doi.org/10.1007/s12161-020-01721-7.

Dugo, P., Piperno, A., Romeo, R., Cambria, M., Russo, M., Carnovale, C., Mondello, L., 2009. Determination of oxygen heterocyclic components in citrus products by HPLC with UV detection. J. Agric. Food Chem. 57 (15), 6543–6551. https://doi.org/10.1021/jf901209r.

Fernández-Ochoa, Á., de la Luz Cádiz-Gurrea, M., Fernández-Moreno, P., Rojas-García, A., Arráez-Román, D., Segura-Carretero, A., 2022. Recent analytical approaches for the study of bioavailability and metabolism of bioactive phenolic compounds. Molecules 27 (3), 777. https://doi.org/10.3390/molecules27030777.

Folin, O., Ciocalteu, V., 1927. On tyrosine and tryptophan determinations in proteins. J. Biol. Chem. 73, 627–650.

Fresco, P., Borges, F., Marques, M., Diniz, C., 2010. The anticancer properties of dietary polyphenols and its relation with apoptosis. Curr. Pharmaceut. Des. 16 (1), 114–134. https://doi.org/10.2174/138161210789941856.

Gackowski, M., Przybylska, A., Kruszewski, S., Koba, M., Mądra-Gackowska, K., Bogacz, A., 2021. Recent applications of capillary electrophoresis in the determination of active compounds in medicinal plants and pharmaceutical formulations. Molecules 26 (14), 4141. https://doi.org/10.3390/molecules26144141.

Gasmi, A., Kumar Mujawdiya, P., Noor, S., Lysiuk, R., Darmohray, R., Piscopo, S., Lenchyk, L., Antonyak, H., Dehtiarova, K., Shanaida, M., Polishchuk, A., Shanaida, V., Peana, M., Bjørklund, G., 2022. Polyphenols in metabolic diseases. Molecules 27 (19), 6280. https://doi.org/10.3390/molecules27196280.

Geeta, H.P., Srinivas, G.P., Champawat, S., 2020. Supercritical fluid extraction of bioactive compounds from bioresource: A review. Int. J. Curr. Microbiol. Appl. Sci. 9, 559–566.

González-Vallinas, M., González-Castejón, M., Rodríguez-Casado, A., Ramírez de Molina, A., 2013. Dietary phytochemicals in cancer prevention and therapy: a complementary approach with promising perspectives. Nutr. Rev. 71 (9), 585–599. https://doi.org/10.1111/nure.12051.

Goto, T., Obara, M., Aoki, S., Okazawa, K., Konisho, K., Osakabe, N., Shoji, T., 2021. Evaluation of polyphenolic content and potential antioxidant activity of Japanese cultivars of peaches, prunes, and plums based on reversed- and normal-phase HPLC and principal component analyses. ACS Food Sci. Technol. 1 (10), 2019–2029. https://doi.org/10.1021/acsfoodscitech.1c00357.

Guo, Q., Li, F., Duan, Y., Wen, C., Wang, W., Zhang, L., Huang, R., Yin, Y., 2020. Oxidative stress, nutritional antioxidants and beyond. Sci. China Life Sci. 63 (6), 866–874. https://doi.org/10.1007/s11427-019-9591-5.

Haddad, P., Eid, H., Nachar, A., et al., 2015. The molecular basis of the antidiabetic action of quercetin in cultured skeletal muscle cells and hepatocytes. Pharmacogn. Mag. 11 (41), 74. https://doi.org/10.4103/0973-1296.149708.

Hernández-Hernández, C., Fernández-Cabanás, V.M., Rodríguez-Gutiérrez, G., Bermúdez-Oria, A., Morales-Sillero, A., 2021. Viability of near infrared spectroscopy for a rapid analysis of the bioactive compounds in intact cocoa bean husk. Food Cont. 120, 107526. https://doi.org/10.1016/j.foodcont.2020.107526.

Herrero, M., 2023. Towards green extraction of bioactive natural compounds. Anal. Bioanal. Chem. https://doi.org/10.1007/s00216-023-04969-0.

Hong, Y.H., Jung, E.Y., Shin, K.S., Yu, K.W., Chang, U.J., Suh, H.J., 2013. Tannase-converted green tea catechins and their anti-wrinkle activity in humans. J. Cosmet. Dermatol. 12 (2), 137–143. https://doi.org/10.1111/jocd.12038.

Hsiao, Y.H., Ho, C.T., Pan, M.H., 2020. Bioavailability and health benefits of major isoflavone aglycones and their metabolites. J. Func. Foods 74. https://doi.org/10.1016/j.jff.2020.104164.

Hsieh, Y.H., Li, Y., Pan, Z., Chen, Z., Lu, J., Yuan, J., et al., 2019. Ultrasonication-assisted synthesis of alcohol-based deep eutectic solvents for extraction of active compounds from ginger. Ultrason. Sonochem. 63, 104915. https://doi.org/10.1016/j.ultsonch.2019.104915.

Huang, D.-D., Shi, G., Jiang, Y., Yao, C., Zhu, C., 2020. A review on the potential of Resveratrol in prevention and therapy of diabetes and diabetic complications. Biomed. Pharmacother. 125, 109767. https://doi.org/10.1016/j.biopha.2019.109767.

Ignat, I., Volf, I., Popa, V.I., 2011. A critical review of methods for characterisation of polyphenolic compounds in fruits and vegetables. Food Chem. 126, 1821–1835. https://doi.org/10.1016/j.foodchem.2010.12.026.

Jafari, S.M., McClements, D.J., 2017. Nanotechnology approaches for increasing nutrient bioavailability. Adv. Food. Nutr. Res. 81, 1–30. https://doi.org/10.1016/bs.afnr.2016.12.008.

Jantan, I., Haque, M.A., Arshad, L., Harikrishnan, H., Septama, A.W., Mohamed-Hussein, Z.A., 2021. Dietary polyphenols suppress chronic inflammation by modulation of multiple inflammation-associated cell signaling pathways. JNB (J. Nutr. Biochem.) 93. https://doi.org/10.1016/j.jnutbio.2021.108634.

Johnson, J.B., Walsh, K.B., Naiker, M., Ameer, K., 2023. The use of infrared spectroscopy for the quantification of bioactive compounds in food: a review. Molecules 28 (7). https://doi.org/10.3390/molecules28073215.

Johnson, J., Mani, J., Ashwath, N., Naiker, M., 2020. Potential for Fourier transform infrared (FTIR) spectroscopy toward predicting antioxidant and phenolic contents in powdered plant matrices. Spectrochim. Acta Mol. Biomol. Spectrosc. 233, 118228. https://doi.org/10.1016/j.saa.2020.118228.

Kalyniukova, A., Holuša, J., Musiolek, D., Sedlakova-Kadukova, J., Płotka-Wasylka, J., Andruch, V., 2021. Application of deep eutectic solvents for separation and determination of bioactive compounds in medicinal plants. Ind. Crop. Prod. 172, 114047. https://doi.org/10.1016/j.indcrop.2021.114047.

Khangholi, S., Majid, F.A., 2016. The mechanisms of inhibition of advanced glycation end products formation through polyphenols in hyperglycemic condition. Planta. Med. 82, 32–45. https://doi.org/10.1055/s-0035-1558086.

Khoddami, A., Wilkes, M.A., Roberts, T.H., 2013. Techniques for analysis of plant phenolic compounds. Molecules 18 (2), 2328–2375. https://doi.org/10.3390/molecules18022328.

Kumar, A., Nirmal, P., Kumar, M., Jose, A., Tomer, V., Oz, E., Proestos, C., Zeng, M., Elobeid, T., Sneha, K., Oz, F., 2023. Major phytochemicals: recent advances in health benefits and extraction method. Molecules 28 (2), 887. https://doi.org/10.3390/molecules28020887.

Kumar, K., Srivastav, S., Sharanagat, V.S., 2021. Ultrasound assisted extraction (UAE) of bioactive compounds from fruit and vegetable processing by-products: A review. Ultrason. Sonochem. 70, 105325. https://doi.org/10.1016/j.ultsonch.2020.105325.

Lama-Muñoz, A., Contreras, M.D.M., 2022. Extraction systems and analytical techniques for food phenolic compounds: A review. Foods 11, 3671. https://doi.org/10.3390/foods11223671.

Leiherer, A., Mündlein, A., Drexel, H., 2013. Phytochemicals and their impact on adipose tissue inflammation and diabetes. Vascul. Pharmacol. 58, 3–20. https://doi.org/10.1016/j.vph.2012.09.002.

Liguori, I., Russo, G., Curcio, F., Bulli, G., Aran, L., Della-Morte, D., Gargiulo, G., Testa, G., Cacciatore, F., Bonaduce, D., Abete, P., 2018. Oxidative stress, aging, and diseases. Clin. Interv. Aging 13, 757–772. https://doi.org/10.2147/CIA.S158513.

Lucarini, M., Durazzo, A., Kiefer, J., Santini, A., Lombardi-Boccia, G., Souto, E.B., Romani, A., Lampe, A., Nicoli, S.F., Gabrielli, P., Bevilacqua, N., Campo, M., Morassut, M., Cecchini, F., 2020. Grape seeds: chromatographic profile of fatty acids and phenolic compounds and qualitative analysis by FTIR-ATR spectroscopy. Foods 9 (1). https://doi.org/10.3390/foods9010010.

Malini, P., Kanchana, G., 2011. Antibiabetic efficacy of ellagic acid in streptozotocin induced diabetes mellitus in albino wistar rats. Asian J. Pharm. Clin. Res. 4 (3), 124–128.

Malireddy, S., Kotha, S.R., Secor, J.D., Gurney, T.O., Abbott, J.L., Maulik, G., Maddipati, K.R., Parinandi, N.L., 2012. Phytochemical antioxidants modulate mammalian cellular epigenome: implications in health and disease. Antiox. Redox Signal. 17 (2), 327–339. https://doi.org/10.1089/ars.2012.4600.

Manach, C., Scalbert, A., Morand, C., Rémésy, C., Jiménez, L., 2004. Polyphenols: food sources and bioavailability. Am. J. Clin. Nutr. 79 (5), 727–747. https://doi.org/10.1093/ajcn/79.5.727.

Maran, J.P., Priya, B., Al-Dhabi, N.A., Ponmurugan, K., Moorthy, I.G., Sivarajasekar, N., 2017. Ultrasound assisted citric acid mediated pectin extraction from industrial waste of Musa balbisiana. Ultrason. Sonochem. 35, 204–209. https://doi.org/10.1016/j.ultsonch.2016.09.019.

Marković, Z., Milenković, D., Đorović, J., Dimitrić Marković, J.M., Lučić, B., Amić, D., 2013. A DFT and PM6 study of free radical scavenging activity of ellagic acid. Monatshefte für Chemie - Chem. Monthly 144 (6), 803–812. https://doi.org/10.1007/s00706-013-0949-z.

Martillanes, S., Rocha-Pimienta, J., Cabrera-Bañegil, M., Martín-Vertedor, D., Delgado-Adámez, J., 2017. Application of Phenolic Compounds for Food Preservation: Food Additive and Active Packaging. InTech. https://doi.org/10.5772/66885.

Menezes, R., Rodriguez-Mateos, A., Kaltsatou, A., González-Sarrías, A., Greyling, A., Giannaki, C., Andres-Lacueva, C., Milenkovic, D., Gibney, E.R., Dumont, J., Schär, M., Garcia-Aloy, M., Palma-Duran, S.A., Ruskovska, T., Maksimova, V., Combet, E., Pinto, P., 2017. Impact of flavonols on cardiometabolic biomarkers: a meta-analysis of randomized controlled human trials to explore the role of inter-individual variability. Nutrients 9 (2). https://doi.org/10.3390/nu9020117.

Mir-Cerdà, A., Nuñez, O., Granados, M., Sentellas, S., Saurina, J., 2023. An overview of the extraction and characterization of bioactive phenolic compounds from agri-food waste within the framework of circular bioeconomy. TrAC, Trends Anal. Chem. 161, 116994. https://doi.org/10.1016/j.trac.2023.116994.

Munteanu, I.G., Apetrei, C., 2021. Analytical methods used in determining antioxidant activity: A review. Int. J. Mol. Sci. 22, 3380. https://doi.org/10.3390/ijms22073380.

Nadar, S.S., Rao, P., Rathod, V.K., 2018. Enzyme assisted extraction of biomolecules as an approach to novel extraction technology: A review. Food Res. Int. 108, 309–330. https://doi.org/10.1016/j.foodres.2018.03.006.

Ozkan, G., Kostka, T., Esatbeyoglu, T., Capanoglu, E., 2020. Effects of lipid-based encapsulation on the bioaccessibility and bioavailability of phenolic compounds. Molecules 25 (23), 5545. https://doi.org/10.3390/molecules25235545.

Pallauf, K., Duckstein, N., Rimbach, G., 2017. A literature review of flavonoids and lifespan in model organisms. Proc. Nutr. Soc. 76, 145–162.

Pandey, K.B., Rizvi, S.I., 2009. Plant polyphenols as dietary antioxidants in human health and disease. Oxid. Med. Cell. Longev. 2 (5), 270–278. https://doi.org/10.4161/oxim.2.5.9498.

Pannucci, E., Spagnuolo, L., De Gara, L., Santi, L., Dugo, L., 2023. Phenolic compounds as preventive and therapeutic agents in diabetes-related oxidative stress, inflammation, advanced glycation end-products production and insulin sensitivity. Discov. Med. 35 (178), 715. https://doi.org/10.24976/discov.med.202335178.68.

Pereira-Coelho, M., Haas, I.C.D.S., Katrin Reinke, C., Dognini, J., Amboni, R.D.D.M.C., Vitali, L., Madureira, L.A.D.S., 2023. A green analytical method for the determination of polyphenols in wine by dispersive pipette extraction and LC-MS/MS. Food Chem. 405. https://doi.org/10.1016/j.foodchem.2022.134860.

Pivari, F., Mingione, A., Brasacchio, C., Soldati, L., 2019. Curcumin and type 2 diabetes mellitus: prevention and treatment. Nutrients 11 (8), 1837. https://doi.org/10.3390/nu11081837.

Przybylska, A., Gackowski, M., Koba, M., 2021. Application of capillary electrophoresis to the analysis of bioactive compounds in herbal raw materials. Molecules 26, 2135. https://doi.org/10.3390/molecules26082135.

Radzali, S.A., Markom, M., Saleh, N.M., 2020. Co-solvent selection for supercritical fluid extraction (SFE) of phenolic compounds from *Labisia pumila*. Molecules 25, 5859. https://doi.org/10.3390/molecules25245859.

Ramlagan, P., Rondeau, P., Planesse, C., Neergheen-Bhujun, V.S., Bourdon, E., Bahorun, T., 2017. Comparative suppressing effects of black and green teas on the formation of advanced glycation end products (AGEs) and AGE-induced oxidative stress. Food Funct. 8 (11), 4194–4209. https://doi.org/10.1039/c7fo01038a.

Rao, H., Ahmad, S., Aati, H.Y., Basit, A., Ahmad, I., Ahmad Ghalloo, B., Nadeem Shehzad, M., Nazar, R., Zeeshan, M., Nasim, M.J., ur Rehman Khan, K., 2023. Phytochemical screening, biological evaluation, and molecular docking studies of aerial parts of Trigonella hamosa (branched Fenugreek). Arab. J. Chem. 16 (7). https://doi.org/10.1016/j.arabjc.2023.104795.

Rasouli, H., Farzaei, M.H., Khodarahmi, R., 2017. Polyphenols and their benefits: A review. Int. J. Food Prop. 20, 1−42. https://doi.org/10.1080/10942912.2017.1354017.

Rivière, C., Pawlus, A.D., Mérillon, J.M., 2012. Natural stilbenoids: distribution in the plant kingdom and chemotaxonomic interest in Vitaceae. Nat. Prod. Rep. 29 (11), 1317−1333. https://doi.org/10.1039/c2np20049j.

Russo, M., Bonaccorsi, I.L., Cacciola, F., Dugo, L., De Gara, L., Dugo, P., Mondello, L., 2021. Distribution of bioactives in entire mill chain from the drupe to the oil and wastes. Nat. Prod. Res. 35 (21), 4182−4187. https://doi.org/10.1080/14786419.2020.1752208.

Sadeghi, A., Rostamirad, A., Seyyedebrahimi, S., Meshkani, R., 2018. Curcumin ameliorates palmitate-induced inflammation in skeletal muscle cells by regulating JNK/NF-kB pathway and ROS production. Inflammopharmacology 26 (5), 1265−1272. https://doi.org/10.1007/s10787-018-0466-0.

Sagbo, I.J., Orock, A.E., Kola, E., Otang-Mbeng, W., 2020. Phytochemical screening and gas chromatography-mass spectrometry analysis of ethanol extract of Scambiosa Columbabria L. Pharmacogn. Res. 12 (1), 35. https://doi.org/10.4103/pr.pr_59_19.

Salah, H.B., Allouche, N., 2019. Plant-based chemicals extraction and isolation. Plant based "green chemistry 2.0". Springer, Singapore, pp. 89−117.

Sampath, C., Sang, S, Ahmedna, M., 2016. In vitro and in vivo inhibition of aldose reductase and advanced glycation end products by phloretin, epigallocatechin 3-gallate and [6]-gingerol. Biomed. Pharmacother. 84, 502−513. https://doi.org/10.1016/j.biopha.2016.09.073.

Sanches-Silva, A., Testai, L., Nabavi, S.F., Battino, M., Pandima Devi, K., Tejada, S., Sureda, A., Xu, S., Yousefi, B., Majidinia, M., Russo, G.L., Efferth, T., Nabavi, S.M., Farzaei, M.H., 2020. Therapeutic potential of polyphenols in cardiovascular diseases: regulation of mTOR signaling pathway. Pharmacol. Res. 152. https://doi.org/10.1016/j.phrs.2019.104626.

Schofield, P., Mbugua, D.M., Pell, A.N., 2001. Analysis of condensed tannins: A review. Anim. Feed Sci. Technol. 91, 21−40. https://doi.org/10.1016/S0377-8401(01)00228-0.

Sharma, M., Kaushik, P., 2021. Vegetable phytochemicals: An update on extraction and analysis techniques. Biocatal. Agric. Biotechnol. 36, 102149. https://doi.org/10.1016/j.bcab.2021.102149.

Shi, L., Zhao, W., Yang, Z., Subbiah, V., Suleria, H.A.R., 2022. Extraction and characterization of phenolic compounds and their potential antioxidant activities. Environ. Sci. Pollut. Control Ser. 29 (54), 81112−81129. https://doi.org/10.1007/s11356-022-23337-6.

Shraim, A.M., Ahmed, T.A., Rahman, M.M., Hijji, Y.M., 2021. Determination of total flavonoid content by aluminum chloride assay: a critical evaluation. LWT 150. https://doi.org/10.1016/j.lwt.2021.111932.

Spagnuolo, L., Della Posta, S., et al., 2023. Chemical composition of hazelnut skin food waste and protective role against advanced glycation end-products (AGEs) damage in THP-1-derived macrophages. Molecules 28, 2680. https://doi.org/10.3390/molecules28062680.

Sun, L., Warren, F.J., Gidley, M.J., 2019. Natural products for glycaemic control: Polyphenols as inhibitors of alpha-amylase. Trends. Food. Sci. 91, 262−273. https://doi.org/10.1016/j.tifs.2019.07.009.

Tebbi, S.O., Debbache-Benaida, N., Moulaoui, K., Zaidi, S., Kadi, R., 2022. Optimized ultrasonic-assisted deep eutectic solvents extraction of Clematis flammula L. leaves, phytochemical screening, biological activities and the characterization of its volatile compounds. Biomass Convers. Biorefin. https://doi.org/10.1007/s13399-022-03585-9.

Thornalley, P.J., 2003. Use of aminoguanidine (Pimagedine) to prevent the formation of advanced glycation endproducts. Arch. Biochem. Biophys. 419 (1), 31−40. https://doi.org/10.1016/j.abb.2003.08.013.

Tsao, R., 2010. Chemistry and biochemistry of dietary polyphenols. Nutrients 2 (12), 1231–1246. https://doi.org/10.3390/nu2121231.

Tuomi, T., Santoro, N., Caprio, S., 2014. The many faces of diabetes: a disease with increasing heterogeneity. Lancet 383 (9922), 1084–1094. https://doi.org/10.1016/S0140-6736(13)62219-9.

Ullah, A., Munir, S., Badshah, S.L., Khan, N., Ghani, L., Poulson, B.G., Emwas, A.H., Jaremko, M., 2020. Important flavonoids and their role as a therapeutic agent. Molecules 25 (22). https://doi.org/10.3390/molecules25225243.

Velderrain-Rodríguez, G.R., Palafox-Carlos, H., Wall-Medrano, A., Ayala-Zavala, J.F., Chen, C.-Y.O., Robles-Sánchez, M., Astiazaran-García, H., Alvarez-Parrilla, E., González-Aguilar, G.A., 2014. Phenolic compounds: their journey after intake. Food Funct. 5 (2), 189–197. https://doi.org/10.1039/c3fo60361j.

Verma, S., Thakur, D., Pandey, C.M., Kumar, D., 2023. Recent prospects of carbonaceous nanomaterials-based laccase biosensor for electrochemical detection of phenolic compounds. Biosensors 13 (3). https://doi.org/10.3390/bios13030305.

Wu, C.H., Huang, S.M., Lin, J.A., Yen, G.C., 2011. Inhibition of advanced glycation endproduct formation by foodstuffs. Food Funct. 2 (5), 224–234. https://doi.org/10.1039/c1fo10026b.

Yahfoufi, N., Alsadi, N., Jambi, M., Matar, C., 2018. The immunomodulatory and anti-inflammatory role of polyphenols. Nutrients 10 (11), 1618. https://doi.org/10.3390/nu10111618.

Yang, B., Choi, E., Shim, S.M., 2017. Inhibitory activities of kaempferol against methylglyoxal formation, intermediate of advanced glycation end products. Appl. Biol. Chem. 57–62. https://doi.org/10.1007/s13765-016-0251-y.

Zálešák, F., Bon, D.J.D., Pospíšil, J., 2019. Lignans and Neolignans: Plant secondary metabolites as a reservoir of biologically active substances. Pharmacol. Res. 146. https://doi.org/10.1016/j.phrs.2019.104284.

Zhang, Q.W., Lin, L.-G., Ye, W.-C., 2018. Techniques for extraction and isolation of natural products: A comprehensive review. Chin. Med. 13, 20. https://doi.org/10.1186/s13020-018-0177-x.

Zhang, H., Tsao, R., 2016. Dietary polyphenols, oxidative stress and antioxidant and anti-inflammatory effects. Curr. Opin. Food Sci. Vol. 8, 33–42. https://doi.org/10.1016/j.cofs.2016.02.002.

Zhang, L., Wang, S., Yang, R., Jin, M., Jiang, J., Wang, X., Wen, Z., Qi, Z., Li, P., 2019. Simultaneous determination of tocopherols, carotenoids and phytosterols in edible vegetable oil by ultrasound-assisted saponification, LLE and LC-MS/MS. Food Chem. 289, 313–319. https://doi.org/10.1016/j.foodchem.2019.03.067.

Zhong, X., Liu, M., Yao, W., Du, K., He, M., Jin, X., Jiao, L., Ma, G., Wei, B., Wei, M., 2019. Epigallocatechin-3-Gallate attenuates microglial inflammation and neurotoxicity by suppressing the activation of canonical and noncanonical inflammasome via TLR4/NF-κB pathway. Mol. Nutr. Food Res. 63 (21). https://doi.org/10.1002/mnfr.201801230.

PART 4

Inorganic compounds state-of-the-art

CHAPTER

16 Nonessential pollutant metals in the food chain: state of art and particular focus on thallium and platinum group metals (PGMs)

Dora Melucci[1], Marcello Locatelli[2] and Clinio Locatelli[1]

[1]*Department of Chemistry, University of Bologna, Bologna, Italy;* [2]*Department of Pharmacy, University "G. d'Annunzio" of Chieti-Pescara, Chieti, Italy*

1. Introduction

Metals are an integral part of the food chain. Even if present in the environment with different origins, both natural and, very often, dependent on criminal acts, a significant number of these metals are truly dangerous, often showing harmful and irreversible effects on human health.

Toxic substances coming exclusively from human activities tend to accumulate in all environmental matrices, evidently causing a strong decrease in food safety (Munir et al., 2022; Briffa et al., 2020; Ferré-Huguet et al., 2008; Merian et al., 2004; Ebdon and Pitts, 2002; Reilly, 2002).

Among these substances, toxic metals together with persistent organic pollutants (POPs) are certainly the most dangerous for human health and wildlife (Alharbi et al., 2018; Vasseghian et al., 2021; Mishra et al., 2022).

The literature reports many works regarding the presence of both toxic and nontoxic metals in the food matrices and all the procedures for their analytical determination (Butler et al., 2018; Mir-Marqués et al., 2016; Komendova, 2020).

The present work therefore wants to underline the state of the art, i.e., the photograph of the current situation regarding the problem aforementioned. As for the three absolutely nonessential toxic metals—Pb, Cd, and Hg—in the literature, there is a very extensive and in-depth discussion and a very large number of works, of which we will however only note the most recent and most interesting. On the contrary, we will pay particular attention, therefore a particular focus, on some emerging pollutant metals—Platinum Group Metals (PGMs: Pt, Pd, Rh, Os, Ru, Ir) (Zhitenko et al., 2016; Pawlak et al., 2014; Wang and Li, 2012; Dubiella-Jackowska et al., 2009; Moldovan, 2007; Ravindra et al., 2004; Rao and Reddi, 2000)and thallium (Belzile and Chen, 2017; Karbowska, 2016; Antonia López Antón et al., 2013; Rodríguez-Mercado and Altamirano-Lozano, 2013; Taylor et al., 2006), since these metals have shown their unstoppable and increasingly growing accumulation and presence in the environment.

Phytochemicals for Health. https://doi.org/10.1016/B978-0-443-15366-2.00016-2

In any case, all the metals considered present a very high risk to human health even at trace and ultratrace concentration levels (Komendova, 2020; Blain, 2022; Liu et al., 2020; Doulgeridou et al., 2020; Savignan et al., 2021; Liu et al., 2022).

It should be underlined, however, that, as regards PGMs and thallium, but, above all, PGMs, many studies still need to be carried out to define with absolute safety their danger and toxicity for humans.

It should also be emphasized that, among the matrices discussed, there have been taken into consideration also those that are not strictly food, but are still involved in the food chain: the so-called food supplements.

Such matrices are nutrients or other substances with a physiological or nutritional effect, whose purpose is to also be supplement to normal diet.

In this work, the different procedures for the determination of the metals listed earlier in food—sea food, vegetables, meals, meat, milk, and milk derivatives—and food supplements—wine, tea, and herbal medicines—are critically discussed and reported.

Finally, a brief discussion is reported regarding a matrix that is not strictly part of the food chain, but which is employed daily to care our body: cosmetics. In this context, particular attention is paid to the use of lipsticks, massively used by women. Lipsticks can be a dangerous source of metal ingestion. Indeed, colored metal salts, even very toxic, are used to give different colors to the lipsticks themselves.

2. Essential, semiessential, and nonessential (toxic) metals

Indeed, it is very difficult to make a distinction between essential and nonessential or toxic metals, considering the fact that probably all metals are toxic depending on their concentration. In this sense, we must partly agree with Philippus Aureolus Theophrastus Bombastus von Hohenheim, known as Paracelsus (1493–1541), who stated « Omnia venenum sunt: nec sine veneno quicquam existit. Dosis sola facit, ut venenum non fit » («Everything is poison: nothing exists that is not poisonous. Only the dose ensures that the poison does not have an effect». In extreme summary: "Nothing is poison in itself, everything is poison itself, it is the dose that makes the poison").

In any case, it should be underlined that the classification between essentiality and nonessentiality of metals is strongly linked to the biological role that each metal plays in the human body (Jomova et al., 2022; Cannas et al., 2020; Genchi et al., 2021b; Voica et al., 2021; Kiran and Sharma, 2022; Sanaei et al., 2021; Zhu and Costa, 2020; Abitew and Ymam, 2023; Syed et al., 2023; Gohar and Shakeel, 2021; Zoroddu et al., 2019).

At the moment, and at least to our knowledge, it is believed that the essential elements for the biochemical processes of mammals are 24: H, C, N, O, F, Na, Mg, Si, P, S, Cl, K, Ca, V, Mn, Fe, Co, Ni, Cu, Zn, Se, Mo, Sn, and I.

It should also be underlined that the specific biochemical role of some of them—F, Si, V, Ni, and Sn—is, to date, still poorly understood, while for others, such as B and Cr, the experimental evidence of their essentiality is still not completely clear.

It is equally true that this classification, essentiality versus nonessentiality, is applicable to almost all metals, as there are some of them—lead, cadmium, and mercury—which are always toxic, whatever their concentration and on which, as previously mentioned, there is in literature a very large number of works concerning their presence in food matrices. In this work, and for these metals, we will therefore only highlight the most recent and, above all, the most significant studies.

For this reason, other "new generation" toxic metals will be taken into consideration and discussed in the various food matrices, because they are consistently present in the environment only since approximately 3 decades. These metals are the platinum group metals (PGMs: Pt, Pd, Rh, Os, Ru, Ir) and thallium. Their growing presence in the environment and therefore, ultimately, in the food chain, is due, respectively, to the introduction of catalytic converters in the construction of vehicles as regards the former (PGMs) and to the introduction of new pesticide preparations for the agriculture as regards the second (thallium). In fact, it should be underlined that the equation suggested by numerous authors remains valid, i.e., that strong environmental pollution implies, with absolute certainty, strong pollution of the matrices belonging to the food chain.

The types of food matrices taken into consideration are: sea food—mussels, clams, oysters, fishes and algae—vegetables, meals, meat, milk, and milk derivatives, and the so-called "food supplements—wine, tea, and herbal medicines.

3. Pollution sources of Tl and PGMs

In all cases, the input of thallium and PGMs into the environment (Schuster et al., 2000) always occurs with the same scheme: the various sources emit into the atmosphere PGMs and thallium in the form of dust, vapors, or liquids.

Once in the atmosphere, the subsequent relapse to the ground strongly causes contamination of water and soil. Finally, bioaccumulation of these metals occurs in all plants, in terrestrial and aquatic organisms, which thus enter the food chain of which humans are the final consumers (Liu et al., 2020; Savignan et al., 2021; D'Orazio et al., 2020; Migaszewski and Gałuszka, 2021; Luo et al., 2020; Hughes et al., 2021; Dey and Mehta, 2020).

As for thallium, the sources are both natural—soil erosion and volcanic emissions—and anthropogenic. The latter concern the entire mining industries sector, but also the ever-increasing use of phytosanitary products for agriculture and also their frequent incorrect disposal in water bodies (canals, rivers, etc.).

As thallium, PGMs have the same natural sources, while as regards the anthropic responsibility for pollution from these metals, the main and most important one concerns vehicular traffic, largely related to the use of autocatalytic converters. Other important anthropic sources are also those linked to mining and metal production.

4. Toxicity of Tl and PGMs

Although widely reported in the literature, it is useful to highlight the aspect regarding the toxicity of the elements taken into consideration.

4.1 *Thallium* (Genchi et al., 2021a; Azucena Morales-Zarco et al., 2023)

As is well known, thallium is extremely toxic by inhalation and ingestion. Unfortunately, poisoning is very difficult to diagnose and causes damage to the stomach and nervous system, leading in many cases to coma and death of the person.

4.2 *PGMs* (Komendova, 2020; Merget and Rosner, 2001; Gebel, 2000)

PGMs in metallic form are to date considered to be biologically inert and nonallergic. Really, in environmental matrices, significant quantities of PGMs undergo numerous chemical processes, so converting them into bioavailable forms, mainly as chlorine- or organic-complexes, and then potentially extremely dangerous for human health.

In their bioavailable forms, PGMs are all toxic, even if with a different degree of toxicity. In particular, as widely known, being reported and described in numerous treatises:

- *Platinum*: DNA alterations, damage to intestines and kidneys with possible onset of cancer.
- *Palladium*: very high toxicity, carcinogenic and irritant of the mucous membranes.
- *Rhodium*: potentially toxic and carcinogenic, irritating to the skin.
- *Osmium*: highly toxic as tetroxide (OsO_4), causing congestion of the respiratory tract and damage to skin and eyes.
- *Ruthenium*: highly toxic and carcinogenic, sometimes irreversible damage to the bones.
- *Iridium*: irritating to the eyes and the digestive tract.

5. Food sample typologies

5.1 *Seafood* (filtering organisms: Mussels, clams, and oysters)

Mussels, clams, and oysters are sessile, filter-feeding organisms. They are in fact natural filters: adult mussels and clams filter up to 100–120 L/day, while adult oysters can filter up to 150–200 L/day, depending on weight, size, and water temperature.

It is clear that, alongside the exceptional cleaning properties of water, they also simultaneously have the enormous disadvantage of being contaminated by any polluting agents present in the aquatic ecosystem in which they live.

For this reason, these species, before being placed on the market for human consumption, must absolutely be examined and analyzed from both a biological and chemical-physical point of view.

Specifically regarding toxic metals, alongside the "traditional ones"—lead, cadmium, and mercury—which have been extensively investigated and for which there is an extensive literature and references therein (Gomez-Delgado et al., 2023; Zhelyazkov et al., 2018; Mehouel and Fowler, 2022; José Tamele, 2020; Wang et al., 2022; Barchiesi et al., 2020; Sun et al., 2023), in recent years, some metals of new generation have also been found to be present in these organisms: Thallium and Platinum Group Metals (PGMs).

Contamination of these metals occurs due to fallout in the water from atmosphere of fine particles containing thallium and platinum group metals (PGMs). As regards thallium, there is a further problem: its presence in many phytosanitary products used in agriculture. Improper management of their use, including the discharge of residues at the end of processing into waterways, can produce major pollution problems.

Although there is no regulation on the maximum permissible limits for these metals in food matrices, this chapter aims to highlight their danger in all food matrices, and in filter-feeding organisms in particular, the latter of which in nature show to be of the exceptional concentrators of the polluting species present in the aquatic environment in which they live.

In this context, over the last 20 years, some interesting works have appeared, in particular by two research groups, which have proposed and developed the analytical determination of PGMs in filter-feeding organisms using both spectroscopic and voltammetric techniques.

Zimmerman and coworkers (Erasmus et al., 2020; Sures et al., 2001; Zimmermann et al., 2004, 2015; Zimmermann and Sures, 2018) proposed the comparison in the determination of PGMs in mussels by electrothermal atomic absorption spectrometry (ET-AAS), adsorptive cathodic stripping voltammetry (AdCSV), and total-reflection-X ray fluorescence analysis (TXRF).

Also our research group has been dealing for some time with the problem of the growing presence of PGMs and Tl in the environment with consequent and evident consequences of great danger to human health on the entire food chain (Locatelli, 2000; Locatelli et al., 2018; Locatelli and Fagioli, 1997; Locatelli and Melucci, 2010, 2014a; Melucci, Casolari, et al., 2017).

Lastly, still with regard to filter-feeding marine organisms, their ever-increasing role as excellent "bio-monitors" of the aquatic ecosystems in which they live should not be forgotten.

In fact, being sessile species, they can provide precious and important information on the long-term trend of pollution of the aquatic ecosystem where they live.

In any case, it should be underlined that "over a long period of time" means that they do not provide precise analytical data on the concentration of a pollutant from a spatial and temporal point of view.

As bio-monitors, they therefore provide information on the long-term trend of the polluting load of a specific aquatic ecosystem (Locatelli et al., 2018; Azizi et al., 2018; Saleh et al., 2021; Melucci et al., 2018; Locatelli, 2003; Locatelli et al., 2001; Locatelli and Melucci, 2016a).

Therefore, by combining these two pieces of information, precise concentration data and data provided by bio-monitoring, a complete vision of the state of an aquatic ecosystem from the point of view of pollution load is obtained, simultaneously highlighting both the critical points and the possible responsible sources.

Finally, as regards thallium, only very recently have extremely interesting works appeared in the literature on its determination in filter-feeding organisms (Mijosek et al., 2020; Esposito et al., 2021; Pearson and Ashmore, 2020; Li et al., 2021), and they all underline the great concern for the growing concentration of this metal in aquatic ecosystems, especially considering its very high danger for human health.

5.2 Seafood (fishes)

Also with regard to this matrix, the contamination occurs due to the fallout in the water from atmosphere of fine particles containing PGMs, while as regards thallium, the contamination occurs both in the way described in the case of PGMs and due to the reckless and criminal introduction of phytosanitary aids used during and at the end of processing in agriculture.

Only very recently, in the past decade, evidently attracting the attention of the scientific community in consideration of their strong toxicity and danger, several works have appeared in the literature regarding the determination of both PGMs and Tl in sea food (Essumang et al., 2008; Erasmus et al., 2022; Moreira et al., 2021; Li et al., 2023; Kim et al., 2021; Nagel et al., 2023).

5.3 Sea food (algae)

Algae are autotrophic organisms, capable of producing energy through photosynthesis. Therefore, taking advantage of the sunlight, they are capable of producing nourishment for their survival. In symbiosis with bacteria, generally cyanobacteria, they also feed on organic waste, which is transformed into oxygen and biomass (bio-purification).

Algae have been eaten everywhere for a long time. In fact, they have very important nutritional values including mineral salts and trace elements, vitamins, amino acids, essential fatty acids.

Evidently, their property of feeding and growing in aquatic ecosystems requires strong attention to their food consumption, and in this sense, a growing number of works have recently appeared in the literature (Tofan et al., 2022; Batley and Campbell, 2022).

Alongside the food aspect, the use of algae as species that can also be used in the aquatic ecosystems bio-remediation should absolutely not be forgotten.

5.4 Wine

Toxic metals are present as pollutants in wine in variable concentrations (Coetzee et al., 2014). This evidently depends on various factors: on the polluting load present in the area where the vineyard is planted, on the agricultural practice followed during the growth and ripening of the grape bunches, and only marginally on the winery processing.

Regarding the usual elemental profile of wines, very interesting are the works of Hopfer et al. (2015) and Melucci et al. (2017b). It should also be underlined that the multielement composition of wine, coupled with chemometric procedures (Serapinas et al., 2008; Arduini et al., 2021), can be usefully exploited as a fingerprint of wine origin, in order to prevent and avoid any criminal acts of counterfeiting (Almeida and Vasconcelos, 2003).

Regarding PGMs in wines, only recently have works appeared in the literature. They all involve spectroscopic and electroanalytical techniques for their determination (Cvetkoviá et al., 2002; Gajek et al., 2021; Amorello et al., 2017; Kilmartin, 2016; Aceto, 2003).

It should be underlined that thallium is always determined when it is necessary to establish the elemental profile of a wine suitable for classifying the wine or establishing its origin, if combined with chemometric tools (Canizo et al., 2018).

5.5 Vegetables

The presence of toxic species in plants placed on the market for human consumption is mainly due to the polluted particulate matter in the atmosphere, to the agricultural practices used before harvesting and only marginally to the absorption by the plants themselves of toxic metals present in the soils.

In the past decade, many research groups have proposed and developed voltammetric and spectroscopic methods for the determination of toxic metals in plant matrices (Zwolak et al., 2019; Esposito et al., 2019; Biddau and Cidu, 2017; Ghanjaoui et al., 2011; Falandysz et al., 2017, 2001; Locatelli and Melucci, 2013; Landi et al., 1990; Locatelli et al., 1984), with particular attention to new emerging pollutants: PGMs (Djingova et al., 2003; Alonso Castillo et al., 2012; Niemelä et al., 2004; Qvarforth et al., 2022; Nischkauer et al., 2013; Bernardino et al., 2019) and Tl (Urbánková and

Sommer, 2008; Patočka et al., 2017; Sasmaz et al., 2016; Yu et al., 2018; Jia et al., 2018; Greda et al., 2016; Ghaderi et al., 2018; Kim et al., 2021, Meeravali et al., 2023).

Over the past decade, our research group has also addressed the problem of the presence of PGMs in vegetable matrices for human consumption, developing analytical methods for their determination using voltammetric and spectroscopic techniques (Locatelli, 2013, 2014; Locatelli et al., 2005; Locatelli and Melucci, 2014b, 2015).

Finally, both for thallium and for PGMs, many really interesting works have been present in the literature for some time, all employing spectroscopic methods in its various versions (Biddau and Cidu, 2017; Ghanjaoui et al., 2011; Falandysz et al., 2017; Locatelli et al., 1984; Patočka et al., 2017; Sasmaz et al., 2016; Yu et al., 2018; Jia et al., 2018; Lesniewska et al., 2004; Parsons and Farago, 1987, Vaezzadeh et al., 2010).

5.6 Meals (wholemeal, wheatmeal, and cornmeal)

As regards the presence of toxic metals in meals—Cu, Pb, Cd, Zn, Hg, As, Se, Mn, Fe, Sn, Sb, Cr—our research group has developed methods for their determination voltammetric and spectroscopic (Locatelli and Melucci, 2014b; Locatelli, 2004, 2005, 2007; Locatelli et al., 2004; Locatelli and Casolari, 2004; Locatelli and Melucci, 2012, 2016b; Locatelli and Torsi, 2003, 2004; Melucci et al., 2007; Melucci and Locatelli, 2007).

Regarding PGMs and Tl, only very recently have works appeared in the literature, in truth in very small numbers. Among these, the precious work of Kollander and coworkers (Kollander et al., 2023) for PGMs and of Garry and coworkers (Garry et al., 2021) for thallium should be mentioned.

5.7 Meat

Evidently, the presence of toxic metals in animal meat for human consumption is closely linked to what the animals eat or drink and, therefore, both to the level of pollution in the place where the animals themselves live and to the quality of the food given to them by the farmers.

As regards the presence of the three nonessential and toxic metals in the meat—Pb, Cd, Hg—the literature reports numerous works, all extremely interesting (Gerofke et al., 2019; Njoga et al., 2021; Wang et al., 2019; Unguryanu et al., 2023; Morshdy et al., 2022; Darwish et al., 2023).

As regards Tl and PGMs, only recently have some works appeared in the literature, many of which addressed to the risks to human health (Filippini et al., 2020; Sabbioni et al., 2018; Rees et al., 2016; Sun and Xing, 2016; Cwynar et al., 2014; Roselli et al., 2016; Jarzyń'ska & Falandysz, 2011; Voica et al., 2023; Angelone et al., 2006; Kim et al., 2021), rather than to innovative instrumental and sample preparation techniques for their determination.

5.8 Milk and milk derivatives

As regards these matrices, their possible contamination essentially depends on the position of the stable where the various milk-producing animals are kept and live, be they cows, sheep, goats, a position that must obviously be far from possible sources of pollution.

A further cause of possible contamination concerns the quality of the food given to the animals by those responsible for their care.

Even for the concentration of toxic metals allowed in milk and milk derivatives, the legislation is very incomplete and takes into consideration only very few metals (EC Regulation 466/2001 Dated March 8, 2001 and Subsequent Amendments: EC Regulation 2375/2001 Dated November 29, 2001 and EC Regulation 221/2002 Dated February 6, 2002, 2001).

Obviously, what was said earlier applies to milk placed on the market for human consumption, but it also applies to mothers' milk taken by newborns during breastfeeding (Parsaei et al., 2019; Bontempo et al., 2019; Sanal et al., 2011; Mohd-Taufek et al., 2016; Rossipal and Krachler, 1998; Krachler et al., 1998; Dawry, 1993; Coni et al., 1996; Oliveira et al., 2020). In this sense, as reported in the literature (Frazzoli et al., 2006; Zhou et al., 2017; Sager and Hobegger, 2013; Ikem et al., 2002; Krachler and Rossipal, 2000; Saribal, 2020; Kılıç Altun and Aydemir, 2021; Bousbia et al., 2019; Mandiá et al., 2021), it should be underlined that living in particularly polluted places also leads to the presence of high concentrations of toxic metals in human milk.

5.9 Tea (*Camellia sinensis*)

Certainly, it is integral part of the food chain everything is taken and ingested by humans, and in this sense the so-called "nutritional or food supplements" are certainly an important part of it.

Also for this matrix, in the literature, there is a vast number of works regarding the presence of nonessential toxic metals and their determination mostly with spectroscopic methods and more rarely with voltammetric methods (Zhang et al., 2020; Salmani et al., 2024; Sun et al., 2019; Schunk et al., 2016; Melucci et al., 2013).

Even if we repeat ourselves, it is worth emphasizing once again that the concentrations of toxic metals in this, as in all previously described food matrices, should be under the maximum limits allowed by law. Unfortunately, in the European Legislation, as in all other International Legislation, laws specifically concerning the tea do not exist.

As regards the "Tea" matrix and the subsequent "Herbal Medicines" the focus on PGMs and TL is very important. This is due to two factors. The first concerns the fact that a large part of the tea leaves for obtaining infusions is imported from countries in which there is no legislation for the problem of "maximum permissible concentrations of toxic metals in food matrices," but above all, and this is the second factor, there are not restrictive regulations regarding the release of PGMs and Tl into the environment due to vehicular traffic (PGMs) or the mismanagement of plant protection products in agriculture (TL).

As regards PGMs in the *Camellia sinensis* matrix, the literature reports very few works (Ni et al., 2017) (Vaezzadeh et al., 2010), among which the work of Melucci and coworkers (Melucci et al., 2019) is very important. As regards Tl, however, the literature reports more works, all very recent and very interesting, which propose almost exclusively spectroscopic methods for its identification and determination (Kim et al., 2021; Kowalczyk et al., 2022; Tish et al., 2023).

5.10 Herbal medicines

The present discussion wants to highlight, possibly in the more exhaustive way, what is the current state of the art in all the very numerous problems relevant to herbal medicines, considering their enormous consumption and, above all, the very dangerous superficiality with which they are taken, a very great risk to human health.

In reality, herbal medicines have a thousand-year history, especially in Eastern countries. It should be underlined that for several years, the use of herbal medicines in the Western world has increased greatly, as they are erroneously considered to be better and/or less dangerous remedies for human health compared with conventional medicine, since they are natural products.

As reported in the discussion on the "tea" matrix, for the "herbal medicines" matrix, the problems are perhaps the same. In fact, there are problems regarding the countries from which the majority of herbal medicines are imported: legislative problems both of an environmental nature and of a nature exquisitely linked to the management and safety from the point of view of human health of the plants themselves from which, generally by infusion, herbal tisanes are obtained and consumed throughout the world (Guidelines on Safety Monitoring of Herbal Medicines in Pharmacovigilance Systems, 2004; Kanfer and Patnala, 2021; Mosihuzzaman and Choudhary, 2008; Supplementary Guidelines on Good Manufacturing Practices for the Manufacture of Herbal Medicine, 2006; Guideline on Specifications: Test Procedures and Acceptance Criteria for Herbal Substances, Herbal Preparations and Herbal Medicinal Products/Traditional Herbal Products CPMP/QWP/2820/00 Rev. 1. London EMEA, 2006; Sahoo et al., 2010).

Regarding the presence and determination of toxic metals in herbal medicines, the literature reports a huge amount of works—for brevity's sake, only a few, even very recent ones, are reported here (Xu et al., 2021; Diuzheva et al., 2020; Kumar and Kumar Bose, 2020; White, 2020; Woo et al., 2012; Zhang et al., 2012; Yuan et al., 2011; Rao and Meena, 2011; Kalny et al., 2007; Başgel & Erdemoğlu, 2006; Chizzola et al., 2003; Zhao et al., 2017; Wang and Liu, 2010; Ma et al., 2018; Esteki et al., 2017)—but, as far as PGMs are concerned, there are very few works dealing with these problems (Chen et al., 2022). Our research group has addressed this topic, trying to develop and optimize analytical methods for the determination of these elements also at the level of trace and ultratrace concentrations of both conventional toxic metals (Locatelli et al., 2014; Melucci and Laurentiis Zappi Locatelli et al., 2018) and of PGMs and thallium (Melucci et al., 2017b, 2021, 2022).

As regards thallium, the literature proposes a greater quantity of works (Quds et al., 2021; Chaudhary, 2021; Yelmate et al., 2023; White, 2020), highlighting the growing concentration in all environmental matrices, and in particular those of the vegetal sector, because more subject to possible pollution deriving, as previously mentioned, from the poor and superficial management of agricultural pesticides used in agriculture.

5.11 Cosmetics (lipstick)

Even if completely unrelated to the food chain, a very brief consideration must be made to highlight the problem relating to the presence of toxic metals in cosmetics (Motswetla1 et al., 2023), with particular attention to lipsticks. Problem that concerns the female world linked to the colorants of the lipsticks themselves, which, once applied to the lips, can slowly be ingested.

The scientific community has been attentive to this problem for some time, and many works—only the most recent ones are reported here—have appeared in the literature (Abed et al., 2024; Dinake et al., 2023; Ullah et al., 2023; Gyamfi et al., 2024; Ali Raza-Naqvi et al., 2022; Arshad et al., 2020; Asantewah Nkansah et al., 2018).

Indeed, even if the EC regulation number 1223/2009 (Regulation (EC) No.1223/2009 of the European Parliament and of the European Council of November 30, 2009), has established the rules, harmonized at European level, for the production, distribution, evaluation, and control of any adverse effects relating to cosmetics, many authors report worrying considerations regarding the presence of metals, many of which are toxic or toxic at high concentrations, in cosmetics placed on the market. Everyone hopes for greater control, especially considering the fact that on the market there are many cosmetics for sale, and lipsticks in particular, originating from illicit and criminal activities, and therefore not traceable or controlled.

Our research group has also addressed this problem (Melucci et al., 2015, 2017c), strongly supporting the need for an "International Regulation for Safe", very punctual and, above all, very restrictive on the presence of certain compounds or chemical elements in cosmetics, which could be particularly dangerous to human health.

References

Abed, M.S., Moosa, A.A., Alzuhairi, M.A., 2024. Heavy metals in cosmetics and tattoos: a review of historical background, health impact, and regulatory limits. J. Hazard. Mater. Adv. 13. https://doi.org/10.1016/j.hazadv.2023.100390.

Abitew, A.D., Ymam, B.B., 2023. Levels of selected essential and non-essential metals in rice and the underlying soil in major rice-growing areas of Ethiopia. J. Environ. Analyt. Chem. (1).

Aceto, M., 2003. 2003 Review in Food and Nutrition Toxicity CRC Press Metals in Wine 1.

Alharbi, O.M.L., Basheer, A.A., Khattab, R.A., Ali, I., 2018. Health and environmental effects of persistent organic pollutants. J. Mol. Liq. 263, 442–453. https://doi.org/10.1016/j.molliq.2018.05.029.

Ali Raza-Naqvi, S., Idrees, F., Sherazi, T.A., Anjum-Shahzad, S., Ul-Hassan, S., Ashraf, N., 2022. Toxicology of heavy metals used in cosmetics. J. Chil. Chem. Soc. 67 (3), 5615–5622. https://doi.org/10.4067/s0717-97072022000305615.

Almeida, C.M.R., Vasconcelos, M.T.S.D., 2003. Multielement composition of wines and their precursors including provenance soil and their potentialities as fingerprints of wine origin. J. Agric. Food Chem. 51 (16), 4788–4798. https://doi.org/10.1021/jf034145b.

Alonso Castillo, M.L., García De Torres, A., Vereda Alonso, E., Siles Cordero, M.T., Cano Pavón, J.M., 2012. Multi-element determination of Pt, Pd and Ir traces in environmental samples by ICP-MS after preconcentration. Talanta 99, 853–858. https://doi.org/10.1016/j.talanta.2012.07.042.

Amorello, D., Barreca, S., Gulli, E., Orecchio, S., 2017. Platinum and rhodium in wine samples by using voltammetric techniques. Microchem. J. 130, 229–235. https://doi.org/10.1016/j.microc.2016.09.010.

Angelone, M., Nardi, E., Pinto, V., Cremisini, C., 2006. Analytical Methods to Determine Palladium in Environmental Matrices: A Review Palladium Emissions in the Environment: Analytical Methods, Environmental Assessment and Health Effects. Springer Berlin Heidelberg, Italy, pp. 245–291. https://doi.org/10.1007/3-540-29220-9_18.

Antonia López Antón, M., Alan Spears, D., Somoano, M.D., Rosa Martínez Tarazona, M., 2013. Thallium in coal: analysis and environmental implications. Fuel 105, 13–18. https://doi.org/10.1016/j.fuel.2012.08.004.

Arduini, S., Zappi, A., Locatelli, M., Sgrò, S., Melucci, D., 2021. An authentication study on grappa spirit: the use of chemometrics to detect a food fraud. Analytica 2 (3), 84–92. https://doi.org/10.3390/analytica2030010.

Arshad, H., Mehmood, M.Z., Shah, M.H., Abbasi, A.M., 2020. Evaluation of heavy metals in cosmetic products and their health risk assessment. Saudi Pharmaceut. J. 28 (7), 779–790. https://doi.org/10.1016/j.jsps.2020.05.006.

Asantewah Nkansah, M., Owusu-Afriyie, E., Opoku, F., 2018. Determination of lead and cadmium contents in lipstick and their potential health risks to consumers. J. Consumer Protect. Food Safety 13 (4), 367–373. https://doi.org/10.1007/s00003-018-1180-y.

Azizi, G., Akodad, M., Baghour, M., Layachi, M., Moumen, A., 2018. The use of Mytilus spp. mussels as bio-indicators of heavy metal pollution in the coastal environment. A review. J. Mater. Environ. Sci. 9 (4), 1170–1181.

Azucena Morales-Zarco, M., Osorio-Rico, L., Aschner, M., Galván-Arzate, S., Santamaría, A., 2023. Thallium neurotoxicity. Springer Sci. Business Media LLC 3, 2331–2357. https://doi.org/10.1007/978-3-031-15080-7_232.

Barchiesi, F., Branciari, R., Latini, M., Roila, R., Lediani, G., Filippini, G., Scortichini, G., Piersanti, A., Rocchegiani, E., Ranucci, D., 2020. Heavy metals contamination in shellfish: benefit-risk evaluation in Central Italy. Foods 9 (11), 1720. https://doi.org/10.3390/foods9111720.

Başgel, S., Erdemoğlu, S.B., 2006. Determination of mineral and trace elements in some medicinal herbs and their infusions consumed in Turkey. Sci. Total Environ. 359 (1–3), 82–89. https://doi.org/10.1016/j.scitotenv.2005.04.016.

Batley, G.E., Campbell, P.G.C., 2022. Metal contaminants of emerging concern in aquatic systems. Environ. Chem. 19 (1), 23–40. https://doi.org/10.1071/EN22030.

Belzile, N., Chen, Y.W., 2017. Thallium in the environment: a critical review focused on natural waters, soils, sediments and airborne particles. Appl. Geochem. 84, 218–243. https://doi.org/10.1016/j.apgeochem.2017.06.013.

Bernardino, C.A.R., Mahler, C.F., Santelli, R.E., Freire, A.S., Braz, B.F., Novo, L.A.B., 2019. Metal accumulation in roadside soils of Rio de Janeiro, Brazil: impact of traffic volume, road age, and urbanization level. Environ. Monit. Assess. 191 (3). https://doi.org/10.1007/s10661-019-7265-y.

Biddau, R., Cidu, R., 2017. Metals and metalloids in wild asparagus at uncontaminated and mining-contaminated sites. J. Environ. Qual. 46 (2), 320–329. https://doi.org/10.2134/jeq2016.09.0354.

Blain, T., 2022. Handbook on the Toxicology of Metals. Elsevier Academic Press Pub, II, pp. 795–806.

Bontempo, L., Barbero, A., Bertoldi, D., Camin, F., Larcher, R., Perini, M., Sepulcri, A., Zicarelli, L., Piasentier, E., 2019. Isotopic and elemental profiles of Mediterranean buffalo milk and cheese and authentication of Mozzarella di Bufala Campana PDO: An initial exploratory study. Food Chem. 285, 316–323. https://doi.org/10.1016/j.foodchem.2019.01.160.

Bousbia, A., Boudalia, S., Gueroui, Y., Ghebache, R., Amrouchi, M., Belase, B., Meguelati, S., Belkheir, B., Benidir, M., Chelaghmia, M.L., 2019. Heavy metals concentrations in raw cow milk produced in the different livestock farming types in guelma province (Algeria): contamination and risk assessment of consumption. J. Animal Plant Sci. 29 (2), 386–395.

Briffa, J., Sinagra, E., Blundell, R., 2020. Heavy metal pollution in the environment and their toxicological effects on humans. Heliyon 6 (9), e04691. https://doi.org/10.1016/j.heliyon.2020.e04691.

Butler, O.T., Cairns, W.R.L., Cook, J.M., Davidson, C.M., Mertz-Kraus, R., 2018. Atomic spectrometry update-a review of advances in environmental analysis. J. Anal. Atomic Spectrom. 33 (1), 8–56. https://doi.org/10.1039/c7ja90059g.

Canizo, B.V., Escudero, L.B., Pérez, M.B., Pellerano, R.G., Wuilloud, R.G., 2018. Intra-regional classification of grape seeds produced in Mendoza province (Argentina) by multi-elemental analysis and chemometrics tools. Food Chem. 242, 272–278. https://doi.org/10.1016/j.foodchem.2017.09.062.

Cannas, D., Loi, E., Serra, M., Firinu, D., Valera, P., Zavattari, P., 2020. Relevance of essential trace elements in nutrition and drinking water for human health and autoimmune disease risk. Nutrients 12 (7), 2074. https://doi.org/10.3390/nu12072074.

Chaudhary, P.H., 2021. Need and scope of standardization of herbal medicines. A review. J. Res. Pharmaceut. Sci. 7 (8), 26–31.

Chen, W., Yang, Y., Fu, K., Zhang, D., Wang, Z., 2022. Progress in ICP-ms analysis of minerals and heavy metals in traditional medicine. Front. Pharmacol. 13. https://doi.org/10.3389/fphar.2022.891273.

Chizzola, R., Michitsch, H., Franz, C., 2003. Monitoring of metallic micronutrients and heavy metals in herbs, spices and medicinal plants from Austria. Eur. Food Res. Technol. 216 (5), 407–411. https://doi.org/10.1007/s00217-003-0675-6.

Coetzee, P.P., Van Jaarsveld, F.P., Vanhaecke, F., 2014. Intraregional classification of wine via ICP-MS elemental fingerprinting. Food Chem. 164, 485–492. https://doi.org/10.1016/j.foodchem.2014.05.027.

Coni, E., Alimonti, A., Bocca, A., La Torre, F., Menghetti, E., Miraglia, E., Caroli, S., 1996. Speciation of trace elements in human milk by high performance liquid chromatography combined with inductively coupled plasma-mass spectrometry. Trace Elem. Electrocyt. 13 (1), 26–32.

Cvetkoviá, J., Arpadjan, S., Karadjova, I., Stafilov, T., 2002. Determination of thallium in wine by electrothermal atomic absorption spectrometry after extraction preconcentration. Spectrochim. Acta B Atom Spectrosc. 57 (6), 1101–1106. https://doi.org/10.1016/S0584-8547(02)00032-0.

Cwynar, P., Kołacz, R., Grudnik, T., 2014. The influence of chronic thallium intoxication on laying hens, including its cumulation in tissues, Organs, and Eggs. Pol. J. Environ. Stud. 23 (3), 949–954.

D'Orazio, M., Campanella, B., Bramanti, E., Ghezzi, L., Onor, M., Vianello, G., Vittori-Antisari, L., Petrini, R., 2020. Thallium pollution in water, soils and plants from a past-mining site of Tuscany: sources, transfer processes and toxicity. J. Geochem. Explor. 209, 106434. https://doi.org/10.1016/j.gexplo.2019.106434.

Darwish, W.S., El Bayoumi, R.M., Elgaffry, O., Hussein, M.A., 2023. Occurrence and sources of heavy metal contamination of meat products with a focus on the associated health risks: a review . Assiut university. J. Adv. Vet. Res. 13 (8), 1721–1725.

Dawry, F.P., 1993. Tl concentration in breast milk. Health Phys. 64 (3).

Dey, S., Mehta, N.S., 2020. Automobile pollution control using catalysis. Res. Environ. Sustain. 2, 100006. https://doi.org/10.1016/j.resenv.2020.100006.

Dinake, P., Motswetla, O., Kereeditse, T.T., Kelebemang, R., 2023. Assessment of level of heavy metals in cosmetics. Toxicol. Res. Appl. 7. https://doi.org/10.1177/23978473231156.

Diuzheva, A., Locatelli, M., Tartaglia, A., Goga, M., Ferrone, V., Carlucci, G., Andruch, V., 2020. Application of liquid-phase microextraction to the analysis of plant and herbal samples. Phytochem. Anal. 31 (6), 687–699. https://doi.org/10.1002/pca.2939.

Djingova, R., Kovacheva, P., Wagner, G., Markert, B., 2003. Distribution of platinum group elements and other traffic related elements among different plants along some highways in Germany. Sci. Total Environ. 308 (1–3), 235–246. https://doi.org/10.1016/S0048-9697(02)00677-0.

Doulgeridou, A., Amlund, H., Sloth, J.J., Hansen, M., 2020. Review of potentially toxic rare earth elements, thallium and tellurium in plant-based foods. EFSA J. 18 (1). https://doi.org/10.2903/j.efsa.2020.e181101.

Dubiella-Jackowska, A., Polkowska, Z., Namieńnik, J., 2009. Platinum group elements in the environment: emissions and exposure. Rev. Environ. Contam. Toxicol. 199, 111–135. https://doi.org/10.1007/978-0-387-09808-1_3.

Ebdon, L., Pitts, L., 2002. In: Cornelis, R. (Ed.), Trace Element Speciation for Environment, Food and Health.

Erasmus, J.H., Malherbe, W., Zimmermann, S., Lorenz, A.W., Nachev, M., Wepener, V., Sures, B., Smit, N.J., 2020. Metal accumulation in riverine macroinvertebrates from a platinum mining region. Sci. Total Environ. 703, 134738. https://doi.org/10.1016/j.scitotenv.2019.134738.

Erasmus, J.H., Zimmermann, S., Smit, N.J., Malherbe, W., Nachev, M., Sures, B., Wepener, V., 2022. Human health risks associated with consumption of fish contaminated with trace elements from intensive mining activities in a peri-urban region. Sci. Total Environ. 825, 154011. https://doi.org/10.1016/j.scitotenv.2022.154011.

Esposito, M., De Roma, A., Cavallo, S., Miedico, O., Chiaravalle, E., Soprano, V., Baldi, L., Gallo, P., 2019. Trace elements in vegetables and fruits cultivated in Southern Italy. J. Food Compos. Anal. 84, 103302. https://doi.org/10.1016/j.jfca.2019.103302.

Esposito, G., Mudadu, A.G., Abete, M.C., Pederiva, S., Griglione, A., Stella, C., Ortu, S., Bazzoni, A.M., Meloni, D., Squadrone, S., 2021. Seasonal accumulation of trace elements in native Mediterranean mussels (*Mytilus galloprovincialis* Lamarck, 1819) collected in the Calich Lagoon (Sardinia, Italy). Environ. Sci. Pollut. Control Ser. 28 (20), 25770–25781. https://doi.org/10.1007/s11356-021-12380-4.

Essumang, D.K., Dodoo, D.K., Adokoh, C.K., Sam, A., Doe, N.G., 2008. Bioaccumulation of platinum group metals on some fish species (*Oreochromis niloticus*, Penaeus laspisulcates, Scylla serrate, Galaxias brevipinnis and Mollusc) in the Pra estuary of Ghana. Toxicol. Environ. Chem. 90 (3), 625–638. https://doi.org/10.1080/02772240701649869.

Esteki, M., Vander Heyden, Y., Farajmand, B., Kolahderazi, Y., 2017. Qualitative and quantitative analysis of peanut adulteration in almond powder samples using multi-elemental fingerprinting combined with multivariate data analysis methods. Food Control 82, 31–41. https://doi.org/10.1016/j.foodcont.2017.06.014.

Falandysz, J., Szymczyk, K., Ichihashi, H., Bielawski, L., Gucia, M., Frankowska, A., Yamasaki, S.I., 2001. ICP/MS and ICP/AES elemental analysis (38 elements) of edible wild mushrooms growing in Poland. Food Addit. Contamin. 18 (6), 503–513. https://doi.org/10.1080/02652030119625.

Falandysz, J., Drewnowska, M., Chudzińska, M., Barałkiewicz, D., 2017. Accumulation and distribution of metallic elements and metalloids in edible Amanita fulva mushrooms. Ecotoxicol. Environ. Saf. 137, 265–271. https://doi.org/10.1016/j.ecoenv.2016.12.014.

Ferré-Huguet, N., Martí-Cid, R., Schuhmacher, M., Domingo, J.L., 2008. Risk assessment of metals from consuming vegetables, fruits and rice grown on soils irrigated with waters of the Ebro River in Catalonia, Spain. Biol. Trace Elem. Res. 123 (1–3), 66–79. https://doi.org/10.1007/s12011-008-8113-z.

Filippini, T., Tancredi, S., Malagoli, C., Malavolti, M., Bargellini, A., Vescovi, L., Nicolini, F., Vinceti, M., 2020. Dietary estimated intake of trace elements: risk assessment in an Italian population. Expos. Health 12 (4), 641–655. https://doi.org/10.1007/s12403-019-00324-w.

Frazzoli, C., Cammarone, R., Caroli S., 2006. 1 2006/01 Pure and Applied Chemistry 10.1351/pac200678010069 00334545 1 69 78 Italy Uptake of platinum-group elements with the diet: a preliminary investigation 78.

Gajek, M., Pawlaczyk, A., Szynkowska-Jozwik, M.I., 2021. Multi-elemental analysis of wine samples in relation to their type, origin, and grape variety. Molecules 26 (1). https://doi.org/10.3390/MOLECULES26010214.

Garry, M.R., Shock, S.S., Salatas, J., 2021. Human health risk assessment of metals exposure through subsistence foods consumption and subsistence harvest activities near a mining transport road in northwest Alaska. Hum. Ecol. Risk Assess. 27 (1), 227–257. https://doi.org/10.1080/10807039.2019.1706151.

Gebel, T., 2000. Toxicology of Platinum, Palladium, Rhodium, and Their Compounds. Springer Science and Business Media LLC, pp. 245–255. https://doi.org/10.1007/978-3-642-59678-0_25.

Genchi, G., Carocci, A., Lauria, G., Stefania Sinicropi, M., Catalano, A., Use, T., 2021a. Toxicity, and detoxification therapy: an overview. Appl. Sci. 11 (18), 8322. https://doi.org/10.3390/app11188322.

Genchi, G., Lauria, G., Catalano, A., Carocci, A., Stefania Sinicropi, M., 2021b. The double face of metals: the intriguing case of chromium. Appl. Sci. 11 (2), 638. https://doi.org/10.3390/app11020638.

Gerofke, A., Martin, A., Schlichting, D., Gremse, C., Müller-Graf, C., 2019. Heavy Metals in Game Meat. Brill, pp. 585–609. https://doi.org/10.3920/978-90-8686-877-3_24.

Ghaderi, A., NasehGhafoori, P., Rasouli-Azad, M., Sehat, M., Mehrzad, F., Nekuei, M., Aaseth, J., Banafshe, H.R., Mehrpour, O., 2018. Examining of thallium in cigarette smokers. Biol. Trace Elem. Res. 182 (2), 224–230. https://doi.org/10.1007/s12011-017-1107-y.

Ghanjaoui, M.E., Cervera, M.L., Rhazi, M.E., Guardia, M.d.l., 2011. Validated fast procedure for trace element determination in basil powder. Food Chem. 125 (4), 1309–1313. https://doi.org/10.1016/j.foodchem.2010.09.091.

Gohar, A., Shakeel, M., 2021. Assessment of environmental impact on essential and toxic elements composition in natural honeys by using inductively coupled plasma mass spectrometry. Environ. Sci. Pollut. Control Ser. 28 (13), 15794–15805. https://doi.org/10.1007/s11356-020-11688-x.

Gomez-Delgado, A.I., Tibon, J., Silva, M.S., Lundebye, A.K., Agüera, A., Rasinger, J.D., Strohmeier, T., Sele, V., 2023. Seasonal variations in mercury, cadmium, lead and arsenic species in Norwegian blue mussels (*Mytilus edulis* L.) — assessing the influence of biological and environmental factors. J. Trace Elem. Med. Biol. 76. https://doi.org/10.1016/j.jtemb.2022.127110.

Greda, K., Swiderski, K., Jamroz, P., Pohl, P., 2016. Flowing liquid anode atmospheric pressure glow discharge as an excitation source for optical emission spectrometry with the improved detectability of Ag, Cd, Hg, Pb, Tl, and Zn. Anal. Chem. 88 (17), 8812—8820. https://doi.org/10.1021/acs.analchem.6b02250.

Guideline on Specifications: Test Procedures and Acceptance Criteria for Herbal Substances, Herbal Preparations and Herbal Medicinal Products/Traditional Herbal Products CPMP/QWP/2820/00 Rev. 1. London EMEA, 2006. European Medicines Agency (EMEA.

Gyamfi, O., Aboko, J., Ankapong, E., Marfo, J.T., Awuah-Boateng, N.Y., Sarpong, K., Dartey, E., 2024. A systematic review of heavy metals contamination in cosmetics. Cutan. Ocul. Toxicol. 43 (1), 5—12. https://doi.org/10.1080/15569527.2023.2268197.

Hopfer, H., Nelson, J., Collins, T.S., Heymann, H., Ebeler, S.E., 2015. The combined impact of vineyard origin and processing winery on the elemental profile of red wines. Food Chem. 172, 486—496. https://doi.org/10.1016/j.foodchem.2014.09.113.

Hughes, A.E., Haque, N., Northey, S.A., Giddey, S., 2021. Platinum group metals: a review of resources, production and usage with a focus on catalysts. Resources 10 (9). https://doi.org/10.3390/resources10090093.

Ikem, A., Nwankwoala, A., Odueyungbo, S., Nyavor, K., Egiebor, N., 2002. Levels of 26 elements in infant formula from USA, UK, and Nigeria by microwave digestion and ICP-OES. Food Chem. 77 (4), 439—447. https://doi.org/10.1016/S0308-8146(01)00378-8.

Jarzyń'ska, G., Falandysz, J., 2011. Selenium and 17 other largely essential and toxic metals in muscle and organ meats of Red Deer (*Cervus elaphus*) - consequences to human health. Environ. Int. 37 (5), 882—888. https://doi.org/10.1016/j.envint.2011.02.017.

Jia, Y., Xiao, T., Sun, J., Yang, F., Baveye, P.C., 2018. Microcolumn-based speciation analysis of thallium in soil and green cabbage. Sci. Total Environ. 630, 146—153. https://doi.org/10.1016/j.scitotenv.2018.02.147.

Jomova, K., Makova, M., Alomar, S.Y., Alwasel, S.H., Nepovimova, E., Kuca, K., Rhodes, C.J., Valko, M., 2022. Essential metals in health and disease. Chem. Biol. Interact. 367. https://doi.org/10.1016/j.cbi.2022.110173.

José Tamele, I., 2020. Patricia Vázquez Loureiro, lead, mercury and cadmium in fish and shellfish from the Indian ocean and red sea (african countries): public health challenges. J. Mar. Sci. Eng. 8 (5), 344. https://doi.org/10.3390/jmse8050344.

Kalny, P., Fijałek, Z., Daszczuk, A., Ostapczuk, P., 2007. Determination of selected microelements in polish herbs and their infusions. Sci. Total Environ. 381 (1—3), 99—104. https://doi.org/10.1016/j.scitotenv.2007.03.026.

Kanfer, I., Patnala, S., 2021. Regulations for the Use of Herbal Remedies Herbal Medicine in Andrology. Elsevier, South Africa, pp. 189—206. https://doi.org/10.1016/B978-0-12-815565-3.00007-2.

Karbowska, B., 2016. Presence of thallium in the environment: sources of contaminations, distribution and monitoring methods. Environ. Monit. Assess. 188 (11). https://doi.org/10.1007/s10661-016-5647-y.

Kilmartin, P.A., 2016. Electrochemistry applied to the analysis of wine: a mini-review. Electrochem. Commun. 67, 39—42. https://doi.org/10.1016/j.elecom.2016.03.011.

Kim, Y.H., Ra, W.J., Cho, S., Choi, S., Soh, B., Joo, Y., Lee, K.W., 2021. Method validation for determination of thallium by inductively coupled plasma mass spectrometry and monitoring of various foods in South Korea. Molecules 26 (21). https://doi.org/10.3390/molecules26216729.

Kılıç Altun, S., Emin Aydemir, M., 2021. Determination of some minerals and heavy metal levels in Urfa cheese and cow's milk. Food and Health 7 (3), 185—193. https://doi.org/10.3153/fh21020.

Kiran, R.B., Sharma, R., 2022. Effect of heavy metals: an overview. Mater. Today: Proc. 51, 880—885. https://doi.org/10.1016/j.matpr.2021.06.278.

Kollander, B., Rodushkin, I., Sundström, B., 2023. Multi-Element assessment of potentially toxic and essential elements in new and traditional food varieties in Sweden. Foods 12 (9), 1831. https://doi.org/10.3390/foods12091831.

Komendova, R., 2020. Recent advances in the preconcentration and determination of platinum group metals in environmental and biological samples. TrAC, Trends Anal. Chem. 122, 115708. https://doi.org/10.1016/j.trac.2019.115708.

Kowalczyk, E., Givelet, L., Amlund, H., Sloth, J.J., Hansen, M., 2022. Risk assessment of rare earth elements, antimony, barium, boron, lithium, tellurium, thallium and vanadium in teas. EFSA J. 20 (1). https://doi.org/10.2903/j.efsa.2022.e200410.

Krachler, M., Rossipal, E., 2000. Concentrations of trace elements in extensively hydrolysed infant formulae and their estimated daily intakes. Ann. Nutr. Metabol. 44 (2), 68–74. https://doi.org/10.1159/000012823.

Krachler, M., Shi Li, F., Rossipal, E., Irgolic, K.J., 1998. Changes in the concentrations of trace elements in human milk during lactation. J. Trace Elem. Med. Biol. 12 (3), 159–176. https://doi.org/10.1016/S0946-672X(98)80005-9.

Kumar, B., Kumar Bose, S., 2020. Toxic Contaminants in Herbal Medicines. Informa UK Limited, pp. 53–63. https://doi.org/10.1201/9780429275852-4.

Landi, S., Fagioli, F., Locatelli, C., Vecchietti, R., 1990. Digestion method for the determination of mercury in vegetable matrices by cold vapour atomic absorption spectrometry. Analyst 115 (2), 173–177. https://doi.org/10.1039/an9901500173.

Lesniewska, B.A., Messerschmidt, J., Jakubowski, N., Hulanicki, A., 2004. Bioaccumulation of platinum group elements and characterization of their species in Lolium multiflorum by size-exclusion chromatography coupled with ICP-MS. Sci. Total Environ. 322 (1–3), 95–108. https://doi.org/10.1016/j.scitotenv.2003.09.019.

Li, C., Wang, H., Liao, X., Xiao, R., Liu, K., Bai, J., Li, B., He, Q., 2021. Heavy metal pollution in coastal wetlands: a systematic review of studies globally over the past three decades. J. Hazard Mater. 424.

Li, H., Zeng, Y., Wang, C., Chen, W., Zou, M., 2023. Variation in the burden and chemical forms of thallium in non-detoxified tissues of tilapia fish (*Oreochromis niloticus*) from waterborne exposure. Chemosphere 333, 138884. https://doi.org/10.1016/j.chemosphere.2023.138884.

Liu, J., Wei, X., Zhou, Y., Tsang, D.C.W., Bao, Z., Yin, M., Lippold, H., Yuan, W., Wang, J., Feng, Y., Chen, D., 2020. Thallium contamination, health risk assessment and source apportionment in common vegetables. Sci. Total Environ. 703. https://doi.org/10.1016/j.scitotenv.2019.135547.

Liu, Z.M., Zeng, Y., Li, Y.Q., Mu, J.L., Wu, M.X., Cao, W.Z., 2022. Review on the speciation analysis methods of platinum group metals in environmental media. Chin. J. Appl. Ecol. 33 (12), 3448–3456. https://doi.org/10.13287/j.1001-9332.202212.032.

Locatelli, C., 2000. Proposal of new analytical procedures for heavy metal determinations in mussels, clams and fishes. Food Addit. Contam. 17 (9), 769–774. https://doi.org/10.1080/026520300415318.

Locatelli, C., 2003. Heavy metal determinations in algae, mussels and clams. Their possible employment for assessing the sea water quality criteria. J. Phys. IV 107 (II), 785–788. https://doi.org/10.1051/jp4:20030418.

Locatelli, C., 2004. Heavy metals in matrices of food interest: sequential voltammetric determination at trace and ultratrace level of copper, lead, cadmium, zinc, arsenic, selenium, manganese and iron in meals. Electroanalysis 16 (18), 1478–1486. https://doi.org/10.1002/elan.200302986.

Locatelli, C., 2005. Overlapping voltammetric peaks - an analytical procedure for simultaneous determination of trace metals. Application to food and environmental matrices. Anal. Bioanal. Chem. 381 (5), 1073–1081. https://doi.org/10.1007/s00216-004-3008-0.

Locatelli, C., 2007. Voltammetric methods for the simultaneous determination of trace metals in foods, plant tissues and soils. J. Sci. Food Agric. 87 (2), 305–312. https://doi.org/10.1002/jsfa.2722.

Locatelli, C., 2013. Iridium and lead as vehicle emission pollutants: their sequential voltammetric determination in vegetable environmental bio-monitors. Microchem. J. 106, 282–288. https://doi.org/10.1016/j.microc.2012.08.008.

Locatelli, C., 2014. Square wave catalytic adsorptive voltammetric determination of osmium, ruthenium and lead in vegetable environmental bio-monitors. Int. J. Environ. Anal. Chem. 94 (3), 277–290. https://doi.org/10.1080/03067319.2013.791980.

Locatelli, C., Casolari, S., 2004. Determination of heavy metal in matrices of food interest. Voltammetry or spectroscopy? critical comparison in the case of meals. Curr. Top. Anal. Chem. 4, 57–65.

Locatelli, C., Fagioli, F., 1997. Determination at trace level of heavy metals in mussels and clams. Ann. Chim. 87 (7–8), 535–538.

Locatelli, C., Melucci, D., 2010. Sequential voltammetric determination of mercury(II) and toxic metals in environmental bio-monitors: application to mussels and clams. Int. J. Environ. Anal. Chem. 90 (1), 49–63. https://doi.org/10.1080/03067310802706611.

Locatelli, C., Melucci, D., 2012. Voltammetric determination of ultra-trace total mercury and toxic metals in meals. Food Chem. 130 (2), 460–466. https://doi.org/10.1016/j.foodchem.2011.07.070.

Locatelli, C., Melucci, D., 2013. Voltammetric method for ultra-trace determination of total mercury and toxic metals in vegetables. Comparison with spectroscopy. Cent. Eur. J. Chem. 11 (5), 790–800. https://doi.org/10.2478/s11532-013-0221-8.

Locatelli, C., Melucci, 2014a. 2014 Voltammetry: Theory, Types and Applications 225 248 Nova Science Publ. Inc Voltammetric Determination of Metals as Food Contaminants- an Excellent Alternative to Spectroscopic Measurements: Application to Meals, Vegetable, Mussels, Clams and Fishes.

Locatelli, C., Melucci, D., 2014b. Voltammetric and Spectroscopic Methods for the Ruthenium Determination in the Environment at Ultra-trace Concentration Level: Critical Comparison and Application to Airborne Particulate Matter, Vegetables, Superficial Waters, Mussels, Clams and Soils Ruthenium: Synthesis, Physicochemical Properties and Applications. Nova Science Publishers, Inc, Italy, pp. 207–233. https://www.novapublishers.com/catalog/product_info.php?products_id=50859.

Locatelli, C., Melucci, D., 2015. 2015 Advances in Environmental Research 97 128 Nova Science Publ. Inc Iridium, the New and Recent Polluting Metal Linked to Vehicular Traffic. Its Determination in Airborne Particulate Matter, Superficial Waters, Soils/sediments, Vegetables, Mussels, Clams and Algae by Voltammetry and Spectroscopy: Critical Comparison.

Locatelli, C., Melucci, D., 2016a. Heavy Metals and Platinum Group Metals (PGMS) Determination in Oysters, Mussels and Clams as Bio-Monitors of the Aquatic System Pollution Oysters and Clams: Cultivation, Habitat Threats and Ecological Impact. Nova Science Publishers, Inc, Italy, pp. 145–188. https://novapublishers.com/shop/oysters-and-clams-cultivation-habitat-threats-and-ecological-impact/.

Locatelli, C., Melucci, D., 2016b. 2016 Chemistry Research Summaries 229 230 Nova Science Publ. Inc Voltammetric Determination of Metals as Food Contaminants – an Excellent Alternative to Spectroscopic Measurements: Application to Meals, Vegetables, Mussels, Clams and Fishes 17.

Locatelli, C., Torsi, G., 2003. Analytical procedures for the simultaneous voltammetric determination of heavy metals in meals. Microchem. J. 75 (3), 233–240. https://doi.org/10.1016/S0026-265X(03)00100-0.

Locatelli, C., Torsi, G., 2004. Simultaneous square wave anodic stripping voltammetric determination of Cr, Pb, Sn, Sb, Cu, Zn in presence of reciprocal interference: application to meal matrices. Microchem. J. 78 (2), 175–180. https://doi.org/10.1016/j.microc.2004.04.003.

Locatelli, C., Fagioli, F., Bighi, C., Scanavini, L., 1984. Simultaneous Determination of Trace Metals in Vegetable Materials by Alternating Current Anodic Stripping Voltammetry and Atomic Absorption Spectroscopy. Walter de Gruyter GmbH, pp. 529–536. https://doi.org/10.1515/9783112417188-059.

Locatelli, C., Fabbri, D., Torsi, G., 2001. Heavy metal determinations in algae and clams and their possible employment for assessing the sea water quality criteria. Ann. Chim. 91 (7–8), 425–434.

Locatelli, C., Casolari, S., Torsi, G., 2004. New analytical procedure for the spectroscopic determination of copper, lead, cadmium, zinc and mercury in meals. Trends Appl. Spectrosc. 5, 159–168.

Locatelli, C., Melucci, D., Torsi, G., 2005. Determination of platinum-group metals and lead in vegetable environmental bio-monitors by voltammetric and spectroscopic techniques: critical comparison. Anal. Bioanal. Chem. 382 (7), 1567–1573. https://doi.org/10.1007/s00216-005-3356-4.

Locatelli, C., Melucci, D., Locatelli, M., 2014. Toxic metals in herbal medicines. A review. Curr. Bioact. Compd. 10 (3), 181–188. https://doi.org/10.2174/1573407210666140716164321.

Locatelli, C., Melucci, D., de Laurentiis, F., Zappi, A., 2018. Determination of Emerging Metal Pollutants and Toxic Metals in Mussels and Bivalve Mollusks, Very Important Food and Environmental Bio-Monitoring Species Mussels: Characteristics, Biology and Conservation. Nova Science Publishers, Inc, Italy, pp. 33–84. https://www.novapublishers.com/catalog/product_info.php?products_id=64372.

Luo, Z., Kayiranga, A., Uwiringiyimana, E., Zhang, Q., Yan, C., Guo, J., Xing, B., 2020. Thallium contamination in agricultural soils and associated potential remediation via biochar utilization. Biochar 2 (1), 33–46. https://doi.org/10.1007/s42773-020-00042-6.

Ma, X., Fan, L., Mao, F., Zhao, Y., Yan, Y., Tian, H., Xu, R., Peng, Y., Sui, H., 2018. Discrimination of three Ephedra species and their geographical origins based on multi-element fingerprinting by inductively coupled plasma mass spectrometry. Sci. Rep. 8 (1). https://doi.org/10.1038/s41598-018-28558-9.

Mandiá, N., Bermejo-Barrera, P., Herbello, P., López-Suárez, O., Fraga, J.M., Fernández-Pérez, C., Couce, M.L., 2021. Human milk concentrations of minerals, essential and toxic trace elements and association with selective medical, social, demographic and environmental factors. Nutrients 13 (6). https://doi.org/10.3390/nu13061885.

Meeravali, N.N., Madhavi, K., Sahayam, A.C., 2023. Determination of thallium in vegetative plant leaves near industrial areas by high-resolution continuum source electrothermal atomic absorption spectrometry after salt induced could point extraction. Spectrochim. Acta B Atom Spectrosc. 200.

Mehouel, F., Fowler, S.W., 2022. Review of the toxic trace elements arsenic, cadmium, lead and mercury in seafood species from Algeria and contiguous waters in the Southwestern Mediterranean Sea. Algeria Environ. Sci. Pollut. Res. 29 (3), 3288–3301. https://doi.org/10.1007/s11356-021-17130-0.

Melucci, D., Locatelli, C., 2007. Sequential voltammetric determination of trace metals in meals. Microchem. J. 85 (2), 321–328. https://doi.org/10.1016/j.microc.2006.07.009.

Melucci, D., Laurentiis Zappi Locatelli, C., 2018. 2018 Advances in Environmental Research 83 129 Nova Science Publ. Inc Toxic Metal Bio-Monitoring: An Innovative Procedure to Get Complete Information on the Long-Term Trend of the Metal Pollution Load in Environmental Ecosystems 61.

Melucci, D., Torsi, G., Locatelli, C., 2007. Analytical procedure for the simultaneous voltammetric determination of trace metals in food and environmental matrices. Critical comparison with atomic absorption spectroscopic measurements. Ann. Chim. 97 (3–4), 141–151. https://doi.org/10.1002/adic.200790000.

Melucci, D., Locatelli, M., Locatelli, C., 2013. Trace level voltammetric determination of heavy metals and total mercury in tea matrices (Camellia sinensis). Food Chem. Toxicol. 62, 901–907. https://doi.org/10.1016/j.fct.2013.10.029.

Melucci, D., Locatelli, C., D'Annunzio, M., Rendina, M., Locatelli, M., 2015. Analytical Methods for Organic and Inorganic Toxics Determination in Cosmetics Products New Developments in Analytical Chemistry Research. Nova Science Publishers, Inc., Italy, pp. 1–39. https://www.novapublishers.com/catalog/product_info.php?products_id=52364.

Melucci, D., Casolari, S., Laurentiis, F.D., Locatelli, C., 2017a. Determination of toxic metals in marine biomonitors by multivariate analysis of voltammetric and spectroscopic data. application to mussels, algae and fishes of the northern adriatic sea. Fresenius Environ. Bull. 26 (6), 3756–3763.

Melucci, D., de Laurentiis, F., Zappi, A., Locatelli, C., 2017b. Voltammetry and Atomic Absorption Spectroscopy: A Critical Comparison of Two Techniques for Toxic Metal Determination in Seafood and Food Supplements

Analytical Chemistry: Developments, Applications and Challenges in Food Analysis. Nova Science Publishers, Inc, Italy, pp. 297–338. https://www.novapublishers.com/catalog/product_info.php?products_id=62688.

Melucci, D., De Laurentiis, F., Zappi, A., Locatelli, M., Locatelli, C., 2017c. Toxic Metals in Cosmetics: A Review of Analytical Determination, Human Health Risk and International Regulation for Safe Use Advances in Health and Disease, vol. 1. Nova Science Publishers, Inc, Italy, pp. 173–207. https://www.novapublishers.com/catalog/product_info.php?products_id=62192.

Melucci, D., Locatelli, M., De Laurentiis, F., Zengin, G., Locatelli, C., 2018. Herbal medicines: application of a sequential voltammetric procedure to the determination of mercury, copper, lead, cadmium and zinc at trace level. Lett. Drug Des. Discov. 15 (3), 270–280. https://doi.org/10.2174/1570180814666170412124634.

Melucci, D., Casolari Laurentiis Zappi Locatelli, M., Locatelli, C., 2019. 2019 Agricultural Research Updates, Gorawala, P. And Mandhatri 99 146 Nova Science Publ. Inc Toxic Metals in Camellia Sinensis: Analytical Methods, Human Health Risk and Regulations 26.

Melucci, D., Casolari, S., Locatelli, M., Locatelli, C., 2021. Thallium: a polluting metal of new generation. Its voltammetric determination in herbal medicines in presence of metal interferences. Analytica 2 (3), 76–83. https://doi.org/10.3390/analytica2030009.

Melucci, D., Locatelli, M., Casolari, S., Locatelli, C., polluting metals, N., 2022. Quantification in herbal medicines by voltammetric and spectroscopic analytical methods. J. Pharmaceut. Biomed. Anal. 211, 114599. https://doi.org/10.1016/j.jpba.2022.114599.

Merget, R., Rosner, G., 2001. Evaluation of the health risk of platinum group metals emitted from automotive catalytic converters. Sci. Total Environ. 270 (1–3), 165–173. https://doi.org/10.1016/s0048-9697(00)00788-9.

Merian, E., Anke, M., Ihnat, M., Stoeppler, M., 2004. Elements and their compounds in the environment – occurrence. Analy. Biol. Relev.

Migaszewski, Z.M., Gałuszka, A., 2021. Abundance and fate of thallium and its stable isotopes in the environment. Rev. Environ. Sci. Biotechnol. 20 (1), 5–30. https://doi.org/10.1007/s11157-020-09564-8.

Mijosek, T., Marijic, Dragun, Z., Ivankovic, D., 2020. Thallium accumulation in different organisms from karst and lowland rivers of Croatia under wastewater impact. Environ. Chem. 17 (2).

Mir-Marqués, A., Cervera, M.L., Guardia, M.de la, 2016. Mineral analysis of human diets by spectrometry methods. Trends Anal. Chem. 82, 457–467. https://doi.org/10.1016/j.trac.2016.07.007.

Mishra, A., Kumari, M., Swati, R.K., Iqbal, K., Thakur, I.S., 2022. Persistent organic pollutants in the environment: risk assessment, hazards, and mitigation strategies. Bioresour. Technol. Rep. 19, 101143. https://doi.org/10.1016/j.biteb.2022.101143.

Mohd-Taufek, N., Cartwright, D., Davies, M., Hewavitharana, A.K., Koorts, P., Shaw, P.N., Sumner, R., Lee, E., Whitfield, K., 2016. The simultaneous analysis of eight essential trace elements in human milk by ICP-ms. Food Anal. Methods 9 (7), 2068–2075. https://doi.org/10.1007/s12161-015-0396-z.

Moldovan, M., 2007. Origin and fate of platinum group elements in the environment. Anal. Bioanal. Chem. 388 (3), 537–540. https://doi.org/10.1007/s00216-007-1234-y.

Moreira, L.S., Costa, F.S., Machado, R.C., Nogueira, A.R.A., Gonzalez, M.H., da Silva, E.G.P., Amaral, C.D.B., 2021. Self-organizing map applied to the choice of internal standards for the determination of Cd, Pb, Sn, and platinum group elements by inductively coupled plasma mass spectrometry. Talanta 233. https://doi.org/10.1016/j.talanta.2021.122534.

Morshdy, A.E.M.A., El Bayomi, R.M., Khalifa, S.M., El-Dien, W.M.S., Darwish, W.S., Mahmoud, A.F.A., 2022. Heavy metal content in chicken meat products: a health risk assessment study. J. Adv. Vet. Res. 12 (4), 451–455.

Mosihuzzaman, M., Choudhary, M.I., 2008. 10 2008/10 Pure and Applied Chemistry 10.1351/pac200880102195 00334545 10 2195 2230 Pakistan Protocols on Safety, Efficacy, Standardization, and Documentation of Herbal Medicine (IUPAC Technical Report) 80.

Motswetla1, Kereeditse, Kelebemang, T.T., 2023. Assessment of level of heavy metals in cosmetics. Toxicol. Res. Appl. 7.

Munir, N., Jahangeer, M., Bouyahya, A., El Omari, N., Ghchime, R., Balahbib, A., Aboulaghras, S., Mahmood, Z., Akram, M., Shah, S.M.A., Mikolaychik, I.N., Derkho, M., Rebezov, M., Venkidasamy, B., Thiruvengadam, M., Ali Shariati, M., 2022. Heavy metal contamination of natural foods is a serious health issue: a review. Sustainability 14 (1), 161. https://doi.org/10.3390/su14010161.

Nagel, A., Cuss, C.W., Goss, G.G., Shotyk, W., Glover, C.N., 2023. Accumulation of thallium in rainbow trout (*Oncorhynchus mykiss*) following acute and subchronic waterborne exposure. John Wiley and Sons Inc, Canada Environ. Toxicol. Chem. 42 (7), 1553−1563. https://doi.org/10.1002/etc.5637.

Ni, Z., Ren, C., Cheng, J., Tang, F., 2017. Determination of rare earth elements in some flower herb teas and their infusions. J. Braz. Chem. Soc. 28 (10), 1960−1965. https://doi.org/10.21577/0103-5053.20170039.

Niemelä, M., Perämäki, P., Piispanen, J., Poikolainen, J., 2004. Determination of platinum and rhodium in dust and plant samples using microwave-assisted sample digestion and ICP-MS. Anal. Chim. Acta 521 (2), 137−142. https://doi.org/10.1016/j.aca.2004.05.075.

Nischkauer, W., Herincs, E., Puschenreiter, M., Wenzel, W., Limbeck, A., 2013. Determination of Pt, Pd and Rh in Brassica Napus using solid sampling electrothermal vaporization inductively coupled plasma optical emission spectrometry. Spectrochim. Acta B Atom Spectrosc. 89, 60−65. https://doi.org/10.1016/j.sab.2013.08.013.

Njoga, E.O., Ezenduka, E.V., Ogbodo, C.G., Ogbonna, C.U., Jaja, I.F., Ofomatah, A.C., Okpala, C.O.R., 2021. Detection, distribution and health risk assessment of toxic heavy metals/metalloids, arsenic, cadmium, and lead in goat carcasses processed for human consumption in south-eastern Nigeria. Foods 10 (4). https://doi.org/10.3390/foods10040798.

Oliveira, M.M., Trevilato, T.M.B., Segura-Muñoz, S.I., Aragon, D.C., Alves, L.G., Nadal, M., Marquès, M., Domingo, J.L., Sierra, J., Camelo, J.S., 2020. Essential and toxic elements in human milk concentrate with human milk lyophilizate: a preclinical study. Environ. Res. 188. https://doi.org/10.1016/j.envres.2020.109733.

Parsaei, P., Rahimi, E., Shakerian, A., 2019. Concentrations of cadmium, lead and mercury in raw bovine, ovine, caprine, buffalo and camel milk. Pol. J. Environ. Stud. 28 (6), 4311−4318. https://doi.org/10.15244/pjoes/94809.

Parsons, P.J., Farago, M.E., 1987. X-ray microanalysis of plants containing the platinum group metals. J. Electron. Microsc. Tech. 5 (4), 291−301. https://doi.org/10.1002/jemt.1060050402.

Patočka, J., Bendakovská, L., Krejčová, A., Černohorský, T., Resano, M., Bělina, P., 2017. Thallium in spruce needles: a comparison of the analytical capabilities of spectrochemical methods. Republ. Analy. Methods 9 (4), 705−715. https://doi.org/10.1039/c6ay02760a.

Pawlak, J., Lodyga-Chruścińska, E., Chrustowicz, J., 2014. Fate of platinum metals in the environment. J. Trace Elem. Med. Biol. 28 (3), 247−254. https://doi.org/10.1016/j.jtemb.2014.03.005.

Pearson, A.J., Ashmore, E., 2020. Risk assessment of antimony, barium, beryllium, boron, bromine, lithium, nickel, strontium, thallium and uranium concentrations in the New Zealand diet. Food Addit. Contam. 37 (3), 451−464. https://doi.org/10.1080/19440049.2019.1704445.

Quds, T., Ahmed, M., Shakeel, S., Jalbani, N., Mazhar, F., Azhar, I., 2021. Determination of the heavy metal contents of frequently used herbal products in Pakistan. Trop. J. Pharmaceut. Res. 20 (2), 377−382. https://doi.org/10.4314/tjpr.v20i2.23.

Qvarforth, A., Lundgren, M., Rodushkin, I., Engström, E., Paulukat, C., Hough, R.L., Moreno-Jiménez, E., Beesley, L., Trakal, L., Augustsson, A., 2022. Future food contaminants: an assessment of the plant uptake of Technology-critical elements versus traditional metal contaminants. Environ. Int. 169, 107504. https://doi.org/10.1016/j.envint.2022.107504.

Rao, M.M., Meena, A.K., 2011. Galib, Detection of toxic heavy metals and pesticide residue in herbal plants which are commonly used in the herbal formulations. Environ. Monit. Assess. 181 (1−4), 267−271. https://doi.org/10.1007/s10661-010-1828-2.

Rao, C.R.M., Reddi, G.S., 2000. Platinum group metals (PGM); occurrence, use and recent trends in their determination. TrAC, Trends Anal. Chem. 19 (9), 565–586. https://doi.org/10.1016/S0165-9936(00)00031-5.

Ravindra, K., Bencs, L., Van Grieken, R., 2004. Platinum group elements in the environment and their health risk. Sci. Total Environ. 318 (1–3), 1–43. https://doi.org/10.1016/S0048-9697(03)00372-3.

Rees, G., Kelly, S.D., Cairns, P., Ueckermann, H., Hoelzl, S., Rossmann, A., Scotter, M.J., 2016. Verifying the geographical origin of poultry: the application of stable isotope and trace element (SITE) analysis. Food Control 67, 144–154. https://doi.org/10.1016/j.foodcont.2016.02.018.

Regulation (EC) no.1223/2009 of the European Parliament and of the European Council of November 30, 2009.

Reilly, C., 2002. Metal Contamination of Food – its Significance for Food Quality and Human Health. Blackwell Science Ltd.

Rodríguez-Mercado, J.J., Altamirano-Lozano, M.A., 2013. Genetic toxicology of thallium: a review. Drug Chem. Toxicol. 36 (3), 369–383. https://doi.org/10.3109/01480545.2012.710633.

Roselli, C., Desideri, D., Meli, M.A., Fagiolino, I., Feduzi, L., 2016. Essential and toxic elements in meat of wild birds. J. Toxicol. Environ. Health 79 (21), 1008–1014. https://doi.org/10.1080/15287394.2016.1216490.

Rossipal, E., Krachler, M., 1998. Pattern of trace elements in human milk during the course of lactation. Nutr. Res. 18 (1), 11–24. https://doi.org/10.1016/S0271-5317(97)00196-6.

Sabbioni, E., Di Gioacchino, M., Farina, M., Groppi, F., Manenti, S., 2018. Radioanalytical and nuclear techniques in trace metal toxicology research. J. Radioanal. Nucl. Chem. 318 (3), 1749–1763. https://doi.org/10.1007/s10967-018-6321-3.

Sager, M., Hobegger, M., 2013. Elementgehalte in Rohmilch aus drei Regionen Niederösterreichs. Ernahrung 37 (7–8), 277–290.

Sahoo, N., Manchikanti, P., Dey, S., 2010. Herbal drugs: standards and regulation. Fitoterapia 81 (6), 462–471. https://doi.org/10.1016/j.fitote.2010.02.001.

Saleh, I., Syamsir, S., Pramaningsih, V., Hansen, H., 2021. The use of green mussel as bioindicator of heavy metal pollution in Indonesia: a review. Environ. Analy. Health Toxicol. 36 (4). https://doi.org/10.5620/eaht.2021026Indonesia.

Salmani, M.H., Gholami, M., Ranjbar, M.J., Mokhberi, F., 2024. Comparison of essential and toxic metals levels in some herbal teas: a systematic review. Biol. Trace Elem. Res. 202 (2), 615–623. https://doi.org/10.1007/s12011-023-03698-w.

Sanaei, F., Amin, M.M., Alavijeh, Z.P., Esfahani, R.A., Sadeghi, M., Bandarrig, N.S., Fatehizadeh, A., Taheri, E., Rezakazemi, M., 2021. Health risk assessment of potentially toxic elements intake via food crops consumption: Monte Carlo simulation-based probabilistic and heavy metal pollution index. Environ. Sci. Pollut. Control Ser. 28 (2), 1479–1490. https://doi.org/10.1007/s11356-020-10450-7.

Sanal, H., Güler, Z., Park, Y.W., 2011. Profiles of non-essential trace elements in Ewe and goat milk and their yoghurt, Torba yoghurt and whey. Food Addit. Contam. Part B Surveill . 4 (4), 275–281. https://doi.org/10.1080/19393210.2011.617520.

Saribal, D., 2020. ICP-MS analysis of trace element concentrations in cow's milk samples from supermarkets in istanbul. Biol. Trace Elem. Res. 193 (1), 166–173. https://doi.org/10.1007/s12011-019-01708-4.

Sasmaz, M., Akgul, B., Yıldırım, D., Sasmaz, A., 2016. Bioaccumulation of thallium by the wild plants grown in soils of mining area. Int. J. Phytoremediation 18 (11), 1164–1170. https://doi.org/10.1080/15226514.2016.1183582.

Savignan, L., Faucher, S., Chéry, P., Lespes, G., 2021. Platinum group elements contamination in soils: review of the current state. Chemosphere 271, 129517. https://doi.org/10.1016/j.chemosphere.2020.129517.

Schunk, P.F.T., Kalil, I.C., Pimentel-Schmitt, E.F., Lenz, D., de Andrade, T.U., Ribeiro, J.S., Endringer, D.C., 2016. ICP-OES and micronucleus test to evaluate heavy metal contamination in commercially available Brazilian herbal teas. Biol. Trace Elem. Res. 172 (1), 258–265. https://doi.org/10.1007/s12011-015-0566-2.

Schuster, M., Schwarzer, M., Risse, G., 2000. Determination of Palladium in Environmental Samples. Springer Science and Business Media LLC, pp. 173–182. https://doi.org/10.1007/978-3-642-59678-0_18.

Serapinas, P., Venskutonis, P.R., Aninkevičius, V., Ežerinskis, Z., Galdikas, A., Juzikiene, V., 2008. Step by step approach to multi-element data analysis in testing the provenance of wines. Food Chem. 107 (4), 1652–1660. https://doi.org/10.1016/j.foodchem.2007.09.003.

Sun, B., Xing, M., 2016. Evaluated the twenty-six elements in the pectoral muscle of as-treated chicken by inductively coupled plasma mass spectrometry. Biol. Trace Elem. Res. 169 (2), 359–364. https://doi.org/10.1007/s12011-015-0418-0.

Sun, J., Hu, G., Liu, K., Yu, R., Lu, Q., Zhang, Y., 2019. Potential exposure to metals and health risks of metal intake from Tieguanyin tea production in Anxi, China. Environ. Geochem. Health 43 (3), 1291–1302. https://doi.org/10.1007/s10653-018-0212-y.

Sun, T., Ji, C., Li, F., Wu, H., 2023. Bioaccumulation and human health implications of trace metals in oysters from coastal areas of China. Mar. Environ. Res. 184, 105872. https://doi.org/10.1016/j.marenvres.2022.105872.

Supplementary Guidelines on Good Manufacturing Practices for the Manufacture of Herbal Medicine, 2006.

Sures, B., Zimmermann, S., Messerschmidt, J., Von Bohlen, A., Alt, F., 2001. First report on the uptake of automobile catalyst emitted palladium by European eels (*Anguilla anguilla*) following experimental exposure to road dust. Environ. Pollut. 113 (3), 341–345. https://doi.org/10.1016/S0269-7491(00)00185-8.

Syed, S.A., Bari, A., Kaimkhani, Z.A., Ali, E.A., 2023. Inductively coupled plasma mass spectrometry assessment of essential and toxic trace elements in traditional spices consumed by the population of the Middle Eastern region in their recipes. Open Chem. 21 (1). https://doi.org/10.1515/chem-2022-0352.

Taylor, A., Branch, S., Day, M.P., Patriarca, M., White, M., 2006. Atomic spectrometry update. Clinical and biological materials, foods and beverages. J. Anal. Atomic Spectrom. 21 (4), 439–491. https://doi.org/10.1039/b601116k.

Tish, B., Unutkan, T., Bakirdere, S., 2023. Determination of thallium in tea with preconcentration by microwave-assisted Synthesized molybdenum disulfide nanoparticles and flame atomic absorption spectrometry (FAAS) analysis. Anal. Lett.

Tofan, L., Bojoaga, C.N., Paduraru, C., 2022. Biosorption for the recovery and analysis of rare earth elements and platinum group metals from real samples. A review . Springer Science and Business Media Deutschland GmbH. Roman. Environ. Chem. Lett. 20 (2), 1225–1248. https://doi.org/10.1007/s10311-021-01330-8.

Ullah, H., Aslam, S., Mustafa, G., Waseem, A., de Freitas Marques, M.B., Gul, Z., Usman Alvi, M., Anwar, S., Sabir, M., Hamid, A., Ibrahim, M., 2023. Potential toxicity of heavy metals in cosmetics: fake or fact: a review. Int. J. Environ. Anal. Chem. https://doi.org/10.1080/03067319.2023.2217406.

Unguryanu, T.N., Lyzhina, A.V., Mitrokhin, O.V., Polibin, R.V., 2023. Human health risk assessment of heavy metals from meat and offal of reindeer and cow in the Far North of European Russia. Feder. Scandinavian J. Publ. Health 51 (7), 1009–1015. https://doi.org/10.1177/14034948221096243.

Urbánková, K., Sommer, L., 2008. The microdetermination of thallium by ICP-AES after previous preconcentration on modified silica. Comparison with integrated-platform graphite atomic absorption spectrometry. Microchim. Acta 162 (1–2), 127–132. https://doi.org/10.1007/s00604-007-0838-4.

Vaezzadeh, M., Shemirani, F., Majidi, B., 2010. Microextraction technique based on ionic liquid for preconcentration and determination of palladium in food additive, sea water, tea and biological samples. Food Chem. Toxicol. 48 (6), 1455–1460. https://doi.org/10.1016/j.fct.2010.03.005.

Vasseghian, Y., Hosseinzadeh, S., Khataee, A., Dragoi, E.-N., 2021. The concentration of persistent organic pollutants in water resources: a global systematic review, meta-analysis and probabilistic risk assessment. Sci. Total Environ. 796, 149000. https://doi.org/10.1016/j.scitotenv.2021.149000.

Voica, C., Nechita, C., Iordache, A.M., Roba, C., Zgavarogea, R., Ionete, R.E., 2021. Icp-ms assessment of essential and toxic trace elements in foodstuffs with different geographic origins available in Romanian supermarkets. Molecules 26 (23). https://doi.org/10.3390/molecules26237081.

Voica, C., Cristea, G., Iordache, A.M., Roba, C., Curean, V., 2023. Elemental profile in chicken egg components and associated human health risk assessment. Toxics 11 (11), 900. https://doi.org/10.3390/toxics11110900.

Wang, Y., Li, X., 2012. 1 2012/01 Procedia Engineering 10.1016/j.proeng.2012.08.273 18777058 1004 1009 Elsevier Ltd China Health risk of platinum group elements from automobile catalysts, 45. http://www.sciencedirect.com/science/journal/18777058.

Wang, H.W., Liu, Y.Q., 2010. Evaluation of trace and toxic element concentrations in Paris polyphylla from China with empirical and chemometric approaches. Food Chem. 121 (3), 887–892. https://doi.org/10.1016/j.foodchem.2010.01.012.

Wang, X., Zhang, Y., Geng, Z., Liu, Y., Guo, L., Xiao, G., 2019. Spatial analysis of heavy metals in meat products in China during 2015-2017. Food Control 104, 174–180. https://doi.org/10.1016/j.foodcont.2019.04.033.

Wang, L., Wang, X., Chen, H., Wang, Z., Jia, X., 2022. Oyster arsenic, cadmium, copper, mercury, lead and zinc levels in the northern South China Sea: long-term spatiotemporal distributions, combined effects, and risk assessment to human health. Environ. Sci. Pollut. Control Ser. 29 (9), 12706–12719. https://doi.org/10.1007/s11356-021-18150-6.

White, C.M., 2020. Dietary supplements pose real dangers to patients. Ann. Pharmacother. 54 (8), 815–819. https://doi.org/10.1177/1060028019900504.

Woo, C.S.J., Lau, J.S.H., El-Nezami, H., Medicine, H., 2012. Toxicity and recent trends in assessing their potential toxic effects. Adv. Bot. Res. 62, 365–384. https://doi.org/10.1016/B978-0-12-394591-4.00009-X.

Xu, Y., Wang, N., Feng, Y., 2021. An Overview of Heavy Metal Research in Traditional Chinese Medicine Metal Toxicology Handbook. CRC Press.

Yelmate, A.A., Thonte, S.S., Satpute, K.L., 2023. Trace Element Determination in Medicinal Plant Samples by ED-XRF Analysis New Advances. IntechOpen.

Yu, H.Y., Chang, C., Li, F., Wang, Q., Chen, M., Zhang, J., 2018. Thallium in flowering cabbage and lettuce: potential health risks for local residents of the Pearl River Delta, South China. Environ. Pollut. 241, 626–635. https://doi.org/10.1016/j.envpol.2018.05.090.

Yuan, X., Chapman, R.L., Wu, Z., 2011. Analytical methods for heavy metals in herbal medicines. Phytochem. Anal. 22 (3), 189–198. https://doi.org/10.1002/pca.1287.

Zhang, J., Wider, B., Shang, H., Li, X., Ernst, E., 2012. Quality of herbal medicines: challenges and solutions. Compl. Ther. Med. 20 (1–2), 100–106. https://doi.org/10.1016/j.ctim.2011.09.004.

Zhang, J., Yang, R., Li, Y.C., Peng, Y., Wen, X., Ni, X., 2020. Distribution, accumulation, and potential risks of heavy metals in soil and tea leaves from geologically different plantations. Ecotoxicol. Environ. Saf. 195, 110475. https://doi.org/10.1016/j.ecoenv.2020.110475.

Zhao, Y., Ma, X., Fan, L., Mao, F., Tian, H., Xu, R., Cao, Z., Zhang, X., Fu, X., Sui, H., 2017. Discrimination of geographical origin of cultivated Polygala tenuifolia based on multi-element fingerprinting by inductively coupled plasma mass spectrometry. Sci. Rep. 7 (1). https://doi.org/10.1038/s41598-017-12933-z.

Zhelyazkov, G., Yankovska-Stefanova, T., Mineva, E., Stratev, D., Vashin, I., Dospatliev, L., Valkova, E., Popova, T., 2018. Risk assessment of some heavy metals in mussels (*Mytilus galloprovincialis*) and veined rapa whelks (Rapana venosa) for human health. Mar. Pollut. Bull. 128, 197–201. https://doi.org/10.1016/j.marpolbul.2018.01.024.

Zhitenko, L.P., Kiseleva, I.N., Khomutova, E.G., 2016. Methods for studying anthropogenic environmental pollution with platinum group metals (review) . Maik Nauka Publishing/Springer SBM. Russ. Feder. Inorg. Mater. 52 (14), 1370–1376. https://doi.org/10.1134/S0020168516140156.

Zhou, X., Qu, X., Zhao, S., Wang, J., Li, S., Zheng, N., 2017. Analysis of 22 elements in milk, feed, and water of dairy cow, goat, and Buffalo from different regions of China. Biol. Trace Elem. Res. 176 (1), 120–129. https://doi.org/10.1007/s12011-016-0819-8.

Zhu, Y., Costa, M., 2020. Metals and molecular carcinogenesis. Carcinogenesis 41 (9), 1161–1172. https://doi.org/10.1093/carcin/bgaa076.

Zimmermann, S., Sures, B., 2018. Lessons learned from studies with the freshwater mussel *Dreissena polymorpha* exposed to platinum, palladium and rhodium. Sci. Total Environ. 615, 1396–1405. https://doi.org/10.1016/j.scitotenv.2017.09.204.

Zimmermann, S., Baumann, U., Taraschewski, H., Sures, B., 2004. Accumulation and distribution of platinum and rhodium in the European eel *Anguilla anguilla* following aqueous exposure to metal salts. Environ. Pollut. 127 (2), 195–202. https://doi.org/10.1016/j.envpol.2003.08.006.

Zimmermann, S., Sures, B., Ruchter, N., 2015. Laboratory Studies on the Uptake and Bioaccumulation of PGE by Aquatic Plants and Animals. Springer Berlin Heidelberg, Germany Environmental Science and Engineering (Subseries: Environmental Science), pp. 361–381. https://doi.org/10.1007/978-3-662-44559-4_23.

Zoroddu, M.A., Aaseth, J., Crisponi, G., Medici, S., Peana, M., Nurchi, V.M., 2019. The essential metals for humans: a brief overview. J. Inorg. Biochem. 195, 120–129. https://doi.org/10.1016/j.jinorgbio.2019.03.013.

Zwolak, A., Sarzyńska, M., Szpyrka, E., Stawarczyk, K., 2019. Sources of soil pollution by heavy metals and their accumulation in vegetables: a review. Water Air Soil Pollut. 230 (7). https://doi.org/10.1007/s11270-019-4221-y.

2001 2001 and EC Regulation EC Regulation 466/2001 Dated March 8, 2001 and Subsequent Amendments: EC Regulation 2375/2001 Dated November 29, 2001 and EC Regulation 221/2002 Dated February 6, 2002. 466.

2004 World Health Organization (WHO) World Health Organization (WHO) Unpublished Content Guidelines on Safety Monitoring of Herbal Medicines in Pharmacovigilance Systems.

Conclusion and future perspectives

Marcello Locatelli[1], Michał Tomczyk[2], Laura Dugo[3], Marina Russo[4]

[1] *Department of Pharmacy, University of Chieti-Pescara "G. d'Annunzio", Chieti, Italy*
[2] *Department of Pharmacognosy, Faculty of Pharmacy with the Division of Laboratory Medicine, Medical University of Bialystok, Bialystok, Poland*
[3] *Department of Science and Technologies for Sustainable Development and One Health, University Campus Bio-Medico di Roma, Rome, Italy*
[4] *Messina Institute of Technology c/o Department of Chemical, Biological, Pharmaceutical and Environmental Sciences, Former Veterinary School, University of Messina, Messina, Italy*

This book presents the state of the art in the field of phytochemicals. In particular, it aims to highlight how, through the interactions of plants with their environment, the potential of an analytical approach for standardization and quality control can be fully exploited. Furthermore, it emphasizes the importance of investigating the correlation between chemical profiles and biological activities to obtain a product that is safe for human health and has proven beneficial activities.

This book comprises chapters that focus on the main problems related to natural products (plants, extraction, quantitative analysis, and relationships with the surrounding environment) and sections that identify and report the main classes of organic and inorganic compounds. It also focuses on natural compounds that have become active principle ingredients (APIs) to highlight the ongoing necessity of studying nature to develop innovative drugs. Specifically, it explores how we can learn from nature to derive new active ingredients for human health, developed through a scientific and analytical approach essential for both standardization and product quality control.

This book contains chapters written by experts in their respective fields to provide a detailed and up-to-date resource for phytochemical investigations. In particular, after an introductory section, it addresses the role of active ingredients in human health. Important chapters follow that discuss the issues of quantitative analysis and standardization of extracts, including innovative configurations that are increasingly used today and the challenges associated with the quantitative analysis of chiral compounds.

The text continues with chapters dedicated to some of the main classes of organic bioactives and their potential uses in developing new formulations and functionalized ingredients. Additionally, there is a chapter highlighting the role of abiotic stress factors from both natural and anthropogenic sources in plant–environment interactions. An important chapter concludes the book, addressing the problem of nonessential pollutant metals in the food chain, a topic often overlooked in this field. These metals, present in extracts and potentially influencing biological activity, are often neglected in industrial quality control due to legislative gaps.

Therefore, the primary audience for this book includes professionals in the food, food supplements, and pharmaceutical fields, as well as researchers and students in pharmacognosy, phytotherapy, food chemistry, medicinal chemistry, chemistry (analytical), phytochemistry, and plant biochemistry.

As made clear in this text, investigations in this field require an interdisciplinary approach in which botanists, chemists, pharmacologists, technologists, and experts in human health and nutrition must interact to integrate their specific knowledge and develop a safe product that demonstrates biological efficacy. The "garden" mentality must now be overcome; we need to focus increasingly on integrated research and knowledge, where everyone brings their own expertise without encroaching on areas outside their responsibility.

A major challenge for the future lies in achieving more integrated research among these professionals on plant systems that are "controlled" from both environmental and qualitative–quantitative perspectives. This is important because, as highlighted, we often work with extracts that are not fully standardized or characterized. Moreover, if the biological activity (or observable effect) does not fully support the literature data (or expectations), this phenomenon is associated with the "infamous" phytocomplex.

The development of telematic platforms, including those utilizing artificial intelligence (AI), which allow for the comparison of large amounts of data (which will need to be standardized) across these various aspects, is becoming increasingly necessary. Furthermore, there is a pressing need to establish national and international regulations that ensure extracts and raw materials are safe for health and suitable for use in the food and food supplement sectors.

Recent developments in AI and chemometrics can strongly boost this process, but a considerable amount of work remains.

An essential element to consider and begin developing immediately is an international, open-access database with "standardized" information and data sheets on analysis and biological activities. Such a resource, potentially integrated with AI approaches, would facilitate the study, development, and marketing of increasingly safe and effective products.

1. Introduction (general aspects on phytochemicals, definitions, classes, chemical structures, properties)
2. Biosynthesis of primary and secondary metabolites
3. Role of active ingredients in human health
4. From Plants to extracts
5. From extracts to quantitative analyses (method validation, analytical approaches (chromatographic, spectroscopic and electrochemical procedures), innovative instrumentations LCxLC)
6. From extracts to quantitative analyses (Chiral separation of natural products)
7. Antidiabetic potential of capsaicinoids
8. Qualitative and quantitative determination of the main components of *Capparis cartilaginea* Decne. infusion by HPLC-UV
9. Potentials of naturally occurring coumarins and derivatives for the treatment of Alzheimer's disease
10. Increasing of biologically active metabolites production using plant biotechnology techniques in {i| Nasturtium officinale} (watercress) microshoot cultures
11. Current knowledge on selected stilbenes and their promising applications
12. Role of abiotic stress factors from natural and anthropogenic sources in plant–environment interaction
13. New active principle ingredient (API) (considering new formulations and new functionalized ingredients)

14. Alkaloids' wide world (extraction, quantification, biological activities)
15. Terpenoids (extraction, quantification, biological activities)
16. Phenolic compounds (extraction, quantification, biological activities)
17. Nonessential pollutant metals in the food chain: state of the art and particular focus on Thallium and Platinum group metals (PGMs)

Index

'*Note:* Page numbers followed by "f" indicate figures and "t" indicate tables.'

D

E

Q

U

V

9780443153662